PROGRESS IN CLINICAL AND BIOLOGICAL RESEARCH

RECENT TITLES

Vol 227: **Advances in Chronobiology**, John E. Pauly, Lawrence E. Scheving, *Editors*. Published in two volumes.

Vol 228: **Environmental Toxicity and the Aging Processes,** Scott R. Baker, Marvin Rogul, *Editors*

Vol 229: **Animal Models: Assessing the Scope of Their Use in Biomedical Research,** Junichi Kawamata, Edward C. Melby, Jr., *Editors*

Vol 230: **Cardiac Electrophysiology and Pharmacology of Adenosine and ATP: Basic and Clinical Aspects,** Amir Pelleg, Eric L. Michelson, Leonard S. Dreifus, *Editors*

Vol 231: **Detection of Bacterial Endotoxins With the Limulus Amebocyte Lysate Test,** Stanley W. Watson, Jack Levin, Thomas J. Novitsky, *Editors*

Vol 232: **Enzymology and Molecular Biology of Carbonyl Metabolism: Aldehyde Dehydrogenase, Aldo-Keto Reductase, and Alcohol Dehydrogenase**, Henry Weiner, T. Geoffrey Flynn, *Editors*

Vol 233: **Developmental and Comparative Immunology,** Edwin L. Cooper, Claude Langlet, Jacques Bierne, *Editors*

Vol 234: **The Hepatitis Delta Virus and Its Infection**, Mario Rizzetto, John L. Gerin, Robert H. Purcell, *Editors*

Vol 235: **Preclinical Safety of Biotechnology Products Intended for Human Use,** Charles E. Graham, *Editor*

Vol 236: **First Vienna Shock Forum,** Günther Schlag, Heinz Redl, *Editors*. Published in two volumes: Part A: *Pathophysiological Role of Mediators and Mediator Inhibitors in Shock.* Part B: *Monitoring and Treatment of Shock.*

Vol 237: **The Use of Transrectal Ultrasound in the Diagnosis and Management of Prostate Cancer,** Fred Lee, Richard McLeary, *Editors*

Vol 238: **Avian Immunology,** W.T. Weber, D.L. Ewert, *Editors*

Vol 239: **Current Concepts and Approaches to the Study of Prostate Cancer,** Donald S. Coffey, Nicholas Bruchovsky, William A. Gardner, Jr., Martin I. Resnick, James P. Karr, *Editors*

Vol 240: **Pathophysiological Aspects of Sickle Cell Vaso-Occlusion,** Ronald L. Nagel, *Editor*

Vol 241: **Genetics and Alcoholism,** H. Werner Goedde, Dharam P. Agarwal, *Editors*

Vol 242: **Prostaglandins in Clinical Research,** Helmut Sinzinger, Karsten Schrör, *Editors*

Vol 243: **Prostate Cancer,** Gerald P. Murphy, Saad Khoury, Réne Küss, Christian Chatelain, Louis Denis, *Editors*. Published in two volumes: Part A: *Research, Endocrine Treatment, and Histopathology.* Part B: *Imaging Techniques, Radiotherapy, Chemotherapy, and Management Issues*

Vol 244: **Cellular Immunotherapy of Cancer,** Robert L. Truitt, Robert P. Gale, Mortimer M. Bortin, *Editors*

Vol 245: **Regulation and Contraction of Smooth Muscle,** Marion J. Siegman, Andrew P. Somlyo, Newman L. Stephens, *Editors*

Vol 246: **Oncology and Immunology of Down Syndrome,** Ernest E. McCoy, Charles J. Epstein, *Editors*

Vol 247: **Degenerative Retinal Disorders: Clinical and Laboratory Investigations,** Joe G. Hollyfield, Robert E. Anderson, Matthew M.LaVail, *Editors*

Vol 248: **Advances in Cancer Control: The War on Cancer—15 Years of Progress,** Paul F. Engstrom, Lee E. Mortenson, Paul N. Anderson, *Editors*

Vol 249: **Mechanisms of Signal Transduction by Hormones and Growth Factors,** Myles C. Cabot, Wallace L. McKeehan, *Editors*

Vol 250: **Kawasaki Disease,** Stanford T. Shulman, *Editor*

Vol 251: **Developmental Control of Globin Gene Expression,** George Stamatoyannopoulos, Arthur W. Nienhuis, *Editors*

Vol 252: **Cellular Calcium and Phosphate Transport in Health and Disease,** Felix Bronner, Meinrad Peterlik, *Editors*

Vol 253: **Model Systems in Neurotoxicology: Alternative Approaches to Animal Testing,** Abraham Shahar, Alan M. Goldberg, *Editors*

Vol 254: **Genetics and Epithelial Cell Dysfunction in Cystic Fibrosis,** John R. Riordan, Manuel Buchwald, *Editors*

Vol 255: **Recent Aspects of Diagnosis and Treatment of Lipoprotein Disorders: Impact on Prevention of Atherosclerotic Diseases,** Kurt Widhalm, Herbert K. Naito, *Editors*

Vol 256: **Advances in Pigment Cell Research,** Joseph T. Bagnara, *Editor*

Vol 257: **Electromagnetic Fields and Neurobehavioral Function,** Mary Ellen O'Connor, Richard H. Lovely, *Editors*

Vol 258: **Membrane Biophysics III: Biological Transport,** Mumtaz A. Dinno, William McD. Armstrong, *Editors*

Vol 259: **Nutrition, Growth, and Cancer,** George P. Tryfiates, Kedar N. Prasad, *Editors*

Vol 260: **EORTC Genitourinary Group Monograph 4: Management of Advanced Cancer of Prostate and Bladder,** Philip H. Smith, Michele Pavone-Macaluso, *Editors*

Vol 261: **Nicotine Replacement: A Critical Evaluation,** Ovide F. Pomerleau, Cynthia S. Pomerleau, *Editors*

Vol 262: **Hormones, Cell Biology, and Cancer: Perspectives and Potentials,** W. David Hankins, David Puett, *Editors*

Vol 263: **Mechanisms in Asthma: Pharmacology, Physiology, and Management,** Carol L. Armour, Judith L. Black, *Editors*

Vol 264: **Perspectives in Shock Research,** Robert F. Bond, *Editor*

Vol 265: **Pathogenesis and New Approaches to the Study of Noninsulin-Dependent Diabetes Mellitus,** Albert Y. Chang, Arthur R. Diani, *Editors*

Vol 266: **Growth Factors and Other Aspects of Wound Healing: Biological and Clinical Implications,** Adrian Barbul, Eli Pines, Michael Caldwell, Thomas K. Hunt, *Editors*

Vol 267: **Meiotic Inhibition: Molecular Control of Meiosis,** Florence P. Haseltine, *Editor*

Vol 268: **The Na^+,K^+-Pump,** Jens C. Skou, Jens G. Nørby, Arvid B. Maunsbach, Mikael Esmann, *Editors*. Published in two volumes: Part A: *Molecular Aspects*. Part B: *Cellular Aspects*.

Vol 269: **EORTC Genitourinary Group Monograph 5: Progress and Controversies in Oncological Urology II,** Fritz H. Schröder, Jan G.M. Klijn, Karl H. Kurth, Herbert M. Pinedo, Ted A.W. Splinter, Herman J. de Voogt, *Editors*

Vol 270: **Cell-Free Analysis of Membrane Traffic,** D. James Morré, Kathryn E. Howell, Geoffrey M.W. Cook, W. Howard Evans, *Editors*

Vol 271: **Advances in Neuroblastoma Research 2,** Audrey E. Evans, Giulio J. D'Angio, Alfred G. Knudson, Robert C. Seeger, *Editors*

Vol 272: **Bacterial Endotoxins: Pathophysiological Effects, Clinical Significance, and Pharmacological Control,** Jack Levin, Harry R. Büller, Jan W. ten Cate, Sander J.H. van Deventer, Augueste Sturk, *Editors*

Vol 273: **The Ion Pumps: Structure, Function, and Regulation,** Wilfred D. Stein, *Editor*

Vol 274: **Oxidases and Related Redox Systems,** Tsoo E. King, Howard S. Mason, Martin Morrison, *Editors*

Vol 275: **Electrophysiology of the Sinoatrial and Atrioventricular Nodes,** Todor N. Mazgalev, Leonard S. Dreifus, Eric L. Michelson, *Editors*

Please contact the publisher for information about previous titles in this series.

Oxidases and Related Redox Systems

OXIDASES AND RELATED REDOX SYSTEMS

Proceedings of the 4th International Symposium on Oxidases and Related Redox Systems held in Portland, Oregon, October 4–8, 1987

Editors

Tsoo E. King

Institute for Structural and Functional Studies
University City Science Center
Philadelphia, Pennsylvania

Howard S. Mason

Department of Biochemistry
Oregon Health Sciences University
Portland, Oregon

Martin Morrison

Department of Biochemistry
St. Jude Children's Research Hospital
Memphis, Tennessee

ALAN R. LISS, INC. • NEW YORK

Address all Inquiries to the Publisher
Alan R. Liss, Inc., 41 East 11th Street, New York, NY 10003

Printed in the United States of America

Library of Congress Cataloging-in-Publication Data

Oxidases and related redox systems.

(Progress in clinical and biological research; v. 274)
"Proceedings of the Fourth International Symposium on Oxidases and Related Redox Systems, held in Portland, Oregon, October 4-8, 1987"—
Includes bibliographies and index.
1. Oxidases—Congresses. I. King, Tsoo E., II. Mason, Howard S. III. Morrison, Martin. IV. International Symposium on Oxidases and Related Redox Systems (4th : 1987 : Portland, Or.) V. Series.
QP603.08095 1988 574.19'258 88-6822
ISBN 0-8451-5124-X

Contents

Contributors and Participants

Robert F. Anderson, Gray Laboratory, Cancer Research Campaign, Mount Vernon Hospital, Northwood, Middlesex HA6 2RN, England **[147,167]***

Laura Andersson, Department of Chemical and Biological Sciences, Oregon Graduate Center, Beaverton, OR 97006*

Helen Anni, Department of Biochemistry and Biophysics, University of Pennsylvania School of Medicine, Philadelphia, PA 19104-6089 **[437]***

G.E. Avaria-Neisser, Department of Chemistry, University of Michigan, Ann Arbor, MI 48109 **[509]**

Angelo Azzi, Institut für Biochemie und Molekularbiologie, Universität Bern, CH-3012 Bern, Switzerland **[591]***

David Ballou, Department of Biological Chemistry, University of Michigan, Ann Arbor, MI 48109 **[211]***

Paul D. Barker, Inorganic Chemistry Laboratory, Oxford University, Oxford OX1 3QR, England **[419]***

Christopher Batie, Department of Biological Chemistry, University of Michigan, Ann Arbor, MI 48109 **[211]**

Ninian Blackburn, Department of Chemical and Biological Sciences, Oregon Graduate Center, Beaverton, OR 97006*

David F. Blair, Department of Biology, Harvard University, Boston, MA 02138 **[731]**

Barbara Bolgiano, Department of Microbiology, University of Pennsylvania School of Medicine, Philadelphia, PA 19104-6076 **[619]**

Martino Bolognesi, University of Pavia, Pavia, Italy **[285]**

Mitch Brenner, Department of Chemistry, University of California, Berkeley, CA 94720 **[227]**

T.C. Bruice, Department of Chemistry, University of California at Santa Barbara, Santa Barbara, CA 93106 **[369]***

G. Buse, Abteilung Physiologische Chemie, Rheinisch-Westfälische Technische Hochschule, 5100 Aachen, Federal Republic of Germany **[687]**

Samuel H.P. Chan, Department of Biology, Syracuse University, Syracuse, NY 13244 **[637]**

Sunney I. Chan, Department of Chemistry, California Institute of Technology, Pasadena, CA 91125 **[731]***

The numbers in brackets are the opening page numbers of the contributors' articles.
*Participants are indicated by an asterisk.

Karen Taylor Conklin, Department of Chemistry, University of Rochester, Rochester, NY 14627 **[387]**

Minor J. Coon, Department of Biological Chemistry, Medical School, University of Michigan, Ann Arbor, MI 48109-0606 **[497]***

Chris E. Cooper, Department of Biological Sciences, Brock University, St. Catharines, Ontario L2S 3A1, Canada **[637]**

M.A. Cusanovich, Department of Biochemistry, University of Arizona, Tucson, AZ 85721 **[401]***

Helen C. Davies, Department of Microbiology, University of Pennsylvania School of Medicine, Philadelphia, PA 19104-6076 **[619]***

E. De Boer, Laboratory of Biochemistry and Biotechnological Centre, University of Amsterdam, 1000 HD Amsterdam, The Netherlands **[477]**

James J. Driscoll, Department of Biochemistry, State University of New York at Buffalo, Buffalo, NY 14214 **[251]**

H. Brian Dunford, Department of Chemistry, University of Alberta, Edmonton, Alberta T6G 2G2, Canada*

Steven L. Edwards, Department of Chemistry, University of California/San Diego, La Jolla, CA 92093 **[463]***

Ronald W. Estabrook, Department of Biochemistry and the Cecil H. and Ida Green Center for Reproductive Biology Sciences, University of Texas Southwestern Medical Center, Dallas, TX 75235 **[525]***

Ole Farver, Department of Chemical Immunology, The Weizmann Institute of Science, Rehovot 76100, Israel **[269]**

Alessandro Finazzi-Agro, University of Rome "La Sapienza", Rome, Italy **[285]**

Kenneth M. Fish, Department of Chemistry, University of Michigan, Ann Arbor, MI 48109 **[509]**

Laurence A. Fishel, Department of Chemistry, University of California/San Diego, La Jolla, CA **[463]**

Jo A. Freedman, Department of Biological Sciences, Brock University, St. Catharines, Ontario L2S 3A1, Canada; present address: Department of Biology, Syracuse University, Syracuse, NY 13244 **[637]**

Yutaka Fujii, Hayaishi Bioinformation Transfer Project, Kyoto Laboratory, Research Development Corporation of Japan, Kyoto 601, Japan **[577]**

Eric A.E. Garber, Department of Biochemistry, Molecular Biology and Cellular Biology, Northwestern University, Evanston, IL 60208 **[749]***

Jeff Gelles, Department of Cell Biology and Physiology, Washington University Medical School, St. Louis, MO 63110 **[731]**

U.A. Germann, Biochemisches Institut, Universität Zürich, CH-8057 Zürich, Switzerland; present address: Laboratory of Molecular Biology, National Cancer Institute, National Institutes of Health, Bethesda, MD 20892 **[331]**

Michael H. Gold, Department of Chemical and Biological Sciences, Oregon Graduate Center, Beaverton, OR 97006*

John T. Groves, Department of Chemistry, Princeton University, Princeton, NJ 08544 **[509]***

Osamu Hayaishi, Hayaishi Bioinformation Transfer Project, Kyoto Laboratory, Research Development Corporation of Japan, Kyoto 601, Japan; present address: President Office, Osaka Medical College, Takatsuki-shi 569, Japan **[577]***

J.T. Hazzard, Department of Biochemistry, University of Arizona, Tucson, AZ 85721 **[401]***

H. Allen O. Hill, Inorganic Chemistry Laboratory, Oxford University, Oxford OX1 3QR, England **[419]**

Wim G.J. Hol, Laboratory of Chemical Physics, Department of Chemistry, University of Groningen, 9747 AG Groningen, The Netherlands **[291]**

Shigeo Horie, Department of Biochemistry, School of Medicine, Kitasato University, Kanagawa 228, Japan*

Robert Huber, Strukturforschung II, Max-Planck-Institut für Biochemie, D-8033 Martinsried, Federal Republic of Germany **[285]**

James K. Hurst, Department of Chemical and Biological Sciences, Oregon Graduate Center, Beaverton, OR 97006*

Fritz-Joachim Hüther, Fachbereich Chemie, Biochemie, Philipps-Universität, D-3550 Marburg, Federal Republic of Germany **[653]**

Tetsutaro Iizuka, Department of Biochemistry, School of Medicine, Keio University, Tokyo 160, Japan **[179]**

M. Imachi, Department of Chemistry, University of Michigan, Ann Arbor, MI 48109 **[509]**

Yuzuru Ishimura, Department of Biochemistry, School of Medicine, Keio University, Tokyo 160, Japan **[179]***

Dean P. Jones, Department of Biochemistry, Emory University School of Medicine, Atlanta, GA 30322 **[127]***

John E. Jones, Laboratory of Developmental Pharmacology, NICHHD, National Institutes of Health, Bethesda, MD 20892 **[557]**

J.A. Jongejan, Department of Microbiology and Enzymology, Delft University of Technology, 2628 BC Delft, The Netherlands*

Bernhard Kadenbach, Fachbereich Chemie, Biochemie, Philipps-Universität, D-3550 Marburg, Federal Republic of Germany **[653]***

Hiroyuki Kagamiyama, Department of Medical Chemistry, Osaka Medical College, Takatsuki 569, Japan **[577]**

Shiro Kanegasaki, Department of Cell Chemistry, Institute of Medical Science, University of Tokyo, Tokyo 108, Japan **[179]**

Kenneth D. Karlin, Department of Chemistry, State University of New York, Albany, NY 12222*

Tsoo E. King, Institute for Structural and Functional Studies, University City Science Center, Philadelphia, PA 19104 **[xxi,xxiii,3]***

Judith P. Klinman, Department of Chemistry, University of California, Berkeley, CA 94720 **[227]***

W.H. Koppenol, Department of Chemistry and Biodynamics Institute, Louisiana State University, Baton Rouge, LA 70803 **[93]***

Daniel J. Kosman, Department of Biochemistry, State University of New York at Buffalo, Buffalo, NY 14214 **[251]***

Joseph Kraut, Department of Chemistry, University of California/San Diego, La Jolla, CA **[463]**

B.E. Krenn, Laboratory of Biochemistry and Biotechnological Centre, University of Amsterdam, 1000 HD Amsterdam, The Netherlands **[477]**

Robert L. Kuczkowski, Department of Chemistry, University of Michigan, Ann Arbor, MI 48109 **[509]**

Seiki Kuramitsu, Department of Medical Chemistry, Osaka Medical College, Takatsuki 569, Japan **[577]**

Rudolf Ladenstein, Strukturforschung II, Max-Planck-Institut für Biochemie, D-8033 Martinsried, Federal Republic of Germany **[285]**

C.P. Lee, Department of Biochemistry, Wayne State University School of Medicine, Detroit, MI 48201*

Bryan Leece, Department of Biological Sciences, Brock University, St. Catharines, Ontario L2S 3A1, Canada **[637]**

K. Lerch, Biochemisches Institut der Universität Zürich, CH-8057 Zürich, Switzerland **[331]***

Joel S. Levine, Atmospheric Sciences Division, NASA Langley Research Center, Hampton, Virginia 23665 **[111]***

Peter Mark Li, Department of Chemistry, California Institute of Technology, Pasadena, CA 91125 **[731]**

Thomas M. Loehr, Department of Chemical and Biological Sciences, Oregon Graduate Center, Beaverton, OR 97006*

Bernd Ludwig, Institute of Biochemistry, Medical University of Lübeck, D-2400 Lübeck, Federal Republic of Germany **[605]***

Münime Lundeen, Atatürk Eğitim Fakültesi Kimya Bölümü, Marmara Üniversitesi, Istanbul, Turkey **[671]***

Thomas L. Luntz, Department of Biochemistry, Molecular Biology and Cellular Biology, Northwestern University, Evanston, IL 60208 **[749]**

Ed Magner, Department of Chemistry, University of Rochester, Rochester, NY 14627 **[387]**

Ryu Makino, Department of Biochemistry, School of Medicine, Keio University, Tokyo 160, Japan **[179]**

Thaddeus Mann, University of Cambridge, Cambridge CB4 2EE, England **[29]***

Agostino Marchesini, Instituto Nutrizione delle Piante, Torino, Italy **[285]**

E. Margoliash, Department of Biochemistry, Molecular Biology and Cell Biology, Northwestern University, Evanston, IL 60208 **[79,749]***

Craig T. Martin, Department of Molecular Biophysics and Biochemistry, Yale University, New Haven, CT 06520 **[731]**

Cheryl Martin-Wixtrom, Departments of Biochemistry, Obstetrics and Gynecology and the Cecil H. and Ida Green Center for Reproductive Biology, University of Texas Southwestern Medical Center, Dallas, TX 75235 **[525]**

Howard S. Mason, Department of Biochemistry, Oregon Health Sciences University, Portland, OR 97201 **[xxi,xxiii,xxv]***

J. Ian Mason, Departments of Biochemistry, Obstetrics and Gynecology, and the Cecil H. and Ida Green Center for Reproductive Biology Sciences, University of Texas Southwestern Medical Center, Dallas, TX 75235 **[525,543]**

Vincent Massey, Department of Biological Chemistry, University of Michigan Medical School, Ann Arbor, MI 48109-0606 **[147]***

J. Matthew Mauro, Department of Chemistry, University of California/San Diego, La Jolla, CA **[463]**

George McLendon, Department of Chemistry, University of Rochester, Rochester, NY 14627 **[387]***

Albrecht Messerschmidt, Strukturforschung II, Max-Planck-Institut für Biochemie, D-8033 Martinsried, Federal Republic of Germany **[285]***

T.E. Meyer, Department of Biochemistry, University of Arizona, Tucson, AZ 85721 **[401]**

Mark A. Miller, Department of Chemistry, University of California/San Diego, La Jolla, CA **[463]**

Anton O. Muijsers, Laboratory of Biochemistry, University of Amsterdam, 1000 HD Amsterdam, The Netherlands*

Michele Müller, Institut für Biochemie und Molekularbiologie, Universität Bern, CH-3012 Bern, Switzerland **[591]***

Ryo Nakajima, Biophysics Division, Research Institute of Applied Electricity, Hokkaido University, Sapporo 060, Japan **[451]**

Shigetada Nakanishi, Institute for Immunology, Kyoto University Faculty of Medicine, Kyoto 606, Japan **[577]**

Kazuhisa Nakayama, Institute for Immunology, Kyoto University Faculty of Medicine, Kyoto 606, Japan **[577]**

Daniel W. Nebert, Laboratory of Developmental Pharmacology, NICHHD, National Institutes of Health, Bethesda, MD 20892 **[557]***

Peter Nicholls, Department of Biological Sciences, Brock University, St. Catharines, Ontario L2S 3A1, Canada **[65,637]***

Thomas Nilsson, Department of Biochemistry and Biophysics, Chalmers Tekniska Högskola, S-412 96 Göteborg, Sweden **[731]**

H. Offenberg, Laboratory of Biochemistry and Biotechnological Centre, University of Amsterdam, 1000 HD Amsterdam, The Netherlands **[477]**

Hiroaki Ohkubo, Institute for Immunology, Kyoto University Faculty of Medicine, Kyoto 606, Japan **[577]**

John Owens, Laboratory of Mathematical Biology, NCI, Frederick Cancer Research Facility, Frederick, MD 21701 **[557]**

Gunnar Panskus, Institute of Biochemistry, Medical University of Lübeck, D-2400 Lübeck, Federal Republic of Germany **[605]**

Sergio Papa, Institute of Medical Biochemistry and Chemistry, University of Bari, 70124 Bari, Italy **[707]***

Israel Pecht, Department of Chemical Immunology, The Weizmann Institute of Science, Rehovot 76100, Israel **[269]***

Julian A. Peterson, Department of Biochemistry, University of Texas, Health Science Center, Dallas, TX 75235*

Raffaelo Petruzelli, University of Rome "La Sapienza", Rome, Italy **[285]**

H. Plat, Laboratory of Biochemistry and Biotechnological Centre, University of Amsterdam, 1000 HD Amsterdam, The Netherlands **[477]**

M.F. Powell, Institute of Pharmaceutical Sciences, Syntex Research, Palo Alto, CA 94304 **[369]**

Alvaro Puga, Laboratory of Developmental Pharmacology, NICHHD, National Institutes of Health, Bethesda, MD 20892 **[557]**

Achim Reimann, Fachbereich Chemie, Biochemie, Philipps-Universität, D-3550 Marburg, Federal Republic of Germany **[653]**

Jill Short Rogalsky, Department of Chemistry, University of Rochester, Rochester, NY 14627 **[387]**

Antonello Rossi, Strukturforschung II, Max-Planck-Institut für Biochemie, D-8033 Martinsried, Federal Republic of Germany **[285]**

Lars Rydén, Department of Biochemistry, Uppsala University, Biomedical Center, S-75123 Uppsala, Sweden **[349]***

Margareta Sahlin, Department of Molecular Biology, University of Stockholm, S-10691 Stockholm, Sweden*

Joann Sanders-Loehr, Department of Chemical and Biological Sciences, Oregon Graduate Center, Beaverton, Oregon 97006-1999 **[193]***

Beatrice Schläpfer, Institut für Biochemie und Molekularbiologie, Universität Bern, CH-3012 Bern, Switzerland **[591]**

Lawrence M. Schopfer, Department of Biological Chemistry, University of Michigan Medical School, Ann Arbor, MI 48109 **[147]**

Daniel Shih, Department of Biochemistry, Oregon Health Sciences University, Portland, OR 97201*

Larry Sieker, Department of Biological Structure, University of Washington School of Medicine, Seattle, WA 98195*

Evan R. Simpson, Departments of Biochemistry, Obstetrics and Gynecology and the Cecil H. and Ida Green Center for Reproductive Biology Sciences, University of Texas Health Sciences Center, Dallas, TX 75235 **[543]**

E.C. Slater, Department of Biochemistry, University of Southampton, Southampton SO9 3TU, England **[51]***

Lucile Smith, Department of Biochemistry, Dartmouth Medical School, Hanover, NH 03755 **[619]***

Edward I. Solomon, Department of Chemistry, Stanford University, Stanford, CA 94305 **[309]***

G.C.M. Steffens, Abteilung Physiologische Chemie, Rheinisch-Westfälische Technische Hochschule, 5100 Aachen, Federal Republic of Germany **[687]***

Peter Steinrücke, Institute of Biochemistry, Medical University of Lübeck, D-2400 Lübeck, Federal Republic of Germany **[605]**

Annemarie Stroh, Fachbereich Chemie, Biochemie, Philipps-Universität, D-3550 Marburg, Federal Republic of Germany **[653]**

Toshihiro Sugiyama, Department of Biochemistry, Osaka University Medical School, Osaka 530, Japan*

Torahiko Tanaka, Department of Biochemistry, School of Medicine, Keio University, Tokyo 160, Japan **[179]**

G. Tollin, Department of Biochemistry, University of Arizona, Tucson, AZ 85721 **[401]**

Olivier Toussaint, Laboratoire de Recherches Organiques, ESPCI, 75231 Paris Cedex 05, France*

C.L. Tsou, Laboratory of Molecular Enzymology, Institute of Biophysics, Academia Sinica, Beijing 100080, China **[85]***

Alfin D.N. Vaz, Department of Biological Chemistry, Medical School, University of Michigan, Ann Arbor, MI 48109-0606 **[497]**

Anne Volbeda, Laboratory of Chemical Physics, Department of Chemistry, University of Groningen, 9747 AG Groningen, The Netherlands **[291]***

Jimin Wang, Department of Chemistry, University of California/San Diego, La Jolla, CA **[463]**

Kikuko Watanabe, Hayaishi Bioinformation Transfer Project, Kyoto Laboratory, Research Development Corporation of Japan, Kyoto 601, Japan **[577]**

Michael R. Waterman, Department of Biochemistry and the Cecil H. and Ida Green Center for Reproductive Biology Sciences, University of Texas Southwestern Medical Center, Dallas, TX 75235 **[525,543]***

R. Wever, Laboratory of Biochemistry and Biotechnological Centre, University of Amsterdam, 1000 HD Amsterdam, The Netherlands **[477]***

Nguyen huu Xuong, Department of Chemistry, University of California/San Diego, La Jolla, CA **[463]**

Isao Yamazaki, Biophysics Division, Research Institute of Applied Electricity, Hokkaido University, Sapporo 060, Japan **[451]***

Takashi Yonetani, Department of Biochemistry and Biophysics, University of Pennsylvania School of Medicine, Philadelphia, PA 19104-6089 **[437]***

Mauricio Zuber, Departments of Biochemistry, Obstetrics and Gynecology, and the Cecil H. and Ida Green Center for Reproductive Biology Sciences, University of Texas Southwestern Medical Center, Dallas, TX 75235; present address: Howard Hughes Medical Institute, Massachusetts General Hospital, Boston, MA 02114 **[525]**

Preface

This volume contains the collected papers and discussion of the 4th International Symposium on Oxidases and Related Redox Systems, held in Portland, Oregon, October 4–8, 1987, at the Oregon Health Sciences University.

The field of oxygen biochemistry has given rise to a number of daughter fields, each with great vitality but all somewhat isolated from one another by the constraints of the disciplines involved. The goal of the 4th International Symposium on Oxidases and Related Redox Systems (informally known as "ISOX IV") was to attempt a synthesis of the biochemistry of oxygen from its various component aspects, chemistry, enzymology, molecular biology, molecular physiology, historical geology, and evolution. Because the number of participants had to be limited for economic reasons, important aspects of oxygen biology, such as model systems, oxygen as a regulator and inducer of gene expression, oxygen toxicity in its many aspects, and structure/function of the mitochondrial respiratory chain, could not be represented properly. The synthesis we aimed at by requesting participants to take a broad comparative point of view in their papers, was only a partial synthesis, but many of the major frontiers of oxygen biochemistry are represented.

As a result of a suggestion by Professor Peter Nicholls of Brock University, St. Catharines, Ontario, the centennial of the birth of Professor David Keilin was celebrated at ISOX IV, in the belief that it was an appropriate context. Professor Keilin made major discoveries that contributed to the founding of the modern fields of respiratory biology and bioenergetics. A number of former associates from abroad and from the United States gave perspectives and reminiscences about Professor Keilin and the Molteno Institute, Cambridge University, which he headed. These detailed and affectionate papers comprise the first section of this volume.

The second section was organized in a format ascribed to the Faraday Society, a style employed by the preceding three ISOX symposia. Preprints of the symposium papers were collected and duplicated. One month before the meeting they were distributed with the hope of catalyzing highly informed discussion among ISOX IV participants. Those discussion sessions were recorded at the conference. They have been edited and are included here as part of the conference proceedings.

The proceedings start with a consideration of thermodynamic properties of oxygen and the evolution of the contemporary atmosphere. The subsequent papers focus on shared and unique properties of iron-, copper-, -heme, and flavoprotein oxidases, and peroxidases. All of these compounds are considered with respect to three-dimensional structures, novel groupings of prosthetic atoms, transfer of reducing equivalents among redox domains, and in general, relationships between structures, mechanisms, and functions and their regulation among the oxygen- and peroxide-activating proteins.

Tsoo E. King
Howard S. Mason
Martin Morrison

Acknowledgments

We acknowledge with gratitude the support ISOX IV received from the following organizations: International Union of Biochemistry; International Union of Pure and Applied Biophysics; Medical Research Foundation of Oregon; National Institute of Arthritis, Diabetes, and Digestive Diseases; National Institute of General Medicine Sciences; Oregon Health Sciences University (OHSU); School of Medicine of the Oregon Health Sciences University; Sommer Fund; Bristol Myers Company; E. I. du Pont de Nemours & Company; and Lilly Research Laboratories.

We owe the following individuals heartfelt thanks for making ISOX IV possible: former President Leonard Laster of OHSU, for unflagging confidence in the meeting and for the kind hospitality he and his wife extended to Professor Keilin's former associates, Dean John Kendall of the School of Medicine, and Professor Richard Jones, Chairman of the Department of Biochemistry, for their essential advice and support; and Consul General Shusaku Tanaka of the Japanese Consulate for extending hospitality to the Japanese participants.

We want to express our gratitude to the Molteno Family participants for making the Keilin Centennial celebration such a special occasion, and for chairing the scientific sessions: Ronald Estabrook, Tsoo King, Thaddeus Mann, Emanuel Margoliash, Peter Nicholls, E.C. Slater, Lucile Smith, and Chen-Lu Tsou. We thank Professor Mann for the photograph of Professor Keilin. Coincidentally, this was the centennial year of the Oregon Health Sciences University School of Medicine.

We thank especially the staff that ran this meeting with great competence: Pat Stange, who managed the meeting itself and was an irreplaceable associate during the organization of the symposium and the editing of the proceedings, deserves special commendation. Caryn Ruch, University Communications, made all of many interactions with University resources run very smoothly. The two are perfectionists. Stephen Gepford operated the audiovisual equipment with high professionalism. We also thank Margaret Mason (women's program); Joyce Morrison; and Mary Ann Asson-Batres, Robin Hall, and Dr. Sarah Barkowski of the Department of Biochemistry, OHSU; Marlene Jacks (Memphis), and Mary Kravitz (Albany) for the valuable contributions.

Tsoo E. King
Howard S. Mason
Martin Morrison

In Memoriam
DR. MARTIN MORRISON

Our friend and ISOX co-organizer, Martin Morrison, died of cancer in his home on August 22, 1987, in Memphis, Tennessee. He was a very good friend, a fine scientist, and an invaluable administrator and co-editor for the ISOX series of meetings. This brief notice is an appreciation and farewell.

Martin (Marty) Morrison was born (1921) and raised in Detroit, Michigan. During World War II he was a 2nd lieutenant in the 8th Army Air Corps but was wounded and so came home early, with decorations and the Purple Heart. He graduated with a B.S. in chemistry from the University of Michigan (1947) and with a Ph.D. in biochemistry from Wayne State University with J.M. Orten (1952). He then joined the Department of Biochemistry at the University of Rochester whose chairman, Professor Elmer Stotz, was an authority on cytochromes and biological oxidations. Morrison stayed at Rochester until 1961, when he took a sabbatical with Professor Keilin in Cambridge. He returned to the States in 1962 to become Head of the Section of Respiratory Enzymology at the City of Hope Medical Center in Duarte, California. Finally, in 1967, he became Chairman of the Department of Biochemistry at St. Jude Children's Research Hospital in Memphis, and affiliated Professor of Biochemistry at the University of Tennessee Center for Health Sciences, where he was until his death.

During the Rochester years, Morrison focused his research on hemoprotein structure-function, later on biological roles of mammalian peroxidases, and on protein arrangement relative to function in normal and defective membranes. In the 34 years of his professional career he published more than 145 research papers, book chapters, and reviews by himself and with a large number of gifted students and co-workers.* He was Ron Estabrook's first research tutor; they published on the purification of cytochrome c in 1954.

Morrison began his studies with purification and characterization of hemoproteins and iron-porphyrin prosthetic groups, especially those of cytochrome c, cytochrome c oxidase, hemoglobins, and peroxidases from many tissues and species. One outcome of the work on lactoperoxidase was the discovery that it catalyzes iodination of exposed tyrosyl and histidyl groups of proteins. He and his co-workers developed this into an important probe for exposed proteins on either surface of membranes. Many membranes and their functional protein components were studied in this way after the appearance of the pioneering paper by Phillips and Morrison (1971) on disposition of human red cell membrane proteins. As a result, Morrison's work and that of others in this field greatly enriched the concept of fluid mosaic membranes, functions of which are regulated by specific protein placement in phospholipid bilayers. This concept and its molecular biological extension is the foundation of the modern science of biological membranes. Morrison sought to elucidate membranal defects in disease states by comparing the membranes, for example, of normal and cancer cells, or of young and old cells, seeking correlations with functional changes. His last paper (in press, 1988) with T.J. Mueller and C.W. Jackson, is on "Membrane skeletal alterations during in vivo mouse red cell aging: increase in the band 4.1a:4.1b ratio." Martin was a member of the editorial boards of *Membrane Biochemistry* and *Archives of Biochemistry and Biophysics.*

His wife, Joyce (Pollard) and their young family accompanied him on his sabbatical leave in 1961, and they fell in love with Cambridge and all. Marty recently recalled with pleasure the wide-ranging and comfortable after-hours conversations with the Professor that other old boys mention in their Centennial papers in this volume.

*Among them, Ernest Beutler, Stanley Cohen, Jerald Connelly, Dean Dawson, Alexander Dounce, Brian Dunford, Ron Estabrook, Shiela Ferguson-Miller, Victor Fried, Tom Hollocher, Shigeo Horie, Toichiro Hosoya, C.W. Jackson, Emanuel Margoliash, G.V. Marinetti, Kenneth Monty, T.J. Mueller, Susumu Ohno, Dave Phillips, George Rouser, Greg Schonbaum, Bill Schroeder, Isao Yamazaki, and (last and least) H.S. Mason.

As a person, Marty managed to disconnect himself from the over-intensity and over-competitiveness of present-day biochemical research, yet temperamentally he was a very hard and productive worker to whom research was a great adventure. He left a charming and affectionate family, the members of which gathered protectively and lovingly around him at home during his last illness. Marty was a happy, warm-hearted man, as his frequent deep, very hearty laughter and brightness of spirit showed, even, remarkably, during his illness, but certainly then to comfort his family and friends. A familiar poem by Byron sums it up:

So we'll go no more a-roving
So late into the night
Though the heart will be as loving
And the moon be still as bright.

For the sword outwears the sheath
And the soul outwears the breast
And the heart must pause to breathe
And love itself have rest.

H. S. Mason
January 12, 1988

PART ONE
David Keilin Centennial Session

Chairman: Tsoo E. King

David Keilin
(March 21, 1887—February 27, 1963)

PART ONE
David Keilin Centennial Session

Chairman: Tsoo E. King

David Keilin
(March 21, 1887—February 27, 1963)

Oxidases and Related Redox Systems, pages 3–27

INTRODUCTION -- MY PERSONAL ENCOMIUM TO PROFESSOR DAVID KEILIN

Tsoo E. King

Institute for Structural and Functional Studies
University City Science Center
3401 Market Street
Philadelphia, PA 19104

First, to express a token of my gratefulness to the Professor, I must say something here about what I learned from him. But, because of my poor command of the English language, I'll not be able to express how the wisdom received from the Professor has benefited and etched my mind so deeply. When my first series of experiments in the Molteno was completed, the Professor and I had a chance to talk more, always at the end of the day under semidarkness in the library among stacks of bound journals. At first, these talks were mostly on drafts of a preliminary report; after many revisions the final paper was published in Nature.

Later, our evening talk turned more to conceptual or even philosophical aspects of reconstitution, biology, and natural phenomena, which led to conversations even in philosophy itself not just those related to science. These discussions were quite diffuse without focus. Naturally, we discussed Herman von Hemholtz, William Harvey, and even from Jean Jacques Rousseau, George Wilhelm Friedrich Hegel, and others up to contemporaries. Soon, I found out that Henri Bergson, a well-known philosopher held a professorship at College de France, a very prestigious post, taught Professor Keilin. Biology lured Professor Keilin, away from philosophy. His tutelage under Maurice Caullery at Sorbonne eventually made him one of the few most innovative and successful biologists in the century. Sorbonne awarded Professor a

D.Sc. degree. He went to Cambridge in 1915 upon invitation by G. H. F. Nutall, then the Quick Professor, prompted by Stanley Gardiner, a powerful professor of zoology, worked first as a "studenship" and evenutally succeeded Nutall as the director of Molteno in 1931.

I consider these conversations quite critical, because they have subsequently exerted profound influence on the ways and tactics in which I attacked scientific problems. In short, I should record my direct feeling here. The professor in the most general form was affected by the thinking of Rousseau and von Helmoltz, but perhaps subconsciously to whatever extent also by the Hegelian thought. Many people may vigorously disagree with me. But is is my true feeling regardless of how much the Professor agreed with Henri Bergson. Just as Jacques Derrida commented as shown in the first transparency. "In its result, Bergson's view is in accord with Hegel's thesis that space "is" time, in spite of the very different reasons they have given. Bergson merely says the reverse: that time (temps) is space. Bergson's view of time too has obviously arisen from an Interpretation of the Aristotelian essay on time. That a treatise of Bergson with the title "Quid Aristoteles de loco senserit" should have appeared at the same time as his "Essai sur les donnes immediates de la conscience" ...J. Derrida.

In my view, the Professor started his work, very approximately to say, by the strategy of structural and then functional studies as reflected in most of his publications including those not directly on cytochrome. But his intuition and "mental image" (a technical term colloquially, approximates to "prepared mind") played a certain role. To me, this aspect precisely embodies the Hegelian idea as Transparency 2 shows. "It [this intuition] is an image, which has received [in sich empfangen hat] as its soul and meaning an "independent" mental representation, its "signification" (Bedeutung). This intuition is the "Sign"." ...G. W. F. Hagel (translated by Alan Bass).

In his first paper of 1925 to then actual isolation of cytochrome <u>c</u> from yeast, and later with Ted Hartree, from mammalian heart, already showed this trend. Professor's methods were most aptly applied in the extremely careful digestion with a suspicious eye toward his own results. To use a cliche, he never jumped to conclusions without verification by all feasible means available to him. Subsequently, by comparison with those reported by others, he would then either unreservedly praise or brutally criticize them, but never was perfunctory. The book review on Otto Warburg's Heavy Metal epitomized his seriousness (cf. the bibliography appended at the end). Some comments were even spiced with a faint sarcasm. Such as Professor once wrote, as shown in Transparency 3. "...the statement made by [certain investigators] that the first indication of the flavoprotein nature of succinic dehydrogenase is found in a paper by Keilin and Hartree (1939) is too generous since no definite suggestion was made in this paper that the faint absorption bands at 495 and 455 mμ in the heart muscle preparation belong to this enzyme [succinic dehydrogenase]...[these bands might be actually] originated from pigments, flavin and carotenoid, present in the retina...of the observer himself...".

Take the first paper for example. He wrote, as transparency 4 "Thus...it is important to mention beforehand that cytochrome (= myohaematin = histohaematin) is a pigment distinct both from blood haemoglobin and muscle haemoglobin (=myochrome of Morner = myoglobin of Gunther) or their derivatives. In many cells cytochrome may, however, coexist with heamoglobin". This first paper was reported before the Royal Society when he was a Beit Memorial Fellow at the Molteno in 1925. Professor always observed four bands, in those scores, eighty-six to be exact, of different species he worked. The wave lengths varied slightly in different organisms. Since he felt these ubiquitous occurrences of such phenomena, why were no previous reports noted--he asked himself. After exhaustive search of the literature, he found that nearly 40 years prior to his observation work was done by Charles MacMunn who noted very similar compounds and coined the words myohaematin (in muscle) and histohaematin (in other tissues).

The term, cytochrome, was proposed by the Professor, as Transparency 5 says "...I have found that the pigment myo- or histohaematin not only exists but has much wider distribution...there is no evidence that is is simple haematin in the proper sense of the term...[myohaematin and histohaematin] are misleading... I propose therefore to describe it under the name cytochrome signifying merely cellular pigment...". Different cytochromes were denoted as components. More significant experimental design included demonstration of the oxidation and reduction of cytochrome in vivo, as beautifully shown by aeration and exhaustion of oxygen of a yeast suspension. Even more brilliantly, in a very ingenous experiment in which a live moth was gummed on a microscopic slide, the moth would intermittently struggle to fly off and then pause for rest periods. The professor examined the insect's wing muscle to note the band change in activity and at rest. His use of narcotics and cyanide, reagents first introduced by Otto Warburg while studying biological oxidation, cleverly enabled his seminal formulation of the first model of the respiratory chain concept. The successive transport of reducing equivalent from substrate to oxygen.

In my opinion, the respiratory chain concept is more valuable than the discovery of cytochrome, although these two facets are so intimately interrelated. Incidentally, his contribution was richly rewarded by election to the Royal Society in about three years after his first paper. The concept settled the anguish and bitter dispute between Otto Warburg and Hendrich Wieland.

The respiratory chain was eventually established by the Professor in 1931 as shown in Transparency 6: "substrate → dehydrogenases → cytochrome oxidase → O_2. Narcotics acted on the dehydrogenase and cyanide on oxidase. After discovery in 1938 and 1939, of cytochrome $\underline{a}$ and $\underline{a}_3$, the oxidase was actually cytochrome $\underline{a}_3$ identical with Warburg's Atmungferment. At that time, nonheme iron proteins and Q cycle/Q-proteins were not known. But no modern version of the respiratory chain has been changed in principle since then. Peter Nicholls succinctly summarized its evolution up to the time of his article in The Enzymes edited by P.D. Boyer in 1963. What I want to emphasize is that without Professor's enunciation of the respiratory chain concept, bioenergetics could never have reached its present stage in this 60 years or so. Without the respiratory chain concept Warburg's Atmungferment was just another one of many hundreds enzymes and his theory, I dare say, would remain as a theory without true evidence of biological importance.

Inevitably, in the library, we quite often talked about some history of the Molteno. He was very proud of Robin Hill's discovery of the Hill reaction first demonstrated in 1939. Indeed Robin worked independently on heme compounds before his formal collaboration with

Professor and they jointly published papers in the Proceedings and other journals. Other discoveries included diaphorase by F. Bruno Straub of Szeged from Albert Szent-Gyorgyi's lab. I vividly remember the professor was not surprised but instead was quite satisfied when I told him that Vincent Massey just reported at the Biochemical Society meeting in the next building that diaphorase actually was lipoate dehydrogenase. Many others at that time used diaphorase (at least the name) as the dehydrogenase for the other branch of the respiratory chain parallel to succinate dehydrogenase. It is too bad that the discovery of superoxide dismutase was indeed the same protein he and Thaddeus Mann isolated, purified, and crystallized in the nineteen thirties was made after the death of the Professor.

He read widely in many subjects. His mastery of linguistics together with his early training, especially in the Forski School in Warsaw and Paris, gave him special advantage. Nevertheless, the Professor considered it critical to know the historical development and perspective of any discipline of science; no one could become truly successful without this knowledge together with curiosity and thinking. Regarding the latter, one could find him sitting in a chair, eyes closed, not resting, but in a deep meditation. On the other hand, the professor somewhat looked down on "professional" scientific historians. In his book published posthumously where some of the edges were polished off, one could still find such passages as Transparency 7 "...those [errors] concerning historical development are perpetuated through sheer laziness and may remain unchallenged for more than a century. Such mistakes are very seldom corrected by the scientific historian...". In contrast it is ironic that nowadays at least biochemists rarely read papers, say, 10 years old. Those conversations gave me not only the insight of some great discoveries in the history of biochemistry but immeasurable knowledge which could never be obtained by just reading the publications. I truly relish and have reflected on these dialogues over the past many years and I am sure I will continue in the future especially when I am depressed. Because of my English, I was and am never able to convey my real joy and the great benefits I obtained from this period in spite of staying only a brief time by the Molteno standard.

A PRELIMINARY COMPILATION OF SOME SELECTED
BIOCHEMICAL PUBLICATIONS OF PROFESSOR DAVID KEILIN

A. The Respiratory Chain, its Components and Related

A- 1. On cytochrome, a respiratory pigment, common to animals, yeast, and higher plants (1925) Proc. Soc. B 98, 312 -339.

Modern bioenergetics is born!

A- 2. A comparative study of turacin and haematin and its bearing on cytochrome (1926) Proc. Roy. Soc. 100, 129-151.

A- 3. The influence of carbon monoxide and light on indophenol oxidase of yeast cells (1927) Nature, Lond. 119, 670-671.

Experiments show in yeast that indophenol oxidase or polyphenol oxidase displays characteristics of Warburg's respiratory ferment (Atmungferment) and is essential for the oxygen consumption by the living cell.

A- 4. Le Cytochrome pigment respiratoire intracellulaire commun aux micro-organisms, aux plantes et aux animaux (1927) C.R.
Seanc Soc. Biol. Paris (Reunion Pleniere), 96 S.P. 39-68.

This paper with questions and answers presented before the Society of Biology in Paris, analyzes the detail of intracellular respiration. Cytochrome is oxidized by indo-phenol oxidase and reduced by reducing substances. The results also afford to explain the effects of narcotics and cyanide on oxidation and reduction of cytochrome. It consists also of 80 references and discussions.

A- 5. Cytochrome and respiratory enzymes (1929) Proc. Roy. Soc. B 104, 206-252.

This is the first paper published (submitted in November 21, 1928) in the Proceedings since his election to the Royal Society, approximately three years after the publication of the first cytochrome paper. One of the important findings is on the results of insoluble "true" indophenol oxidase in muscle. It is inhibited by cyanide, sulfide, CO in dark. Under the light CO-oxidase dissociates and the active enzyme is liberated liberated. The thermo liability of the respiratory chain is also found.

A- 6. Cytochrome and intracellular oxidase (1930) Proc. Roy. Soc. B. 106, 418-444.

Extracted cytochrome c from yeast cells has been studied in rather detail. Cytochrome c, itself, is shown to be not autoxidizable and inactive for the oxidation of cystein. In the presence of indophenol oxidase (actually is a kind of modern version of the Keilin-Hartree preparation from sheep heart), c promotes the cystein oxidation. For oxidase in the presence of c for cystein oxidation, partition constant of about 8 is changed to about 63 under strong light. Once again it has been confirmed that narcotics do not act on the oxidase system.

*Originally, hematin, hemochromogen, etc., is spelled haematin, haemochromogen, etc various cytochromes as components, and nm as m .

A- 7. The porphyrin of component c of cytochrome and its

relationship to other porphyrins (1930) Proc. Roy. Soc. B 107, 282-292. (With R. Hill)

Cytochrome c yields a hematoporphyrin by its reaction of HBr7 and acetic acid. This is the first report for preparation of the protoporphyrin (called as "unmodified porphyrin c") from cytochrome c. Protoporphyrin is obtained by the action of hydrochloric acid and a reducing agent such as SO_2.

A- 8. The absorption spectrum of the component c of cytochrome (1931) Proc. Roy. Soc. B 109, 29-34. (With M. Dixon and R. Hill)

The first absorption spectra are presented for other hematin-free ferri- and ferro-cytochrome c which is prepared from Delft yeast. The ferrocytochrome c shows alpha maximum at 550, beta-peak at 520 and Soret at 418 nm.

A- 10. Discussion on recent advances in the study of enzymes and their action (1932) Proc. Roy. Soc. B 111, 280-297.A- 11. Cytochrome and intracellular respiratory enzymes (1933) Ergebnisse der Enzymforschung 2, 239-271.

A- 13. Supposed direct spectroscopic observation of the oxygen-transporting ferment (1933) Nature, Lond. 132, 783.

A- 15. Estimation of haematin iron and the oxidation-reduction equivalent of cytochrome c. (1933) Proc. Roy. Soc. B. 114, 104-109. (With R. Hill).

A- 16. Cytochrome and the supposed direct spectroscopic observation of oxidase (1934) Nature, Lond. 133, 290-291.

A- 20. The action of sodium azide on cellular respiration and on some catalytic oxidation reactions (1936) Proc. Roy. Soc. B 121, 165-173.

A- 21. Le mecanisme de la respiration intracellulaire (1936) Bull. Soc. Chim. Biol. 18, 96-130.

A- 23. Biological oxidations and reductions (1936) Ann. Rev.

. 5, 1-24. (With D. E. Green)

A- 27. Structure chimiuque et propriete de quelques ferments (ou diastases) intracellulaires (1937) Congres du Palais de la Decouverte, Paris, pp.357-363.

A- 29. Preparation of pure cytochrome c from heart muscle and some of its properties (1937) Proc. Roy. Soc. B 122, 298-308. (With E. F. Hartree)

Trichloroacetic acid, 0.15 N, is used to extract heart mince. After a simple procedure of purification, cytochrome c with 0.34% iron is obtained with 1.65 and 2.64 g from 10 kg of ox and horse heart respectively The method is much simpler than reported by Thorell in 1935 and 1936. Some properties and rudimental amino acid analysis are also presented.

A- 31. Cytochrome a and cytochrome oxidase (1938) Nature, Lond. 141, 870-8871. (With E. F. Hartree)

A- 33. Cytochrome oxidase (1938) Proc. Roy. Soc. B 125, 171-186. (With E. F. Hartree)

This is the first full-length paper to equate indophenol oxidase as cytochrome oxidase and a preface on more definitive enouncement of cytochrome (or component) a and a_3. It has examined its catalytic activity with relation to cytochrome. Not infrequently but straight forwardly the paper has emphasized, "although a great deal of work has been devoted to the study of this enzyme, the literature on the subject still contains a great bulk of inaccurate and controversial statements ... [they are] naturally responsible for the introduction of several unsound theories...." Indeed all the proposals by the schools such as Remeson, Bigwood, Harrison and Shibata have been rejected.

A- 40. Cytochrome and cytochrome oxidase (1939) Proc. Roy. Soc. B 127, 167-191. (With E. F. Hartree)

Numerous experiments show that cytochrome a is not autoxidizable while a_3 is. a^{2+} and a_3^{2+}

show alpha maxima at 605 and 600 nm whereas Soret peaks at 452 and 448 nm, respectively. The alpha bands are rather not sharp but slightly diffuse. Bands of a and a_3 are fused together. The heme groups of a and a_3 are identical. Cytochrome b exhibits maxima at 564, 530 and 432 nm in the reduced form. The now well known Keilin-Hartree preparation (HMP) has been used for identification of the existence of cytochromesa, a_3, b, and c in addition to succinate dehydrogenase mainly by studies with inhibitors and spectroscopy. The identity of a_3 and cytochrome oxidase is proved. a_3^{2+} can react with O_2, CO; whereas a_3^{3+} can react with H_2S, azide, hydroxlamine, but not CO. Cytochrome oxidase or a_3 is identical with the Warburg's respiratory ferment. However, very slow or practically no reduction of a_3^{3+} in usual HMP or in sucrose can be demonstrated by c^{2+} under strictly anaerobic condition. Possibility of cytochrome oxidase as a copper protein is indicated; actually copper is found in HMP to be 0.0129%.

A- 44. Cytochrome system in Bacterium coli commune. (1940) Biochem. J. 35, 688-692. (With C. H. Harpley)

A- 45. Properties of cytochrome c (1940) Nature, Lond. 145, 934. (With E. F. Hartree)

A- 47. Succinic dehydrogenase-cytochrome system of cells. Intracellular respiratory system catalyzing aerobic oxidation of succinic acid (1940) Proc. Roy. Soc. B 129, 277-306. (With E. F. Hartree). The Keilin-Hartree preparation (HMP) actually is one of the best has been traced to the formation of oxalacetate, a potent inhibitor of SDH, through residual fumarase and malate dehydrogenase in HMP in the oxidation of succinate.

A- 53. A new soluble cytochrome component from yeast (1942) Nature, Lond. 149, 21. (With S. J. Bach and M. Dixon)

A- 66. Comparative study of spores and vegetative forms of Bacillus subtilis (1947) Antonie van Leeuwenhoek 12, 115-128. Jubilee Volume, Albert J. Kluyver.

(With E. F. Hartree).

A- 67. Activity of the cytochrome system in heart muscle preparations (1947) Biochem. J. 41, 500-502. (With E. F. Hartree)

The heart muscle preparation (HMP) contains all intracellular respiratory components. Their experimental results prompt them to reject the claim made by Hess for splitting of cytochrome oxidase into two components and actually the Hess method of preparation of cytochrome oxidase is not suitable because the enzyme is much less active than HMP.

A- 68. Effect of haematin on the oxidation of succinic acid by tissue preparations (1947) Biochem. J. 41, 503-506. (With E. F. Hartree)

Free hematin or proto-, hemato-, and deuterohematin inhibits succinate oxidation by O_2 as well as anaerobically by methylene blue. It does not affect the cytochrome oxidase system.

A- 73. Activity of the succinic dehydrogenase-cytochrome system in different tissue preparations (1949) Biochem. J. 44, 205-218. (With E. F. Hartree)

A succinate oxidase chain is formulated which is practically the same as the modern linear form.

A- 74. Cytochrome c (1949) Biochemical Preparations 2, 1-4. (With E. F. Hartree)

A- 75. Effect of low temperature on the absorption spectra of haemoproteins; with observations on the absorption spectrum of oxygen (1949) Nature, Lond. 164, 254-259. (With E. F. Hartree)

A new cytochrome, called e, with an alpha-absorption maximum of about 552 nm at liquid air temperature is observed (see No. A-96 for renaming it as c_1 to conform the prior observation in 1941 by the Japanese workers.)

A- 76. Metal catalysis in cellular metabolism (1950) Nature, Lond. 165, 4-5.

This is a short (less than 2 pages of 140x245 mm) book review of Otto Warburg's Schwermetalle als Wirkungs gruppen von Fermenten, 1949 (pP 195 + 4 plates. Freibrug i. Br.; Editio Cantor G.m.b.H.). It is an objective, courteous but critical review. The topics are dealt with respect to Warburg's view of Cl. Bernard's "venous blood," charcoal iron models ("The reader ... must be reminded that charcoals as they are prepared with hematin compounds from blood or from hemin have no more in common with hematin compounds than charcoal--iron prepared from Bismarck brown or safranin"), (italics added) [misinterpretation of] MacMunn's work by using other people's quotations, the dialogues between Pasteur and Berthelot, among others. Professor Keilin has expressed: "It is the author's [Warburg] approach to the development of our knowledge in this subject ... in this respect the book is disappointing ... the author revives his old controversies, applying to them all the vigour and skill of an experienced polemist. However, once this attitude is adopted, the desire to score a few points over his supposed opponents prevents the author from appreciating fully the real significance of their work."

A- 88. Cytochrome oxidase and the 'Pasteur enzyme' (1953) Nature, Lond. 171, 413. (With E. F. Hartree)

A- 90. Cytochrome (1953) Brit. Med. Bull. 9, 89-97. (With E. C. Slater)

A review contains concise but succinct evaluation of historic development, cytochrome(s), the respiratory chain and very brief disclosures of oxidative phosphorylation and photosynthesis. Also was formally presented a respiratory chain with two branches, one from succinate and other, NADH, which converge at the "factor."

A- 96. Relationship between certain components of the cytochrome system (1955) Nature, Lond. 176, 200-206. (With E. F. Hartree)

"Since cytochromes $\underline{c}_1$ and $\underline{e}$ appear to be

Identical, the term cytochrome e, proposed in 1949 [by the authors], should be replaced by cytochrome c_1, which was introduced in 1940," [by Yakaushiji and Okunuki], the authors have characteristically stated. The instrument used for measurements is a Hilger Uvispec spectrophotometer under special arrangement, to minimize a slight effect of turbidity, for determination of the

heart muscle preparation in 20% sucrose so that a band width equivalent to 0.5 nm is achieved at 550 nm.

A- 99. Effect of cyanide on succinic dehydrogenase (1958) Biochem. J. 69, 32P. (With Tsoo E. King)

A-100. Reconstitution of the succinic oxidase system from soluble succinic dehydrogenase and a particulate cytochrome system preparation (1958) Nature, Lond. 181, 1520-1522. (With Tsoo E. King)

A soluble succinate dehydrogenase (SDH) prepared with a butanol method without cyanide preincubation, which is very critical, structurally recombining with a particulate system containing all respiratory components except SDH. Succinate oxidase is thus reconstituted which is identical with all properties tested of the original oxidase.

A-107. Effect of inhibitors on the activity of soluble succinic dehydrogenase and on the reconstitution of the succinic dehydrogenase-cytochrome system from its components (1960) Proc. Roy. Soc. B 152, 163-187. (With Tsoo E. King)

A-111. Absorption spectra and some other properties of cytochrome c and of its compounds with ligands (1962) Proc. Roy. Soc. B, 156, 429-458. (With W. D. Butt)

B. Hemoglobin, Myoglobin and their Derivatives

B- 12. On the combination of methaemoglobin with H_2S (1933) Proc. Roy. Soc. B 113, 393-404.

Sulfo-hemoglobin (Hb) or sulfo-methamoglobin with an absorption maximum of 618 nm, which has been described by several investigators since middle of last century, by reaction of H_2S with HbO_2 in oxygen, is proved incapable to be formed by a simple action between H_2S and hemoglobin. On the other hand, H_2S-methemoglobin, formed from H_2S and methemoglobin, shows diffuse faint -bands at about 578 nm and stronger-band at 545 nm.

B- 19. The combination between methaemoglobin and peroxides: hydrogen peroxide and ethyl hydroperoxide (1935) Proc. Roy. Soc. B. 117, 1-15. (With E. F. Hartree)

B- 20. The action of sodium azide on cellular respiration and on some catalytic oxidation reactions (1936) Proc. Roy. Soc. B 121, 165-173.

The brown solution of acid methaemogen on addition of NaN_3 turned red and its absorption spectrum is replaced by two bands at 575 nm and 542.5 nm. It required one molecule of azide per atom of iron of methaemoglobin. Like KCN, H_2S, sodium azide inhibited the activities of indophenol oxidase, catechol oxidase, catalase and peroxidase.

B- 27. Structure chimique et propriete de quelques ferments (ou diastases) intracellulaires (1937) Congres du Palais de la Decouverte, Paris, pp. 357-363.

B- 28. Reaction of nitric oxide with haemoglobin and methaemoglobin (1937) Nature, Lond. 139, 548. (With E. F. Hartree)

B- 51. Activity of carbonic anahydrase within red blood corpuscles (1941) Nature, Lond. 148, 493-496. (With T. Mann)

B- 52. Absorption spectra of haemoglobin in solution and in red blood corpuscles (1941) Nature, Lond. 148, 75-77.

B- 56. Reactions of haemoglobin and its derivatives with phenyl hydroxylamine and nitrosobenzene (1943) Nature, Lond. 151, 390-391. (With P. Tate)

B- 62. Haemoglobin in the root nodules of leguminous plants (1945) Nature, Lond. 155, 227-229. (With Y. L. Wang)

This is first time in finding hemoglobin in (the root nodules of leguminous) plants.

B- 63. Spectroscopic study of the permeability and lysis of red blood corpuscles (1946) Nature, Lond. 157, 210-213. (With E. F. Hartree)

B- 65. Haemoglobin of Gastrophilus larvae. Purification and properties (1946) Biochem. J. 40, 855-866. (With Y. L. Wang)

A hemoglobin from larvae of a parasite that lives in the horse stomach has been isolated and crystallized.

It differs from the horse hemoglobin in several aspects such as low affinity with CO, position of absorption band, the number of subunits, magnitude of the shift of -band of HbO_2 to alpha-band of HbCO. The latter is very similar to root nodules hemoglobin—(100 A) vs. (95 A).

B- 69. Haemoglobin and nitrogen fixation in the root nodules of leguminous plants (1947) Nature, Lond. 159, 692-694. (With J. D. Smith)

Within the root nodules of leguminous plants, hemoglobin exhibits usual reversible oxygenation. No evidence has been found that during nitrogen fixation, oxidation of the hemoglobin takes place.

B- 70. Stability of haemoglobin and of certain endoerythrocytic enzymes in vitro (1947) Biochem. J. 41, 491-500. (With Y. L. Wang)

B- 78. Reaction of methaemoglobin with hydrogen peroxide (1950) Nature, Lond. 166, 513-514. (With E. F. Hartree)

B- 83. Relationship between haemoglobin and erythrocruorin (1951) Nature, Lond. 168, 266-269. (With E. F. Hartree)

B- 84. Effect of drying upon the absorption spectra of haemoglobin and its derivatives (1952) Nature, Lond. 170, 161-162.

B- 86. Haemoglobin in fungi. Occurrence of haemoglobin in east and the supposed stabilization of the oxygenated cytochrome oxidase (1953) Nature, Lond. 172, 390-393.

B- 87. Stability of biological materials and its bearing upon the problem of anabiosis (1953) Sci. Progr. 41, 577-592.

B- 89. Haemoglobin in Protozoa (1953) Nature, Lond. 172, 451. (With F. J. Ryley)

B- 91. Haemoglobin in moulds: Neurospora crassa and Penicillium notatum (1953) Nature, Lond. 172, 393-394. (With A. Tissieres)

B- 92. Catalase, peroxidase and metmyoglobin as catalysts of coupled peroxidatic reactions (1954) Biochem. J. 60, 310-325. (With E. F. Hartree)

B- 94. Haemoglobin in certain strains of yeast (Saccharomyces cerevisiae) (1954) Proc. Biochem. Soc. Biochem. J. 57, Pxxix. (With A. Tissieres)

B- 95. Cyanide compounds of ferroperoxidase and myoglobin and their reversible photo-dissociation (1955) Biochem. J. 61, 153-171. (With E. F. Hartree)

B- 97. Distribution and diversity of haemoglobin (1956) Acta Biochim. Polonica 3, 439-457.

B-109. Reactions of haemoproteins with hydrogen peroxide and the supposed formation of hydrogen peroxide during the autoxidation of haemoglobin (1961) Nature, Lond. 191, 769-770.

The dissociation constants between H_2O_2 have been found to be 1.3 mM and 2 M for reduced and oxidized peroxidase (horseradish), respectively. The absorption maxima for the reduced CN peroxidase are 565.5, 535, and 432 nm identical to the cyano compounds of hemoglobin and myoglobin. The reduced cyanoperoxidase compound is found to be light sensitive and is stronger than the cyanomyoglobin compound. The quantum efficiency is 0.8 and the partition ratio of [(ferroperoxidase.CO)(HCN)]/[(ferroperoxidase.CN)(CO)] = about 2500.

C. Catalase, Peroxidase and their Derivatives

C- 10. Discussion on recent advances in the study of enzymes and their action (1932) Proc. Roy. Soc. B 111, 280-297.

C- 17. The haematin content of horseradish peroxidase (1934) Proc. Roy. Soc. B 114, 210-222. (With K.A.C. Elliott)

C- 18. Inhibitors of catalase reaction (1934) Nature, Lond. 134, 933-934. (With E. F. Hartree)

C- 20. The action of sodium azide on cellular respiration and on some catalytic oxidation reactions (1936) Proc. Roy. Soc. B 121, 165-173.

C- 25. Coupled oxidation of alcohol (1936) Proc. Roy. Soc. B. 119, 141-159. (With E. F. Hartree)

C- 26. On some properties of catalase haematin (1936) Proc. Roy. Soc. B 121, 173-191. (With E. F. Hartree)

C- 27. Structure chimique et propriete de quelques ferments (ou diastases) intracellulaires (1937) Congres du Palais de la Decouverte, Paris, pp. 357-363.

C- 30. On the haematin compound of peroxidase (1937) Proc. Roy. Soc. B, 122, 119-133. (With T. Mann)

C- 32. On the mechanism of the decomposition of hydrogen peroxide by catalase (1938) Proc. Roy. Soc. B 124, 397-405. (With E. F. Hartree)

C- 38. Mechanism of decomposition of hydrogen peroxide by catalase (1939) Nature, Lond. 144, 787-788. (With E. F. Hartree)

C- 39. Decomposition of hydrogen peroxide by catalase (1939) Nature, Lond. 144, 1091. (With E. F. Hartree)

C- 54. Decomposition of hydrogen peroxide by catalase (1943) Nature, Lond. 152, 626. (With E. F. Hartree)

C- 59. Properties of azide-catalase (1945) Biochem. J. 39, 148-157. (With E. F. Hartree)

C- 61. Properties of catalase. Catalysis of coupled oxidation of alcohols (1945) Biochem. J. 39, 293-301. (With E.

F. Hartree)

C- 80. Purification of horseradish peroxidase and comparison of its properties with those of catalase and methaemoglobin (1951) Biochem. J. 49, 88-106. (With E. F. Hartree)

C- 92. Catalase, peroxidase and metmyoglobin as catalysts of coupled peroxidatic reactions (1954) Biochem. J. 60, 310-325. (With E. F. Hartree)

C- 93. Reactions of methaemoglobin and catalase with peroxides and hydrogen donors (1954) Nature, Lond. 173, 720-723. (With E. F. Hartree)

C- 95. Cyanide compounds of ferroperoxidase and myoglobin and their reversible photo-dissociation (1955) Biochem. J. 61, 153-171. (With E. F. Hartree)

C-101. On the supposed catalytic oxidation of thiol groups by catalase (1958) Biochim. Biophys. Acta 28, 225. (With P. Nicholls)

C-102. Reactions of catalase with hydrogen peroxide and hydrogen donors (1958) Biochim. Biophys. Acta 29, 302-307. (With P. Nicholls)

C-104. Effect of azide on free haematins (1959) Biochim. Biophys. Acta 36, 257-259. (With P. Nicholls)

C-108. Precautions in the use of ferri- or ferro-cyanide for the study of haemoproteins (1961) Nature, Lond. 190, 717-718.

C-109. Reactions of haemoproteins with hydrogen peroxide and the supposed formation of hydrogen peroxide during the autoxidation of haemoglobin (1961) Nature, Lond. 191, 769-770.

See abstract B-109.

D. Other Metalloproteins and Derivatives

D- 2. A comparative study of turacin and haematin and its bearing on cytochrome (1926) Proc. Roy. Soc. B 100, 129-151.

D- 21. Le mecanisme de la respiration intracellulaire see

abstract A-2 (1936) Bull. Soc. Chim. Biol. 18, 96-130.

D- 22. The action of cyanide and other respiratory inhibitors on xanthine oxidase (1936) Proc. Roy. Soc. B 119, 159-190. (With M. Dixon)

They show that cyanide inhibition on xanthine oxidase is different from that on the respiratory chain. The reaction is relatively slow and irreversible. Purines protect the enzyme aerobically.

D- 24. Uricase, amino acid oxidase, and xanthine oxidase (1936) Proc. Roy. Soc. B 119, 114-140. (With E. F. Hartree)

D- 25. Coupled oxidation of alcohol (1936) Proc. Roy. Soc. B 119, 141-159. (With E. F. Hartree)

D- 34. Polyphenol oxidase: purification, nature and properties (1938) Proc. Roy. Soc. B 125, 187-204. (With T. Mann)

A relatively simple method for the preparation of polyphenol oxidase has been worked out giving 6000-7000 times more active than the starting material, mushroom. The method includes purification with ammonium sulfate fractionation of the aqueous extract by lead acetate, calcium phosphate gel and acetone. Ten kg of the mushroom produces about 10 mg of the purest enzyme of 0.3% copper with the activity in Q_{O_2} (1 of O_2 consumed per mg per hour at 20) = 1.16 x 10^6. The enzyme is sulfide cyanide and CO sensitive.

D- 35. Haemocuprein, a copper-protein compound of red blood corpuscles (1938) Nature, Lond. 142, 148. (With T. Mann)

D- 36. Haemocuprein and hepatocuprein, copper-protein compounds of blood and liver in mammals (1938) Proc. Roy. Soc. B 126, 303-315. (With T. Mann)

A blue protein with 0.34% Cu, haemocuprein, has been isolated and crystallized from ox red blood corpuscles. It also exists in serum and blood corpuscle of other animals. A practically colorless Cu (0.34%) protein, hepatocuprein, has been also isolated and highly purified from liver by the same method as haemocuprein. Neither of them combines oxygen or exhibits activity of peroxidase, catalase, polyphenol and cytochrome oxidase or carbonic anhydrase.

D- 37. Laccase, a blue copper-protein oxidase from the latex of Rhus succedanea (1938) Nature, Lond. 143, 23-24. (With T. Mann)

A deep blue copper protein isolated from latex of Rhus succedanea has been purified up to 0.154% Cu and 55% polysaccarides. The enzyme oxidizes ascorbate, dihydroquinone, phenylene diamine and at the same time the enzyme is bleached. The activity is sensitive to cyanide thiocarbamate, azide, sulfide, but not CO.

D- 41. Carbonic anhydrase (1939) Nature, Lond. 144, 422-423. (With T. Mann)

D- 43. Activity of carbonic anhydrase in relation to gastric secretion (1940) Nature, Lond. 146, 651-652.

D- 48. Carbonic anhydrase (1940) Biochem. J. 34, 1163-1776. (With T. Mann)

Carbonic anhydrase has been extensively purified from ox or sheep erythrocytes. The purification includes the removal of hemoglobin by alcohol-$CHCl_3$, fractionation with lead acetate, calcium phosphate gel adsorption, alumina C , or by fractionation with ammonium sulfate and alumina C with a yield of about 200 mg per liter of blood. It has been estimated 100 g erythrocytes contains 0.21 g of carbonic anhydrase. The purest preparation contains 0.33% and is claimed to contain (from evidence of sedimentation) 2 Zn atom per mol of protein. Authors states: "the presence of Zn in carbonic anhydrase establishes for the first time the physiological function of this metal in organisms." The enzyme is free from Fe, Cu, Mn, Mg, and Pb and sensitive to cyanide, sulfide, azide. The enzymic activity parallels to the Zn content. It has also purified by gastric mucosa up to Zn content of about 0.13%.

D- 49. Some properties of laccase from the latex of lacquer trees (1940) Nature, Lond. 145, 304. (With T. Mann)

D- 50. Sulphanilamide as a specific inhibitor of carbonic anhydrase (1940) Nature, Lond. 146, 164-165. (With T. Mann)

D- 51. Activity of carbonic anhydrase within red blood corpuscles (1941) Nature, Lond. 148, 493-496. (With T. Mann)

D- 57. Activity of purified carbonic anahydrase (1944) Nature, Lond. 153, 107-108. (With T. Mann)

D- 58. Trace elements in relation to physiological function and enzyme systems (1944) Proc. Nutr. Soc. 1, 189. (With T. Mann)

D-105. Haemerythrin (1960) Acta Biochim. Polonica 7, 415-424.

E. Non-metalloenzymes

E- 64. Prosthetic group of glucose oxidase (notatin) (1946) Nature, Lond. 157, 801. (With E. F. Hartree)

E- 70. Stability of haemoglobin and of certain endoerythrocytic enzymes in vitro (1947) Biochem. J. 41, 491-500. (With Y. L. Wang)

E- 71. Properties of glucose oxidase (notatin) (1948) Biochem. J. 42, 221-229. (With E. F. Hartree)

Glucose oxidase is identical with penicillin B and notatin, catalyzes the oxidation of D-glucose to gluconic acid and hydrogen peroxide. Samples of 80-90% purity determined by physical methods show absorption maxima at 377 and 455 nm in the oxidized form at pH 6.5. Glucose bleaches the color and exhibit a spectrum without peaks. FAD is shown to be the prosthetic group. The enzyme is rather specific and practically only active to D-glucose among 50 or so sugars tested. Its action is not affected by narcotics or respiratory inhibitors. The specificity does not change at all after some 20 times of repeated cultivations of Penicillin notatin in a fructose medium.

E- 72. The use of glucose oxidase (notatin) for the determination of glucose in biological material and for the study of glucose-producing systems of manometric methods (1948) Biochem. J. 42, 230-238. (With E. F. Hartree)

Taking advantage of the specificity of glucose oxidase, a simple manometric method has been worked out for glucose determination in the presence of catalase.

E- 81. Specificity of glucose oxidase (notatin) (1951) Biochem. J. 50, 331-341. (With E. F. Hartree)

E- 82. Biological catalysis of mutarotation of glucose (1951) Biochem. J. 50, 341-348. (With E. F. Hartree)

E- 85. Position of haems in the haemoglobin molecule (1953) Nature, Lond. 171, 922-925.

E- 86. Haemoglobin in fungi. Occurrence of haemoglobin in yeast and the supposed stabilization of the oxygenated cytochrome oxidase (1953) Nature, Lond. 172, 390-393.

F. Methodology

F- 14. An improved method for the measurement of tissue respiration (1933) Biochem. J. 27, 86-95. (With M. Dixon)

F- 15. Estimation of haematin iron and the oxidation-reduction equivalent of cytochrome c (1933) Proc. Roy. Soc. B 114, 104-109. (With R. Hill)

F- 42. Direct perception of pigment in the nerve tissue of human retina (1939) Nature, Lond. 143, 333. (With E. L. Smith)

F- 72. The use of glucose oxidase (notatin) for the determination of glucose in biological material and for the study of glucose-producing systems of manometric methods (1948) Biochem. J. 42, 230-238. (With E. F. Hartree)

F- 75. Effect of low temperature on the absorption spectra of haemoproteins; with observations on the absorption spectrum of oxygen (1949) Nature, Lond. 164, 254-259. (With E. F. Hartree)

F- 98. Spectrophotometric study of suspensions of pigmented particles (1958) Biochim. Biophys. Acta 27, 173-184. (With E. F. Hartree)

F-108. Precautions in the use of ferri- or ferro-cyanide for the study of haemoproteins (1961) Nature, Lond. 190, 717-718.

G. General and Other Subjects

G- 9. Uteroverdin (1931) Nature, Lond. 128, 967-968. (With R. Lemberg and J. Barcroft)

A tetrapyrrol compound related to bilirubin.

G- 10. Discussion on recent advances in the study of enzymes and their action (1932) Proc. Roy. Soc. B 111, 280-297.

G- 26. On some properties of catalase haematin (1936) Proc. Roy. Soc. B 121, 173-191. (With E. F. Hartree)

G- 55. Properties and function of natural tetrapyrrolic compounds (1943) Biochem. J. 37, xxi-xxii.

G- 58. Trace elements in relation to physiological function and enzyme systems (1944) Proc. Nutr. Soc. 1, 189. (With T. Mann)

G- 77. Further observations on absorption spectra at low temperatures (1950) Nature, Lond. 165, 504-505. (With E. F. Hartree)

G- 78. Absorption spectrum of oxygen (1950) Nature, Lond. 165, 543-544. (With E. F. Hartree)

G- 87. Stability of biological materials and its bearing upon The problem of anabiosis (1953) Sci. Progr. 41, 577-592.

G-103. The Leeuwenhoek Lecture—The problem of anabiosis or latent life: history and current concept (1959) Proc. Roy. Soc. B 150, 149-191.

Given on May 11, 1958 before the Royal Society in the Burlington house.

G-106. The problem of language in publications of works of scientific and other fields of learning (1960) Nature, Lond. 186, 912.

H. Index of Coauthors

S. J. Bach, 53(A)

J. Barcroft, 9(G)

W. D. Butt, 110(A)

M. Dixon, 8(A), 14(F), 22(D), 53(A)

K. A. C. Elliott, 17(C)

W. Feldeberg, 46(B)

D. E. Green, 23(A)

C. H. Harpley, 44(A)

E. F. Hartree, 18(C), 19(B), 24(D), 25(C, D), 26(C, G), 28(B), 29(A), 31(A), 32(C), 33(A), 38(C), 39(C), 40(A), 45(A), 46(B), 47(A), 54(C), 59(C), 61(C), 63(B), 64(E), 66(A), 67(A), 68(A), 71(E), 72(E, F), 73(A), 74(A), 75(A, F), 78(B, G), 80(C), 81(E), 82(E), 83(B), 92(B, C), 93(0), 95(B, C), 96(A), 98(F), 99(A)

R. Hill, 7(A), 8(A), 15(A, F)

T. E. King, 99(A), 100(A), 107(A)

R. Lemberg, 9(G)

T. Mann, 30(C), 34(D), 35(D), 36(D), 37(D), 39(D), 46(B), 48(D), 49(D), 50(D), 51(B), 51(D), 57(D), 58(D, G)

P. Nicholls, 102(C), 104(C)

F. J. Ryley, 89(B)

E. C. Slater, 90 (A)

E. L. Smith, 42(F)

J. D. Smith, 69(B)

P. Tate, 56(B)

A. Tissieres, 91(B), 94(B)

Y. L. Wang, 62(B), 65(B), 70(B, E)

Oxidases and Related Redox Systems, pages 29–49

AN EXERCISE IN NOSTALGIA ON THE THEME OF DAVID KEILIN

Thaddeus Mann

University of Cambridge, Cambridge, England.

Like many of my scientific friends, I have been fortunate in my time to inhabit several hospitable shells. In the early 1930's, prior to coming to Cambridge, I had the good fortune of spending several years under the wings of Professor Jacob Parnas, when grappling youthfully with intricacies of the intermediary metabolism of carbohydrate and phosphorylations in muscle and in yeast. The School of Parnas was at that time one of Europe's, let alone Poland's leading centres of biochemical research. It was in 1935, shortly after I managed to secure both a M.D. degree and a Rockefeller Foundation Research Fellowship, that I was able to move to the Molteno Institute in Cambridge, England, and meet David Keilin for the first time. In later years, though for short periods only, I had also the good fortune of inhabiting some other hospitable shells, among them several provided by American friends. However, by far the longest period of my life, 52 years in all, was spent at the University of Cambridge, all of it filled with happy recollections, not least those of the years 1935 - 1941, when I actually worked together with Keilin on several research projects. My present account, confined largely to these particular years, I can best describe as an exercise in nostalgia on the theme of David Keilin.

Reflecting, as I often do, on my memories of that period, I realize that my close friendship with the man whose centenary of birth we are commemorating, owes much to two, rather exceptional circumstances. First, both he and I had been brought up in Poland, though in different parts of that country, he in Warsaw, at that time still under Russian rule, and I in the southern region of Galicia which formed part of the Austro-Hungarian Empire. Consequently, he and I could freely converse in Polish, and frequently did, not only when working at the bench, but even more so at home, in the company of our wives. His wife, Dr Anna Keilin, was a medical practitioner. To her he remained absolutely devoted ever since they studied at the University of Paris, and subsequently throughout married life in Cambridge, where their daughter, Joan, was born. He often told me that to his wife he owed not only his great domestic happiness, but life as well, since it was her wisdom and medical advice, that made possible for him to cope with asthma, a burden that he bravely carried on his shoulders from the time of early childhood. The confidence that he placed in me by talking about his private life, I valued as an exceptional privilege, often naturally responding by relating to my happy married life with a wife who shared with me also lifelong comradeship in research.

Secondly, equally important, was the circumstance that my partnership with Professor Keilin extended over a period longer than that enjoyed by most of his biochemical friends, with the exception of Edward Hartree, with whom he collaborated closely and published extensively on many aspects of cellular metabolism, the function of cytochrome and cytochrome oxidase specially, and also on the role of a wide range of enzymes such as succinic dehydrogenase, catalase, uricase, and glucose oxidase. As scientist, teacher and man, Keilin was truly extraordinary and unforgettable. His versatility was expressed in many directions, foremost perhaps in a thorough knowledge of biology as a whole, a characteristically comparative approach to biological problems, great personal devotion to experimental work, and the ability to guide and advise others.

As rightly emphasized by Hartree in the Obituary Notice in the Biochemical Journal (1963), all biological sciences, and not merely cellular metabolism and enzymes, came within Keilin's orbit, so that he could speak freely and with authority on many of them. "His zest for the day ahead, his eagerness to get to the bench and the spectroscope, his confidence that his plans for the day's work contained the key to one of his current problems, all these were epitomized in the sound of his lively, purposeful step along the corridor to his room as he arrived at the Laboratory in the morning".

My task of recording recollections of Keilin's work on the present occasion has been also made easier by the perusal of the Biographical Memoir which I prepared for the Royal Society of London (Mann 1964). In it I included a photograph bearing Keilin's own signature, and a detailed account of his studies in Paris, prior to the move to Cambridge in 1915, and his research apprenticeship at the Laboratoire d'Évolution des Êtres Organisés (Fig.3), located together with its secluded garden near the imposing building of the Panthéon (Keilin 1915). In later years he delighted in telling the story of how giving a slight twist to his microscope's mirror he could see the reflection of the imposing building of the Panthéon instead of the image of insect eggs and larvae on which he was working at that time. Work of similar kind, which soon earned him the recognition of a widely acknowledged authority in entomology, continued in Cambridge after 1915, but in its course his attention turned gradually from morphology to cellular physiology and biochemistry as means of advancing the knowledge of respiratory processes and adaptations in insect larvae. It was the work on insects that led ultimately to the discovery of cytochrome in 1925, and enabled Keilin to lay the foundations for new concepts and our present knowledge of mechanisms underlying cellular respiration in general (Keilin 1925).

The larva on which Keilin's attention was chiefly concentrated in the years preceding the discovery of cytochrome, was that of the horse bot-fly *Gasterophilus* which inhabits the stomach of horses and is characterized by the presence of haemoglobin, different however, in composition and function from that of the horse-blood. Pursuing the fate of the larval haemoglobin in the adult bot-fly, he soon discovered that unlike the larva, the adult insect, that is the bot-fly as such, showed spectroscopically no signs of any absorption bands of either haemoglobin or oxyhaemoglobin, but exhibited instead, in its thoracic muscles, another, most unusual four-banded spectrum. The same spectrum he was also able to detect in the thoracic muscles of other insects, none of them containing haemoglobin during the larval stage of development, as well as in a variety of animal tissues and micro-organisms, yeast in particular. Baker's yeast turned out to be the best material for demonstration of yet another surprising phenomenon, namely, that shaking of the yeast suspension in air causes the bands to disappear, but they re-appear after the yeast suspension has been left standing without shaking. A similar phenomenon was observed by Keilin in live insects: the four spectral bands were much stronger in an insect actively vibrating its wings than in one at rest.

From these and other similar observations on a variety of cells and tissues, Keilin was able to conclude that the newly discovered cellular constituent which he proposed to name cytochrome, occurs in two clearly distinguishable, yet easily interconvertible states, a reduced and an oxidized form. In the wing muscles of a bee, for example, he found the reduced cytochrome to give a spectrum with four distinct absorption bands (in the visible part of the spectrum) with maximum intensity: *a*, 604.6; *b*, 566.5; *c*, 550.2; and *d*, 521.0 mμ, and the oxidized cytochrome to exhibit a much more diffuse spectrum with two bands only: α, 566.5 and β, 528.7 mμ.

The events that preceded the discovery of cytochrome, and the major subsequent developments in studies on the relationships between the different cytochrome components and cytochrome oxidase (previously called indophenol oxidase), and their role in cellular respiration, have been admirably recounted by Keilin himself in his monumental monograph on "The History of Cell Respiration and Cytochrome" published posthumously, in 1966, by the Cambridge University Press. In this monograph, Keilin described in detail not only the spectroscopic, chemical and functional properties of the various components of the cytochrome system, but their distribution and relation to respiratory requirements in a multitude of cells, the mechanism of cytochrome oxidation and reduction, and the effects of the different respiratory inhibitors, such as carbon monoxide, potassium cyanide, hydrogen sulphide, azide and hydroxylamine (Fig.1). The book contains a wealth of information about the early history of cellular respiration, and a detailed account of the events that led to the discovery of the relationship between cytochrome a_3 and cytochrome oxidase. It includes also a characteristically generous tribute to the memory of Charles MacMunn, the discoverer of histohaematin. Keilin's account of the ideas and work developed by him and Hartree in later years has been supplemented in a meticulous manner by Joan Keilin-Whiteley in the process of updating and editing her father's monograph

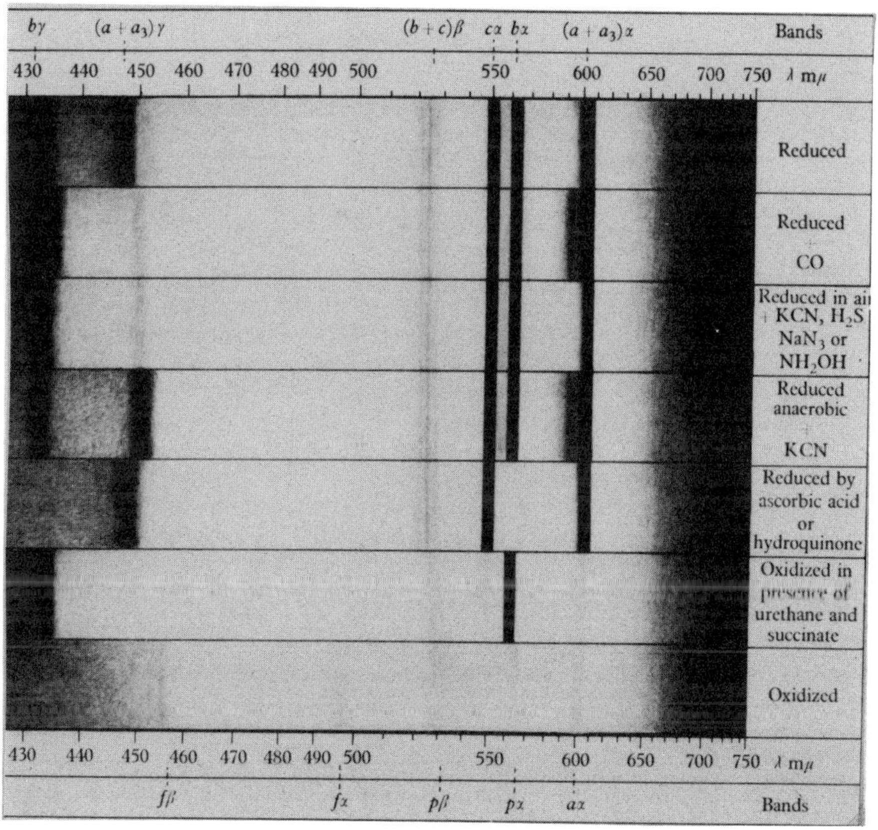

Figure 1. Absorption spectra of reduced and oxidized cytochromes in heart-muscle preparation, untreated and treated with different respiratory inhibitors (Keilin 1966).

prior to publication by the Cambridge University Press. In the current issue of the Biochemical Society's Bulletin (August 1987), I was glad to see a statement that Keilin's monograph is again available for purchase.

A most characteristic and illuminating trend in Keilin's outlook and ideas, which I never ceased to admire, was the comparative approach to biochemical studies. This is evident in the publications by him and Hartree that led to the identification of cytochrome oxidase with cytochrome a_3 (Keilin and Hartree 1939), the work with Tsoo King on the reconstruction of the respiratory chain (Keilin and King 1958), and above all, the Leeuwenhoek Lecture on anabiosis (Keilin 1959). A similar trend is characteristic of the studies on the coupled oxidation of alcohols by catalase (Keilin and Hartree 1936, 1945), and the work with Wang on haemoglobins in both the larvae of *Gasterophilus* and in the root nodules of leguminous plants (Keilin and Wang 1945, 1946).

Why a man who pioneered so much of modern biochemistry, should not have been awarded a Nobel Prize, even for the discovery of cytochrome alone, still remains beyond my comprehension. In 1955, when that Prize was awarded to Theorell alone, for work on the nature and mode of action of oxidation enzymes, I felt so appalled by the decision not to include Keilin in the award, that on the day of the official announcement I went to see the Professor in his office so as to express my dismay and disappointment. He, however, having taken me by the elbow, reacted by pointing out to me that on the day when he will have to enter the Heaven, Saint Peter is unlikely to ask him if he is bringing with him a Nobel Prize. Great modesty, bordering on humility, was indeed yet another outstanding attribute of David Keilin.

In addition to possessing an original and enquiring mind, and a boundless enthusiasm for research, he was also a superb teacher, at first, in the years 1925-1931, in his capacity as University Lecturer in Parasitology, and thereafter as Professor of Biology. His wide and profound knowledge of comparative cellular physiology as a whole, and equally so, the elegance and accuracy of his lecture demonstrations, continued for many years to attract to the lecture room at the Molteno Institute not merely undergraduates, but graduates and scientists from various University Departments, including Zoology, Botany, Physiology and Biochemistry. The meticulously prepared demonstrations of experiments, on many occasions involving the projection of cytochrome and haemoglobin spectra directly from the spectroscope to a screen, were indeed a most enjoyable feature of his lecturers. When demonstrating, he would also frequently remind the audience that there is serious trouble in store for any biologist who surrenders prematurely to the temptation of mistaking a working hypothesis for a certainty of the kind that only a convincing and conclusive experiment can furnish. Standing in front of the audience assembled in the lecture room, Keilin must have given the impression of a man with a particularly small build. This is also manifested by a photograph (Fig.2.) taken in front of the Molteno Institute on June 7th, 1937, that is 50 years ago, on the occasion of my Ph.D. degree; in addition to the Professor and myself, it features also Dr Parr Tate, the parasitologist who in due course, succeeded Keilin as director of the Institute, and Dr Malcolm Dixon, the author of the well known textbook on "Enzymes", who though a member of the University Department of Biochemistry, at that time directed by Frederick Gowland Hopkins, was a constant visitor to the Molteno and a dear friend of all who worked there. Though in company of others, Keilin may appear to have been of small stature, he was a man of an enormous abundance of heart, so much so as to earn him a well deserved compliment from another friend, Albert Szent-Györgyi, of "a small man with a big heart".

The building of the Molteno Institute itself is shown in another photograph (Fig.3.), taken at the same time as the previous one; among the panoramic windows of the first floor, the second on the right is of the laboratory room in which I was privileged to work with Keilin for several years, and the small window at the right corner is the one of the professorial office, by visitors usually referred to as the "Citizens' Advice Bureau", thus named after the British public institution which offers legal advice to citizens at large. Advice, in fact, was never withheld by the Professor even when he was not in his office, but working at the bench in the room nearby.

As might be expected, I myself was often the one to seek advice. One such occasion, which I will never forget, occurred shortly after the 2nd World War, when I received a letter from one of the Swiss Universities, enquiring if I

Figure 2. Standing in front of the Molteno Institute, on June 7, 1937. From left to right: Dr. Keilin, T. Mann, M. Dixon and P. Tate.

Figure 3. The garden of the Laboratoire d'Évolution des Êtres Organisés in Paris, and below, the Molteno Institute in Cambridge.

might be interested in applying for a recently vacated chair of biochemistry. This was at the time when I was already in the midst of my explorations of the biochemistry of mammalian semen, and shortly after the identification of the seminal sugar as fructose. My wife's research moreover, in the Department of Biochemistry, was proceeding equally well so that both of us felt reluctant to leave Cambridge. Nonetheless, the first obvious step was to show the letter to the Professor. He, having read it, simply remarked that since I am obviously unable to reach a decision by myself, this can only mean that I am not seriously interested in the proposal.

On the subject of Keilin's advisory talents alone, I could continue with many more examples, but though I find it difficult to desist the temptation of recounting further examples, I feel, without wishing to appear chatty or egocentric, that I should say something now about the experimental projects that I have been actually privileged to tackle while collaborating with Keilin directly. This, I feel, will give me the best chance of illustrating the Professor's attitude to his junior colleagues, his approach to joint collaborative work, and provide examples of his true love of work at the bench. He used to divide the scientific community into two types of people: those who work at the bench, and those who merely talked about "research with a capital R". He despised the latter type, and he himself practiced at all times what he preached.

The first enzyme that Keilin and I investigated jointly was peroxidase which we purified from horseradish roots and identified as a methhaemoglobin-like haematoprotein endowed with the ability to form spectroscopically well defined compounds with both the substrate, hydrogen peroxide, and several specific inhibitors, including carbon monoxide, cyanide, fluoride and hydrogen sulphide (Keilin and Mann 1937). Next to be purified and investigated were three copper proteins: haemocuprein from mammalian blood corpuscles (Mann and Keilin 1938a, b), polyphenol oxidase from mushrooms (Keilin and Mann 1938), and laccase from the latex of lacquer trees (Keilin and Mann 1939a, 1940a). The first of these three copper-proteins, haemocuprein, nowadays also commonly called erythrocuprein, was the first compound of this kind ever isolated and crystallized. An upsurge of new interest in this copper-protein resulted from recent observations on the relationship of haemocuprein to the enzyme superoxide dismutase, which catalyses the dismutation of a superoxide radical to hydrogen peroxide and oxygen. I beg your forgiveness for quoting from the extensive recent literature on this subject just one sentence in a paper by Gärtner and Weser (1984). It reads as follows: "We would like to express our admiration to Professors T. Mann and the late D. Keilin for their intriguing work on haemocuprein and their far reaching conclusions drawn already in 1939." I wish, Professor Keilin could have seen this gracious comment and I wonder what his reaction to it would have been this time.

The last metalloprotein to be jointly purified and investigated was carbonic anhydrase from both blood and the gastric mucosa (Keilin and Mann 1939b, 1940b). It turned out to be a zinc-protein, the first enzyme ever to be identified as one containing zinc; I shall refer to it in more detail later. But first, may I tell the story of our work on peroxidase, the first enzyme that Keilin and I purified together, known already from an earlier study by Elliott and Keilin (1934) to contain haematin in some, as yet undetermined state. Just before embarking on the purification of peroxidase from the horseradish roots, the Professor said to me that we must first obtain more information about the distribution and content of haematin in plants generally, that is, not merely in roots such as horseradish, but in leaves, stems, fruit, seed, and similar plant material. This task he allocated to me, insisting that I should tackle it independently so as to provide the material for my future Ph.D. dissertation (Mann 1937). The technique which I used for this purpose, much of it based on Keilin's previous experience, was simple enough. It depended on preparing sections from plant material and soaking them in pyridine and a hyposulphite solution so as to convert all of the plant-haematin to pyridine haemochromogen.

In order to carry out the determination, I would place the material on the stage of Keilin's favourite instrument, a microscope fitted with the Zeiss double-spectrum microspectroscope at the top, and a platform carrying a calibrated scale and a double-wedge trough filled with a standard solution of pyridine haemochromogen on one side, and water on the other side (Fig.4.). In this way, two spectra could be observed simultaneously, one derived from the plant material on the stage of the microscope, and the other given by the haemochromogen standard on the platform. Next, by moving the trough along the scale, one was able to bring the two

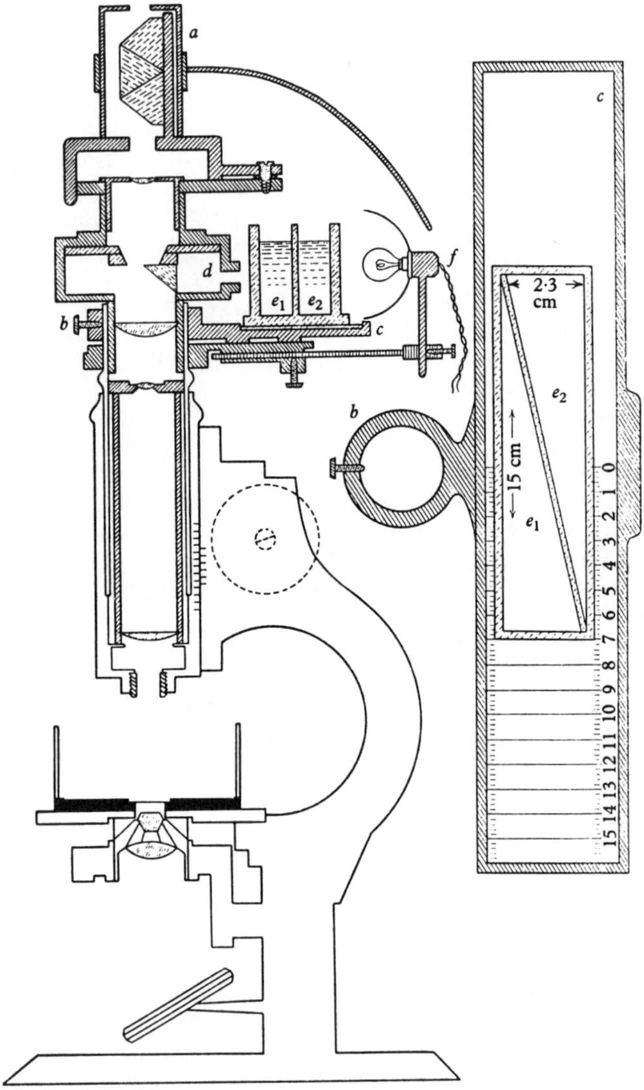

Figure 4. The microspectroscope: *a*, the spectroscope ocular; *b*, ring, holding *c*, a calibrated platform; *d*, side aperture; e_1 and e_2, the compartments of the double-wedge glass trough, illuminated by *f*, lamp.

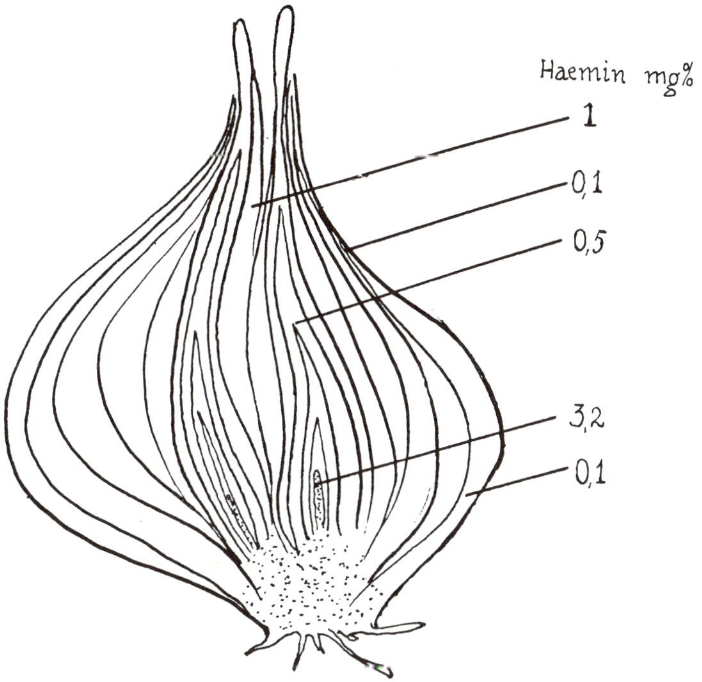

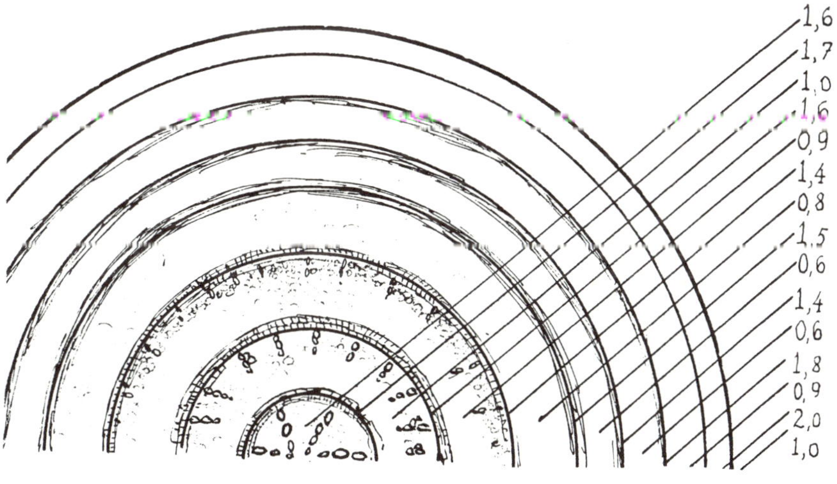

Figure 5. Haematin (mg/100 g fresh weight) in sections from onion (top) and sugar beetroot (bottom).

spectra into a position where they matched precisely each other in the intensity of the pyridine haemochromogen absorption bands, and then take the reading on the calibration scale. This made possible for me to determine the concentration of total haematin not merely in the plant section as a whole, but in different parts of it. In this manner, I was able to demonstrate that in a sliced onion, for example, the bulk of haematin was concentrated in the young, immature parenchymatic tissue of the cambium, the phloëm containing much less haematin, and xylem least. Similarly, in a section from the beetroot, I found that all the cambia corresponded to rings with characteristically high concentration of haematin; and in the trunk of a 5-year old pine tree, I found the haematin to be concentrated more in the bark portion than in other parts (Fig.5). Not unexpectedly, of all the plant materials that I examined, the horseradish turned out after all to be the richest source of haematin, the total content being about 10mg/100g dry weight. It was at this point of the investigation, that the Professor agreed to join forces with me and carry out the first large-scale preparation of peroxidase. The quantity of horseradish roots that we used for this purpose consisted of about 15 Kg of material which had to be patiently sliced, then minced and pressed out, and finally the extract filtered through Kieselguhr on a huge Buchner funnel mounted on a vacuum flask with similarly impressive dimensions. No amount of persuasion on my part, would deter the Professor to carry out all these manipulations himself instead of watching me doing it, and this inspite of the fact that in order to reach the top of the Buchner funnel high above the level of the bench, he had to climb and mount a special stool.

Peroxidase, after the preparation of several extracts, and extensive purification and concentration, turned out to be a deep brown coloured enzyme with a spectrum very similar, but not entirely identical, to that of blood-methaemoglobin. On addition of a drop of hydrogen peroxide, the brown colour of the concentrated enzyme preparation turned to red, and at the same time the spectrum changed from four bands to two entirely different bands characteristic of the H_2O_2 - peroxidase complex. Naturally, what delighted us most was that for the first time in the history of enzymology, we could demonstrate the existence of a well defined enzyme-substrate complex, and our delight was further enhanced as soon as we discovered that not only the substrate of peroxidase, but its various inhibitors formed spectroscopically distinct complexes with the pure enzyme. Among these inhibitors were some, such as cyanide, azide and H_2S, which combined with peroxidase in its natural, that is methaemoglobin-like oxidized form, while others, such as carbon monoxide, would form a complex only after the iron in the haematin had been reduced with hyposulphite from the trivalent to a divalent state, that is after a reduction of the methaemoglobin-like form to a haemoglobin-like state.

The routine which we adopted for following the formation of a H_2O_2 complex of peroxidase, depended on placing a solution of the highly concentrated enzyme on the stage of the microscope, and while the Professor was looking at it through the Zeiss microspectroscope, I, standing on his side, would add a drop of hydrogen peroxide to the sample on the stage, so as to enable him to read the position of the two absorption bands of the enzyme-substrate complex. Next, he would step over to the side so as to give me a chance of confirming his reading. However, on several occasions the position of the absorption bands, as I read them, was somehwat different. Far from suggesting that my reading may be faulty, he proposed that I should be the one to look at the sample first, while he, standing on my side, would add H_2O_2 to the enzyme thus enabling me to take the first reading. When adopting this procedure, we realized immediately that peroxidase forms more than one compound with its substrate, the compound formed first differing in respect of the position of the spectral bands from one produced a moment later.

Of the copper-proteins which we purified and investigated together, laccase presented the greatest challenge, mainly because of the extraordinarily high toxicity of the starting material, that is the latex from either the Indochinese or the Japanese lacquer tree. Laccase, it must be recalled, was the enzyme first accorded the name of an "oxidase" by Gabriel Bertrand in 1894, and it was also Bertrand who first used the latex of the Indochinese lacquer tree for a partial purification of this enzyme. Having found some manganese in his preparations, he actually concluded that this metal may be involved in the oxidizing action of laccase on its phenolic substrates which include laccol, the natural compound present in the crude latex. Though fully aware in advance, of the health hazards to be faced when handling crude latex, and in spite of equipping ourselves with gas masks and protective clothing, both the

Professor and I soon developed a painful rash on the skin of our faces and hands. Moreover, we experienced some breathing difficulty of the kind that would have undoubtedly alarmed any asthmatic. Yet in spite of the obvious risk, Keilin insisted on participating in several subsequent preparations of laccase, thus setting to his young associate once more an example of courage and devotion to experimental work. Pure laccase turned out to be a beautifully deep coloured protein with copper in its prosthetic group, but totally free from manganese.

Carbonic anhydrase was the last enzyme that we purified together, first from animal and human blood corpuscles, and subsequently from the pig's gastric mucosa. Its extraction presented no risk, except from spilling occasionally some blood which we had to use in huge quantities. Purification of carbonic anhydrase involved several simple and at that time conventional steps, such as precipitation of the dialysed ethanol-chloroform extract with ammonium sulphate and lead acetate, and frational adsorption on calcium phosphate and alumina gels. Right from the beginning we suspected carbonic anhydrase to be yet another metalloprotein in view of the sensitivity of enzymatic activity to metal-binding agents, such as cyanide or azide, and in fact, as the purity of our preparations increased so did their copper content and blue coloration. However, further steps of purification, involving mainly the alumina gel cγ, yielded two distinct fractions, one bluish coloured and rich in copper, but with little carbonic anhydrase, and the other colourless, with little copper but plentiful carbonic anhydrase. On further purification, the first fraction yielded ultimately pure, crystalline haemocuprein, while the other produced a copper-free preparation of highly active carbonic anhydrase with a characteristically high content of zinc.

Diphenylthiocarbazone was the chemical reagent which we used mainly for identifying zinc as the prosthetic group of carbonic anhydrase and for proving that all our preparations of the enzyme, both crude and partially purified, contained zinc in amounts strictly proportional to the level of enzymatic activity. In spite of this unequivocal evidence, we felt that it might be advisable if we asked a well-known expert on mass spectrography in London, for an independent analysis of metals, without telling him of course, that we are already aware of the presence of zinc in our purified enzyme preparation. When weeks later, the report finally arrived, it amused us greatly to read in it that our preparations contained minute traces of lead, calcium and aluminium which, as I mentioned before, we used during the purification procedure. Somehow, however, our expert missed the presence of zinc altogether, a sad reflection I suppose, on what the state of mass spectrometry must have been 50 years ago. The distribution of carbonic anhydrase and its physiological role have been subsequently studied in a variety of tissues, including the female reproductive tract where my late wife discovered high concentrations of this enzyme and demonstrated the dependence of its activity on progesterone levels (Lutwak-Mann 1955). In 1983, a special conference commemorating the discovery and subsequent studies of carbonic anhydrase was held by the New York Academy of Sciences; the proceedings were published shortly thereafter (Tashian and Hewett-Emmett 1984).

Whenever he embarked on the purification of any metalloprotein, be it haemoglobin, cytochrome or an enzyme, Keilin was always motivated not so much by the mere desire to obtain these compounds in a pure state, as by the prospect of shedding subsequently better light on the mechanism of their action and their physiological role inside the cells. As an example may serve our last joint study on the subject of "Activity of carbonic anhydrase within red blood corpuscles" which we published in 'Nature' (1941). Once more, we used in this study Keilin's beloved instrument, the microspectroscope, this time for determining the changes in activity of carbonic anhydrase inside intact erythrocytes under the influence of sulphanilamide, a substance that we showed one year previously to be a powerful and highly specific inhibitor of this enzyme (Mann and Keilin 1940).

The main steps in the procedure for following the activity of carbonic anhydrase in the blood corpuscles were as follows. First, a suspension of the washed erythrocytes was treated with sodium nitrite so as to convert their haemoglobin to acid methaemoglobin. Next, a small sample of this suspension was placed on the stage of the microscope, and the rest divided into two halves. One half of it served as standard of cell-bound acid methaemoglobin, but the other was treated with sodium carbonate in order to convert methaeglobin from an acid into an alkaline form and thus prepare a standard of cell-bound alkaline methaemoglobin. The two standards were then placed in the two sides of the double-wedge trough on the calibrated platform of the microspectroscope, identical in kind to the one which I described earlier when discussing the work on plant haematin (Fig.4). Now, the sample of the acid-methaemoglobin corpuscles was also treated with sodium carbonate, but immediately thereafter the trough with the standards had to be moved repidly along the calibrated platform until a position was reached where the intensity of the spectral bands in the standard matched exactly that in the sample on the stage, now undergoing rapid change from acid to alkaline methaemoglobin. Readings would be taken every 10 seconds over a period of altogether 2 minutes in order to register quantitatively the velocity with which the acid-methaemoglobin erythrocytes in the sample on the stage of the microscope were converted into alkaline-methaemoglobin erythrocytes. This experiment was then followed by a second one, in which the sample of the acid-methaemoglobin corpuscles on the stage was treated with sulphanilamide prior to the addition of sodium carbonate. In this way we could demonstrate that the conversion of acid methaemoglobin to alkaline methaemoglobin within the erythrocytes was markedly delayed in the presence of a carbonic anhydrase inhibitor.

Fifty years have passed since I was able to join Keilin in his studies. Next year, a quarter of a century will pass since his sudden death on 27 February 1963, at the moment when he was attending a discussion in the University of Cambridge on the future of the Molteno Institute. But the memory of the great man whose birth centenary we are commemorating today, remains much alive as can be seen from many recent developments in science. Even within the last few months there have been several significant events well illustrating the impact that Keilin's work continues to exert. Last May, in an article under the title "The birth of molecular biology" (New Scientist, 1987), Max Perutz reminded us once more that Keilin was among the first to appreciate fully the potentialities of the molecular approach to biological problems, and it was Keilin, in fact, whose personal initiative led 40 years ago to the creation of the first Unit of Molecular Biology in the University of Cambridge. Equally significant events, which would have pleased Keilin most particularly, concern the unbroken interest in the nature and function of the cytochrome system. The most recent examples include: the Royal Society Croonian Lecture by Peter Mitchell (1987) on the subject of "Proton-motive coenzyme mechanisms in cytochrome systems: variations on a theme by David Keilin", in which the lecturer reminded us once more of the key role currently ascribed to the membrane-bound ubiquinol-cytochrome *c* oxidoreductase complex in energy-transducing mechanisms; the paper by Daldal et al. (1987) on the genes for the cytochromes *b* and c_1 and other subunits of the ubiquinol-cytochrome oxidoreductase complex; the announcement of no less than two forthcoming books devoted specifically to the cytochromes *c* and the mammalian cytochromes P-450 (Pettigrew and Moore 1987; Geungerich 1987); and finally, the impressive number of papers on various aspects of cytochrome function delivered in the course of the "Fourth international symposium on oxidases and related redox systems" (King, Mason, 1988).

The great esteem in which the name of David Keilin is nowadays held among scientists, will continue and it is bound, I believe, to acquire in time the dimensions of an academic saga. In addition, those of us who actually worked together with Keilin, realize that his greatness extended far beyond the traditional limits of scientists, and that it was expressed above all, in his outstanding qualities as a man. All of us will cherish in particular the memory of a man with a heart of gold and a degree of modesty that only the greatest scientists can hope to achieve in life. We must also not forget that in science, as in other fields of human endeavour, knowledge and devotion to duty are no less important to a beginner than the ways in which he or she was briefed and initiated into research. It is in this respect that I owe to David Keilin my special and lasting gratitude.

REFERENCES

Bertrand G (1894). Sur le latex de l'arbe a lacque. C R Acad Sci Paris 118:1215-1218.

Daldal F, Davidson E, Cheng S (1987). Isolation of the structural genes for the Rieske Fe-S protein, cytochrome *b* and cytochrome c_1, all components of the ubiquinol:cytochrome c_2 oxidoreductase complex of *Rhodopseudomonas capsulata*. J Mol Biol 195:1-12.

Elliott KC, Keilin D (1934). The haematin content of horseradish peroxidase. Proc Roy Soc B 114:210-222.

Gärtner A, Weser U (1984). Aqueous isolation of erythrocyte copper proteins: formation of [(Heme b) copper] protein. Inorg Chim Acta 93:89-92.

Guengerich FP, ed. (1987). Mammalian cytochromes P-450. London:Wolfe Medical Publications, 484 pages in two volumes.
Hartree EF (1963). Obituary Notice David Keilin (1887-1963). Biochem J 80:1-5.
Keilin D (1915). Recherches sur les larves de Diptères cyclorhaphes:I. Cycle évolutif de *Pollenia rudis* Fabr., parasite d'*Allolobophora chlorotica* Sav. II. Biologie comparée des larves de Diptères. Bull Sci Fr Belg 49: 15-198.
Keilin D (1925). On cytochrome, a respiratory pigment, common to animals, yeast and higher plants. Proc Roy Soc B 98:312-339.
Keilin D (1959). The Leeuwenhoek Lecture - The problem of anabiosis or latent life: history and current concept. Proc Roy Soc B 150:149-191.
Keilin D (1966). "The History of Cell Respiration and Cytochrome." Cambridge University Press, 416 pages.
Keilin D, Hartree EF (1936). Coupled oxidation of alcohol. Proc Roy Soc B 119:141-159.
Keilin D, Hartree EF (1939). Cytochrome and cytochrome oxidase. Proc Roy Soc B 127:167-191.
Keilin D, Hartree EF (1945). Properties of catalase. Catalysis of coupled oxidation of alcohols. Biochem J 39:293-301.
Keilin D, King TE (1958). Reconstitution of the succinic oxidase system from soluble succinic dehydrogenase and a particulate cytochrome system preparation. Nature 181: 1520-1522.
Keilin D, Mann T (1937). On the haematin compound of peroxidase. Proc Roy Soc B 122:119-133.
Keilin D, Mann T (1938). Polyphenol oxidase. Purification, nature and compounds. Proc Roy Soc B 125:187-204.
Keilin D, Mann T (1939a). Laccase, a blue copper-protein oxidase from the latex of *Rhus succedanea*. Nature 143: 23-24.
Keilin D, Mann T (1939b). Carbonic anhydrase. Nature 144: 442-443.
Keilin D, Mann T (1940a). Some properties of laccase from the latex of lacquer trees. Nature 145-304.
Keilin D, Mann T (1940b). Carbonic anhydrase. Purification and nature of the enzyme. Biochem J 34:1163-1176.
Keilin D, Mann T (1941). Activity of carbonic anhydrase within blood corpuscles. Nature 148:493-496.
Keilin D, Wang YL (1945). Haemoglobin in root nodules of leguminous plants. Nature 155:227-229.

Keilin D, Wang YL (1946). Haemoglobin of *Gastrophilus* larvae. Purification and properties. Biochem J 40:855-866.

Lutwak-Mann C (1955). Carbonic anhydrase in the female reproductive tract. Occurrence, distribution and hormonal dependence. J Endocrol 13:26-38.

Mann T (1937). The haematin compounds in plants and their relation to peroxidase. Ph.D.Dissertation, University of Cambridge.

Mann T (1964). David Keilin 1887-1963. Biographical Memoirs of Fellows of the Royal Society. 10:183-205.

Mann T, Keilin D (1938a). Haemocuprein, a copper-protein compound of red blood corpuscles. Nature 142:148.

Mann T, Keilin D (1938b). Haemocuprein and hepatocuprein, copper-protein compounds of blood and liver in mammals. Proc Roy Soc B 126:303-315.

Mann T, Keilin D (1940). Sulphanilamide as a specific inhibitor of carbonic anhydrase. Nature 146:164-165.

Mitchell P (1987). Proton-motive osmoenzyme mechanisms in cytochrome systems: variation on a theme by Keilin. Proc Roy Soc B, in press.

Perutz M (1987). The birth of molecular biology. New Scientist No. 1561:38-41.

Pettigrew GW, Moore GR (1987). "Cytochromes c. Biological Aspects". Berlin Heidelberg London: Springer-Verlag, in press.

Tashian RE, Hewett-Emmett D (eds) (1984). "Biology and chemistry of carbonic anhydrases". Ann N Y Acad Sci 429, 640 pages.

Oxidases and Related Redox Systems, pages 51-64

DAVID KEILIN AND THE MOLTENO INSTITUTE FAMILY

E. C. Slater

Department of Biochemistry, University of Southampton, Southampton SO9 3TU, United Kingdom

Since Tsoo King and Thaddeus Mann have spoken mainly about Keilin's scientific work and its influence on the development of biochemistry and Peter Nicholls will also be doing this, I shall try to trace an also important although less direct effect that Keilin had through his influence on those who worked for shorter or longer periods at the Molteno Institute and thereby became members of what became known variously as the Molteno Family or Old Boys (and Girls) Club.

When I joined the Molteno Institute in October 1946, one year after the end of the World War, Professor Keilin (it would never have entered our heads in those days to call him David even among ourselves - maybe Keilin or simply DK, but more often the Professor) was a famous parasitologist as well as biochemist. His discovery of cytochrome had been made more than twenty years earlier, and this had been followed in the 15 years preceding the War by a series of papers on cytochrome and the other haemoproteins - haemoglobin, myoglobin, peroxidase and catalase and on other metal-containing enzymes, such as polyphenol oxidase, laccase and carbonic anhydrase. He was then 59 years old, a generation older than the post-War generation of students, even the more ancient among us whose post-graduate training had been delayed by the War. To me, coming from Australia, he looked just like a famous European scientist should look like and I approached my first interview with him with considerable fear and trepidation. However, although he was essentially a shy man, he looked kind and was kind and had the great facility of putting people at their ease with a gentle smile or even just a twinkle in the eyes.

Those were heady days immediately after the War. Although life in England was very austere with shortages of food and fuel in the exceptionally cold winter of 1946-47, the laboratories were often quite primitively equipped some in Nissen huts, and the English people as a whole were tired after the trials of the War, morale among the scientists, particularly in those returning to academic science after service in the forces or other War work, was high, in contrast, I must say, to the general atmosphere one finds today. In Cambridge particularly, an intellectual climate was being developed that was to lead, before the end of the first decade after the War, to the first elucidation of the amino acid sequence of a protein, to the first determination of the three-dimensional structure of a protein, to establishing the chemical structure of the gene, to the first chemical synthesis of a nucleotide, to the discovery of the basic mechanisms of nerve conduction and muscle contraction - just to mention the discoveries honoured with the Nobel Prize.

The Molteno Institute (Fig. 1) was quite small. It had been built in 1920 with money provided by a South African couple - Mr and Mrs Percy Molteno - to support research in parasitology by G.H.E.Nuttall, an immunologist as well

Figure 1. Molteno Institute in the winter of 1947-8, probably taken on a Sunday morning.

as parasitologist, whom Keilin joined from Paris and later succeeded as Director. Keilin's discovery of cytochrome made the Institute a centre for biochemical research as well as for parasitology and by 1946 it had, in fact, been renamed the Molteno Institute for Parasitology and Biology, which is a bit like the name of the hotel my wife and I stayed at during our first visit to Paris - Hotel of Portugal and the Universe.

However, less than one half of this quite small building was devoted to Keilin's biochemical research. An Agricultural Research Unit on Plant Viruses was housed in two tiny rooms at one corner of the top floor. This unit, then under the leadership of Kenneth Smith, was extremely active and successful. Second-in-command was Roy Markham and John Smith (known mostly by his initials,J.D.,at that time) who had just joined as a young research student - and beginning research students are very young in Britain - was the Benjamin of the Molteno. Shortly afterwards, a New Zealander - Dick Matthews, who had been engaged in the military campaign in Italy, showed up. Kenneth Smith was already a Fellow of the Royal Society and all three of the others named - Markham, Smith and Matthews - later became Fellows. This small unit attracted a number of distinguished visitors from overseas, of whom I shall now mention Leon Heppel from this country and Gerry Wyatt from Canada who discovered a new base in nucleic acids.

I should mention that no piece of apparatus in the Molteno was the special reserve of any individual or group, so that even those who knew nothing about plant viruses, or nucleic acids for that matter, could not fail to be aware of the fact that, when we received our first Beckman in 1949,J.D. Smith wanted to use it for measuring absorbance at 260 nm of the fluid eluted from pieces of paper in test tubes, in competition with my wish to measure the NADH oxidase activity at 340 nm. At least, I got to know that there was a new way of measuring nucleotides and that the old idea that nucleic acids contain equal amounts of the four nucleotides is not correct. I think that Roy Markham's contribution to the general know-how in Cambridge about nucleic acids in the late forties and early fifties, freely communicated to those inbibing at the local pub - called the Bunshop - has not been generally recognised.

Most of the top floor was reserved for a practical class in parasitology, but it was also used by Keilin for a course of eight lectures with demonstrations on respiratory enzymes. This course was designed specifically for a handful of third-year students in physiology, but despite the fact that it was given at noon on Saturdays the room was generally packed. Later during my stay in the Molteno, these lectures were shared with Ted Hartree and myself. Max Perutz also gave a course on X-ray diffraction of proteins in this room.

At one side there was a small museum and, at the other end, two tiny rooms. I was allocated one of these where to my surprise I found a schoolmate from Australia, Jack Legge, who was busy with the proofs of Lemberg and Legge, the first and for quite a long time only comprehensive treatise on haem compounds since Fischer. Jack and I had been in the same classes at preparatory school, lower and middle school in Australia and at University and our war work had also brought us into contact. The other room was occupied by Leo Levenbook, who was studying the biochemistry of the parasite Gastrophilis, the organism in which Keilin had first seen cytochrome in 1924. His room was directly above Keilin's office, which was unfortunate, since Leo was somewhat accident-prone and one night the rubber tubing to the cooling tube of a distillation set-up became disconnected, with the result that by morning a waterfall was flowing down the stairs and streaming under Keilin's door into his office. In fact, Jack Legge shared the blame, since he had forgotten to turn off the water when he left the Laboratory as Leo had asked him to do, so that when I arrived in the Molteno there were dark mutterings about visitors from drought-ridden Australian.

Legge returned to Australia to a lectureship in the University of Melbourne, and his place was taken by Claude Liébecq from Belgium, who has since its inception been Managing Editor of the *European Journal of Biochemistry*. The good relations that have always existed between this journal and its competitor, *Biochimica et Biophysica Acta*, is a good example of the effectiveness of the Molteno Club. Claude contributed to the reputation of the top floor by, with the help of a water pump, evaporating an acetone-containing solution - he was isolating $NADP^+$ from yeast - while the manometer bath that I was using was lit. [You might not understand this phrase - lighting a manometer bath - but at that time gas lights were widely used, not only for street lighting but also, until a few years previously, for lighting the old Chemistry Laboratory.The Molteno was, it is true, more modern with electrical lighting, but water baths, including those for the immersion of manometers for measuring respiratory activity, were heated by gas. It was then customary to do such experiments at blood temperature, which so happens is not far from the boiling point of ether. So the temperature of these gas-heated manometer baths could be controlled very closely by a simple but ingenious thermostat consisting of a column of mercury on top of a litle ether.] In any case, the acetone vapour over the sink came into contact with the gas flame under my bath. The inevitable explosion blew out my window and scattered burning acetone over the room and on Claude's hair. I was just outside the room and, with the help of a fire extinguisher, was able to help Claude subdue the fire. Fortunately, Claude was not badly burnt.

Early in 1948, Chen-lu Tsou arrived from China after a short sojourn in Birmingham and took Liébecq's place in the top room. Chen-lu claims that he taught me how to use chop-sticks by using two glass rods to remove Barcroft manometer flasks from a bath of chromic acid - sulphuric acid cleaning mixture. He also gives me the credit of teaching him the proper purpose of these manometer flasks, which was perhaps the more significant contribution, since Chen-lu published a series of remarkable papers on cytochrome *c* and succinate dehydrogenase. At a time when we knew virtually nothing about the structure of the mitochondria, he showed that cytochrome *c* is reversibly bound by electrostatic interactions to the particles of the Keilin and Hartree heart-muscle preparation (now known as sub-mitochondrial particles) and that the bound cytochrome becomes inaccessible to cyanide. He proved also the non-identity of succinate dehydrogenase with cytochrome *b* , which had seemed a possibility at that time. When he returned to China to join the Institute in Shanghai directed by Wang, Yin-lai, also a member of the Molteno Family, he and his colleagues succeeded in solubilizing and purifying succinate dehydrogenase and showing that it is a flavoprotein. Later, he played a vital role in the first chemical synthesis of a protein - insulin - by a team led by Wang. After the Cultural Revolution, Chen-lu became Deputy Director of the Institute of Biophysics of the Chinese Academy of Science in Beijing and still leads an active research group there, although,in the spirit of the four modernizations, China has seen fit to follow the Western European countries in introducing retirement from leading positions at much too early an age. In the tradition of the Molteno Family, I was pleased to supervise for a Ph.D. in Amsterdam one of Chen-lu's students and a second is at present working under Simon Albracht for her degree in Amsterdam.

A second Chinese - Chin - also arrived at about this time to work on bacterial cytochromes. His English was not as good as Chen-lu's, and I was appointed, together with Robin Hill, as his Ph.D. examiner, because as Keilin put it, I could understand "Chinnese". Sadly, Chin, who moved to Cleveland, Ohio, was killed there by a hit-and-run driver.

In the autumn of 1949, when I went to Severo Ochoa's laboratory in New York, my place in the top room was taken by David Morell, an earlier associate of mine from my Canberra period, who worked on xanthine oxidase and leucoperoxidase. After obtaining his Ph.D., David Morell returned to Australia to become the chief associate of Rudi Lemberg,the leading biochemist in Australia, who had collaborated with Keilin in the 1930's. Later Morell became Principal of the College of Higher Education at Wagga Wagga.

Before leaving the Molteno, Morell moved to the second floor and his place was taken by Walter Bonner, who firmly established that phosphate is not necessary *per se* for the operation of the respiratory chain, a point that needed to be established at that time and repeatedly told to the organic chemists for many years. Walter and I collaborated in one of my most cited papers, simply because we showed that Briggs-Haldane rather than Michaelis-Menten kinetics applies to succinate dehyrogenase, a finding that greatly pleased Briggs, whom, unknown to me, was then Professor of Botany in Cambridge and who sent me by personal messenger over a distance of about 100 yards a series of hand-written letters asking for clarification of certain points.

Walter returned to the United States and joined the Johnson Foundation where he became the great proponent of plant mitochondria.

Chen-lu was succeeded in the top room by a Swiss mountain climber - Alfred Tissieres - who, like Chin, studied bacterial cytochromes, in particular those in Azotobacter. Alfred remained in Cambridge for a number of years, becoming a Fellow of Kings College. He had a wide circle of interesting friends - including the Cambridge Himalaya climbers - who frequently visited him in this room.

When I returned from America in the summer of 1950, I moved into Leo Levenbook's room on the other side of the top floor. Leo had left for USA where he became a leading insect biochemist and is now at the National Institutes of Health. I stayed in this room until 1955, being joined at different times by a young technician Ann Searle, a research student Francis Holton and post-docs Ken Cleland from Australia and Paul Greengard from America. Alfred Tissieres, with whom I collaborated in studying oxidative phosphorylation in Azotobacter, Ann Searle and Paul Greengard all worked with me for short periods in Amsterdam. Francis became Lecturer in Biochemistry at the Royal Veterinary College in London and Ken Cleland Professor of Histology and Embryology in Sydney. Ann later joined Ef Racker and her name appears (not directly recognisable as Ann Searle) on some classical papers of his on oxidative phoshorylation. Paul Greengard has made his name as one of the founders of a whole new area of biology - cyclic AMP.

Among the large group of Alfred Tissieres friends were Francis Crick and a young American who never seemed to do any work but had time to talk endlessly with Alfred, disturbing me in my work with what I considered their pointless chatter. How wrong can you be? I never imagined that I was listening to the greatest advance in biology since Darwin. Alfred later joined Jim Watson in Harvard and then returned to his native country, to take his part in developments that was to make Switzerland one of the leading centres of molecular biology.

Before leaving the top floor, I should mention a small room at the head of the stairs where, as long as I was in the Molteno, Parr Tate kept a colony of blowflies (*Calliphora erythrocephala*). I am sorry to say that I never discovered what he did with them, but Stan Lewis and I made thankful use of them when we were able to demonstrate oxidative phosphorylation in their thoracic muscle mitochondria - not surprising, you might think, but it had been claimed that they got along without it. Also I must not forget the roof on which the Kipps apparatus was placed, an essential piece of apparatus in the days before Boehringer *et al* . when one had to make one's own NAD^+, ATP and glutathione, all of which involved precipitation as the salts of heavy metals, which then had to removed as the sulphide. The roof was also the cold room - in winter only of course.

Up to now, I have mentioned only the periphery of the Molteno - the top floor. Below was the nerve centre of the biochemical activity. At one corner was Keilin's office where I learned whatever I know about the philosophy of scientific research. My fondest and most thankful memories are of the long discussions with Keilin, particularly in my first years in Cambridge. He did not have many visitors and usually responded with a "come-in" to my knock towards the end of the day, when I was excited to tell him about our latest experiment. The subsequent discussion ranged widely, sometimes into reminiscences of his controversy with Warburg, and I would not get home until much later than the usual Australian dinner time, much to the despair of my poor wife, since the longest discussions with Keilin seemed to be on the days when she had decided to cook an accumulated meat ration (one shilling and threepence worth per person per week). However, "I was talking with the Professor" was an accepted alibi and it was true also. I think that it is the shared experience of having these discussions with a great scientist and a kind human being that welds together those who have worked at the Molteno Institute into a family.

The tearoom at the opposite end of this floor played its role. In my time it was presided over by the parasitologists - Parr Tate and Ann Bishop - with Ted Hartree making the tea. It was a real family group with discussions about the most trivial matters, such as the weather, housing, trains, births, marriages, tennis, stories for the newcomers about Molteno Old Boys, never science. Perhaps that is why these teas played a role in creating the family atmosphere. There was a genuinely kindly but not prying interest in one's doings and one made lifelong friends.

Outside Keilin's door was an International Centrifuge and in a separate little room - just under the fly room of the top floor - was a monster of a centrifuge of French manufacture that could spin four 1-litre flasks at 3000 rev./min, which was unique in those days. It was used mainly for spinning down the cytochrome *c* after precipitation with trichloroacetic acid. If one wanted to centrifuge cold in this or the International Centrifuge, one just added sufficient ice to the tubes to keep the contents cold during centrifugation. A rapid (then called "ultra") Sorvall centrifuge (about 14,000 rev./min) did not arrive until 1950. This was placed in a deep-freeze cabinet and, by carefully controlling the time in which the head was placed in the deep-freeze before centrifuging, it was possible to spin a mitochondrial suspension while keeping the temperature between 0^0 and 5^0. A few years later, Keilin obtained a grant from the Royal Society to purchase one of the the first generation of M.S.E. refrigerated centrifuges which was brought up the main staircase to the room on the top floor, badly scratching the wall on the way, for which I was blamed and reminded of whenever I visited the Molteno Institute, as I did regularly long after I had moved to Amsterdam.

Next to Keilin's room was the Instrument Room, which housed a large Hilger visual spectrometer, the Hartridge reversion spectroscope that Keilin had used to measure the position of the absorption bands of cytochrome, as well as a photoelectric absorptiometer, balances, pH meters. The addition of a Beckman photoelectric spectophotometer in 1948 opened up new possibilities.

Opposite the Instrument Room was what was known as the Chemical Laboratory, the largest room in the Institute. This is where Ted Hartree worked and where many of us received basic instruction in such things as making the Keilin and Hartree heart-muscle preparation and, from Keilin himself, in the use of the microspectroscope which had pride-of-place in this room.

Ted Hartree was Keilin's longest associate, having joined him in the early thirties, after obtaining his Ph.D. from London. He published no less than 45 papers with Keilin. He remained with Keilin for many years after the latter's retirement in 1952 and then joined Thaddeus Mann's laboratory. Ted Hartree, who is now retired and living in Cambridge,is quite well,but he did not feel that his health is sufficiently stable to warrant the tiring trip to Portland. His selfless devotion and skilful experimentation was a great help to Keilin over the years of their association, and he is a dear friend to the members of the Molteno Club. His name has an honoured place in the history of biochemistry as the co-author of a series of classical papers.

Opposite the Chemical Laboratory was the room of Thaddeus Mann, who, after collaborating with Keilin in the work that he has just described, turned to the study of biochemistry of spermatozoa. Thaddeus and his wife Cecilia became the great authorities on the biochemistry of reproduction, culminating in the appearance a few years ago now of their book on this subject. Already in 1951, Thaddeus became the third Molteno Institute F.R.S., after Keilin and Kenneth Smith. He usually had one visitor working with him, among whom the memory of the late Enzio Leone, who became Professor of Biochemistry at the University of Naples, evokes fond recollections.

In the room next to Thaddeus was Hans Laser, who had worked in Meyerhof's laboratory in Heidelberg, before having to flee from Hitler. Hans was a colourful character with a warm personality, full of enthusiasm and convinced that every research project would uncover the philosopher's stone. He worked in or near Cambridge right up to his death only a few years ago. We collaborated on the effects of heavy water on the activity of the respiratory chain and oxidative phosphorylation, research that was completed during a visit of Hans to Amsterdam.

The next room was occupied by a succession of visitors who stayed for one or two years. The ground floor was largely, but at that time not exclusively, occupied by the parasitologists : Parr Tate, the University Lecturer in Parasitology, and Ann Bishop, head of a M.R.C. Unit in malariology, who became a F.R.S. in 1959. There was also a large preparation room, where the first stages of the large-scale isolation of enzymes, such as cytochrome *c*, catalase, horse-radish peroxidase and later hexokinase, and of coenzymes, such as ATP, NAD^+ and glutathione, were carried out - usually in a large and vulnerable glass jar which often took a part of a scientist with it when it gave up the ghost - in my case the top of my index (and pipetting) finger. The whole Institute wept for days during the first days of the isolation of horse-radish peroxidase, not for emotional reasons, but because of the lachrymatory mustard oil (not to be confused with mustard gas!) liberated into the atmosphere.

The large belt-driven mortar and pestle used for making the Keilin and Hartree heart-muscle preparation was in the basement, as was the workshop and chemical store, presided over by the most important member of the Institute after the Professor - Charles (surname Harpley), the chief technician, as well as James (Corley), Percy (Larkin) and Baker (whose first name I never knew).

There was a large bird room for Ann Bishop's malaria work, occupied by a bevy of young girls, as well as the feathered variety. When I arrived in the Molteno, there were a number of non-parasitologists also working on this floor - a young man recently discharged from the services was to be seen preparing myoglobin from whale meat which was not rationed. The crystals were examined by X-ray crystallography in the Cavendish Laboratory, with results that finally led John Kendrew as well as Max Perutz to a handshake from the King of Sweden. Next door (or was it in the same room as John Kendrew? - I am not sure), Philip George, who gave the opening talk on the fitness of oxygen at the first ISOX Symposium and who identified the chemical nature of the Michaelis intermediate in peroxidase activity, was then working on catalase.

Also on this floor I think, but I am not sure, Dan Arnon, who was already well known for his work on trace elements in plant growth, was taking his first steps in photosynthesis. Later, Dan made one of the most fundamental discoveries in this field - photophosphorylation - which has not, in my opinion, received the recognition that it deserves. In 1947, we stood in awe of his ability to talk, rather slowly, but without a pause, for the entire tea break.

There was also a small office, occupied by the secretary Miss Seamer, on this floor, and a spare room, which in the summer of 1947 was completely filled by what looked at the time like apparatus from a science-fiction novel. In fact, it was the 1947 version of Britton Chance's spectrophotometer, which Brit had brought over from Stockholm to do some experiments on catalase, which, like all Brit's experiments, were duly published. Unfortunately, I do not have a photograph of Keilin's face when he first saw this apparatus.

I find that I have concentrated unduly on the 1940's, but my memory is the clearer the further back it goes. I have not mentioned my good friend Jean Rosenberg from Paris or two more Australians, Max Clark, who became Head of a new School of Biological Sciences at Flinders University, Adelaide and George Humphrey, who became Director of the Fisheries Research Laboratory near Sydney. There was a continuous stream of distinguished visitors in the 1950's and 1960's. A most important one was Emanuel Margoliash, who tried to persuade us that every experiment had a simple explanation if only one knew what it was, and who became the great expert on cytochrome *c* and has used this protein as a model for incursions into many fields of biology; of course, I recall Lucile Smith, the investigator of cytochrome *c* oxidase for longer than it is polite to remind a lady, growing Rhodospirillum in the museum. Ron Estabrook, Tsoo King, the late Martin Morrison, David Shappirio, Geoff Schatz and Bob Burris arrived just as or after I moved to Amsterdam. And, of course, I must not forget the next speaker in this programme, who was Keilin's last student and is now Chairman of the Department of Biological Sciences at Brock University in Canada.

Frequent visitors to the Molteno from other laboratories were Malcolm Dixon, Bob Morton, Robin Hill and Joan Keilin from the Biochemistry Department; Lord Rothschild from Zoology; Professor Roughton from Colloid Science; and Max Perutz, the most distinguished Molteno Old Boy, who was given space in the Molteno in 1938 and in the 1940's was working at the Cavendish in the Medical Research Council Unit for the Study of the Molecular Structure of Biological Systems, the forerunner of the MRC Laboratory of Molecular Biology. On the precise 100th Anniversary of Keilin's birth on March 21 last, Max gave the Keilin Memorial Lecture at Magdalen College, Cambridge, Keilin's old college [much of the content of this lecture is to be found in (Perutz,1987a)]. He has also written a delightful article entitled "Keilin and the Molteno" [Perutz (1986) and Perutz(1987b)], in which he emphasises the great debt he and Kendrew owed to Keilin.

The strength of the ties holding together the Molteno Family were revealed when one visited other countries or attended international meetings and met previous generations of the family, such as André Lwoff in Paris, Bruno Straub in Budapest, Emil Smith in U.S.A. or Wang, Yin-lai at the International Congress of Biochemistry in Brussels and at the FEBS Meeting in Warsaw and in Shanghai. Membership of the Old Boy's Club also meant easy acceptance by Keilin's contempories, including Otto Warburg in Copenhagen.

An example is the way these ties helped to restore scientific contacts with China. Even before the end of the Cultural Revolution but after the visit of President Nixon, Emil Smith was a member of a party of American scientists who visited China. Emil insisted that he visit Chen-lu Tsou in his laboratory in Beijing. In fact, because of the Cultural Revolution, Chen-lu had no laboratory or room to work in, not even a chair to sit on, but he was duly provided with a room furnished with a chair and a shelf of bottles which he filled with tap water to impress the American visitor. He even kept the laboratory after the visit was over, perhaps because of the difficulty of reversing decisions taken by powerful political figures. The discussions that Bill Whelan and I had with various officials in China in 1979 that finally led to the re-entry of China into the International Union of Biochemistry and, on the basis of our agreement, into the other scientific unions and eventually into the International Council of Scientific Unions, were greatly helped by the friendship started at the Molteno Institute with Chen-lu and the Old Boy connections with Wang, Yin-lai.

Re-reading what I have just written, I think that I can say that the Molteno Institute Family constitutes a pretty distinguished group, present company except for the speaker, not excepted, for what was essentially a tiny

Institute. However, it is good to remember, especially now that the Molteno Institute has been abolished, that the Molteno Institute Family was not the creation of a building, but of a very special man - David Keilin - whose memory we honour today.

I would like to conclude by quoting a passage from the Magdalen College Magazine written by a Fellow of the College, Francis Turner (1963), that has also been quoted by Max Perutz (1987b), that I believe expresses beautifully how the members of the Molteno Institute Family saw Keilin.

"David Keilin wore the mantle of his world-wide distinction as an invisible garment: what was visible was a disarming modesty, true friendliness, and a mind so quick that he immediately understood all that his neighbour meant, however imperfectly expressed. But this understanding was not only quickness of wit: it arose from the deeply charitable view he had of human nature, which, together with his wide experience gave him an unusual insight into those with whom he came in contact. He was a most remarkable and lovable person, patient and wise, with so great a spirit in so small a frame."

REFERENCES

Perutz M F (1986). Keilin and the Molteno. Cambridge Review October 1986: 152-156.
Perutz M F (1987a). Physics and the Riddle of Life. Nature 326: 555-558.
Perutz M F (1987b) Keilin and the Molteno. Biochem. Soc. Bull. 9: 4-8.
Turner F (1963). Magdalen College Magazine and Record I,13.

Oxidases and Related Redox Systems, pages 65–78

'INTEGRITY OF STRUCTURE' AND 'LATENT LIFE' - THE RESPIRATORY CHAIN, CLAUDE BERNARD, AND DAVID KEILIN

Peter Nicholls

Department of Biological Sciences
Brock University
St. Catharines, Ont. L2S 3Al Canada

INTRODUCTION

While we are looking at Keilin's photograph, I want to begin by telling you a dream. In the dream, I was standing in front of a door and looking at the handle on the door, and I said to myself 'there is no handle on the other side', and I opened the door and walked into the room. There, sitting at a desk, was David Keilin, and he got up and came to me and said "Peter, we have been waiting for you" in that unmistakable tone of mild reproof. Then I knew where I was, I became afraid, and I awoke.

Centenaries -- births and deaths -- have different ambiences. A lot depends not only on when one is born but also upon when one dies. David Keilin's dates are 1887 - 1963. He shared a birth year with a great variety of people. I noted some of them recently as they came along -- Edith Sitwell, Artur Rubinstein, Rupert Brooke; and it may be worth thinking about the fact that while many people doubtless dream of Rubinstein - we all have records of his piano performances because he died only a couple of years ago - nobody, I suspect, dreams of Brooke, who died during the First World War -- another place in time.

KEILIN AS A BIOLOGIST

Keilin has been emphasized as a biologist. What is biology? Biology is decay balanced by some kind of restoration of a previous state, and this maintenance that we all depend on was, of course, pointed out in the famous phrase of Claude Bernard that a "Free life depends upon the fixity of the interior milieu." Keilin had an enormous respect for Bernard and thought of himself as coming from that French School of experimental biology, and in his background there was undoubtedly a very strong French connection.

Maurice Caullery was his supervisor and mentor, and Alfred Giard, Caullery's supervisor, was the first occupant of the chair at the laboratory of the Evolution des Etres Organises, the evolution of organized beings, which was established in the 1880's just after Bernard's death. Bernard was teaching general physiology at the Jardin des Plantes, a few years previously.

And Giard was a French evolutionist. He thought of himself as a Darwinian but his biography said that he was 'Darwin matine de Renan', that is, made more optimistic. The French School of biology has never, to this day, perhaps, fully accepted the implications of Darwinian thought. Giard's research programme was called transformism. It was based upon his work in parasitology and developmental biology, and one of his main themes was that life, unlike the inanimate world, depends upon the past. As he said, on the other hand in the world of mechanics 'la connaissance du passe est inutile'. So biology has something to do with time, as Tsoo King pointed out at the beginning. In my view, many of Keilin's ideas and approaches become more comprehensible if he is seen as an experimental biologist working with cell extracts and not as some kind of protoenzymologist or protobioenergeticist. It was from that point of view - that angle, that Keilin, in contrast to some of the biochemists or chemists of the time, was able to develop ideas such as the idea of the steady state in the respiratory chain. This idea seems to us almost self evident but was one which in fact Warburg and the Japanese school had enormous difficulties with. The appearance of the cytochrome system as seen in vivo was regarded by them as evidence for some kind of special state of oxygen carriage -- a different state from either the ferrous or the ferric. And it took a biologist who was working from the ideas of the experimental physiology school of the previous century to explain what we now take for granted, that the in vivo condition is a steady state. That is part of the French connection.

At the same time, and in Keilin's own lifetime, there was the development of experimental physiology in the United Kingdom and in this country, and this development of ideas followed Bernard, but adopted a set of Anglo-Saxon attitudes to that experimental point of view. A prime example is Lawrence Henderson, who was the first to publish Bernard's 'Introduction to the Study of Experimental Medicine' in English, and who introduced the ideas of homeostasis in the blood and the fitness of the environment, the latter being something which also reflected the ideas of Alfred Giard.

BIOLOGY IN THE EARLY 20TH CENTURY

Joseph Barcroft, who used the ideas of Claude Bernard very extensively in his analysis of physiology and blood biochemistry, entitled the first three chapters of his book 'Features in the Architecture of Physiological Function', after the quotation from Bernard on the fixity of the 'milieu interieur'. And he at the same time discusses Keilin's work. In that period of the 1920's and 1930's, Barcroft had an enormous influence in Cambridge on people's ideas about control and physiology. He was also a coauthor, as noted at the end of the abstracts, of one paper with David Keilin at the time and mentor in a number of others, particularly those which concerned the effects of gases on the respiratory chain. His idea of 'Life as flux' was another French idea that was previously exemplified by Bergson - there is that something about life that we can't quite picture. And someone like J. S. Haldane, who was an extraordinarily skilled analyst of physiological systems, and able to explain nearly all the physiological control mechanisms of respiration and circulation which he looked at, in what we would regard as a materialistic or a mechanistic way, was at the same time a kind of vitalist. He developed a theory which he called 'organicism'. This was a bit like Newton who developed the theory of gravitation and then invoked the hand of God in putting right the little things that went wrong when the planetary orbits didn't quite work out. J. S. Haldane did the same thing when he developed the vitalistic theory of organicism.

Gowland Hopkins, the great originator of the Biochemistry School at Cambridge, was a much more thoroughgoing materialist than Haldane and when he lectured on organicism, attempted to take Haldane to task . But Hopkins was also responsible for the idea that life was a "dynamic equilibrium in a polyphasic system" - his version of 'Life as flux'. This led to the extreme version of this school of thought, which was expressed at the end of the war and in the middle 1940's by Schoenheimer in "The Dynamic State of the Body Constituents", which postulated that even macromolecules were in a continuous steady state. Proteins could be steady state entities. And that kind of viewpoint went too far, because it meant that there was a chemistry which was not approachable. Warburg called Hopkins a romanticist who preferred the unknown to the known. Yet Hopkins was the one who was criticising Haldane's drift into vitalism. There were also others at that time who issued warnings. Pirie, in Hopkins' 75th birthday book, wrote a chapter which now seems to us self evident but was very unusual at the time, entitled "The meaninglessness of the terms 'life' and 'living'".

It was during this period that Keilin had transformed himself from a parasitologist to an experimental biologist or biological enzymologist, although he was still interested in morphology and structure, and he remained for a long time editor of "Parasitology". I remember that on one occasion, in a moment of youthful indiscretion, I made some disparaging remarks about traditional biology, and Keilin told me that if he heard such comments again, he would include on my research programme a study of the anatomy of a fly's brain, which would teach me respect for that kind of work. So after that, I was a little more cautious in expressing my views.

THE 'COLLOIDAL' SYSTEM

It was at that point that Keilin produced what I think is one of his most significant papers; one that is not usually considered as such because it was published in the era when cytochrome oxidase was being defined. This was the 1940 paper in the Proceedings of the Royal Society on the colloidal structure of the succinate oxidase system. In that paper, Keilin and Hartree showed that the activity of a heart muscle preparation, a submitochondrial particle preparation as we now know it,depended on both the integrity of its components and the integrity of the colloidal structure. Keilin and Hartree were at pains to point out that the action of inhibitors may be due either to an effect on the components of a system or on the colloidal structure which sustains them. For example, succinate dehydrogenase and cytochrome oxidase can both be perfectly active in a given preparation but not form a succinate oxidase system. As they said, "The efficiency of the succinate dehydrogenase - cytochrome system depends not only on the integrity of its components, but also upon that of the colloidal structure that supports them and ensures their mutual accessibility". And when over the next ten years or so, where various people claimed to have taken the succinate oxidase system apart and put it back together again, Keilin and Hartree were in the position of critics who showed in many cases up to the late 1950's that those separations and reconstitutions were not in fact valid, that they were due to activations of one sort or another or to the removal of inhibitors, and were not true reconstitutions. They developed a position in the mid and late 1940's in which they envisaged the respiratory chain or respiratory complexes as being arranged in some way on a colloidal surface (as Dr. Slater has pointed out).

The actual model which was in their minds at that time is hard to reconstruct, and unlike the cavalier style of today when with Macintoshes and MacPaint programs we quickly draw what we think is there, you have to look a long way through the primary articles of the period to find anything that we would regard as a picture. But they pointed out, for example, in 1947-1949, that drying a Keilin-Hartree preparation inactivated it, and then you could reactivate it with an appropriate buffer, with appropriate proteins, or even things like calcium or aluminum whose hydroxides formed colloidal systems; and they envisaged those colloidal precipitates as providing 'new surfaces for orientation'. These were a kind of forerunner of the proteoliposome but an image of the respiratory chain which perhaps now, if we think about it, we would find a little strange. It was only in the late 1950's that the first true reconstitution was achieved, and of course, that was in the hands of Tsoo King at the Molteno, when for the first time, the succinate dehydrogenase was removed and replaced. But it is rather interesting to note that the exciting part of this was in the letter to Nature and the lengthy paper in the Proceedings of the Royal Society put a tremendous emphasis upon the action of inhibitors and things which seemed important to Keilin. It is only recently that I realized the continuity of those ideas because one of the important points was that cyanide in that system was acting as a blocker of reconstitution and not, despite its known effects elsewhere, as a blocker of one or other of the individual components of the system before reconstitution.

LATENT LIFE.

So where does this theme intersect with the ideas of Claude Bernard on latent life? According to Bernard, there are three forms of life. This is in his book on the "Phenomena of Life common to Animals and Plants", which is the one Bernard work which is referenced by Keilin in his Leeuwenhoek lecture.

According to Bernard, there are three forms of life -- and only the third form is dynamic. The first form, according to Bernard, is pure structure. What, according to Bernard, is life? "Life," says Bernard, at one point, "is memory". Memory -- what is memory? - pure structure. The three forms of life are 'latent', 'oscillating' and 'constant'. It is only the last one which maintains its internal milieu in the way that became so important for the Cambridge physiologists of the 20's and 30's. The first form, latent life, is not characterized by any metabolism; it is characterized by structure alone. It is the form of life shown by mummy wheat, dried eel worms, dried rotifers and tardigrades -- the creatures which take their opportunity one year or another if the right environment comes by, the parasites that Giard and Caullery worked on. Giard was the one who introduced the term 'anhydrobiosis'

And what was Keilin's contribution to the idea of the importance of structure? It was to take it to the molecular level. In 1947, Keilin and Wang wrote a paper about the stability of proteins, and this is a paper which when I first read it, my thought was - 'it was amusing, but why do it?' It takes a while to understand why a thing is done. It was a paper which looked at George Henry Falkiner Nuttall's samples of haemoglobin; Nuttall, the godfather of immunochemistry, had collected haemoglobin and blood samples from all over the world from the most exotic animals. They used Himalayan bears', sea-lions', jackals', and ant-bears' blood. They had all been in a drawer sealed in ampoules in the Molteno, and they were taken out after 40 years and an experiment was done. This experiment involved looking at the stability of the haemoglobin and the other proteins in the blood samples. What they found was that if you keep out the light and the oxygen and the bacteria, structure -- molecular structure -- is maintained. The comparison between a 24-year-old sample of horse haemoglobin and a fresh sample showed that the sigmoid oxygen binding curve is preserved -- one of the characteristic molecular features of haemoglobin will last over 20 years.

Their other idea was that when the molecules were in the Bernardian latent state, they were in a structurally different form. When dried, haemoglobin existed in a different form and when frozen haemoglobin existed in a different form. The apotheosis of the latent life phenomenon was provided by low temperature freezing of organisms or of molecules, because at the temperature of liquid air or as Luyet showed soon after, liquid helium, it's hard to imagine, as many of the opponents of this idea suggested, that there was anything going on of the steady state variety.

LIFE AT LOW TEMPERATURES.

Total latent life or complete anabiosis could be proved if something could be frozen to the temperature of liquid helium and then restored to its functional state. The original use of low temperature was, as I understand it, to store the Keilin-Hartree preparation or submitochondrial particles, and it was made possible, despite the absence of cold rooms in the Molteno, by the presence of the Low Temperature Station next door, which had a plentiful supply of things like liquid air and eventually other liquid gases. But when the material that was being stored was frozen, of course, Keilin took an opportunity to look at it in the microspectroscope and realized that the freezing technique could be used for other purposes. The low temperature technique became a method not only for storing, but examining material, and it, of course, then developed in the hands of the photosynthetic biophysicists into a way of working with photosynthetic systems. It was a state in which at the temperature of liquid air everything had frozen out - structure alone was left .

In 1959, Keilin presented the Leeuwenhoek lecture to the Royal Society, which was the only formal lecture which I ever heard him give, and as a student at that time, I was puzzled by his choice of topic. His topic was "Anabiosis or Latent Life", and he went into the history of the subject , discussing the work of Leeuwenhoek himself, of Needham and Baker and the 19th century French workers who called the phenomenon of freezing organisms and thawing them out in a viable form - "resurrection", which, of course, caused some raised eyebrows even when the term was used by the Abbé Spallanzani. Keilin also quoted the work of John Hunter who attempted to do something on the large scale - the California scale, by freezing carp and reviving them, and it may be worthwhile reminding ourselves that this idea of a kind of immortality is not just a phenomenon of the southern part of the United States in the present era, but it was also Hunter's idea in 1776. "Until this time", Hunter remarked, "I had imagined that it might be possible to prolong life to any period by freezing a person in the frigid zone, as I thought all action and waste would cease until the body was thawed. I thought that if a man would give up the last ten years of his life to this kind of alternate oblivion and action, it might be prolonged for a thousand years, and by getting himself thawed every hundred years, he might learn what had happened during his frozen condition. Like other schemers, I thought I should make my fortune by it, but the experiment on the carp undeceived me." Well, we've improved things since then, but whether we shall make our fortunes or become immortal is yet to be seen.

There are various kinds of immortality. There is the immortality of the Greeks who believed that there was a kind of soul which escapes from the body at death, and there is the immortality of that branch of the Christian faith which was held by Milton and Priestley who believed that the soul sleeps or dies with the body to be resurrected later on. What does Keilin say in the Leeuwenhoek lecture? He says that the concept of life is synonymous with that of the structure which supports all the components of its catalytic systems.

TABLE 1

active life = normal metabolism

hypobiosis (Monterosso 1934)	dormancy (hypometabolism)	hibernation aestivation diapause quiescence
	(ametabolism) latent life = cryptobiosis (D.K.) = anabiosis (Preyer 1891) = abiosis (Schmidt 1948)	

(from Keilin, 1959)

TABLE 2

	states*		conditions
hypobiosis (Monterosso, 1934)	(1) anhydrobiosis (Giard 1894)	(1)	dehydration
	(2) cryobiosis (D.K.)	(2)	cooling
	(3) anoxybiosis (D.K.)	(3)	lack of oxygen
	(4) osmobiosis (D.K.)	(4)	high salt concentration
	(5) combinations of (1) to (4)		

*New terms in tables 1 and 2 are initialled D.K.

(from Keilin, 1959)

And he tried to pull together all this work on different states of latent life attained by dehydration or by low temperature or by anoxia or even by high salt concentration by introducing a whole group of new terms (see Tables 1 and 2), the first one in Table 2 having been created by his academic grandfather in 1894, when he discovered the phenomenon of anhydrobiosis. And Keilin also introduced the term 'cryptobiosis', which did not, as far as I know, take off, although it was used by Crowe and Cooper in their Scientific American article.

CONSEQUENCES.

What were the consequences of this idea? I have listed three kinds of consequence - the technical, the theoretical and the philosophical. The technical consequence of the work on low temperature was its use to study photosynthesis and to trap steady states -- to take the system in Bernard's third living state and to trap it by reducing the temperature, -- the 'triple trapping' method of Britton Chance. But I think it is also important to remember that X-ray crystallography of proteins would not be possible unless we thought that they were static molecular entities whose crystal structure is relevant to their biological functioning, which was not the idea of Schoenheimer and some of his followers. And in that respect, Keilin could be regarded as a protomolecular biologist rather than a proto-biochemist, as Dr. Mann commented.

The theoretical consequence is that the idea of the steady state - the chemical consequence of Bernardian 'constant life' - is based upon an underlying structure, as Pirie claimed without much evidence in 1937, and Keilin was at pains to point out later on, and not on some living force or entelechy of the kind that Henri Bergson and Haldane in Britain espoused, which has been and remains in some quarters a very persistent idea. I think it's just as well that Keilin didn't follow Bergson into philosophy and potential oblivion,

And the last point, the philosophical point -- is that if life is equal to structure, then the disappearance of that structure means the loss of life, because structure and life are 'non-conserved entities' , and in this respect completely different from matter and energy. And so when I stand in front of that door with a handle on my side but none on the other side, what I shall remember as far as possible is not to turn that handle -- because Keilin taught me that he is not on the other side -- there is no other side.

BIBLIOGRAPHY

Baldwin E (1949) 'Dynamic Aspects of Biochemistry' Cambridge Univ. Press, Cambridge.

Barcroft J (1934) Features in the Architecture of Physiological Function (Cambridge Comparative Physiology Series) Cambridge Univ. Press.

Bergson H (1911) 'Creative Evolution' (trans. A. Mitchell) MacMillan & Co., London, U.K.

Bernard C (1878) 'Lecons sur les phenomenes de la vie communs aux auinaux et aux vegetaux', Paris (trans. R. P. & M. A. Cook (1974) Burns & Harris, Dundee).

Caullery M (1909) 'L'oeuvre scientifique d'Alfred Giard' Bull. Scient. de la France et de la Belg. 42, xv-lxiii.

Crowe JH & Cooper AF (1971) 'Cryptobiosis' Scient. Amer. 225, no. 6, 30-36.

Franklin KJ (1953) 'Joseph Barcroft, 1872-1947' Blackwells, Oxford.

Haldane JS (1917) 'Organism and Environment as illustrated by the Physiology of Breathing' (Silliman Memorial Lectures) Yale Univ. Press (New Haven).

Haldane JS (1932) 'Materialism' Hodder & Stoughton, London.

Hartree EF (1973) 'The Discovery of Cytochrome'. Biochemical Education, 1, 69-74.

Henderson LJ (1913) 'The Fitness of the Environment' MacMillan Co., New York.

Hopkins FG (1949) 'A lecture on organicism, delivered upon an unknown occasion, 1927' in Hopkins and Biochemistry (ed. J. Needham & E. Baldwin) W. Heffer & Sons, Cambridge.

Keilin D (1953) 'Stability of Biological Materials and its bearing upon the problem of anabiosis' Science Prog. 41, 577-592.

Keilin D (1959) 'The Problem of anabiosis or latent life: history and current concept' (The Leeuwenhoek Lecture) Proc. Roy. Soc. B150, 149-191.

Keilin D (1960) 'Maurice Jules Gaston Corneille Caullery, 1868-1958' Biog. Mem. Fell. Roy. Soc. 6, 13-31.

Keilin D & Hartree EF (1940) 'Succinic dehydrogenase-cytochrome system of cells. Intracellular respiratory system catalysing aerobic oxidation of succinic acid'. Proc. Roy. Soc. B129, 277-306.

Keilin D & Hartree EF (1947). 'Activity of the Cytochrome System in Heart Muscle Preparations' Biochem. J. 41, 500-502.

Keilin D & Hartree EF (1947) 'Effect of Haematin on the oxidation of Succinic Acid by Tissue Preparations' Biochem. J. 41, 503-506.

Keilin D & Hartree EF (1947) 'Comparative study of spores and vegetative forms of Bacillus subtilis' Ant. van Leeuw. Jour. Microbiol. Serol. (A. J. Kluyver Jubilee Vol.) 12, 115-128.

Keilin D & Hartree EF (1949) 'Activity of the Succinic Dehydrogenase-cytochrome System in Different Tissue Preparations'. Biochem. J. 44, 205-218.

Keilin D & Hartree EF (1949) 'Effect of low temperature on the absorption spectra of haemoproteins; with observations on the absorption spectrum of oxygen' Nature 164, 254-259.

Keilin D & Hartree EF (1950) 'Further observations on Absorption Spectra at Low Temperatures' Nature 165, 504.

Keilin D & Hartree EF (1952) 'Effect of Drying upon the Absorption Spectra of Haemoglobin and its Derivatives' Nature 170, 161.

Keilin D & Hartree EF (1953) 'Cytochrome' Brit. Med. Bull. 9, 89-96.

Keilin D & Hartree EF (1955) 'Relationship between certain components of the cytochrome system'. Nature 176, 200-206.

Keilin D & King TE (1958) 'Reconstitution of the succinic oxidase system from soluble succinic dehydrogenase and a particulate cytochrome system preparation' Nature 181, 1520-1522.

Keilin D & King TE (1960) 'Effect of inhibitors on the activity of soluble succinic dehydrogenase and on the reconstitution of the succinic dehydrogenase-cytochrome system from its components' Proc. Roy. Soc. B152, 163-187.

Keilin D & Wang YL (1947) 'Stability of Haemoglobin and of certain Endoerythrocytic Enzymes in vitro' Biochem. J. 41, 491-500.

Le Dantec F (1909) 'Alfred Giard (1846-1908)' Bull. Scient. de la France et de la Belg. 42, ii-xiii.

Mann, TRR (1964) 'David Keilin, 1887-1963' Biog. Mem. Fell. Roy. Soc. 10, 183-205.

Nicholls P (1963) 'Cytochromes - a survey' in The Enzymes (2nd ed.), Vol. 8, 3-40 (ed. P. Boyer, H. Lardy & K. Myrback), Academic Press, N. Y.

Nicholls P (1984) Cytochromes and Cell Respiration (Carolina Biology Reader No. 66) Carolina Biological Supply Co., Burlington, N. C.

Pirie NW (1937) 'The Meaninglessness of the Terms Life and Living' in Perspectives in Biochemistry (ed. J. Needham & D. E. Green), pp. 11-22, Cambridge Univ. Press, Cambridge.

Schoenheimer R (1941) 'The Dynamic State of Body Constituents' Harvard Univ. Press, Cambridge, Mass.

Oxidases and Related Redox Systems, pages 79–84

A PREPARED MIND, INFINITE PAINS AND GENIUS

E. Margoliash

Department of Biochemistry, Molecular Biology and and Cellular Biology, Northwestern University, Evanston, Illinois 60208

It is a great honor for me to have been asked to contribute to the Keilin Centennial Session of this meeting, because the two years I spent at the Moleteno in 1951 to 1953, and the shorter stays thereafter, have moulded my life in a way that no other scientific experience could have done.

I hope, therefore, that I may be allowed to start on a personal note. When in early September 1952 my young family and I arrived in Cambridge, I was an escapee from the medical profession, six years out of medical school. I was hoping to be able to work on the mechanism by which large doses of thiourea, injected into rats, dramatically decreased the catalase activity of their livers, in a matter of hours. After patiently listening to my story, and it must have lasted well over half an hour, occasionally nodding his head, Professor Keilin finally said: "This is a very nice idea, but you know it is rather complicated. Why don't you start by doing something simpler? Why don't you make some cytochrome *c*? It is such a pleasant preparation." And these words--I can remember them as clearly as if they had been said this morning--decided the rest of my life. It took some weeks to find out what it was he wanted me to do about cytochrome *c*, a material I had no acquaintance with whatsoever at the time, a circumstance that has since been somewhat remedied. Professor Keilin really wanted me to find out why the preparation of cytochrome *c* from Delft yeast, that had worked extremely well between 1930 and 1936, when the source material had been switched to horse hearts--in my time perilously carried on a bicycle by Baker--why the yeast preparation had simply refused to work, during the war with British yeast, and after the war even with Delft yeast. But that is another story.

Professor Keilin was a gentle, compassionate man, in speech, demeanor and feeling, towards the many people, particularly the young scientists, who came within his orbit. But intellectually he was an overpowering giant. He never imposed on you any idea at all. He just waited patiently until you came around to his way of conceiving a scientific situation. And you nearly always did. In any case, nobody will deny that it was David Keilin's spirit that made the Molteno Institute the intellectual 'home away from home' for the most remarkable list of scientists who had been there, as inmates or visitors, from many lands and circumstances, and over many years.

When Tsoo King, just two weeks ago, asked me to say a few words about David Keilin on this occasion, he mentioned that it would be nice if I could address the question of how is it that the professor had been, throughout his life, scientifically so relatively infallible. Indeed, essentially every significant experiment he undertook had ended up being part of a worthwhile, and often seminal, publication, and he regularly made such major strides forward, none of which have had to be retraced.

The story of Keilin's discovery of cytochrome, in the mid-1920s (Keilin, 1925), has been told so often that it hardly bears repeating to this audience. Nevetheless, it does carry in it the three prime characteristics that underlie so much of his work.

First, a prepared mind. When he first saw the four-banded spectrum of cytochrome in the thoracic muscles of the adult bot fly, Keilin was already a very successful field naturalist and experimental parasitologist. He had discovered the intricate life cycles of several parasitic insects, then totally unknown, had in fact identified some 13 new species and 8 genera, and had realized full well the complexities of evolutionary adaptations. In some of the cases, he had discovered how these strange life cycles were largely driven by the need of respiration for growth and activity, and he obviously had a very good grasp on the finely tuned, involved, mechanisms of the real biological world, in all its variegated splendor. One can count 62 papers before the epoch making "On Cytochrome, a Respiratory Pigment common to Animals, Yeast, and Higher Plants" (Keilin 1925). Most of the giants he came in contact with at Cambridge, Hopkins, Barcroft, Roughton, and others, were either chemists, by training and feeling, or physiologists, with necessarily a narrower view of how life operated. It would be so effective, if our younger people today, who are trying to unravel macromolecular mechanisms of function, had some of that same appreciation for the subtleties of evolutionary adaptations to function, which, rarely if ever, seem to follow our generally naive initial ideas. In the long run, they may have no choice but to become familiar with evolutionary concepts.

Second, infinite pains. The four-banded spectrum observation in one tissue, was immediately followed by the exmination of a range of species and tissues that is truly remarkable. It included vertebrates and invertebrates in enormous profusion, plants as well as animals, even some prokaryotes, such as *B. subtilus*, which showed a cytochrome spectrum, while the anaerobe, *B. sporagenes*, had none. What biochemist today, or even in the 1920s, observing a new phenomenon, would go to the enormous trouble of carefully examining 86 species before the first publication? The 1925 paper does not include most of the bacteria examined because of editorial comments about a too lengthy manuscript. As you see, with the passage of time, editors do not change, and sometimes their decisions are not the best. Bacterial cytochromes were shortly thereafter described by other workers, and Keilin decided to forego any publication at that time.

As a biologist, well aware of how evolution is likely to adapt to selection pressures by very subtle mechanisms, Keilin must obviously have been fascinated by how stable the cytochrome system was, presumably a strong indication of its antiquity, and of its fundamental importance to some major phenomenology of aerobic life processes.

Last, but not least, of these factors is genius. The prepared mind, having done many observations, under as many conditions as seem to be applicable to the situation at hand, still does not necessarily produce a great scientific advance. Somewhere along the line there must be a leap forward, across the abyss of ignorance, connecting a new idea with a new phenomenon. In the case of cytochrome there were two such leaps. One, was the result of observations on the ways three of the bands appeared and disappeared under various conditions, not simultaneously, but one after the other. Thus, the four-banded spectrum was identified as the α-bands of three hemochromes, termed components, and named *a*, *b* and *c* from left to right, as viewed in the low dispersion optical microspectroscope, the instrument that had made all these observations possible. The fourth band was considered to represent the combined β-bands.

The other leap, came from observations made on living organisms, the most remrkable being those on the living wax moth, *Galleria mellonela*. The disappearance of the bands when the moth moved its wings, corresponding to the oxidation of the cytochrome components, their reappearance when the moth was quiet, corresponding to the reduction of the hemochromes, showed that the level of reduction in a living, completely intact tissue--the muscle of the living moth--was what we call today a steady state, established between the rate of reduction and the rate of oxidation of a series of components operating in a bucket brigade fashion.

Thus, in one shot, Keilin swept aside all the problems that had bedevilled McMunn's observations of cytochrome, (see Keilin, 1966), which had been termed myohematin and histohematin nearly 40 years earlier, and which had been argued to result from artifactual degradation products of myoglobin. This could no longer be the case, since the material was present in intact tissues, operating under physiological conditions. Second, even more important, he had bridged the two main theories that were disputing the mechanism of cellular respiration, with Warburg insisting on O_2 activation, at the oxidase end of the respiratory chain, and Thunberg insisting on H activation, at the dehydrogenase end of the respiratory chain. One could now go ahead and attempt to prepare and examine the different components of the bridging cytochrome chain, cytochrome *c*, as you know, being the first to yield to a direct attack.

This pattern can, I believe, be discerned again and again in David Keilin's work. Thus, my answer to the question Tsoo King posed, as to why Keilin was so remarkably effective as a scientist, is simply that he nearly always had a prepared mind, well aware of biological reality, that he took infinite pains and these were often illuminated by flashes of genius.

I may add that in regard to the matter of taking pains, he expected the same devotion from others. I remember, particularly vividly, the day I excitedly ran into the Professor's office, one morning early in 1952, and showed him the result of an iron content determination on a horse cytochrome *c* preparation that had been absorbed on and eluted from a column of the old zeolite cation exchanger used for water softening, Permutit. How I had come to be doing that is another story. Anyhow, the number at the bottom of the page on my notebook was 0.457%. The professor hesitated, was silent, and after what seemed an unusually long time, took out of the breast pocket of his laboratory coat the very well sharpened pencil that he usually carried there--the sort of pencil you would expect a draftsman to use--and carefully added three zeros after my number, so as to make it into 0.457000%. He then said "Maybe your value is not very accurate. Maybe you should do it again."

His skepticism was of course well founded, but that is also another story. All I want to point out is that "doing it again" became about eight months of work. Not only was it deemed necessary to show that the preparation with 0.45% iron was, per heme, precisely as active as the more usual preparation containing 0.34% iron, but also to identify what had been removed from the 0.34% iron preparation. It eventually turned out to be myoglobin globin, which precipitated from near saturated ammonium sulfate solutions containing cytochrome *c*, using trichloracetic acid, in a quasi-stoichiometric complex of one molecule of myoglobin globin to 4 molecules of cytochrome *c*, corresponding to 0.34% iron. Further, the 0.34% iron preparation had to be reconstituted from the 0.45% iron preparation, and it was found that only myoglobin or myoglobin globin would do that, not any other protein tested. It was also shown that the protein segment of the optical spectrum fitted the assumed composition of the putative complex, etc... To me it certainly looked like infinite pains, but it taught me an invaluable lesson in experimental life sciences.

REFERENCES

Keilin D (1925). On cytochrome, a respiratory pigment, common to animals, yeast and higher plants. Proc Roy Soc B 98:31-339.

Keilin D (1966). The history of cell respiration and cytochrome. Cambridge: Cambridge University Press.

The following obituary notices and historical reviews contain excellent accounts and complete bibliographies of David Keilin's life and work:

Hartree EF (1963). David Keilin (1887-1963). Biochem J 89:1-5.

Slater EC (1964). In Memoriam David Keilin, F.R.S. Enzymologia 26:313-320.

Mann T (1964). David Keilin, 1887-1963. Biographical Memoirs of Fellows of the Royal Society 10:183-205.

Tate P (1966). David Keilin, 1887-1963. Parasitology 55:1-28.

Dixon M, Tate P (1966). David Keilin, 1887-1963. J Gen Microbiol 45:159-185.

Hartree EF (1973). The discovery of cytochrome. Biochem Educ 1:69-74.

Hartree EF (1981). Keilin, cyotchrome, and the concept of a respiratory chain. In Semenza G (ed) "Of Oxygen, Fuels, and Living Matter, Volume 1, Part 1," New York: John Wiley and Sons, pp 161-226.

Oxidases and Related Redox Systems, pages 85–90

MY DAYS WITH DAVID KEILIN AND DOING SCIENCE IN CHINA

C. L. Tsou

Institute of Biophysics, Academia Sinica,

Beijing 100080. China

To begin where Peter Nicholls left off, about the immortality of life under low temperatures, the Prof. usually had with him a thermoflask of liquid nitrogen and a piece of very soft rubber tubing while lecturing on cryobiology in the top floor lecture room of the Molteno. As a demonstration during the lecture, he dipped one end of the rubber tubing into the thermoflask for a few minutes and after the boiling of the liquid nitrogen had come to a stop, took it out and banged it on the bench. It was then as hard as a metallic rod and we were all very much impressed by the remarkable change in properties of rubber at low temperatures. I asked Prof. Keilin once how he became interested in activities of life at low temperatures. He said that his study for the immortality of life at low temperatures was all for a thesis to be submitted to the University of Cambridge for the degree of Doctor of Divinity, because during a University ceremony, the D. D.'s always marched at the head of the procession followed by the LL. D.'s and D. Litt.'s. The D. Sc.'s were always the last and of course, the Ph. D.'s were no where to be seen.

Now allow me to give you three stories of my days in Cambridge each followed by some episodes of my scientific activities in China.

Professor Keilin made his great contributions to

science with very limited equipment. Apart from the hand spectroscope with which the Prof made his discovery of the cytochromes, people at the Molteno had at their disposal only the Barcroft manometers, Thunberg tubes, and an old swing-out centrifuge. A later addition of a Beckman DU spectrophotometer was a luxury. The MSE refrigerated centrifuge which Bill mentioned came only a few weeks before my work for the Ph. D. thesis had finished. When we needed to centrifuge in the cold, we had to put a few ice cubes made from distilled water to each of the centrifuge tubes when dilution was no objection. The Prof. told us a story that once he had a rich Professor from a rich country as a visitor who asked his advice on what to work on with all kinds of fancy equipment available and enough money to buy anything you can think of. I am sorry to say that this kind of thing does not happen only in rich countries. In recent years, many Chinese Universities had generous loans from the World Bank to buy scientific instruments and I had the same experience more than once with my visitors asking not only what to do but also what to buy with all the money made available to them! In China as it is anywhere else, it is not always the people who had the best equipment who are doing the best work.

My training with Prof. Keilin in Cambridge made it much easier for me to start working in my own group when I went back to Shanghai in 1951. The Institute of Biochemistry of Academia Sinica had then similar equipment as the Molteno except that we had to use a model A stone mill, (A for ancient, this kind of hand operated stone mill must have been in use in China for making flour from wheat since time immemorial) instead of a mechanical mortar for grinding up heart muscle mince with sand to make Keilin-Hartree preparations of succinic dehydrogenase-cytochrome oxidase. During our work on the purification of succinic dehydrogenase, we used the same trick of adding ice cubes made from distilled water to the centrifuge tubes to keep them cool in some of the steps where dilution didn't matter.

With Professor Keilin, I made all the enzymes that I used including the proteases for the study on the effect of proteolytic digestion of cytochrome c. One day I was discussing with the Prof. my work on the reactions of cyanide with cytochrome c and succinic dehydrogenase and he suggested to me to try the effect of carbylamine on the respiratory enzymes; it is after all similar to cyanide in structure. It wasn't available commercially, and as I had some training in organic chemistry, I tried to make it myself. On account of its smell, the Prof. suggested that it would be better to work with as little material as was convenient. The Molteno wasn't a chemical lab and no hood was available for such experiments. I therefore worked on the roof which served not only as a cold room for Bill but also as a hood for me. The Prof. commended me upon my dexterity for making the final fractional distillation step with 0.5 ml of material collecting fractions with a drop each and determining the boiling point of each of the fractions. However, I was an unwelcome person for sometime because of the stink which remained with my lab coat for quite a long time.

When I was back in China in the early fifties, it became apparent that my training in organic chemistry and the habit of making myself some of the things I needed stood me in good stead. During my early days in the Shanghai Institute of Biochemistry, we had to synthesize even dimercaptopropanol ourselves. For our work on the resynthesis of insulin from its chains, we made sodium tetrathionate ourselves for the splitting of the insulin into the S-thiosulfonate derivatives of the chains.

Among those who are engaged in fundamental research in China today, we are fortunate in being generously supported by Academia Sinica and in having a Grant from the National Institutes of Health of the U. S. so that we are not only moderately well equipped but also have some hard currency at our disposal. Nevertheless, I still ask my graduate students to make their own enzymes and to synthesize whatever reagents that were good for their training and above everything to make them remember always that worthwhile contributions in fundamental research can be made from very limited resources.

I got into the bad habit of smoking when I was very young and in England those days, cigaretes were difficult to obtain for which one had to know the tobacconist and got the cigaretes "under the counter". I therefore learnt to smoke a pipe as pipe tobacco was easier to obtain. Unfortunately, my technique with the pipe wasn't very good and someone in the Molteno once remarked that my pipe seemed to work most of the times in anaerobic conditions. It is for this reason that I kept the pilot flame of a gas burner always on so as to light my pipe whenever it went anaerobic which happened quite frequently. One day Prof. Keilin walked into my room in the corner of the third floor and saw the pilot flame of the burner on. He asked me to turn it off and use the match for lighting my pipe, he said that if I always left it on I may regret it if I had an accident.

It was a few days after, when I went home after work, my wife, who was then working in the Goldsmith laboratory of Metallurgy downstairs to the Chemistry Department in Pembroke site, told me that Professor A. R. Todd dismissed a gradute student because of an accident. The student was doing a distillation the previous evening and he forgot to turn the cooling water off when he finished the experiment and left for home. Because of the increased water pressure after working hours when everyone turned their taps off, the connection between the rubber tubing and the condenser burst and he had a flood in his room which was right above the office of Professor Todd. The water dripped through the floor and soiled some important papers of the Professor. This made Prof. Todd very angry and dismissed him on the spot when he came to his office in the morning. After my wife told me this story, we had our dinner, talked a little and then went to bed. It happened that I couldn't go to sleep and sometime around midnight it suddenly occurred to me that I left the pilot flame of the gas burner in my lab on. The more I thought of it the more I was sure that I didn't turn it off when I left. I hurriedly put on my Mackintosh over my pyjamas and rode my bicycle to the lab. I had to climed over the locked iron gate of Downing site, went into my lab on the top floor of the

Figure 1. The back of Professor Keilin's house, facing the garden. The photograph shows the new owner, Richard Perham, and son, and Chen Lu Tsou.

Figure 2. The door behind Perham and Tsou leads to the sitting room, a corner of which is shown in this figure with Tsou. Old boys and girls and their spouses would recognize this room where they had many memorable evenings, sometimes with M. Dixon playing the piano, and once with W. Bonner playing the clarinet.

Molteno and found to my great relief that the pilot flame was off after all.

Very recently. before I came to the States in September, something very similar to the accident in Prof. Todd's lab happened to one of my graduate student. Although his room was not directly above my office, but he did make a mess of a few laboratories. However, I had no say on what to do with him, the Security Officer of the Institute came, inspected the damage done, talked to my student and asked him to write a "self-criticism" on the accident. Depending on how well he did with his essay of self-criticism, he may get away with the accident, pay a fine or be dismissed as a graduate student. In case the security officer decided on a fine, as his supervisor, I should pay 80% of the fine for "failure to teach the students to observe the security regulations"! It is only common sense that since I have more money than the poor student I should pay the lion's share of the fine. There you have the difference in Chinese and Western philosophies. Fortunately for him and for me, he apparently did his essay well enough so as to satisfy the Security Officer that he had learnt the lesson and further punishment was not necessary.

I was able to visit Cambridge several times in recent years and was invited by Professor Hans Kornberg and Dr. Richard Perham to spend a few days in Cambridge in 1984. Dr. Perham kindly asked me to dinner at St. John's and to have a drink before dinner at his home. He didn't say where his house was but drove me there after we finished in the lab. When we came out of the car, he asked me whether I recognized the house. I stood there silent for may be a minute and then I suddenly realized that, of course, it was Keilin's house!

PART TWO

Oxidases and Related Redox Systems

Session I: General Considerations

Chairman: Thaddeus Mann

Oxidases and Related Redox Systems, pages 93–109

THE PARADOX OF OXYGEN: THERMODYNAMICS VERSUS TOXICITY

W. H. Koppenol
Department of Chemistry and Biodynamics Institute
Louisiana State University
Baton Rouge, Louisiana 70803

INTRODUCTION

After hydrogen and helium, oxygen is the most abundant element in the solar system (1). It is formed through a sequence of nuclear reactions which starts with hydrogen. In a process called "hydrogen burning" helium is produced (2). Two α-particles react then to form an unstable ^{8}Be nucleus. Interaction with another He nucleus yields ^{12}C and oxygen is formed similarly. These processes are known as "helium burning" and lead to an enhanced presence of isotopes with masses which are multiples of 4. Of the metals, ^{56}Fe is by far the most abundant. On earth there is very little hydrogen (H_2) and helium, and oxygen accounts for approximately 30 % of the weight of our planet (1). On a molar basis oxygen is the most abundant element. Yet the amounts of iron, silicon, magnesium and nickel, the other major constituents of the earth, are so large that when the earth was formed all oxygen combined with these elements to form their oxides. Interestingly, metallic iron and nickel make up the core of the earth and silicon, magnesium and iron are found in the mantle and crust as various silicates (1). Thermodynamically, this is not surprising if one considers the Gibbs energies of formation of the various oxides. The relevant values are: MgO, -135 kcal; 1/2 SiO_2, -102 kcal, FeO, -58.5 kcal and NiO, -50.7 kcal per mole of oxygen (3). The oxides are lighter than the elements and "float" on the core. The formation of the oxides of the naturally occurring radioactive elements uranium and thorium is also more favourable than the formation of iron(II) oxide, as shown in Equation 1 for uranium:

$$2\ FeO + U \rightarrow UO_2 + 2\ Fe \qquad \Delta_{rxn}G^o = -130\ kcal \qquad (1)$$

For thorium dioxide $\Delta_{rxn}G^o$ is -162 kcal. These radioactive elements are therefore expected to be mainly present in the earth's crust. Their decay with that of ^{40}K releases heat which is thought to be responsible for the geothermal gradient, which is on average 30 $^oC/km$ on the surface of the earth. The observation that meteorites, which might have a composition close that of the mantle or the core, contain very little of these radioactive elements can be taken as evidence for concentration of these radioactive elements in the crust. The release of heat upon decay is also likely to be the driving force behind the uplifting of land above sea level, continental drift and, secondarily, orogenesis and vulcanism. It would seem that oxygen is indirectly responsible for the emergence of life on land (4). While the crust is especially rich in oxygen, the atmosphere of the young earth did not contain dioxygen, and probably consisted mainly of dinitrogen, some carbon dioxide, water vapour and possibly some sulphur dioxide from vulcanic eruptions. There are no physical or chemical mechanisms in operation by which dioxygen is liberated into the atmosphere on a large scale[1], and as such the occurrence of this gas is a geochemical anomaly (6).

It is possible that oxygen was introduced by photosynthetic organisms, probably blue-green algae, as early as 3.3 - 3.5×10^9 years before present (7). For a long time, until approximately 1.7×10^9 years before present, the atmosphere stayed essentially anoxic, while photosynthetically produced oxygen was consumed in the oxidation of ferrous compounds and other oxidizable materials. Geological evidence from this time consists of banded iron-formations - indicating presence of oxygen in the hydrosphere (5) - and kerogens depleted in ^{13}C (8). Isotopic fractionation occurs when carbon dioxide is fed into the Calvin cycle by the enzyme ribulose-1,5-biphosphate carboxylase. ^{13}C values of up to -50 $^o/oo$ vs. the Peedee belemnite standard suggest that "light" methane was incorporated into the biomass, a process that is thought to require oxygen (8). It is very difficult to derive precisely what the partial oxygen pressures were

---

[1] Ultraviolet radiation from the sun would photolyze water in the absence of oxygen and ozone. This could conceivably lead to the production of oxygen via hydroxyl radical and hydrogen peroxide (5).

at various stages of the earth's history. For the mid-Precambrian a value of 3×10^{-5} atm with an estimated uncertainty of one order of magnitude has been suggested (9). During the Phanerozoic the partial oxygen pressure might have been higher than at present at some times (10), but was probably never lower than 10 % (9).

Compared to these geological times it is only very recent that we realize that oxygen is made by green plants at the expense of energy and used for respiration to drive unfavourable reactions. In his fascinating book "The History of Cell Respiration and Cytochrome", Keilin (11) gives an account of the discoveries which led to the current concepts of oxygen uptake and electron transfer. Limiting ourselves to oxygen, we notice that the circulation of the blood was discovered by Harvey in 1628, and that it was Mayow who realized in 1668 that only a component of air is taken up by blood in the lungs (11). The importance of this concept could not be realized until after the discovery of oxygen by Priestley, Scheele and Lavoisier around 1774. The latter performed the first quantitative studies of respiration. The ultimate destination of nearly all oxygen used by organisms is cytochrome *c* oxidase, an enzyme previously known as indophenol oxidase. It was Keilin who identified this enzyme as the terminal link in the electron transport chain (11, 12). His rediscovery of the cytochromes after MacMunn's work was forgotten (13, 15) forms the basis for the current views on energy conservation in oxidative phosphorylation. On the subject of oxygen and related species Keilin made a number of other important contributions. Together with Mann he discovered haemocuprein (16, 17). McCord and Fridovich (18) later showed that the enzymatic function of this enzyme is to scavenge superoxide radicals. With Hartree he studied reactions and spectroscopic properties of the haemoproteins horseradish peroxidase, methaemoglobin and catalase (11, 19).

Oxygen is eminently suitable for respiration because of the energy released upon reduction to water (20). However, it is becoming increasingly evident that partially reduced oxygen species - superoxide, hydrogen peroxide and the hydroxyl radical - are detrimental. Reactive oxyradicals initiate lipid peroxidation and are formed during phagocytosis, the metabolism of toxic chemicals and microsomal oxidation (21-25). They are implicated in pathological conditions such as rheumatoid arthritis (26) and reperfusion

TABLE 1

Gibbs energies of formation and solution, pK_a's and affinities of various oxygen species at 25 °C and 1 atm (101.3 kPa) (34)

Species	State	$\Delta_fG°$	$\Delta_fG°'$ (pH 7)	$\Delta_sG°$*	pK_a	Electron Affinity
		kcal/mol	kcal/mol	kcal/mol		volt
O_2 ($^3\Sigma_g^-$)	g	0				0.44
	aq	3.9	3.9	3.9		
($^1\Delta_g$)	g	22.6				1.42
($^1\Sigma_g^+$)	g	37				2.04
O_3	g	39.1				2.14
	aq	41.6	41.6	2.5		
O	g	55.4				1.46
$O_3^{\cdot-}$	aq	18.5	18.5	-74		
$HO_2\cdot$	g	6.4				1.19
	aq	1.2	10.7	-5.2	4.7	
$O_2^{\cdot-}$	aq	7.6	7.6	-85		
H_2O_2	g	-25.3				
	aq	-33.1	-14.1	-7.80#	11.7	
HO_2^-	aq	-17.1	-7.6	-98		
·OH	g	8.2				1.83
	aq	6.2	15.7	-2.0	11.9	
$O^{\cdot-}$	aq	22.4	19.4	-102		
H_2O	g	-54.7				
	liq	-56.7	-37.6		14.0	
	aq+	-59.1		-4.4		
OH^-	aq	-37.6	-28.1	-106		

* $\Delta_sG°$ for ionic species are with respect to $\Delta_fG°(H^+)$=-102.5 kcal/mol (39).

\# The difference with the accepted Gibbs energy of the solution of H_2O_2 of -6.6 kcal reflects our more negative value of ΔfG° of H_2O_2 (aq).

\+ H_2O_{aq} refers to a hypothetical 1 M solution of water in water.

injury (27), and play a role in the toxicity of asbestos and cigarette smoke (28, 29). At a conference of which a session was devoted to radicals in clinical medicine (30) many other disorders were presented in which oxyradicals are believed to be involved. Damage can often be prevented by superoxide dismutase, catalase, removal of iron by desferrioxamine, or by compounds which react with the hydroxyl radical and form stable products (21). It is therefore widely accepted that a reduced metal ion reacts with hydrogen peroxide to form the reactive hydroxyl radical (21, 31). The superoxide anion is a precursor of hydrogen peroxide and can serve as a reducing agent for the metal and thereby mobilize it, as is the case for iron (32). Superoxide is formed in small amounts by mitochondria (33) and possibly from oxyhaemoglobin and oxymyoglobin (34). It would seem that harmful reactions of oxyradicals are a trade-off with the large amount of energy that is available through the four-electron reduction of dioxygen (20).

The fitness of oxygen has been examined by George in 1965 from a thermodynamic point of view (20). This paper will reassess the possibility of other elements as terminal oxidants and present recent thermodynamic data for oxy-radicals. Such data allows one to discuss the feasibility of various interconversion reactions.

THERMODYNAMICS

The usefulness of oxygen as an oxidant will be compared to that of carbon, sulphur, nitrogen and chlorine with the help of oxidation state diagrams. These elements were chosen because they might have been available during the evolution of the earth's hydrosphere and atmosphere. The diagrams were calculated for a pH of 7, which seems reasonable since geological evidence suggests that the pH of the oceans has been more or less constant. At present the pH is 8.1 and this value might have been some what lower when the pCO_2 was higher. A hundred fold increase in this pressure, as present in the soil (9), reduces the pH of seawater by about one unit (35).

In an oxidation state diagram the slope of a line joining A and B is equal to the reduction potential $E^{o\prime}(B/A)$. Potentials refer to 1 molal concentrations or a pressure of one atmosphere (101.3 kPa), as indicated, at 25

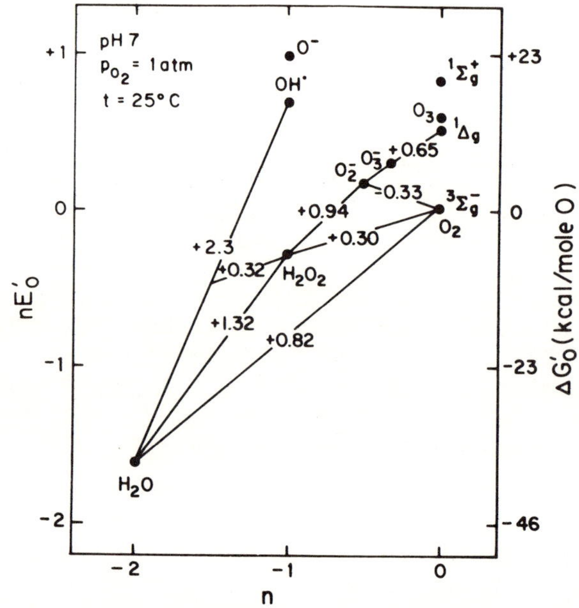

Figure 1. Oxidation state diagram of oxygen. Conditions: pH 7, 298 K, 1 m concentrations, and partial pressures of 1 atm (101.3 kPa) for ozone, singlet and triplet oxygen. Abscissa, oxidation state, n, of the element in the compound shown. Ordinate, left: n times the reduction potential in volts, relative to NHE; ordinate, right: $\Delta_f G^{o\prime}$ per mole of oxygen. The numbers ($E^{o\prime}$) refer to the slopes of the lines (34, 36).

°C and follow the Stockholm Convention with respect to sign. Values for all potentials are relative to the Normal Hydrogen Electrode. Construction of such diagrams requires Gibbs energies of formation ($\Delta_f G^o$), reaction ($\Delta_{rxn} G^o$) or experimentally determined reduction potentials. Data were taken from Refs. 3, 20, 34, 36-39. Table 1 lists the thermodynamic data used for the calculation of reduction potentials of O and H containing species. An oxidation state diagram for oxygen is shown in Fig 1. Such a diagram has an advantage over a Latimer diagram in that it shows graphically which species - O_2^- and H_2O_2 - are unstable with respect to disproportionation. The dismutation of superoxide does not (34) appear to lead to $^1\Delta_g O_2$ although this is thermodynamically possible (40). Singlet oxygen as a disproportionation product from hydrogen peroxide is possible and has been reported (41). The reduction potential of the hydroxyl radical in Fig. 1 is based on two recent experimental values (42, 43) of 1.89 V and 1.91 V for $E^{o\prime}(\cdot OH/OH^-)$ which are slightly higher than an earlier calculated value of 1.77 V (44). The positions of O^- and O_3^-, which depend on this reduction potential (45), have likewise been adjusted. The line from H_2O_2 to the centre of the line $\cdot OH$ - H_2O represents the one-electron reduction of hydrogen peroxide. The value of 0.32 V (36) is obtained by subtracting $E^{o\prime}(\cdot OH/H_2O)$ from $2E^{o\prime}(H_2O_2/H_2O)$.

Oxidation state diagrams for carbon (46), sulphur, nitrogen and chlorine[2] at pH 7 are shown in Figures 2A-2D, respectively. The radicals $\cdot SO_3^-$, $\cdot SO_4^-$ and $\cdot SO_5^-$ have been studied (47) and appear to be rather oxidizing, with a value of 0.63 V for $E(\cdot SO_3^-/SO_3^{2-})$ and higher values for the two other radicals. Especially $\cdot SO_4^-$ mimicks the hydroxyl radical in many respects (48). The value of 0.63 V for $E(\cdot SO_3^-/SO_3^{2-})$ implies a value of -1.7 V for $E^{o\prime}(SO_4^-/\cdot SO_3^-)$ at pH 7, showing that the one-electron reduction of sulphate is extremely unfavourable. The same holds for carbonate (or bicarbonate), and presumably perchlorate, where no data on one-electron reduction are available. A general phenomenon illustrated by the "saw-tooth" appearance of part of these diagrams is that one-electron transfer processes are less feasible than two-electron transfers. Of the elements shown, only nitrogen and chlorine show that reductions of the elements releases energy. Although hydrochloric acid has a more negative Gibbs energy of formation, water has unique properties as a solvent. It is these qualities that make oxygen the best terminal oxidant, as discussed by George (20).

Thermodynamic data regarding the intermediates in the reduction of oxygen are shown in Table 2. The first part of this Table is based on an earlier compilation (37) and the entries have been adjusted for the "best" value for the

---

2 W.H. Koppenol, unpublished

A

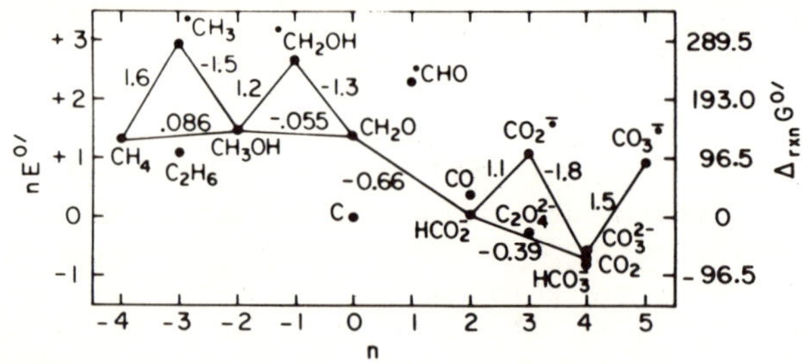
•CH3
•CH2OH
•CHO
1.6
-1.5
1.2
-1.3
.086
-.055
CH4
CH3OH
CH2O
C2H6
C
-0.66
CO
CO2•
1.1
-1.8
C2O4 2-
HCO2-
-0.39
HCO3-
CO2
CO3•
1.5
CO3 2-
nE°′
ΔrxnG°′
289.5
193.0
96.5
0
-96.5
n

B

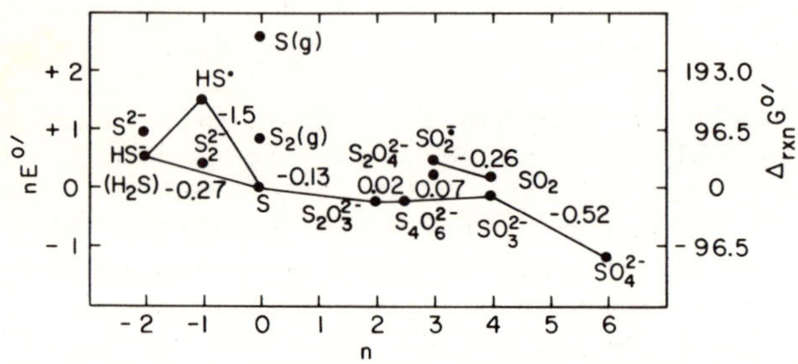
S(g)
HS•
-1.5
S2-
HS-
S2 2-
S2(g)
(H2S) -0.27
S
-0.13
S2O4 2-
SO2•
-0.26
SO2
0.02
0.07
S2O3 2-
S4O6 2-
SO3 2-
-0.52
SO4 2-
nE°′
ΔrxnG°′
193.0
96.5
0
-96.5
n

C

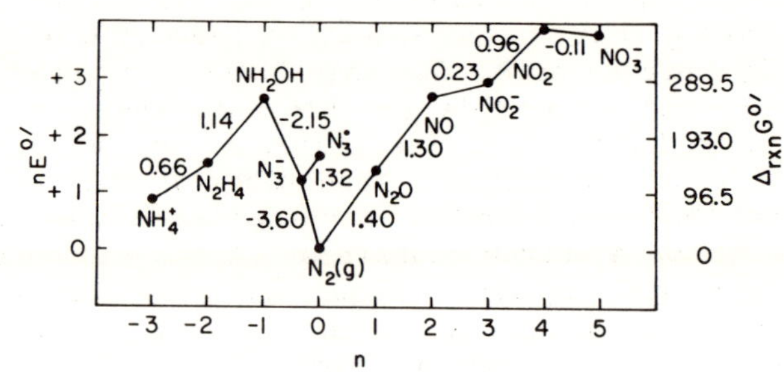
NH2OH
1.14
-2.15
0.66
N2H4
NH4+
N3-
N3•
1.32
-3.60
N2(g)
1.40
N2O
1.30
NO
0.23
NO2-
0.96
NO2
-0.11
NO3-
nE°′
ΔrxnG°′
289.5
193.0
96.5
0
n

D

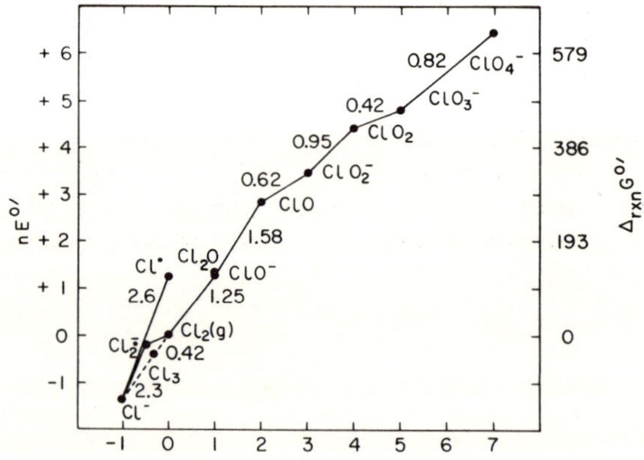
0.82
ClO4-
ClO3-
0.42
ClO2
0.95
ClO2-
0.62
ClO
1.58
Cl•
Cl2O
ClO-
2.6
1.25
Cl2-
Cl2(g)
0.42
Cl3
2.3
Cl-
nE°′
ΔrxnG°′
579
386
193
0
n

Figure 2. Oxidation state diagrams of carbon (A), sulphur (B), nitrogen (C) and chlorine (D). Conditions: pH 7, 298 K and 1 m concentrations, except for the elements. Abscissa, oxidation state, n, of the element in the compound shown. Ordinate, left: n times the reduction potential in volts, relative to NHE; ordinate, right: $\Delta_{rxn}G^{o\prime}$ per mole of the element, to form the compound of interest from the element, water and/or hydrogen. The numbers ($E^{o\prime}$) refer to the slopes of the lines. Gibbs energies of formation at pH 7 are calculated from standard values (3) by adding 9.53 kcal per hydrogen atom. Solvation energies, if unknown, are estimated according to Ref. 42 or taken to be identical to those of similar molecules. Thus, SO, NO and ClO were assigned the same Gibbs hydration energy as CO, +17.3 kJ/mol, and ClO_2, N_2O and NO_2 that of CO_2, +8.4 kJ/mol (3). Data regarding the species SO_2^-, $N_3\cdot$ and ClO_2 are from Refs. 49, 50 and 43, respectively. The dashed line in Figure 2D represents the reduction potential of the Cl_2/Cl^- couple, 1.36 V.

TABLE 2

Thermodynamic data for $O_2^{\cdot-}$ and $\cdot OH$ at 25°C and 101.3 kPa

Reaction	ΔH^o kcal/mol	ΔS^o cal/degree-mol	ΔG^o kcal/mol
$1/2\ H_2(g) + O_2(g) \rightarrow HO_2\cdot(g)$	3.5	-10	6.4
$1/2\ H_2(g) + O_2(g) \rightarrow HO_2\cdot(aq)$	-8.5	-32	1.2
$HO_2\cdot(g) \rightarrow HO_2\cdot(aq)$	-12.0	-22	-5.2
$1/2\ H_2(g) + O_2(g) \rightarrow O_2^{\cdot-}(aq) + H^+(aq)$	-7.5	-50	7.6
$HO_2\cdot(aq) \rightarrow O_2^{\cdot-}(aq) + H^+(aq)$	1.0	-18	6.4
$O_2^{\cdot-}(g) \rightarrow O_2^{\cdot-}(aq)$	-95	-33	-85
$1/2\ H_2(g) + 1/2\ O_2(g) \rightarrow \cdot OH(g)$	9.3	+4	8.2
$1/2\ H_2(g) + 1/2\ O_2(g) \rightarrow \cdot OH(aq)$	3.2	-10	6.2
$\cdot OH(g) \rightarrow \cdot OH(aq)$	-6.2	-14	-2.0
$1/2\ H_2(g) + 1/2\ O_2(g) \rightarrow O^{\cdot-}(aq) + H^+(aq)$	13.4	-30	22.4
$\cdot OH(aq) \rightarrow O^{\cdot-}(aq) + H^+(aq)$	10.3	20	16.2
$O^{\cdot-}(g) \rightarrow O^{\cdot-}(aq)$	-111	-30	-102

enthalpy of formation of the hydrodioxyl radical, $HO_2\cdot$[3], which is 3.5 kcal/mol (51). This change led to a more negative Gibbs energy of solvation, which is now more in agreement with that predicted by the model of Schwarz and Dodson (42). The second part of Table 2, on the hydroxyl radical, is based on data from Refs. 3, 42, 43 and the assumption that the enthalpy of solvation of $\cdot OH$ is half of that of hydrogen peroxide.

DISCUSSION

As mentioned in the Introduction, there is a risk attached to the use of oxygen, which is the possible release of superoxide, hydrogen peroxide and hydroxyl radical. The superoxide anion is not overly reactive but might serve to reduce ferric ions and thereby mobilize them from ferritin (32). The hydroxyl radical shows an indiscriminate reactivity towards biomolecules, with rate constants at, or close to, the diffusion limit. The product is an organic radical ($R\cdot$) that will react rapidly with oxygen to form a peroxyl radical ($ROO\cdot$) which itself is an oxidizing species. Thus, oxygen not only makes the formation of an oxidizing radical possible, it also exacerbates damage through chain reactions. Sometimes these harmful reaction are used for a good purpose, as by leucocytes to combat infections (52).

One might ask how aerobic organisms defend themselves. The partial oxygen pressure in the cell is kept low by cytochrome c oxidase (53). Superoxide dismutases scavenge $O_2^{\cdot-}$, and catalase, gluthatione peroxidases and cytochrome c peroxidase dispose of H_2O_2. Furthermore, ascorbic acid and tocopherol might repair organic radicals formed after one-electron oxidation. The source of the damage is generally considered to be the hydroxyl radical from the one-electron reduction of hydrogen peroxide by a transition metal complex as in Reactions 2-4 or through the interaction of two hydrogen peroxide molecules with the metal complex, as shown for iron EDTA, Reaction 6. Subsequent reactions of the

---

[3] In the literature one finds for $HO_2\cdot$ also the names perhydroxyl and hydroperoxyl radical. IUPAC rules for nomenclature allow the latter, but not the former; the name hydrodioxyl is the recommended one (J.G. Traynham, 1987, personal communication)

hydroxyl and other radical are also listed (36):

Reactants		Products	$\Delta G^{o\prime}$ (pH 7) kcal/mol	
$HO_2 + O_2^{\cdot-} + 2H^+$	---->	$H_2O_2 + O_2$ (triplet)	-32.1	(2)
$Fe(III)EDTA^- + O_2^-$	---->	$Fe(II)EDTA^{2-} + O_2$	-10.4	(3)
$Fe(II)EDTA^{2-} + H_2O_2$	---->	$Fe(III)EDTA^- + \cdot OH + H_2O$	-4.8	(4)
$Fe(III)EDTA^- + H_2O_2$	---->	$Fe(II)EDTA^{2-} + O_2^{\cdot-} + 2H^+$	+18.9	(5)
$Fe(III)EDTA^- + 2H_2O_2$	---->	$Fe(II)EDTA^{2-} + \cdot OH + O_2 + H_2O$	+3.6	(6)
$\cdot OH + O_2^{\cdot-} + H^+$	---->	$H_2O + O_2$	-60.6	(7)
$\cdot OH + H_2O_2$	---->	$H_2O + H^+ + O_2^{\cdot-}$	-28.6	(8)
$\cdot OH + Fe(II)EDTA^{2-} + H^+$	-->	$Fe(III)EDTA^- + H_2O$	-50.3	(9)
$\cdot OH + CH_3CH_2OH$	---->	$CH_3C\cdot HOH + H_2O$	-30	(10)
$CH_3C\cdot HOH + Fe(III)EDTA^-$	-->	$CH_3CHO + Fe(II)EDTA^{2-} + H^+$	-35	(11)
$2CH_3C\cdot HOH$	---->	$CH_3CHOHCHOHCH_3$	-?	(12)
$CH_3C\cdot HOH + O_2$	---->	$CH_3CHO + O_2^{\cdot-}$	-29	(13)

Formation of the hydroxyl radical is thermodynamically allowed, as shown in Reaction 4, but regeneration of ferrous EDTA by a single hydrogen peroxide, Reaction 5, seems unlikely at first. It is possible that this Reaction does take place because the superoxide anion rapidly disappears. The role of the iron complex is that of a catalyst; the direct reduction of hydrogen peroxide by superoxide or its protonated form, the hydrodioxyl radical, proceeds very slowly, if at all (See Ref. 55 for review). In principle, it should be possible for a metal complex to catalyze superoxide dismutation and not the one-electron reduction of hydrogen peroxide, as the thermodynamic requirements under standard conditions (pH 7) are different: any metal complex with a reduction potential between 0.94 V ($O_2^{\cdot-}/H_2O_2$) and -0.33 V ($O_2/O_2^{\cdot-}$) can be reduced and oxidized by the superoxide anion. To be able to reduce hydrogen peroxide the metal complex must have a reduction complex below 0.32 V. However, when one makes the necessary corrections for *in vivo* and *in vitro* conditions one finds that nearly every metal complex can thermodynamically catalyze both the dismutation reaction and the one-electron reduction of hydrogen peroxide (36).

Evidence that the hydroxyl radical is formed is mainly

based on observations that scavengers inhibit damage. Such assays are complex and it is not surprising that results obtained with different scavengers are often inconsistent (56). It might be possible that the reaction of hydrogen peroxide with a ferrous complex does not yield a hydroxyl radical but a higher oxidation state of iron (FeO^{2+}), similar to Compound II in haems (53). The following observations support this suggestion. Ferrocytochrome c was not damaged by Fe(II)EDTA and hydrogen peroxide as it is by radiolytically generated hydroxyl radicals (57). Stopped-flow studies of the reaction of various ferrous amino-polycarboxylates (apc) with hydrogen peroxide show that the hydroxyl radical scavengers ethanol, sodium formate, t-butanol and sodium benzoate are less efficient then one would expect from radiolysis experiments (58-60). In the presence of sodium formate the reactions of ferrous penta- and quadridentate apc complexes with hydrogen peroxide yield short-lived intermediates with absorption peaks near 300 nm (ϵ = 2,600 $M^{-1}cm^{-1}$) and 420 nm (ϵ = 620 $M^{-1}cm^{-1}$) which have been ascribed to a Fe(II)apc ligand radicals (58, 59). A study of the steady-state decomposition of hydrogen peroxide by quadridentate ferrous apc complexes in the presence of the scavengers formate and bromide led to the conclusion that the reduction potential of the intermediate/ferric apc couple is greater than $E^{o\prime}(CO_2^{\cdot-}/HCO_2^-)$, 1.1 V (46) but less than that of the $Br^{\cdot}/Br^-$ couple, 1.93 V (42). The reduction potential of the couple $^{\cdot}OH/H_2O$ is 2.32 V at pH 7. If indeed a higher oxidation state of iron is formed that is less oxidizing than the hydroxyl radical by about one volt, then the equivalent versions of Reactions 7-10 are still thermodynamically possible. Ferrous complexes appear not to be alone in that they might be able to form a higher oxidation state. The reduction of hydrogen peroxide by copper(I) phenanthroline also appears to involve copper(III), rather than the hydroxyl radical (61, 62).

A higher oxidation state of iron could be more selective in its reactions, because it less oxidizing than the hydroxyl radical. This would make such an intermediate more dangerous than a randomly reacting hydroxyl radical. In principle, the formation of a metal complex in a higher oxidation state can be prevented by sequestering the metal.

Acknowledgement. I thank Dr. J.D. Rush and Dr. P.L. Bounds for helpful discussions. Supported by Grant GM 33883 from the National Institutes of Health.

REFERENCES

1. Henderson, P. (1982) Inorganic Geochemistry, Pergamon Press, Oxford, Chapters 2 and 4.
2. Burbidge, E.M., Burbidge, G.R., Fowler, W.A. and Hoyle, F. (1957) Rev. Mod. Phys. **29**, 547-640.
3. Wagman, D.D., Evans, W.H., Parker, V.B., Schumm, R.H., Halow, I., Bailey, S.M., Churney, K.L. and Nuttall, R.L. (1982) J. Phys. Chem. Ref. Data **11**, Suppl. 2.
4. Schuiling, R.D. (1970) Chemie en Techniek **25**, 215-218.
5. Walker, J.C.G, Klein, C., Schidlowski, M., Schopf, J.W., Stevenson, D.J. and Walter, M.R.(1983) in Earth's Earliest Biosphere (Schopf, J.W., ed.) Chapter 11, Princeton University Press, Princeton, New Jersey.
6. Chapman, D.J. and Schopf, (1983) in Ref. 5, Chapter 13.
7. Schopf, J.W. and Packer, B.M. (1987) Science **237**, 70-73.
8. Hayes, J.M. (1983) in Ref. 5, Chapter 12.
9. Holland, H.D., Lazar, B. and McCaffrey, M. (1986) Nature **320**, 27-33.
10. Holland, H.D. (1984) The Chemical Evolution of the Atmosphere and Oceans, Princeton University Press, Princeton, New Jersey, pp. 420, 421 and 525.
11. Keilin, D. (1970) The History of Cell Respiration and Cytochrome, Cambridge University Press, Cambridge.
12. Keilin, D. (1930) Proc. Roy. Soc. B **106**, 418-444.
13. Keilin, D. (1925) Proc. Roy. Soc. B **98**, 312-339.
14. Hartree, E. F. (1981) in Of Oxygen, Fuels and Living Matter (Semenza, G., ed.) Vol. 1, Part 1. Wiley, Chichester, U. K. pp. 161-173.
15. Margoliash, E. and Schejter, A. (1984) TIBS **9**, 364-366.
16. Mann, T. and Keilin, D. (1938) Proc. Roy. Soc. B **126**, 303-315.
17. Mann, T. and Keilin, D. (1938) Nature **142**, 148.
18. McCord, J.M. and Fridovich, I. (1969) J. Biol. Chem. **244**, 6049-6055.
19. Keilin, D. and Hartree, E.F. (1951) Biochem. J. **49**, 88-104.
20. George, P. (1965) in Oxidases and Related Redox Systems (King, T.E., Mason, H.S. and Morrison, M., eds.) Wiley and Sons, New York, pp. 3-36.
21. Halliwell, B. and Gutteridge, J.M.C. (1984) Biochem. J. **219**, 1-14.
22. Babior,B.M., Kipnes, R.S. and Curnutte, J.T. (1974) J. Clin. Invest. **47**, 741-744.
23. Mason, R.P. (1982) in Free Radicals in Biology (Pryor,

W.A., ed.) Vol. 5, Academic Press, New York, pp. 161-222.
24. Cohen, G. and Cederbaum, A.I. (1979) Science **204**, 66-68.
25. Pryor, W.A. (1978) Photochem. Photobiol. **28**, 787-801.
26. McCord, J.M. (1974) Science **185**, 529-531.
27. Koyama, I., Bulkley, G.B., Williams, G.M. and Im, M.J. (1985) Transplantation **40**, 590-595.
28. Weitzman, S.A. and Graceffa, P. (1984) Arch. Biochem. Biophys. **228**, 373-376.
29. Church, D.F. and Pryor, W.A. (1985) Env. Health Perspect. **64**, 111-126.
30. Proceedings of the Third Biennial General Meeting of the Society for Free Radical Research (1987) Free Radical Res. Commun. **3** (1-5).
31. Aust, S.D., Morehouse, L.A. and Thomas, C.E. (1985) J. Free Radicals Biol. Med. **1**, 3-25.
32. Thomas, C.E., Morehouse L.A. and Aust, S.D. (1985) J. Biol. Chem. **260**, 3275-3280.
33. Loschen, G., Azzi, A., Richter C. and Flohe, L. (1974) FEBS Lett. **42**, 68-72.
34. Koppenol, W.H. and Butler, J. (1985) Adv. Free Radical Biol. Med. **1**, 91-131.
35. Koppenol, W.H., Arnoldus, R., Kreulen, R. and Schuiling, R.D. (1979) in: Proceedings of the VI Colloquium on the Geology of the Aegean Region, Vol. III (Kallergis, G., ed.) Institute of Geological and Mining Research, Athens, pp. 985-994.
36. Koppenol, W.H. (1987) Bioelectrochem. Biophys., in press.
37. Koppenol, W.H. (1983) in Oxyradicals and their Scavenger Systems, Vol. 1 (Cohen, G. and Greenwald, R.A.) Elsevier Biomedical, Amsterdam, pp. 274-276, and references therein.
38. Milazzo, G. and Caroli, S. (1978) Tables of Standard Electrode Potentials, Wiley and Sons, Chichester, U.K.
39. Rosseinsky, D.R. (1965) Chem. Rev. **65**, 467-490.
40. Koppenol, W.H. (1976) Nature **262**, 420-421.
41. Smith, L.L. and Kulig, M.J. (1976) J. Amer. Chem. Soc. **98**, 1027-1029.
42. Schwarz, H.A. and Dodson, R.W. (1984) J. Phys. Chem. **88**, 3643-3647.
43. Klaening, U.K, Sehested, K. and Holcman, J. (1985) J. Phys. Chem. **89**, 760-763.
44. Koppenol, W.H. and Liebman, J. (1984) J. Phys. Chem. **88**, 99-101.
45. Koppenol, W.H. (1982) FEBS Lett. **140**, 169-172.

46. Koppenol, W.H. and Rush, J.D. (1987) J. Phys. Chem. **91**, 4429-4430.
47. Hayon, E., Treinin, A. and Wilf, J. (1972) J. Am. Chem. Soc. **94**, 47-57.
48. Neta, P. and Huie, R.E. (1985) Env. Health Perspect. **64**, 209-217.
49. Stanbury, D.M. and Lednicky, L.A. (1984) J. Am. Chem. Soc. **106**, 2847-2853.
50. Alfassi, Z.B., Harriman, A., Huie, R.E., Mosseri, S. and Neta, P. (1987) J. Phys. Chem. **91**, 2120-2122.
51. Shum, L.G.S. and Benson, S.W. (1983) J. Phys. Chem. **87**, 3479-3482.
52. Badwey, J.A. and Karnovsky, M.L. (1980) Ann. Rev. Biochem. **49**, 695-726.
53. Chance, B., Sies, H. and Boveris, A. (1979) Phys. Rev. **59**, 527-605.
54. Fridovich, I. and Freeman, B. (1980) Ann. Rev. Physiol. **48**, 693-702.
55. Koppenol, W.H. (1982) in Oxidases and Related Redox Systems (King, T., Mason, H.S. and Morrison, M., eds) Pergamon Press, Oxford, pp. 127-136.
56. Winterbourn, C.C. (1987) Free Radical Biol. Med. **3**, 33-39.
57. Koppenol, W.H. (1985) J. Free Radicals Biol. Med. **1**, 281-285.
58. Rush, J.D. and Koppenol, W.H. (1987) J. Inorg. Biochem. **29**, 199-215.
59. Rush, J.D. and Koppenol, W.H. (1987), submitted for publication.
60. Rush, J.D. and Koppenol, W.H. (1986) J. Biol. Chem. **261**, 6730-6733.
61. Johnson, G.R.A., Nazhat, N.B. and Saadalla-Nazhat, R.A. (1985) J. Chem. Soc. Chem. Commun. 407-408.
62. Goldstein, S. and Czapski, G. (1985) J. Free Radcials Biol. Med. **1**, 373-380.

DISCUSSION

ESTABROOK: Oxygen radical species have the potential of initiating the degradation of proteins for turnover. Would you comment on the relative reactivity of amino acids in proteins such as histidine, tryptophan, etc., with various species of oxygen radicals? Further, do you consider such reactions a significant use of oxygen radicals?

KOPPENOL: Superoxide radicals are rather unreactive, hydroxyl radicals react randomly at diffusion-controlled rates, and metal binding to proteins lead to site specific damage after reaction with hydrogen peroxide. Histidine especially is destroyed in this reaction. Generally, these reactions are considered harmful, however, when they take place in a phagocyte they serve a useful purpose, namely, the destruction of invading bacteria or foreign materials.

SLATER: Although reduction of O_2 to water via cytochrome oxidase proceeds without any formation of free O_2^-, a side reaction caused by reduction of O_2 by a species of ubiquinone can lead to O_2^- formation under certain conditions in which superoxide can accumulate, for example, when the respiratory chain is highly reduced during steady state respiration. This is illustrated by C.P. Lee's studies of cytochrome oxidase-deficient patients from whom mitochondria can be isolated showing normal respiratory activity but with a highly reduced cytochrome c. Superoxide formed in this way may well be the cause of myopathy.

KOPPENOL: I agree. Production of superoxide by mitochondria has been shown by Loschen et al. (1974) FEBS Lett 42: 68-72.

AZZI: The presence of cytochrome c oxidase, superoxide dismutase and other scavenging mechanisms will keep the concentration of O_2 and its radicals vanishingly small. How can such small radical concentrations lead to toxological consequences?

KOPPENOL: Under normal conditions the defense you mention is quite adequate. We calculated some time ago that the steady state concentration of superoxide in a red blood cell is very small, about 10^{-15}M [(1985) Adv Free Radical Biol Med 1: 91-131]. However, under certain conditions (reperfusion injury, hyperbaric oxygen) it is believed that there is an increased flux of superoxide which exceeds the capacities of the defense.

SANDERS-LOEHR: In your mechanism for the reaction of an Fe(II) complex with H_2O_2, you suggest an intermediate with a side-on bound peroxide. I would like to comment that we have recently found direct evidence for the side-on binding of peroxide from resonance Raman experiments on Fe(II)EDTA plus peroxide using a mixed oxygen isotope probe.

KOPPENOL: We also obtained evidence for side-on binding of peroxide from the fact that the Fe(II)HEDTA reacts more rapidly than Fe(II)EDTA and that the kinetics are consistent with the formation of a dimeric iron intermediate [(1987) J Inorg Biochem 29: 199-215].

Oxidases and Related Redox Systems, pages 111–126

THE ORIGIN AND EVOLUTION OF ATMOSPHERIC OXYGEN

Joel S. Levine

Atmospheric Sciences Division, NASA Langley Research Center, Hampton, Virginia 23665

INTRODUCTION

A major event in the history of the atmosphere and in the development of life on our planet was the evolution of atmospheric oxygen. Not since the first appearance of living systems on our planet did an event have such an important impact on the evolution and development of life. The presence of atmospheric oxygen directly impacted biological evolution from a metabolic and biochemical point of view, e.g., respiration replaced fermentation as the process for energy generation in living systems. Of comparable importance was the appearance and evolution of atmospheric ozone, which accompanied the rise of atmospheric oxygen. Ozone (O_3), an allotropic form of oxygen, is the only atmospheric species in sufficient abundance capable of absorbing biologically lethal solar ultraviolet radiation (200-300 nm). (Solar radiation of wavelength less than 200 nm is absorbed by molecular (O_2) and atomic (O) oxygen and molecular (N_2) and atomic (N) nitrogen in the middle and upper atmosphere.) The absorption of solar ultraviolet radiation in the 200-300 nm range by ozone shielded the Earth's surface from this energetic radiation and permitted life to leave the safety of the ocean and go ashore for the first time in history. Once life could safely survive on land, protected by the ozone layer, a new and very important biological niche was opened.

ORIGIN OF THE ATMOSPHERE

It is generally believed that the Earth and the rest of the solar system condensed out of an interstellar cloud of gas and dust, called the "primordial solar nebula," about 4.6 billion years ago. The atmospheres of the Earth and the other terrestrial planets (Venus and Mars) are thought to have formed as a result of volatile outgassing; the release of trapped volatiles from the solid planet (Walker, 1977; Lewis and Prinn, 1984; Levine, 1985a). By contrast, the atmospheres of the giant planets Jupiter, Saturn, Uranus, and Neptune are believed to be the captured remnants of the primordial solar nebula. Some volatile outgassing and formation of the Earth's atmosphere may have been associated with the impact heating during the final stages of the formation of the Earth.

For many years it was believed that the early atmosphere of the Earth was a strongly reducing chemical mixture composed of methane (CH_4), ammonia (NH_3), and molecular hydrogen (H_2). The more recent picture envisions the early atmosphere as a mildly reducing mixture of carbon dioxide (CO_2), molecular nitrogen (N_2), and water vapor (H_2O), with only trace amounts of hydrogen (Levine, 1985b). This mixture is not unlike that emitted by present-day volcanoes (Table 1). There is very little question but that the Earth

Table 1. Average Composition of Hawaiian Volcanic Gas (Walker, 1977)

Gas	% Volume
Water vapor (H_2O)	79.31
Carbon dioxide (CO_2)	11.61
Sulfur dioxide (SO_2)	6.48
Nitrogen (N_2)	1.29
Hydrogen (H_2)	0.58
Carbon monoxide (CO)	0.37
Sulfur (S_2)	0.24
Chlorine (Cl_2)	0.05
Argon (Ar)	0.04

outgassed tremendous quantities of H_2O, CO_2, and N_2 over geological time. The question is whether outgassing of CH_4, NH_3, and H_2 occurred for a short while prior to the longer period of extensive outgassing of H_2O, CO_2, and N_2. Most researchers now believe that the answer to this question is no. Clearly, the composition of the present atmosphere (Table 2) bears very little resemblance to the composition

Table 2. Composition of the Present Atmosphere (Levine, 1985b)

Gas	% Volume
Nitrogen (N_2)	78.08
Oxygen (O_2)	20.95
Argon (Ar)	0.93
Water vapor (H_2O)	Variable: 0 to few %
Carbon dioxide (CO_2)	0.034

of the early atmosphere. The bulk of the H_2O that outgassed from the interior condensed out of the atmosphere, forming the Earth's vast oceans. Only small amounts of H_2O remained in the atmosphere, with almost all of it confined to the troposphere (the lowest region of the atmosphere which extends from the surface to about 10 km). At the surface, the H_2O concentration is variable, ranging from a fraction of a percent to a maximum of several percent by volume. At the top of the troposphere, H_2O has a mixing ratio in the parts per million by volume (ppmv = 10^{-6}) range. Most of the CO_2 that outgassed over the Earth's history formed sedimentary carbonate rocks [calcite, $CaCO_3$, and dolomite, $CaMg(CO_3)_2$] after dissolution in the ocean. The mixing ratio of CO_2 in the present atmosphere is about 340 ppmv (0.034% by volume). It has been estimated that the preindustrial (around the year 1860) level of atmospheric CO_2 was about 280 ppmv (0.028%). For each CO_2 molecule presently in the atmosphere, there are about 10^5 CO_2 molecules incorporated as carbonates in sedimentary rocks. All of the carbon presently in sedimentary rocks outgassed from the interior of the Earth and was at one time in the atmosphere in the form of CO_2. Hence, the early atmosphere may have contained orders of magnitude more CO_2 than it presently contains. Molecular nitrogen is chemically inert, non water-soluble (as is CO_2) and noncondensable (as in H_2O). Hence, most of the outgassed nitrogen accumulated in the atmosphere over geological time to become the most abundant constituent (78% by volume). It is important to note that oxygen, the second most abundant constituent of the atmosphere (21% by volume), is not released via volatile or volcanic activity. However, oxygen may be produced from volcanic H_2O and CO_2 via photochemical processes.

THE ABIOGENIC PRODUCTION OF OXYGEN

The photolysis of H_2O and CO_2 served as an abiogenic source of O_2 in the early, prebiological atmosphere. The photolysis of H_2O leads to the photochemical production of O_2 via the following reactions:

(1) $H_2O + h\nu \rightarrow OH + H$, $\lambda \leq 240$ nm (ν = frequency).

A small percentage (<10%) of the atomic hydrogen (H) produced by the photolysis of H_2O will eventually escape into space. The hydroxyl radical (OH) formed by the photolysis of H_2O forms atomic oxygen (O) via the reaction:

(2) $OH + OH \rightarrow O + H_2O$

Similarly, the photolysis of CO_2 leads to the formation of O via:

(3) $CO_2 + h\nu \rightarrow CO + O$, $\lambda \leq 230$ nm

The atomic oxygen produced in reactions (2) and (3) forms O_2 via:

(4) $O + O + M \rightarrow O_2 + M$

where M is any third body. O_2 may also be formed from the products of reactions (1) - (3) via:

(5) $O + OH \rightarrow O_2 + H$

Some of the O_2 produced by reactions (1) - (5) was lost via the oxidation of minerals exposed to the atmosphere during the course of weathering. O_2 was also lost by direct photolysis (reaction (6)) and by atmospheric reaction with H_2 (reaction (7)), which led to the reformation of H_2O:

(6) $O_2 + h\nu \rightarrow O + O, \lambda \leqslant 242$ nm

(7) $2H_2 + O_2 \rightarrow 2H_2O$ (Net cycle)

Levels of O_2 in the early prebiological atmosphere were very sensitive to atmospheric levels of H_2O, CO_2, and H_2, and to the flux of incoming solar radiation, which initiates the photolysis of H_2O, CO_2, and O_2. There is reason to believe that all of these parameters may have varied significantly over geological time (Hart, 1978; Canuto et al., 1982; 1983). The H_2O distribution in the troposphere is controlled by the saturation vapor pressure, which is regulated by the tropospheric temperature profile. Even though very large amounts of H_2O may have outgassed in the early history of our planet, it seems unlikely that the early atmosphere contained significantly more H_2O than the present atmosphere contains, since any outgassed H_2O in excess of its saturation vapor pressure would have simply condensed into cloud droplets and then precipitated out of the atmosphere. CO_2 is a different story. Prior to the formation of carbonates, the early atmosphere may have contained significantly higher levels of CO_2, perhaps orders of magnitude more CO_2 than presently found in the atmosphere (Hart, 1978). To assess the importance of CO_2 on early prebiological levels of O_2, we have performed photochemical calculations for the preindustrial level (280 ppmv) and for 100 times that value, a mixing ratio of 0.028 (2.8% by volume) (Levine, 1982). While very little is known about levels of H_2 in the early atmosphere, previous studies suggested an H_2 mixing ratio of between 1.7×10^{-5} (17 ppmv) (Kasting and Walker, 1981) and 10^{-3} (Pinto et al., 1980). To assess the importance of H_2 on early atmospheric levels of O_2, we performed photochemical calculations over a wide range of H_2 concentrations for mixing ratios ranging from 10^{-6} to 10^{-1}.

A key parameter in the photochemical production of O_2, which is initiated by the photolysis of H_2O and CO_2, is the level of solar ultraviolet (UV) flux (see reactions (1) and (3)). Measurements of young, Sun-like stars obtained with the International Ultraviolet Explorer (IUE) satellite suggest that the young Sun may have emitted considerably more UV radiation than it presently emits (Canuto et al., 1982; 1983; and Zahnle and Walker, 1982), although the total visible luminosity of the young Sun was only about 75% of its present value (Hart, 1978). The variation in UV radiation emitted by a Sun-like star over its history is summarized in Table 3 (Canuto et al., 1982). To assess the role

Table 3. Solar Ultraviolet Radiation as a Function of Sun's Age (Canuto et al., 1982)

Age (Years)	Ultraviolet Enhancement
10^6	10^4
10^7	500
5×10^7	100
10^8	32
5×10^8	8
10^9	4
4.5×10^9	1

of enhanced solar UV radiation on early atmospheric levels of O_2, we performed calculations over a wide range of UV levels (Levine, 1982).

The results of these photochemical calculations are shown in Figures 1-4. Figure 1 shows the vertical distribution of O_2 in the early atmosphere for three different sets of calculation parameters. Profile B represents the "standard" case: CO_2 = 1 (280 ppmv), H_2 mixing ratio = 1.7×10^{-5}, and solar UV = 1 (present level). Profile A represents a combination of parameters to give a minimum O_2 profile: CO_2 = 1, H_2 mixing ratio = 10^{-1}, and UV = 1, while profile C represents a combination of parameters to yield a maximum O_2 profile: CO_2 = 100, H_2 mixing ratio = 1×10^{-6}, and UV = 100. All three calculations exhibit a similar distribution with altitude. The O_2 maximum occurs above 40 km, the region of maximum photolysis, and the O_2 minimum occurs close to the surface, away from the region of maximum photolysis.

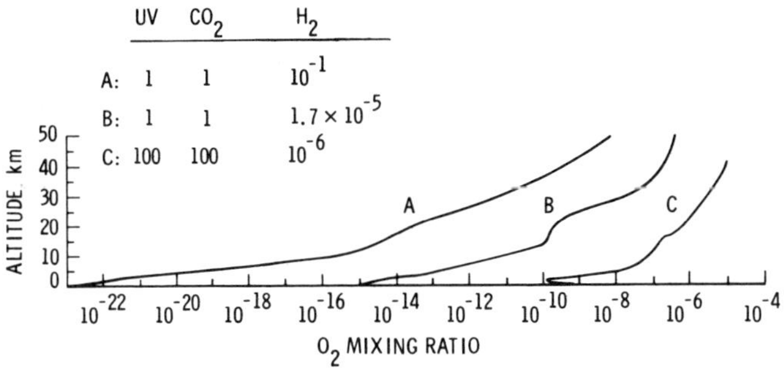

Figure 1. Vertical distribution of O_2 in the prebiological early atmosphere for three different sets of values for CO_2, H_2, and solar ultraviolet (UV) radiation (Levine, 1985b).

The variation of surface O_2 mixing ratio of a function of CO_2 and H_2 for different solar UV levels is shown in Figures 2-4. Figure 2 shows the surface O_2 mixing ratio as a function of CO_2 (for CO_2 = 1, 10, and 100) and H_2 (for H_2 mixing ratio varying from 10^{-6} to 10^{-1}) for solar UV = 1. Figures 3 and 4 give similar calculations for solar UV = 10 and 200, respectively. The important conclusion from these calculations is that in the prebiological early atmosphere, photochemical processes could not account for surface O_2 mixing ratios in excess of the parts per billion by volume level (ppbv = 10^{-9}). Hence, photochemical processes were not responsible for transforming the early atmosphere from a mildly reducing mixture to a strongly oxidizing mixture.

THE RISE OF OXYGEN AND OZONE

The calculations described in the previous section clearly indicate that photochemical processes were not responsible for transforming the atmosphere from a mildly reducing mixture to a strongly oxidizing mixture. The production of O_2 as a by-product of photosynthetic activity was the overwhelming source of atmospheric oxygen. The production of oxygen as a result of photosynthesis may be represented as:

$$\text{(8)} \qquad nH_2O + mCO_2 \xrightarrow{\text{chlorophyll}} C_m(H_2O)_n + mO_2$$

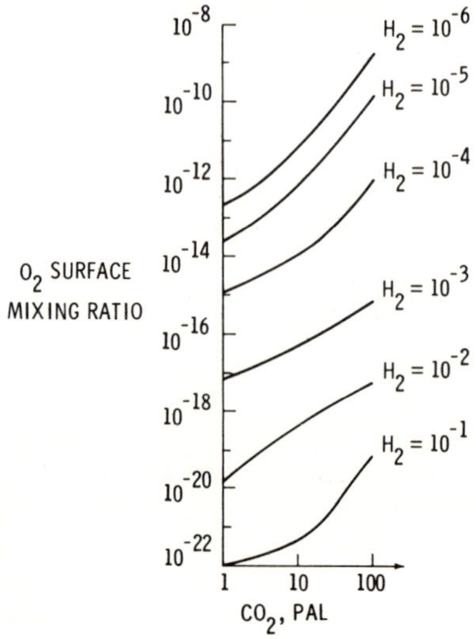

Figure 2

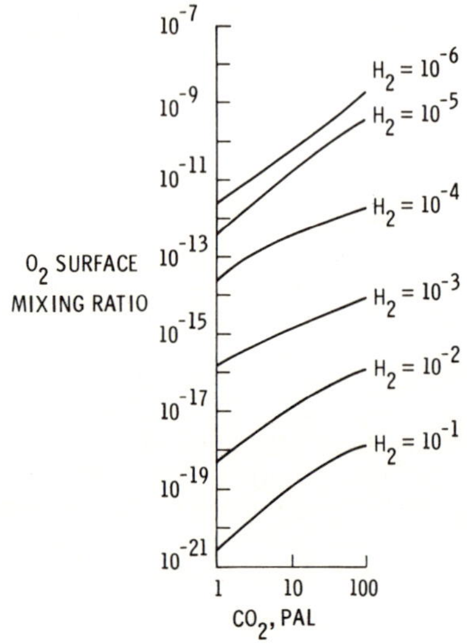

Figure 3

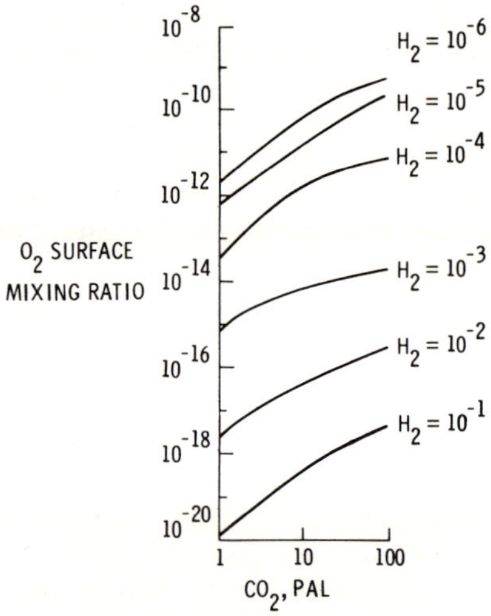

Figure 4

Figures 2-4. Variation of surface O_2 mixing ratio as a function of CO_2 and H_2 for solar ultraviolet (UV) of 1 (Figure 2), 10 (Figure 3), and 100 (Figure 4) (Levine, 1985b).

A possible timetable for the rise of atmospheric oxygen is summarized in Table 4 (Cloud, 1983).

Table 4. The Evolution of Atmospheric Oxygen (Cloud, 1983)

Time	Oxygen Level (% of present level)
2 billion years ago	1%
1 billion years ago	5%
670 million years ago	7%
550 million years ago	10%
400 million years ago	100%

The sources, sinks, and transfer rates of oxygen in the present atmosphere are summarized in Table 5. If all photosynthetic activity, the overwhelming source of atmospheric oxygen, were to cease, respiration and decay would continue to consume atmospheric oxygen. In the absence of photosynthesis, organic carbon (responsible for oxygen consumption via respiration and decay) would no longer be added to the reservoirs of surface organics. In the absence of photosynthetic production, the reservoir of surface organic carbon would be completely exhausted after about 20 years, at which time the amount of oxygen in the atmosphere would have decreased by less than 1% (Walker, 1977). With no carbon left in the surface organic reservoir, the burial of organic carbon in sediments would cease, but weathering would continue. It would take approximately 4 million years for weathering to consume all of the oxygen in the atmosphere (Walker, 1977).

As a result of photosynthetic activity, O_2 became a major constituent of that atmosphere. Accompanying and directly controlled by the buildup of O_2 was the evolution of O_3, which is formed photochemically from O_2. The photochemical production of O_3 was initiated by the photolysis of O_2 (reaction (9)), followed by the three-body recombination of O, O_2, and M (reaction (10)):

Table 5. Rates of Oxygen Production, Destruction and Transfer (Walker, 1977)
(Units of Moles O_2/year; the atmosphere contains 3.8×10^{19} moles O_2)

1. <u>Production</u>	
Photosynthesis	10^{16}
Photolysis of water/escape of hydrogen	7×10^{9}
2. <u>Destruction</u>	
Respiration and decay	10^{16}
Combustion of fossil fuels	3×10^{14}
Weathering of sedimentary rocks	10^{13}
Reaction with volcanic hydrogen	5×10^{10}
3. <u>Transfer</u>	
Burial of surface organic matter to sedimentary rocks	10^{13}

(9) $O_2 + h\nu \rightarrow O + O, \lambda \leqslant 242$ nm

(10) $O + O_2 + M \rightarrow O_3 + M$

There are a number of photochemical and chemical processes that lead to the destruction of O_3, including

(11) $O_3 + h\nu \rightarrow O + O_2, \lambda \leqslant 1100$ nm, and

(12) $O_3 + O \rightarrow 2O_2$

In addition, O_3 is chemically destroyed through a series of catalytic cycles involving the oxides of nitrogen (nitric oxide, NO; and nitrogen dioxide, NO_2), hydrogen (hydroxyl, OH; and hydroperoxyl radical, HO_2) and chlorine (atomic chlorine, Cℓ; and chlorine oxide, CℓO). Sources of nitrogen oxides in the early atmosphere included atmospheric lightning, biogenic production, and the oxidation of nitrous oxide (N_2O). The oxides of hydrogen were produced photochemically and chemically from water vapor. Volcanic emissions and sea salt spray were sources of chlorine in the early atmosphere. The catalytic cycles leading to the chemical destruction of O_3 are summarized here:

(13) $O_3 + NO \rightarrow NO_2 + O_2$

(14) $NO_2 + O \rightarrow NO + O_2$

Net cycle: $O_3 + O \rightarrow 2O_2$

(15) $O_3 + OH \rightarrow HO_2 + O_2$

(16) $HO_2 + O_3 \rightarrow OH + 2O_2$

Net cycle: $2O_3 \rightarrow 3O_2$

(17) $O_3 + C\ell \rightarrow C\ell O + O_2$

(18) $C\ell O + O \rightarrow C\ell + O_2$

Net cycle: $O_3 + O \rightarrow 2O_2$

We have investigated the origin and evolution of atmospheric O_3 as a function of the buildup of O_2 by solving reactions (9) to (18) for O_3. The results of these calculations are given in Figure 5, where the vertical profiles of O_3 with and without the inclusion of chlorine species chemistry are given in terms of present atmospheric level (P.A.L.) of O_2, ranging from 10^{-4} P.A.L. to the present atmospheric level (1 P.A.L.)

The absorption of solar UV radiation is controlled by the total atmospheric burden or column density of O_3. The total atmospheric burden of O_3 corresponding to the five O_3 profiles shown in Figure 5 is given in Table 6. It has been suggested that biological shielding of the Earth's surface was achieved when the total atmospheric burden of O_3 reached about 6×10^{18} O_3 molec. cm^{-2} (Berkner and Marshall, 1965), which is approximately half of the total O_3 burden in the present atmosphere. According to the calculations presented

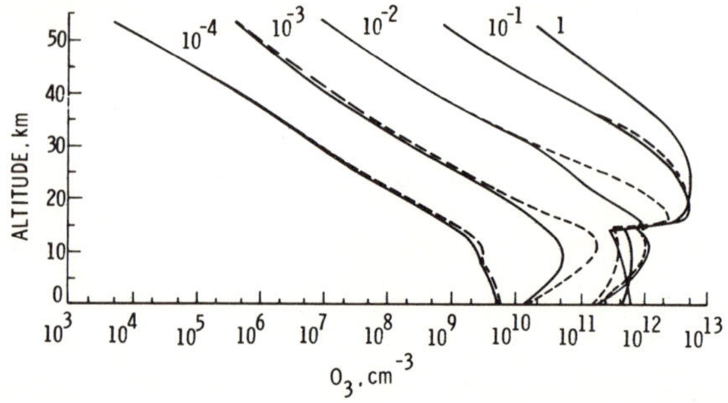

Figure 5. Vertical distribution of ozone (O_3) with (——) and without (---) the inclusion of chlorine species chemistry, as a function of O_2 level (in terms of present atmospheric level, P.A.L.) (Levine, 1982).

in Table 6, this atmospheric burden of O_3 was reached when O_2 reached 10^{-1} P.A.L.

Once atmospheric oxygen reached about 10% of its present atmospheric level, our calculations indicate that there was sufficient ozone in the atmosphere to shield the surface from biologically lethal solar ultraviolet radiation (200-300 nm). At this point in the Earth's history, life could leave the safety of the ocean and go ashore for the first time. The land, shielded from solar ultraviolet radiation for the first time, provided a major new niche for life. Once on land, life "exploded" both in numbers and diversification.

Table 6. Evolution of Ozone as a Function of Increasing Oxygen Levels (Levine, 1982)

O_2 Level (PAL)	O_3 Column Density (cm^{-2})	Height of O_3 Peak (km)	O_3 Density at Peak (cm^{-3})
Without chlorine-species chemistry			
1	9.93 (18)*	20.5	5.53 (12)
10^{-1}	6.07 (18)	19	4.57 (12)
10^{-2}	2.47 (18)	16	2.48 (12)
10^{-3}	1.88 (17)	11.5	1.92 (11)
10^{-4}	5.58 (15)	0	5.63 (09)
Without chlorine-species chemistry			
1	9.70 (18)	20.5	5.40 (12)
10^{-1}	5.94 (18)	19	4.62 (12)
10^{-2}	1.59 (18)	10	1.16 (12)
10^{-3}	6.98 (16)	9	5.72 (10)
10^{-4}	5.18 (15)	0	5.42 (09)

*9.93 (18) = 9.93×10^{18}.

REFERENCES

Berkner, LV, Marshall, LC (1965). On the origin and rise of oxygen concentration in the Earth's atmosphere. Journal of Atmospheric Sciences 22: 225-261.

Canuto, VM, Levine, JS, Augustsson, TR, Imhoff, CL (1982). UV radiation from the young Sun and oxygen and ozone levels in the prebiological paleoatmosphere. Nature 296: 816-820.

Canuto, VM, Levine, JS, Augustsson, TR, Imhoff, CL, Giampapa, MS (1983). The young Sun and the atmosphere and photochemistry of the early Earth. Nature 305: 281-286.

Cloud, P. (1983). The biosphere. Scientific American 249: 176-189.

Hart, MH (1978). The evolution of the atmosphere of the Earth. Icarus 33: 23-29.

Kasting, JF, Walker, JCG (1981). Limits on oxygen concentration in the prebiological atmosphere and the rate of abiotic fixation of nitrogen. Journal of Geophysical Research 86: 1147-1158.

Levine, JS (1982). The photochemistry of the paleoatmosphere. Journal of Molecular Evolution 18: 161-172.

Levine, JS (1985a). "The Photochemistry of Atmospheres: Earth, the Other Planets, and Comets": Orlando: Academic Press, Inc., 518 pages.

Levine, JS (1985b). The photochemistry of the early atmosphere. In Levine JS (ed): "The Photochemistry of Atmospheres: Earth, the Other Planets, and Comets," Orlando: Academic Press, Inc., pp. 3-38.

Lewis, J, Prinn, R (1984). "Planets and Their Atmospheres: Origin and Evolution" New York: Academic Press, Inc., 470 pages.

Pinto, JP, Gladstone, GR, Yung, YL (1980). Photochemical production of formaldehyde in the Earth's primitive atmosphere. Science 210: 183-185.

Walker, JCG (1977). "Evolution of the Atmosphere." New York: MacMillan, 318 pages.

Zahnle, KJ and Walker, JCG (1982). The evolution of solar ultraviolet luminosity. Reviews of Geophysics and Space Physics 20: 280-292.

DISCUSSION

<u>MASON</u>: It has been proposed that photosynthetic oxygen production started about 3.5 billion years ago based upon the discovery of microfossils having some resemblance to blue-green algae. What is your thinking about the probability that the proposal is correct?

<u>LEVINE</u>: The idea that there may have been blue-green algae-like photosynthetic organisms producing oxygen as

early as 3.5 billion years ago is based on the study of microfossils in some of the oldest rocks ever found. I believe that this hypothesis is plausible although it is only a hypothesis.

KOPPENOL: One of your equations leading to the production of oxygen in the atmosphere reads: $HO\cdot + HO\cdot \rightarrow O + H_2O$. Did you also consider the alternative reaction: $HO\cdot + HO\cdot \rightarrow H_2O_2$, which then decomposes: $2H_2O_2 \rightarrow 2H_2O + O_2$?

LEVINE: In our photochemical calculations we consider many chemical reactions involving $HO\cdot$ including the reaction you mention. We also consider the reactions involving $HO\cdot$ and other species in the atmosphere, such as N_2, H_2, and H_2O.

PAPA: Would you comment on the evolutionary relationship between anaerobic and aerobic life?

LEVINE: The earth's early atmosphere was a mildly reducing mixture of gases with no free oxygen. Hence, the first microorganisms were most probably anaerobic. As oxygen built up in the atmosphere, the by-product of photosynthetic activity, the atmosphere was transformed from a reducing mixture to a highly oxidizing mixture. This transformation was probably the major "air pollution" event in the history of our planet.

Once oxygen built up in the atmosphere, aerobic microorganisms evolved and flourished. However, even today there are anaerobic environments where anaerobic microorganisms survive and thrive, such as wetlands and water saturated soils.

MANN: How serious is the risk to life at the present time and in the foreseeable future, from "holes" and thinning of the ozone layer?

LEVINE: The possible depletion of ozone in the stratosphere which shields the surface of our planet from biologically lethal solar ultraviolet radiation, is of real concern. During the month of October, the concentration of ozone in the stratosphere over the Antarctic has decreased by more than 50%. The most likely culprit for this significant decrease in stratospheric ozone is a group of man-made chemicals called chlorofluorocarbons (CFCs). CFCs relelease chlorine in the stratosphere, which catalytically destroys ozone. Accompanying the decrease in stratospheric ozone is a significant increase in the flux of solar ultraviolet radiation (200-300 nm) reaching the earth's surface.

RYDEN: Your diagram on oxygen concentration vs. time was a log-log plot. If it is remade into a linear plot, one can see a dramatic increase in O_2 concentration at about 500 million years ago up to today's concentration when it stays very constant. Have you an explanation for how the concentration is kept constant?

LEVINE: Over the recent geological past, it appears that the global production of oxygen (photosynthetic activity) has been approximately balanced by its global destruction (respiration and decay), and the atmospheric concentration of oxygen is believed to have been fairly constant. The atmospheric oxygen increase about 500 million years ago may have been resulted from a significant increase in the rate of photosynthetic activity by the world's vegetation.

AZZI: How could it be possible that O_2 evolved from living organisms under conditions where no ozone being present, no form of life would be expected to survive due to the very high UV irradiation?

LEVINE: This is indeed a strange paradox! At the very time that life was forming on our planet, the flux of biologically lethal solar ultraviolet radiation was greater than the present value and there was no ozone in the atmosphere to shield the surface from this radiation. For these reasons, I believe that early life was restricted to the oceans, where seawater could absorb the biologically lethal solar ultraviolet radiation in the absence of atmospheric ozone. Several meters of ocean water provides the same ultraviolet radiation absorption as the present ozone layer (or about 10 meters of clear non-turbid water). Hence, several meters below the ocean surface, evolving life could have been protected from solar ultraviolet radiation, and yet at the same time still receive sufficient levels of solar visible radiation.

Oxidases and Related Redox Systems, pages 127–144

NEW CONCEPTS OF THE MOLECULAR PATHOGENESIS ARISING FROM HYPOXIA

Dean P. Jones

Department of Biochemistry
Emory University School of Medicine
Atlanta, Georgia 30322 USA

INTRODUCTION

Inadequate supply of O_2 to cells is a major cause of human morbidity and mortality. Despite decades of intensive research, the molecular basis of irreversible injury is not known. In recent years, studies of freshly isolated cells, primary cultures and cultured cell lines have begun to connect the extensive knowledge of oxidase and oxygenase functions with the complex pathophysiological processes that occur in intact tissues in vivo.

The purpose of this presentation is to summarize studies of cellular hypoxia, focusing on emerging new concepts of the molecular pathogenesis arising from hypoxia. This includes discussions of the molecular definition of hypoxia, the central role of mitochondrial distribution in determining cellular O_2 dependence, and the importance of cellular regulatory mechanisms in modulating the selective vulnerability of cells to hypoxic and post-ischemic reperfusion injury.

DEFINITION OF HYPOXIA

Studies of the O_2 dependence of physiological and biochemical changes have frequently addressed the earliest and most sensitive responses to O_2 deficiency. Hence, efforts were made to define the *critical O_2 tension*, i.e., the O_2 tension at which the first detectable change from normoxia occurs. A variety of studies of recent years have shown that this concept of a critical O_2 tension is not meaningful at the molecular level and of limited utility relative to understanding the mechanism(s) of irreversible injury. The primary event of hypoxia is a limitation of O_2 to enzymes that utilize O_2 as substrate, namely, the oxidases and oxygenases (Keevil and Mason, 1978). Over 100 mammalian oxidases and oxygenases are known and many of these have high Km_{O_2} values relative to physiological O_2 concentrations. Thus some enzymes are O_2 dependent in cells even at normoxic O_2 concentrations (Aw et al, 1985; Jones, 1986a; Jones and Mason, 1978a; Jones et al, 1985).

The O_2 dependence of cytochrome P-450 catalyzed reactions has been extensively studied and serves as an illustrative example. While the Km_{O_2} values for metabolism of substrates by the liver microsomal enzymes are typically in the range of 4 to 10 μM, the Km_{O_2} is dependent upon the nature of the oxidizable substrate and the form of cytochrome P-450 involved (Jones et al, 1978b). In studies of the O_2 dependence of biosynthetic forms of cytochrome P-450 (Aw et al,1985), we found that the aromatization reactions of androstenedione and testosterone by a choriocarcinoma cell line have Km_{O_2} values in the physiological range of O_2 concentrations that occur *in vivo* (Fig. 1). Thus, aromatization is expected to be O_2 dependent *in vivo*, and this O_2 dependence is likely to be important in the variety of complications associated with hypoxia during pregnancy.

The O_2 dependence of cytochrome P-450-dependent oxidations is also dependent upon the concentration of the oxidizable substrate (Jones and Mason, 1978b), and recent studies have revealed that this substrate-dependent modulation of Km_{O_2} can be important in regulation of cellular functions. In studies of the O_2 dependence of cytochrome $P\text{-}450_{scc}$, the rate-determining step of adrenal steroidogenesis, we found that the Km_{O_2} is highly dependent upon cholesterol concentration (Stevens et al, 1974). Under nonstimulated conditions, the cholesterol concentration is low and the Km_{O_2} is high relative to the physiological O_2 concentration (Fig. 2); thus the reaction is O_2 limited. Following stimulation by ACTH, the cholesterol concentration is increased, and at high cholesterol concentrations, the Km_{O_2} is low relative to physiological O_2 concentrations (Fig. 2). Thus, the

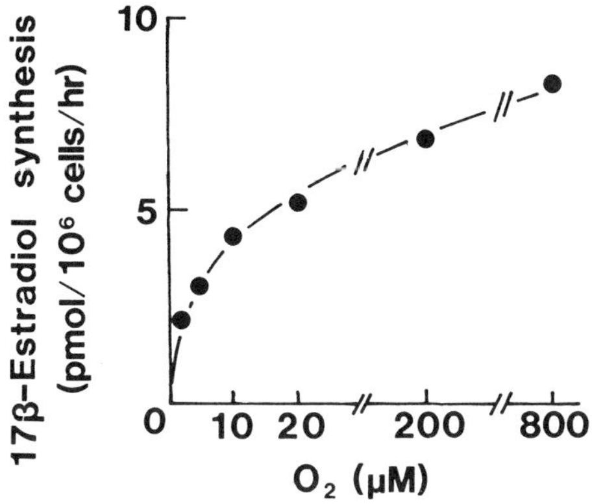

Figure 1 O_2 dependence of 17β-estradiol production in BeWo choriocarcinoma cell line with testosterone as substrate. Data from Aw et al. (1985).

modulation of Km_{O_2} from a high value where O_2 is limiting to a low value where O_2 is not limiting greatly enhances the rate increase effected by an increase in cholesterol concentration.

The results of these studies and others (Aw and Jones, 1982, 1984, 1985) of drug metabolism and intermediary metabolism have provided clear evidence that hypoxia must be considered a continuum with normoxia, especially since some reactions are O_2 dependent *in vivo* during normoxia. From this perspective, "critical P_{O_2}" is more a measure of the sensitivity of analytical methods to detect changes than a measure of an important physiological or pathological parameter.

MITOCHONDRIAL O_2 DEPENDENCE AND INTRACELLULAR O_2 GRADIENTS

Although certain aspects of metabolism are O_2 dependent under normoxic conditions, pathological processes are only known to occur in cells under conditions where mitochondrial function is impaired. The Km_{O_2} for cytochrome oxidase function in isolated mitochondria is very low, i.e., in the range of 0.02 to 0.2 μM (Oshino

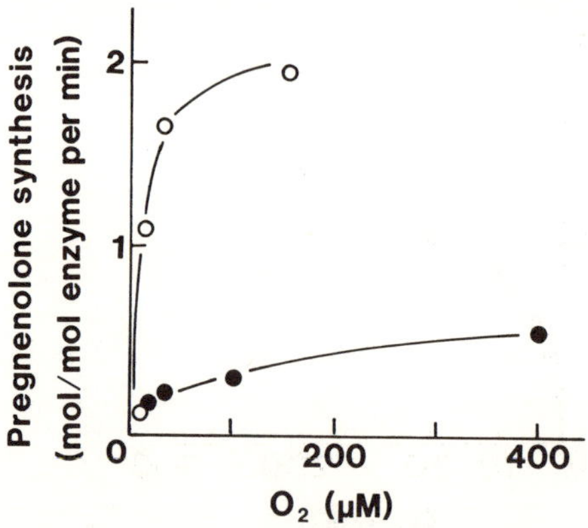

Figure 2 O_2 dependence of pregnenolone synthesis by purified, reconstituted cytochrome P-450$_{scc}$ in the presence of low (•) and high (o) cholesterol. Data Stevens et al, 1984.

et al, 1974; Sugano et al, 1974). Physiological effects of hypoxia are seen *in vivo* when arterial P_{O_2} drops below about 75 torr (105 μM; Hillis and Braunwald, 1977). Thus a substantial O_2 gradient must occur from the circulation to the mitochondria. The location and basis for this gradient has been a matter of controversy and speculation for decades. However, studies with isolated cells revealed that much of the gradient is at the cellular level (Jones, 1986b; Jones and Mason, 1978b; Jones et al, 1985; Wilson et al, 1979) and is due to the spatial distribution and density of mitochondria in the cells (Jones, 1984, 1986b).

A digitized illustration of a portion of a hepatocyte (Fig 3A) with an associated O_2 gradient map (Fig. 3B) that was generated from mathematical modeling of O_2 diffusion into mitochondrial clusters reveals that the O_2 concentration is heterogeneous within cells. This heterogeneity in concentration explains why mitochondria in some cells require higher O_2 concentrations than isolated mitochondria and also why proteins of the endoplasmic reticulum, peroxisomes and cytoplasm do not show the same differences in O_2 sensitivity as do mitochondrial oxidases when the function of isolated form is compared to that in intact cells (Jones, 1986b; Jones and Mason, 1978a and b).

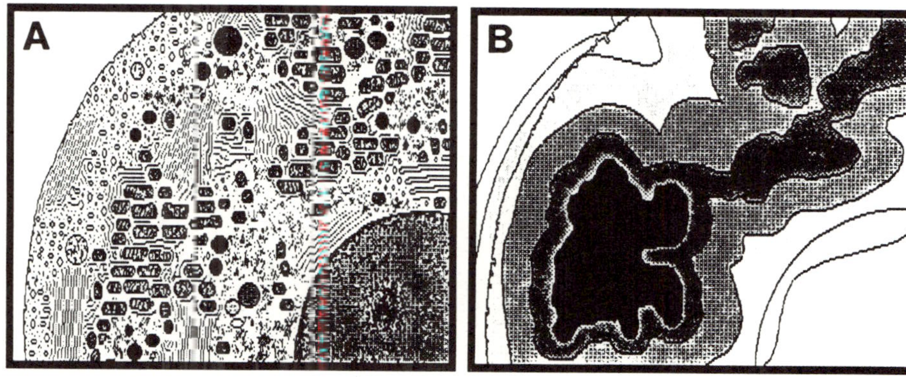

Figure 3 Computer-assisted drawing of a portion of hepatocyte (A) showing the distribution of mitochondria. In B, the O_2 concentration profile was constructed for this drawing from computer analyses of the O_2 gradient in the vicinity of clusters of mitochondria in cells that are exposed to 6 μM O_2 as described in Ref 14. O_2 concentrations are scaled from black (0-1 μM O_2) to white (5-6 μM) with each interval representing 1 μM O_2.

Fetal and neonatal hepatocytes have a relatively low density of mitochondria and little clustering. Studies of the O_2 dependence of mitochondrial function in these cells show that the P_{50} values are comparable to those for isolated mitochondria (Aw and Jones, 1987a). Thus the fetal and neonatal hepatocytes can function with substantially lower blood P_{O_2} than adult cells. This may be an important factor in the relative resistance of these cells to hypoxic injury. Similarly, primary cultures of adult hepatocytes are O_2 dependent over the same range as neonatal cells, i.e., the O_2 dependence is very similar to that for isolated mitochondria (Aw and Jones, 1987b). Studies of the change in O_2 dependence as a function of culture time show that cells undergo a dramatic decrease in P_{O_2} during the first 24 h. Although these cells lose mitochondrial cytochromes during this time, it is not known whether the change in O_2 dependence is coincident with the decrease in density or a change in clustering of mitochondria.

Studies with isolated cardiac myocytes and kidney cells from adult rat show that these cells also require much higher O_2 concentrations for mitochondrial function than do isolated mitochondria from these tissues (Kennedy and Jones, 1987; Aw et al, 1987). In both cases, the high densities and clustering of mitochondria contribute to the high values. Studies with these cells provide insight as to how variations in functional state affect the cellular mitochondrial O_2 dependence. Isolated cardiac myocytes from adult rat ventricle are quiescent upon isolation and have an O_2 consumption rate comparable to an arrested intact heart. Upon electrical stimulation of these cells, the O_2 consumption rate increases up to 6-fold (Jones et al, 1987). Associated with this increase is an increase in the O_2 concentration required for function: half-maximal oxidation of cytochrome $c+c_1$ occurred at 5.8 μM O_2 in quiescent cells but increased to 11.6 μM in electrically stimulated heart cells. Thus, the effective Km_{O_2} for mitochondrial function is increased as O_2 consumption is increased.

Similarly, isolated kidney cells are active in Na^+ pumping by the Na^+-K^+-ATPase, but this activity can be modulated by inhibiting the pump with ouabain or stimulating the pump by increasing the Na^+ leak with the ionophore nystatin. Comparison of the P_{50} values for cytochrome *c+c₁* oxidation and the cellular O_2 consumption rates for these conditions shows that as the ATP-requiring function is increased, the P_{50} value is increased (Aw et al, 1987c).

These studies show that both the spatial distribution and O_2 consumption rate of mitochondria in cells affects the required O_2 concentration for cellular functions. Perhaps of most importance is that structural defects that lead to clustering of mitochondria (see D'Agostino et al, 1968; Nishizaw et al, 1987; Tandler et al, 1968) or increases in functional demand that increase O_2 consumption increase the O_2 concentrations required for optimal ATP production and other O_2-dependent processes.

Knowledge of the factors that affect the cellular O_2 dependence provides a basis to describe the variations in sensitivity to hypoxia that occur in different tissues *in vivo* (Jones et al, 1985). With this knowledge, one can address important questions concerning cellular oxygenation and function during developmental and pathological processes.

SUPPRESSION OF TRANSPORT FUNCTIONS DURING ANOXIA TO PRESERVE A VIABLE STATE

Mammalian tissues vary considerably in their ability to tolerate exposure to anoxia without irreversible loss of function. There are many possible explanations for this difference, e.g., cells with the highest energy demand die most rapidly, cells with least energy reserve die most rapidly, cells with lowest glycolytic rate die most rapidly, etc. Although factors such as these are doubtlessly of some importance, they have not provided an acceptable general explanation for cellular susceptibility or resistance to hypoxic injury.

We recently studied the effect of anoxia on the mitochondrial protonmotive force in hepatocytes (Andersson et al, 1987) and found that these cells maintain over 80% of the protonmotive force even following 30 min of anoxia. These results suggest that cells have control mechanisms to protect against irreversible injury and form the basis for the suggestion that irreversible loss of function is a consequence of failure of protective mechanisms rather than a failure of cell function *per se*. Although the concept that cells contain mechanisms to protect against anoxic injury is well recognized in the example of stimulation of glycolysis during anaerobiosis, the possibilities that the early events of anoxia are protective and that many systems are specifically regulated during anoxia have not been fully explored.

We initially examined the effect of anoxia on the protonmotive force in freshly isolated hepatocytes from adult rat, calculating the protonmotive force as the sum of the components of the electrochemical proton gradient ($\Delta\mu_{H^+} = \Delta\psi - Z\Delta pH$), where the membrane potential ($\Delta\psi$) and pH gradient (ΔpH) were measured independently to obtain the $\Delta\mu_{H^+}$. To obtain $\Delta\psi$, triphenylmethylphosphonium ion ($TPMP^+$) was allowed to equilibrate in cell compartments under aerobic or anaerobic conditions. Cells were rapidly removed from the suspending medium by centrifugation through a silicone oil:mineral oil layer to allow measurement of total cellular uptake. To obtain mitochondrial uptake, cells were briefly exposed to digitonin prior to centrifugation. Following corrections for non-specific binding, adventitious water, and a reoxygenation artefact (Andersson and Jones, 1985), the results showed that $\Delta\psi$ was decreased only 19% by 30 min anoxia (Table 1). Studies with Rhodamine B distribution confirmed that anoxia has no major effect on $\Delta\psi$ since accumulation in mitochondria of aerobic cells was indistinguishable from that for anoxic cells. With both approaches, FCCP or valinomycin plus K^+ eliminated accumulation of the indicator, showing that the methods are suitable for measuring a collapse of $\Delta\psi$.

Measurement of ΔpH was obtained from the distribution of dimethadione (DMO) between mitochondrial, cytosolic and extracellular compartments and showed that anoxia has no significant effect on ΔpH (Andersson and Jones, 1985). This result was confirmed by independent measures of pyruvate distribution between the compartments (Aw et al, 1987a). Thus, the results show that the protonmotive force is largely preserved during 30 min anoxia. FCCP eliminated these pH-dependent distributions, showing that the methods can detect a collapse of the pH gradient. Calculations of the free energy necessary for ATP synthesis in the mitochondria under these

Table 1. Mitochondrial membrane potentials and ΔpH in hepatocytes after 30 min anoxia[a]

Conditions	Δψ (mV)	ΔpH
Aerobic	-163	0.78
+FCCP (1 μM)	- 60	0.20
+valinomycin (40 μg/ml)	- 40	--
Anaerobic	-133	0.76

[a]Data from Andersson et al, 1987.

conditions (ΔG′ = 10.3 kCal/mol) and the amount available from the protonmotive force (ΔG′ = -12.3 kCal/mol; 3 H^+/ATP), reveals that sufficient energy is available for ATP synthesis (Andersson et al, 1987). Thus, energy is available from the protonmotive force that is either not used or used very slowly for ATP synthesis during anoxia.

Maintenance of $\Delta\mu_{H^+}$ in the absence of O_2 could, in principle, result from an alternate source of energy input, such as from utilization of ATP from glycolysis, or from selective inhibition of ion conductance so that $\Delta\mu_{H^+}$ is only slowly dissipated. To utilize ATP from glycolysis would require that ATP be taken up by the adenine nucleotide carrier and hydrolyzed by the ATP synthase (F_1 ATPase). However, such a mechanism would place an additional demand for ATP on the already ATP-limited cell. Addition of atractyloside to inhibit the carrier or oligomycin to inhibit the ATPase did not result in loss of $\Delta\mu_{H^+}$ under anaerobic conditions (Andersson et al, 1987). Therefore, ATP from glycolysis is not utilized to maintain $\Delta\mu_{H^+}$ in hepatocytes during hypoxia.

Because of these results, we focussed our efforts on examining the possibility that ion transport (conductance) across the inner membrane is inhibited during anoxia. Only a small number of electrophoretic carriers function in the inner membrane, and inhibition of these systems eliminates a major source of ion movement. Thus, if these systems are inhibited, a relatively slow rate of ion movement is sufficient to maintain Δψ for a prolonged period.

Studies of ion transport rates across the mitochondrial inner membrane in intact cells is not feasible with current technologies. Consequently, we initially compared mitochondrial transmembranal distributions of ions in anoxic cells to those of aerobic cells (Aw et al, 1987a). The solutes transported by the adenine nucleotide carrier and the phosphate carrier are more displaced from equilibrium than under aerobic conditions, showing that transport of these solutes is inhibited, even though the rates are not directly measurable.

In a survey of activities, we found that the ATP synthase (Aw et al, 1987b) and transport systems for malate and glutamate are also inhibited (Aw et al, 1987a). In contrast, pyruvate, citrate or K^+ transport is not inhibited during anoxia (Aw et al, 1987a). Thus, there is a selective regulation of ion transport across the mitochondrial inner membrane during anoxia.

To determine the mechanisms involved in this anoxic regulation, we have studied regulation of the ATP synthase and the phosphate carrier. The ATP synthase is regulated by an endogenous inhibitor that binds at low mitochondrial Ca^{2+} (Yamada et al, 1980). Under anoxia, the 19% decrease in $\Delta\psi$ results in decreased mitochondrial Ca^{2+} (Aw et al, 1987b) and this change is associated with both a decrease in O_2 consumption rate and a decrease in measured F_1-ATPase activity (Aw et al, 1987b). Thus, it appears that the decrease in $\Delta\psi$ during anoxia results in decreased mitochondrial Ca^{2+} and inhibition of the ATP synthesis. In this way, the cell can preserve the protonmotive force even though there is sufficient $\Delta\mu_{H^+}$ for ATP synthesis.

The effect of anoxia on the electroneutral phosphate uptake system is also of critical importance because the cytosolic phosphate increases to over 10 mM as a result of ATP hydrolysis. With the ΔpH that is retained during anoxia (Andersson et al, 1987), phosphate would accumulate to over 50 mM if the uptake were not regulated. Studies with hepatocytes showed that such accumulation does not occur (Aw et al, 1987a), so we examined transport in isolated mitochondria.

Under steady-state anoxic conditions, phosphate accumulation by isolated liver mitochondria is 2- to 3-fold greater than that occurring in hepatocytes (Aw et al, 1987d). Thus, cells contain some control mechanism that is not completely preserved in isolated mitochondria. This indicates that a factor extrinsic to mitochondria is involved in regulation, but the nature of this factor is unknown.

Elucidation of this regulatory mechanism remains elusive because studies with isolated mitochondria show that high phosphate inhibits phosphate uptake and another intrinsic mechanism of regulation is involved. Thus, although we have not yet unraveled the details of these regulatory mechanisms, available data clearly show that regulatory mechanisms exist to preserve ion distribution during anoxia and that these mechanisms function to prolong the period during which cells can recover from anoxia.

NEAHYPOXIA

We use the term "neahypoxia" to refer to those conditions where cellular functions are altered by O_2 deficiency but where there is no irreversible cell injury (Jones et al, 1985). Under these conditions, cellular regulatory mechanisms preserve a viable state. Irreversible injury can thus be considered a result of the inability of control mechanisms to preserve the viable state. This is a departure from the usual consideration that the earliest changes are the initiators of the pathophysiological sequence of events. Thus, the challenge before us is to understand the neahypoxic state, the protective mechanisms that are involved, and the conditions under which these regulatory mechanisms fail.

ISCHEMIA, REOXYGENATION INJURY AND THE SELECTIVE VULNERABILITY OF THE NEAHYPOXIC STATE

Studies during the last several years have provided substantial evidence that oxidative injury can occur during the reoxygenation phase following ischemia and can be an important component of the overall pathogenic process (Granger et al, 1981; McCord, 1985; Parks et al, 1983). Antioxidant enzymes, such as superoxide dismutase or catalase, protect against reoxygenation injury.

Table 2. Characteristics of the neahypoxic state that enhance vulnerability to oxidative injury

Reduced respiratory capacity
Altered Ca^{2+} homeostasis
Imbalances of electrolytes, particularly Na^+ and K^+
Osmotic imbalances, mitochondrial and cellular swelling
Decreased membrane potentials
Decreased adenylate pools
Plasma membrane abnormalities, blebbing
Decreased NADPH supply
Decreased GSH synthesis

Allopurinol, an inhibitor of xanthine oxidase, protects against injury, and this indicates that this enzyme is involved in generation of toxic oxygen species. Breakdown of ATP during anoxia or ischemia results in accumulation of hypoxanthine and xanthine which serve as substrates for the this enzyme and the generation of H_2O_2 and O_2^-. Moreover, ischemia promotes the conversion of xanthine dehydrogenase to xanthine oxidase. Thus, several lines of evidence indicate that ischemia can create conditions suitable for generation of a burst of reactive oxygen species upon reoxygenation.

The suppression of metabolic and transport functions during hypoxia and ischemia suggests a further consideration that may be of general importance in pathophysiological processes associated with O_2 deficiency, namely, that the neahypoxic state has a reduced ability to tolerate oxidative stress (see Table 2) and therefore may be vulnerable to injury by oxidants at concentrations that would normally be nontoxic (Jones, 1985). In recent experiments (Tribble and Jones, 1986, 1987), we found that hypoxic and anoxic hepatocytes are more susceptible to damage from *t*-butylhydroperoxide than aerobic cells. There is no difference in the rate of *t*-butylhydroperoxide elimination or the rate of *t*-butylhydroperoxide-dependent free radical production. Thus, the hypoxic cells appear to have an increased susceptibility to oxidative stress.

CONCLUSIONS

1. Numerous enzymes are O_2 dependent in cells at normoxic O_2 concentrations; thus, hypoxia is a continuum with normoxia when one considers enzymatic O_2 dependence at the molecular level. In the case of cytochrome P-450_{scc}, cholesterol concentration modulates the effective Km_{O_2} over the physiological range of O_2 concentrations. Thus, O_2 concentration is an important component of the regulation of adrenal steroidogenesis.

2. Substantial intracellular O_2 concentration gradients can occur in hypoxic cells due to the non-uniform, non-random distribution of mitochondria. The presence of large aggregates or clusters of mitochondria dramatically increases the O_2 concentrations required for optimal cell function, and such clusters are a major determinant of the location and magnitude of O_2 gradients in tissues. Further, control of cluster size provides a mechanism to regulate cellular O_2 dependence.

3. Cells contain regulatory mechanisms to suppress trans-membranal transport and metabolic reactions during anoxia. Knowledge of the function and failure of these mechanisms is central to understanding the selective vulnerability of tissues to anoxia and to the development of improved therapies for treating and preventing anoxic cell injury.

4. The numerous metabolic and bioenergetic changes that occur during anoxia render the cells less capable of tolerating oxidative injury. Thus, the oxidative stress of reperfusion injury may simply reflect a reduced functional capacity of the neahypoxic state rather than a substantial increase in generation of toxic O_2 species.

ACKNOWLEDGEMENTS

The research upon which this manuscript is based was supported by NIH Grants GM-36538, HL-30286 and GM-28176 and by American Heart Association Grants-in-Aid 80-902 and 83-976, with funds contributed in part by the local affiliate. I also thank the numerous collaborators and colleagues who have contributed to and supported this research and Marion Little who prepared the manuscript for publication.

REFERENCES

Andersson, BS, and Jones, DP (1985). Use of digitonin fractionation to determine mitochondrial transmembrane ion distribution during anoxia. Analyt Biochem 146, 164-172.

Andersson, BS, Aw, TY, and Jones, DP (1987). Mitochondrial transmembrane potential and pH gradient during anoxia. Am J Physiol 252, C349-C355.

Aw, TY, and Jones, DP (1982). Secondary bioenergetic hypoxia. Inhibition of sulfation and glucuronication reactions in isolated hepatocytes at low O_2 concentration. J Biol Chem 257, 8997-9004.

Aw, TY, and Jones, DP (1984). Control of glucuronidation during hypoxia. Limitation by UDP-glucose pyrophosphorylase. Biochem J 219, 707-712.

Aw, TY, and Jones, DP (1985). ATP concentration gradients in the cytosol of liver cells during hypoxia. Am J Physiol 249, C385-C392.

Aw, TY, Jones, DP, O'Shannessy, DJ, Priest, JH, and Priest, R.E. (1985). Oxygen dependence of estrogen production by human placental microsomes and cultured choriocarcinoma cells. J. Steroid Biochem. 22, 753-758.

Aw, TY, and Jones, DP (1987). Respiratory characteristics of neonatal rat hepatocytes. Pediat Res 21, 492-496.

Aw, TY, and Jones, DP (1987). Microzonation of ATP and pH in the aqueous cytoplasm of mammalian cells. In *Microcompartmentation* (Jones, DP, Ed) CRC Press, Boca Raton, Florida, in press.

Aw, TY, Wilson, E, Hagen, TM and Jones, DP (1987). Determinants of mitochondrial O_2 dependence in kidney. Am J Physiol, in press.

Aw, TY, Andersson, BS, and Jones, DP (1987). Mitochondrial transmembrane ion distribution during anoxia. Am J Physiol 252, C356-C361.

Aw, TY, Andersson, BS, and Jones, DP (1987). Suppression of mitochondrial respiratory function following short-term anoxia. Am J Physiol 252, C362-C368.

Aw, TY, Shan, X, Andersson, BS, and Jones, DP (1987). Mitochondrial membrane potential and pH gradient in hypoxic hepatocytes: Characteristics of the neahypoxic state. In: *Integration of Mitochondrial Function*

(Lemasters, JJ, Hackenbrock, C.R., Thurman, R.G., and Westerhoff, HV, Eds) Plenum Press, in press.

D'Agostino, AN, Ziter, FA, Rallison, ML, and Bray, PF (1968). Familial myopathy with abnormal muscle mitochondria. Arch Neurol 18, 388-401.

Granger, DN, Rutili, G, and McCord, J (1981). Superoxide radicals in feline intestinal ischemia. Gastroenterology 81, 22-29.

Hillis, LD, and Braunwald, E (1977). Myocardial ischemia. New Engl J Med 296, 971-978.

Jones, DP, and Mason, HS (1978). Metabolic Hypoxia: Accumulation of Tyrosine Metabolites in Hepatocytes at Low pO_2. Biochem Biophys Res Commun 80, 477-483.

Jones, DP, and Mason, HS (1978). Gradients of O_2 Concentration in Hepatocytes. J Biol Chem 253, 4874-4880.

Jones, DP (1984). Effect of mitochondrial clustering on O_2 supply in hepatocytes. Am J Physiol 247, C83-C89.

Jones, DP, Kennedy, FG, Andersson, BS, Aw, TY, and Wilson, E (1985). When is a mammalian cell hypoxic? Insights from studies of cells vs. mitochondria. Mol Physiol 8, 473-482.

Jones, DP (1985). The role of oxygen concentration in oxidative stress: hypoxic and hyperoxic models. In *Oxidative Stress* (Sies, H, Ed). Academic Press, New York, pp 151-195.

Jones, DP (1986). Renal metabolism during normoxia, hypoxia and ischemic injury. Annu Rev Physiol 48, 33-50.

Jones, DP (1986). Intracellular diffusion gradients of O_2 amd ATP. Am J Physiol 250, C663-C675.

Jones, DP, Kennedy, FG, and Aw, TY (1987). Intracellular O_2 gradients and the distribution of mitochondria. In: *Hypoxia: The Tolerable Limits* (Houston, C, and Sutton, J, Eds) Benchmark Press, in press.

Keevil, T, and Mason, HS (1978). Molecular oxygen in biological oxidations--an overview. Meth Enzymol 52, 3-40.

McCord, JM (1985). Oxygen-derived free radicals in post-ischemic tissue injury. N Engl J Med 312, 159-163.

Nishizawa, M, Tanaka, K, Shinozawa, K, Kuwabara, T, Atsumi, T, Miyatake, T, and Ohama, E (1987). A mitochondrial encephalomyopathy with cardiomyopathy. A case revealing a defect in complex I in the respiratory chain. J Neurolog Sci 78, 189-201.

Oshino, N, Sugano, T, Oshino, R, and Chance, B (1974). Mitochondrial function under hypoxic conditions: the steady states of cytochrome a+a3 and their relation to mitochondrial energy states. Biochim Biophys Acta 368, 298-310.

Parks, DA, Bulkley, GB, and Granger, DN (1983). Role of oxygen free radicals in shock, ischemia and organ presentation. Surgery 94, 428-432.

Stevens, VL, Aw, TY, Jones, DP, and Lambeth, JD (1984). Oxygen dependence of adrenal cortex cholesterol side chain cleavage: Implications in the rate-limiting step(s) in steroidogenesis. J Biol Chem 259, 1174-1179.

Sugano, T, Oshino, N, and Chance, B (1974). Mitochondrial function under hypoxic conditions. the steady states of cytochrome $_c$ reduction and of energy metabolism. Biochim Biophys Acta 374, 340-358.

Tandler, B, Erlandson, RA, and Wynder, EL (1968). Riboflavin and mouse hepatic cell structure and function. I. Ultrastructural alterations in simple deficiency. Am J Pathol 52, 69-95.

Tribble, DL and Jones, DP (1986). The oxygen dependence of t-butylhydroperoxide toxicity. Federat Proc 45, 1725 (abst).

Tribble, DL and Jones, DP (1987). Anoxia impairs the ability of hepatocytes to withstand oxidative stress. Federat Proc 46, 1942 (abst).

Wilson, DF, Erecinska, M, Drown, C, and Silver, IA (1979). The oxygen dependence of cellular metabolism. Arch Biochem Biophys 195, 485-493.

Yamada, EW, Shiffman, FH, and Huzel, NJ (1980). Ca^{2+}-regulated release of an ATPase inhibitor protein from submitochondrial particles derived from skeletal muscles of the rat. J Biol Chem 255, 267-273.

DISCUSSION

SLATER: There is another intrinsic control phenomenon in mitochondria that has to be taken into account. Whatever the mechanism, the experimental finding is a very steep relationship between the rate of ATP synthesis and Δ**p**. ATP synthesis can be almost completely inhibited by concentrations of respiratory inhibitors or uncouplers that have a very small effect on Δ**p**.

JONES: We have evidence that at least three mechanisms function to suppress mitochondrial function during anoxia. One of them may be a voltage-dependent gating of ion channels or carriers to decrease ion conductance. It would not be surprising for multiple control mechanisms to be involved because many enzyme systems are regulated by more than one allosteric effector.

ESTABROOK: I question the interpretation of your data suggesting the "O_2 dependence of physiological and biochemical changes" where you define "the critical O_2 tension". A tissue such as rat liver contains about 15 μM cytochrome oxidase and 30-100 μM cytochrome P-450 (assuming uniform distributions). Is there any meaning to speak of critical O_2 concentrations of 1 μM or less? It seems to me that only second order rates of reaction are applicable. This criticism is more severe if one accepts the hypothesis that mitochondria are aggregated as you describe.

JONES: The problem of relating enzyme kinetics obtained with dilute enzyme and excess substrates to *in vivo* conditions in which enzyme concentrations are in excess of substrate concentrations is difficult to solve. Our data comparing the O_2 dependence of cytochrome functions in isolated mitochondria (state 3) and in cells suspended with the same amount of cytochrome c per ml illustrate this. Isolated mitochondria can function maximally at 10% of the O_2 concentration required for intact cells. Clearly the packing of high concentrations of enzymes in a small volume can create conditions wherein solution kinetics do not apply. A further complication is that we don't know the solubility coefficient for O_2 in the vicinity of the active site. Consequently, at this physiological level we can only discuss O_2 dependence in terms of solution O_2 concentrations which we can measure. The values that we obtain for O_2 dependence are for near steady state conditions in cells and, if we are correct in our interpretation that intracellular diffusion limits O_2 delivery to the oxidase, it becomes apparent that the spatial geometry of capillaries and clusters of mitochondria are just as important in determining intracellular function as the kinetic characteristics of the enzyme.

PAPA: You have clearly stressed in your talk that mitochondria in cells in situ will have to function at extremely low O_2 tensions - close to anaerobiosis. Factors which may contribute to adapt to this are: possible changes in the affinity of cytochrome c oxidase for oxygen, tighter binding of the ATPase inhibitor protein to F_o-F_1 complex, to prevent extensive ATP hydrolysis when aerobic $\Delta\mu_{H^+}$ subsides to low values. There is also evidence for voltage-gated H^+ channels in the mitochondrial membrane whose inhibition may contribute to preserve $\Delta\mu_{H^+}$ in anaerobiosis. Would you comment on these factors?

JONES: Our studies do not eliminate the possibility that changes in O_2 affinity of cytochrome oxidase can occur, but they indicate that such modulation is not necessary to change the effective O_2 dependence of mitochondrial function in cells.

With regard to the mechanisms of controlling mitochondrial functions during anoxia, we have data showing that multiple mechanisms exist. We do not know the details of these regulatory processes but they may include voltage-gating of ion channels and covalent or non-covalent modification of carrier systems. We have recent data showing that mitochondrial Ca^{2+} decreases during anoxia and that the F_1 ATPase activity is decreased. These results are consistent with binding of the endogenous inhibitor to the ATPase under conditions of low mitochondrial Ca^{2+}, which would result in decreased proton conductance and ATP synthesis. We do not suggest that this is the only or most important regulatory event but present this as an example of how the cell can respond to preserve mitochondrial homeostasis during anoxia and allow for facile recovery upon reoxygenation.

Session II: Flavoprotein Oxidases and Oxygenases

Oxidases and Related Redox Systems, pages 147–166

STRUCTURAL DETERMINANTS OF THE OXYGEN REACTIVITY OF DIFFERENT CLASSES OF FLAVOPROTEINS

Vincent Massey*, Lawrence M. Schopfer* and Robert F. Anderson†
*Department of Biological Chemistry, University of Michigan, Ann Arbor, MI 48109, and †Gray Laboratory, Mount Vernon Hospital, Northwood, Middlesex HA6 2RN, England

The reaction of reduced flavins and flavoproteins with molecular oxygen has been a subject of great interest for many years. Free reduced flavins react comparatively slowly with O_2, probably due to the spin conservation problems associated with the reaction of singlet reduced flavin with triplet O_2. The primary event with free flavins, as elucidated by Bruice and colleagues (1,2), appears to be a 1-electron transfer, generating a radical pair, which could undergo either a second 1-e transfer, (yielding oxidized flavin and hydrogen peroxide), or after spin inversion, collapse to form a flavin hydroperoxide:

$$Fl_{red}H^-(\uparrow\downarrow) + O_2(\uparrow\uparrow) \longrightarrow [FlH^{\bullet}\uparrow\uparrow O_2^-] \longrightarrow [FlH^{\bullet}\uparrow\downarrow O_2^-] \longrightarrow FlHOOH \xrightarrow{H^+} FlHOO^-$$

It should be noted that the flavin hydroperoxide in this reaction path has not been observed directly with free unsubstituted flavins, only with N(5)-substituted flavins, where the flavin C(4a)-hydroperoxide is produced (2,3). This species is very unstable in aqueous solution, where the elimination of H_2O_2 and regeneration of oxidized flavin, Flox, occurs (1). Thus, if the hydroperoxide is an intermediate in the aqueous reaction of reduced flavins with O_2, it must break down faster than it is formed. This is entirely reasonable; Kemal et al. calculate the rate of the bimolecular reaction above to be $\sim 250\ M^{-1}\ s^{-1}$ at pH 6.4, 30°. The proposed radical mechanism received direct support from the experiments of Anderson (4) who generated riboflavin semiquinone and O_2^- on a μsec time scale by pulse radiolysis. He was able to follow the rapid formation of the flavin C(4a)-hydroperoxide ($k = 7.8 \times 10^8\ M^{-1}\ s^{-1}$) and its subsequent decay to oxidized flavin and H_2O_2 (k = 260 s^{-1} at pH 6.5, 25°).

The work described above thus provided a satisfying explanation of the chemical pathway for formation of flavin C(4a)-hydroperoxides, which had been observed as intermediates in the catalytic reactions of flavoprotein monooxygenases, and which will be discussed in more detail below. Less understandable, however, is why the reaction is so slow. The calculated rate constant of 250 M^{-1} s^{-1} for hydroperoxide formation from $FlredH^-$ and O_2 is equivalent to a $t_{1/2}$ of formation of ~10 s in air saturated solution. The overall reoxidation of reduced flavin proceeds much faster than this, due to autocatalysis by various products, Flox, $Fl^{\cdot -}$ and O_2^- (5,6,2). Such reactions are unlikely to play an important role in the reaction of protein-bound reduced flavin and O_2, which is why it is important to unravel the details of the primary step(s).

Flavoprotein-Oxygen Reactivities

In terms of their reactions with O_2 there are three well defined groups of flavoproteins: the electron transferases (often referred to misleadingly as "dehydrogenases"), which generate O_2^- and enzyme flavin radical, the oxidases, which yield H_2O_2 and oxidized enzyme as the only observed products, and the monooxygenases, which bring about a splitting of the O-O bond, inserting one oxygen atom into a substrate and reducing the other atom to H_2O. With all enzymes of the monooxygenase class that have been investigated in detail, the flavin C(4a)-hydroperoxide and the C(4a)-hydroxyflavin have been detected as intermediates in the pathway of substrate monooxygenation (see references 7 and 8 for reviews). As the pathway of reoxidation of free reduced flavins by O_2 appears also to proceed via the C(4a)-hydroperoxide, it was of obvious interest to determine whether the C(4a)-hydroperoxide was an intermediate common to all three classes of flavoproteins, with the ultimate products being dictated by the particular interactions between flavin and protein in each class. In support of the concept that reactivity might be controlled by specific flavin-protein interactions was the finding of many characteristics distinctive to each of the classes, as detailed below.

Properties of Flavoprotein Electron Transferases

Typical examples of enzymes of this class are flavodoxin, ferredoxin-NADP reductase, cytochrome P-450 reductase, and electron transfer flavoprotein. Enzymes of this class generally react quite rapidly with one electron oxidants such as ferricyanide or cytochrome c (9). The overall reaction rate of the reduced flavoprotein with O_2 is slow, but the first electron transfer reaction can be quite fast (e.g. with flavodoxin at pH 7.2, 25° the rate of the reaction $EFlredH^- + O_2 \rightarrow EFlH^{\bullet} + O_2^-$ is $3.5 \times 10^4\ M^{-1}\ s^{-1}$ (10)). The subsequent oxidation of flavin radical by either O_2^- or O_2 is very slow (10,11). With most enzymes of this class, the flavin semiquinone which is stabilized by the protein is the blue neutral radical (9).

Solvent accessibility studies have been carried out with a number of proteins of this group, using chemically or environmentally sensitive flavins in place of the native coenzyme. With few exceptions the benzene ring of the flavin is the only part which is exposed to solvent with the N(5)-C(4) locus being buried in the protein (see ref. 12 for a recent review). Most enzymes of this class fail to offer any substantial stabilization of the anionic forms of artificial flavins. With few exceptions, the reaction of O_2 with the reduced flavin forms of enzymes of this class is satisfactorily accounted for by the $1e^-$ oxidation-reduction pathway. The flavin semiquinone and O_2^- produced may fail to form hydroperoxide simply because the O_2^- does not have access to the flavin C(4a)-position. The radical pair must either undergo a second $1e^-$ transfer to yield EFlox and HO_2 (H_2O_2), or dissociate into the uncaged free radical products.

Properties of Flavoprotein Oxidases

Typical enzymes of this class are the D- and L-amino acid oxidases, glucose oxidase, glycolate oxidase and L-lactate oxidase. In general these enzymes react only slowly with one electron oxidants such as ferricyanide and cytochrome c, but react rapidly with O_2 (9). The reduction product of O_2 is H_2O_2, without any flavin semiquinone or O_2^- ever being detected as an intermediate. These enzymes also share many other properties highly suggestive of similar flavin-protein interactions. Without exception they stabilize the red anionic semiquinone when subject to $1e^-$-reduction conditions. They all form flavin N(5)-sulfite adducts, a property unique to this class of flavoproteins

(9). They also stabilize the benzoquinoid anion forms of the synthetic, 6-hydroxy-, 6-mercapto-, 8-hydroxy- and 8-mercaptoflavins (12). These all have the common characteristic of localizing negative charge in the N(1)-C(2)O locus of the flavin

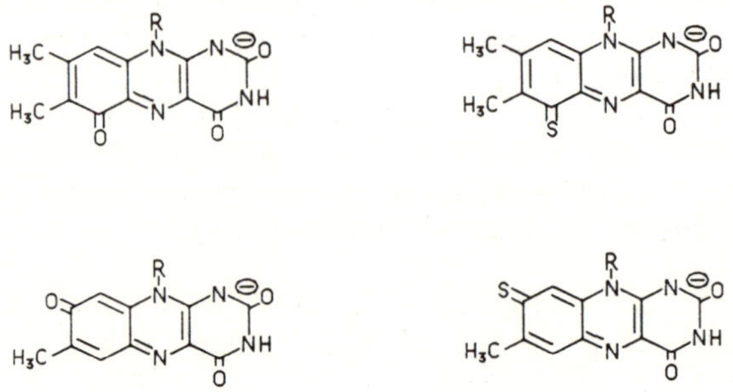

Of these environmentally sensitive flavins, the 8-mercaptoflavins are the most versatile indicators of the protein environment surrounding the bound flavin, as the anion (pK 3.8) can exist in various resonance forms ranging from that where the negative charge is localized on the sulfur atom to that where it is localized at the N(1)-C(2)O locus (13).

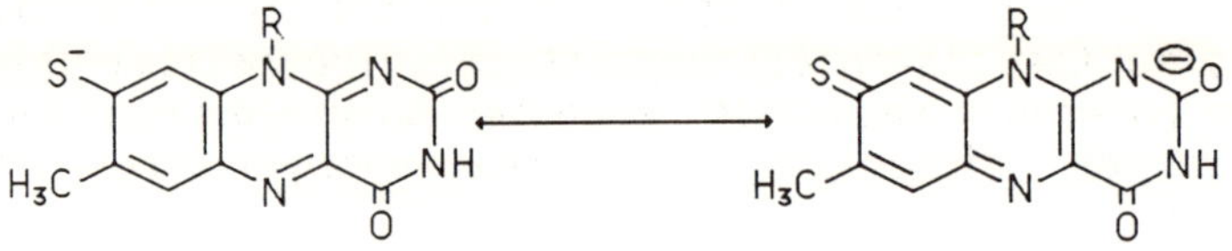

In free solution, and when bound to electron transferases such as flavodoxin and ferredoxin-NADP reductase, the flavin appears to exist predominantly in the thiolate anion form, whereas when bound to typical oxidases such as D-amino acid oxidase or L-lactate oxidase, it is stabilized as the benzoquinoid anion form.

All the properties listed above argue strongly for a positively charged residue of the protein located in the vicinity of the flavin N(1)-position, which by H-bonding and/or Coulombic interaction would stabilize a negative charge in the N(1)-C(2)O locus of the flavins, (including the anion form of fully reduced normal flavin).

This ascription is also supported by the spectra of the reduced forms of 4-thioflavins bound to this class of protein. With native flavin the spectral properties of the neutral and anionic forms of reduced flavin (pK ∿ 6.5) are sufficiently similar that it is often difficult to determine when a protein-bound reduced flavin is neutral or anionic. With 4-thioflavins, however, there is a large spectral difference. For the oxidases examined, the reduced 4-thioflavin was clearly in the anionic form (14).

Solvent accessibility studies have shown that while the flavin 2- and 8- positions show various patterns of reactivity, depending on the particular enzyme, in all cases the N(5)-C(4) region appears to be accessible. Thus, the possibility was clear that this group of enzymes might react with O_2 in the same way as proposed for free flavins, with the initially formed flavin radical-O_2^- pair collapsing to the C(4a)-hydroperoxide, which then rapidly eliminates H_2O_2 to reform oxidized flavin enzyme. This possibility received what appeared to be convincing support by pulse radiolysis studies, where the neutral semiquinone form of glucose oxidase was found to react very rapidly with O_2^- to form a labile intermediate with the spectral properties of a flavin C(4a)-hydroperoxide (15). However, as will be discussed in a later section, our later studies appear to exclude this mechanism.

Properties of Flavoprotein Monooxygenases

Enzymes of this class introduce one atom of the O_2 molecule into substrate and reduce the other atom to H_2O. Among the earliest examples of this type to be recognized were L-lactate monooxygenase (16) and L-lysine monooxygenase (17) which carry out oxidative decarboxylation and oxidative deamination reactions, respectively. With these enzymes, the reducing source for the enzyme flavin is the substrate itself, and so they were referred to as "internal" monooxygenases to distinguish them from the larger group which uses NAD(P)H as reductant, the so called "external" monooxygenases. Rapid reaction studies showed that the oxidative decarboxylation brought about by L-lactate monooxygenase occurred via enzyme bound pyruvate and H_2O_2 being held in juxtaposition at the active site long enough for the reaction to occur (18). Pyruvate and H_2O_2 are the typical products expected for a simple oxidase reaction, and the overall reaction is an adventitious monooxygenation, caused by slow dissociation of the primary oxidase products. These enzymes are then better classified mechanistically as oxidases; as shown in the previous section L-lactate monooxygenase behaves as a typical oxidase and is generally referred to as L-lactate oxidase.

The "external" flavoprotein monooxygenases catalyze reactions which can be formulated in general terms.

$$NAD(P)H + H^+ + S + O_2 \rightarrow NAD(P)^+ + SO + H_2O$$

A large amount of mechanistic work has been done with such enzymes. The function of the reduced pyridine nucleotide is to reduce the enzyme flavin, which then reacts with O_2 to produce the flavin C(4a)-hydroperoxide, in all cases studied. As such hydroperoxides have not been found experimentally in reactions of O_2 with reduced flavoprotein oxidases or electron transferases, the protein probably stabilizes the hydroperoxide, at least transiently, in the case of the monooxygenases. How this is done is not at all clear. With some enzymes, such as bacterial luciferase and the microsomal N,S-monooxygenase, the hydroperoxide is remarkably stable in the absence of the substrate. For example, the luciferase hydroperoxide can be isolated in almost pure state by low temperature chromatography (19). With the N,S-monooxygenase, the hydroperoxide is stable only when $NADP^+$ is bound to the enzyme (20). With these two enzymes, on addition of the third substrate, S, the hydroperoxide rapidly transfers its terminal oxygen to the substrate, giving rise to a C(4a)-hydroxyflavin intermediate, which then loses H_2O to generate the oxidized flavoprotein, ready for the next catalytic cycle. It would appear likely that in the case of the N,S-monooxygenase the oxygenation reaction occurs via a simple nucleophilic attack of the nitrogen or sulfur electron pair of the substrate on the terminal oxygen of the hydroperoxide, since model studies of Ball and Bruice (21) have shown that N(5)-substituted flavin C(4a)-hydroperoxides behave like other organic peracids in carrying out such reactions in free solution.

What factors could contribute to the remarkable stabilization of the hydroperoxides in these enzymes? One obvious way would be to form a strong H-bridge from the flavin N(5)H to some protein residue, in a fashion analogous to that found in the crystal structure of flavodoxin neutral semiquinone (10).

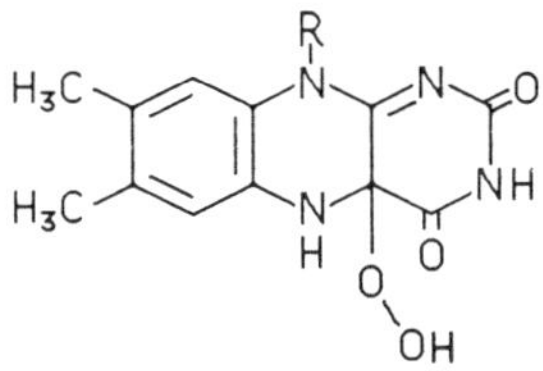

Other possibilities would be to form a H-bridge from the peroxide itself, or simply to surround the flavin in a hydrophobic environment. The model studies of Bruice and coworkers have shown that N(5)-substituted flavin C(4a)-hydroperoxides are rather stable in apolar solvents, and eliminate H_2O_2 rapidly only in aqueous media (1). With luciferase, H-bonding from the N(5)H appears an attractive possibility, since the enzyme has been shown also to stabilize strongly the neutral flavin radical (22).

The largest known group of flavoprotein monooxygenases is that of enzymes hydroxylating phenolic compounds. These enzymes are usually referred to as aromatic or phenolic hydroxylases, and differ significantly from the enzymes discussed above. With the phenolic hydroxylases, the phenolic substrate plays two distinctive roles. The first is to modulate the reactivity of the enzyme flavin with NAD(P)H; the EFlox.S complex is often reduced at rates 10^3-10^5 greater than the substrate-free enzyme. This change in reactivity has to be due to some critical difference in flavin-protein interactions on binding of substrate. In this role the substrate serves as an "effector", since the same phenomenon can often be elicited by binding of substrate analogs which cannot be hydroxylated. The second important role played by the substrate is to modulate the reaction of the reduced enzyme with O_2. In the presence of substrate, the reduced flavin reacts with O_2 to form the C(4a)-hydroperoxide. Such reactions have been found with all enzymes investigated by rapid reaction techniques. For purposes of discussion, we will restrict ourselves largely to what is known about p-hydroxybenzoate hydroxylase, since this is the enzyme which has been studied most; it is where the first demonstration of a flavin C(4a)-hydroperoxide was made (23,24), and the structure of the enzyme, in some of its forms, has been determined from X-ray crystallography (25,26,27). In the presence of 2,4-dihydroxybenzoate (a substrate which is converted to 2,3,4-trihydroxybenzoate) a series of three intermediates is formed (24). The first is the flavin C(4a)-hydroperoxide. This is converted rapidly to a high extinction intermediate, referred to as Intermediate II. Rapid quench experiments have shown that the formation of this intermediate occurs concomitantly with the transfer of one oxygen atom from the hydroperoxide to the substrate (28). The structure of this intermediate is still unknown, but the rapid quench experiments provide the essential information that oxygen atom transfer has already occurred at this step. Intermediate II then decays to the C(4a)-hydroxyflavin, which subsequently dehydrates to give oxidized flavin enzyme ready for the next cycle of catalysis. Many proposals have been made for the structure of Intermediate II. These have been summarized in reference 29. A recent proposal has been made that II may be a complex of an oxyflavin radical species and a hydroxycyclohexadienyl radical species (30 a,b). This proposal awaits further experimental tests. In the presence of a non-substrate effector, or in the presence of high concentrations of azide, variable amounts of the C(4a)-hydroperoxide can be detected by rapid reaction spectrophotometry. Under these conditions the hydroperoxide decays rapidly ($k = 3\text{-}11\ s^{-1}$ at pH 6.55, 4° (24)) to oxidized enzyme and H_2O_2. The rate of formation of hydroperoxide is in the range $2\text{-}5 \times 10^5\ M^{-1}\ s^{-1}$ for all substrates and effectors studied (24). In the absence of substrate, no intermediate was observed; reduced enzyme reacted with O_2 with a rate of $2.6 \times 10^4\ M^{-1}\ s^{-1}$. It has been commonly assumed that the hydroperoxide is indeed formed in this reaction, but decays more rapidly than it is formed. We will show in the next section that this assumption is not correct. Whatever the actual mechanism of hydroperoxide formation, the differences in stability of the hydroperoxide between the phenolic hydroxylases and enzymes such as bacterial luciferase and the microsomal N,S-monooxygenase are quite dramatic. While strong flavin protein interactions must be invoked to explain the hydroperoxide stability of the latter enzymes, only minimal stabilization is required for the phenolic hydroxylases.

There are several structural features which are also common to the phenolic hydroxylases. Work with artificial flavins has shown that in all cases the flavin N(5)-C(4) region is accessible to solvent, as is also the flavin-8 position (12). These findings are in agreement with the X-ray crystal structure of p-hydroxybenzoate hydroxylase (25,26,27). It has also been shown that in all cases NAD(P)H gains access to the re-face of the flavin, presumably by a channel starting at the exposed benzene ring portion of the flavin (31). The pyridine nucleotide pocket of PHBH has not been defined directly in the crystal structure but recent model building by Drenth and colleagues has led to the same conclusion (26,27).

On the Mechanism of Reaction of Reduced Flavoproteins with O_2

From the body of work already discussed, a reasonable explanation of the reactions with O_2 of reduced flavoenzymes could be formulated as shown in Scheme 1. Following the formation of the radical pair (II), several alternative routes are possible.

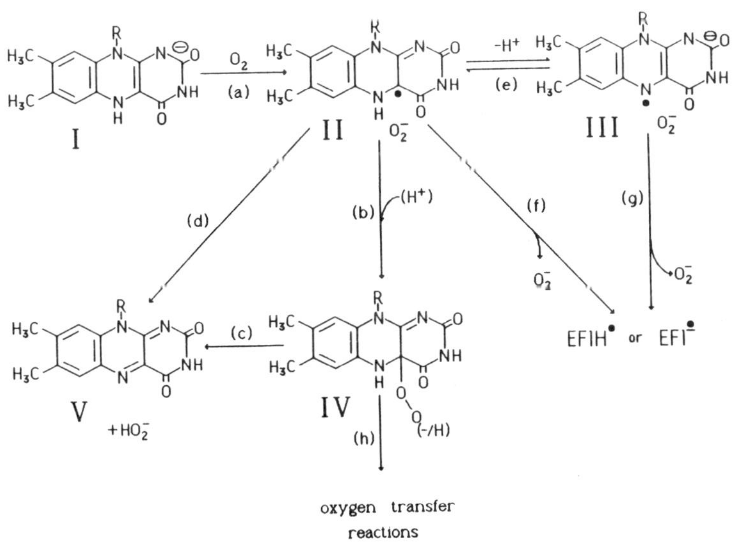

Scheme 1

In the case of the monooxygenases, after spin inversion of the initially formed radical pair, the C(4a)-hydroperoxide

(IV) would be formed, and either undergo productive oxygen atom transfer (step h) or eliminate hydrogen peroxide (step c). As quantitative formation of the hydroperoxide is generally observed with these enzymes the rate of step b must be considerably greater than those of the competing reactions shown.

With electron transferases, such as flavodoxin, the initial reaction (a) would appear to be followed largely by step f, with perhaps a small amount of second $1e^-$ transfer through step (d) (10).

With oxidases, neither flavin semiquinone nor O_2^- have ever been detected. If the radical pair II is formed, it must undergo further reactions before any dissociation can occur. The rapid reoxidation of enzyme with generation of H_2O_2 would then have to occur either by a direct second $1e^-$-transfer (step d) or via the hydroperoxide (steps b and c).

Comparison of Results from Stopped Flow Monitoring of the Reaction of Reduced Glucose Oxidase with O_2 and Pulse Radiolysis Reaction of Glucose Oxidase Semiquinone and O_2^-

As a direct test of the possible pathways outlined in Scheme 1 we subjected glucose oxidase at pH 6.0, 20° to pulses from the 4 MeV Van der Graaf generator at the Gray Laboratory, using conditions similar to those described previously (4,15). Reactions were carried out in 0.1 M formate, 0.002 M phosphate, pH 6, using a 50 µM solution of enzyme equilibrated with N_2O/O_2 (4:1). The electron pulse (in 0.5 µs) results in the following radical yield:

$$H_2O \rightsquigarrow \underset{(0.3\ \mu M/Gy)}{e^-_{aq}} + \underset{(0.3\ \mu M/Gy)}{OH^{\bullet}} + \underset{(0.05\ \mu M/Gy)}{H^{\bullet}} + H_3O^+,\ OH^-,\ H_2O_2$$

Secondary reactions are:

$$e^-_{aq} + O_2 \rightarrow O_2^{\overset{-}{\bullet}} \quad k = 2 \times 10^9\ M^{-1}\ s^{-1}$$
$$e^-_{aq} + N_2O \rightarrow OH^{\bullet} + OH^- + N_2 \quad k \sim 8 \times 10^8\ M^{-1}\ s^{-1}$$
$$OH^{\bullet} + HCOO^- \rightarrow CO_2^- + H_2O \quad k = 2 \times 10^9\ M^{-1}\ s^{-1}$$
$$CO_2^- + O_2 \rightarrow CO_2 + O_2^- \quad k = 2 \times 10^9\ M^{-1}\ s^{-1}$$

The rate of formation of the neutral flavoprotein semiquinone is too rapid to measure; it is probably formed partially from scavenging some e^-_{aq} in competition with O_2

and N_2O, and partially from reaction with CO_2^-. At the time of measurement after the end of the pulse (~2 µs), there is thus a mixture of glucose oxidase neutral radical (Fig. 1) and an excess of O_2^-. Over the next 100 µs the neutral semiquinone disappears, but without the concomitant formation of oxidized enzyme. In the particular experiment of Fig. 1 bovine superoxide dismutase was present at a concentration of 15 µM in order to eliminate secondary long term effects of O_2^- (15). Hence O_2^- was consumed before complete reaction of the flavin semiquinone with O_2^- could be achieved.

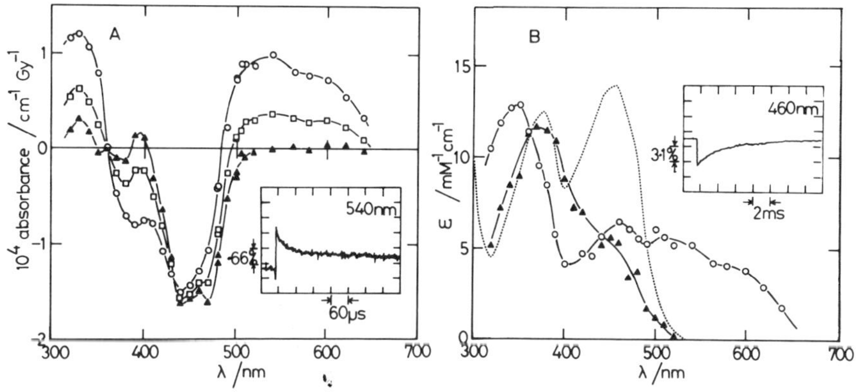

Figure 1. A. Change in absorbance following pulse radiolysis (36 Gy in 0.5 µs) of a N_2O/O_2 (4/1) saturated solution buffered at pH 6.0 (2 mM phosphate) containing glucose oxidase (50 µM), sodium formate (0.1 M) and superoxide dismutase (ca. 15 µM). Spectra measured (O), 2 µs and (□), 150 µs after the pulse were used to calculate the change in absorbance due to the presence of the intermediate formed at 150 µs (▲), as described in the text. Insert: oscilloscope trace of absorbance vs time following a dose of 33 Gy.

B. Absorption spectra of the initially formed species at 2 µs (O), and the intermediate present at 150 µs (▲), corrected for the loss in absorption due to reacted oxidized glucose oxidase. Insert: oscilloscope trace of absorbance vs time for the decay fo the intermediate formed by 150 µs following a dose of 53 Gy.

The difference spectra at 2 µs and 150 µs are shown in Fig. 1A, as well as that due to the conversion of the flavin semiquinone to the new species (corrected for the remaining 38% flavin semiquinone). The intermediate decays on a ms time scale as shown in the inset of Fig. 1B, with a rate of 350 s^{-1}. As the absorption spectrum and extinction coefficient of the glucose oxidase semiquinone is known (ε 570 = 4140 M^{-1} cm^{-1} (32)) the absorption spectrum of the intermediate can be related to those of the native enzyme and the neutral flavoprotein radical (Fig. 1B). It can be seen that the intermediate has an absorption maximum around 380 nm and extinction coefficient ($\sim$11,500 M^{-1} cm^{-1}) similar to those of other known flavoprotein C(4a)-hydroperoxides. The small distortion in the 450 nm region would indicate the formation of at most 10-20% EFlox in a competing reaction to that of formation of the hydroperoxide. Thus starting from the radical pair (Species II), the formation of hydroperoxide (step b), must proceed at a rate approximately 5-10 x faster than the second $1e^-$-pathway of step d. The concentration of O_2^- is readily varied in the pulse experiments by varying the pulse dose. Experiments of this sort, carried out in the presence or absence of small concentrations of superoxide dismutase yielded a linear relationship of kobs vs [O_2^-] with a second order rate constant of 1 x 10^9 M^{-1} s^{-1} for the formation of the hydroperoxide. The decay of the hydroperoxide was independent of O_2^-, yielding a value of 350 s^{-1} for the elimination of H_2O_2 (step c).

From the pulse radiolysis results, it is clear that if the reaction of reduced enzyme with O_2 proceeded via the radical mechanism of Scheme 1, we would expect to find a limiting rate of 350 s^{-1} for the formation of EFlox. If the bimolecular reaction of step (a) was fast enough, we would also expect to find detectable concentrations of IV as an intermediate. Previous stopped flow studies of the reaction of reduced enzyme with O_2 (32,33) had failed to detect any intermediate species. Since these had been done under different conditions, we repeated the stopped flow studies under conditions identical to those employed in the pulse radiolysis experiments. Enzyme was reduced with a small molar excess of mannose. This is a very poor substrate which reduces the enzyme very slowly, so that re-reduction of the enzyme in the O_2-half reaction is negligible (33). The reduced enzyme was reacted with buffer equilibrated with different O_2 concentrations, and the progress of the reaction monitored at 10-15 nm intervals over the range 320-570 nm. At all wavelengths and at all concentrations of O_2, the return of absorbance to that of the oxidized enzyme was exponential, i.e. we were unable to detect any intermediate in the reaction. The kobs values were directly proportional to [O_2], giving a second order rate constant of $2.2 \times 10^6\ M^{-1}\ s^{-1}$. There was no indication of saturation with O_2, and kobs values of 1000 s^{-1} could be measured experimentally, just about at the limit of the stopped flow technique.

The most obvious interpretation of these results is that the reduced enzyme-O_2 reaction does not proceed via the radical mechanism of Scheme 1. A possible way to reconcile the stopped flow results with a radical mechanism would be to have the initial caged radical pair of Species II, with its electron orbital spins in the same direction, react faster (step d) than spin inversion and subsequent formation of the hydroperoxide (step b). However, this explanation is still difficult to reconcile with the pulse radiolysis experiments, since the O_2^- produced there should be in a random spin configuration with respect to the flavoprotein radical, yet it reacts preferentially (> 5:1) to form the hydroperoxide rather than through a second 1e$^-$ transfer (step d).

An alternative explanation is that in the reduced enzyme the protein conformation is such that O_2 does not have access to the flavin C(4a)-position, and so proceeds through steps a) and d), without being able to form the C(4a)-hydroperoxide. Studies of glucose oxidase in the oxidized state, coupled with artificial flavins, have shown that the N(5)-C(4)O locus is exposed to solvent, as is also to some extent the C(2) position. The benzene ring of the flavin is quite inaccessible (34). Unfortunately we have as yet no information regarding accessibility in the reduced state, so we cannot comment with any confidence on this possibility. Clearly, from the pulse radiolysis experiments, the flavin C(4a) position of the semiquinoid enzyme is quite accessible to O_2^-. Hence, if there is a conformational change on reduction, it must be exhibited mainly at the fully reduced flavin level. Thus neither of these explanations is entirely satisfactory. We will present a third alternative later.

Concerning flavoprotein oxidases in general, we should comment on some other aspects of Scheme 1. With most flavoprotein oxidases, the protein stabilizes the anionic flavin radical species (III). When such anionic radical species are produced in conjunction with O_2^- in pulse radiolysis experiments (e.g. glucose oxidase at pH 9), no formation of hydroperoxide could be detected. In reaction of reduced flavoprotein oxidases with O_2, (e.g. D-amino acid oxidase, L-lactate oxidase, glucose oxidase at pH 9), no flavoprotein radical species has ever been detected (9, 18, 32). Thus, if the radical pathway of Scheme 1 is operational, the ionization step (e) must be much slower than steps d) or b).

Is a Flavin C(4a)-Hydroperoxide an Intermediate in the Reaction of O_2 with Substrate-Free p-Hydroxybenzoate Hydroxylase?

The reaction of O_2 with reduced phenolic hydroxylases in the absence of substrate or effector is generally assumed to involve an unstable flavin C(4a)-hydroperoxide (decaying faster than it is formed) simply because the hydroperoxide is an easily demonstrated intermediate in the presence of substrate or substrate-like effectors. We decided to test this assumption directly, using the same combination of pulse radiolysis and stopped flow approaches as employed with glucose oxidase. The results were quite similar to those obtained with glucose oxidase, merely differing in reaction rates. Reaction conditions were 50 mM phosphate, 0.1 M formate, pH 6.4, 20° for both the stopped flow and pulse radiolysis experiments. In the pulse radiolysis experiments an intermediate with spectral properties expected for a C(4a)-hydroxperoxide was produced rapidly and decayed with a rate constant independent of $[O_2^-]$ of ~70 s^{-1}. In the stopped flow studies, where dithionite-reduced substrate-free enzyme was mixed with different concentrations of O_2, the reformation of oxidized enzyme was exponential at all wavelengths, i.e., there was no spectroscopically detectable intermediate. The kobs values were strictly proportional to $[O_2]$, yielding a second order rate constant of 9×10^4 M^{-1} s^{-1}. This is appreciably smaller than that found with glucose oxidase, but again there was no sign of saturation with respect to $[O_2]$. The highest $[O_2]$ experimentally possible (1.08 mM) gave a kobs value of 91 s^{-1}, which was greater than the limiting rate expected from the pulse radiolysis data.

The same discussion that was made for glucose oxidase can be made here. It is quite clear that a flavin C(4a)-hydroperoxide is <u>not</u> involved in the reaction of the substrate-free reduced enzyme, even though it clearly is formed when substrate or substrate-like effectors are bound to the reduced enzyme. This suggests that the reduced enzyme substrate complex has a different conformation from that of the substrate-free reduced enzyme. This conclusion is supported by the substantial differences in reactivity which exist between the substrate-free oxidized enzyme and the EFlox.S complex, and between substrate-free oxidized enzyme and substrate-free reduced enzyme. Thus, the substrate-free enzyme is reduced by NADPH some 3-5 orders of magnitude more slowly than complexes of the enzyme with various substrates (24). Similarly, while free oxidized enzyme binds and releases substrate very rapidly, the free reduced enzyme binds and releases substrate very slowly. For example, at pH 6.55, 4° the binding of p-hydroxybenzoate to oxidized enzyme has a k_{on} of > 5 x 10^6 M^{-1} s^{-1} and k_{off} >200 s^{-1}, while the binding to reduced enzyme is characterized by k_{on} = 160 M^{-1} s^{-1} and k_{off} = 3.4 x 10^{-3} s^{-1} (24).

Solvent accessibility studies on the substrate-free oxidized enzyme substituted with artificial flavins show that the flavin 2,4 and 8 positions are all quite accessible to solvent. In the EFlox.S complex, the 8-position is still quite accessible, but the 2- and 4-positions are significantly less exposed to solvent. These conclusions are in agreement with the X-ray crystal structure, which has been solved to high resolution in the case of the complex of the oxidized enzyme with p-hydroxybenzoate (25). This structure shows the benzene ring of the flavin at the edge of the protein, and a narrow channel leading toward the pyrimidine ring of the flavin. The substrate is positioned with its aromatic ring almost perpendicular to the flavin face, with the substrate 3-position near the flavin 4a-position. The substrate is buried within the protein folds, so that it is clear that a conformational change must occur on substrate binding, since the substrate could not find its active site location without the protein folding being different before its addition. Recently several more crystal forms of the enzyme have been solved, although the results have not yet been published in detail (26,27). The reduced enzyme-substrate complex is reported to be essentially identical in structure to EFlox.S. The substrate-free EFlox form (in the presence of high [Cl^-]), has also been solved, but to comparatively low resolution. Significant differences from EFlox.S are obvious, particularly a 0.8 A° movement of the flavin N(5) toward the position occupied by the substrate in the EFlox.S complex (27). The crystal structure of the substrate-free reduced enzyme has not been solved, but is probably somewhat similar to that of the EFlox.S and EFlredH$^-$.S complexes. This follows from the very slow binding (and release) of substrate to the free reduced enzyme, clearly implying that substrate has difficulty in binding in its correct position when the flavin is reduced first.

These results imply that major changes in flavin-protein interaction occur, a), between EFlox and EFlox.S and b) between EFlox and EFlred. In both the EFlred and EFlred.S forms there would appear, however, to be limited direct access to the flavin C(4a)-position. The crystal structures of the EFlox.S and EFlred.S forms show that O_2 might gain access to the flavin via a channel oriented toward the flavin 2-position, or via the proposed pyridine nucleotide channel on the re-face of the flavin. Direct access to the flavin C(4a)-position would appear difficult.

In terms of the radical mechanism of Scheme 1, the initial radical pair II would have to react by different routes depending on whether or not substrate was bound; by step d) in the absence of substrate and by step b) in its presence. Alternatively, in the absence of substrate the radical pair might react to form an adduct at another position of the flavin ring, such as the flavin C(10a)-hydroperoxide, which could perhaps eliminate H_2O_2 faster than it is formed, and so escape detection in the stopped flow experiments.

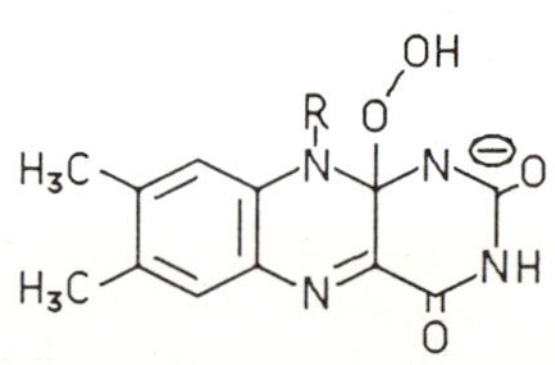

Such a possibility could of course also explain the results obtained with glucose oxidase, but again the results from pulse radiolysis experiments are difficult to accommodate. In view of these difficulties we think that the possibility of a non-radical mechanism for the oxygen reaction of reduced flavoproteins should again receive serious consideration. The fact that a flavin neutral radical, where the unpaired spin is localized chiefly at the

C(4a)-position (35,36) should react efficiently with O_2^- to produce a flavin C(4a)-hydroperoxide is chemically very reasonable, but the fact that such a reaction occurs is obviously no proof of the radical pathway, particularly in those cases where the expected C(4a)-hydroperoxide cannot be found in the reoxidation pathway starting with the reduced flavoprotein and O_2.

Early suggestions for overcoming the spin conservation dilemma of triplet O_2 reacting with singlet reduced flavin, without involving a radical mechanism, were made by Orf and Dolphin (37) and Hemmerich and Wessiak (38). The latter authors state "The center pyrazine subnucleus of this species ($FlredH^-$) exhibits 8 π-electrons and is, therefore, antiaromatic in the flat state. Hence the assumption of a low-lying triplet state, which might be thermally populated in contact with a paramagnetic species, seems to be plausible". Both authors propose the primary formation of a dioxetane structure and subsequent rearrangement to yield hydroperoxide. In this way, either the C(4a)- or the C(10a)-hydroperoxide could be formed readily, as shown in Scheme 2.

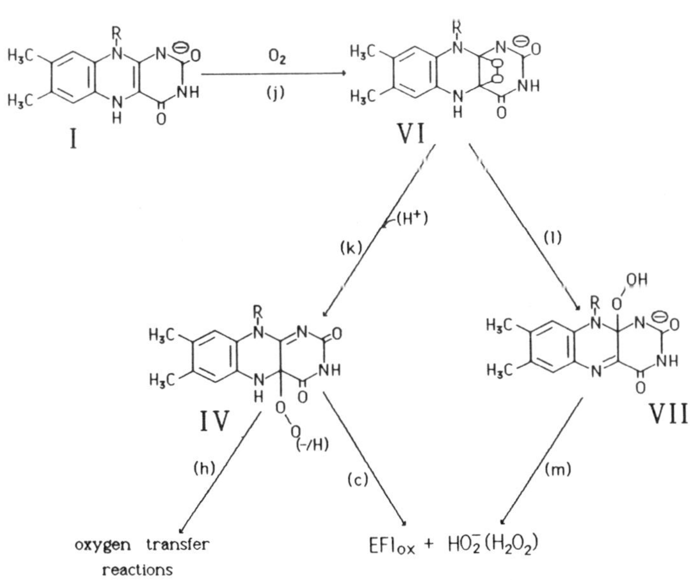

Scheme 2

If the C(10a)-hydroperoxide flavin anion state VII were favored by the protein, peroxide elimination through step (m) would be expected to be rapid, and so VII would not be expected to accumulate to a detectable extent during the reaction. In fact, if Scheme 2 is correct, the rate limiting step in the overall reaction of reduced p-hydroxybenzoate hydroxylase with O_2 has to be the bimolecular reaction of step (j), since both the formation of the C(4a)-hydroperoxide IV in the presence of substrate, and the overall reformation of EFlox in the absence of substrate, are strictly second order reactions.

In the crystal structure of the EFlox.S and EFlred.S complexes of p-hydroxybenzoate hydroxylase, helix H10 has its N-terminal end pointed toward the flavin N(1) position, and has been suggested by Drenth and colleagues to stabilize anionic flavin species bearing a negative charge in the N(1) region, by Coulomb interaction with the partial positive charge of the helix (27). If the protein structural change accompanying reduction of the substrate-free enzyme changes the orientation of this helix relative to the flavin, then any stabilization of VII would be removed, and peroxide elimination would be expected to be rapid. This possibility must await further crystallographic studies for proper evaluation.

A final point which we would like to make regarding p-hydroxybenzoate hydroxylase is that in all forms investigated the flavin positions most accessible to solvent are those of the benzene ring. If the radical mechanism were operational, it is difficult to conceive why O_2 does not react preferentially in this exposed area, as it appears to do readily with flavodoxin. It is almost as though the flavoprotein monooxygenases possess an active mechanism for bringing O_2 to the C(4a)- and perhaps also to the C(10a)-center.

ACKNOWLEDGEMENTS. This work was supported by a grant from the U.S. Public Health Service, GM 11106.

REFERENCES

1. Kemal, C. and Bruice, T.C. (1976) Proc. Natl. Acad. Sci. USA 73, 995-999
2. Kemal, C., Chan, T.W. and Bruice, T.C. (1977) J. Am. Chem. Soc. 99, 7272-7286

3. Eberlein, G. and Bruice, T.C. (1982) J. Am. Chem. Soc. 104, 1449-1452
4. Anderson, R.F. (1982) in Flavins and Flavoproteins (Massey, V. and Williams, C.H., eds) Elsevier North Holland, New York, pp. 278-283
5. Gibson, Q.H. and Hastings, J.W. (1962) Biochem. J. 83, 368-377
6. Massey, V., Palmer, G. and Ballou, D.P. (1973) in Oxidases and Related Systems (King, T.E., Mason, H.S. and Morrison, M., eds) University Park Press, Baltimore, pp. 25-43
7. Massey, V. and Hemmerich, P. (1975) in The Enzymes 3rd Ed. (Boyer, P.D., ed) Academic Press, New York, Vol. XII, pp. 191-252
8. Ballou, D.P. (1984) in Flavins and Flavoproteins (Bray, R.C., Engel, P.C. and Mayhew, S.G., eds) de Gruyter, Berlin, pp. 605-618
9. Massey, V., Müller, F., Feldberg, R., Schuman, M., Sullivan, P.A., Howell, L.G., Mayhew, S.G., Matthews, R.G. and Foust, G.P. (1969) J. Biol. Chem. 244, 3999-4006
10. Mayhew, S.G. and Ludwig, M.L. (1975) in The Enzymes 3rd Ed. (Boyer, P.D., ed) Academic Press, New York, pp. 57-118
11. Mayhew, S.G. Foust, G.P. and Massey, V. (1969) J. Biol. Chem. 244, 803-810
12. Ghisla, S. and Massey, V. (1986) Biochem. J. 239, 1-12
13. Massey, V., Ghisla, S. and Moore, E.G. (1979) J. Biol. Chem. 254, 9640-9650
14. Massey, V., Claiborne, A., Biemann, M. and Ghisla, S. (1984) J. Biol. Chem. 259, 9667-9678
15. Anderson, R.F., Massey, V. and Schopfer, L.M. (1984) in Flavins and Flavoproteins (Bray, R.C., Engel, P.C. and Mayhew, S.G., eds) de Gruyter, Berlin, pp. 573-576
16. Hayaishi, O. and Sutton, W.B. (1957) J. Am. Chem. Soc. 79, 4809
17. Takeda, H. and Hayaishi, O. (1966) J. Biol. Chem. 241, 2733-2736
18. Lockridge, O., Massey, V. and Sullivan, P.A. (1972) J. Biol. Chem. 247, 8097-8106
19. Hastings, J.W., Balny, C., LePeuch, C. and Douzou, P. (1973) Proc. Natl. Acad. Sci. USA 70, 3468-3472
20. Beaty, N.B. and Ballou, D.P. (1981) J. Biol. Chem. 256, 4619-4625
21. Ball, S. and Bruice, T.C. (1980) J. Am. Chem. Soc. 102, 6498-6503; 103, 5494-5503

22. Kurfürst, M., Ghisla, S., Presswood, R. and Hastings, J.W. (1982) Eur. J. Biochem. 123, 355-361
23. Spector, T. and Massey, V. (1972) J. Biol. Chem. 247, 7123-7127
24. Entsch, B., Ballou, D.P. and Massey, V. (1976) J. Biol. Chem. 251, 2550-2563
25. Wierenga, R.K., de Jong, R.J., Kalk, K.H., Hol, W.G.J. and Drenth, J. (1979) J. Mol. Biol. 131. 55-73
26. Van der Laan, J.M. (1986) Thesis, University of Groningen
27. Schreuder, H.A., van der Laan, J.M., Thunnissen, M.M.G.M., Kalk, K.H., Swarte, M.B.A., Hol, W.G.J. and Drenth, J. (1987) in Flavins and Flavoproteins (McCormick, D.B. and Edmondson, D.E., eds) de Gruyter, Berlin, in press
28. Entsch, B., Massey, V. and Ballou, D.P. (1974) Biochem. Biophys. Res. Commun. 57, 1018-1025
29. Wessiak, A., Schopfer, L.M. and Massey, V. (1984) J. Biol. Chem. 259, 12547-12556
30a. Anderson, R.F., Patel, K.B. and Stratford, M.R.L. J. Biol. Chem., in press.
30b. Anderson, R.F., this volume.
31. Manstein, D.J., Pai, E.F., Schopfer, L.M. and Massey, V. (1986) Biochemistry 25, 6807-6816
32. Stankovich, M.T., Schopfer, L.M. and Massey, V. (1978) J. Biol. Chem. 253, 4971-4979
33. Gibson, Q.H., Swoboda, B.E.P. and Massey, V. (1964) J. Biol. Chem. 239, 3927-3934
34. Schopfer, L.M., Massey, V. and Claiborne, A. (1981) J. Biol. Chem. 256. 7329-7337
35. Müller, F., Hemmerich, P., Ehrenberg, A., Palmer, G. and Massey, V. (1970) Eur. J. Biochem. 14, 185-196
36. Palmer, M.H., Simpson, I. and Platenkamp, R.J. (1980) J. Mol. Struc. 66, 243-263
37. Orf, H.W. and Dolphin, D. (1974) Proc. Natl. Acad. Sci USA 71, 2646-2650
38. Hemmerich, P. and Wessiak, A. (1976) in Flavins and Flavoproteins (Singer, T.P., ed) Elsevier, Amsterdam, pp. 9-22

Oxidases and Related Redox Systems, pages 167–178

A POSSIBLE RADICAL MECHANISM FOR THE FUNCTIONING OF FLAVOPROTEIN PHENOLIC HYDROXYLASES

Robert F. Anderson

Cancer Research Campaign Gray Laboratory,
Mount Vernon Hospital, Northwood,
Middlesex, HA6 2RN, England

The mechanism by which certain flavoprotein monooxygenases insert a hydroxy group into aromatic substrates has been the subject of intensive study for some time (see Wessiak *et al.*, 1984 for a listing). The most chemically favored mechanism has involved a ring-opened form of the flavin at the C(4a)-N(5) position (Entsch *et al.*, 1976) resulting from activation of the precursor flavin C(4a)-hydroperoxide (Intermediate I) for electrophilic attack on a substrate. Rapid quench experiments with *p*-hydroxybenzoate hydroxylase have shown that the formation of the high extinction Intermediate II is coincident with the transfer of one oxygen atom from the hydroperoxide to the substrate (Entsch *et al.*, 1974). As no chemical models have been found to support the ring-opened mechanism (Wessiak and Bruice, 1983; Bruice, 1984) its validity is questionable.

We have initiated a novel approach to determine the spectra of intermediates formed upon ·OH radical addition to substrates using pulse radiolysis to see if the resulting radicals can partially account for the spectral properties of Intermediate II. A possible parallel for the enzymatic reaction would be a concerted homolytic fission of the O-O bond of the flavin C(4a)-hydroperoxide (Intermediate I), to form a flavin C(4a)-alkoxyl radical, and a substituted hydroxycyclohexadienyl radical form of the substrate.

Radical Spectra of Hydroxylated Substrates

The radiolysis of aqueous solutions containing solutes is a convenient method to generate and quantitatively study the chemistry of radical species. Details of the Gray Laboratory's 4-MeV Van de Graaff accelerator pulse radiolysis radical detection system, ^{60}Co γ-ray steady-state irradiation facility, methods of dosimetry and sample handling have been published (Vojnovic, 1983; Anderson *et al.*, 1986; Anderson *et al.*, 1987a).

The radiolysis of water results in the production of radicals, molecular products and ionic species in well defined yields.

$$H_2O \rightarrow e_{aq}^-, H^\cdot, {}^\cdot OH, H_2O_2, H_2, H_3O^+, OH^- \quad (1)$$

The reducing e_{aq}^- species can be conveniently converted to the oxidizing $^\cdot OH$ radical in N_2O-saturated solution.

$$e_{aq}^- + N_2O \rightarrow {}^\cdot OH + OH^- + N_2 \quad (2)$$

giving rise to a $\cdot OH$ radical concentration (*G*) of 0.62 $\mu M\ Gy^{-1}$ (Klein and Schuler, 1978) and a $H^\cdot$ atom concentration of 0.06 $\mu M\ Gy^{-1} < 9\%$ of the total radical yield which slightly perturbs the spectra of $\cdot OH$ radical adducts in the low wavelength region).

As the enzyme *p*-hydroxybenzoate hydroxylase has been the most studied, using a variety of substrates which give rise to spectrally distinct intermediates (Entsch *et al.*, 1976; Spector and Massey, 1972), we have investigated these substrates first. 4-Hydroxybenzoic acid (*p*HBA) serves as a good model to describe the approach taken and has been extensively studied (Anderson *et al.*, 1987a). The $\cdot OH$ radical reacts fast with *p*HBA ($k = 6 \times 10^9\ M^{-1}\ s^{-1}$) to form isomeric dihydroxycyclohexadienyl radicals as shown in scheme I.

These initial radicals can undergo unimolecular dehydration and also a variety of bimolecular reactions. A method to determine the initial distribution of $\cdot OH$ radical addition to substrates formed upon steady-state irradiation is to oxidize the initial radicals through electron transfer to quinones of relatively high potential (e.g. benzoquinone, BQ and methylbenzoquinone, MBQ) to form identifiable hydroxylated products (Steenken and Raghavan, 1979; Raghavan and Steenken, 1980). From similar experiments with *p*HBA it was concluded that the fractions of $\cdot OH$ radical addition to the 1:2:3:4: positions were 0.16:0.04:0.65:0.15: respectively (Anderson *et al.*, 1987a). Adjunctive pulse radiolysis experiments showed that the $\cdot OH$ radical adduct ipso to the hydroxy group undergoes fast dehydration ($k = 2 \times 10^7\ s^{-1}$) while dehydration of the major radical intermediate (addition to the 3 position) is slower (e.g. $k = 6 \times 10^3\ s^{-1}$ at pH 6) but undergoes a complex acid-base catalysis reaction. The spectrum produced following pulse radiolysis (*ca.* 5 Gy in 30 ns) of a N_2O-saturated solution containing *p*HBA (5 mM) at pH 6.0 (phosphate, 2 mM) is centered at 365 nm (O), figure 1. This spectrum decays to form a new spectrum centered at 420 nm. As this transformation is relatively slow at pH 6, leading to some loss of radical yield

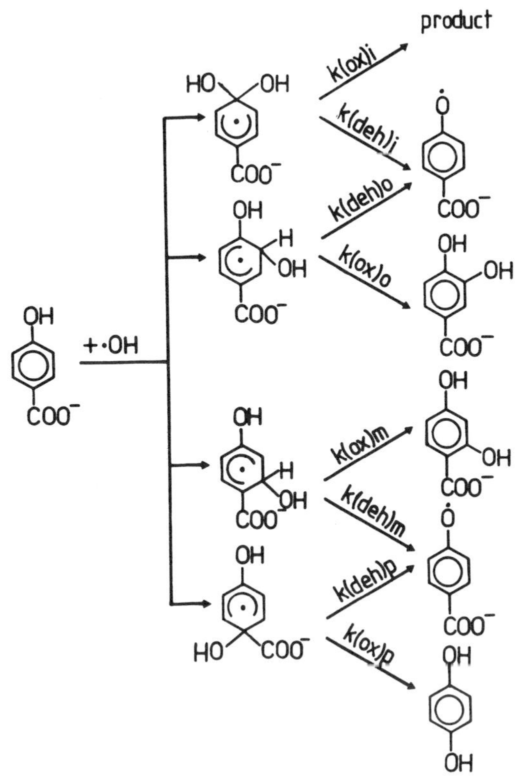

Scheme I. Reactions induced by ·OH radical attack on *p*HDA.

through bimolecular reactions, the spectrum of the second intermediate was determined at pH $\geq$ 8 (□) where it is rapidly formed through base catalysis. The second intermediate spectrum is similar to that known for phenoxyl radicals (Habersbergerova *et al.*, 1972). A small proportion of the phenoxyl radical spectrum was formed on the nanosecond timescale consistent with fast dehydration following ·OH radical attack at the 4 position (Anderson *et al.*, 1987a). As an increase in absorption was observed at all measured wavelengths in this nanosecond time period it was concluded that the initial ·OH radical adduct to the substituted 4 position absorbs at $\leq$ 320 nm. A similar conclusion can be made that ipso attack to the carboxylate group forms a radical which absorbs at $\lambda \leq$ 320 nm.

By considering both the steady-state radiolysis and pulse radiolysis results it is thus possible to determine the spectrum of the hydroxycyclohexadienyl radical formed by ·OH radical addition to the 3 posi-

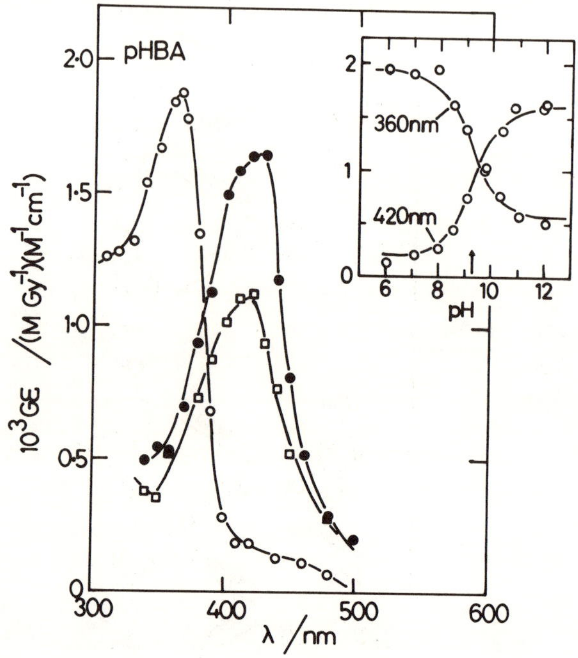

Figure 1. Absorption spectra following pulse radiolysis (5 Gy in 30 ns) of a N_2O-saturated solution containing *p*HBA (5 mM): (○) pH 6, 0.5 μs; (□) pH8, 40 μs; (●) pH 12, 100 ns. **Insert**: The dependence of the initial absorbance on pH.

tion of *p*HBA (Anderson *et al.*, 1987b), which is the position of enzymatic hydroxylation. The same procedure has been employed to obtain the equivalent radical spectra for the substrates 2,4-dihydroxybenzoic acid (2,4-diHBA) and 4-aminobenzoic acid (*p*ABA) (Anderson *et al.*, 1987b.) (We have recently redetermined the spectrum of the phenoxyl radical of 2,4-diHBA at pH 10 for use in calculating the initial ·OH radical adduct spectrum). The three radicals have absorption maxima in the 360-410 nm region with ϵ 3700-5300 M^{-1} cm^{-1} and are presented as part of figure 2.

Previous work (Anderson *et al.*, 1987a) has shown that the rate of conversion of the hydroxycyclohexadienyl radical of *p*HBA to the phenoxyl radical does not increase linearly with the concentration of base but is described kinetically by the above equations. We have extended this study to the high base region (pH 12) where k_f $[OH^-] \gg k_r$ and k_f $[OH^-] \gg k_e$ and measured the absorption spectrum of the deprotonated

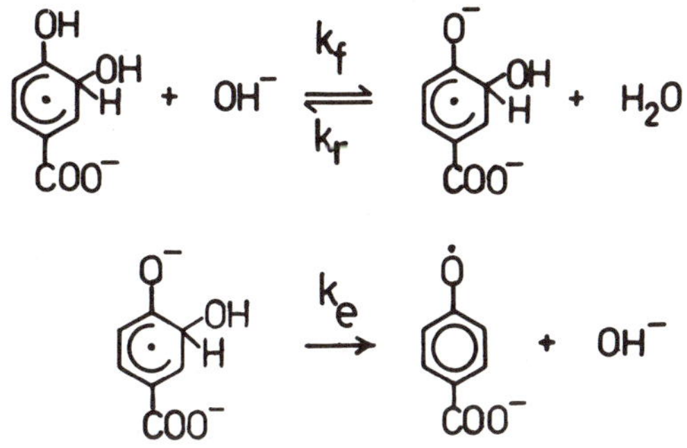

form of the hydroxycyclohexadienyl radical of *p*HBA, figure 1 spectrum (●). The absorption maximum shifts from 365 nm to 420 nm with a $pK_a = 9.3 \pm 0.1$ for the two radical forms (not 8.4 ± 0.2 as estimated from the preliminary data, Anderson *et al.*, 1987a). No change in initial absorption spectrum (up to pH 12) was seen upon ·OH radical addition to *p*ABA but complex spectral and kinetic changes were seen with di-HBA, most probably related to the several possible acid-base forms of this substrate.

Combined Spectra and Proposed Mechanism

The possible formation of the ·OH radical adduct species of the substrates upon interaction with the flavin C(4a)-hydroperoxide (Intermediate I) necessitates the postulate of O-O bond scission. It has been proposed from recent crystallographic studies (Schreuder *et al.*, 1987) that polarization of the O-O bond of the flavin C(4a)-hydroperoxide occurs through dipole moment interactions and hydrogen bonding of the flavin ring with the protein. Also deprotonation of the hydroxy group in the 4 position of *p*HBA and 2,4-diHBA would lead to increased activation of the 3 position to electrophilic attack. Both these factors working together may render scission of the hydroperoxy bond and formation of ·OH radical adducts of the substrates favorable. The resulting peroxy radical on the reduced flavin ring would be expected to have the chemical properties of an alkoxyl-type radical rather than a phenoxyl-type radical. Absorption bands of alkoxyl radicals are typically < 300 nm whereas phenoxyl radicals also absorb in the 400-450 nm region (Habersbergerova *et al.*, 1972). As the uv/vis absorption spectra of both the flavin C(4a)-hydroperoxy (Intermediate I) and the flavin C(4a)-hydroxy (Intermediate III) are similar (Entsch *et al.*, 1976), it can be assumed that such a flavin C(4a)-alkoxyl species would give rise to a similar band associated with the C(4a)-substituted reduced flavin ring. Assuming hydroxycyclohexadienyl substrate radicals are relatively stable in the active site of the enzyme and do not undergo a rapid dehydration reaction as observed in

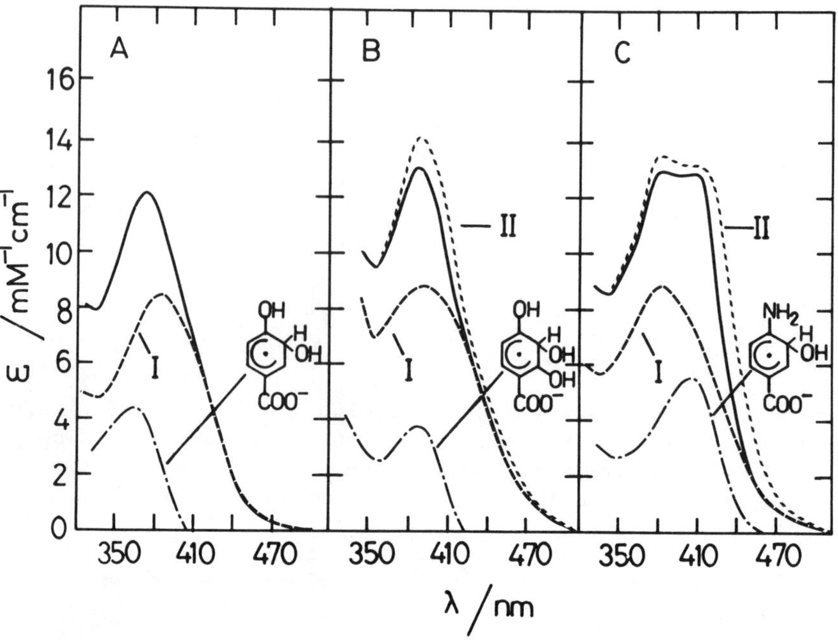

Figure 2. Combined spectra. The spectra of the substituted hydroxy-cyclohexadienyl radicals are combined with the spectra of Intermediate I (from Entsch *et al.*, 1976) to yield the solid curves. The spectra of Intermediate II (from Entsch *et al.*, 1976) are also drawn for comparison.

free solution, it is of interest to combine these spectra with the spectra of Intermediate I, as shown in figure 2. The spectra of Intermediate II, which have been measured for the substrates 2,4-diHBA and *p*ABA are closely matched by the combined spectra. A blue spectral shift of Intermediate II, only observed for the substrate 2,4-diHBA (Wessiak *et al.*, 1984; Entsch *et al.*, 1976) at high pH, might well be related to the deprotonation of the hydroxy radical as described earlier for *p*HBA. (Intermediate II for *p*HBA is invisible, probably for kinetic reasons).

The proposed radical mechanism, using a notation similar to that of Entsch *et al.*, 1976), is the same as the previous scheme in respect of the binding of substrate, reduction of the flavin ring by NADPH and the formation of Intermediate I upon reaction with molecular oxygen. The formation of the radical pair then occurs, (Intermediate II) followed by the formation of the pseudobase Intermediate III (as described below) which then decays to yield the reoxidized flavin and eliminated product, Scheme II.

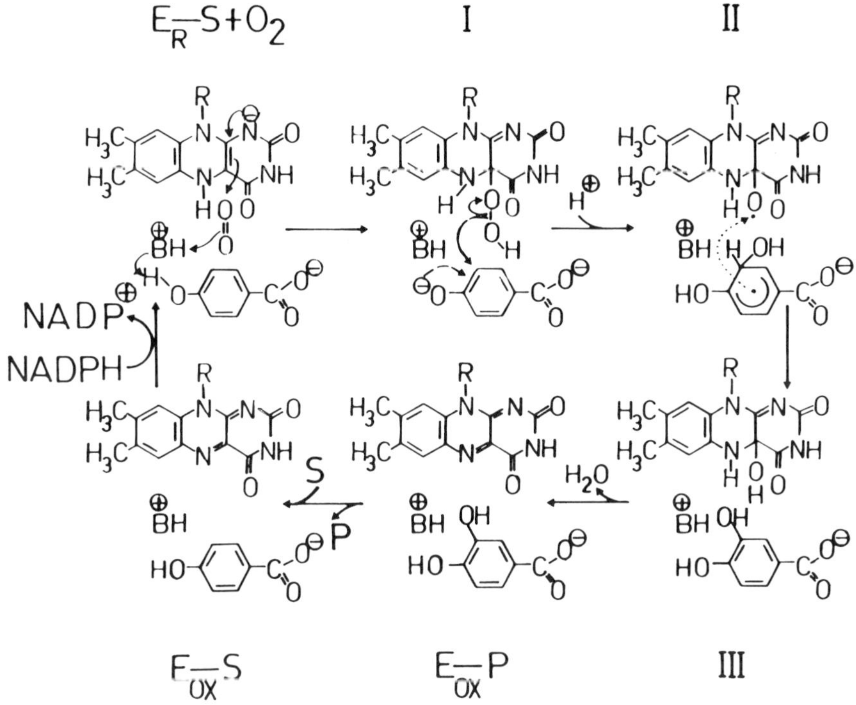

Scheme II. Proposed scheme for the functioning of the flavoprotein *p*-hydroxybenzoate hydroxylase. Notation for Intermediates I, II, III and E_R (reduced enzyme), E_{OX} (oxidized enzyme), S (substrate), P (product) are the same as that presented in Entsch *et al.*, 1976. A full description of the scheme is given in the text.

Conversion of Proposed Intermediate II to Intermediate III

Essential to the proposed radical mechanism is some discussion on the conversion of Intermediate II to Intermediate III. Three possible pathways need to be considered: these are presented in Scheme III.

(i) The flavin C(4a)-alkoxyl radical could abstract the H atom on position C(3) of the hydroxycyclohexadienyl radical to give directly Intermediate III and product. This pathway may be active in the functioning of phenol hydroxylase since when resorcinol was used as the substrate with this enzyme, a large kinetic isotope effect, $k_H/k_D = 8.5$ was measured with deuterated resorcinol (Detmer and Massey, 1985). No such kinetic isotope effect was seen for *p*-hydroxybenzoate hydroxylase using deuterated 2,4-diHBA as substrate (Entsch *et al.*, 1976).

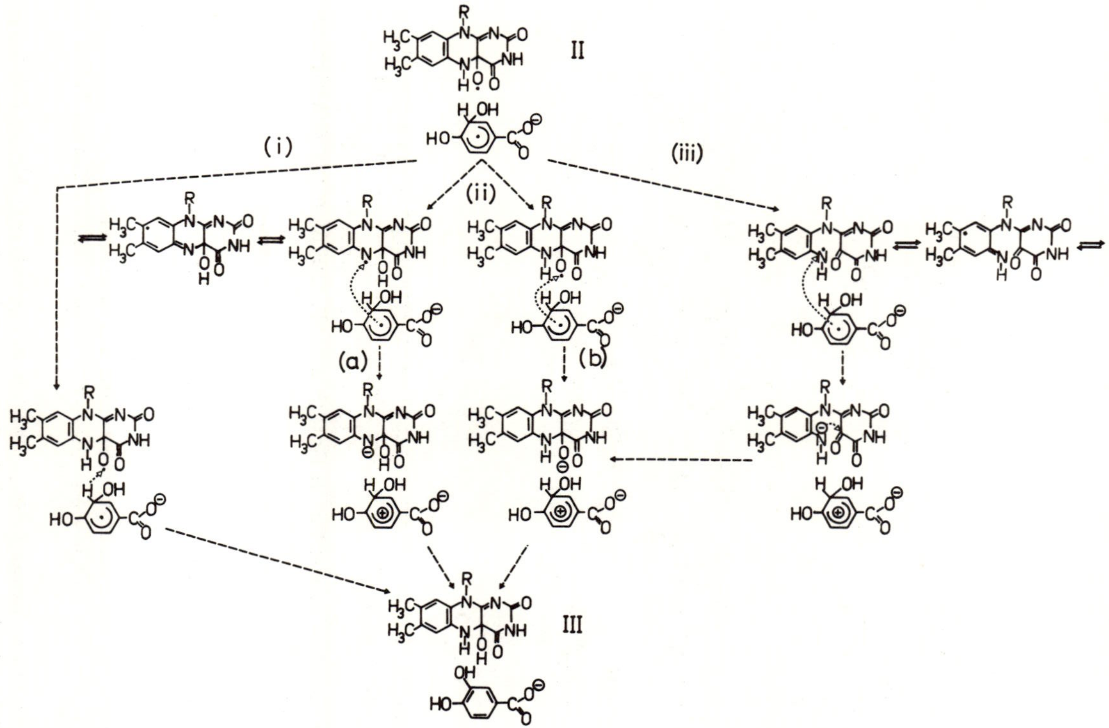

Scheme III. Possible pathways for the conversion of Intermediate II to III. A full description is given in the text.

(ii) Electron transfer from the initial hydroxycyclohexadienyl radical to the flavin C(4a)-alkoxy radical could occur to form an ionic pair, pathway (ii)b. The cationic form of the product would undergo rapid deprotonation while the flavin C(4a)-hydroxy group would exhibit acid-base properties. A pK_a of 9.2 has been measured for a C(4a)-hydroxy flavin with N(5)-substituted (Bruice, 1984). A more likely electron transfer pathway would be intramolecular H atom transfer from N(5) to the C(4a)-alkoxyl radical to form a radical which could be delocalized in the benzene ring, pathway (ii)a. Electron transfer could then occur to form a similar ionic pair as before. Since no intermediary species have been found between Intermediate II and Intermediate III such an electron transfer pathway to occur must be by a concerted mechanism.

(iii) Tertiary alkoxyl radicals are able to undergo β-fragmentation reactions, the rate of which increases with increasing polarity of the solvent (Walling and Wagner, 1964). Fragmentation of the ring form of D-glucose upon alkoxyl radical formation has been shown to occur (Schuchmann and von Sonntag, 1977) and a similar reaction can be postulated for the flavin C(4a)-alkoxyl radical, pathway (iii). The fragmented flavin ring would delocalise the radical center over the benzenoid ring and could undergo an electron transfer reaction followed by ring closure through nucleophilic attack to form the ionic species in pathway (ii)b. Such a mechanism implies that for steric reasons the quinonoid and benzonoid fragments of the flavin ring would need to be twisted away from each other. No evidence has been found from crystallography studies (Schreuder *et al.*, 1987) that this is possible.

In the course of a study with phenol hydroxylase (Detmer and Massey, 1985) it was found that the rate constants for the conversion of Intermediate I to II plotted against the Hammett σ^+ parameter for different substrates containing substituents in the *meta*-position yielded a straight line plot of slope -0.5. The negative slope is consistent with an electrophilic aromatic substitution reaction and the small magnitude supports the suggestion of a radical mechanism (Wells, 1963).

A recent study on *p*-hydroxybenzoate hydroxylase using the modified flavin 6-hydroxy-FAD and *p*ABA as substrate was found to have very similar spectra for Intermediates I and III and an unusual spectrum for Intermediate II in that a satellite band was observed centered at 445 nm (Entsch *et al.*, 1987). Such a spectrum might be accounted for by the proposed mechanism through the formation of the ·OH radical adduct on the substrate together with an aromatic phenoxyl radical in position C(6). Such a radical on the benzenoid ring of the flavin would be relatively stable and could be formed upon the shift of the initial peroxyl radical center in position C(4a) via N(5).

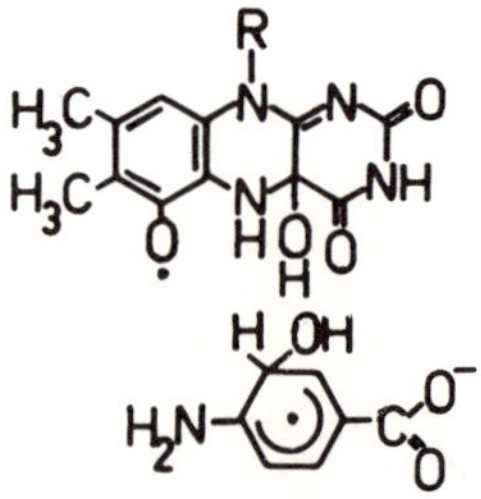

The radical mechanism proposed in this paper can account for some of the observed spectral changes observed for *p*-hydroxybenzoate hydroxylase. Extension of the method to investigate other enzymes and substrates together with electron spin resonance experiments are in hand to study further the proposed mechanism.

Acknowledgements

The author wishes to thank colleagues Mr. K. B. Patel and Dr. M. R. L. Stratford for technical assistance and discussion, Professor V. Massey and his group at the University of Michigan for helpful comments, and also Mrs. M. Hance for typing the manuscript. This work was financially supported by the Cancer Research Campaign.

REFERENCES

Anderson R F, Vojnovic B, Patel K B, Michael B D, (1986). Reaction of alcohol radicals with cyclic disulphides. An optical and conductimetric pulse radiolysis study. *Radiat Phys Chem* **28**: 219-222.

Anderson R F, Patel K B, Stratford M R L (1987a). Radical spectra and product distribution following electrophilic attack by the ·OH radical on 4-hydroxybenzoic acid and subsequent oxidation. *J Chem Soc Faraday Trans* **I 83**: 3177-3188.

Anderson R F, Patel K B, Stratford M R L (1987b). Absorption spectra of radicals of substrates for *p*-hydroxybenzoate hydroxylase following electrophilic attack of the OH· radical in the 3 position. *J Biol Chem* **262**: 17475-17479.

Bruice T C, (1984). Flavin oxygen chemistry brought to date. in *Eighth International Symposium on Flavins and Flavoproteins* (Bray R C, Engel P C and Mayhew S G, eds) pp 45-55, Walter de Gruyter, Berlin.

Detmer K, Massey V (1985). Effect of substrate and pH on the oxidative half-reaction of phenol hydroxylase. *J Biol Chem* **260**: 5998-6005.

Entsch B, Massey V, Ballou D P (1974). Intermediates in flavoprotein catalyzed hydroxylations. *Biochem Biophys Res Commun* **57**: 1018-1025.

Entsch B, Ballou D P, Massey V (1976). Flavin-oxygen derivatives involved in hydroxylation by *p*-hydroxybenzoate hydroxylase. *J Biol Chem* **251**: 2550-2563.

Entsch B, Massey V, Claiborne A (1987). *para*-Hydoxybenzoate hydroxylase containing 6-hydroxy-FAD is an effective enzyme with modified reaction mechanisms. *J Biol Chem* **262**: 6060-6068.

Habersbergerova A, Janovsky I, Kourin P (1972). Absorption spectra of intermediates formed during radiolysis and photolysis. Part II. *Radiat Res Rev* 4: 123-231.

Klein G W, Schuler R H (1978). Oxidation of benzene by radiolytically produced OH radicals. *Radiat Phys Chem* **11**: 167-171.

Raghavan N V, Steenken S (1980). Electrophilic reaction of the OH radical with phenol. Determination of the distribution of isomeric dihydroxycyclohexadienyl radicals. *J Am Chem Soc* **102** 3495-3499.

Schreuder H A, van der Laan J M, Thunnissen M M G M, Kalk K H, Swarte M B A, Hol W G J, Drenth J (1987). Crystallographic studies on the mechanism of *p*-hydroxybenzoate hydroxylase from *Pseudomonas fluorescens*. In *Flavins and Flavoproteins* (McCormick D B, Edmondson D E eds) Walter de Gruyter, Berlin, in press.

Schuchmann M N, von Sonntag C (1977). Radiation chemistry of carbohydrates. Part 14. Hydroxyl radical-induced oxidation of D-glucose in oxygenated solution. *J Chem Soc Perkin Trans* **II**: 1958-1963.

Spector T, Massey V (1972). *p*-Hydroxybenzoate hydroxylase from *Pseudomonas fluorescens*. Evidence for an oxygenated flavin intermediate. *J Biol Chem* **247**: 5632-5636.

Steenken S, Raghavan N V (1979). Oxidation of methoxyhydroxycyclo hexadienyl radicals by quinones. The influence of the isomeric structure of the radical on the rate constant for electron transfer. *J Phys Chem* **83**: 3101-3107.

Vojnovic B, (1983). The measurement of radiation-induced transient conductivity in aqueous solutions on a nanosecond timescale. PhD Thesis, The University of London.

Walling C, Wagner P J (1964). Positive halogen compounds. X. Solvent effects in the reactions of *t*-butoxy radicals. *J Am Chem Soc* **86**: 3368-3375.

Wells P R (1963). Linear free energy relationships. *Chem Rev* **63**: 171-219.

Wessiak A, Bruice T C (1983). Synthesis and study of a 6-amino-5-oxo-3*H*-,5*H*-uracil and derivatives. The structure of an intermediate proposed in mechanisms of flavin and pterin oxygenases. *J Am Chem Soc* **105**: 4809-4825

Wessiak H, Schopfer L M, Massey V (1984). pH Dependence of the reoxidation of *p*-hydroxybenzoate hydroxylase ·2,4-dihydroxybenzoate complex. *J Biol Chem* **259**: 12546-12556.

DISCUSSION

KLINMAN: Hydroperoxide homolysis is an uphill process (51 kcal/mol). What is the driving force for the flavin hydroperoxide homolysis in your reaction scheme?

ANDERSON: Two factors are important: (1) the O-O bond of the flavin C(4a)-hydroperoxide is polarized in the enzyme-substrate complex, and (2) the 3-position of the substrates is activated to electrophilic attack. These two factors may assist a concerted homolytic fission of the O-O bond and formation of the radical pair.

Oxidases and Related Redox Systems, pages 179–190

NADPH OXIDASE IN HUMAN AND PORCINE NEUTROPHILS: STOICHIOMETRY BETWEEN O_2 CONSUMPTION AND O_2^- FORMATION

Yuzuru Ishimura, Ryu Makino, Torahiko Tanaka, Tetsutaro Iizuka, and Shiro Kanegasaki
Department of Biochemistry (Y.I., R.M., T.T., T.I.), School of Medicine, Keio University, Shinjuku-ku, Tokyo 160 and The Institute of Medical Science (S.K.), The University of Tokyo, Minato-ku, Tokyo 108, JAPAN

INTRODUCTION

Upon ingestion of particles or exposure to soluble stimuli such as phorbol myristate acetate (PMA[1/]) and a chemoatractant, fMLP(formylmethionylleucylphenylalanine), neutrophils markedly increase the oxygen consumption and produce a variety of reactive oxygen species such as superoxide anion (O_2^-), hydrogen peroxide (H_2O_2) and•OH radical (Babior, 1978; Badwey and Karnovsky, 1980; Prince, 1987). This process has been called as the respiratory burst, and the reactive oxygen species produced are believed to serve as bactericidal agents. It has been long suggested that an initial biochemical event in the production of such reactive oxygen species is a one electron reduction of molecular oxygen (O_2) to O_2^- catalyzed by a membrane-bound enzyme or enzyme system designated as NADPH oxidase (Babior, 1978; Babior et al., 1973). The rest of reactive oxygen species are assumed to be derived secondarily from O_2^-. There is no definite evidence, however, that H_2O_2 is not formed directly from molecular oxygen. In fact, a considerable amount of H_2O_2 formation has been always observed during the respiratory burst.

1/ The abbreviation used are: PMA, phorbol 12-myristate 13-acetate; O_2^-, superoxide anion; HRP, horseradish peroxidase; diacetyl HRP, diacetyldeuteroheme-substituted horseradish peroxidase; SOD, superoxide dismutase.

The difficulties in determining the primary product(s) come from the fact that O_2^- easily dismutates into molecular oxygen and H_2O_2 through either an enzymatic or nonenzymatic route (McCord and Fridovich, 1969), and further reactions among these species yield many oxygen derivatives including OH radical (Green et al., 1979). In this regard, most, if not all, previous experiments on the stoichiometry between O_2 consumption and product formation were inadequate. For example, when O_2 consumption is measured by an oxygen electrode, the values must be corrected for the amounts of O_2 evolved from dismutation of O_2^- and/or from decomposition of H_2O_2 (Dri et al., 1979). Such corrections are not precise, since both O_2^- and H_2O_2 have been shown to accumulate during the respiratory burst and unpredictable amounts of other oxygen metabolites are produced from the reaction between O_2^- and H_2O_2. Even when all O_2^- was converted to H_2O_2 and then quantified (Root and Metcalf, 1977; Nabi et al., 1979), an accurate stoichiometry could not be established unless a direct release of H_2O_2 from stimulated neutrophils is eliminated.

In this paper, we present direct evidence that the primary product in the respiratory burst is O_2^- and not H_2O_2 by using the diacetyldeuteroheme-substituted horseradish peroxidase (diacetyl HRP) as the trapping agent for both O_2^- and H_2O_2 (Makino et al., 1986).

MATERIALS AND METHODS

Porcine and human neutrophils were isolated as described previously (Makino et al., 1986; Tanaka et al., 1987) with minor modifications from heparinized venous bloods (Iizuka et al., 1985). The cells obtained were washed and suspended in a Ca^{2+}-enriched or Ca^{2+}-free HBSS medium at pH 7.3 to give a concentration of about 10^8 cells/ml. For the preparation of a cell-free system, the suspension was frozen in a glass culture tube at liquid N_2 temperature. At use, the suspension was quickly thawed at 37°C in a water bath with vigorous shaking and was cooled immediately in an ice bath. It was sonicated for 30 times with 0.2 sec-pulses by using a microprobe of Branson sonicator set at 20% of the full power in ice. When required, subcellular fractionation was carried out by conventional centrifugal methods.

O_2 consumption, O_2^- and H_2O_2 formations were measured simultaneously at 25°C in a single cuvette using the diacetyl HRP as described previously (Makino et al., 1986). HRP was purchased from Toyobo Co. Ltd., Osaka, Japan, and the diacetyl HRP was prepared as described earlier (Makino and Yamazaki, 1972). Other reagents and chemicals were of the highest grade commercially available.

RESULTS

Formation of compound III of diacetyl HRP during the respiratory burst induced by PMA

Fig. 1 shows changes in spectrum of diacetyl HRP which occured during the respiratory burst of porcine neutrophils stimulated by PMA (left) and a simultaneous record of O_2 consumption (right). The addition of diacetyl HRP alone had no effect on the rate of O_2 consumption of the neutrophils. Upon addition of PMA, the spectrum of ferric diacetyl HRP with absorption maxima at 507 and 640 nm changed to another spectral species with peaks at 550 and 583 nm, showing a set of isosbestic points. The absorption maxima of the new spectral species agreed with the previously reported values for compound III of diacetyl HRP (Makino and Yamazaki, 1972). During the conversion, spectrum characteristic of compound I or II of diacetyl HRP was not detectable. The result indicates that O_2^-, but not H_2O_2, is the primary product in the respiratory burst of neutrophils when induced by PMA. If the same reaction was carried out in the presence of an excess amount of SOD, formation of compound II instead of compound III was observed having clear isosbestic points with the spectrum of ferric diacetyl HRP (data not shown). The result is consistent with the view that O_2^- is converted to H_2O_2 and O_2 by SOD, and the resultant H_2O_2 reacts with the ferric from of diacetyl HRP to give the compound II. The same stoichiometric relationship between the O_2 consumption and O_2^- formation was obtained also with human neutrophils.

When SOD was added during the PMA-stimulated burst, the resultant spectrum was that of a mixture of compound II and III as judged from the shifts of the peak positions as well as the shift of an isosbestic point at around 610 nm. This means that if H_2O_2 was generated simultaneously with O_2^-, corresponding spectrum should be detectable. On the basis of these results, we concluded that the PMA-stimulated neutrophils released exclusively O_2^-, but not H_2O_2, into the extracellular medium. The conclusion was further supported by the analyses on the rates of O_2 consumption and compound III formation, as described below.

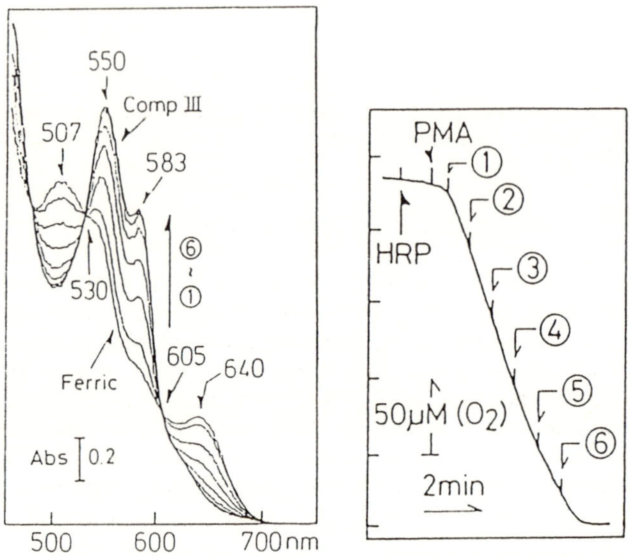

Figure 1. Formation of compound III of diacetyl HRP during PMA-stimulated respiratory burst of porcine neutrophils. Spectral changes from the ferric state to compound III of diacetyl HRP (left) and time course of oxygen consumption measured polarographycally (right). The numerals in the circle in the left indicate the time when each spectrum was recorded corresponding to those in the time course of the reaction shown in the right.

Rates of O_2 consumption and compound II and III formation during the respiratory burst induced by PMA

A simultaneous record of O_2 consumption and optical absorbance change at 585 nm due to compound III formation was shown in Fig. 2A. Upon addition of PMA, formation of compound III began after a short lag period less than 1 min in paralell with the time course of O_2 consumption. The rates calculated from the steepest of both traces (dotted lines) coincided well with each other, being 57 and 54 µM/min for the O_2 uptake and compound III formation, respectively. On the other hand, the addition of SOD significantly slowed down the rate of O_2 consumption to 38 µM/min, accompaning with a concomitant formation of compound II at a rate of 36 µM/min (Fig. 2B). The results indicate that a mole of O_2^- formed by neutrophils converted to a half mole each of H_2O_2 and O_2 molecule, and the latter is reutilized in the system. Thus the amounts of O_2 consumed and H_2O_2 produced was stoichiometric. The situation was again the same with human neutrophils.

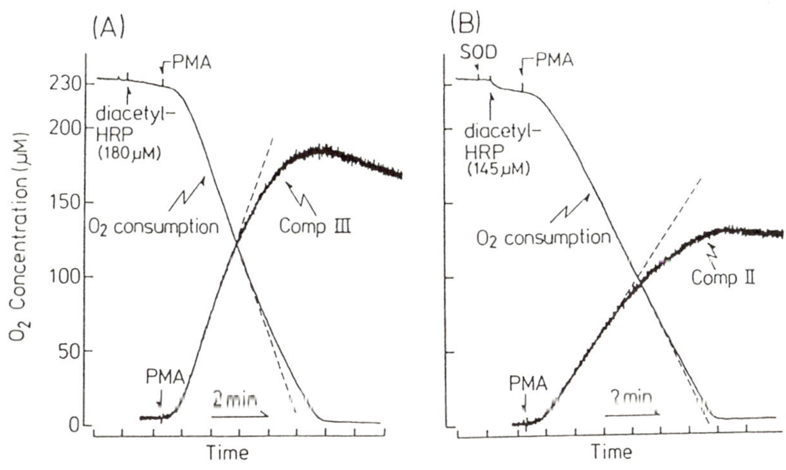

Figure 2. Simultaneous measurements of oxygen consumption, and the formation of compounds III and II during the PMA-stimulated respiratory burst of porcine neutropuils. (A), a simultaneous record of oxygen consumption and compound III formation as measured at 585 nm; (B), a simultaneous record of oxygen consumption and compound II formation in the presence of an excess amount of superoxide dismutase.

Stoichiometry between O_2 consumption and O_2^- formation in a cell-free preparation from porcine neutrophils

We recently found that both saturated and unsaturated long chain fatty acids are capable of activating a cell-free preparation of porcine neutrophils (Tanaka et al., 1987) A typical experiment where laurate (2.5 mM, $C_{12:0}$) was used as the stimulus is shown in Fig. 3. As seen, the cell-free system produced O_2^- upon addition of NADPH. Presence of both NADPH and a fatty acid with an appropriate chain length were absolutely required for the reaction. The initial rates of O_2 consumption and O_2^- formation in this experiment were found to be 32.7 and 31.9 nmoles/min, respectively, and the total amounts of oxygen consumed and O_2^- generated in the presence of 40 nmoles NADPH were 58.4 and 56.4 nmoles, respectively. Accordingly, the O_2^- producing systems both in intact cells and cell-free preparations convert O_2 stoichiometrically to O_2^- under our experimental conditions. The details for the experiments with the cell-free system will be described elsewhere (Tanaka et al., 1988)

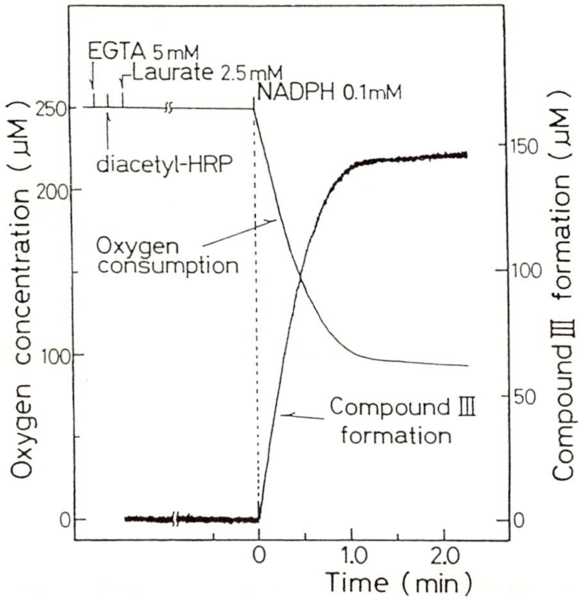

Figure 3. Oxygen consumption and formation of compound III of diacetyl HRP during laurate-stimulated O_2^- formation in a cell-free system from porcine neutrophils under the Ca^{2+}-free conditions.

Stoichiometry between O_2 consumption and O_2^- formation during the particle ingestion

When neutrophils were exposed to opsonized zymosan, a

particulate stimulus, the rates of compound III formation and O_2 consumption did not coincide with each other as shown in Fig. 4A. The O_2 consumption rate was 32 μM/min, while the rate of compound III formation was only 14 μM/min. It was also noted that compound III formation occured only in the initial stage of O_2 consumption and ceased thereafter. Similar results were obtained also when cytochrome c was employed as an indicator of O_2^- production (Fig. 4B). Thus, only 20 % of the O_2 consumed was released into the extra-cellular medium as the form of O_2^- in zymosan-stimulated neutrophils. These findings are in contrast with those of foregoing experiments by PMA-stimulation (Fig. 2A), where formation of compound III was stoichiometric to the amount of O_2 consumed. Again no formation of compound II nor I was observed (Fig. 5A). When azide (1.5 mM) was included in the reaction mixture, however, the formation of compound II was evident (Fig. 5B) with an increased O_2 consumption rate of 46 μM/min.

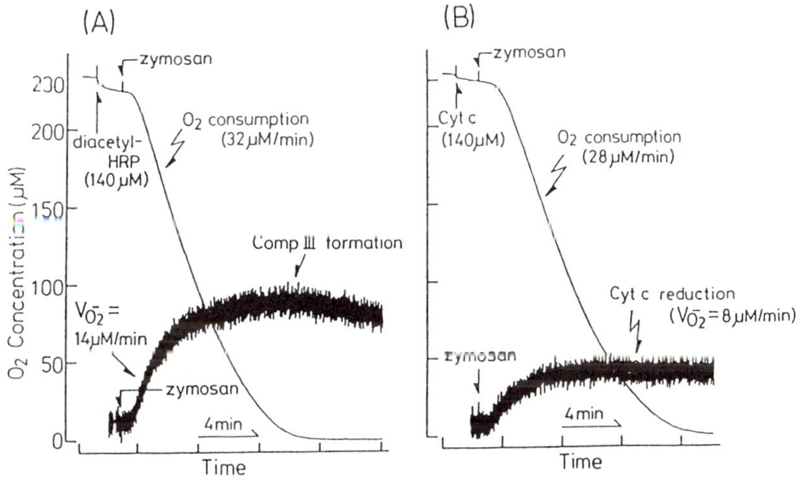

Figure 4. Simultaneous measurements of oxygen consumption, and compound III formation and cytochrome c reduction by porcine neutrophils during phagocytosis of opsonized zymosan at 28°C. (A), oxygen consumption and compound III formation as measured at 585 nm; (B), oxygen consumption and cyto-chrome c reduction as measured at 570 nm. $V_{O_2^-}$ denotes the rate of O_2^- formation.

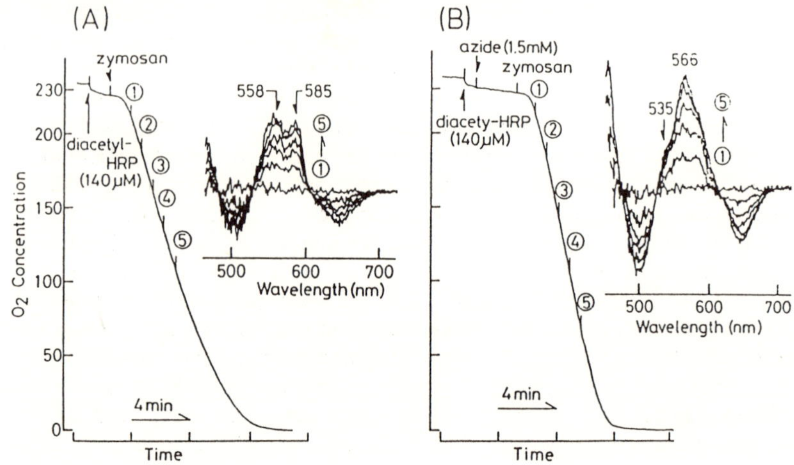

Figure 5. Formation of O_2^- and H_2O_2 during the particle ingestion of intact neutrophils. (A), Compound III formation in the absence of azide. (B), Compound II formation in the presence of azide.

DISCUSSION AND CONCLUSIONS

In this study, aforementioned difficulties in measuring quantitatively the oxygen metabolites were circumbented by the use of diacetyldeuteroheme-substituted horseradish peroxidase (diacetyl HRP) as the trapping agent for both O_2^- and H_2O_2. The heme-substituted HRP reacts stoichiometrically with O_2^- to form a stable complex called as compound III, which, once formed, scarcely dissociates to regenerate oxygen derivaties. It also reacts with H_2O_2 and forms another stable entity, compound II, which is spectrally distinguishable from compound III. Thus both of oxygen metabolites can be trapped in identifiable HRP derivatives which do not release O_2^- or H_2O_2 into the medium, so that conversion to other reactive oxygen species is prevented. We have also found that this method is useful to study the oxygen metabolism in a cell-free preparation from porcine neutrophils (Tanaka et al., 1987). The stoichiometric relationship thus established between the amounts of molecular oxygen consumed and O_2^- formed in the respiratory burst of porcine and human neutrophils indicates that the process is attributable to a single enzymatic reaction that carries out effectively a one electron reduction of molecular oxygen. When intact cells are stimulated by a soluble reagent such as PMA, all of O_2^- thus produced is excreted to extracelluar medium, while the O_2^- is mostly converted to H_2O_2 and then is consumed in the cells when stimulated by phagocytozable particles such as zymosan. The H_2O_2 is excreted from neutrophils only when its consumption within the cells is exceeded by the formation. Further studies on the basic properties of neutrophil NADPH-oxidase system are in progress.

ACKNOWLEDGEMENT

This work was supported in part by Grants-in-Aid for Scientific Research from the Ministry of Education, Science and Culture, Japan and by grants from Keio University. One of the authors (Y.I.) is grateful to Nissan Science Foundation (Tokyo, Japan) for the kind support of travel expense. Thanks are also to Mrs. T. Ichikawa for the assistance in preparing the manuscript.

REFERENCES

Babior BM (1978). Medical progress - oxygen-dependent microbial killing by phagocytes (first of two parts). N Engl J Med 298: 659-668.

Babior BM, Kipnes RS, Curnutte JT (1973). Biological defense mechanisms: The production by leukocytes of superoxide, a potential bactericidal agent. J Clin Invest 52: 741-744.

Badwey JA, Karnovsky ML (1980). Activative oxygen species and the functions of phagocytic leukocyte. Annu Rev Biochem 49: 695-726.

Dri P, Bellavite P, Berton G, Rossi F (1979). Interrelationship between oxygen consumption, superoxide anion and hydrogen peroxide formation in phagocytosing guinea pig polymorphonuclear leucocytes. Mol cell Biochem 23: 109-122.

Green MR, Hill HAO, Okolow-Zurkowska MJ, Segal AW (1979). The production of hydroxyl and superoxide radicals by stimulated human neutrophils: Measurements by EPR spectroscopy. FEBS Lett 100: 23-26.

Iizuka T, Kanegasaki S, Makino R, Tanaka T, Ishimura Y (1985). Studies on neutrophil b-type cytochrome <u>in situ</u>

by low temperature absorption spectroscopy. J Biol Chem 260: 12049-12053.

Iizuka T, Kanegasaki S, Makino R, Tanaka T, Ishimura Y (1985). Pyridine and imidazole reversibly inhibit the respiratory burst in porcine and human neutrophils: Evidence for the involvement of cytochrome b558 in the reaction. Biochem Biophys Res Commun 130: 621-626.

Makino R, Tanaka T, Iizuka T, Ishimura Y, Kanegasaki S (1986). Stoichiometric conversion of oxygen to superoxide anion during the respiratory burst in neutrophils: Direct evidence by a new method for measurement of superoxide anion with diacetyldeuteroheme-substituted horseradish peroxidase. J Biol Chem 261: 11444-11447.

Makino R, Yamazaki I (1972). Effects of 2,4-substituents of deuterohemin upon peroxidase functions. I. Preparation and some properties of artificial enzymes. J Biochem (Tokyo) 72: 655-664.

McCord JM, Fridovich I (1969). Superoxide dismutase: An enzymatic function for erythrocuprein (hemocuprein). J Biol Chem 244: 6049-6055.

Nabi ZF, Takeshige K, Hatae T, Minakami S (1979). Hydrogen peroxide formation of polymorphonuclear leukocytes stimulated with cytochalasin D. Exp Cell Res 124: 293-300.

Prince RC (1987). Superoxide production by neutrophils. Trends Biochem Sci 12: 86-87.

Root RK, Metcalf JA (1977). H_2O_2 release from human granulocytes during phagocytosis. J Clin Invest 60: 1266-1279.

Tanaka T, Makino R, Iizuka T, Ishimura Y, Kanegasaki S (1988). Activation by saturated and trans-unsaturated fatty acids of superoxide generating system in a cell-free preparation from porcine neutrophils. Submitted for publication to J Biol Chem.

Tanaka T, Kanegasaki S, Makino R, Iizuka T, Ishimura Y (1987). Saturated and trans-unsaturated fatty acids elicit high levels of superoxide generation in intact and cell-free preparations of neutrophils. Biochem Biophys Res Commun 144: 606-612.

DISCUSSION

MASSEY: I gather that most workers believe that the neutrophil respiratory chain burst oxidase is a flavoprotein. In that case it is somewhat of a paradox that this enzyme should produce O_2^- as product, if indeed the "killing" reaction requires H_2O_2, and therefore needs superoxide dismutase to convert O_2^- to H_2O_2. There are very many flavoprotein oxidases which produce H_2O_2 very rapidly and directly. It is, therefore, strange that the respiratory burst enzyme should be tailored to produce O_2^-!

ISHIMURA: Thank you for the comment.

PECHT: Is anything known concerning the steps preceeding the formation of O_2^- radicals? (i.e. What are the coupling steps between PMA activation of kinase C and the oxidase burst?).

ISHIMURA: I don't think the signal transduction pathway after the activation of protein kinase C in this system has been elucidated.

KOOMAN. The O_2^- you are measuring is that released into the extracellular medium. All killing occurs in the phagosome. Is it possible that the O_2^- released into the phagosome is substrate for bacterial SOD which converts it to H_2O_2 within the pathogen, leading to killing?

ISHIMURA: Yes. The O_2^- I am measuring, after PMA stimulation, is that released into the extracellular medium. You are correct that in phagocytosis O_2^- is mainly released into phagosomes.

DUNFORD: I agree with your assessment that superoxide produced exclusively by NADPH oxidase, and with your schematic neutrophil diagram which shows myeloperoxidase (MPO) poised for the killing process. What is missing, I feel, is the role of superoxide dismutase (SOD). It was originally proposed that SOD is present to protect the neutrophil from oxidative damage by superoxide. My proposal (in press, Free Radical Biology and Medicine) is that SOD is present as part of the attack mechanism, to produce H_2O_2 in a hurry for MPO so that the latter enzyme can produce hypochlorous acid most efficiently. Thus, the chemical sequence is: O_2 -> O_2^- -> H_2O_2 -> HOCl. My question is what happens to compound I of the peroxidase in your assay procedure conducted in the presence of H_2O_2?

ISHIMURA: As you know, compound I is so unstable that we could only see the accumulation of compound II when H_2O_2 is produced.

NICHOLLS: Does myeloperoxidase need H_2O_2 as a substrate? Could it not work with superoxide itself via compound III (as with the horseradish enzyme used in the assay)?

DUNFORD: Myeloperoxidase needs a hydroperoxide or peracid. Native myeloperoxidase plus superoxide would form compound III which is relatively inert. The reaction to form compound III of a peroxidase from superoxide was first demonstrated by Professor Yamazaki.

AZZI: Have you tried stimulants other than PMA to activate neutrophils? Is the stoichiometry between O_2 and O_2^- always one-to-one?

ISHIMURA: Yes, we did. We found that neutrophils activated by long chain fatty acids, such as laurate and oleate, gave exactly the same stoichiometry. However, the stoichiometry apparently changes depending on the kind of stimulants employed, as I described in the text. That is, a stimulant such as zymosan gave a very small amount of O_2^- compared to the amount of oxygen consumed.

HURST: Comment. It is recognized that neutrophils are capable of several distinct types of microbiological reactions, both oxidative and non-oxidative in character. Myeloperoxidase-dependent and -independent mechanisms exist, the latter presumably involving intracellular Fenton chemistry. The non-oxidative toxins include a set of cationic polypeptides and proteins which appear to be capable, at least in part, of disrupting the normal functioning of membrane-localized metabolism systems [see Elsbach and Weiss (1983) Rev Infect Dis 5: 843-853 for a discussion of bacterial permeability-increasing proteins, and Sestead, Skzlarek and Lehrer (1985) Infect. in Immunity 45: 150-154 for "defensins"].

Session III: Iron- and Copper-Oxidases, Oxygenases, and O_2 Transporters

Chairman: Emanuel Margoliash

Oxidases and Related Redox Systems, pages 193–209

INVOLVEMENT OF OXO-BRIDGED BINUCLEAR IRON CENTERS IN OXYGEN TRANSPORT, OXYGEN REDUCTION, AND OXYGENATION

Joann Sanders-Loehr

Department of Chemical and Biological Sciences, Oregon Graduate Center, Beaverton, Oregon 97006-1999

ABSTRACT

The occurrence of an oxo-bridged binuclear iron site is well-established for the oxygen transport protein, hemerythrin, and strongly implicated in ribonucleotide reductase, purple acid phosphatase, ferritin, and methane monooxygenase. Key identifying characteristics are an antiferromagnetic interaction between the two iron atoms, an Fe-O-Fe vibrational mode in the resonance Raman spectrum, and an S = 1/2 EPR signal upon one-electron reduction. In hemerythrin the oxo bridge serves as a hydrogen bond acceptor which stabilizes the bound hydroperoxide. In ribonucleotide reductase both the binuclear iron center and a protein tyrosine undergo oxidation in the presence of molecular oxygen, whereas in methane monooxygenase a binuclear iron moiety may activate O_2 for substrate oxygenation.

INTRODUCTION

Oxo-bridged binuclear Fe(III) centers are found in a number of proteins which interact with molecular oxygen. The structure of this site, as elucidated for the oxygen transport protein hemerythrin, is shown in Figure 1. The two iron atoms are situated in a face-sharing bioctahedron, bridged by a solvent-derived oxo group (O^{2-}) and two protein carboxylates (1). Of the remaining coordination sites, five are occupied by histidines and one is available for exogenous ligands such as azide (which binds to the met form) and dioxygen (which binds to the deoxy form) (2). Monobridged

Fe-O-Fe complexes have long been known for Fe(III) in aqueous solution under alkaline conditions, whereas the more recently discovered tribridged complexes (with two acetates in addition to the oxo group) form readily in the presence of tridentate capping groups (4). An example of the latter occurs with tris(1-pyrazolylborate) in which the three pyrazole ligands are analogues for the three imidazole ligands in hemerythrin. However, no models have yet been synthesized with a ligand-exchangeable site. The important features of the oxo-bridged binuclear iron center from the viewpoint of protein function are (i) its ability to undergo either a one- or a two-electron reduction, (ii) its ability to interact with exogenous ligands such as molecular oxygen in the precursor ferrous state, and (iii) the potential for functional variability that may be achieved by altering amino acid residues in the bridging or terminal ligand positions.

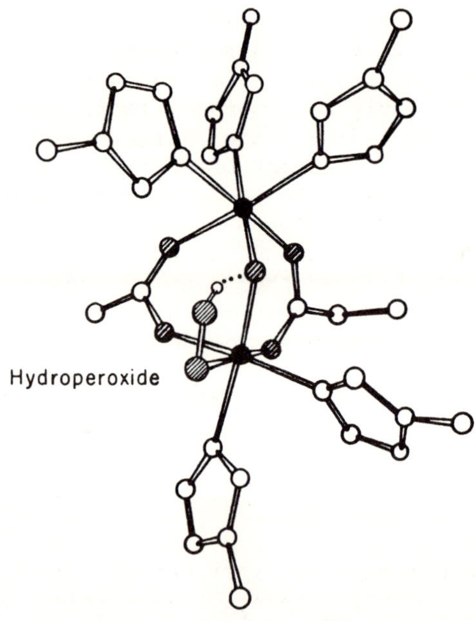

Figure 1. Binuclear iron center of oxyhemerythrin. Disposition of protein ligands based on crystal structure of azidomet form at 2.0 Å resolution (1). Iron (●), carbon and nitrogen (○), oxygen (⊜). Protein ligands are Glu 58, Asp 106, His 25, His 54, His 73, His 77, and His 101. The hydroperoxide moiety is bound end-on, similar to azide in methemerythrin, but with additional stabilization from hydrogen bonding to the oxo bridge (2, 3).

Model compounds containing oxo-bridged binuclear iron centers exhibit a number of properties which are characteristic of the Fe-O-Fe moiety and independent of the number of additional bridging groups present (5, 6). These include intense absorption in the near-uv region ($\epsilon > 5,000$ M^{-1} cm^{-1}), strong antiferromagnetic coupling between the two iron atoms ($-J > 90$ cm^{-1}), short Fe-O bond lengths (< 1.82 Å), and a resonance Raman spectrum dominated by the Fe-O-Fe symmetric stretch (< 550 cm^{-1}) whose frequency is dependent on the Fe-O-Fe bond angle and the oxygen isotopic composition. All of these characteristics are observed for the known Fe-O-Fe site in oxyhemerythrin (Table 1). The hemerythrin protein shows one additional feature not yet routinely detected in the model compounds: the ability to form a stable, mixed-valence species upon one-electron reduction which is characterized by an $S = 1/2$ EPR spectrum arising from the magnetic interaction of a pair of high spin Fe(II) and Fe(III) ions. The existence of an Fe-O-Fe center in ribonucleotide reductase seems unquestionable on the basis of its similarity to hemerythrin and the tribridged model compounds, particularly the Fe-O-Fe stretch from Raman spectroscopy and the short Fe-O distance from EXAFS (Table 1). For purple acid phosphatase and methane monooxygenase, the EPR evidence for a mixed valence species is indicative of an oxo- or hydroxo-bridged binuclear iron site, but the definitive identification from Raman spectroscopy or EXAFS is still lacking.

The binuclear iron site of hemerythrin clearly functions in the reversible binding of dioxygen. The two Fe(II)'s in the deoxy form are connected by a hydroxo bridge which is converted to an oxo bridge in conjunction with the two-electron reduction of O_2 to peroxide (Table 2). This coupling by a hydroxo bridge in the reduced state and an oxo bridge in the oxidized state appears to be a common characteristic of binuclear iron proteins. Although the binuclear iron sites in ribonucleotide reductase and purple acid phosphatase also undergo reactions with dixoygen, these processes are more related to the regulation of enzymatic activity than to catalysis. The conversion of ribonucleotides to deoxyribonucleotides, for example, requires the participation of a tyrosyl radical which is generated during the reaction of the Fe(II)-containing protein with O_2 (11). Conversely, with acid phosphatase (also known as uteroferrin), it is the mixed-valence reduced form which catalyzes phosphate ester hydrolysis, and the enzyme is inactivated by reacting with O_2 in the presence of excess

TABLE 1. Properties of Oxo-Bridged Binuclear Iron Proteins[a]

	Oxy-hemerythrin	Ribonucleotide reductase	Purple acid phosphatase	Methane monooxygenase
Absorption max (nm)	330, 360	325, 370	320	410
Fe-O bond length (Å)	1.83	1.78	?	?
Magnetic susceptibility (-J) in cm^{-1}	~90	110	>150	?
EPR g values after one electron red'n	1.96, 1.88, and 1.66	?	1.92, 1.77, and 1.63	1.95, 1.88, and 1.78
Raman ν(Fe-O-Fe) in cm^{-1} for ^{16}O (^{18}O)	486 (472)	492 (479)	?	?

[a]Data sources: oxyhemerythrin (7-10), ribonucleotide reductase (11-14), purple acid phosphatase (15, 16), and methane monooxygenase (17, 18).

TABLE 2. Proteins with Oxo-Bridged Binuclear Iron Centers

	Function	Binuclear Iron Site
Hemerythrin	Oxygen transport	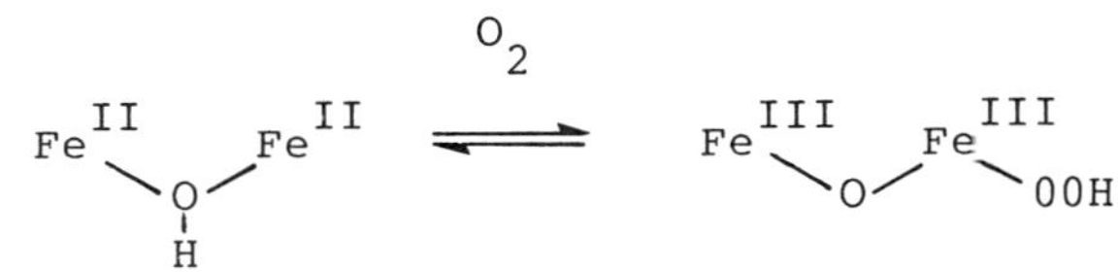
Ribonucleotide reductase	$NDP + 2H \longrightarrow dNDP + H_2O$	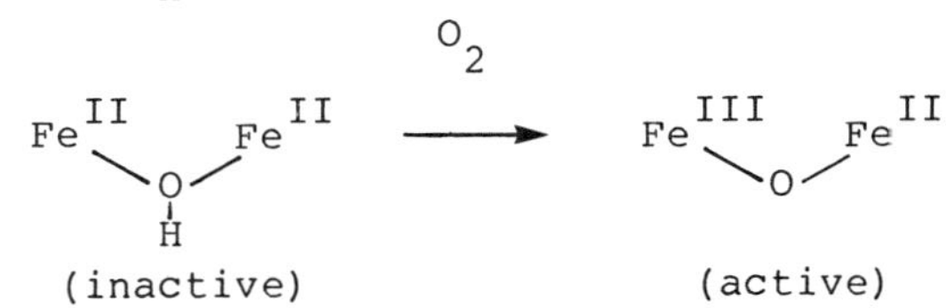
Purple acid phosphatase	$RO\text{-}P \xrightarrow{+H_2O} ROH + P_i$	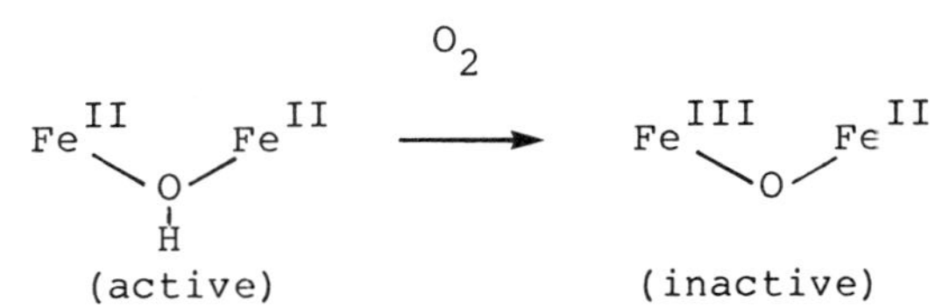
Methane monooxygenase	$CH_4 + 2H \xrightarrow{O_2} CH_3OH + H_2O$	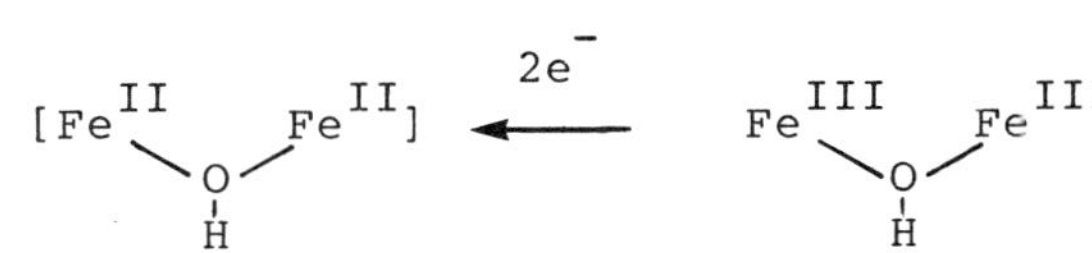

phosphate (15). Methane monooxygenase presents a case where the binuclear iron site is first reduced to the ferrous state and then reacts with dioxygen to produce a species capable of inserting oxygen into a hydrocarbon (17).

OXYGEN TRANSPORT

The two binuclear iron proteins known to be involved in reversible binding of dioxygen are the blood protein, hemerythrin, and the muscle protein, myohemerythrin. Both of these occur in several phyla of marine invertebrates and have closely related tertiary structures but differ in that the muscle form is monomeric whereas the blood form is generally an octamer (7). Although this distribution is reminiscent of myoglobin and hemoglobin, there is only a single case where cooperative oxygen binding has been observed in a hemerythrin (19). As in the case of hemoglobin (20) and hemocyanin (21), the binding of dioxygen by hemerythrin can be described as a reversible, oxidative addition reaction (Table 2).

Although the favored reaction of oxyhemerythrin is the release of dioxygen as O_2, there is a slow, but significant, rate of dissociation of oxygen in the form of peroxide with the resultant essentially irreversible oxidation of the binuclear iron site to the met state (7). In another analogy to the vertebrate hemoglobin system, a cytochrome b_5 reductase capable of reducing methemerythrin has been isolated from invertebrate erythrocytes, and it may be responsible for regenerating oxygen-reactive deoxyhemerythrin <u>in vivo</u> (22).

The remarkable ability of the respiratory proteins to release oxygen as O_2 rather than as O_2^{2-} is believed to be a consequence of the hydrophobic character of the oxygen binding site. This would favor the dissociation of a neutral ligand such as O_2 over a charged ligand such as HO_2^- (23). For example, the conversion of oxyhemerythrin to deoxyhemerythrin (i.e., transport) has a first-order rate constant of 50 s^{-1} (24), whereas the conversion of oxyhemerythrin to methemerythrin (i.e., oxidation) has a first-order rate constant of 10^{-5} s^{-1} (7). Rate constants of 10^{-4} to 10^{-5} s^{-1} are also observed for the dissociation of other anionic ligands such as Cl^-, SCN^-, and N_3^- from methemerythrin (25).

Another feature which appears to be important in stabilizing the oxygenated complex with respect to release of either O_2 or O_2^{2-} is the strong interaction of the bound dioxygen with the protein. In hemocyanin, both ends of the peroxide ion are coordinated to Cu(II) ions (21) whereas in hemoglobin, the bound superoxide ion is hydrogen bonded to the imidazole of the distal histidine (20, 26). In oxyhemerythrin, the peroxide ion is coordinated in an end-on fashion (Fig. 1), as was originally suggested by single-crystal polarized spectroscopy (27) and the oxygen isotope dependence of the peroxide O-O vibration in the resonance Raman spectrum (28). This has been verified by x-ray crystallography (2). A further resonance Raman study of the deuterium isotope dependence of the peroxide ion vibrations, ν(O-O) and ν(Fe-O_2), indicated that the bound peroxide ion is protonated (29). Furthermore, this protonated moiety appears to be hydrogen bonded to the oxo group of the Fe-O-Fe center.

The hypothesized role of the oxo bridge as a hydrogen bond acceptor for the bound hydroperoxide ion in oxyhemerythrin is based primarily on resonance Raman spectroscopy (Fig. 2). The Fe-O-Fe symmetric stretch at 486 cm^{-1} undergoes a 4 cm^{-1} shift to higher energy when the protein is equilibrated in D_2O (3). Deuterium isotope dependence is observed with other protonated ligands such as hydroxide which are also capable of hydrogen bonding to the oxo bridge, but not with unprotonated ligands such as azide, cyanide, or thiocyanate. Thus, it is unlikely that the observed deuterium effect can be ascribed merely to a protein conformational change. Furthermore, the only possible hydrogen bond donor within 6 Å of the oxo bridge in oxyhemerythrin is the coordinated hydroperoxide group. The shift of the Fe-O-Fe vibration to higher energy in D_2O indicates that the hydrogen bond strength is actually greater with hydrogen than with deuterium. The proposed hydrogen-bonding interaction between the hydroperoxide ligand and the oxo bridge is also supported by x-ray difference density maps which show a merging of electron density between the terminal peroxide oxygen and the Fe-O-Fe oxygen in oxyhemerythrin relative to the azide electron density in methemerythrin (2).

The following reaction mechanism for the interconversion of deoxy and oxyhemerythrin incorporates currently known structural and kinetic details.

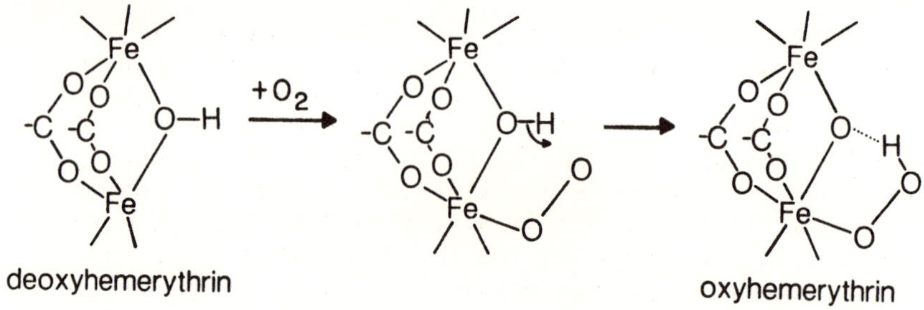

The Fe(II)-OH-Fe(II) formulation for deoxyhemerythrin is consistent with the observed -J of 15 cm^{-1}, a value indicative of weak antiferromagnetic coupling between the iron atoms (9, 30). This is the expected behavior for a hydroxo-bridged species where bridge protonation is accompanied by an increase in Fe-O bond length to ~2.0Å (compared to the abnormally short Fe-O distance of ~1.8Å in oxo-bridged complexes). The deoxyhemerythrin structure is also depicted

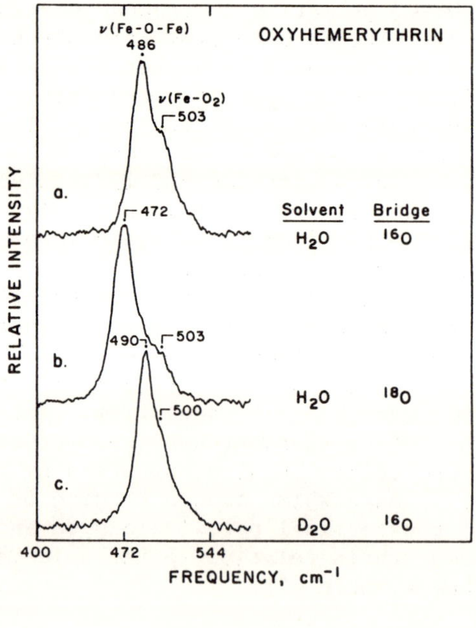

Figure 2. Resonance Raman spectra of oxyhemerythrin in H_2O or D_2O and with ^{16}O or ^{18}O in the oxo bridge. From Ref. 3.

with a vacant site on the lower Fe which is in agreement with the spectroscopic evidence for both 5- and 6-coordinate iron atoms (30) and with the kinetic evidence that the rate of O_2 binding is too rapid to allow time for ligand displacement (24). The bimolecular reaction with dioxygen appears to occur in a single step, and the second-order rate constant of 7 x 10^6 M^{-1} s^{-1} is close to being diffusion controlled. It is likely that the reaction is limited by the diffusion and binding of dioxygen and that the subsequent reduction of dioxygen occurs rapidly. Other neutral species such as HN_3, HOCN, and NO have similarly rapid rates of binding to deoxyhemerythrin (i.e., 10^4 to 10^5 M^{-1} s^{-1}) (31). In contrast, the binding of charged anions such as N_3^- or OCN^- to methemerythrin occurs several orders of magnitude more slowly (i.e., 10 to 10^2 M^{-1} s^{-1}) (25). The two-electron reduction of dioxygen is presumably accompanied by a simultaneous proton transfer from the bridging oxygen to the peroxide oxygen, thereby accounting for the known pH-independence of the reaction (24). In the slower reverse reaction, the release of dioxygen is dependent on proton as well as electron transfer, and the rate constant is ~15% lower for the deuterium-substituted protein (32).

OXYGEN REDUCTION

In ribonucleotide reductase, reduction of dioxygen in the presence of Fe(II) results in the production of a tyrosyl radical, which is required for the enzymatic activity of the protein (33, 34). Evidence for the oxo-bridged binuclear iron site in the resulting oxidized form (Table 2) has been obtained by a number of different techniques (Table 1). Among the most unequivocal is resonance Raman spectroscopy. The Fe-O-Fe vibration at 492 cm^{-1} undergoes a 13-cm^{-1} shift to lower energy when the protein is equilibrated with $H_2{}^{18}O$ (Fig. 3). Normal coordinate analysis calculations for these vibrational frequencies suggest an Fe-O-Fe angle of 138° (14), similar to the 135° angle observed for azidomethemerythrin (1). An Fe-O-Fe angle in this range is also derived from EXAFS analyses of ribonucleotide reductase (12, 13). An additional ^{18}O-dependent feature in the resonance Raman spectrum occurs at 598 cm^{-1} (Fig. 3). This peak was originally assigned to an iron-coordinated hydroxide (14), but is more likely due to a photochemically induced structural rearrangement of the Fe-O-Fe center.

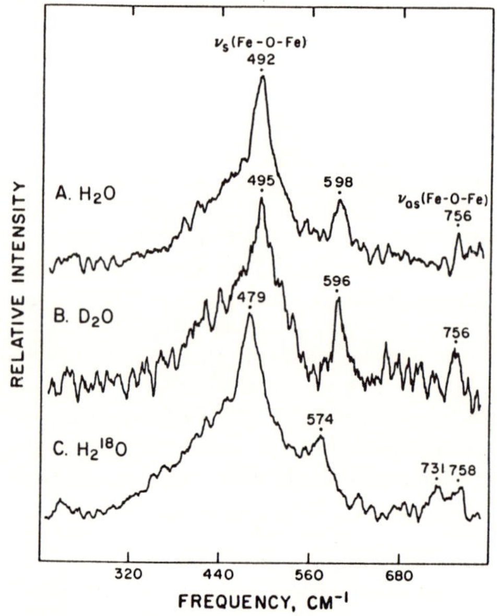

Figure 3. Resonance Raman spectra of ribonucleotide reductase (radical-free B2 subunit) in H_2O, D_2O, and $H_2{}^{18}O$. From Ref. 14.

It is likely that the binuclear iron center in ribonucleotide reductase has a tribridged geometry, in view of its small Fe-O-Fe angle and the close correspondence of its physical properties with those of hemerythrin and tribridged model complexes. At least one histidine ligand is implicated from NMR contact-shifted resonances (35). This may be His 118 which is one of two evolutionarily conserved histidine residues (36) and which is in close proximity to Tyr 122. The latter has been identified as the radical-bearing tyrosine by site-directed mutagenesis (37). The location of this His trans to the oxo bridge is likely from the strong resonance Raman enhancement of the Fe-O-Fe vibration (38). The EXAFS and NMR data suggest that additional non-bridging ligands are more likely to be protein carboxylates or exogenous hydroxide (12, 13, 35).

Since the physical properties of the Fe-O-Fe site are essentially unaffected by destruction of the tyrosyl radical (11), it is unlikely that Tyr 122 is itself an iron ligand.

Regeneration of the radical and the accompanying enzymatic activity requires that the binuclear iron center be in the ferrous state. This can be accomplished either by reconstituting apoprotein with anaerobic Fe(II) (11) or by exposing the radical-free protein to a cell-derived reduction system which includes the enzyme superoxide dismutase (39). Subsequent exposure to O_2 leads to the appearance of the oxo-bridged binuclear Fe(III) structure and a single tyrosyl radical. A suggested scheme for this reaction is shown in Figure 4. The Fe(II) form of the protein (left) is depicted as having a hydroxo-bridged structure similar to deoxyhemerythrin with an empty exogenous ligand site. A single molecule of dioxygen then coordinates to one of the

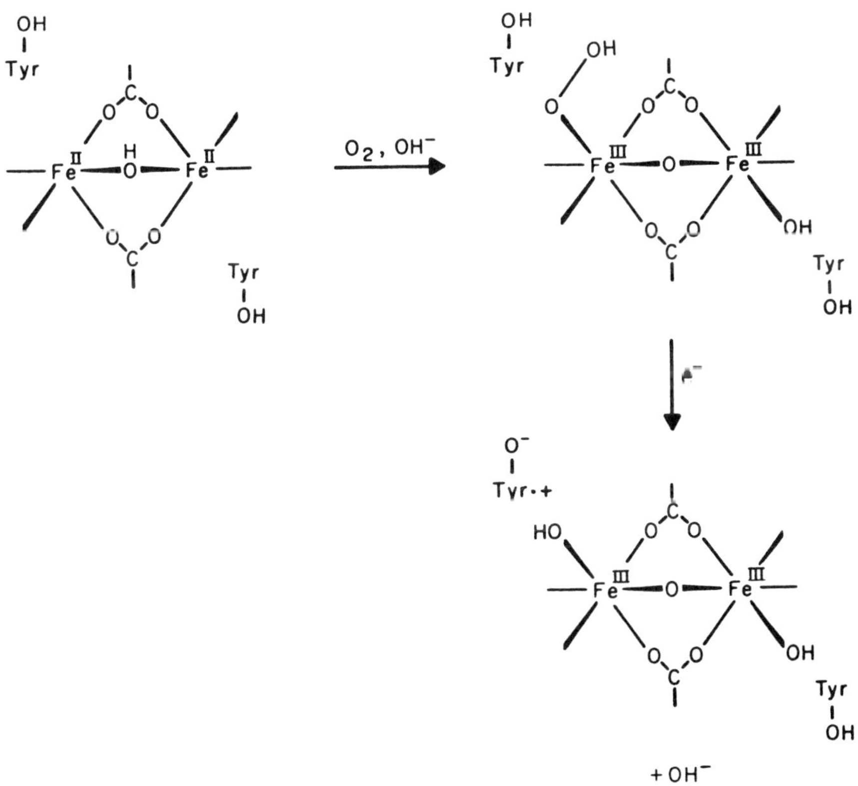

Figure 4. Proposed mechanism for the regeneration of the tyrosyl radical in ribonucleotide reductase.

iron atoms, followed by reduction to peroxide and proton transfer yielding a structure analogous to oxyhemerythrin (Fig. 4, middle). The subsequent oxidation of the adjacent tyrosine to a tyrosyl radical requires the input of an additional reducing equivalent and results in the reduction of peroxide to water.

There are several different pathways by which the tyrosine oxidation in Figure 4 could proceed. The reaction is most likely initiated by the decomposition of peroxide to yield a ferryl intermediate, either Fe(IV)=O or Fe(V)=O, depending on the availability of the additional reducing equivalent. This would be analogous to the proposed reaction pathways for heme enzymes such as cytochrome P-450 and the peroxidases (40) as well as non-heme-containing monooxygenases (41). High-valent oxo species involving formal iron oxidation states of IV and V have been observed in the reaction of iron EDTA with peroxides (42, 43). These types of compounds should be capable of abstracting an electron from a nearby phenolate group, thereby generating a tyrosyl radical. Facile oxidation of phenols to phenoxy radicals has been observed with ferryl-containing enzymes such as horseradish peroxidase (40) and also with simple cobalt-dioxygen complexes (44).

The binuclear iron center of acid phosphatase undergoes a seemingly more straightforward oxidation in the presence of O_2. The enzymatically active pink form is converted to an inactive purple form in the presence of phosphate ions and dioxygen. The reaction ostensibly involves the one electron oxidation of a mixed valence binuclear Fe(II)/Fe(III) cluster to an oxo-bridged binuclear Fe(III) cluster (Table 2).

The existence of a hydroxo-bridged structure in the pink enzyme seems likely from the weak antiferromagnetic coupling with a $-J$ of $\sim 10\ cm^{-1}$ (45) and from the similarity of the EPR spectrum to that of the one-electron-reduced form of hemerythrin (Fig. 5). Tyrosine and histidine ligands to the binuclear iron site in pink acid phosphatase are well-documented from NMR contact-shifted resonances (45). For the oxidized purple acid phosphatase, an oxo bridge is indicated by the strong antiferromagnetic coupling (Table 1) and a tribridged structure seems likely from the short Fe-Fe distance of 3.0 Å determined by EXAFS (47). However, it has not yet been possible to detect the expected Fe-O-Fe vibration by resonance Raman spectroscopy (16). Studies of tribridged model compounds indicate that resonance enhancement of the Fe-O-Fe stretch can be significantly weaker when the structure lacks histidine-type ligands in a position trans to the oxo bridge (38).

SUBSTRATE OXYGENATION

Methane monooxygenase is a multienzyme complex which utilizes reducing equivalents from NADH to catalyze the oxidation of methane to methanol, with the alcoholic oxygen

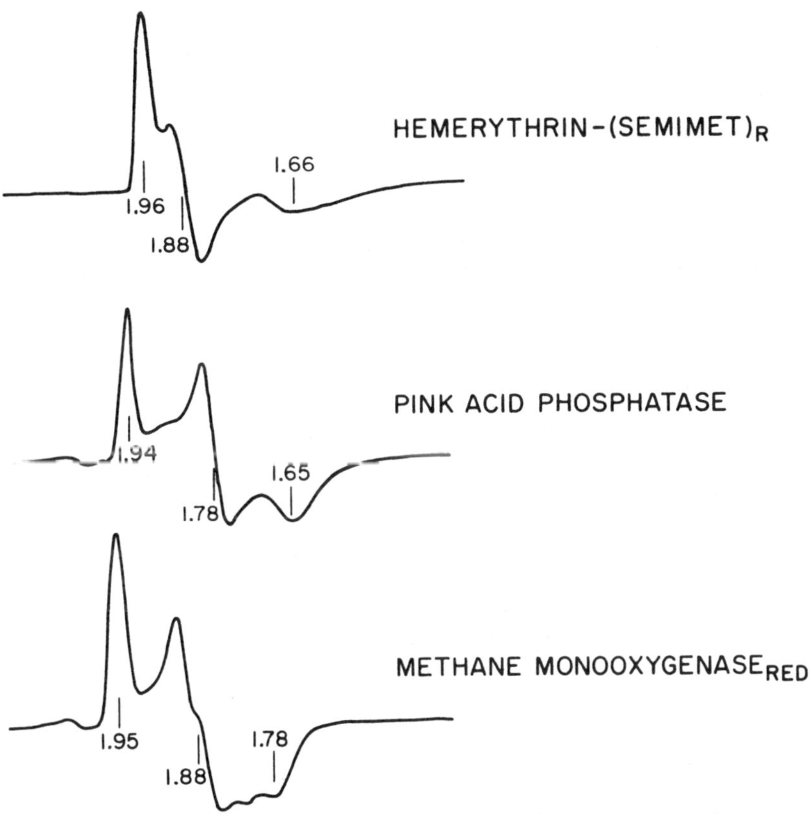

Figure 5. EPR spectra of one-electron reduced hemerythrin, pink acid phosphatase, and one-electron reduced methane monooxygenase. Data taken from Refs. 46, 16, and 18, respectively.

being derived from O_2 (17, 48, 49). Component A, which carries out the reaction between CH_4 and O_2, is likely to contain a binuclear iron center. In its one-electron-reduced form it exhibits the characteristic S = 1/2 EPR signal for a mixed valence Fe(II)/Fe(III) cluster (Fig. 5). Further identification of the binuclear iron site must await the preparation of sufficient quantities of enzyme for other types of physical characterization. Should an Fe-O-Fe center prove to be present in the oxidized form of component A, an oxygen activation sequence similar to that proposed for ribonucleotide reductase (Fig. 4) can be visualized. This would involve an initial NADH-dependent reduction of component A. The resulting deoxyhemerythrin-like binuclear ferrous iron center could then convert O_2 to peroxide. Decomposition of this intermediate could then lead to oxygen insertion into the CH_4 substrate.

ACKNOWLEDGEMENT

The author acknowledges the valuable contributions of students and coworkers whose work is cited herein and the support of the National Institutes of Health (GM 18865).

REFERENCES

1. Stenkamp, R. E., Sieker. L. C., and Jensen, L. H. (1984) J. Am. Chem. Soc. 106, 618.
2. Stenkamp, R. E., Sieker, L. C., Jensen, L. H., McCallum, J. D., and Sanders-Loehr, J. (1985) Proc. Natl. Acad. Sci. USA 82, 713.
3. Shiemke, A. K., Loehr, T. M., and Sanders-Loehr, J. (1986) J. Am. Chem. Soc. 108, 2437.
4. Lippard, S. J. (1986) Chemistry in Britain, 222.
5. Murray, K. S. (1974) Coord. Chem. Rev. 12, 1.
6. Armstrong, W. H., Spool, A., Papaefthymiou, G. C., Frankel, R. B., and Lippard, S. J. (1984) J. Am. Chem. Soc. 106, 3653.
7. Wilkins, R. G., and Harrington, P. C. (1983) Adv. Inorg. Biochem. 5, 52.
8. Elam, W. T., Stern, E. A., McCallum, J. D., and Sanders-Loehr, J. (1982) J. Am. Chem. Soc. 104, 6369.
9. Maroney, M. J., Kurtz, D. M., Jr., Nocek, J. M., Pearce, L. L., and Que, L., Jr. (1986) J. Am. Chem. Soc. 108, 6871.

10. Loehr, T. M., and Shiemke, A. K. (1988) in Biological Applications of Raman Spectroscopy (Spiro, T. G., Ed.) Volume 3, in press, Wiley, New York.
11. Sjöberg, B.-M., and Gräslund, A. (1983) Adv. Inorg. Biochem. 5, 87.
12. Scarrow, R. G., Maroney, M. J., Palmer, S. M., and Que, L., Jr. (1986) J. Am. Chem. Soc. 108, 6832.
13. Bunker, G., Petersson, L., Sjöberg, B.-M., Sahlin, M., Chance, M., Chance, B., and Ehrenberg, A. (1987) Biochemistry 26, 4708.
14. Sjöberg, B.-M., Sanders-Loehr, J., and Loehr, T. M. (1987) Biochemistry 26, 4242.
15. Antanaitis, B. C., and Aisen, P. (1983) Adv. Inorg. Biochem. 5, 111.
16. Averill, B. A., Davis, J. C., Burman, S., Zirino, T., Sanders-Loehr, J., Loehr, T. M., Sage, J. T., and Debrunner, P. G. (1987) J. Am. Chem. Soc. 109, 3760.
17. Woodland, M. P., and Dalton, H. (1984) J. Biol. Chem. 259, 53.
18. Woodland, M. P., Patil, D. S., Cammack, R., and Dalton, H. (1986) Biochim. Biophys. Acta 873, 237.
19. Richardson, D. E., Emad, M., Reem, R. C., and Solomon, E. I. (1987) Biochemistry 26, 1003.
20. Phillips, S. E. V., and Schoenborn, B. P. (1981) Nature 292, 81.
21. Reed, C. A. (1985) in Biological Inorganic Copper Chemistry (Karlin, K. D., and Zubieta, J., Eds.) pp. 61-73, Adenine Press, New York.
22. Utecht, R. E., and Kurtz, D. M., Jr. (1985) Inorg. Chem. 24, 4459.
23. Loehr, T. M. (1988) in Oxygen Complexes and Oxygen Activation by Metal Complexes (Martell, A. E., Ed.) in press, Plenum, New York.
24. deWaal, D. J., and Wilkins, R. G. (1976) J. Biol. Chem. 251, 2339.
25. Meloon, D. R., and Wilkins, R. G. (1976) Biochemistry 15, 1284.
26. Jameson, G. B., and Drago, R. S. (1985) J. Am. Chem. Soc. 107, 3017.
27. Gay, R. R., and Solomon, E. I. (1978) J. Am. Chem. Soc. 100, 1972.
28. Kurtz, D. M., Jr., Shriver, D. F., and Klotz, I. M. (1977) Coord. Chem. Rev. 24, 145.
29. Shiemke, A. K., Loehr, T. M., and Sanders-Loehr, J. (1984) J. Am. Chem. Soc. 106, 4951.

30. Reem, R. C., and Solomon, E. I. (1987) J. Am. Chem. Soc. 109, 1216.
31. Wilkins, P. C., and Wilkins, R. G. (1987) Biochim. Biophys. Acta 912, 48.
32. Armstrong, G. D., and Sykes, A. G. (1986) Inorg. Chem. 25, 3135.
33. Reichard, P., and Ehrenberg, A. (1983) Science 221, 514.
34. Ashley, G. W. and Stubbe, J. (1985) in Pharmacol. Ther. 30, 301.
35. Sahlin, M., Ehrenberg, A., Gräslund, A., and Sjöberg, B.-M. (1986) J. Biol. Chem. 261, 2778.
36. Sjöberg, B.-M., Eklund, H., Fuchs, J. A., Carlson, J., Standart, N. M., Ruderman, J. V., Bray, S. J. and Hunt, T. (1985) FEBS Lett. 183, 99.
37. Larsson, Å., and Sjöberg, B.-M. (1986) EMBO J. 5, 2037.
38. Wheeler, W. D., Shiemke, A. K., Averill, B. A., Loehr, T. M., and Sanders-Loehr, J., unpublished results.
39. Eliasson, R., Jörnvall, H., and Reichard, P. (1986) Proc. Natl. Acad. Sci. USA 83, 2373.
40. Ortiz de Montellano, P. R. (1987) Acc. Chem. Res. 20, 289.
41. Valentine, J. S., Burstyn, J. N., and Margerum, L. D. (1988) in Oxygen Complexes and Oxygen Activation by Metal Complexes (Martell, A. E., Ed.) in press, Plenum, New York.
42. Rush, J. D., and Koppenol, W. H. (1987) J. Inorg. Biochem. 29, 199.
43. Balasubramanian, P. N., and Bruice, T. C. (1987) Proc. Natl. Acad. Sci. USA 84, 1734.
44. Bedell, S. A., and Martell, A. E. (1983) Inorg. Chem. 22, 364.
45. Lauffer, R. B., Antanaitis, B. C., Aisen, P., Que, L., Jr. (1983) J. Biol. Chem. 258, 14212.
46. Muhoberac, B. B., Wharton, D. C., Babcock, L. M., Harrington, P. C., and Wilkins, R. G. (1980) Biochim. Biophys. Acta 626, 337.
47. Kauzlarich, S. M., Teo, B. K., Zirino, T., Burman, S., Davis, J. C., and Averill, B. A. (1986) Inorg. Chem. 25, 2781.
48. Green, J., and Dalton, H. (1986) J. Biol. Chem. 236, 155.
49. Dalton, H. (1980) Adv. Appl. Microbiol. 26, 71.

DISCUSSION

KOPPENOL: How stable is the tyrosine radical?

SANDERS-LOEHR: Very. It persists for days upon storage of the enzyme at ice temperature.

BALLOU: The tyrosine radical in ribonucleotide reductase is required for activity, presumably to abstract a hydrogen atom from ribonucleotide reductase. I have collaborated with Joanne Stubbe and her colleagues to try and detect changes in the tyrosine radical during turnover. None could be found - perhaps because the hydrogen abstraction and replacement occur in very rapid steps which are not rate limiting in the overall reaction.

Determination of the stoichiometry of the formation of the tyrosine radical may be difficult since this is a repair or maintenance function on ribonucleotide reductase. Thus, it occurs only rarely.

SANDERS-LOEHR: But one would still like to be able to account for all of the reducing equivalents during an *in vitro* reconstitution from adding Fe(II) and O_2 to apoenzyme. Is it possible that some excess Fe(II) serves as the additional reductant? The alternative of hydroxyl radical production is not very attractive.

SAHLIN: No hydroxyl radical has been observed during reconstitution experiments. We have used different spin traps but this could, of course, mean that the hydroxyl radical is not formed or that the geometry of the "active site cleft" is narrow and/or wrongly charged.

Oxidases and Related Redox Systems, pages 211–226

PHTHALATE OXYGENASE, A RIESKE IRON-SULFUR PROTEIN FROM *Pseudomonas cepacia*

David Ballou and Christopher Batie

Department of Biological Chemistry, University of Michigan, Ann Arbor, MI 48109

ABSTRACT

Phthalate oxygenase catalyzes the oxygenation of phthalate to form a *cis*-dihydrodiol. It is comprised of two proteins: a flavo-iron-sulfur protein with NADH-dependent oxidoreductase activity (POR) and a nonheme iron protein with oxygenase activity (PO). The latter is a tetramer of 48 kDa units and contains a Rieske [2Fe-2S] center and one mononuclear iron per monomer. The mononuclear iron is likely the site of oxygenation. This system can be isolated in large quantities and is sufficiently stable for detailed mechanistic studies. We briefly describe some of our studies on this system which include kinetics, and visible, magnetic resonance, EXAFS, and ENDOR spectroscopies.

INTRODUCTION

The aerobic degradation of aromatic compounds in the environment is an important and widespread activity of soil microorganisms . This process, which is critical to the overall carbon cycle in nature, usually involves a series of oxygenative steps coupled with some rearrangements which lead to activated catecholic derivatives (Dagley, 1975). The catechols formed are then cleaved, thus breaking the aromaticity and permitting entrance of the products into the mainstream of intermediary metabolism. Figure 1 shows several typical pathways by which aromatic compounds are degraded. The first step in the degradation of unactivated aromatic compounds is typified by the benzene path. It begins with the formation of a *cis*-dihydrodiol, a class of reaction catalyzed by specific dioxygenases (Batie et al., 1987 and references therein) that utilize NADH and O_2. The *cis*-dihydrodiol is then rearomatized by a *cis*-dihydrodiol dehydrogenase to produce a catechol and to regenerate NADH. This remarkable two step process forms a highly activated aromatic compound from a non-activated compound at the mere expense of "a breath of fresh air" (O_2) and no undesirable side products are released.

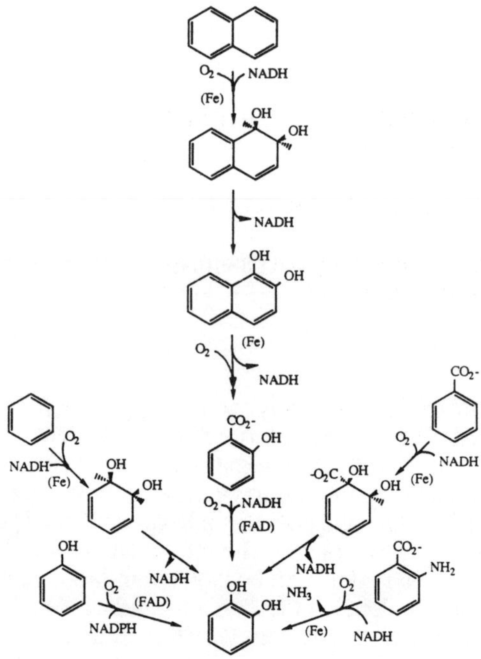

Figure 1. Pathways of degradation of aromatic compounds. Primary cofactors involved in the oxygenations are shown in parentheses.

More highly activated aromatic substances such as phenols are often converted to catechols by flavoprotein monooxygenases. These utilize NAD(P)H and oxygen to hydroxylate the substrate while converting the other atom of O_2 to water. These flavoenzymes, which carry out less difficult oxygenations than those involving unactivated aromatic compounds, are discussed in some detail by Massey et al. in this volume.

The catechols which are formed in these reactions are cleaved to form *cis*, *cis*-muconates or mucono-semialdehydes by iron-containing dioxygenases such as catechol dioxygenase and protocatechuate dioxygenase (Nozaki, 1974). The muconates are routed to β-ketoadipate and acetate wherein they enter mainstream metabolism. The overall scheme (Figure 1) shows how a wide variety of aromatic compounds are routed via oxygenative steps to form central metabolites.

cis-DIHYDRODIOL FORMING DIOXYGENASES

The formation of *cis*-dihydrodiols is catalyzed by various specific dioxygenases. The nature of this class of reactions, including the determination of the product stereochemistry, has been clearly shown by Gibson et al., (1970). Table 1 shows a partial list of enzyme systems that catalyze this type of reaction

Table 1. Enzymes Catalyzing Formation of *cis*-Dihydrodiols

Enzyme	Reference
Benzene Dioxygenase	Axcell and Geary, 1975
Naphthalene Dioxygenase	Ensley and Gibson, 1983
Benzoate Dioxygenase	Yamaguchi and Fujisawa, 1982
Toluene Dioxygenase	Subramanian et al., 1979
Phthalate Dioxygenase	Batie et al., 1987

In all cases examined these enzyme systems are composed of two or three protein components and involve a flavin, a ferredoxin-like [2Fe-2S] center, a Rieske [2Fe-2S] center, and a mononuclear Fe^{2+} center. The flavin and the ferredoxin centers are apparently involved in electron transfer from NAD(P)H to the mononuclear iron center which is thought to be the site of oxygenation. The electron transfer chain is thus quite similar to those found with various cytochrome P-450 systems; the Rieske center and the mononuclear iron replace the cytochrome P-450 moiety at the site of oxygenation.

With some of the *cis*-dihydrodiol-forming systems the flavin and the ferredoxin centers are found on a single protein component (eg phthalate dioxygenase and benzoate dioxygenase) while with others they are found on separate small proteins (eg. naphthalene dioxygenase and benzene dioxygenase). The oxygenase proteins have been found to utilize either multiples of α subunits or an equimolar combination of α and β subunits. The quarternary structures are varied, with molecular weights ranging from 70-220 thousand. In all known cases the oxygenase component contains both a Rieske [2Fe-2S] center and a mononuclear iron center.

One other such system has been studied which, although containing completely analogous components, catalyzes a monooxygenation, the hydroxylation of the methyl group of 4-methoxybenzoate (Bernhardt et al., 1975). The product, basically a gem diol, hydrolyzes to yield 4-hydroxybenzoate and formaldehyde. This indicates the versatility of this type of oxygenase .

GENERAL DESCRIPTION OF THE PHTHALATE OXYGENASE SYSTEM

Phthalate is an unactivated aromatic chemical that is released into the environment by industrial as well as by natural processes (Keyser et al., 1976). Ribbons and his colleagues (Keyser et al., 1976, Pujar and Ribbons, 1985) have isolated bacteria that will utilize phthalate as sole carbon source and have characterized the pathways and the biochemical components. In ***Pseudomonas cepacia*** phthalate is first dihydroxylated by phthalate oxygenase to form phthalate 4,5-dihydrodiol; this product is then dehydrogenated to give 4,5-dihydroxyphthalate and NADH. Decarboxylation yields protocatechuate (3,4-dihydroxybenzoate) which is oxygenatively cleaved to form β-carboxy-*cis, cis*-muconate. The structure of the phthalate oxygenase system from ***P. cepacia*** is illustrated in Scheme 1 (Batie et al., 1987).

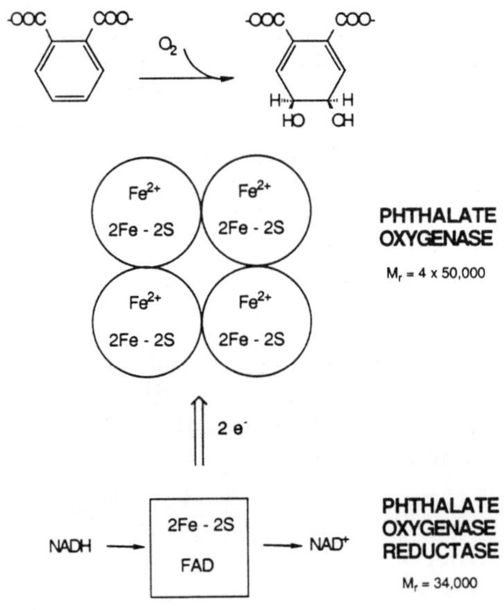

Scheme 1. The Phthalate Oxygenase System

Two protein components are involved in the catalysis: phthalate oxygenase (PO) and phthalate oxygenase reductase (POR). The reductase is a single polypeptide with a MW of 34,000 and contains both a FMN cofactor and a ferredoxin-like [2Fe-2S] center. This protein is capable of catalyzing the reduction of the oxygenase as well as a number of other electron acceptors such as cytochrome c, ferricyanide, and dichloroindophenol. The oxygenase is a tetramer of 48 kDa units and harbors one Rieske [2Fe-2S] center and one mononuclear iron per subunit. The general flow of reducing equivalents and the stoichiometry of the catalyzed reaction are shown in Scheme 1. We have determined that in the absence of reductase and NADH the oxygenase can effect oxygenation of phthalate if it is supplied with two reducing equivalents from sources such as dithionite and then exposed to O_2 (Batie et al., 1987). However, this process is sluggish and inefficient when compared to normal turnover in the presence of the reductase and NADH. Thus in normal turnover at 25° it is found that the rate is 23 s^{-1}. In contrast, when the oxygenase is reduced with dithionite, the reoxidation rate is only 0.01 s^{-1}. Moreover, in that reoxidation only about 0.2 mol of product/ mol of oxygenase is formed. In the normal turnover system coupling is essentially complete.

PHTHALATE OXYGENASE (PO)

Phthalate oxygenase is produced by ***P. cepacia*** DB01 in copious quantities. More than 5% of the cell protein appears to be PO when the cells are grown with phthalate as the sole carbon source. Preparations from 500 g of frozen cells yield up to 1.5 g of homogeneous PO (Batie et al., 1987). Moreover, this enzyme, by contrast to many of the other systems listed in Table 1, is quite stable to normal biochemical procedures. Thus it is an excellent system for investigating properties which are general to this class of enzymes. The fully active PO contains three iron atoms: two as part of the Rieske center and one ferrous atom. The latter can be removed and replaced reversibly. It is required for activity (Batie et al., 1987).

Rieske Center Properties

The visible spectrum of PO is dominated by the Rieske center. The absorbance spectra of oxidized and reduced PO (Figure 2) are nearly indistinguishable from those of the Rieske protein from ***Thermus Thermophilus*** (Fee et al., 1984). EPR spectra (Figure 3) of reduced PO are also very similar to those of the Rieske protein, suggesting that the [2Fe-2S] centers are nearly the same. In fact, all of the oxygenases of this class have very similar visible and EPR spectra (Geary et al., 1984, Twilfer et al., 1981, Sauber et al., 1977). The EPR spectra, which are diagnostic of this center, have g-values of 2.01, 1.90, and 1.77-1.72.

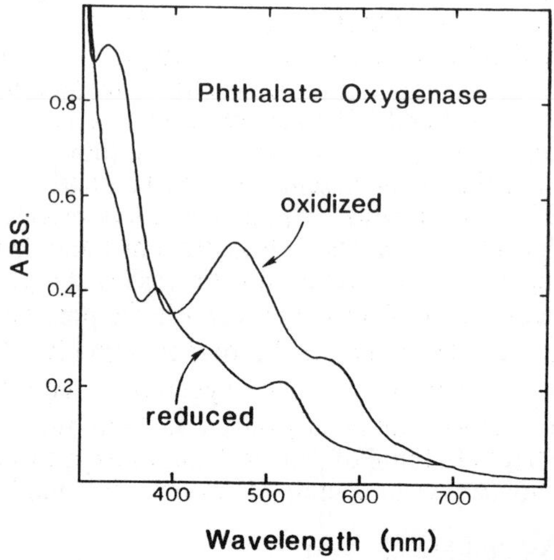

Figure 2. Visible spectra of oxidized and reduced phthalate oxygenase. Concentration = 63 μM, 0.1 M HEPES, 20 mM EDTA, pH 8.

Because of its unique EPR spectrum the Rieske [2Fe-2S] center has been recognized for many years as a distinct type of iron-sulfur center. Although it was known that the center contains two antiferromagnetically coupled Fe^{3+} atoms and two labile sulfurs (ie. it is diamagnetic in the ferric state and ferromagnetic when reduced by one electron/two irons of the Rieske center), the full nature of the other ligands remains uncertain. The Rieske protein from ***T. Thermo-philus*** has been found to have only two cysteine residues per two irons; the other ligands are not sulfurs. Therefore the ligation of the Rieske center must be different from that of ferredoxin [2Fe-2S] centers which have four cysteines centrosymmetrically positioned around the [2Fe-2S] center. This difference in ligands is also consistent with the fact that Rieske [2Fe-2S] centers have redox potentials 250 - 400 mV higher than ferredoxin centers. Since it has not been possible to carry out extensive physical studies on the small quantities of ***T. thermophilus*** Rieske protein available, we have utilized the abundant PO to investigate the nature of the non-sulfur ligands.

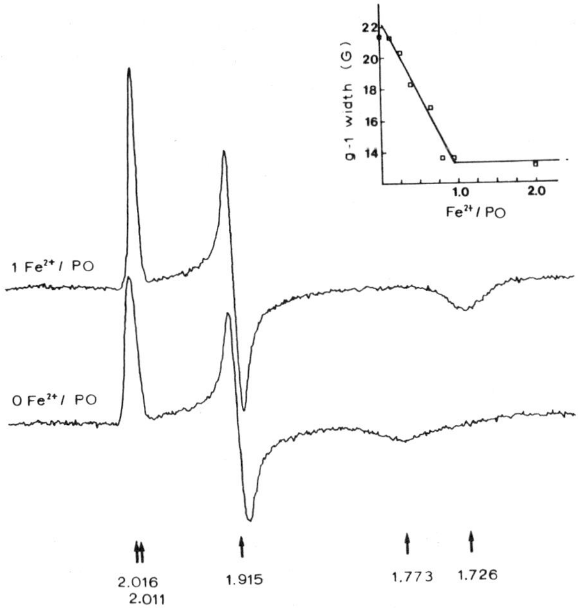

Figure 3. EPR spectra of reduced phthalate oxygenase. PO was depleted of monomer iron as described in (Batie, et al., 1987). Enzyme was reduced by excess dithionite either with or without Fe^{2+} as indicated. EPR spectra were recorded at 24 K, 0.2 mW power, 10 G modulation. Inset: The width at half-height of the low field signal (g-1) was plotted vs. the Fe/PO ratio.

Electron Nuclear Double Resonance (ENDOR) experiments in collaboration with the laboratories of B. Hoffman (Northwestern) and J. Fee (Los Alamos) have indicated the presence of nitrogen ligands in the center (Cline et al., 1985). More recently, we have performed further ENDOR experiments on samples which had been isolated from ***P. cepacia*** **DB01** grown on $^{15}NH_4SO_4$ (Hoffman et al., 1987). These results show quite clearly that two nitrogens are involved in the ligand environment. It is likely that at least one of the nitrogen ligands derives from a histidine residue. Possible configurations of this ligand environment are shown in Scheme 2. Resonance Raman spectroscopy implied that the center was not symmetric (Kuila et al., 1987). Thus the form on the left in Scheme 2 is more likely to be correct. Mossbauer studies of the Rieske protein from ***T. Thermophilus*** have shown that the two irons are inequivalent, in keeping with this idea (Fee et al., 1984). We are now undertaking ENDOR investigations using PO which has been highly enriched in ^{15}N-histidine. This should allow us to establish more fully the identity of the ligands in the Rieske center.

EXAFS studies with J. Penner-Hahn and colleagues have been particularly fruitful (Penner-Hahn et al., 1987). Results with the iron-depleted PO (The two irons of the Rieske center remain) have allowed the determination of Fe-S, Fe-Fe and Fe-N distances for both the oxidized and reduced states of the Rieske center in PO. The Fe-N distances were calculated by assuming a nitrogenous ligand was present. These distances are indicated in Scheme 2.

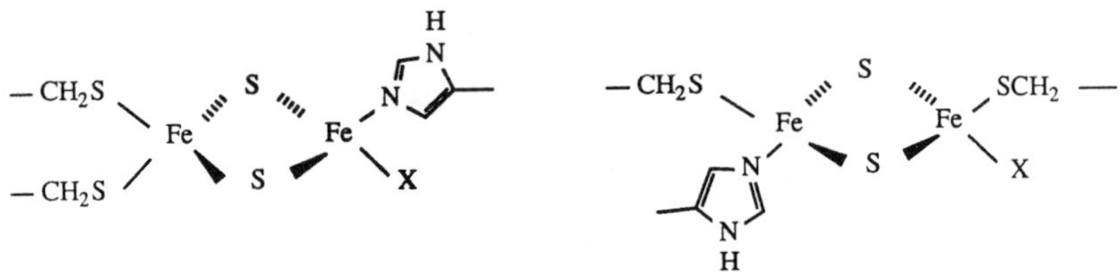

	Fe-S	Fe-N	Fe-Fe
Oxidized	2.20(2)	2.01(5)	2.68(4)
Reduced	2.24(2)	2.08(5)	2.67(4)

Scheme 2. Possible models of the Rieske [2Fe-2S] Centers

Mononuclear Iron Site

As mentioned above, in addition to the stable Rieske center, PO has a dissociable mononuclear Fe^{2+}. This iron was shown to be required for activity by demonstrating that Fe^{2+}-depleted enzyme (inactive) can be reconstituted to full activity by the addition of 1 equivalent of ferrous ion (Batie et al., 1987). The binding constant for Fe^{2+} has been estimated to be <1 m<u>M</u>. The addition of one equivalent of Fe^{2+} to iron-depleted enzyme broadens the EPR spectrum (Figure 3) to yield g-values of 2.01, 1.90, and 1.72. This perturbation does not result from a magnetic interaction, as one equivalent of Zn^{+2}, Co^{+2}, Ni^{+2} or Mn^{+2} also cause nearly identical changes in the EPR spectrum. Moreover, no changes in the relaxation behavior of the center are observed (Dunham, 1987). Thus, the changes in the EPR spectrum due to the binding of divalent metals must result from a slight structural change rather than from a magnetic interaction. The lack of magnetic interaction implies that the Rieske center and the mononuclear iron sites are more than 10 Å apart.

The Fe-EXAFS for PO in which the mononuclear site was populated with either Fe^{2+} or Zn^{2+} established that the bond lengths in the Rieske center were unchanged from those of the iron-depleted enzyme (Penner-Hahn et al., 1987). Thus, twist or other strain may be responsible for the EPR observations (above). With zinc ion in the mononuclear site, Zn-EXAFS could be performed. These studies indicated that ligation to zinc was likely to be hexacoordinate with N and/or O ligands at an average distance of 2.01 A. No backscattering from iron or sulfur could be detected. A perturbation of these distances on binding phthalate suggests that the site of oxygenation is at the mononuclear iron.

PHTHALATE OXYGENASE REDUCTASE (POR)

POR catalyzes the rapid reduction of a variety of electron acceptors by NADH. Its spectral properties are consistent with the presence of FMN and a ferredoxin-like [2Fe-2S] center (Batie et al., 1987). Titrations of this protein with reducing agents (NADH or dithionite) show that the redox potentials of the FMNH·/FMN and of the $[2Fe\text{-}2S]_{red}/[2Fe\text{-}2S]_{ox}$ couples are very nearly the same. The visible absorption spectra of POR are shown in figure 4. The EPR Spectrum is shown in Figure 5.

In the EPR spectrum g-values at 2.041, 1.949, and 1.900 are associated with the [2Fe-2S] center, and that at 2.008 is attributed to a blue flavin semiquinone radical. The latter is badly saturated at this power level. The fact that one equivalent of NADH produces nearly equivalent amounts of each of the species confirms the similarity of the redox couples. The ability to observe both radicals clearly by EPR indicates that the two redox centers are more than 10 A apart (ie. there is no apparent magnetic interaction).

We have examined in some detail the kinetics at 4° of the reaction of the reductase with NADH. Although studies are not yet complete, several features are clear. The reaction is stereospecific, utilizing the pro-R- proton of NADH and a steady state kinetic isotope effect of *ca* . 3-fold has been determined. Several spectral intermediates can be delineated which are consistent with Scheme 3.

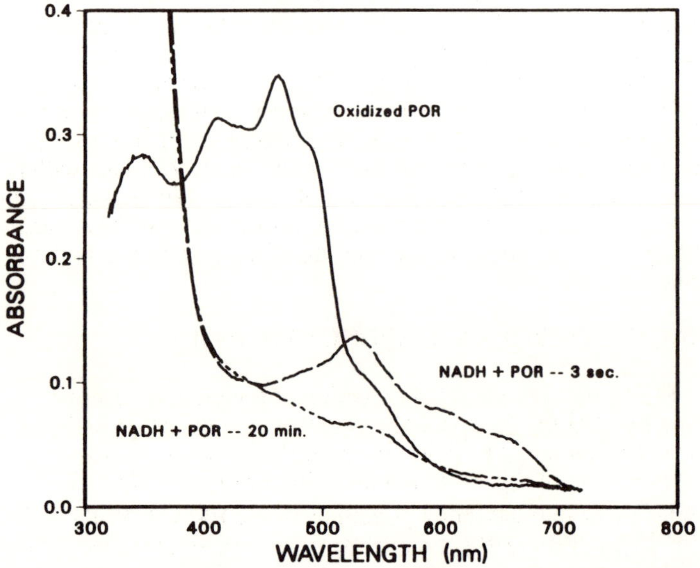

Figure 4. Absorption Spectra of POR. Reduced sample was produced by mixing excess NADH anaerobically with oxidized POR. The spectrum was recorded 3-10 s after mixing. Final spectrum of fully reduced POR was recorded at 20 min. Oxidized spectrum was recorded after mixing POR with buffer. Spectra were recorded in a stopped flow spectrophotometer (Beaty and Ballou, 1981). Conditions: 4° C, 0.1M HEPES, pH 8. POR = 14 mM.

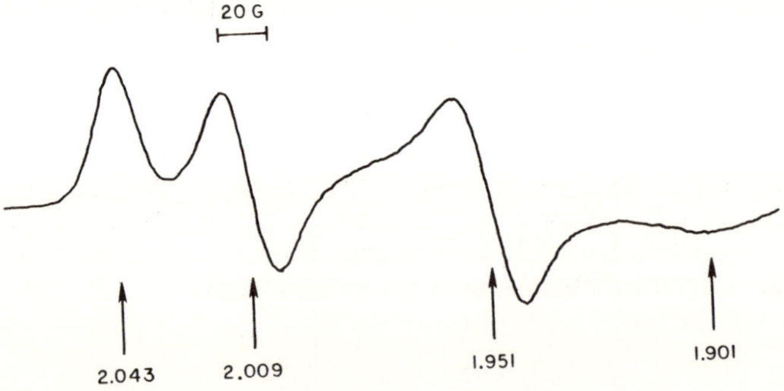

Figure 5. EPR Spectrum of phthalate oxygenase reductase. Spectrum was recorded after reducing anaerobically the POR (0.096 m M) with 1 equivalent of NADH. The spectrum was recorded at 21 K, with 10 G modulation at 5 mW power.

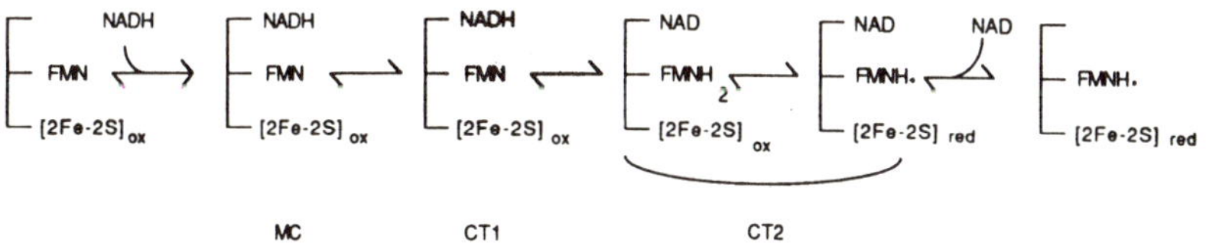

Scheme 3. Steps in the Reduction of POR

After formation of the Michaelis complex (MC), a charge transfer complex involving NADH and FMN develops (CT1). MC1 is included since the development of CT1 is kinetically saturable. CT2 forms after intramolecular electron transfer and involves a previously unknown type of charge-transfer interaction between semiquinone and NAD. It can be observed quite clearly if 2-electron reduced POR is mixed with a high concentration of NAD. Thus, various charge-transfer species appear in the course of reduction which can be assigned to structures consistent with a logical chemical mechanism. All steps are sufficiently rapid (45 - 200 s^{-1}) for consideration in the overall mechanism in the phthalate oxygenase system. It should be noted that intramolecular electron transfer is quite rapid (> 200 s^{-1}) although the distance of transfer is more than 10 Å.

Crystals of POR are available and a determination of the X-ray structure is underway in M. Ludwig's laboratory (Correll et al., 1985). With both centers on the same small protein this should provide us with a unique opportunity to probe how flavin and ferredoxin-like centers interact in enzymatic electron transfer systems. We intend to make site-directed changes to the protein to delineate further the roles of various residues in the electron transfer process.

CATALYSIS

The phthalate dioxygenase system is very efficient in catalyzing the difficult formation of a *cis*-dihydrodiol on an unactivated aromatic ring. Steady state kinetic studies coupled with various rapid reaction studies of the partial reactions have allowed us to propose Scheme 4 to account for the catalysis.

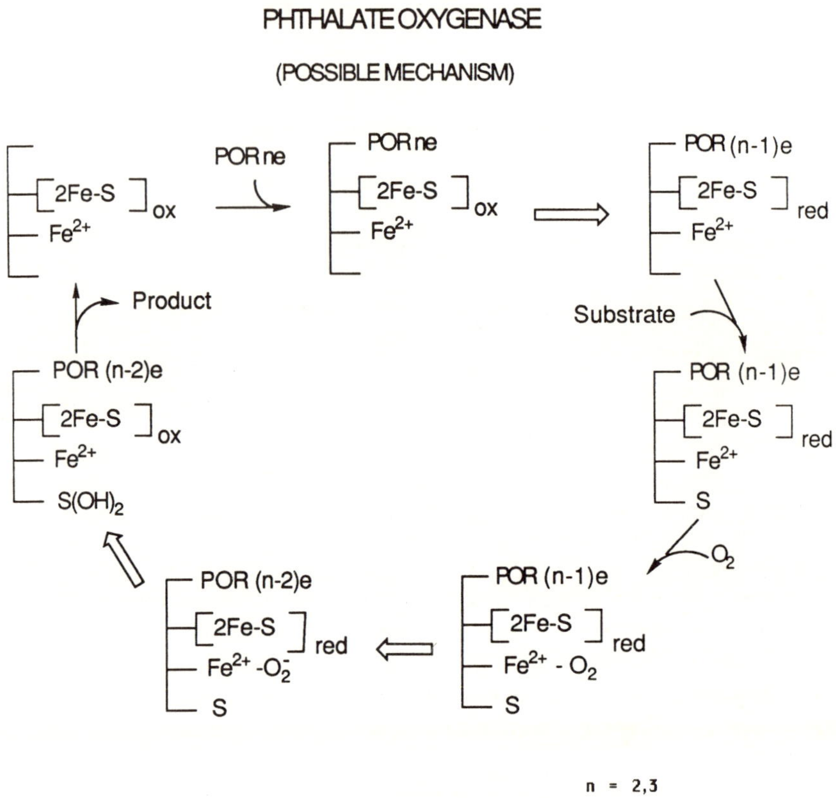

Scheme 4. Reaction Sequence of Phthalate Oxygenase

The order of the scheme is deduced from several observations. Steady state studies employing varying concentrations of reductase, oxygenase, phthalate, oxygen, and NADH yield patterns that indicate the reductase is reduced before interacting with the oxygenase. Excess PO is reduced by NADH and catalytic quantities of POR at 12 s^{-1} /POR. The kinetic patterns indicate there is an irreversible step between association of POR and PO and the addition of either S or O_2 ; this is probably the electron transfer step. PO is reduced by stoichiometric levels of POR_{red} at a rate of 60 s^{-1}. These rates do not depend markedly on substrate concentration and are consistent with the notion that S combines with the complex after PO is reduced. It has also been noted above that PO reacts very slowly with oxygen in the absence of POR and S. Thus, a complex involving PO, POR, S and O_2 must be present for catalysis to occur efficiently.

The steps following combination of substrates and the two proteins are much less certain and are the current focus of our studies. Preliminary rapid kinetics studies of the reaction of reduced phthalate oxygenase (in the presence of all components except NADH) with oxygen reveals at least five distinct phases. Three of them occur at rates consistent with being involved with catalysis.

The dihydroxylation of aromatic compounds may require the development of reasonably powerful oxidizing agents. These could produce undesirable products, perhaps deleterious to the organism. It is tempting to think that the purpose of requiring all components to be present before a rapid reaction with oxygen occurs is to prevent the formation of such free radicals until such time that the reaction can be carried out in a concerted fashion.

It is clear that much remains to be done in order to understand this interesting oxygenation system. This catalysis employs complex electron transfer processes involving four internal redox centers: FMN, ferredoxin-like [2Fe-2S] , Rieske [2Fe-2S] center, and a mononuclear Fe^{2+}. In addition, three substrates: NADH, O_2 , and phthalate are involved.

REFERENCES

Axcell, B. C. and Geary, P. J. (1975). Biochem. J. 146, 173-183.
Batie, C. J. LaHaie, E., and Ballou, D. P. (1987). J. Biol. Chem. 262, 1510-1518.
Beaty, N. B. and Ballou, D. P. (1981). J. Biol. Chem. 256, 4611-4618.
Bernhardt, F. H., Pachowsky, H., and Mabry, T. J. (1975). Eur. J. Biochem. 57,241-256.
Cline, J. F., Hoffman, B. M., Mims, W. B., LaHaie, E., Ballou, D. P., and Fee, J. A. (1985). J. Biol. Chem. 260 3251-3254.
Correll, C. C., Batie, C. J., Ballou, D. P., and Ludwig, M. L. (1985). J. Biol. Chem. 260, 14633-14635.
Dagley, S. (1975). Am. Sci. 63, 681-689.
Dunham, W. R. (1987). Personal Communication.
Ensley, B. D. and Gibson, D. T. (1983). J. Bacteriol., 155, 505-511.
Fee, J. A., Findling, K. L., Yoshida, T., Hille, R., Tarr, G. E., Hearshen, D. O., Dunham, W. R., Day, E. P., Kent, T. A., and Munck, E. (1984). J. Biol. Chem. 259, 124,133.
Geary, P. J., Saboowalla, F., Patil, D. and Cammack, D. (1984). Biochem. J. 217, 667-673.
Gibson, D. T., Hensley, M., Yoshioka, H. and Mabry (1970). Biochemistry 9, 1626-1630.

Hoffman, B. M., Fee, J. A., Batie, C. J., Ballou, D. P. and Kuila, D. (1987). Unpublished results.
Kuila, D., Fee, J. A., Schoonover, J. R., Woodruff, W. H., Batie, C. J. and Ballou, D. P. (1987). J. Am. Chem. Soc. 109, 1559-1562.
Keyser, P. K., Pujar, B. G., Eaton, R. W., and Ribbons, D. W. (1976). Environ. Health Perspect. 18, 159-166.
Ornston, L. N. (1966). J. Biol. Chem. 241, 3787-3794.
Penner-Hahn, J. and Tsang, H. T. (1987). Personal Communication.
Pujar, B. G. and Ribbons, D. W. (1985). Appl. Environ. Microbiol. 49, 374-376.
Sauber, K., Frohner, C., Rosenberg, G. , Eberspcher, J. and Lingens, F. (1977). Eur. J. Biochem., 74, 89-97.
Subramanian, V,, Liu, T. N., and Gibson, D. T. (1979). Biochem. Biophys. Res. Commun. 91, 1131-1139.
Twilfer, H., Bernhardt, F. H. and Gersond, K. (1981). Eur. J. Bichem. 119, 595-601.
Yamaguchi, M. and Fujisawa, H. (1982). J. Biol. Chem. 257, 12497-12502.

ACKNOWLEDGEMENTS

J. A. Fee, D. Kuila, D. (Los Alamos), RAMAN, EPR; B. M. Hoffman, (Northwestern Univ.), ENDOR; J. Penner-Hahn, H. T. Tsang, (Univ. Michigan), EXAFS; R. Olsen, (Univ. Michigan), Microbiology; W. R. Dunham, (Univ. Michigan), EPR

DEDICATION

To Stanley Dagley (1916-1987) who pioneered this field over the past 40 years. It is largely through his insight and, moreover, his honest and fair approach towards science that this field of aerobic metabolism of aromatic compounds has thrived. We miss you, Stan.

DISCUSSION

GROVES: There are two curious things about benzene dioxygenases and related enzymes: The cis-stereochemistry of the glycol product and the unactivated nature of the aromatic ring of the substrate. The latter seems to suggest that a reactive iron-oxygen intermediate such as Fe=O will be a prerequisite for substrate reaction. The cis-stereochemistry could be conveniently accommodated by oxidation of an intermediate hydroxycyclohexadienyl radical as shown below:

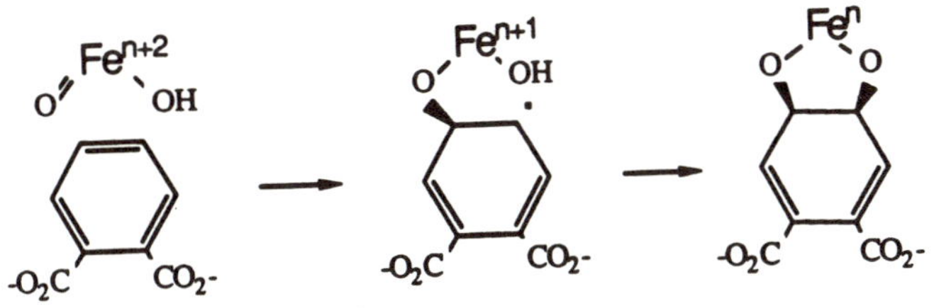

We have described a related process for the formation of diols from cyclohexanol by $Fe(II)/H_2O_2$.

BALLOU: Your suggestion is attractive. Perhaps I can expand upon it in the scheme below. Roman numerals indicate the tentative oxidation state of the iron in the mononuclear site. Species B might allow for the formation of your suggested iron species D. Nonheme iron centers such as the mononuclear iron in phthalate dioxygenase may have available liganding positions to accommodate forms of iron-oxygen as proposed. This is not likely with hemoproteins, which are restricted by ligation to a rigid ring structure. Here the species, C, is at a higher oxidation state and might abstract an electron from the substrate. The addition of the second electron via the Rieske center could lead to species C, which would then be displaced by water or protein ligands to yield product. The poor reactivity of the reduced oxygenase with O_2 in the absence of reduced reductase might be attributed to reversible steps with an equilibrium lying far to the left (since there are no observed intermediates). Thus, species C might only be formed after addition of the second electron to the oxygenase, a process requiring the reductase. Species D is effectively a ferric hydroxide potentially poorly exchangeable. Perhaps this accounts for incorporation of both atoms of O_2 into the product.

Species C might also be capable of attacking a methoxy carbon to effect a monooxygenation. In this case water would be lost and the system would be analogous to P-450 oxygenations. Monooxygenases involving both a Rieske type iron sulfur and a similar mononuclear iron site have been found for substrates such as p-methoxybenzoate.

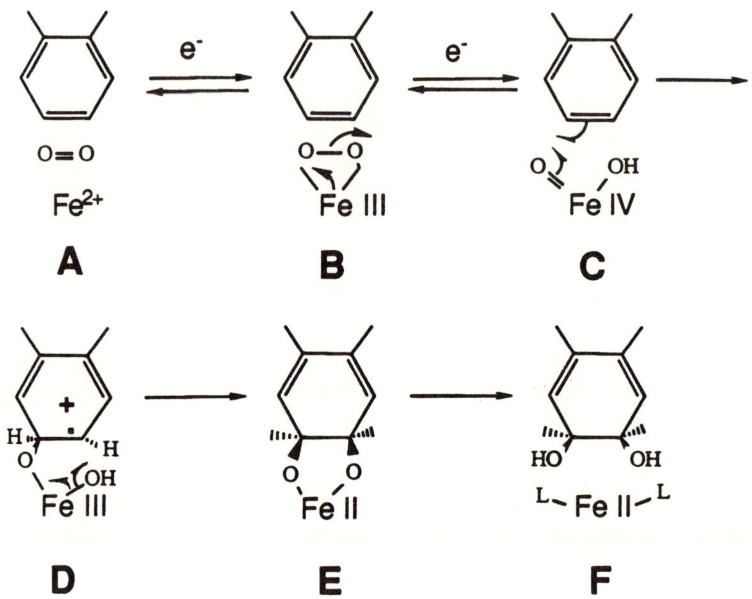

Oxidases and Related Redox Systems, pages 227–248

ROLE OF COPPER AND CATALYTIC MECHANISM IN THE COPPER MONOOXYGENASE, DOPAMINE β-HYDROXYLASE (DβH)

Judith P. Klinman and Mitch Brenner,
Department of Chemistry, University
of California, Berkeley, CA 94720

ABSTRACT

Although classically characterized as a Type II copper protein, recent work has shown that DβH requires two coppers per subunit for optimal activity. A major challenge has been to elucidate the relationship of these copper sites to one another, specifically whether there exists a single binuclear copper site or distinct metal sites catalyzing separate functions. In this paper, copper sites have been investigated by EPR techniques and rapid mixing methods to measure the rates and stoichiometries of copper reduction by ascorbate and subsequent reoxidation by tyramine. As shown, there is little or no evidence by EPR for spin coupling between metal sites in resting enzyme. *Both* coppers per subunit undergo reduction and reoxidation with single exponential values of 185 s^{-1} and 80 s^{-1}, respectively. Significantly, the ratio of copper reoxidation to product formation in a single turnover is close to 2:1, implicating both coppers in the catalytic mechanism and ruling out any significant spin coupling in the enzyme-product complex. In contrast to single turnover experiments, rapid kinetics in the presence of excess ascorbate reveal a very low level of E-Cu(II). This result, which appears to conflict with evidence that the dominant enzyme form in the steady state is E-P, leads us to propose a reduction of enzyme at the level of E-P rather than free enzyme. Taken together with steady state kinetic parameters using alternate reductants, the data support separate binding sites for reductants and product/substrate and hence, separate functions for each copper per subunit. A detailed catalytic mechanism for DβH is discussed in the context of a single metal site catalyzing dopamine hydroxylation.

INTRODUCTION

Type II copper proteins represent a fairly disparate group, catalyzing the functions summarized in Table 1.

Table 1: Type II Copper Proteins

Name	Function	Unique Features
Superoxide Dismutase	One e- transfer to/from O_2^-	Cu/Zn, bridging imidazole
Laccase, ceruloplasmin, ascorbate oxidase	Reduce O_2 to H_2O	Multiple copper sites: I, II, III
Oxidases	Reduce O_2 to H_2O_2	
Galactose	Oxidizes galactose	No additional cofactor
Amine	Oxidizes amines	Contains an organic redox cofactor
Monooxygenases	Reduce O_2 to H_2O	
DβH	Dopamine → Norepinephrine	Contains two coppers per subunit
Peptide amidating enzyme	Generates α-amidated neuropeptides	Same cofactor requirements as DβH
Methane monooxygenase	Methane → Methanol	Found in methanotrophs
Phenylalanine Hydroxylase	Phenylalanine→ Tyrosine	Contains a pterin

As shown these include single electron transfer [superoxide dismutase (Fee, 1981; Valentine & Mota de Freitas, 1985)], multiple electron transfer yielding water from dioxygen [laccase, ceruloplasmin, ascorbate oxidase(Farver & Pecht, 1981; Reinhammer, 1979)], oxidases [galactose oxidase (Kosman,1984), amine oxidases (Knowles & Yadav, 1984; Lobenstein-Verbeek *et al.*, 1984)], and monooxygenases [DβH (Villafranca, 1981; Klinman *et al.*, 1984; Ash *et al.*,1984), peptide amidating enzyme (Eipper, 1983), methane monooxygenase (Tonge *et al.*,1977), phenylalanine hydroxylase (Pember *et al.*,1986)]. Of the proteins in Table I, only the latter two classes catalyze a functionalization of organic substrates. Closer examination indicates the presence of additional redox cofactors in the amine oxidases and phenylalanine hydroxylase, such that a relatively small number of known proteins use a copper center to oxidize substrates. Among the monooxygenases, DβH is by far the best characterized.

Although considerable structural information is now available for the type I electron transport proteins, e.g.,the blue, type I electron transport proteins such as stellacyanin and azurin (Gray & Solomon, 1981) and the binuclear, type III proteins such as hemocyanin and tyrosinase (Solomon, 1983), these proteins have not appeared to bear directly on the structure or function of DβH. Classically, DβH has fallen into the category of a type II copper protein, meaning that it has been found to lack the characteristic charge transfer and magnetic properties of the type I and III proteins. However, until recently, the characterization of DβH has been focussed on a form of enzyme containing only a single copper per subunit. This situation changed dramatically in 1984, with the reports of Klinman *et al.*(1984) and Ash *et al.*(1984), that DβH requires two coppers per subunit for maximal expression of activity. Thus, the question of the possible formation of binuclear complexes within a subunit arises, specifically whether dopamine β-hydroxylase contains a binuclear copper site analogous to tryosinase and hence, should be reclassified as a type III protein.

In this study, the interrelationship of the two copper sites per subunit in DβH has been examined by EPR and rapid mixing techniques, leading to the conclusion of little or no spin coupling between copper sites in either the resting form of enzyme or in the catalytically generated product complex. Although both coppers undergo reoxidation in the course of product formation, supporting a catalytic role for two metals, the currently available data on DβH support a model involving distinct copper sites which may perform the separate functions of electron transfer and substrate hydroxylation, respectively. A detailed catalytic mechanism is presented which involves a single copper center in the substrate hydroxylation step. We suggest that this mechanism may be generally applicable to type II copper monooxygenases and possibly non-heme iron monooxygenases.

RESULTS

EPR spectra for DβH are shown in Figure 1 for enzyme containing either one (1-A) or two, (1-B) coppers per subunit. The EPR spectrum of Cu(II)-

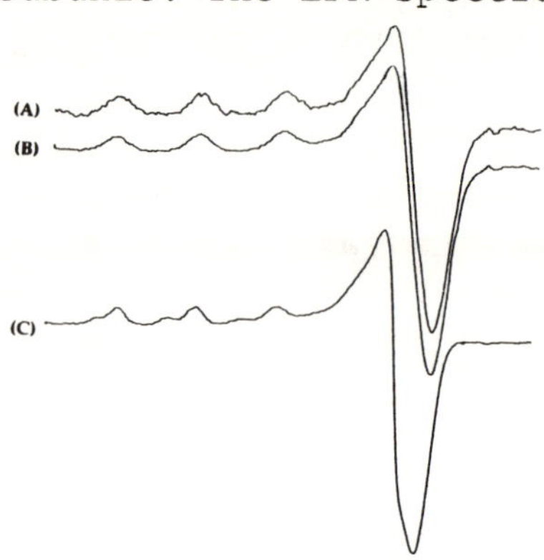

Figure 1: EPR spectra for DβH and Cu(II)-EDTA. 1-A: 57.9 μM DβH subunits (0.1 ml) with one copper per subunit, Gain = 160. 1-B: Enzyme with two coppers per subunit (0.1 ml), Gain = 79. 1-C: 200 μM $CuSO_4$ in 20 mM EDTA, pH 4.5 (0.067 ml). Gain = 20. EPR settings were 0.2 mW microwave power, 10 Gauss modulation amplitude, T = 14.7 K, 2970±500 Gauss field sweep, 57 s/scan X 10 scans, 0.3 s time constant and 9.2073 GHz microwave frequency, 100kHz modulation frequency.

EDTA is given in 1-C, as a frame of reference. The differences between spectra 1-A and 1-B are minor, although some broadening can be ascertained in the g_{11} region on proceeding from one to two coppers per subunit. This small amount of broadening, which differs from a preliminary spectrum of Villafranca (1981), was also reported by Skotland *et al.* (1980) who compared a difference spectrum for DβH at a ratio of eight *vs* four coppers per tetramer to native enzyme. Importantly, double integration of EPR spectra (at 50 K, 1 mW power to avoid a possible saturation of signal) consistently indicates a level for copper which is equivalent to values obtained by atomic absorption spectroscopy. Thus, there is no evidence for significant antiferromagnetic coupling between metal sites in DβH containing two coppers per subunit.

Further characterization of copper sites was attempted by power saturation techniques using DβH with either one or two coppers per subunit and varying temperature. At 50 K no evidence for power saturation was seen with either enzyme form up to the limits of our instrumentation (40 mW). By decreasing the temperature to 11.7 K, an onset of saturation was detectable. Once again, no significant differences in relaxation rates could be detected for enzyme at varying copper stoichiometries. Although not a highly sensitive probe of structure at copper sites, power saturation data support the conclusions of Figure 1 - that the addition of a second copper to DβH leads to an enzyme form which is practically indistinguishable by EPR techniques.

In conjunction with studies of the resting form of DβH, rapid acid and freeze quench studies were considered essential, in an effort to characterize catalytic intermediates formed in the course of enzyme reduction and reoxidation. Steady state kinetic studies of DβH have indicated a pattern of ping pong kinetics (Rosenberg & Lovenberg, 1980), implying the interaction of reductant and substrates with different enzyme forms. The traditional interpretation of this pattern has been a reduction of free enzyme, although, as will be discussed below, alternate interpretations are possible. In any case, it was expected that the

Cu(II) form of DβH would undergo a rapid reduction by ascorbate. The conversion of enzyme bound Cu(II) to Cu(I) was followed by EPR, using freeze quench techniques to monitor reduction on a millisecond time scale.

As shown in Figure 2-A for reduction of 12 μM enzyme subunits by 0.2 mM ascorbate, 80% of reducible copper is lost with a half time of 12 ms. The very slow reduction of the remaining 20% of signal is attributed to inactive enzyme in DβH preparations, a common problem which is difficult to overcome completely since inactive and active enzyme appear to co-chromatograph (Colombo *et al.*, 1987). Significantly, a semi-log plot of the data in Figure 2-A indicated a single exponential process for the reduction of 80% of bound copper (data not shown).

Subsequent studies of enzymic reduction in the range of 0.2 to 2 mM ascorbate revealed a hyperbolic dependence on ascorbate concentration, shown in reciprocal form in Figure 2-B. These data are characterized by a limiting rate constant of $k_R = 185\ s^{-1}$ and an apparent Km = 0.5 mM. The latter is very similar to the steady state Km value for ascorbate (Kaufman & Friedman, 1965). The limiting first order rate is considerably faster than an early report of 20 to 30 s^{-1} (at 0.9 mM ascorbate) by Skotland *et al.*(1980) using enzyme with only one copper per subunit. The magnitude of k_R relative to $k_{cat} = 16\ s^{-1}$ supports the idea that enzyme reduction is a relatively fast process under turnover conditions.

A somewhat surprising aspect of the data in Figure 2 is the appearance of a single exponential for reduction of the bulk of enzyme bound metal, since the two copper sites per subunit may be expected to be differentially accessible to reductant. One possible explanation is that intraprotein electron transfer between copper sites is fast, such that the overall reduction process appears as a first order process. Alternatively, it is difficult to rule out a protein conformational change from resting enzyme as a prerequisite for the reduction of both copper sites. In any case, as will be discussed below, enzyme reduction under

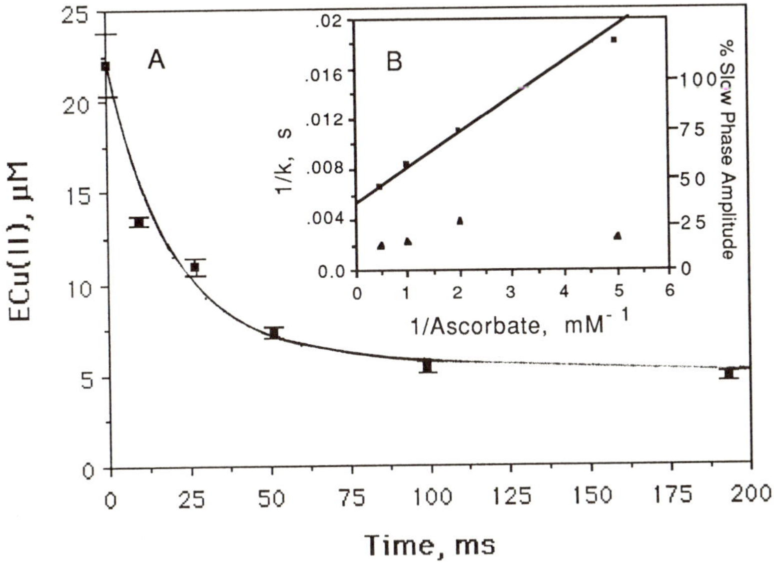

Figure 2: Time and Concentration Dependence of Reduction of Enzyme Bound Copper by Ascorbate. 2-A: After mixing, the concentration of DβH subunits was 12 μM, with 22.1 μM Cu(II) and 1.9 μM Cu(I). Ascorbate was 0.2 mM in buffers containing 25 mM potassium phosphate, pH 6, 10 mM disodium fumarate and 5 mM KCl, T = 35° C. At the indicated times, reactions were quenched by freezing with isopentane at -140° C. The concentration of enzyme bound cupric ion in each quenched sample was calculated from the amplitude of the EPR signal at $g_{\perp} = 2$, relative to a zero time control in which enzyme was mixed against a buffer blank. The solid curve is a simulation, with 80% of copper showing a rate constant of 55 s^{-1} and 20% of copper showing a rate constant of 2 s^{-1}. In this and other samples, ~ 2.5 μM copper cannot be reduced. 2-B: The result of kinetic fits to DβH reduction curves for 0.2 to 2 mM ascorbate are plotted in a reciprocal form, (■-■). Percent of enzymic copper reduced in the slow phase, (▲-▲). EPR conditions were the same as Figure 1, with the exception of 5 mW power, 20 Gauss modulation amplitude and 4 rather than 10 scans per sample.

turnover conditions is likely to occur via a different mode than is occurring in direct reduction studies.

According to current models of DβH, enzyme in a Cu(I) form interacts with substrate and oxygen, leading to hydroxylated product, water and E-Cu(II) (Rosenberg & Lovenberg, 1980). If correct, it should be possible to pre-reduce enzyme in a stoichiometric manner and see a rapid burst of product. Early experiments involving pre-reduction of enzyme for an unstated time followed by incubation with tyramine (0.5 mM) and oxygen indicated a very slow formation of E-Cu(II) ($t_{1/2}$ ~ 20 min), attributed to a requirement for excess ascorbate as an effector or reductant (Ljones *et al.*, 1978). In retrospect, the slow observed rate can be ascribed to copper inadequacy, since experiments were performed with enzyme containing only one copper per subunit.

Our recent experiments using two copper per subunit enzyme and rapid mixing techniques to control times for enzyme incubation with both ascorbate and tyramine have permitted measurement of rate constants for formation of both octopamine and E-Cu(II). As shown in Figure 3, reoxidation of E-Cu(I) and hydroxylation of tyramine parallel one another with rate constants of 31 s^{-1} and 34 s^{-1}, respectively, from semilog plots (data not shown).It should be noted that the data in Figure 3 were conducted at 5 mM tyramine which, in contrast to steady state conditions, did not lead to substrate saturation. Elevation of tyramine to 0.4 M did, however, lead to a limiting rate constant, k_O = 80 s^{-1}.

An initially surprising result was a reduction in k_O from an anticipated value of 590 s^{-1} at saturating tyramine and oxygen (Miller & Klinman, 1985). The approximately 7 fold reduction in rate is disturbing and requires explanation. One possibility is that pre-reduction produces an enzyme form that proceeds from ternary complex to product at a reduced rate. An alternate, favored explanation is that oxygen binds much more weakly to binary than ternary complex [K_D reported to be 0.8 mM (Kruse *et al.*, 1986) from binary complex *vs* 0.1 mM (Miller & Klinman, 1985) from ternary complex], such that oxygen at 0.2 mM is not significantly bound to pre-reduced enzyme. While subsequent addition of tyramine to 0.4 M can produce saturation, oxygen remains non-saturating at 0.2 mM reacting in a second order fashion. Under these conditions, the relevant rate constant for comparison to rapid studies is the steady state

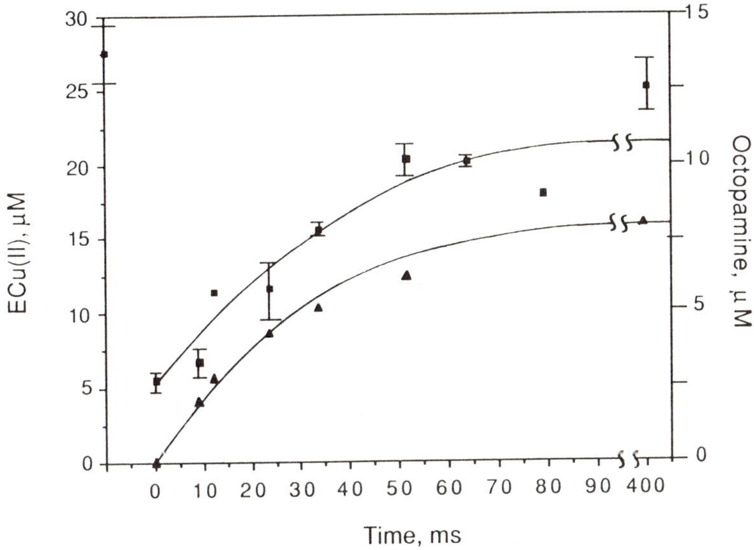

Figure 3: Time Course for Octopamine and Cu(II) Formation from Stoichiometrically Reduced DβH. 30 µM DβH, which had been incubated for 43 sec with one equivalent of ascorbate at 35° C in 50 mM potassium phosphate, pH 6, and 10 mM disodium fumarate was rapidly mixed (1:1) with 10 mM tyramine hydrochloride in the same buffer, giving 15 µM DβH subunits and 5 mM tyramine after mixing. (■-■), EPR amplitude at $g_\perp$ of freeze quench samples. EPR settings were the same as Figure 2. Initial value of 30 µM E-Cu(II) refers to a signal from samples in which ascorbate and tyramine were replaced by a buffer blank. The value of E-Cu(II) at 0 ms refers to pre-reduced enzyme mixed with a buffer blank instead of tyramine. Error bars reflect the range of duplicate shots. (▲-▲), the concentration of octopamine in acid quenched samples, obtained from reaction mixtures identical to those described for freeze quench. Product levels were obtained from HPLC traces of quenched reaction mixtures, relative to octopamine standard curves.

V/K_{O2} for reaction of oxygen with E-tyramine; at 0.21 mM oxygen this value is 70 s^{-1}, which compares very favorably with k_O = 80 s^{-1}. We conclude that the available data for reaction of pre-reduced enzyme with tyramine and oxygen are fully compatible with steady state parameters and that, in contrast to an earlier suggestion (Brenner & Klinman, 1987), the E-Cu(I) form of enzyme is catalytically competent.

The relationship of the rate of Cu(II) to product formation is of considerable interest, in light of current working models invoking two distinct copper sites per subunit. Although resting forms of enzyme fail to show any interaction between copper sites, it is conceivable that binuclear species form in the course of catalysis. In such an instance, intermediate copper species would be expected to be magnetically coupled and to show reduced or possibly no signal in the EPR. However, as shown in Figure 3, the rate of E-Cu(II) formation correlates with octopamine production. While this result cannot provide information on copper species formed prior to the hydroxylation step, it does rule out significant coupling between metal sites in the catalytically produced product complex. Consistent with a direct role for both coppers in the catalytic process, the stoichiometry of copper reoxidation to octopamine formation is consistently in the range of 2:1.

In an early study by Skotland *et al.*(1980), rapid mixing of DβH with reaction components and excess ascorbate led to a very low level of E-Cu(II). This result has been repeated with enzyme containing two coppers per subunit, Figure 4. As shown, only 10% of total enzyme is oxidized in the steady state. This low level of detectable Cu(II) contrasts with the above described studies conducted in the presence of limiting (stoichiometric) ascorbate. Further, both steady state (Miller & Klinman, 1985; Ahn & Klinman, 1983) and pre-steady state (Murray & Klinman, 1985) kinetic studies of DβH have indicated that the dominant enzyme species (>70%) is the enzyme product complex.

A reasonable explanation for the low levels of Cu(II) in Figure 4 is a reduction at the level of E-P rather than free enzyme. Such a phenomenon would be fully consistent with first, ping pong kinetics since the hydroxylation step itself is irreversible and second, the much faster apparent

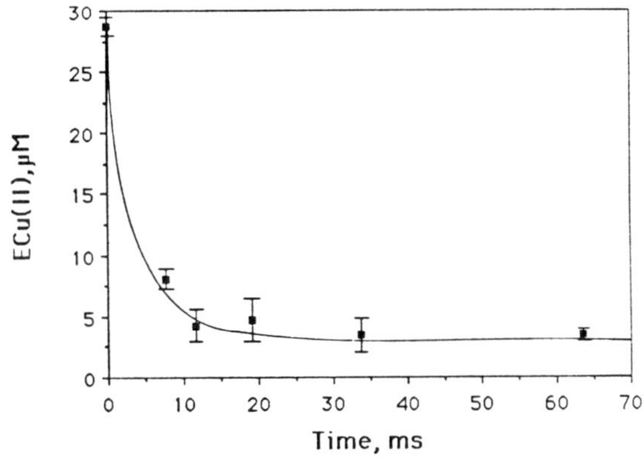

Figure 4: Monotonic Decrease in E-Cu(II) to a Low Steady State Level. The points refer to the amplitude at $g_\perp$ of rapid freeze quench samples from the reaction of 15 μM DβH subunits containing 30 μM copper [28.7 μM Cu(II) and 1.3 μM Cu(I)] with 10 mM ascorbate and 5 mM tyramine hydrochloride at 35° C in 75 mM potassium phosphate, pH 6, and 10 mM disodium fumarate (all concentrations refer to after mixing). EPR conditions are the same as Figure 2.

rate constant for enzyme reduction (185 s^{-1}) than product release (16 s^{-1}) (Miller & Klinman, 1985). Although steady state studies by Villafranca and co-workers have shown that the second order rate constant for reduction of enzyme by ascorbate is substrate independent (Fitzpatrick *et al.*, 1986), reduction of copper in E-P could be fairly insensitive to modifications in product structure. Further, Stewart and Klinman (in press) have recently carried out a full examination of kinetic parameters with varying reductants, finding that ferrocyanide is nearly as competent as ascorbate as reductant whereas dopamine is slow; model building suggests that ferrocyanide and ascorbate possess an unexpected but striking similarity in possible binding interactions with enzymic residues. Taken together, the available data suggest a model for DβH in which distinct sites exist for reductant and product/substrate. In such an instance, separate functions for each copper per subunit, that of electron transfer and substrate hydroxylation, respectively, become a distinct possibility.

DISCUSSION

A key reference point for discussion of mechanism is the nature of the catalytic copper site, specifically whether this site contains mono- or binuclear copper. As shown under Results, no evidence has been forthcoming for a binuclear copper complex in DβH from either EPR or kinetic approaches. Recent EXAFS data on both the E-Cu(II) and E-Cu(I) forms of DβH reconstituted with two coppers per subunit have led to a similar conclusion, since no evidence for back scattering between metal sites could be detected (Scott *et al.*). The EXAFS studies of Scott *et al.* have also provided insight into the identity of ligands to copper in DβH, with best fits obtained for E-Cu(II) using four imidazoles and for E-Cu(I) using two to three imidazoles and a heavy atom such as sulfur or chloride ion. While the results with E-Cu(II) are analogous to an earlier report by Hasnain *et al.*(1984) using enzyme with only one copper per subunit, those with E-Cu(I) are unexpected, requiring a significant ligand rearrangement on copper reduction. It is conceivable, that prolonged incubation of enzyme with ascorbate under the conditions of EXAFS studies leads to an isomerized enzyme form. Although incubation of DβH with stoichiometric ascorbate for 43 seconds leads to a catalytically competent enzyme (*c.f.* discussion under Results), the time course for possible structural changes in the E-Cu(I) form of DβH deserves further attention.

With the background reference of a single copper atom as the site of substrate hydroxylation, the nature of reactive intermediates derived from oxygen and substrate become compelling questions. Unlike cytochrome P-450 chemistry, relatively few model systems have addressed the question of catalysis by mononuclear copper. In this context, a recent study by Karlin and co-workers (1987) on a binuclear model is of particular interest, showing reversible oxygen binding under aprotic conditions, coupled to a rapid catalysis of phosphine and sulfide oxidation following acidification to form a postulated copper hydroperoxide species. An earlier investigation of the DβH catalyzed hydroxylation of dopamine (Ahn & Klinman, 1983) had led to a similar conclusion of the intermediacy of copper hydroperoxide (Eq.1). As described by Ahn and Klinman, detailed pH and deuterium isotope

$$E\begin{matrix}\sim Cu(I)\\ \sim BH\\ \sim Cu(I)\end{matrix} + O_2 \underset{\leftarrow}{\rightarrow} E\begin{matrix}\sim Cu(II)\\ \sim B\\ \sim Cu(II)\text{-}OOH\end{matrix} \qquad (1)$$

effects demonstrate the uptake of a single proton to be essential for the oxygen insertion step; since cleavage of the benzylic C-H bond itself is unlikely to show such a dependence, a role for general acid catalysis in oxygen activation has been implicated.

Turning to the question of a substrate derived intermediate, this point has received considerable attention in the literature. An early measurement of the intrinsic isotope effect in the DβH reaction indicated very large deuterium isotope effects of 9-11 (Miller & Klinman, 1983). In general, proton and hydrogen atom abstraction reactions are characterized by large, maximal primary isotope effects when either ΔG or ΔH for the hydrogen abstraction step is close to zero. By contrast, no evidence for chemical reversibility could be detected on the DβH surface, consistent with the large thermodynamic driving force for the net reaction. This apparent discrepancy led Miller and Klinman (1983) to suggest a two step mechanism for DβH involving first, hydrogen abstraction to form an intermediate (E-I) in a nearly isoenergetic process and second, oxygen insertion. Irreversibility is then due to a very rapid and essentially irreversible collapse of E-I to the E-P complex.

Direct evidence for a chemical intermediate was later provided by secondary isotope effects measured in the DβH catalyzed hydroxylation of dopamine (Miller & Klinman, 1985). From a compilation of fractionation factors, $^{D}K_{\alpha}$ was estimated as 0.84 for the equilibrium conversion of dopamine to norepinephrine (Cleland, 1982), such that the kinetic isotope effect, $^{D}k_{\alpha}$, was expected to be unity or inverse for a direct insertion mechanism. By contrast, $^{D}k_{\alpha}$ was observed to be normal and large (1.19±.06), implicating an intermediate which has undergone substantial bond rehybridization at the benzylic carbon. Support for a chemical intermediate has also come from studies of mechanism based inhibitors (Colombo et al., 1984; Bossard & Klinman, 1986; Wimalasena & May, 1987).

The chemical nature of the intermediate generated following C-H bond cleavage in the DβH reaction has been actively pursued in several laboratories, addressing the question of whether this intermediate arises from proton, hydride or hydrogen atom abstraction. Structure reactivity correlations have played a major role in these investigations beginning with an early study by May and co-workers (1981). A limitation of this early study was the use of a single oxygen concentration such that Vmax could not be distinguished from Vmax/Km parameters; additionally no information was available concerning rate limiting steps. Subsequent studies were conducted by Fitzpatrick *et al.*(1985) and Miller and Klinman (1985) who examined enzyme inhibition by 3-phenylpropenes and turnover with phenethylamines, respectively. In the case of 3-phenylpropene inhibition, a deuterium isotope effect of 2.0 was observed on V/K_{O2} for 3-(*p*-hydroxyphenyl)propene indicating some rate limitation by C-H bond cleavage. Restricting themselves to V/K_{O2}, a free energy correlation indicated a slope of -1.2, effectively ruling out a proton abstraction mechanism. Although the best correlation of rate was obtained with σ+, Fitzpatrick *et al.* favored a radical mechanism, since the generation of a carbocation intermediate was expected to show a rho value considerably more negative than -1.2.

A particularly thorough study was performed by Miller and Klinman (1985), who obtained deuterium isotope effects for each of nine phenethylamine derivatives. As shown, considerable variation in isotope effects on Vmax and Vmax/Km was observed. Using the available intrinsic isotope effect for dopamine, however, intrinsic rate constants could be calculated for the C-H bond cleavage and product release steps. Subsequent multiple regression analysis permitted a separation of electronic from steric and hydrophobic effects. This treatment led to several conclusions, the most important of which were electronic effects of ρ = 1.4 and ρ = -1.5 on the product release and substrate hydroxylation steps, respectively. The large positive rho value on product release was surprising and led to a model in which product dissociation from an inner sphere complex limits turnover. With regard to the negative rho value on hydroxylation, this was found to correlate significantly better with σ than σ+, consistent with a hydrogen atom abstraction mechanism *via* a polarized transition state.

As seen above, two independent studies support the generation of a substrate derived radical as the key intermediate in DβH. This point has been further substantiated by Fitzpatrick and Villafranca from a comparison of partition ratios for enzyme turnover *vs* inactivation with 3-phenylpropenes, 1-benzylcyclopropanes and 3-phenylpropynes (Fitzpatrick *et al.*,1985). *At this point the central question arises as to how a Cu(II)-OOH with no unpaired electrons is competent to abstract a hydrogen atom from substrate.* As has been discussed (Miller & Klinman, 1985), prior formation of high valent copper species, leading to either Cu(III) or Cu(IV) and hydroxyl radical or Cu(IV)=O as the hydroxylating agents appears extremely unlikely, given a more positive redox potential for enzymic copper [310 to 360 mV (Ljones *et al.*, 1978; Walker *et al.*, 1977)] than aqueous copper [160 mV (Weast, 1971)], the absence of oxyanion ligands which have been shown to stabilize Cu(III) complexes (Anson *et al.*, 1987), and the described kinetic properties indicating that C-H bond cleavage occurs in a close to isoenergetic step (Miller & Klinman, 1983).

In this context, Miller and Klinman (1985) have proposed a concerted mechanism for DβH, Figure 5. As depicted, generation of some radical character at oxygen, via partial O-O homolysis of the Cu(II)-OOH, initiates C-H homolysis - leading to a transition state where both the O-O and C-H bonds of reactants are significantly cleaved and the O-H bond of product is significantly formed. In particular, three major features of catalysis are noted. First, in the binding of substrate and oxygen the C-H and O-O bonds are linearly aligned, such that no additional entropy need be overcome to attain maximal orbital overlap at the transition state. Second, the active site copper provides a catalyst for the homolysis of the O-O bond, lowering the energy for complete homolysis of

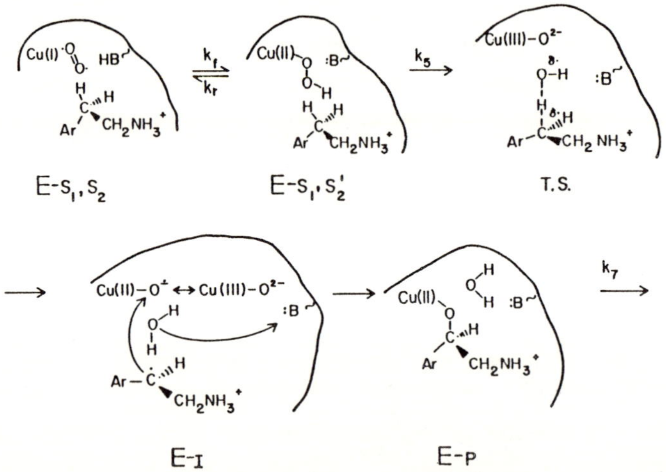

Figure 5: Postulated Chemical Mechanism for the Reaction of DβH. As shown, $E\text{-}S_1,S_2$ is the initial complex, $E\text{-}S_1,S_2'$ involves oxygen reduction to hydrogen peroxide, T.S. is the transition state, E-I is the radical intermediate and E-P is enzyme bound product. Adapted from Miller and Klinman (1985).

peroxide from 51 to an estimated value of +30 kcal/mol. Third, and most significantly, two highly energetic processes - one endoenergetic and the other exoenergetic - are coupled, such that the large energy available from the transfer of hydrogen from substrate to oxygen [-34 kcal/mol, (Weast, 1971)] can be used to drive the highly unfavored O-O homolysis step.

In closing, it is of value to question whether the mechanism shown in Figure 5 can be generalized to other copper and non-heme iron containing monooxygenases. As noted, a key feature of Figure 5 is the coupling of the energy available from hydrogen transfer from carbon to oxygen to the activation of the postulated Cu(II)-OOH intermediate. In principle, one can envisage a spectrum of mechanisms in which the extent of prior Cu(II)-OOH homolysis varies as a function of the reactivity of the organic substrate. This concept has recently been advanced by Valentine *et al.* in the context of cytochrome P-450 reactivity. For organic substrates substantially less activated than the benzylic carbon of phenethylamines, however, e.g., methane in the methane monooxygenase, Table 1, virtually full cleavage of a Cu(II)-OOH may be necessary to initiate C-H bond abstraction.[1] While the intermediacy of such a high valent copper species prior to substrate hydroxylation is incompatible with the known properties of DβH (*vide supra*), it is interesting to speculate that a replacement of ligands to copper [e.g.from imidazoles to oxyanions (Anson *et al.*, 1987)] could achieve the desired stabilization of a Cu(III) intermediate in other protein systems.

[1]Only a single report of a membrane-derived, copper-containing methane monooxygenase has appeared in the literature (Tonge *et al.*, 1977). While consistent with the appearance of a membrane associated methane monooxygenase activity following growth of *M. trichosporium* on a copper rich medium (Davis *et al.*, 1987), further work is needed. The methane monooxygenases which have been well characterized to date contain iron as active site cofactor (cf. Sanders-Loehr,J., this volume).

ACKNOWLEDGEMENTS: We wish to thank Dr. D. Edmondson (Emory University) and Dr. A. Bearden (University of California, Berkeley) for assistance with EPR studies. The National Institutes of Health provided support for these studies.

REFERENCES:

Ahn, N, Klinman, JP (1983) Biochem. 22: 3096.
Anson, FC, Collins, TJ, Richmond, TG, Santarsiero, BD, Toth,JE, Treco, BGRT (1987). J Am Chem Soc 109: 2974.
Ash, DE, Papadopoulos, NJ, Colombo, G, Villafranca, JJ, (1984). J Biol Chem 259: 3395.
Bossard, M, Klinman,J.(1986). J Biol Chem 261: 16421.
Brenner, M, Klinman,JP (1987). Fed Proc 46: 2262.
Cleland, WW (1982). CRC Crit Rev Biochem 13: 385.
Colombo, G, Papadopoulos, N, Ash, DE, Villafranca, JJ (1987). Arch Biochem Biophys 252: 71.
Colombo, G, Rajashekar, B, Giedroc, DP, Villafranca, JJ (1984). Biochem 23: 3590.
Davis, KJ, Cornish, A, Higgins, JI (1987). J Gen Microbiol 133: 291.
Eipper, BA, Mains, RE, Glembotski, CC (1983). Proc Natl Acad Sci USA 80: 5144.
Farver, O, Pecht, I (1981). Metal Ions in Biology, Vol 3, Spiro, TG, Ed, Wiley, New York: pp 151-192.
Fee, J (1981). Metal Ion Activation of Dioxygen, Spiro, TG, Ed, Wiley, New York: pp 209-238.
Fitzpatrick, PF, Flory, DR, Villafranca, JJ (1985). Biochem 24: 2108.
Fitzpatrick, PF, Harpel, MR, Villafranca, JJ (1986). Arch Biochem Biophys 249: 70.
Gray, HG, Solomon, EI (1981). Metal Ions in Biology, Vol 3, Spiro, TG, Ed, Wiley, New York: pp 1-39.
Hasnain, SS, Diakun, GP, Knowles, PF, Binsted, N, Garner, CD, Blackburn, NJ (1984). Biochem J 221: 545.
Karlin, KD, Cruse, RW, Gultneh, Y (1987). J Chem Soc Chem Commun 599.
Kaufman, S, Friedman, S (1965). Pharmacol Rev 17:

71.
Klinman, JP, Krueger, M, Brenner, M, Edmondson, D (1984). J Biol Chem 259: 3399.
Knowles, PF, Yadav, KDS (1984). Copper Proteins and Copper Enzymes, Vol 2, Lontie, R, Ed, CRC Boca Raton, FL, pp 103-130.
Kosman, DJ, (1984). Copper Proteins and Copper Enzymes, Vol 2, Lontie, Ed, CRC, Boca Raton, FL, pp 1-26.
Kruse, LI, De Wolf, Jr., WE, Chambers, PA, Goodhart, PJ, (1986). Biochem 25: 7271.
Ljones, T. Flatmark, T. and Skotland, T., FEBS Letts. 92, 81 (1978).
Lobenstein-Verbeek, CL, Jogejan, JA, Duine, JA, (1984). FEBS Letts 107: 305.
May, SW, Phillips, RS Mueller, PW, Herman, HH (1981). J Biol Chem 256: 8470.
Miller, SM, Klinman, JP (1983). Biochem 22: 3091.
Miller, SM, Klinman, JP (1985). Biochem 24: 2114.
Murray, C, Klinman, JP (1985).Fed Proc 44: 8173.
Pember, SO, Villafranca, JJ, Benkovic, SJ (1986). Biochem 25: 6611.
Reinhammar, B (1979). Adv Inorg Biochem 1: 91.
Rosenberg, RC, Lovenberg, W (1980). Essays in Neurochem and Neuropharm, Vol 4, Youdim, MBH, Lovenberg, W, Sharman, DF, Lagnado, JR, Wiley, New York: pp 163-209.
Scott, R, De Wolf,Jr., WE, Kruse, LI, personal communication.
Skotland, T, Petersson, L, Blackstrom, T, Ljones, T Flatmark, T (1980). Eur J Biochem 103: 5.
Solomon, EI (1983). Copper Coordination Chemistry: Biochemical and Inorganic Perspectives, Karlin, KD, Zubieta, J, (eds), Adenine, Guilderland, New York: pp 1-22.
Stewart, LC, Klinman, JP (1987). Biochem 26: 5302.
Tonge, GM, Harrison, DEF, Higgins, IJ (1977). Biochem J 161: 333.
Valentine, JS, Burstyn, JN, Margerum, LD. Proceeding of the 5th IUCCP Symposium on Oxygen Complexes and Oxygen Activation by Metal Complexes, in press.
Valentine, JS, Mota de Freitas, D (1985). J Chem Educ 62: 992.
Villafranca, JJ, (1981). Metal Ions in Biology, Vol 3, Spiro, TG, Ed Wiley, New York, pp 263-290.

Walker, GA, Kon, H, Lovenberg, W (1977). Biochim Biophys Acta 482: 309.
Weast, RC (1971). (ed) CRC Handbook of Chemistry and Physics, 51st Ed, The Chemical Rubber Co, Cleveland, OH.
Wimalasena, K, May, SW (1987). J Am Chem Soc 109: 4036.

DISCUSSION

PECHT: What is the source of the rather large isotope effect on the activity of the enzyme?

KLINMAN: Hydrogen abstraction from substrate in the hydroxylation step.

BLACKBURN: We have two results from our lab which I think are relevant to the data you presented. (1) Our most recent EXAFS analysis of oxidized 8 Cu dopamine beta hydroxylase indicate the presence of 2-3 imidazoles and a heavy atom scatter at about 2.6 A such as from S or Cl. This suggests a nice mechanism for reduction whereby an axial S atom comes into the coordination sphere upon reduction. (2) Potentiometric titrations using a Cu^{2+} sensitive electrode provide unmabiguous evidence for 2 unequivalent Cu(II) series per subunit with pK values of 8.9 and 7.3 respectively.

KLINMAN: (1) Your ability to detect sulfur in the oxidized form of the enzyme suggests that the ligand rearrangement on enzyme reduction, observed by Scott and co-workers, may not be as extreme as previously thought. As I described, our kinetic data do support the kinetic competence of enzyme in the reduced Cu(I) form.

(2) Your titrimetric results are quite interesting and similar to previous work of Ljones. Although our enzyme reconstitution studies were fit by a statistical binding of copper, this may represent differences in binding between oxidized and reduced copper. In our activity reconstitution studies, ascorbate levels were high and copper should have been completely reduced.

MCLENDON: In light of Dr. Blackburn's observation that the two Cu(II) centers bind differently to apo-dopamine beta hydroxylase (pK_1, 8.9, pK_2, 7.3), do both coppers reduce at the same potential? If both Cu(I) centers bind with the same differential affinity to the Cu(II) centers, a difference in redox potential of about 70 mV between the two sites would be expected.

KLINMAN: At the moment, we only know the redox potential for enzyme with one copper. Determination of the redox potential of the second copper should be an experiment of high priority!

BALLOU: Do you envisage the reduction of the enzyme-product complex as aiding the release of the oxygenated compound from its coordination to the copper?

KLINMAN: From structure/reactivity correlations we believe that product release from an inner sphere complex (rather than a conformational change) controls regeneration of free enzyme. It seems reasonable that reduction of Cu(II) to Cu(I) would facilitate this process.

GROVES: The mechanism you have suggested requires assistance of the CuO-OH bond homolysis by the benzylic C-H bond which is to be broken. The low polarizability of C-H could make such a process unlikely. Perhaps it would suffice to use an electron from the aromatic ring in the following way:

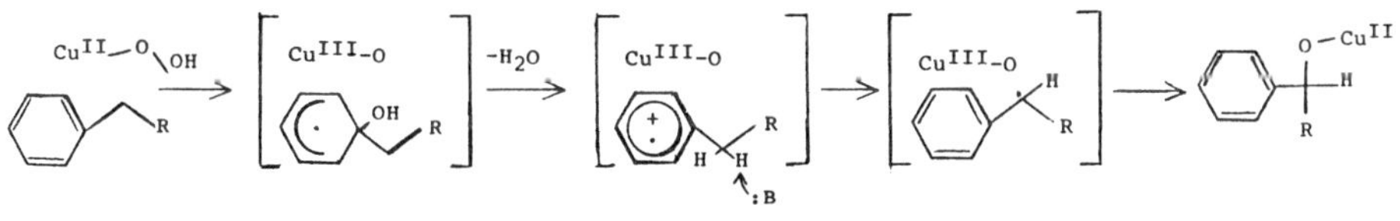

Thus, hydroxyl radical transfer to the aromatic ring and subsequent dehydration would form an aromatic cation radical. The deprotonation of such species is known to have a large isotope effect if the pKa of the base is appropriate.

KLINMAN: A mechanism involving one electron oxidation prior to C-H bond cleavage seems very unlikely since we see a linear free energy correlation for a range of ring substituted phenylethylamines. In particular the p-CF_3 substituent appears to react by the same mechanism as the catechol substrate dopamine. Although I haven't thought about your "ipso" mechanism, it appears to lead (by an additional elimination mechanism) to a radical cation prior to the C-H bond cleavage and, therefore, seems unlikely for the above stated observations. A second feature of your mechanism is the requirement for <u>two</u> protons in the hydroxylation step. At the present time we only have evidence for a single proton in this step.

Session IV: Copper-Proteins, continued

Oxidases and Related Redox Systems, pages 251–267

SOLVENT EXCHANGEABLE PROTONS AND THE ACTIVATION OF MOLECULAR OXYGEN: THE GALACTOSE OXIDASE REACTION

Daniel J. Kosman and James J. Driscoll

Department of Biochemistry, State University of New York at Buffalo, Buffalo, New York 14214

The reduction of O_2 to H_2O_2 requires two protons as well as two electrons. Thus, activation of dioxygen reasonably may involve either general or specific acid catalysis. Consequently, the reduction of O_2 to H_2O_2 could exhibit a kinetic solvent isotope effect (KSIE). The reaction catalyzed by the mononuclear Cu(II) enzyme, galactose oxidase does exhibit a KSIE (+1.55). The pL-rate profile exhibits an alkaline shift in D_2O which can be attributed to the differential partitioning of H^+ versus D^+ between bulk water and a metal-bound H_2O (ΔpK_a = +0.19). A variety of spectral evidence places an equatorial, Cu(II)-liganded water molecule at the active site of galactose oxidase. The analysis of the KSIE data is detailed and the potential generality of the function of such metal-bound H_2O at other type 2 Cu(II) sites is discussed.

INTRODUCTION

Galactose oxidase, hereafter referred to as (GO) is a copper-containing enzyme that catalyzes the conversion of a primary alcohol and molecular oxygen to an aldehyde and hydrogen peroxide (Kosman et al., 1974; Ettinger and Kosman, 1981). GO is the only known mononuclear Cu(II) protein that catalyzes a two-electron redox reaction involving molecular oxygen without the involvement of another metal ion or organic cofactor. This feature makes GO an experimentally attractive choice for the study of the structure and function of Cu(II)-active sites involved in substrate level oxidation.

One of the most fundamental but least well-characterized aspects of metal-dependent oxidation is the coupling of the requisite H^+ and electron transfers to O_2. Molecular oxygen is a better oxidant in the presence of protons due to the increase in its basicity as it is progressively reduced in one-electron steps (Dolphin, 1985).

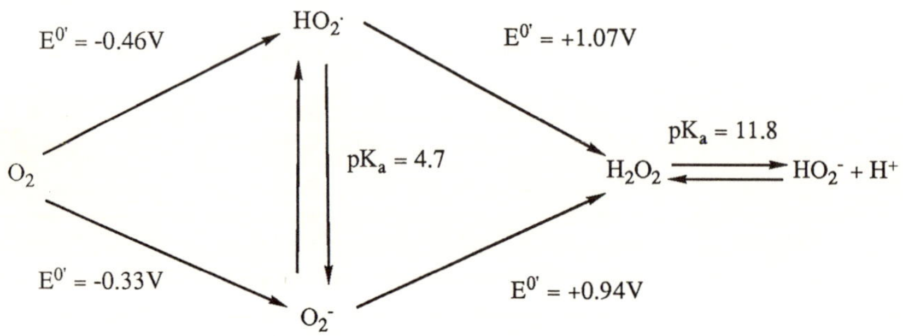

Thus, enzyme catalysis of these electron transfers can be effected by coupled H^+ transfers from suitable donor groups. A donor group serving this function should be evident as a pH dependence of the enzymic reaction in those instances for which the e^- and H^+ transfers are (partially) rate limiting. In addition, such protic species should be in rapid exchange with bulk solvent and thus be manifest as a kinetic solvent isotope effect on reactions run in isotopic waters (Schowen and Schowen, 1982).

GO exhibits a bell-shaped pH dependence for V_{max}. The ascending limb of the pH-rate profile has been assigned to a histidine imidazole (pK_a = 6.3) which may provide base-catalysis of abstraction of a substrate carbinol proton (Kwiatkowski et al., 1977). The origin of the descending, alkaline limb has not been elucidated. However, $H_2{}^{17}O$ ESR measurements show H_2O to coordinate equatorially to the Cu(II) (Melnyk, 1979). This H_2O is displaced by CN^-, known also to coordinate equatorially (Bereman and Kosman, 1977; Marwedel et al., 1981), again via direct ESR observation. In addition, CN^-, kinetically, is a non-competitive inhibitor with respect to alcohol substrate, suggesting a role for this coordination site in turnover (Winkler et al., 1981; Ettinger and Kosman, 1981). The alcohol substrate does not bind at this site (Marwedel and Kurland, 1981). Metal bound solvent H_2O may be a feature common to type 2 Cu(II) sites.

In addition to GO, plasma amine oxidase (Barker et al., 1979) and Cu,Zn superoxide dismutase (Boden et al., 1979; Tainer et al., 1983) appear to have such species as does the type 2 Cu(II) in fungal laccase (Deinum and Vanngard, 1975). Kinetic evidence for a protic (conjugate acid) group whose function could be to protonate an oxygen reduction intermediate is available for the *Rhus vernicifera* laccase (Koudelka et al., 1985), plasma amine oxidase (Farnum et al., 1986), and dopamine β-monooxygenase reactions (Ahn and Klinman, 1983). In addition, the mechanism suggested for superoxide dismutation catalyzed by Cu,Zn SOD is suggested to involve proton transfer from solvent H_2O in the O_2^- *reduction* half-reaction (Tainer et al., 1983). Lastly, general acid catalysis of acetol (monohydroxyacetone) oxidation by O_2 catalyzed by Cu(II)-tetrapyridine has been suggested, based on the pH dependence of the reaction and measured KSIE (Driscoll and Kosman, 1987). Thus, there is accumulating evidence that transition metal ion-bound water can act as an effective Bronsted acid and that general acid catalysis of O_2 reduction does occur. The question whether or not these are characteristics of the H_2O bound at the active site Cu(II) in GO and of the enzymic reaction mechanism is the focus of the experiments discussed herein.

MATERIALS AND METHODS

Galactose oxidase was isolated and purified from cultures of the fungus *Dactylium dendroides* (Tressel and Kosman, 1980). Initial velocity measurements were obtained under air (0.24 mM oxygen). A biological oxygen monitor (YSI 53), equipped with a Clark-type electrode, was used. Indicated experimental pD values of D_2O-containing buffers were calculated by adding 0.40 to the pH meter reading (Glasoe and Long, 1960). Initial velocity data were handled by computer fitting to the Michaelis-Menten equation using the computer program MM-FIT. Proton inventory, slope, and intercept data were analyzed using the linear regression fitting program POLYFIT. The pL-rate data were computer fit to individual rate laws using the computer program PENNZYME, which optimizes kinetic parameters through a two-stage nonlinear weighted least-squares minimization (Kohn et al., 1979).

RESULTS

Figure 1 displays linear transformations of kinetic data fit initially to the Michaelis-Menten equation which were obtained for the GO catalyzed oxidation of β-methylgalactopyranoside. Substitution of D_2O for H_2O altered V_{max} and V_{max}/K_m while K_m for the alcohol substrate remained constant within experimental error. Figure 1 (inset) shows that the ratio of the initial velocities in either solvent was constant at all substrate concentrations examined, consistent with the insensitivity of K_{π} (alcohol) to isotopic substitution. Thus, the kinetic isotope effect observed was on the rate of oxidation of β-methylgalactopyranoside by GO and apparently not an effect of D_2O on enzyme conformation or stability.

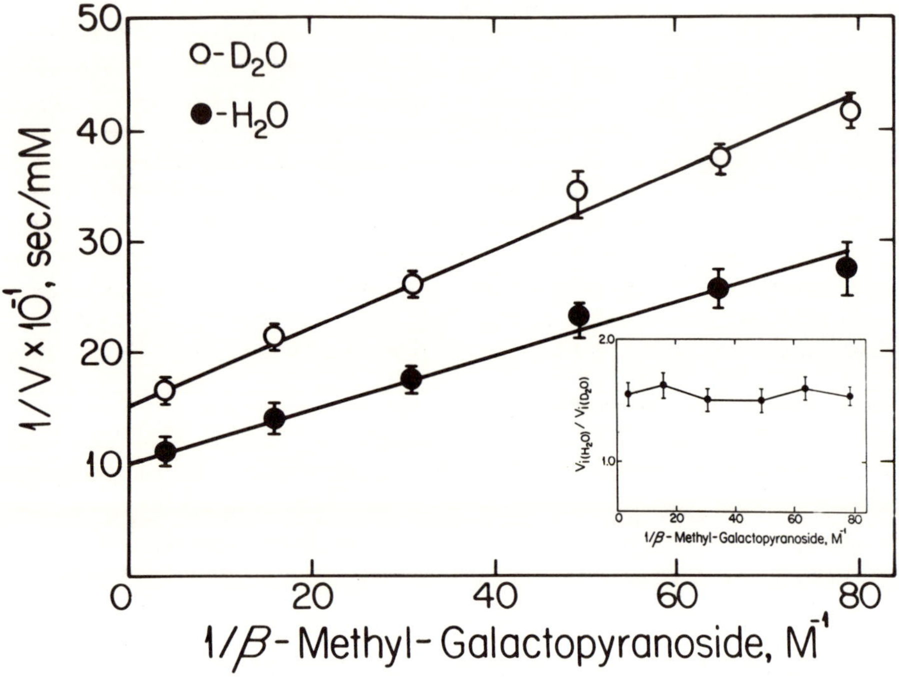

Figure 1: Lineweaver-Burk double-reciprocal plot for the GO-catalyzed oxidation of methyl β-galactopyranoside. Assay buffers were used at either pH 6.70 or pD 7.10 at 0.24 mM oxygen. Inset displays the ratio of the rates in either solvent plotted vs. reciprocal substrate concentration.

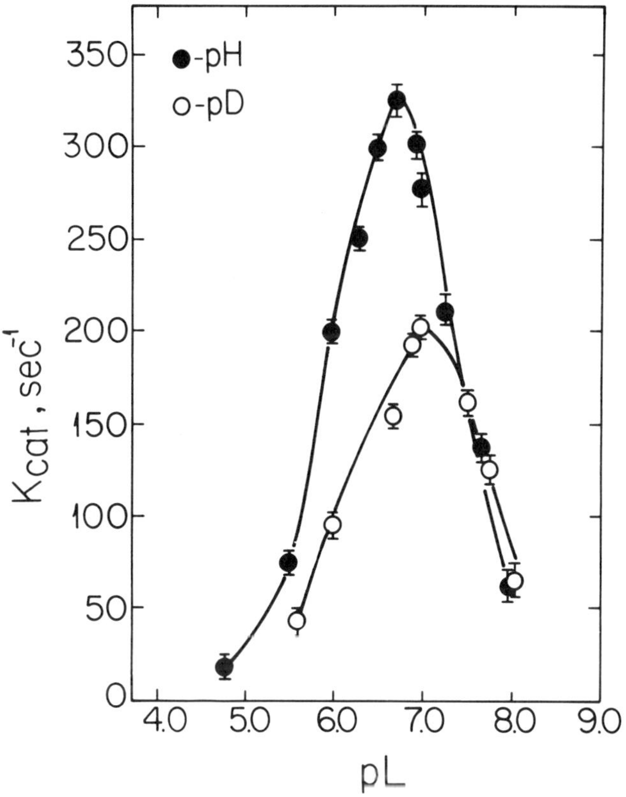

Figure 2: pL-k_{cat} profile for the GO-catalyzed oxidation of methy β-galactopyranoside. Data as in Figure 1 was fit to Equation 1.

Figure 2 displays pL-rate profiles for the GO catalyzed oxidation of β-methylgalactopyranoside. Both curves were bell-shaped with the data points fitting well to Equation 1.

$$k_{cat,\ obs} = k_{cat,\ max}/\{1 + ([H^+]/K_{es1}) + (K_{es2}/[H^+])\} \qquad (1)$$

The term $k_{cat,max}$ is the pH-independent value for the turnover rate of the enzyme, while K_{es1} and K_{es2} represent the dissociation constants for the prototropic groups which modulate the observed k_{cat} values. The pK_{es} values associated with the enzyme central complex (E·ALC·O_2) were pK_{es1} = 6.30±0.05 and pK_{es2} = 7.16±0.05. When D_2O was substituted for H_2O as solvent, a shift in these pK_{es} values (ΔpK_{es}) as well as a decrease in the overall rate of reaction (KSIE) was observed. The computer fit yielded the following dissociation constants: pK_{es1} = 6.80±0.05 and pK_{es2} = 7.35±0.06. Thus, ΔpK_{es1} = +0.50 and ΔpK_{es2} = +0.19.

A proton inventory was conducted in a series of mixed isotopic waters to characterize the protic events responsible for the observed KSIE. The magnitude of the KSIE was 1.55±0.02 (at equivalent pL) and as Figure 3 shows, the rate data were best fit to a linear equation for which n = 1. The linear shape of the k_{cat} versus mole fraction of deuterium plot indicated the existence of a single solvent-derived proton in a rate-determining step(s) of the reaction. That is, the isotope effect seen is generated by a single such hydrogenic site associated with the transition state of the reaction (Schowen, 1972).

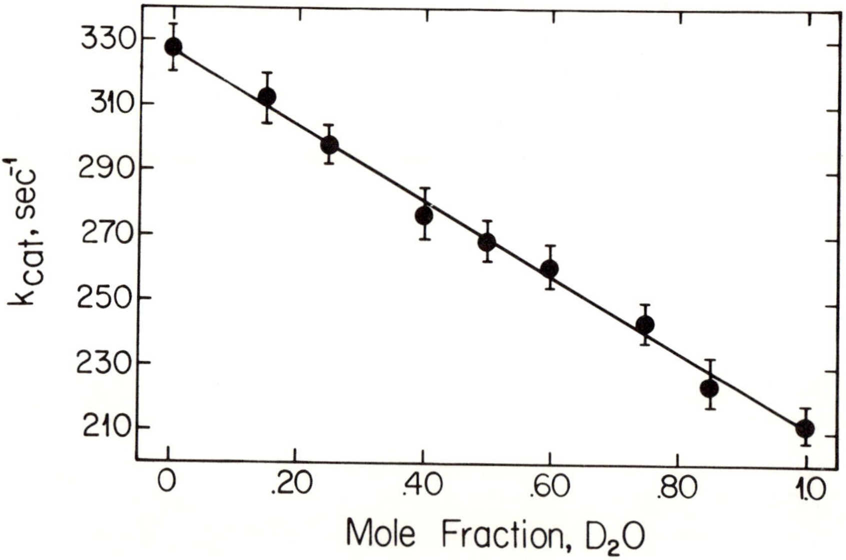

Figure 3: Proton inventory data for the GO-catalyzed oxidation of methyl β-galactopyranoside performed at equivalent pL (pH 6.70, pD 7.10). The simulated fit was determined by using the polynomial regression analysis program POLYFIT.

The pH dependence of CN^- inhibition was examined also to determine whether the H_2O molecule bound at the Cu(II) (Melnyk, 1979) and displaced by CN^- (Melnyk, 1979; Marwedel et al., 1981) might deprotonate at neutral pH during catalytic turnover. Figure 4 illustrates the effect of pH and solvent composition on CN^- inhibition of the GO reaction. Apparent K_i values were determined over the pH range 5.0 to 8.9 and then fit to Equation 2.

$$K_{iapp} = K_i'(1 + K_{ai}/[H^+])/(1 + \beta K_{ai}/[H^+]) \qquad (2)$$

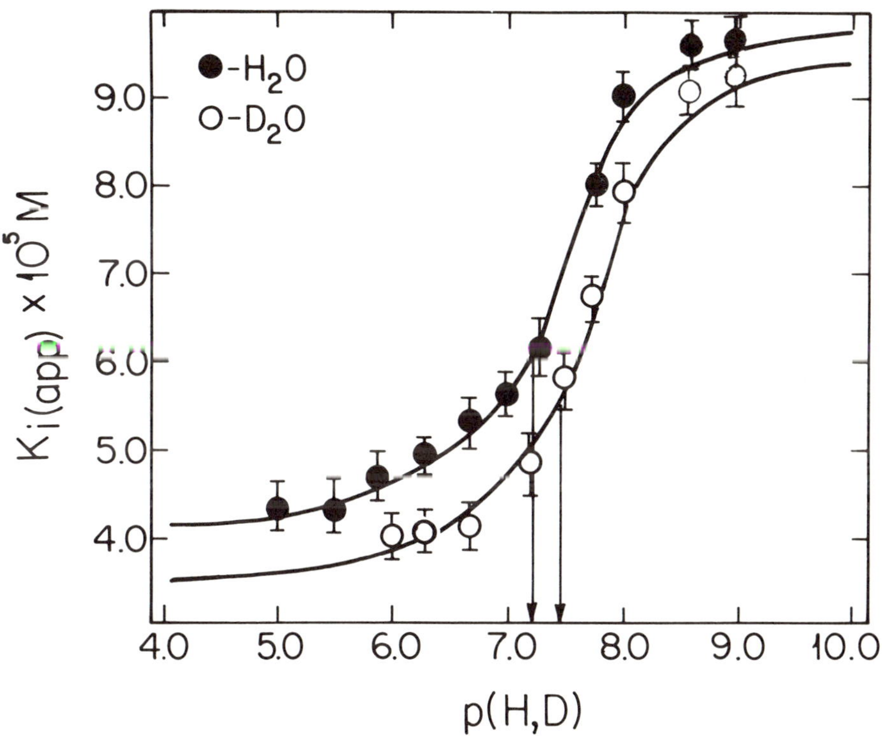

Figure 4: Apparent K_i values for the inhibition of the GO reaction by CN^- plotted as a function of pH or pD. Data were fit to Equation 2; the values for the group ionizations contolling K_i are indicated.

Equation 2 describes the pH-dependent behavior of the apparent CN^- inhibition constant (K_{iapp}). The term K_i' represents the pH-independent value for this inhibition constant, while K_{ai} (fitted pK value, 7.17±0.05) represents the group dissociation constant which modulates the pH-dependent behavior of K_{iapp}. The symbol β (fitted value, 0.12) is a solvent-independent factor which describes the limiting value for K_{iapp} as $[H^+]$--> 0. This limit was demonstrably not due to a term representing the ionization of HCN. Although the pK_a of HCN is 9.3, it is assumed that CN^-, not HCN, is bound to the metal center. This would indicate that K_i should increase with decreasing pH: it doesn't. Possibly, CN^- binding results in endogenous ligand displacement and an associated ligand protonation. Figure 4 also shows the results of experiments in which the effect of D_2O on the inhibition by CN^- was determined. The pK_a value of the observed group ionization was 7.34±0.06, indicating a shift in pK of 0.17 pH units. Note that the pK_a value and the shift in pK_a seen in the CN^- inhibition study were extremely similar to the values determined for the descending limb of the pL-rate profile.

DISCUSSION

A KSIE can yield two types of information: the chemical nature of the exchangeable site(s) whose ionization modulates enzymic activity and the nature of the transition state (in relationship to the reactant state) in which the associated proton transfer occurs. The former derives from consideration of the *shift* in pKa observed in D_2O, the latter from consideration of the *magnitude* of the KSIE itself. However, both types of information are related to the *fractionation factors* for the reactant and transition states, denoted ϕ(RS) and ϕ(TS), respectively.

The ratio of deuterium to protium binding at a particular hydrogenic site will depend upon the binding of hydrogen at that site relative to the binding of hydrogen in bulk water. If hydrogen binding at the exchangeable site of the protic group is weaker relative to bulk water, protium will accumulate at that site. If hydrogen binding is tighter relative to bulk water, deuterium will accumulate at the exchangeable site (Schowen, 1972). A measure of the deuterium preference (and, therefore, the tightness of hydrogen binding) at an exchangeable site is given by the isotopic fractionation factor, ϕ, which represents the deuterium-to-protium ratio at a given site relative to bulk water. Therefore, it is equivalent to the equilibrium constant for an exchange reaction (see below). When ϕ = 1, binding to the hydrogenic site in question is equal to that in bulk water. When ϕ > 1, binding is tighter at the exchangeable site than in bulk water resulting in deuterium accumulation at these sites. When ϕ < 1, binding to the exchangeable site is weaker than in bulk water and, therefore, protium accumulates.

The following scheme illustrates the relationship of ϕ to the kinetic constants for the reaction.

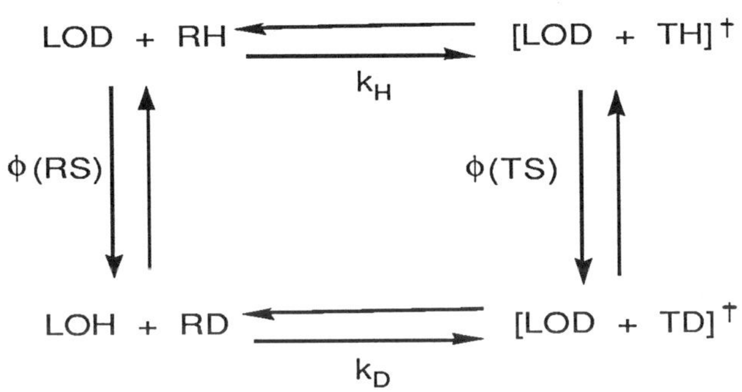

The reactant-state, RS, with a single exchangeable site (RH) yields a transition-state, TS (TH), characterized by rate constant k_H. In D_2O, RH would exhibit an equilibrium exchange constant of ϕ(RS); the transition-state would be associated with rate constant k_D. TH would exhibit an equilibrium exchange constant of ϕ(TS). Since the system is a closed thermodynamic cycle, k_H ϕ(TS) = k_D ϕ(RS) and k_H/k_D = ϕ(RS)/ϕ(TS). The KSIE is, therefore, determined by the ratio of the reactant state and transition state fractionation factors.

The relative magnitudes of ϕ(RS) and ϕ(TS) serve to characterize the nature of the transition state. In a simple way, the discussion above leads to the conclusion that if, for example, ϕ(RS) > ϕ(TS), protium is more tightly bound in the transition state relative to the reactant state. However, since the KSIE, as k_H/k_D, indicates only the ratio of these parameters, their absolute relationship can only be deduced if one of them can be determined. In fact, ϕ(RS) can be assigned based on the *shift* in pK_a observed in D_2O, or ΔpK_a.

Ionization of an acid in a mixture of H_2O and D_2O where L = H or D and LX represents a given acid, is described as:

$$LX + LOL \rightleftharpoons L_3O^+ + X^- \quad \text{with} \quad K_a = [X^-][L_3O^+]/[LX]$$

Since protium accumulates where binding is weaker, e.g., H_3O^+, there will be an accumulation of H_3O^+relative to D_3O^+ for common biologic acids. Thus, the ionization constant (pK_a) for a given acid-base site will increase in a D_2O-containing solvent (Kresge, 1977). The relative tightness of protium (or deuterium) binding at the acid-base site will be reflected in the magnitude of this decrease (or shift) in ionization constant(ΔpK_a). Equation 3 can then be used to calculate the reactant state fractionation factor, ϕ(RH) (Schowen and Schowen, 1982). Note that y measures the polyproticity of the ionizing acid, i.e., y = 1 for acetic acid, 2 for a Cu-bound water molecule, and 3 for H_3O^+. The standard fractionation factor for H_2O is 1.0 and that for H_3O^+ is 0.69.

$$\Delta pK_a = \log \frac{(\phi RH_y)^y \ (\phi H_2O)^2}{(\phi RH_{y-1})^{y-1} \ (\phi H_3O^+)^3} \qquad (3)$$

Table I (Schowen and Schowen, 1982) lists ΔpK_a values for common heteroatom organic acids as well as some hydrated metal atoms [$(H_2O)_n$, n = 6-8]. For a monoprotic acid, ΔpK_a = 0.5 (a "standard" shift) indicates ϕ(RS) = 1. Sulfhydryl acids exhibit smaller shifts; they are weaker hydrogenic sites than bulk water and thus accumulate protium. The effect of polyproticity is readily apparent in the range of shifts exhibited by the three ionizations of phosphoric acid. Similarly, the shifts for hydrated metal ions can be compared only with caution because they, too, may differ with respect to the number of coordinated water molecules. However, in general, tighter protium binding is observed at hydrogenic sites coordinated to metal ions (Schowen and Schowen, 1982). This may arise from a lack of transfer of unit positive charge to the water ligand. Transfer of unit positive charge from a cation to H_2O should induce a fractionation factor equivalent to that exhibited by H_3O^+, e.g., 0.69. With this in mind, it is reasonable to estimate the fractionation as $\phi(RH) = (\phi H_3O^+)^{\delta} = 0.69^{\delta}$ where δ is the amount o charge transferred to the metal bound H_2O (Schowen and Schowen, 1982). The fractionation factor thus calculated can then be compared to the ratio $\phi(RH)/\phi(RH_{y-1})^{y-1}$ determined from Equation 3 placing limits on the possible fractionation factor values for the hydrogenic site contributing to the KSIE.

Evaluation of the pL data obtained for the GO reaction generally supports the hypothesis that the Cu(II)-bound H_2O indicated by the spectral studies is the origin of the decending limb of the pH-rate profile ($pK_a = 7.16$), does exhibit a ΔpK_a (+0.19) and is responsible for the KSIE. First, the pK_a value itself and the shift are within the range exhibited by hydrated Cu(II) (Tables I and II). The estimated values for the fractionation factor terms in Equation 3 for $\delta = +1$, for example, 0.48/0.94 are reasonable, giving the conjugate acid a lyonium-like character and a value to the conjugate base - a copper hydroxo species - consistent with the coordination of HO^- to Cu^{2+}. The Cu(II) in GO is coordinated to two histidine imidazoles and one other ligand; for unit charge to be transferred to the H_2O is a reasonable assumption.

Second, the displacement by CN^- of this H_2O exhibits a pH and solvent dependence which is essentially identical to that exhibited by k_{cat}. The simplest interpretation of these latter data is that both kinetic phenomena are associated with the ionization of the same group.

Table I

Solvent Isotope Effects on the Ionization of Weak Acids

	pK_a^H	$\Delta pK_a = pK_a^D - pK_a^H$
Carboxylic and Phenolic Acids		
Acetic	4.76	0.51
Benzoic	4.20	0.49
p-Nitrophenol	6.98	0.50
Polyprotic Oxygen Acids - Phosphoric		
pK_1	2.15	0.21
pK_2	7.20	0.53
pK_3	12.15	0.67
Sulfhydryl Acids		
Thioacetic	3.20	0.30
Mercaptoethanol	9.61	0.35
p-Nitrothiophenol	4.50	0.31
Hydrated Metal Ions		
Fe^{3+}	2.54	0.29
Cu^{2+}	7.71	0.49
Gd^{3+}	8.20	0.14
La^{3+}	10.04	0.31

Table II

Experimental Values of pKa and Δ pKa for Model Cu(II) Complexes

	pKa	Δ pKa
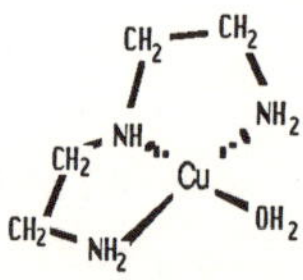 Cu(II)-DET	8.42	+0.38
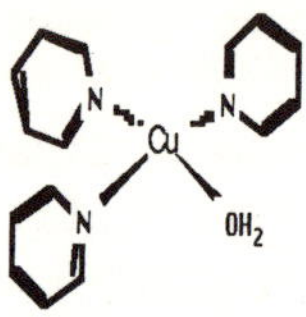 $Cu(II)-Pyr_3$	7.11	+0.34
$pKes_2$ Galactose Oxidase	7.16	+0.19

The questions of whether this group is responsible for the KSIE (not just the shift, ΔpK_{es2}) or not, and that of its function in catalysis are not answered by this analysis alone, however. Simulations of the kinetic data shown in Figure 1 were performed utilizing several different boundary conditions. The data were only consistent with an assignment of the KSIE (+1.55) to pK_{es2}, i.e., to the Cu-bound H_2O. With this assignment, and the value for ϕ(RH), a value for ϕ(TS) was calculated. These values and conclusions are summarized on the following page.

Reactant and Transition State Fractionation Factors and Structures

Theory

$k_H/k_D = \phi(RS)/\phi(TS) = +1.55$ (Experimental Value)

Assignment

$\phi(RS)$ (for Cu^{2+}-OH_2) = +0.69

Calculation

$\phi(TS) = +0.46$, i.e., $\phi(RS) > \phi(TS)$

Conclusion

H^+ is bound more tightly in the transition state than in the reactant state. For example:

Potentially, the relatively small KSIE seen in the GO reaction may reflect a transition state in which the transferred proton is assymetrically displaced towards the donor atom rather than towards the acceptor and/or with this proton's line-of-flight being bent, as in the scheme above (Kresge, 1977). The data are consistent with a reaction pathway involving a transfer of one electron to O_2 in a transition state which is stabilized by a hydrogen bond from the Cu^{2+}-OH_2, i.e., a rate-determining electron transfer that is catalyzed by partial (asymmetric) proton transfer. As noted above, protonation of O_2 makes O_2 a better oxidant by 6.4 kcal/mol vs. $O_2 + e^- \rightarrow O_2^-$ only. This proposal can be illustrated within the context of the elements at the GO active site in the central complex as shown on the following page.

Proposal for the Proton Transfers in the Galactose Oxidase Reaction

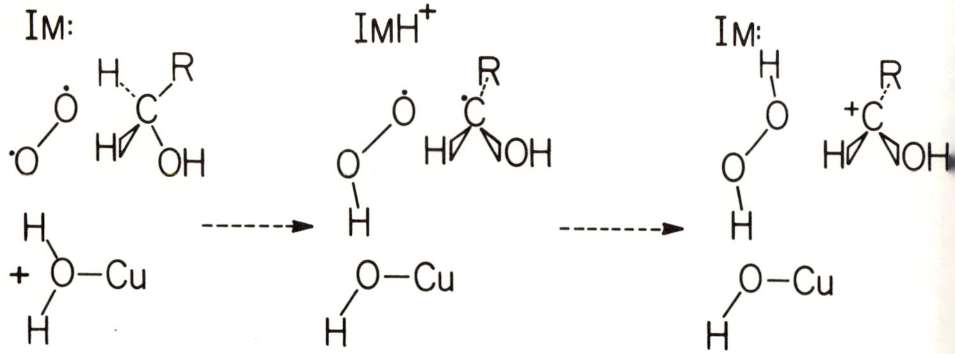

ACKNOWLEDGEMENTS

The authors wish to express their appreciation to Dr. Eric Dahmer for his assistance in the use of PENNZYME and Angela Grys for the preparation of this manuscript. This research was supported in part by NSF Grant DMB 8417792.

REFERENCES

Ahn, N., Klinman, J.P. (1983). Mechanism of modulation of dopamine β-monooxygenase by pH and fumarate as deduced from initial rate and primary deuterium isotope effect studies. *Biochemistry* **22**:3096-3106.

Bereman, R.D., Kosman, D.J. (1977). Stereoelectronic properties of metalloenzymes. 5. Identification and assignment of ligand hyperfine splittings in the electron spin resonance spectrum of galactose oxidase. *J. Am. Chem. Soc.* **99**:7322-7325.

Barker, R., Boden, N., Cayley, G., Charlton, S.C., Henson, R., Holmes, M.C., Kelly, I.D., Knowles, P.F. (1979). Properties of cupric ions in benzylamine oxidase from pig plasma as studied by magnetic-resonance and kinetic methods. *Biochem. J.* **177**:289-302.

Boden, N., Holmes, M.C., Knowles, P.F. (1979). Properties of the cupric sites in bovine superoxide dismutase studied by nuclear-magnetic-relaxation measurements. *Biochem. J.* **177**:303-309.

Deinum, J.S., Vanngard, T. (1975). ^{17}O Hyperfine interaction in the EPR spectrum of fungal laccase A. *Febs. Lett.* **58**:62-65.

Dolphin, D. (1985). Cytochrome P_{450}: substrate and prosthetic-group free radicals generated during the enzymatic cycle. *Phil. Trans. Roy. Soc. Lond.* **B311**:579-591.

Driscoll, J., Kosman, D.J. (1987). Proton transfer in the copper(II)-tetrapyridine catalyzed oxidation of acetol(monohydroxyacetone). *J. Am. Chem. Soc.* **109**:1765-1772.

Ettinger, M.J., Kosman, D.J. (1981). Chemical and catalytic properties of galactose oxidase. In Spiro, TG (ed): *Metal Ions in Biology* , Vol III, New York: Wiley-Interscience, pp 220-261.

Farnum, M., Palcic, M., Klinman, J.P. (1986). pH dependence of deuterium isotope effects and tritium exchange in the bovine plasma amine oxidase reaction: a role for single-base catalysis in amine oxidation and imine exchange. *Biochemistry* **25**:1898-1904.

Glasoe, P.K., Long, F.A. (1960). Use of glass electrodes to measure acidities in deuterium oxide. *J. Phys. Chem.* **64**:188-190.

Kohn, M.C., Menten, L., Garfinkel, D. (1979). Computer programs for enzyme kinetics. *Comp. Biomed. Res.* **12**:461-469.

Kosman, D.J., Ettinger, M.J., Weiner, R., Massaro, E.J. (1974). The molecular properties of the copper enzyme galactose oxidase. *Arch. Biochem. Biophys.* **165**:456-467.

Koudelka, G.B., Hansen, F.B., Ettinger, M.J. (1985). Solvent isotope effects and the pH dependence of laccase activity under steady-state conditions. *J. Biol. Chem.* **260**:15561-15565.

Kresge, A.J. (1977). Magnitude of primary hydrogen isotope effects. In Cleland WW, O'Leary M, Northrup, DB (eds): *Isotope Effects on Enzyme-Catalyzed Reactions* , Baltimore, MD: University Park Press, pp 37-63.

Kwiatkowski, L.D., Siconolfi, L., Weiner, R., Giordano, R.S., Bereman, R.D., Ettinger, M.J., Kosman, D.J. (1977). Histidine as an essential residue in the active site of the copper enzyme galactose oxidase. *Arch. Biochem. Biophys.* **182**:712-722.

Marwedel, B.J., Kosman, D.J., Bereman, R.D., Kurland, R.J. (1981). Magnetic resonance studies of cyanide and fluoride binding to galactose oxidase copper(II): evidence for two exogenous ligand sites. *J. Am. Chem. Soc.* **103**:2842-2847.

Marwedel, B.J., Kurland, R.J. (1981). Fluoride ion as an NMR relaxation probe of galactose oxidase-substrate binding. *Biochim. Biophys. Acta* **657**:495-506.

Melnyk, D. (1979). An investigation of the type 2 Cu(II) enzyme galactose oxidase: determination of the oxidation-reduction potential, identification of an equatorial ligand and determination of pH dependence of the stereoelectronic properties. Ph.D. Dissertation, Buffalo, NY: State University of New York at Buffalo.

Schowen, K.B., Schowen, R.L. (1982). Solvent isotope effects on enzyme systems. *Meth. Enzymol.* **87**:551-606.

Schowen, R.L. (1972). Mechanistic deductions from solvent isotope effects. *Prog. Phys. Org. Chem.* 9:275-332.

Tainer, J.A., Getzoff, E.D., Richardson, J.S., Richardson, D.C. (1983). Structure and mechanism of copper, zinc superoxide dismutase. *Science* **306**:284-287.

Tressel, P., Kosman, D.J. (1980). A simplified purification procedure for galactose oxidase. *Anal. Biochem.* **105**:150-153.

Winkler, M.E., Bereman, R.D., Kurland, R.S. (1981). Kinetic and magnetic resonance studies of substrate binding to galactose oxidase copper(II). *J. Inorg. Biochem.* **14**:223-235.

DISCUSSION

PAPA: Would your mechanism of proton exchange imply a pH dependence of the Cu center?

KOSMAN: Yes. Kinetically, the ionization of this water molecule is what we believe contributes to the alkaline, descending limb of the pH-rate profile (cf. Fig. 2). As to the pH dependence of the properties of the Cu(II) center itself, one would expect the ionization of the metal-bound water to alter the redox and spectral character of the metal. We have not measured the latter, but the spin Hamiltonian parameters of the Cu(II) do change with pH in the direction expected, i.e., as water is replaced by hydroxide (see Melnyk, 1979).

Oxidases and Related Redox Systems, pages 269–283

PREFERRED SITES AND PATHWAYS FOR ELECTRON TRANSFER IN BLUE COPPER PROTEINS

Ole Farver and Israel Pecht

Department of Chemical Immunology
The Weizmann Institute of Science
Rehovot, Israel

ABSTRACT

Long-range electron transfer reactions proceed within and between metalloproteins at relatively fast rates and with marked specificities. The blue single copper proteins are well known electron carriers with their redox center being of limited accessibility to solvent and solutes. The question of where and how electrons are transferred to and from the copper-ion have been investigated. One experimental approach developed in order to pursue these problems is that of reductively labeling several representative, yet structurally distinct blue single copper proteins; azurin, plastocyanin, and stellacyanin with chromium ions. In all three cases, a substitution inert Cr(III)-adduct is formed when the oxidized protein is reduced by $Cr(II)_{aq}$ ions. In azurin, Cr(III) binds to the Glu-91 carboxylate ~10 Å from the copper center. In both plastocyanin and stellacyanin the Cr(III) label is most probably also coordinated to carboxylate groups, present in plastocyanin, and in stellacyanin 12 Å and 6 Å, respectively, from the copper center. The salient feature emerging from examination of the three copper proteins is that a π-facilitated electron transfer (E.T.) pathway may be operative; in azurin, E.T. proceeds via an extended imidazole ring system, and in plastocyanin and stellacyanin via a weakly coupled π-system. Therefore, a case emerges for suggesting that this is the common feature of the long-distance intramolecular E.T. in this class of metalloproteins. These pathways are most probably a regulatory alternative to the E.T. site recognized at the exposed, "Northern" imidazole coordinated to copper in all these proteins.

INTRODUCTION

The blue copper proteins represent an ancient line of redox proteins that have evolved into a diversity of forms participating in biochemical processes ranging from photosynthesis to mitochondrial and bacterial electron transfer networks. With the prosthetic group - a "blue" copper redox site - having limited accessibility to solvent and solutes, several fundamental questions arise:
1) Are there specific site(s) on these proteins' surface for electron transfer to and from the copper center?
2) Which amino acid residues are involved in the electron transfer from the protein/solvent interface?
3) Do different blue single copper proteins share analogous intramolecular electron transfer pathways?

These questions gain further significance because of the accumulating evidence for effective long-range intramolecular electron transfer processes proceeding within proteins between the native redox site, i.e. the copper or iron prosthetic group and extraneously introduced redox centers (Kostic et al., 1983; Liang et al., 1986; Bechtold et al., 1986; Mayo et al., 1986).

NMR studies (Cookson et al., 1980a,b; Armstrong et al., 1986a) using paramagnetic transition metal complexes have identified electrostatic interactions with preferred locations assumed to be involved in electron transfer on the protein surfaces, just as electron spin echo studies have provided further insight into the accessibility of the Cu(II) ion in several blue proteins (Mimms et al., 1984).

In order to try and further resolve the above questions, we have developed a methodology of affinity labeling electron-transfer loci on several, structurally distinct, blue copper proteins: Pseudomonas aeruginosa azurin, spinach plastocyanin, and Rhus vernicifera stellacyanin with chromium ions (Farver and Pecht, 1981a,b; Morpurgo and Pecht, 1982; Farver et al., 1987). These Cr(III) modified proteins have then been valuable in identifying regions on their respective surfaces that react with different redox partners. These studies entailed examination of the reactivity between the physiological partners, e.g. azurin with cytochrome c-551 and Ps. aeruginosa cytochrome c oxidase (Farver et al., 1982a) and plastocyanin with the cytochrome b_6/f complex in broken chloroplast and with isolated P-700 reaction centers (Farver et al., 1982b).

RESULTS AND DISCUSSION

Reduction of the above mentioned blue copper proteins by $Cr(II)_{aq}$ yields stoichiometrically a substitution inert Cr(III)-adduct formally described by the scheme:

$$Cr(II) + P[Cu(I)] \longrightarrow Cr(III) - P[Cu(I)]$$

The salient feature of this reaction is that the coordination sphere of the Cr(II) ion in the activated complex is assumed to be preserved in the substitution inert Cr(III) reaction product. Thus, if the Cu(II) ion in a blue protein is reduced by Cr(II), a substitution inert Cr(III)-protein complex will be formed with the chromium ion attached in the very same position where the electron transfer from Cr(II) to Cu(II) takes place. By proteolytic cleavage it is possible to isolate and identify the peptide fragment to which Cr(II) is coordinated. This procedure has so far been employed for several redox-proteins and allowed identification of the labeling site for spinach plastocyanin, *Pseudomonas aeruginosa* azurin, and *Rhus* stellacyanin. In azurin, Cr(III) is most probably chelated to the side-chains of Lys-85 and Glu-91 (Farver and Pecht, 1981a). In plastocyanin the Cr-label was found to be attached to one or more of the carboxylate side chains of the peptide Asp-42, Glu-43, Asp-44, Glu-45 (Farver and Pecht, 1981b). Finally, following reductive labeling of stellacyanin with Cr(II) the Cr(III) was found to be bound to a short peptide Val-48, Asp-49 and Lys-50 and thus is most probably chelated by the side chains of the latter residues (Farver et al., 1987). Figures 1,2, and 3 show the three dimensional structures of these proteins, and the identified locations of Cr-labeling.

The shortest distance between the Cr-binding location in azurin (Fig. 1) and the Cu-center can be calculated from the crystallographic data (Adman and Jensen, 1981) and is ~ 10 Å. The peptide stretch Lys-85 to Glu-91 defines an opening into the protein's interior exposing N_δ of His-35. Interestingly, this peptide sequence is essentially conserved in all azurins sequenced so far (Ryden and Lundgreen, 1976). The two imidazole rings of His-35 and the Cu-ligating His-46 are virtually co-planar with an inter-plane distance of 3.8 Å; thus the non-coordinating imidazole of His-35 may extend the effective π-system from the copper site towards the above opening and serve as a relay for a resonance transfer of an electron from the surface of the protein to the redox center and vice versa.

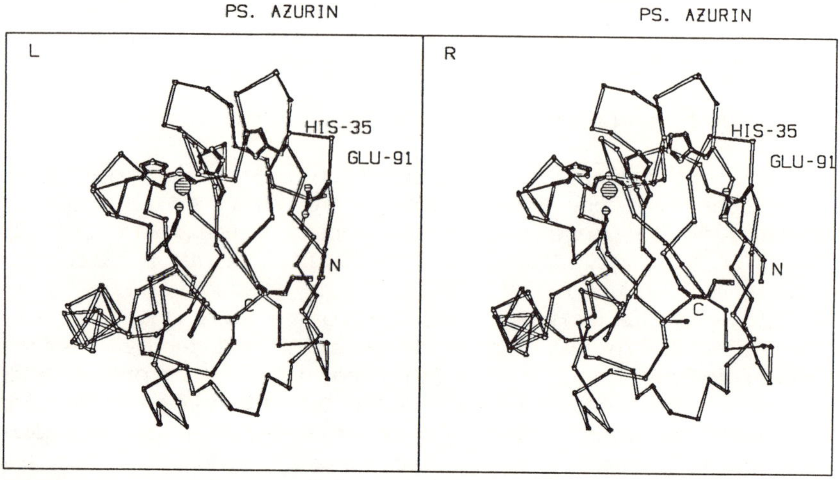

Figure 1. Stereo view pair of the molecular structure (C^{α} atom framework) of Pseudomonas aeruginosa azurin. The side chains of the four copper ligands are shown along with the side chains of His-35 and Glu-91. The coordinates were taken from the Brookhaven data file 1AZU (Adman & Jensen, 1981).

The question now arises whether this site identified by chromium affinity labeling has relevance to the physiological electron transfer function of azurin. The assumed natural partners for azurin are cytochrome c-551 and the bacterial cytochrome c oxidase. Reactions of both these proteins with native and chromium labeled azurin have been investigated. A detailed kinetic comparison between the electron transfer reactions of native and Cr(III) labeled azurin with Cyt c-551 was carried out using the temperature-jump chemical-relaxation methodology. These measurements have shown that the fast electron exchange between Az and Cyt c-551 is perturbed by the presence of Cr(III) on azurin (Farver et al., 1982a), to the extent that the exchange rate decreases from 7.0×10^{6} for native Az to $4.5 \times 10^{6} M^{-1} s^{-1}$ for the labeled one. Since the chromium label is coordinated near His-35, and the heme edge of Cyt c-551 is surrounded by positively charged lysyl side chains, an optimal juxtaposition of the two proteins in a precursor complex is probably perturbed by the presence of Cr(III) on azurin. By contrast, no difference in the rates of the cytochrome c oxidase catalyzed oxidation of Az[Cu(I)] by O_2 was detected whether azurin was in the native state or labeled with Cr(III). Therefore, one may deduce the presence of two physiologically active sites for electron transfer on azurin. One center is engaged in electron exchange with cytochrome c-551 via His-35 and another in electron transfer from Az[Cu(I)] to cytochrome c oxidase. One of the Cu-ligands, His-117, has an edge of the imidazole ring exposed to the solvent, surrounded by hydrophobic amino acid residues. This has been proposed to serve as an electron transfer site already in the structural analysis of azurin (Adman and Jensen, 1981).

Line broadening ^{1}H-NMR studies of Pseudomonas azurin have provided electron self-exchange rates (Groeneveld and Canters, 1985). It was found that the exchange rate is independent of pH in striking contrast with many other redox reactions of azurin, of particular interest the azurin/cytochrome c-551 exchange rates. The pH-independence and the activation parameters of the azurin self-exchange led to the conclusion that the electron transfer takes place via the Cu-ligating His-117. This conclusion is in full agreement with the above notion of two independent sites and pathways for electron transfer in azurin. One is the hydrophobic area around His-117 which constitutes part of the Cu-coordination sphere. The other is the hydrophilic loop around His-35 which is in close contact with the Cu-ligating imidazole of His-46. Thus one may conclude that Cyt c-551 donates electrons to the copper center of azurin from the hydrophilic site around His-35 while the electrons are passed on to cytochrome c oxidase from Cu(I) via the accessible imidazole ring of His-117 at the "Northern" end of the molecule (Fig. 1).

In plastocyanin, the Cr(III)-label was found to be coordinated to carboxylate side chains of the stretch Asp-42 to Glu-45 (Fig. 2) (Farver and Pecht, 1981b) constituting the so called "negative patch" on this protein (Guss and Freeman, 1983). These side chains are positioned next to Tyr-83, the phenol group of which protrudes into solution. The distance between the Cr-binding site (and Tyr-83) and the copper center is ~ 12 Å and the intervening area consists of conserved aromatic amino acids (Guss and Freeman, 1983). It was therefore suggested that the electron transfer take place by electronic delocalization of a weakly coupled π-electron system. It is noteworthy that this electron transfer site is identical with the binding site for positively charged transition metal complexes to reduced Pc[Cu(I)] as identified by NMR (Cookson et al., 1980a,b; Armstrong et al., 1986a). Identification of an electron transfer site and pathway in plastocyanin raises naturally again the question of its physiological relevance; i.e. is it related to the interaction of Pc with its two physiological partners in the photosynthetic system, cytochrome f and the photosystem I reaction centers, P-700. A detailed comparison of the kinetic behavior demonstrated that oxidation of the Cr(III)-labeled Pc[Cu(I)] by P-700 is considerably slower ($k = 7.10^7 M^{-1}s^{-1}$) than oxidation of the native Pc[Cu(I)] ($k = 2.10^8 M^{-1}s^{-1}$) (Farver et al., 1982b). In contrast, the reduction of Pc[Cu(II)] with the membrane bound Cyt b_6/f complex is not affected by the Cr-label ($k = 4.10^6 M^{-1}s^{-1}$) (Farver et al., 1982b). Therefore, the hydrophilic site around the negatively charged carboxylates and Tyr-83 is most likely where intermolecular electron transfer from Pc[Cu(I)] to P-700 reaction centers takes place. The imidazole ring of one of the copper ligands (His-87) is exposed to the solvent and like in azurin, surrounded by hydrophobic amino acid residues (Guss and Freeman, 1983). By analogy to azurin, it was therefore suggested that this hydrophobic area on plastocyanin serve as specific recognition and electron transfer site from the membrane bound Cyt b_6/f complex (Farver et al., 1982b). A favorable activation entropy observed for the latter reaction (Wood, 1974) also supports this notion since ordered water molecules are expected to be displaced from the interface between the two proteins upon formation of a precursor complex governed by hydrophobic interactions.

A third blue copper protein examined by this Cr(II) reductive labeling method is stellacyanin and here the coordination site was identified as Asp-49 and Lys-50 (Farver et al., 1987). Since crystallographic structure determination of stellacyanin has so far not been achieved, we have constructed a tentative computer model by energy minimization based on a combination of spectroscopic and chemical features of stellacyanin. This model is presented in Fig. 3 (Wherland et al., 1987, submitted). According to calculations from this model, the Cr(III) coordinating Asp-49 and Lys-50 in stellacyanin seem close enough to the Cu(II) center to ensure rapid electron transfer.

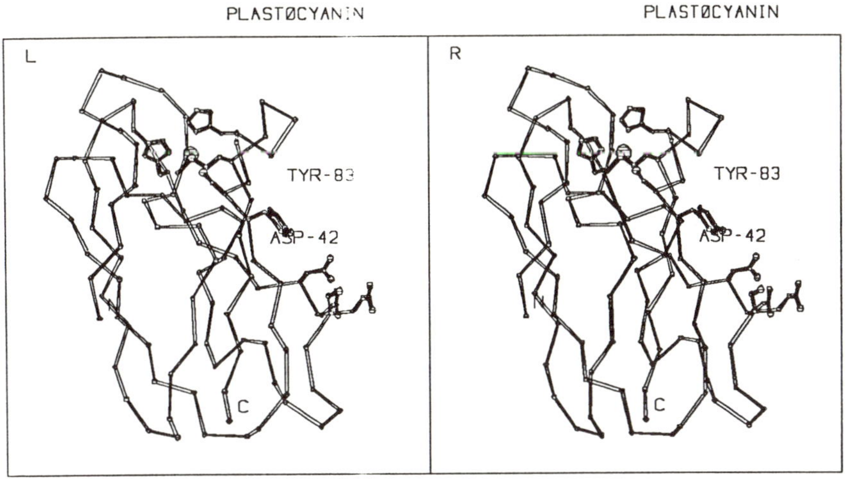

Figure 2. Stereo view pair of the C^{α} atom framework of poplar plastocyanin. The side chains of the four copper ligands are shown along with the side chains of Asp-42 to Glu-45 and Tyr-83. The coordinates were taken from the Brookhaven data file 1PCY (Guss & Freeman, 1983).

The figure also shows that stellacyanin - like plastocyanin - contains a tyrosyl residue (Tyr-85) with a solvent exposed phenol ring. Whereas this tyrosyl ring is most probably not part of the electron transfer pathway from Cr(II) to Cu(II), it may provide a part of a π-electron system available for electron transfer reactions with other proteins, as in the case assumed to be for plastocyanin. Intervening aromatic side chains were indeed observed in electron transfer pairs like cytochrome-c -cytochrome per-oxidase (Cheung et al., 1986) and in the bacterial photo-synthetic reaction centers (Deisenhofer et al., 1985) . From our proposed three dimensional model, one can estimate a 6 Å distance from this Cr(III) binding site to the copper center which is much closer than the Cr-Cu distances calculated from the x-ray crystallographic data of azurin and plastocyanin (10-12 Å, vide supra).

Since the physiological role of stellacyanin is so far unknown, no comparative studies of reactivity as those done with azurin and plastocyanin could be attempted. Computer modeling, however, has shown the existence of a solvent exposed edge of the imidazole ring of His-92 (Wherland et al., 1987) analogous to that found in the two other copper proteins. Thus, this imidazole is most likely also a potential site for electron transfer.

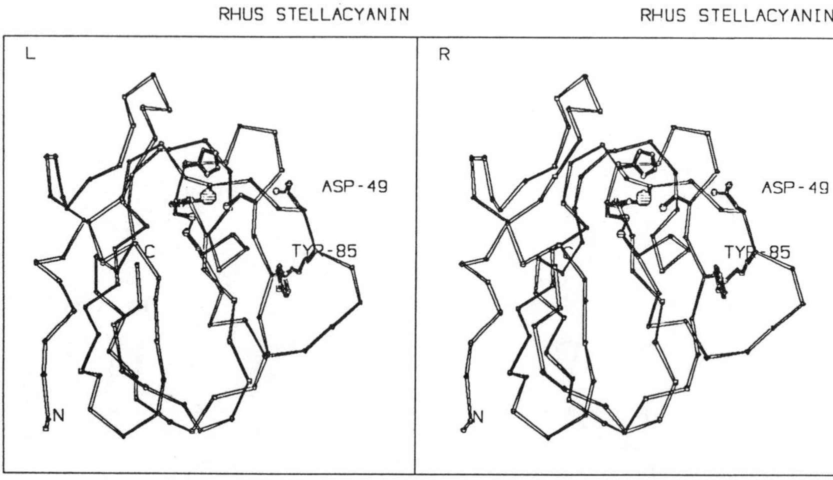

Figure 3. Stereo drawing of the proposed three dimensional C^{α} folding of Rhus vernicifera stellacyanin. The four copper ligands as well as the side chains of Asp-49 and Tyr-85 are also shown. The data were taken from the three dimensional computer model of stellacyanin (Wherland et al., 1987).

Significantly, many electron transfer reactions between transition metal complexes and stellacyanin proceed markedly faster than the corresponding reactions with azurin and plastocyanin. This could now be rationalized in terms of the more favorable distances if the area near Tyr-85 in stellacyanin is also an active site on this protein. Another noteworthy observation is that the Cr-coordinating Asp-49 in stellacyanin is homologous to Asp-42 in plastocyanin, one of the four potential chelating carboxylates for Cr(III) in the latter protein (vide supra) demonstrating a further resemblance between these two copper proteins. Considering this point, we have suggested that electron transfer from these homologous sites to the copper center proceeds via homologous pathways (Farver et al., 1987). However, in plastocyanin the immediate vicinity of Tyr-83 contains four exposed carboxylate side chains (Fig. 1) making this region negatively charged at physiological pH. These charged residues probably also provide a recognition site for its natural partner, the photosystem I reaction center (vide supra). In stellacyanin, which is an extracellular protein, only one carboxylate residue, Asp-49, is conserved out of the above four present in Pc. Moreover, the local negative charge it carries is fully compensated for by its neighboring protonated ε-amino group of Lys-50. Therefore, it does not seem to be due to any advantageous electrostatic effect that stellacyanin is specifically labeled at this particular site. Rather, the specificity for this site is most probably caused by a facile electron transfer to the Cu(II) center via the π-electron system of the Tyr-85 phenol ring and the Cys-87 sulfur atom in stellacyanin. In conclusion, a π^*-electron transfer pathway seems to exist in all three examined types of single blue copper proteins.

The existence of two distinct functional sites and pathways for electron transfer in the blue copper proteins seems to be in marked contrast to cytochrome c, the ubiquitous electron transfer protein where the redox reactions take place at the solvent accessible crevice containing the heme prosthetic group. Already Cr(III)-labeling experiments demonstrated that upon reducing the ferricytochrome c with Cr(II), the Cr(III) ions are quantitatively coordinated inside the heme crevice indicating that the redox site of this protein is directly accessible to electron transfer reagents (Grimes et al., 1974). Other experiments such as differential chemical modification, enabled comparison of the chemical reactivity of each of the 19 lysyl residues of cytochrome c in free state with that bound to either cytochrome c oxidase or to cytochrome c reductase (Rieder and Bosshard, 1980). The binding sites on cytochrome c in terms of particular lysyl residues were found indistinguishable and both redox partners seem to interact at the front of the exposed heme edge. In another study, the electron transfer rates of eight different singly modified cytochrome c derivatives with plastocyanin and stellacyanin, respectively, were determined and compared (Armstrong et al., 1986b). Again, it was found that the preferred electron exchange site is near or at the exposed heme edge, even with these non-physiological protein partners.

In summary, it can be stated that both the cytochromes and the blue single copper proteins make use of an extended π-electron system for the intramolecular electron transfer from the surface of the protein to take the metal center. But while electron transfer process in the c-type cytochromes place predominantly via one pathway, namely the exposed heme edge, in the blue proteins, where the copper site is less accessible to extraneous redox agents, electron exchange occurs from one of two preferred sites on the protein surface. Another question is of course whether there are more than the above two preferred electron transfer pathways in the blue copper proteins. This problem has been addressed by attaching $Ru(NH_3)_5^{3+/2+}$ to solvent exposed histidines of azurin and plastocyanin, respectively. With Pseudomonas azurin it is possible to attach ruthenium specifically to His-83. The intramolecular electron transfer rate from Ru(II) to Cu(II) was found to be 2 s^{-1}, independent of temperature over a wide range (Kostic et al., 1983). Intramolecular electron transfer has also been investigated in a ruthenium labeled derivative of plastocyanin from Anabaena variabilis where the Ru-label is coordinated to His-59. The rate constant for the intramolecular electron transfer from Ru(II) to Cu(II) was found to be ~ 0.3 s^{-1} (Jackman et al., 1987). In both cases the intramolecular processes are far too slow to participate in any physiological redox process of these proteins and therefore these particular histidines cannot be considered as "preferred" sites for electron transfer. Obviously the inherent importance of the observed intramolecular processes lies, however, in the activation-less nature of the electron transfer in azurin (Kostic et al., 1983).

Finally, is it possible to relate the above described physiologically preferred sites and pathways to the spatial orientation of Cu electronic orbitals? From spectroscopic studies on plastocyanin single crystals, the orientation of the $d_{x^2-y^2}$ orbital has been estimated (Penfield et al., 1981). This is the orbital carrying the unpaired electron in the blue Cu(II) proteins and the one being engaged in electron transfer. Ligand field calculations show that the two histidines make the largest contributions to the blue copper ligand field, the cysteine being strong as well albeit somewhat weaker in spite of the short Cu-S distance (Penfield et al., 1981). Thus, it is apparent that very little energy is required to transfer an electron through one of these ligands to or from the $d_{x^2-y^2}$ orbital. The barrier to electron transfer must therefore come from other factors e.g. the distance from the actual redox agent to the edge of one of the copper ligands mentioned above and the nature of intervening medium.

Acknowledgment: O.F. gratefully acknowledges a short-term EMBO fellowship.

REFERENCES

Adman ET, Jensen LH (1981). Structural features of azurin at 2.7 Å resolution. Israel J Chem 21:8-12.

Armstrong FA, Driscoll PC, Hill HAO, Redfield C (1986a). ^{1}H NMR studies of Cr $(NH_3)_6^{3+}$ binding to spinach plastocyanin. J Inorg Biochem 28:171-180.

Armstrong GD, Chapman SK, Sisley MJ, Sykes AG, Aitken A, Osheroff N, Margoliash E (1986). Preferred sites on cytochrome c for electron transfer with two positively charged copper proteins, Anabaena variabilis plastocyanin and stellacyanin. Biochemistry 25:6947-6951.

Bechtold R, Kuehn C, Lepre C, Isied SS (1986). Directional electron transfer in ruthenium-modified horse heart cytochrome c. Nature 332:286-288.

Cheung E, Taylor KJ, Kornblatt JA, English AM, McLendon G, Miller JR (1986). Direct measurement of intramolecular electron transfer rates between cytochrome c and cytochrome c peroxidase: Effects of exothermicity and primary sequence on rate. Proc Natl Acad Sci USA 83:1330-1333.

Cookson, DJ, Hayes MT, Wright PE (1980a). Electron transfer reagent binding sites on plastocyanin. Nature (London) 283:682-683.

Cookson DJ, Hayes MT, Wright PE (1980b). NMR study of the interaction of plastocyanin with chromium (III) analogues of inorganic electron transfer reagents. Biochim Biophys Acta 591:162.

Deisenhofer J, Epp O, Miki K, Huber R, Michel M (1985). Structure of the protein subunits in the photosynthetic reaction centre of Rhodopseudomonas viridis at 3 Å resolution. Nature 318:618-624.

Farver O, Blatt Y, Pecht I (1982a). Resolution of two distinct electron transfer sites on azurin. Biochemistry 21:3556-3561.

Farver O, Licht A, Pecht I (1987). Electron transfer pathways in stellacyanin: A possible homology with plastocyanin. Biochemistry (in press).

Farver O, Pecht I (1981a). Affinity labeling of an electron transfer pathway in azurin by Cr(II) ions. Israel J Chem 21:13-17.

Farver O, Pecht I (1981b). Identification of an electron transfer locus in plastocyanin by chromium (II) affinity labeling. Proc Natl Acad Sci USA 78:4190-4193.

Farver O, Shahak Y, Pecht I (1982b). Electron uptake and delivery sites on plastocyanin in its reactions with the photosynthetic electron transport system. Biochemistry 21:1885-1890.

Grimes CJ, Piszkiewicz D, Fleischer EB (1974). Electron transfer reactions in biological systems: The reduction of ferricytochrome c by chromous ions. Proc Natl Acad Sci USA 71:1408-1412.

Groeneveld CM, Canters GW (1985). The pH dependence of the electron self-exchange rate of azurin from Pseudomonas aeruginosa as studied by ^{1}H-NMR. Eur J Biochem 153:559-564.

Guss JM and Freeman HC (1983). Structure of oxidized poplar plastocyanin at 1.6 Å resolution. J Mol Biol 169:521-563.

Jackman, MP, Sykes, AG, Salmon GA (1987). Preparation and characterization of a penta-ammine-ruthenium (III) derivative of plastocyanin and the kinetics of long distance electron transfer. J Chem Soc, Chem Commun 1987:65-66.

Kostic NM, Margalit R, Che C-M, Gray HB (1983). Kinetics of long distance ruthenium to copper electron transfer in [Pentaammineruthenium histidine-83]azurin. J Am Chem Soc 105:7765-7767.

Liang N, Kang CH, Ho PS, Margoliash E, Hoffman BM (1986). Long range electron transfer from iron(II)-cytochrome c to (zinc-cytochrome c peroxidase) within the 1:1 complex. J Am Chem Soc 108:4665-4666.

Mayo SL, Ellis WR, Crutchley RJ, Gray HB (1986). Long-range electron transfer in heme proteins. Science 233:948-952.

Mimms WB, Davis JL, Peisach J (1984). The accessibility of type 1 Cu(II) centers in laccase, azurin, and stellacyanin to exchangeable hydrogen and ambient water. Biophys J 45:755-766.

Morpurgo, G, Pecht, I (1982). Affinity labeling of stellacyanin by Cr (II) aquo ions. Biochem Biophys Res Commun 104:1592-1596.

Penfield KW, Gat RR, Himmelwright RS, Eichman NC, Norris VA, Freeman HC, Solomon EI (1981). Spectroscopic studies on plastocyanin single crystals: A detailed electronic structure determination of the blue copper active site. J Am Chem Soc 103:4382-4388.

Rieder R, Bosshard HR (1980). Comparison of the binding sites on cytochrome c for cytochrome c oxidase, cytochrome bc_1, and cytochrome c_1. J Biol Chem 255:4732-4739.

Ryden L, Lundgreen J-O (1976). Homology relationships among the small blue proteins. Nature 261:344-346.

Wherland S, Farver O, Pecht I (1987). A three dimensional model of stellacyanin and its implications for electron transfer reactivity. Submitted.

Wood PM (1979). Rate of electron transfer between plastocyanin, cytochrome f, related proteins and artificial redox reagents in solution. Biochim Biophys Acta 357:370-379.

DISCUSSION

HURST: In early studies using this probe the concern was often expressed that Cr^{3+} binding might not actually identify the lead-in group for electron transfer because Cr^{2+} - Cr^{3+} electron exchange is possible. Hence, the binding site observed might actually be thermodynamically stable for Cr^{3+} binding and not accurately reflect the kinetic pathway. What controls have you implemented to ensure that Cr^{3+} has not migrated as a consequence of electron exchange following initial reduction of the protein?

PECHT: The point raised of migration to a thermodynamically more stable site is indeed an important question. However, there are several observations that confirm the Cr(III) -binding location as the site where the electrons are transferred to Cu(II): 1) No binding could be observed upon addition of Cr(III) to either of the Cu-proteins studied. 2) All experiments were carried out adding only one equivalent of Cr(II) to the respective proteins at a very slow rate compared to the rate of electron transfer (about 1 min, $k = 10^4 - 10^7\ M^{-1}\ s^{-1}$). Hence, there should never be any free Cr(II) in solution to be involved in the proposed Cr^{2+} -Cr^{3+} exchange. 3) In stellacyanin we have clearly shown that there is no exchange between the originally coordinated Cr(III) and additionally added Cr(II) (Morpurgo and Pecht, 1982). 4) Finally, the fact that we consistently observed stoichiometric reduction and binding at one Cr(III) per protein and find quantitatively that this Cr(III) is attached to one specific peptide fragment provides at least circumstantial evidence against the hypothesis of migration.

KOPPENOL: We have done kinetic ionic strength studies with plastocyanin similar to those with cytochrome c [(1986) J Am Chem Soc 109: 2679] in which the ionic strength dependence is interpreted as a function of the net charges of the reactants and the dipole moments of cytochrome c and the transition complex. Our results indicate that the site of electron transfer is equatorial with respect to the dipole axis, which fits with a location close to copper but not with the site of the acidic patch.

PECHT: One of our main conclusions from studying electron transfer pathways of the copper proteins is that they employ two distinct pathways. Hence, the question which you raise cannot be answered in a general fashion. Rather it depends on the particular reagent employed. For example, it has been shown that positively charged metal complexes associate preferentially with the "negative patch" (Cookson et al. 1980) and even inhibit electron transfer there (Lappin et al. 1979). This site is, however, an alternative, possibly a regulatory one, to the long time recognized site present at the so-called "Northern" side of plastocyanin, at the edge of His-87 imidazole.

MESSERSCHMIDT: Wouldn't you think that 2D-NMR measurements combined with molecular dynamics calculations should give a better molecular model for stellacyanin?

PECHT: Yes, 2D-NMR measurements would yield a substantial improvement and consolidation for the model, which currently doesn't have a direct structural basis. Even better as you would say, are direct structure determinations which hopefully will be done some day.

LUNDEEN: Which end of the disulfide cys-59 - cys-93 coordinates to copper in your model of stellacyanin? What is the bond distance? Cu(I) forms tetrahedral, usually dimeric complexes with disulfide but Cu(II) is reluctant to coordinate. Do you think a disulfide ligand would allow facile reduction and reoxidation of a copper site?

PECHT: The answer to the first question is that the sulfur of cys-93 is coordinated to the copper. The bond distance between copper and the sulfur is 3.0 A. As to the effect(s) of having disulfide as a ligand, we assume it plays a role similar to that of methionine sulfur in other blue copper proteins. It may be responsible for differences in the spectroscopic, and in some chemical properties, of stellacyanin vis-a-vis azurin and plastocyanin.

Oxidases and Related Redox Systems, pages 285–288

PRELIMINARY X-RAY CRYSTAL STRUCTURE AND PARTIAL cDNA-SEQUENCE OF ASCORBATE OXIDASE FROM ZUCCHINI

Albrecht Messerschmidt, Antonello Rossi, Rudolf Ladenstein, Robert Huber, Martino Bolognesi, Agostino Marchesini, Raffaelo Petruzelli, and Alessandro Finazzi-Agro

Max-Planck-Institut fuer Biochemie, Martinsried, F.R.G. (A.M., A.R., R.L., R.H.),
University of Pavia, Italy (M.B.),
Instituto Nutrizione delle Piante, Torino, Italy (A.Ma.),
University of Rome 'La Sapienza', Italy (R.P., A.F-A)

The blue protein ascorbate oxidase belongs to the group of 'blue' oxidases together with laccase and ceruloplasmin. These are multi-copper enzymes catalysing the 4-electron reduction of molecular oxygen to water with concomitant one-electron oxidation of the substrate [Malkin & Malmstrom (1970), Malmstrom et al. (1975)].

We were able to crystallise ascorbate oxidase (AO) in two different crystal forms [Ladenstein et al. (1979), Bolognesi et al. (1983)], which were both used in the solution of the crystal structure. The structure was determined by multiple isomorphous replacement (MIR) techniques in one crystal form, by Patterson search methods using the uninterpreted MIR-electron density in the other form, and by cyclic averaging about non-crystallographic symmetry in both crystal forms (the subunits of AO are arranged as tetramers with local 222-symmetry in both crystal forms).

The present 2.5 A resolution electron density map allows the trace of the polypeptide chain to be followed partly. Two domains within one monomer with plastocyanin folding could be identified. Each of these 'plastocyanin'-domains contains one copper-ion bound analogous to the blue copper proteins plastocyanin and azurin.

Half-way between these copper-ions at the interface between the 'plastocyanin'-domains bound to both domains is a trinuclear cluster of copper-ions.

A 165 amino acid residue stretch at the C-terminus of AO could be determined by cDNA-sequencing techniques. It shows homology with C-terminal sequences of Neurospora crassa laccase [German & Lerch (1986)] and human ceruloplasmin [Takahashi et al. (1984)].

REFERENCES:

Bolognesi M, Gatti G, Coda A, Avigliano L, Marcozzi G, Finazzi-Agro A (1983). A new crystalline modification of the copper enzyme ascorbate oxidase. J Molecular Biology 169: 351-352.

German U, Lerch K (1986). Isolation and partial nucleotide sequence of the laccase gene from Neurospora crassa: Amino acid sequence homology of the protein to human ceruloplasmin. Proc Natl Acad Sci USA 83: 8854-8858.

Ladenstein R, Marchesini, A, Palmieri S (1979). Preliminary crystallographic study of ascorbic acid oxidase from green Zucchini squash. FEBS Letters 107: 407-408.

Malkin R, Malmstrom BG (1970). The state and function of copper in biological systems. In Advanc Enzymol 33: 177-243.

Malmstrom BG, Andreasson L-E, Reinhammar B (1975). Copper-containing oxidases and superoxide dismutase. In The Enzymes (Boyer PD, ed) XII: 507-579.

Takahashi N, Ortel TL, Putnam FW (1984). Single-chain structure of human ceruloplasmin: The complete amino acid sequence of the whole molecule. Proc Natl Acad Sci USA 81: 390-394.

DISCUSSION

SIEKER: (1) You said there was a copper binding site with 50% occupancy - did Hg go into this site (substitute) also? If so, that's strange since the site seems labile. (2) Where was the trinuclear copper cluster located in the beta-sandwich domains?

MESSERSCHMIDT: (1) Hg did not go into the labile Cu binding site. (2) The trinuclear copper cluster is located on the three strand portion/side.

PAPA: Do your nice structural data on the ligands of the four Cu atoms/monomer allow you to specify or predict their electron accepting capability, the oxygen binding and redox characteristics? Is the dimer essential for the activity of the oxidase?

MESSERSCHMIDT: It is known from redox potentials for the different copper types in ascorbate oxidase that type I copper is the electron acceptor from ascorbate. The binuclear site is supposed to be the oxygen binding site and electron acceptor for the electrons coming in from the type I copper. It is known from the literature that the redox potential of the type III copper is slightly higher than that of the type I copper. How the mechanism of O_2 reduction at the trinuclear copper site is performed is at the present state of knowledge difficult to say. It needs a lot of further investigation, e.g., kinetic and spectroscopic studies.

PECHT: Could you please state the distances between the trinuclear center and the type II copper, and among themselves.

MESSERSCHMIDT: The distance from the trinuclear copper to the type I copper is around 13 A each. I don't have the other distances in mind at the moment.

RYDEN: First of all, congratulations on solving this structure; the achievement is certainly a landmark for the study of blue copper proteins. I have two questions: (1) How do you account for the 8.5 copper atoms in the protein? (2) Will you please point out the corresponding residues in the linear structure you have worked out so far.

MESSERSCHMIDT: (1) Both subunits are, aside from minor conformational differences and perhaps carbohydrate attachment, identical. We find in one subunit 4 fully occupied coppers and that type I copper in domain A is approximately half occupied. This would add up to 4.5 coppers per subunit, or 9 coppers per dimer.

(2) There is a peptide sequence known for ascorbate oxidase which is also present in the first domains of the Neurospora crassa laccase and human ceruloplasmin sequences. It is I H W H G with the marked histidines being ligands of the corrsponding copper (see Domain 1 below). There is a further sequence, H X H, at the end of the first domain with the ligands noted. In the third domain, we have H P W H L H G H..........H C H I E P H L H M G M (See Domain 3 below).

Domain 1						Domain 1		
...I	H	W	H	G...		...H	X	H...
	Cu4		Cu3			Cu2	Cul	Cu3

Domain 3

H	P	W	H	L	H	G	H	--- H	C	H	I	E	P	H	L	H	M	G	M
Cu5			Cu4		Cu3			Cu2	Cu5	Cu3				Cu5			Cu5		

Then:	Cul -	type 1 copper in domain 1
	Cu2 -> Cu4	trinuclear copper site
	Cu2, Cu3	each has 3 histidine ligands, probably type III Cu2
	Cu4	2 histidine ligands, probably type II copper
	Cu5	type I copper in domain 3

Session V: Copper-Proteins, continued

Chairman: Peter Nicholls

Oxidases and Related Redox Systems, pages 291–307

STRUCTURE OF ARTHROPODAN HEMOCYANIN

Anne Volbeda and Wim G.J. Hol

Laboratory of Chemical Physics, Department of Chemistry, University of Groningen, Nijenborgh 16, 9747 AG Groningen, The Netherlands

ABSTRACT

The X-ray structure of spiny lobster deoxy-hemocyanin has been determined at 3.2 Å resolution. This oxygen-carrying type III copper protein is a hexamer with a molecular weight of ~ 460,000. Each subunit is folded into three domains, of which the structure is qualitatively described. The oxygen binding dinuclear copper site is located in the second domain. Each copper is surrounded by three histidines. An analysis of inter subunit contacts shows that the hemocyanin hexamer is best described as a trimer of dimers. Available amino acid sequence and electron microscopy data suggest that the polypeptide architecture of spiny lobster hemocyanin is common to all arthropodan hemocyanins.

INTRODUCTION

Four different classes of oxygen transporting proteins have been developed in nature: hemoglobins, hemerythrins, arthropodan hemocyanins and molluscan hemocyanins. Many X-ray structures are known of hemoglobin and hemerythrin, some of them to a very high resolution. Hemocyanins are much less well characterized at the atomic level. Only of one arthropodan hemocyanin a medium resolution X-ray structure is known. The X-ray studies show very different polypeptide foldings in these oxygen carriers. Only hemoglobin utilises an iron containing heme group for oxygen binding, whereas, in spite of their name, hemerythrin and hemocyanin have found a different solution for this purpose: a dinuclear iron site in the first, a dinuclear copper site in the latter.

Unlike hemoglobins and hemerythrins, which are often found inside specialized cells, hemocyanins are freely dissolved in the hemolymph. They are also much bigger. Whereas in hemerythrin and hemoglobin ~ 120 or ~ 140 residues are used for the binding of one oxygen molecule, this number increases to ~ 410 in molluscan hemocyanins and ~ 620 - 660 in arthropodan hemocyanins. Hemocyanin and hemoglobin show a high degree of cooperativity in oxygen binding, but little if any cooperativity is found in hemerythrin.

Because of extensive spectroscopic similarities between arthropodan and molluscan hemocyanins it is generally believed that their oxygen binding centres are quite similar. They have, however, a completely different quaternary structure, as is shown by a large number of electron microscopy studies (Van Bruggen et al., 1981; Van Holde et al., 1982; Ellerton et al., 1983). Molluscan hemocyanins are cylindrical molecules consisting of 10 or 20 subunits and have molecular weights up to ~ 9,000,000. The dinuclear copper site is located in a so-called functional unit, which has a molecular weight of ~ 50,000 and is repeated 7 or 8 times inside the polypeptide chain of one subunit. Arthropodan hemocyanins, the main subject of this paper, consist of hexamers or multihexamers with subunits of molecular weight ~ 75,000, each subunit containing one dinuclear copper pair. The largest arthropodan hemocyanins are found in horseshoe crabs and have a molecular weight of ~ 3,500,000, made up by 48 subunits.

In the last 4 or 5 years many amino acid sequences of arthropodan hemocyanins have been determined. These sequences all appear to be homologous (Linzen et al., 1985). Conserved residues are found throughout the entire polypeptide chain. Therefore the tertiary structure of different arthropodan hemocyanin subunits must be very similar. Very recently the first amino acid sequence was published of a functional unit of a molluscan hemocyanin, showing practically no homology with arthropodan hemocyanins, except for one interesting section of 42 residues. The very large differences both in the primary and quaternary structure of arthropodan and molluscan hemocyanins justifies their classification into two separate classes of oxygen carriers.

Whereas functionally hemocyanins belong to the oxygen carriers, structurally they belong to the class of type III copper proteins. The characteristic feature of this class of proteins is that they contain one or more dinuclear copper sites which are ESR silent both in the reduced Cu(I)Cu(I) and in the oxidized Cu(II)Cu(II) form. Therefore a strong diamagnetic coupling between the Cu(II) unpaired electrons must exist, which is believed to be mediated by a bridging ligand (Van Holde et al., 1982; Ellerton et al., 1983). Other type III copper proteins are tyrosinase and the blue oxidases ascorbic acid oxidase, ceruloplasmin and laccase. Except for ascorbic acid oxidase, of which a crystallographic structure determination shows much progress (see the contribution by Albrecht Messerschmidt) no crystal structures are known yet of these proteins. It would be interesting to see whether the different dinuclear copper sites have a similar structure. In this aspect tyrosinase probably looks much like hemocyanin, because the spectrum of oxy-hemocyanin is remarkably similar to the spectrum of oxy-tyrosinase. A comparison of amino acid sequences of tyrosinases to arthropodan hemocyanins shows a high degree of homology for a section of about 42 residues which makes up half of the dinuclear copper site. The same section of residues is the only homologous region between arthropodan hemocyanins and the one sequence known of a molluscan hemocyanin (Drexel et al., 1987). This very interesting observation and its consequences is discussed into more detail in Konrad Lerch's contribution.

In our laboratory the first crystal structure of an arthropodan hemocyanin has been determined (Gaykema et al., 1984; Gaykema et al., 1986). The hemocyanin used comes from the spiny lobster Panulirus interruptus. Being a hexamer it is one of the smallest hemocyanins known. Still it has a molecular weight of ~ 460,000, which is quite big for a crystallographic structure determination. Like most arthropodan hemocyanins it displays subunit heterogeneity. Three types of subunits are found, designated as subunit a, b and c. In several multi-hexameric hemocyanins this subunit variation has been linked with the formation of precise connections between hexamers. In Panulirus hemocyanin, however, there must be another reason for the occurrence of subunit heterogeneity. Possibly the different subunits are used in connection with fine-tuning the oxygen-transport properties of the hemolymph.

The crystals used for the X-ray structure determination were grown from a solution of native hemocyanin, containing all three subunit types. Upon dissolving them, however, they appeared to contain only the a and the b subunits, in roughly equal amounts. Amino acid sequence determinations have been completed for subunit a (Bak et al., 1986) and subunit b (Peter Jekel, personal communications), whereas the sequence of subunit c is almost complete (Ben Neuteboom, personal communications). Subunit a and b are very similar, the difference in their sequence being only ~ 3%, but subunit c shows much less similarity. Covalently bound to the a and b subunit is a carbohydrate chain, which function is yet unknown. It is unlikely that it plays a role in the oxygen transport process, because many arthropodan hemocyanins are known without bound carbohydrates.

The crystal structure of Panulirus hemocyanin has been determined at the medium resolution of 3.2 Å. Crystallographic refinement of an atomic model is still in progress. Spectroscopic measurements on hemocyanin crystals have shown that we have determined the structure of deoxy-hemocyanin (Volbeda et al., manuscript in preparation). Recently great progress has been made with the structure determination of a hemocyanin subunit of the horseshoe crab Limulus polyphemus, using the coordinates of the atomic model of Panulirus hemocyanin as a starting point (Karen Magnus, personal communications). Interestingly the Limulus hemocyanin crystals are in the oxygenated state. Much work still has to be done, however, before these two hemocyanin structures can be compared in detail.

In this report the structure of Panulirus hemocyanin is described. Inter subunit contacts are discussed in connection with a possible allosteric mechanism. Special attention is paid to the dinuclear copper site and to the structure of arthropodan hemocyanins in general.

METHODS

Monoclinic crystals of Panulirus hemocyanin were grown for X-ray data collection, containing roughly equal amounts of subunit a and b. They were colourless and diffracted reasonably well to ~ 3.2 Å resolution. The entire hexamer was contained in the asymmetric unit. Its precise location and orientation in the crystallographic unit cell could be determined by using data from two isomorphous heavy atom derivatives, giving useful phase information to ~ 4 Å. The 32 symmetry of the hexamer together with the near identity of the a and b subunit subsequently allowed the application of density averaging procedures to improve and extend the phases to 3.2 Å resolution. This resulted in an electron density map of remarkably high quality, in which the hemocyanin polypeptide chain could be traced without problems. The same density averaging procedures were later succesfully used for phase extension to atomic resolution in the structure determination of many other proteins, amongst them several virus structures.

A nearly complete atomic model of Panulirus hemocyanin was built into the electron density distribution, using the program GUIDE at a computer graphics system and the amino acid sequence of subunit a, which was simultaneously determined by the group of Beintema in Groningen. The crystallographic R-factor of this model, containing more than 32,000 atoms, was 35.2%, including ~ 60,000 reflections between 8.0 and 3.2 Å resolution. This model is being refined now on a Cyber 205 supercomputer. The present R-factor is 24.0%.

THE FOLDING OF A HEMOCYANIN SUBUNIT

Subunit a and b of Panulirus hemocyanin each contain 657 residues (Bak et al., 1986). These are folded into three distinct domains, as became apparent after the polypeptide chain was traced in a six-fold averaged electron density map (Gaykema et al., 1984). The first domain contains ~ 175 residues (figure 1). The first 4 or 5 residues are invisible

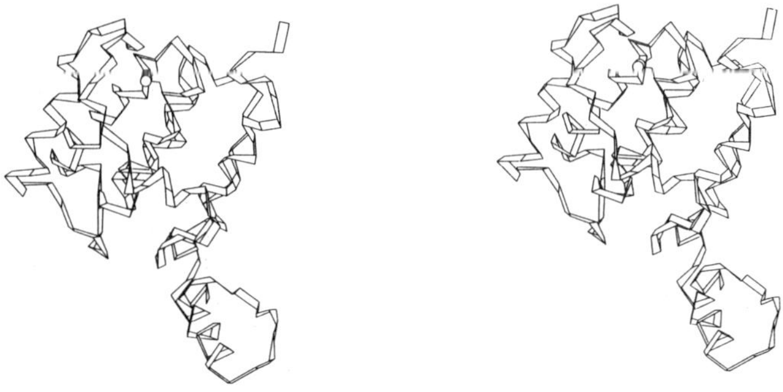

Figure 1. Stereo ribbon diagram of domain 1 of spiny lobster hemocyanin. The disulfide bond is depicted by the open spheres.

or show uninterpretable density in the electron density distribution, indicating that they are disordered. This domain consists of a larger globular part, formed by six α-helices and containing one disulfide bond, and, at the C-terminal end, a small but interesting appendix, formed by one α-helix and one β-strand. Part of this β-strand is asparagine 167, to which the carbohydrate moiety is attached. The appendix is located very close to the molecular two-fold axis.

Functionally the most important domain is domain 2 (figure 2). It contains the oxygen binding dinuclear copper

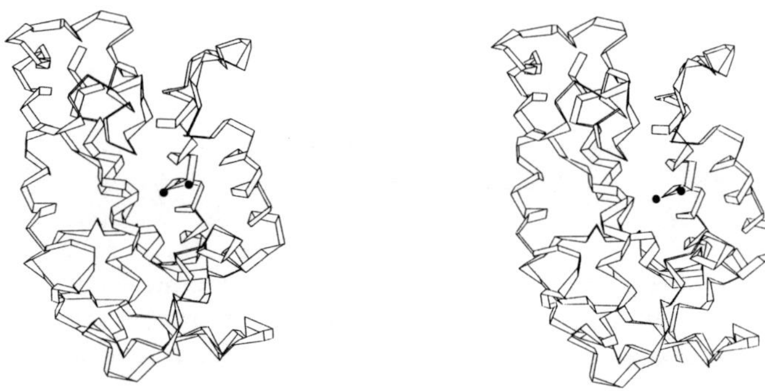

Figure 2. Stereo diagram of domain 2. The black spheres represent the coppers.

site, which will be described below in a special section. There are ~ 220 residues in this domain, which are folded into seven α-helices and five β-strands. The last two α-helices follow each other immediately and therefore they can equally well be described by one α-helix which has a kink in the middle. The β-strands can be divided into one extended chain and two two-stranded anti-parallel β-sheets. Domain 2 is the most central domain in the hexamer. It is almost completely surrounded by domain 1 and 3 and by the polypeptide chains of other subunits.

Figure 3. Stereo diagram of domain 3.

Domain 3 contains ~ 260 residues (figure 3). It is the most irregularly shaped domain. A seven-stranded anti-parallel β-barrel, comprising only ~ 60 residues, is the core. From this β-barrel five loops extend, making up the remaining ~ 200 residues and containing six α-helices and 5 β-strands. Three of these loops extend far away from the β-barrel. They are involved in contacts with the other two domains. The first of them contains two anti-parallel β-strands which form a three-stranded β-sheet with the β-strand located in the appendix of domain 1. Two disulfide bonds occur in the third domain.

Four short sections of residues show weak or uninterpretable densities in a six-fold averaged electron density distribution. The first of these is at the N-terminus, as already mentioned. Also the 3 or 4 C-terminal residues of the hemocyanin subunit are practically invisible. The other 2 sections are located in the two longest loops which extend from the β-barrel in domain 3, close to the disulfide bridge connecting these loops. It is presently investigated whether these weak densities arise from high mobility and disorder or from a different conformation in the six subunits of a hexamer, which could be caused by crystal packing effects.

A nice illustration of the presence of 3 domains in a

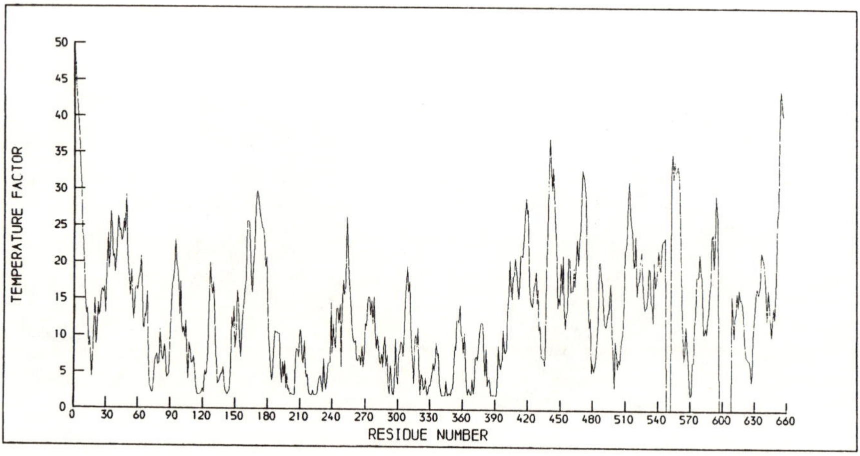

Figure 4. Temperature factor plot of spiny lobster hemocyanin.

hemocyanin subunit is given in figure 4, which shows a temperature factor plot obtained after several cycles of refinement using the constraint of the 32 non-crystallographic symmetry. Clearly the average temperature factor in domain 2 is much lower than in the other two domains. This agrees well with its being the most buried and therefore probably the least mobile domain.

THE DINUCLEAR COPPER SITE

Up to very recently we were not certain about the oxidation state of the coppers in the monoclinic crystals of Panulirus hemocyanin. Oxygen is known to bind as a peroxide ion (Van Holde et al., 1982; Ellerton et al., 1983):

$$\underset{\text{deoxy hemocyanin}}{Cu(I)Cu(I) + O_2} \Leftrightarrow \underset{\text{oxy hemocyanin}}{Cu(II)O_2^{2-}Cu(II)}$$

The oxygenated form could be excluded because the crystals were colourless and did not show an absorption band at 340 nm. Half-met forms containing Cu(I)Cu(II) pairs could also be excluded because no ESR signals could be detected. The remaining possibilities were deoxy hemocyanin, met hemocyanin, which contains Cu(II)Cu(II), or combinations of these two forms. XANES experiments, carried out last year, finally indicated that we have solved the structure of deoxy hemocyanin (Volbeda et al., manuscript in preparation).

Figure 5 shows the structure of the dinuclear copper site. Both coppers are surrounded by three histidines, which are provided by two helical pairs. They are His-194, His-198 and His-224, which interact with CuA, and His-344, His-348 and His 384, interacting with CuB. The first two histidines of each helical pair are separated by one helix turn, i.e. three residues. Similar His-X-X-X-His sequences are present in other type III copper proteins (Lerch, 1978; Takahashi et al., 1984; Germann et al., 1986; Drexel et al., 1987).

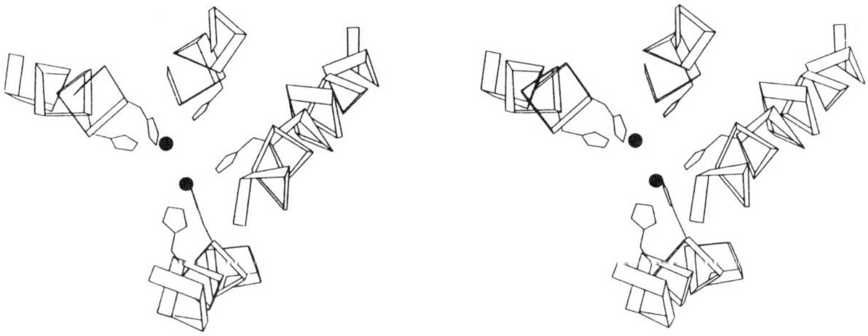

Figure 5. Stereo diagram of the dinuclear copper site.

Many hydrophobic residues are found in the immediate neighbourhood of the dinuclear copper site. The electron density map shows no sign of a protein bridging ligand. The 3,2 Å resolution of the X-ray data unfortunately makes definite statements regarding the presence or absence of a small exogenous ligand, such as a μ-hydroxo group, almost impossible. We hope that further refinement of the hemocyanin model will give more clues. Due to the medium resolution accurate distances between the coppers and their ligands cannot be obtained from our structure. After model building in an unaveraged electron density map, calculated from the last refined model with a crystallographic R-factor of 24.0%, we measure copper to copper distances in the six subunits of the hemocyanin hexamer varying from 3.4 to 3.7 Å. This agrees reasonably well with EXAFS results (Woolery et al., 1984), where a distance of ~ 3.45 Å was found for deoxy hemocyanin.

There is a considerable variation in the distances between the coppers and the NE2 atoms of the surrounding histidines. Some of them, especially His-198 and His-344, appear to be strong ligands, because in all six subunits they show distances smaller than 2.0 Å to CuA and CuB. For His-224 and His 348, however, after careful model building we find distances to the coppers varying between 2.4 and 3.0 Å. For the remaining two histidines we measure in the six independent subunits Cu-N distances in between 1.9 and 2.5 Å. The apparent variation in distances might explain that only 2 ligands per copper are found by EXAFS measurements of deoxy hemocyanin (Woolery et al., 1984), because this technique is known to be rather insensitive to more distant ligands. The structure of the dinuclear copper site in Panulirus hemocyanin will become more accurate if we succeed in further refinement of the hemocyanin model. Better even would be the growth and subsequent X-ray analysis of higher quality hemocyanin crystals. Numerous attempts to do so with several different arthropodan hemocyanins have yielded many well-shaped large size crystals which, however, not diffract beyond 3 Å (See e.g. Gaykema et al., 1983).

THE STRUCTURE OF THE HEXAMER

The 32 symmetry of the hexamer is illustrated in figure 6b. We can distinguish four contact regions. They are located respectively between subunits I and II, I and III, I and VI and I and V. There are no contacts within 4 Å between subunits I and IV. Subunits I and II are related by the molecular two-fold and show the largest number of contacts (see table 1). Together with subunits III-IV and V-VI they form the three "tight" dimers of a hexamer. Subunits I and VI also are related by a molecular twofold, but they make much less contacts. Accordingly we call the pair of them a "loose" dimer. The remaining contacts are related by the molecular threefold. They involve the subunits I-III-V and II-IV-VI, which form the two trimers of the hexamer.

Because the contacts between the members of a tight dimer are much more numerous than the ones between the subunits of a trimer, the hemocyanin hexamer is best described as a trimer of dimers. The contacts in the trimers mainly involve large hydrophilic side chains, whereas in the tight dimers many hydrophilic as well as hydrophobic contacts are found, also involving many smaller side chains. Consequently the trimer contacts appear to be much more flexible. This may be an important observation with respect to possible conformational changes upon cooperative oxygen binding.

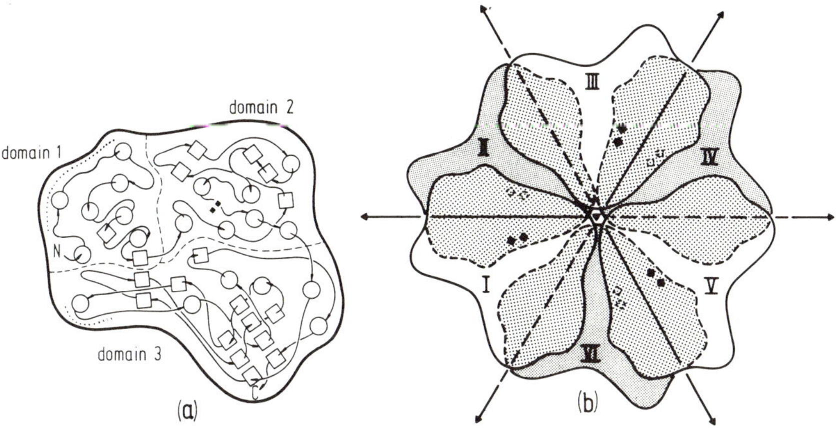

Figure 6. (a) Schematic drawing of the secondary structure in a single subunit of spiny lobster hemocyanin. Circles represent α-helices, squares β-strands. The small black diamonds are the copper positions. The three domains are separated by dashed lines. The dotted lines indicate 2 loops which are absent in Cheliceratan hemocyanins. (b) Schematic representation of the 6 subunits in the hexamer. The lower trimer is shaded. The arrows are the molecular twofold axes, of which the dashed parts indicate the region of loose dimer contacts and the drawn parts the region of tight dimer contacts

TABLE 1. General information about intersubunit contacts in Panulirus interruptus hemocyanin. Not included are contacts mediated by ordered water molecules.

Type of contact[†]	No. of contacts[‡]	No. of atoms involved[§]			No. of amino acids involved[§]		
		Domain 1	Domain 2	Domain 3	Domain 1	Domain 2	Domain 3
I-II	129	19	38	7	8	16	3
I-III	51	10	16	0	4	7	0
I-V	51	0	18	5	0	7	1
I-VI	17	4	0	6	1	0	3
all	248	33	72	18	13	30	7

[†] Subunit numbers as defined in figure 6b

[‡] A contact is defined as a pair of atoms within 4 Å of each other

[§] These numbers apply to atoms or residues from subunit I only

The binding of oxygen induces a change in the charge of the coppers from Cu(I) to Cu(II), which causes a change in the preferred mode of copper coordination, probably a change in Cu to Cu distance and in the electrical field generated. Because of these effects conformational changes are induced which are transmitted to the copper sites of neighbouring subunits where the oxygen affinity is increased. The information about oxygen binding in one subunit is transferred over a remarkable distance, because the distance between the nearest dinuclear copper sites is ~ 46 Å. As the most simple model for the allosteric mechanism we propose that upon oxygenation of the hemocyanin hexamer the conformational changes induced cause a rotation of quasi-rigid tight dimers with respect to each other (Gaykema et al., 1986). In this model the tight dimer contacts essentially remain the same, whereas the more flexible trimer contacts change. This implies that hemocyanin hexamers may occur as partly oxygenated molecules with some dimers in the oxy orientation and others still in the deoxy orientation. This may explain why oxygen binding by arthropodan hemocyanins cannot be described by a simple two-state model (Robert et al., 1987).

THE STRUCTURE OF ARTHROPODAN HEMOCYANINS

The crystal structure of Panulirus interruptus hemocyanin is the only arthropodan hemocyanin structure known at this moment. A comparison of the sequences of seven hemocyanin subunits belonging to five different arthropods however, indicates that the Panulirus structure occurs in all arthropodan hemocyanins (Linzen et al., 1985). Regions with identical and functionally similar residues are found throughout the entire polypeptide chain. Most conserved are the residues which are buried in the Panulirus structure. The functionally important domain 2 shows the highest sequence homology. Very conserved are the residues belonging to the four α-helices which surround the dinuclear copper site. No substitutions are found for the copper surrounding histidines. The conclusion that different arthropodan hemocyanins have very similar structures is recently confirmed by the crystal structure determination of subunit II of Limulus polyphemus hemocyanin (Karen Magnus, personal communications). It appeared to be possible to solve this structure by using Panulirus hemocyanin as a starting model.

The most striking differences between arthropodan hemocyanins appear to be different modes of multi-hexamer formation (Markl, 1986). In the subphylum of the Crustacea, to which Panulirus interruptus belongs, only hexamers and dodecamers are found, whereas in the subphylum of the Chelicerata, to which Limulus polyphemus belongs, aggregation states of two, four and eight hexamers are functional. It appears that in the sequences of Crustacean hemocyanins there is an insertion of 21 residues, including the second α-helix of domain 1, which is absent in the Chelicerata. In addition there is a smaller Crustacean insertion of about 8 residues at the end of the third α-helix in domain 3 (see also figure 6a). In the crystal structure these two insertions are close to each other. Their removal can be accomodated for by relatively minor changes in neighbouring parts of the subunit polypeptide chain (Linzen et al., 1985; Soeter et al., 1987). Electron microscopy studies show that some important Chelicerata inter-hexamer contact regions overlap the Crustacean insertions (Trijntje Wichertjes & Marin van Heel, private communication). The existence of these two insertions therefore may explain why no higher hexameric aggregation states are found in the Crustacea.

CONCLUSION

The crystal structure determination of Panulirus interruptus hemocyanin, combined with amino acid sequence and electron microscopy results, has revealed the basic architecture of all arthropodan hemocyanin hexamers, yielded important details of the dinuclear copper site and is giving clues as to the control of the assembly of the large multihexameric hemocyanins.

REFERENCES

Bak HJ, Neuteboom B, Jekel PA, Soeter NM, Vereijken JM, Beintema JJ (1986). Structure of arthropod hemocyanin. FEBS Lett. 204: 141-144.

Bruggen EFJ van, Schutter WG, Breemen JFL van, Bijlholt MMC, Wichertjes T (1981). Arthropodan and Molluscan Haemocyanins. In "Electron Microscopy of Proteins" (Harris M, ed.) pp. 1-38, Academic Press, New York.

Drexel R, Siegmund S, Schneider HJ, Linzen B, Gielens C, Preaux G, Lontie R, Kellerman J, Lottspeich F (1987). Complete Amino-Acid Sequence of a Functional Unit from a Molluscan Hemocyanin (Helix Pomatia). Biol. Chem. Hoppe-Seyler 368: 617-635.

Ellerton HD, Ellerton NF, Robinson HA (1983). Hemocyanin - a current perspective. Prog. Biophys. Mol. Biol. 41: 143-248.

Gaykema WPJ, Groendijk H, Doorten G, Vereijken JM, Hol WGJ (1983). Crystals Containing a Single Subunit Type of Panulirus interruptus Hemocyanin. J. Mol. Biol. 168: 197-201.

Gaykema WPJ, Hol WGJ, Vereijken JM, Soeter NM, Bak HJ Beintema JJ (1984). 3.2 Å structure of the copper-containing, oxygen carrying protein Panulirus interruptus haemocyanin. Nature (London) 309: 23-29.

Gaykema WPJ, Volbeda A, Hol WGJ (1986). Structure Determination of Panulirus interruptus Haemocyanin at 3.2 Å Resolution. Succesful Phase Extension by Sixfold Density Averaging. J. Mol. Biol. 187: 255-275.

Germann UA, Lerch K (1986). Isolation and partial nucleotide sequence of the laccase gene from Neurospora crassa: Amino acid sequence homology of the protein to human ceruloplasmin. Proc. Natl. Acad. Sci. USA 83: 8854-8858.

Holde KE van, Miller KI (1982). Hemocyanins. Quart. Rev. Biophys. 15: 1-129.

Lerch K (1978). Amino acid sequence of tyrosinase from Neurospora crassa. Proc. Natl. Acad. Sci. USA 75: 3635-3639.

Linzen B, Soeter NM, Riggs AF, Schneider HJ, Schartau W, Moore MD, Yokata E, Behrens PQ, Nakashima H, Takagi T, Nemoto T, Vereijken JM, Bak HJ, Beintema JJ, Volbeda A, Gaykema WPJ, Hol WGJ (1985). The Structure of Arthropod Hemocyanins. Science 229: 519-524.

Markl J (1986). Evolution and function of structurally diverse subunits in the respiratory protein hemocyanin from arthropods. Biol. Bull. 171: 90-115.

Robert CH, Decker H, Richey B, Gill SJ, Wyman J (1987). Nesting: Hierarchies of allosteric interactions. Proc. Natl. Acad. Sci. USA 84: 1891-1895.

Soeter NM, Jekel PA, Beintema JJ, Volbeda A, Hol WGJ (1987). Primary and tertiary structure of the first domain of Panulirus interruptus hemocyanin and comparison of arthropod hemocyanins. Eur. J. Biochem.: accepted.

Takahashi N, Ortel TL, Putnam FW (1984). Single chain structure of human ceruloplasmin: The complete amino acid sequence of the whole molecule. Proc. Natl. Acad. Sci USA 81: 390-394.

Woolery GL, Powers L, Winkler M, Solomon EI, Spiro TG (1984). EXAFS Studies of Binuclear Copper Site of Oxy-, Deoxy, Metaquo, Metfluoro- and Metazide hemocyanin from Arthropods and Molluscs. J. Am. Chem. Soc. 106: 86-92.

DISCUSSION

SANDERS-LOEHR: Your identification of a copper-containing helical pair is somewhat reminiscent of the situation in hemerythrin where the A and B helices provide two histidine ligands to Fe_A and the C and D helices provide three histidine ligands to Fe_B. What is the probability of a sequence homology between hemocyanin and hemerythrin?

VOLBEDA: I would like to speculate that hemocyanin and hemerythrin are indeed evolutionarily related. The Fe_B helical pair of hemerythrin can be structurally superimposed quite well onto the copper-binding helical pairs of hemocyanin in such a way that the histidine ligands are at equivalent positions. There is almost no sequence similarity between the other residues of the helical pair. The structural similarity may equally well be the result of convergent evolution.

SIEKER: (1) Are you suggesting that there may be a μ-oxo bridge between the two copper atoms? The μ-oxo bridge in hemerythrin linking the Fe atoms has an Fe-O-Fe angle of about 120°. (2) Is there an angle between the planes of, or defined by, three histidines of one copper to the histidine plane of the second copper?

VOLBEDA: (1) Yes, it seems possible that there may be a μ-oxo bridge between the two copper atoms.

(2) The resolution of 3.2 A may not be good enough to see the oxygen. I am not sure about the angle between the planes defined by the liganding atoms of Cu_A and Cu_B, but I think that it is small. The copper ions are definitely in the plane defined by their liganding atoms.

KARLIN: I would like to make a suggestion in response to the comment of Larry Sieker that a possible "endogenous" bridging hydroxide or water molecule could be found in the structure of deoxyhemocyanin by considering a tetrahedral structure around Cu(I). A trigonal pyramidal geometry should or could also be considered where Cu(I) is in the plane of a histidine nitrogen and a shared hydroxo bridging ligand, and where a 4th ligand for each Cu(I) is an axial histidine.

I would also like to comment that while the His-Cu-His angles of your structure of deoxyhemocyanin are not consistent with trigonal Cu(I), a bridging H_2O or HO^- ligand seems unlikely on the grounds that Cu(I) is not oxophilic, preferring nitrogenous ligands which are provided.

VOLBEDA: I had not yet considered a distorted trigonal pyramidal geometry and will certainly do this.

KARLIN: Is there anything about the structure of hemocyanin which might suggest why it binds only one carbon monoxide molecule per two coppers, when it would seem from a coordination chemistry perspective that it should be able to bind two carbon monoxide molecules per one Cu(I).

VOLBEDA: This may be related to the asymmetrical distribution of non-liganding residues in the immediate neighborhood of the two copper ions.

RYDEN: You showed the 2-fold symmetry of the peptide chains around the copper domain 2. How much of the second domain showed this symmetry?

VOLBEDA: Only the copper-binding helical pairs are related by the pseudo-2 fold symmetry. So, this symmetry does not extend to other parts of the polypeptide chain.

PAPA: Is the cooperative interaction for oxygen binding among the three dimers? And, if so, is anything known about the residues involved in such interactions?

VOLBEDA: *Panilurus interruptus* hemocyanin binds oxygen cooperatively so there must be a cooperative interaction between the subunits of the hexamer. The residues involved in these interactions are not yet known.

MESSERSCHMIDT: What is the state of structure refinement?

VOLBEDA: Presently the R-factor is 24.0% for about 60,000 reflections between 3.0 and 3.2 A resolution.

NICHOLLS: If the two-copper center evolved by gene duplication, what might have been the role for a one-copper precursor? Do one-copper proteins of this type exist? Do they react with O_2? O_2^-? or what?

VOLBEDA: This is an interesting question. It is conceivable that the presumed one-copper precursor would have had enzymatic or electron-carrying properties, but I do not know any one-copper protein in which the copper is bound by a helical pair as in hemocyanin. It would be interesting if such a protein is discovered.

Oxidases and Related Redox Systems, pages 309–329

COUPLED BINUCLEAR COPPER PROTEINS: CATALYTIC MECHANISMS AND STRUCTURE-REACTIVITY CORRELATIONS

Edward I. Solomon

Department of Chemistry
Stanford Univerity, Stanford, CA 94305

INTRODUCTION

A coupled binuclear copper active site is found in a wide variety of proteins and enzymes involved in various reactions utilizing dioxygen as summarized in Table 1. In the deoxy [Cu(I)Cu(I)] state hemocyanin and tyrosinase react with dixoygen producing stable oxygenated species with similar spectroscopic features (Figure 1). In contrast to hemocyanin, which binds dioxygen reversibly,

Table 1. Summary of Proteins Containing Coupled Binuclear Copper Active Sites and Their Functions.

HEMOCYANIN:	Arthropod	deoxy + O_2 $\rightleftarrows$ oxy
	Mollusc	deoxy + O_2 $\rightleftarrows$ oxy
		$2H_2O_2$ $\rightleftarrows$ O_2 + $2H_2O$
TYROSINASE:		deoxy + O_2 $\rightleftarrows$ oxy
		$2H_2O_2$ $\rightleftarrows$ O_2 + $2H_2O$
		phenol + $2e^-$ + O_2 + $2H^+$ $\rightleftarrows$ _o_-diphenol+H_2O
		2_o_-diphenol + O_2 $\rightleftarrows$ 2_o_-quinone + $2H_2O$
MULTICOPPER OXIDASES:		
	Laccase	
	Ceruloplasmin	4AH + O_2 $\rightarrow$ 4A + $2H_2O$
	Ascorbate Oxidase	

tyrosinase oxygenates monophenols to ortho-diphenols and further oxidizes this product to the o-quinone.[1]

Laccase, the simplest of the multi-copper oxidases contains, in addition to a coupled binuclear site, two other copper centers which together catalyze the irreversible four-electron reduction of dioxygen to water at the binuclear site. These proteins provide an important series of active sites for which one can ultimately hope to correlate geometric and electronic structure with biological function.

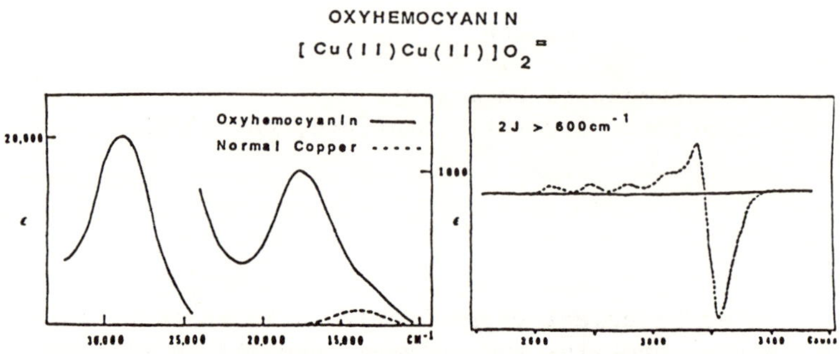

Figure 1. Comparison of absorption (left) and EPR (right) spectra of normal tetragonal copper(II) and oxyhemocyanin.

Resonance Raman spectroscopy has shown that in oxyhemocyanin[2] and oxytyrosinase[3] bound dixoygen has been formally reduced to peroxide; thus the binuclear site exists in the [Cu(II)Cu(II)] oxidation state. However, the spectroscopic features of oxyhemocyanin shown in Figure 1 are quite unusual when compared to those of normal, monomeric, tetragonal cupric complexes. Instead of the normal weak d-d transitions at 600-700 nm, oxyhemocyanin exhibits two intense absorption bands in the visible and near UV, one at 600 nm ($\varepsilon \sim 1000\ M^{-1}\ cm^{-1}$) and a second extremely intense band at 350 nm ($\varepsilon \sim 20{,}000\ M^{-1}\ cm^{-1}$). Further, in contrast to the usual tetragonal Cu(II) EPR spectra, oxyhemocyanin is EPR nondetectable. Since both coppers in the active site are known to be d^9 cupric from X-ray absorption edge studies,[4] they must be strongly antiferromagnetically coupled. SQUID magnetic susceptibility studies[5] have only been able to put a lower limit on the magnitude of this coupling of $-2J > 600\ cm^{-1}$ ($H_{ex} = -2JS_i\ S_j$).

Significant insight into these unique spectral features of oxy hemocyanin has been obtained, and has been used to generate a "spectroscopically effective" model of this active site.[6] This has been accomplished through the generation of a series of protein derivatives which allow the active site to be systematically varied (Figure 2). These derivatives have been used as perturbations for a variety of spectroscopic studies. The simplest derivative is met apo,[7] in which one copper is removed and the remaining one is oxidized thus exhibiting the usual spectral features associated with tetragonal Cu(II). Next in complexity is the mixed valent half-met derivative[8] where the copper(II) can be studied directly. The copper(I), while d^{10} and thus spectroscopically inaccessible by EPR and optical methods, can still be investigated by comparison to the met apo derivative. Finally, two derivatives contain binuclear cupric sites. Dimer has a large EPR signal associated with two weakly (dipolar) interacting tetragonal Cu(II)'s while the met derivative, like oxyhemocyanin, is EPR-nondetectable.[9]

Studies on these derivatives have generated the spectroscopically effective model of the oxy-active site illustrated at the bottom of Figure 2. In the site, two tetragonal cupric ions are bridged by both an endogenous ligand (OR^-) and the exogenous μ-1,2 peroxide. The endogenous bridge is responsible for the strong antiferromagnetic coupling between the coppers and the bridging peroxide gives rise to the intense electronic absorption features (μ-1,2 $O_2^{=}$ --> 2Cu(II) charge transfer transitions).

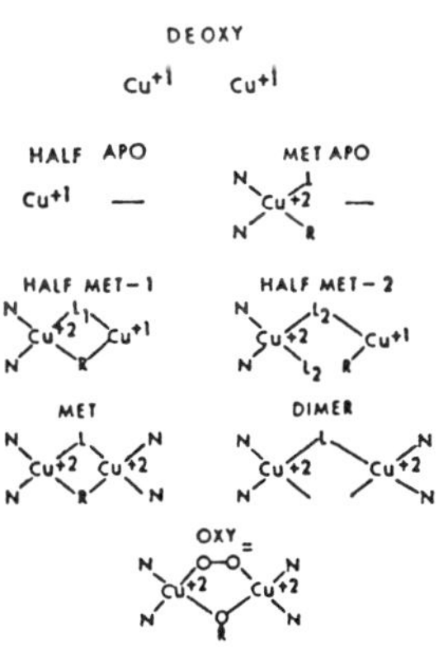

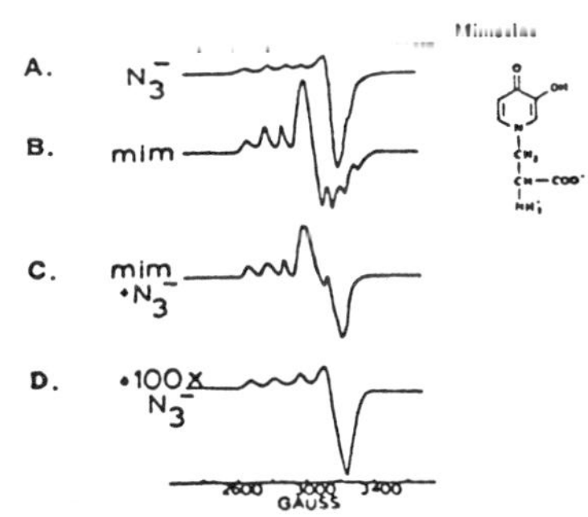

Figure 2. (Left) Derivatives of the coupled binuclear copper active site (L = exogenous ligand, R = endogenous bridge). Figure 3. (Right) Tyrosinase EPR spectra (77K) of (a) half met-N_3^-, (b) half met-mimosine, (c) half met ternary complex of azide and mimosine, and (d) half met-mimosine with excess azide (after 24 hours).

This earlier research has been the subject of several detailed reviews[10] and was presented at ISOX III.[11] Here we summarize our more recent studies which extend this chemical and spectroscopic approach to parallel studies on the coupled binuclear copper site in tyrosinase and to the Type 3 site in laccase.

I. Tyrosinase: Catalytic Mechanism

Half met, met, dimer and oxy derivatives of the coupled binuclear copper active site in tyrosinase have been prepared.[12] Chemical and spectral studies on these derivatives have shown[12b] that the active site in this enzyme is extremely similar to that found in the hemocyanins, and is thus reasonably described by the spectroscopically effective active site model presented at the bottom of Figure 2. Here we consider the involvement of this oxy-site in the catalytic mechanism of tyrosinase.

Differences between the tyrosinase and hemocyanin active sites have been determined from peroxide displacement studies.[12b] Azide displaces the equatorially bound peroxide in an associative displacement mechanism, similar to associative ligand displacement reactions of square-planar Ni and Pt complexes. These reactions are considered to occur through initial axial coordination by the incoming ligand. This is followed by rearrangement of the five-coordinate complex through a trigonal bipyramidal intermediate with subsequent loss of the displaced ligand. The rates of peroxide displacement by azide show that the tyrosinase active site is considerably more accessible to exogenous ligands than the hemocyanin site (k(tyrosinase) = .95 hr^{-1}; k(hemocyanins) < .04 hr^{-1}).

Mimosine, an analogue of the organic substrate (see Figure 3) also displaces peroxide from oxytyrosinase (K=162 hr^{-1}) but not from oxyhemocyanin (k << 10^{-4} hr^{-1}) and is found to bind directly to the coupled binuclear copper active site.[13] This is demonstrated by the appearance of a new absorption band at 425 nm, assigned as a phenolate(mimosine)-to-copper(II) charge transfer transition, upon addition of mimosine to met-tyrosinase. Also, addition of mimosine to half-mettyrosinase results in a new, quite unusual EPR signal with large hyperfine splitting of the lowest g value (Figure 3B, vida infra). Thus, in tyrosinase, both substrate analogues and small molecules are capable of binding at the binuclear copper site with concomitant displacement of peroxide.

Furthermore, ligand competition studies of met- and half-mettyrosinase have shown[13] that the organic substrate analogue (mimosine) and the peroxide analogue (N_3^-) displace each other, competing for the same binding site. Using EPR spectroscopy we have observed during this displacement reaction the formation of a ternary intermediate with mimosine and N_3^- simultaneously bound to the half-mettyrosinase active site (Figure 3C).

These studies have led to a structural model[13] for the reaction mechanism for ortho-hydroxylation of monophenols and oxidation of the resulting diphenols (Figure 4), which is based on associative ligand substitution at the tyrosinase site. Phenolic substrate initially

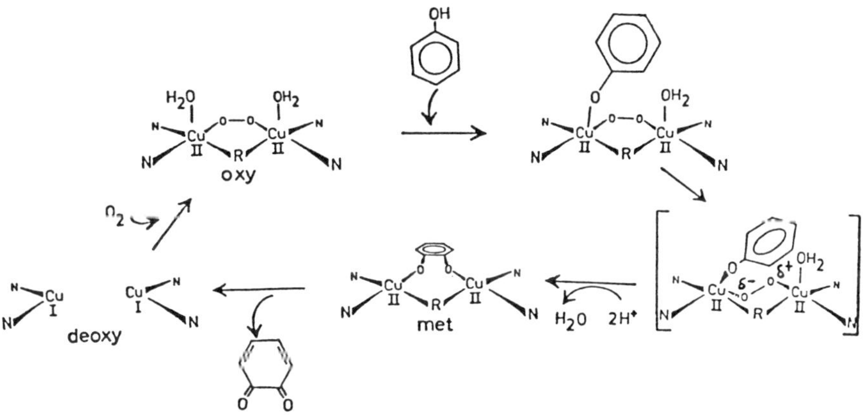

Figure 4. Active site structural mechanism for hydroxylation and oxidation of phenols to form o-quinones by tyrosinase.

coordinates to an axial position of one of the coppers of oxytyrosinase. Rearrangement through a trigonal bipyramidal intermediate leads to ortho-hydroxylation by the bound peroxide, loss of H_2O and coordination of the new diphenol. Subsequent oxidation of the diphenol to a quinone results in a reduced, binuclear cuprous site which releases the quinone and then binds dioxygen to regenerate oxytyrosinase.

A systematic chemical and spectroscopic study of the binding of a series of substrate analogues to the tyrosinase active site was additionally pursued.[14] Carboxylate ligands were used because they are known competitive inhibitors which offer a wide range of structural and electronic variations.

These studies have demonstrated that competitive inhibitors can be divided into two groups: (1) poor inhibitors, which have an equilibrium constant for binding to the enzyme similar to that for binding to aqueous Cu(II) complexes (e.g. phenylacetate: $K_{Cu(II)aq} = 110\ M^{-1}$, $K_{Tyr} = 160\ M^{-1}$, $K_I = 110\ M^{-1}$), and (2) good inhibitors, which are substrate analogues in that the carboxylate is conjugated into the aromatic ring, producing a planar structure, and which bind with an equilibrium constant higher by an order of magnitude relative to aqueous Cu(II) (e.g. benzoate: $K_{Cu(II)aq} = 40\ M^{-1}$, $K_{Tyr} = 800\ M^{-1}$, $K_I = 1400\ M^{-1}$). Thus, it appears that the protein pocket surrounding the copper site contributes to the stabilization of substrate (analogue) binding to the active site. Associated with this increased stability are unusual Cu(II) spectral features which are related to a difference in the geometry of substrate binding to the copper site.

The poor inhibitors produce EPR, absorption, and CD spectral features typical of normal tetragonal Cu(II) (see phenylacetate, Figure 5 and Table 2).

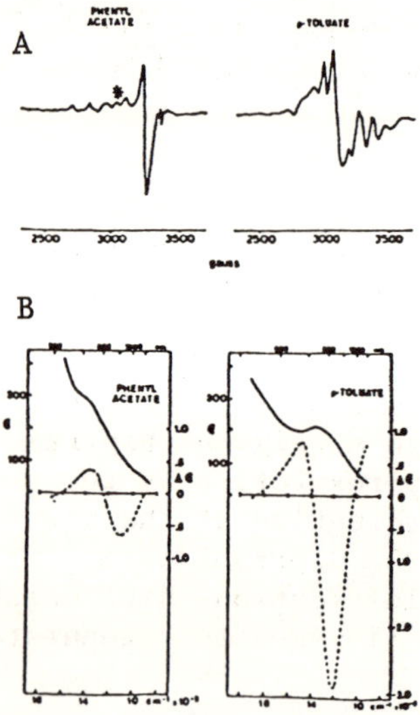

Table 2.

EPR Simulation Parameters for Half-mettyrosinase

g	A, $\times 10^4$ cm^{-1}	line width, G
Phenylacetate		
2.080 ± 0.005	0 ± 5	20
2.080 ± 0.005	0 ± 5	20
2.332 ± 0.005	158 ± 2	20
p-Toluate		
2.014 ± 0.005	83 ± 6	25
2.195 ± 0.010	22 ± 4	30
2.315 ± 0.005	90 ± 2	30

Figure 5. (Left) Comparison of (a) EPR, and (b) absorption (——) and CD (----) spectra of carboxylate inhibitor complexes with half met tyrosinase (after subtraction of the residual half met-NO_2^- signal). The feature labelled * arises from an impurity.

Alternatively, the good competitive inhibitors (p-toluate, Figure 5 and Table 2) exhibit unusual spectral features upon binding to the binuclear copper site in tyrosinase. For the half met derivative, these include a large rhombic splitting of $g_{\perp}$ and a large hyperfine coupling to the lowest g value in the EPR spectrum, and a low lying (< 10,000 cm^{-1}) d-->d transition in the CD spectrum. A ligand field analysis[14] of these spectral changes has shown that these result from a large distortion of the Cu(II) center toward a trigonal bipyramidal geometry.

Thus, the protein packet appears to contribute to the stabilization of substrate analogue binding in a geometry which is at an intermediate position along the reaction coordinate for associative peroxide displacement in Figure 4. This substrate-protein interaction will clearly assist in the catalytic hydroxylation reaction as both the rate of reaction will be increased and the substrate will be positioned in a geometry suited to chemical attack at the ortho aromatic position by the peroxo oxygen atom bound at the adjacent copper. Further, this peroxide, initially forming an unpolarized μ-1,2 bridge between equatorial positions of the Cu(II) ions, is labilized (and perhaps protonated) by this rearrangment of the copper coordination geometry, introducing some polarization into the 0-0 bond and activating the reaction with substrate. The structural and electronic characteristics of the peroxo-substrate-binuclear copper ternary complex in brackets in Figure 4, seem to be most compatible with oxygenation of monophenol via electrophylic attack by peroxide on the aromatic ring of the substrate.

Finally, kinetic results[14] indicate that the oxidation of diphenol substrates to o-quinones by tyrosinase has less geometric and electronic requirements than does monophenol oxidation. Diphenol coordinates to the reduced (deoxy), oxidized (met), and oxygenated (oxy) forms of the binuclear copper active site and is oxidized by the latter two forms. In contrast to monophenol oxidation, bulky substituents do not appear to affect diphenol turnover, indicating that axial-to-equatorial rearrangment at the copper may not be required for a simple two-electron oxidation.

II. Laccase: Structure-Reactivity Correlation

The model for the coupled binuclear copper site in

hemocyanin and tyrosinase (Figure 2) may now be compared to the parallel site in laccase which contains a blue copper (denoted Type 1 or T1), a normal copper (Type 2, T2), and a coupled binuclear copper (Type 3, T3) center. Initial studies have focussed on the simplified Type 2 depleted (T2D) derivative[15] in which the T2 center has been reversibly removed. Native laccase exhibits an absorption band at 600 nm ($\varepsilon = 5700\ M^{-1}\ cm^{-1}$) which is associated with the T1 copper and a peak at 330 nm originally assigned to the T3 center (Figure 6A).[16] Upon removal of the T2 copper the 330 nm absorption disappears, raising concern regarding its assignment as a Type 3 spectral feature. As shown in Figure 6B, both the T1 and T2 sites contribute to the EPR spectrum of native laccase (the T3 cupric ions are coupled and hence EPR nondetectable as in hemocyanin) while the T2 contributions are absent in the T2D derivative.

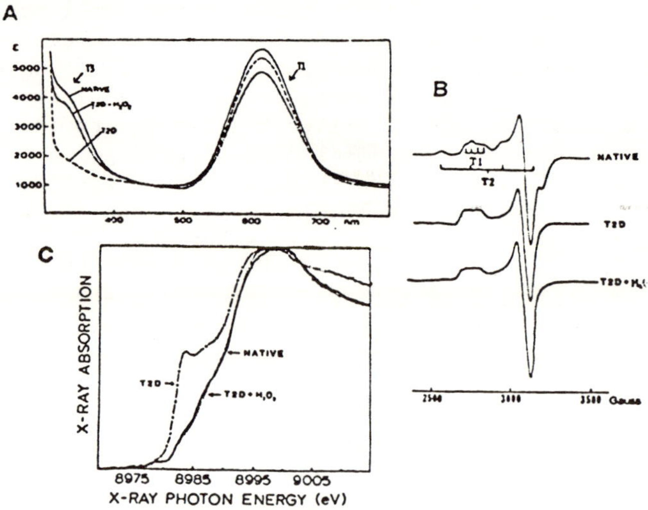

Figure 6. (a) Optical absorption, (b) EPR, and (c) X-ray absorption edge spectra for native and T2D laccase and after reaction of T2D with peroxide.

When T2D laccase is reacted with peroxide the 330 nm band reappears.[17] This observation suggested the possibility that the T3 site might be reduced in the T2D derivative but could be oxidized by H_2O_2. Alternatively, other researchers assigned the 330 nm band as an O_2^{2-} --> Cu(II) charge transfer transition in analogy to oxyhemocyanin in Figure 1 and therefore associated this spectral feature with peroxide binding to a binuclear cupric T3 site.[18] An absorption feature at 330 nm, however, cannot directly distinguish between these possibilites. Also, the EPR spectrum is not diagnostic of the T3 oxidation state since both a reduced site and an antiferromagnetically coupled oxidized binuclear cupric site would be EPR nondetectable. A direct probe of oxidation state is, however, available through X-ray absorption edge spectroscopy (Figure 6C). The T2D form exhibits a peak at 8984 eV which is eliminated by reaction with peroxide, qualitatively indicating that Cu(I) is being oxidized to Cu(II).[17] Such conclusions must be drawn cautiously however, since covalent cupric complexes can also exhibit X-ray absorption edge intensity at low energy.[19]

The oxidation state dependence of the X-ray absorption edge feature was defined and quantitated by a systematic study of a series of twenty Cu(I) and forty Cu(II) complexes.[4b,4c,20] The Cu K edge band energies and shapes depend on oxidation state, ligation, and coordination geometry. Figure 7 (top) presents representative edges for two, three, and four coordinate Cu(I),

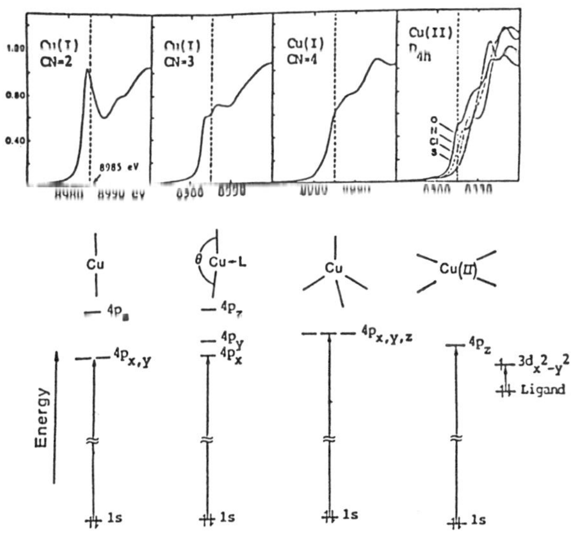

Figure 7. Representative X-ray absorption edge data for Cu(I) and Cu(II) complexes in a variety of coordination environments. Also shown are ligand field splitting diagrams for the complexes.

and tetragonal Cu(II) complexes with a variety of ligand sets. Two or three coordinate Cu(I) complexes always exhibit a peak at energies below 8984 eV. In addition, the three coordinate complexes display a double peak with lower intensity. No Cu(II) complex studied shows a peak below 8986 eV. There is however a low energy tail in the 8984 eV region which is associated with a peak at higher energy.

These spectral differences can be understood through ligand field theory.[4b,4c] For two coordinate Cu(I) with an approximately linear geometry, the electric dipole allowed 1s --> 4p transition is predicted to split into 1s --> $4p_z$ and 1s --> $4p_{x,y}$ components. Further, the 1s --> $4p_z$ transition is expected to be highest in energy due to repulsive interactions with the axial ligands. Polarized single crystal X-ray edge spectroscopy confirms the assignment of the lowest energy peak as 1s --> $4p_{x,y}$. As shown at the bottom of Figure 7, increasing the ligand field strength by adding a third ligand along the y axis leads to an additional repulsive interaction which raises the energy of the $4p_y$ level relative to the $4p_x$. This accounts for the experimentally observed edge splitting in the three coordinate Cu(I) complexes. For a tetrahedral geometry, all p orbitals are equally destabilized by the ligand field and the 1s --> 4p transitions are all above 8985 eV.

For the tetragonal Cu(II) complexes, both the low energy tail and its associated higher energy peak shift to lower energy as the covalent interaction with the equatorial ligand set increases. Since this band is z polarized with an energy shift dependent on equatorial ligand ionization energy, this transition can be assigned as the 1s --> $4p_z$ combined with a ligand --> Cu(II) charge transfer shake-up, which is also supported by final state calculations.[21]

Turning to the X-ray absorption edge spectrum of the T3 site in T2D laccase (Figure 8), a peak is observed below 8984 eV. This energy is characteristic of Cu(I) while the shape of the peak suggests that it is due to three coordinate copper. The amount of reduced copper present can then be quantitated from the normalized edge intensities of copper model complexes with the appropriate coordination number. For T2D laccase, the 8984 eV peak indicates that 98 $\pm$ 22% of the T3 sites are reduced. Peroxide reacted T2D laccase shows no 8984 eV feature and corresponds quantitatively to a fully oxidized T3 site.

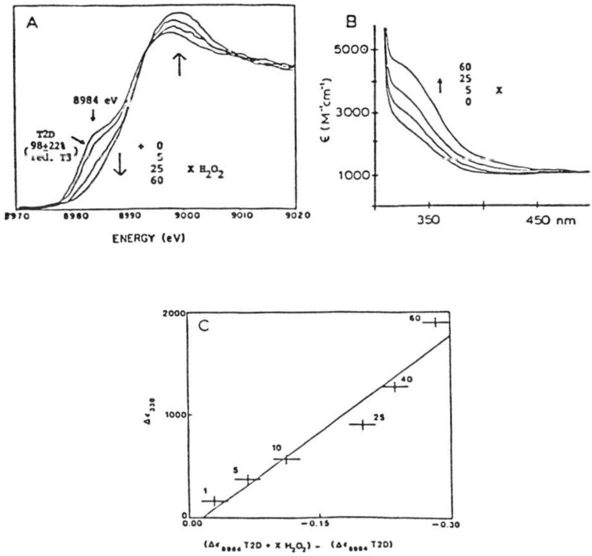

Figure 8. Correlation of X-ray edge and optical absorption data for the reaction of T2D laccase with peroxide.

The possibility of the 330 nm absorption band being a O_2^{2-} --> Cu(II) charge transfer transition was investigated through a correlation of the 330 nm absorption intensity increase with the loss of the 8984 eV Cu(I) peak as T2D was titrated with H_2O_2. The linear correlation in Figure 8C demonstrates that all change at 330 nm is associated with oxidation of the T3 site with no evidence for additional peroxide binding to the oxidized Cu(II). Thus, the T3 site in T2D laccase is a deoxy [Cu(I)Cu(I)] form which, in strong contrast to deoxyhemocyanin and deoxytyrosinase, does not react with dioxygen. It is, however, oxidized by peroxide to give the binuclear cupric met derivative, [Cu(II)Cu(II)].[22]

In comparing the spectroscopic properties of coupled binuclear copper site in met T2D laccase with that of methemocyanin, it is noted that both have no EPR signal. Since both sites contain two Cu(II)'s, the sites must be antiferromagnetically coupled. SQUID magnetic susceptibility measurements place a lower limit on the singlet-triplet splitting of $-2J > 400$ cm^{-1} for methemocyanin and $-2J > 200$ cm^{-1} for met T2D laccase.[23] By lowering the pH and adding azide a new broad EPR signal is produced in both met derivatives (Figure 9).[22,24] This signal accounts for ∿15% of the T3 sites and can be simulated based on two dipolar interacting cupric ions (dashed spectra at bottom of Figure 9). In met T2D laccase this signal clearly shows a high field component at g_{eff} = 1.86 associated with a larger zero field splitting and thus shorter Cu-Cu distance relative to methemocyanin. These results are interpreted in terms of an endogenous bridge being present in both sites, providing the superexchange pathway for strong antiferromagnetic coupling and the lack of an EPR signal. Protonation of this bridge uncouples the sites and a quantitative study of the pH dependence of the broad EPR signal indicates an intrinsic pK_a of > 7.0

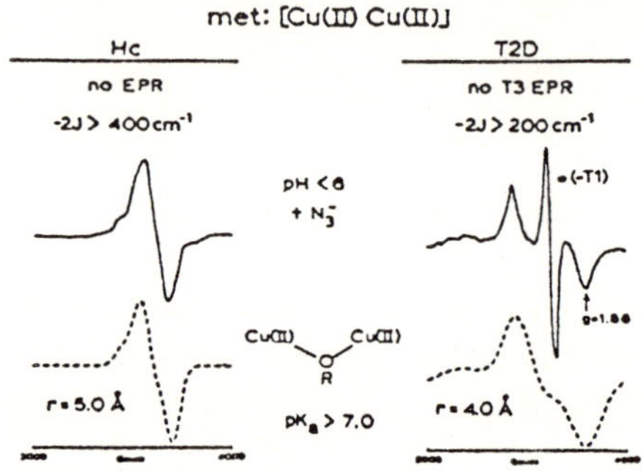

Figure 9. Comparison of EPR spectra of the uncoupled met derivatives of hemocyanin and T2D laccase with simulations for dipolar interacting copper(II's) at distances indicated. (Program supplied by Professor John Pilbrow, Monash University). The peak indicated by the asterisk results from subtraction of the T1 signal.

for the endogenous bridge in both proteins.[22,24] Thus, met T2D laccase and methemocyanin are similar with respect to exchange coupling through the endogenous bridging ligand.

Reduction of the met derivatives by one electron produces the mixed-valent 1/2-met [Cu(II)Cu(I)] derivative. A comparison of the chemical and spectroscopic properties of 1/2-met hemocyanin and 1/2-met T2D laccase reveals a number of important differences[22] (Figure 10). In particular, the 1/2-met T3 site in T2D has EPR spectra and equilibrium binding constants for exogenous ligands which are similar to those observed for normal tetragonal Cu(II) complexes. In 1/2-met hemocyanin, however, binding constants of exogenous ligands are, in general, two orders of magnitude greater and the spectroscopic features are very unusual relative to those of normal cupric complexes.

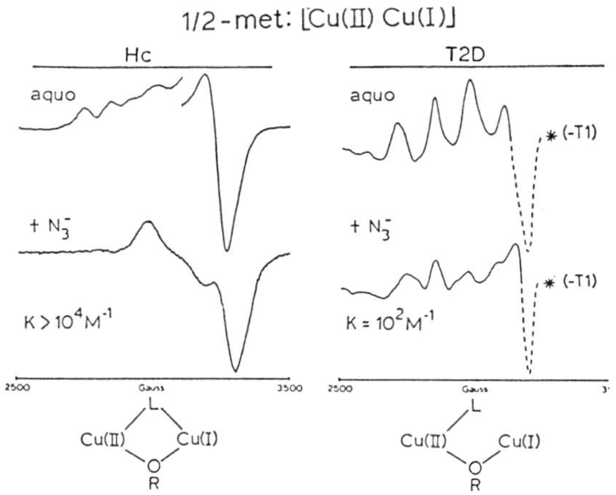

Figure 10. Comparison of EPR spectra of half met hemocyanin and T2D laccase in the absence and presence of azide. The T1 copper signal has been subtracted from the T2D laccase data.

We have studied this unusual chemistry and spectroscopy in detail as summarized in reference 10, and find that these result from exogenous ligand bridging between the Cu(II) and Cu(I) centers of the 1/2 met hemocyanin active site of Figure 10, left). In contrast, the lack of both high affinity binding and mixed-valent spectral features in the 1/2-met T2D laccase strongly indicates that exogenous ligands bind only to one copper at the T3 site (Figure 10, right).

Native laccase catalyzes the reduction of dioxygen to water but the reduced T3 site in T2D laccase does not react with dioxygen. Thus, the T2 copper must play an important role in exogenous ligand interactions with the T3 site. Low temperature MCD spectroscopy was employed to probe exogenous ligand interactions with the T2 and T3 centers since the optical features associated with the paramagnetic T2 site can be distinguished from those associated with the antiferromagnetic T3 centers.[25] As illustrated in Figure 11 (left) application of a magnetic field to the T2 (S=1/2) center splits both the ground and the excited state energies by $g\beta H$. The selection rules for MCD transitions are $\Delta m = +1$ for left circularly polarized (LCP) light and $\Delta m = -1$ for right circularly polarized (RCP) light. Thus, one predicts for the T2 center two transitions with equal intensity and opposite

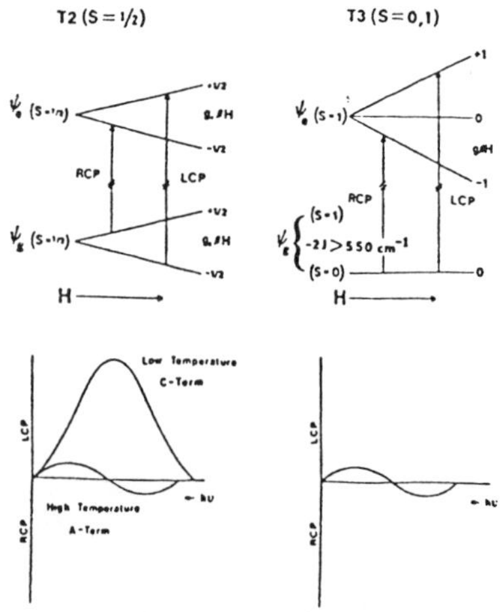

Figure 11. Origin of low temperature MCD features for the T2 (S=1/2) and T3 (S=0) centers in native laccase.

sign. Since the Zeeman splitting is on the order of 10 cm^{-1} and the absorption bands are several thousand cm^{-1} wide, the transitions will cancel except for a broad, weak derivative-shaped signal at high temperatures (an A term). However, as the temperature is lowered the Boltzmann population of the m_s=+1/2 level is reduced relative to that of the m_s=-1/2 level. This leads to a large increase in absorption of LCP light and results in a temperature dependent C term. At very low temperatures ($\sim$ 4K), these C terms will be two to three orders of magnitude more intense than the A terms.

In contrast, the strong antiferromagnetic coupling in the T3 center leads to an S=0 ground state and a thermally inaccessible S=1 excited state (-2J > 550 cm^{-1} for native laccase). Since the S=0 level is nondegenerate, it cannot be split by a magnetic field and there are no changes in ground state population with temperature (Figure 11, right). Thus, there are no C terms associated with the T3 center and the low temperature MCD spectrum of native laccase should be dominated by the contributions from the paramagnetic T2 and T1 centers where only the T2 center is considered to be capable of binding exogenous ligands.

The binding of azide to the T3 and T2 sites in native laccase has been probed via absorption and low temperature MCD spectroscopic studies of the N_3^- --> Cu(II) charge transfer transitions in the 400-500 nm region (Figure 12A). The low temperature MCD and absorption behavior

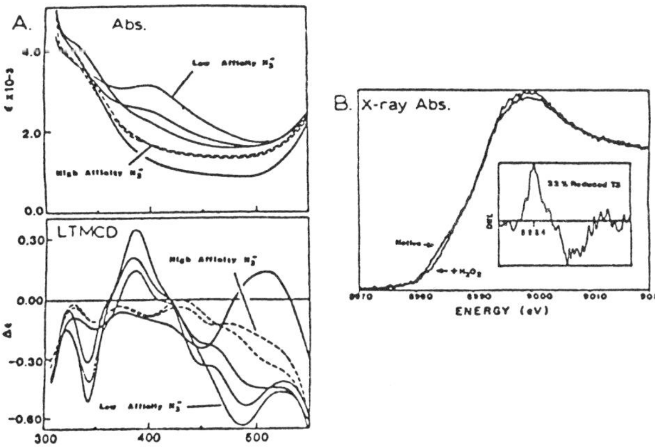

Figure 12. (a) Absorption and low temperature MCD spectra of azide binding to native laccase; (b) X-ray absorption edge data for reaction to native laccase with peroxide.

in this region is quite complex and consists of two contributions. The spectral changes associated with high affinity (HA) binding are completed with the addition of less than one equivalent of azide ($K_{HA} \geq 10^4$ M^{-1}). Prior addition of H_2O_2 eliminates these features, suggesting that they may be associated with ligand binding to a fraction of native laccase molecules in which the T3 site is reduced. This interpretation has been confirmed through X-ray absorption edge studies, shown in Figure 12B, in which the edge features of H_2O_2 treated laccase have been subtracted from those of native enzyme.[4b] A peak is present at 8984 eV with an intensity that corresponds to ∿22% reduced T3 sites in the native enzyme.

The features associated with low affinity azide binding to fully oxidized laccase sites can then be obtained by subtraction of the high affinity azide spectral contributions. The low affinity absorption and low temperature MCD spectra are shown in Figure 13 for increasing [N_3^-]. Two N_3^- -> Cu(II) charge tranfer bands are observed in absorption, one at 500 nm and a more intense band at 400 nm. Both increase with azide concentration; the 400 nm absorption intensity is plotted as a function of [N_3^-] in Figure 13B. In the low temperature MCD spectrum a negative peak is observed at 485 nm which increases in magnitude with increasing azide concentration. It is thus clear that the corresponding 500 nm absorption band is associated with a paramagnetic ground state and must correspond to azide binding to the T2 center. In the region of the intense 400 nm absorption, a positive feature at 385 nm is observed in the low temperature MCD spectrum. The intensity of this feature first increases, then decreases with increasing azide concentration. As shown in Figure 13B, the intensity of this low temperature MCD feature does not correlate with the 400 nm absorption intensity, but in fact parallels the appearance of the g=1.86 EPR signal which is associated with the small fraction of uncoupled T3 sites (Figure 9, Right). Thus, no low temperature MCD intensity is associated with the intense 400 nm absorption band and it must be associated with azide binding to the antiferromagnetically coupled T3 site.

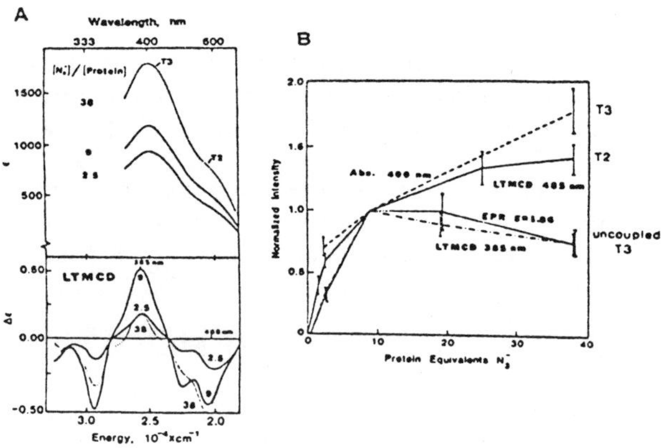

Figure 13. (a) Absorption and low temperature MCD spectra for binding of low affinity azide to native laccase. (b) Correlation of the LT MCD signals at 485 and 385 nm to the 400 nm absorption and the g=1.86 EPR signal with increasing $[N_3^-]$.

From the low temperature MCD data it has been found that azide binds to both the T2 and T3 centers with similar binding constants. The stoichiometry of this binding has been determined through a series of ligand competition experiments, one of which is summarized in Figure 14. The appearance of fluorine superhyperfine splitting in the EPR spectrum (I_n=1/2 for ^{19}F) indicates that one fluoride binds to the T2 copper center (Figure 14A). In the absorption and low temperature MCD spectra, the low affinity N_3^- -> Cu(II) charge transfer transitions associated with the T2 (485 nm MCD) and the T3 (400 nm abs.) centers are eliminated with the addition of increasing concentrations of fluoride (Figure 14B). The quantitative loss of these features parallel each other as shown by the solid line in Figure 14C (right). Fitting these data to models for competitive binding of one or two fluorides clearly demonstrates that only one fluoride

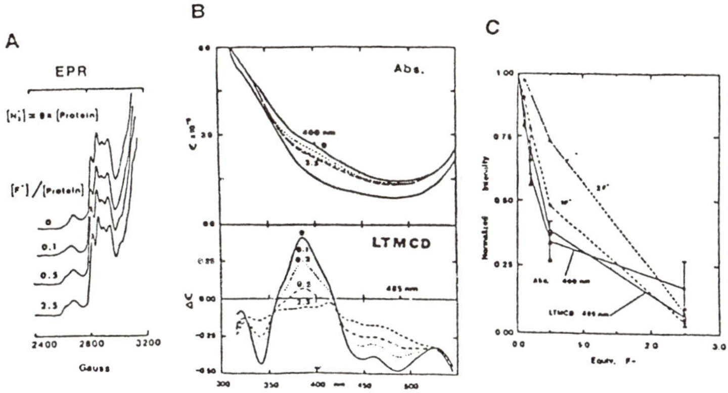

Figure 14. Results of N_3^-/F^- binding competition experiments for native laccase.

binds to the T2 center and is involved in displacing azide from both the T2 and T3 centers. Therefore one exogenous azide ligand must bridge between the T2 and T3 centers in native laccase.[25]

Figure 15 summarizes the differences between exogenous azide binding to the coupled binuclear copper site

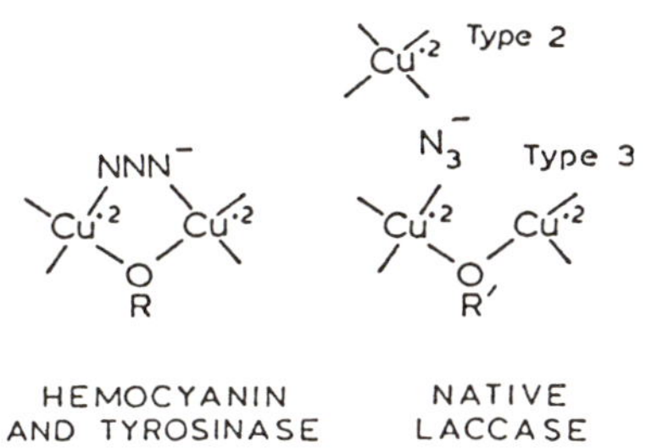

Figure 15. Comparison of the spectroscopically effective models for azide binding at the binuclear copper active site in hemocyanin and the trinuclear copper cluster in laccase.

in hemocyanin and tyrosinase as compared to laccase. For hemocyanin and tyrosinase exogenous ligands bridge the two coppers at the active site while in laccase azide binds to only one copper of the T3 center and bridges to the T2 center forming a trinuclear copper cluster. It is the focus of current research to define in detail the role of the μ-1,2 peroxy bridge in reversible O_2 binding and activation and the T2-T3 trinuclear copper cluster with respect to its role in the multi-electron reduction of dioxygen to water.

Acknowledgements: NIH grant #DK31450 is gratefully acknowledged for support of this research.

REFERENCES:
1. Mason, H.S.; Fowlks, W.B.; Peterson, E.W. J. Am. Chem. Soc. 1955, 77, 2914.
2. Freedman, T.B.: Loehr, J.S.; Loehr, T.M. J. Am. Chem. Soc. 1976, 98, 2809.
3. Eickman, N.C.; Solomon, E.I.; Larrabee, J.A.; Spiro, T.G.; Lerch, K. J. Am. Chem. Soc. 1978, 100, 6529.
4. (a) Brown, J.M.; Powers, L.; Kincaid, B.; Larrabee, J.A.; Spiro, T.G. J. Am. Chem. Soc. 1980, 102, 4210. (b) Kau, L.S.; Spiro, D.J.; Penner-Hahn, J.E.; Hodgson, K.O.; Solomon, E.I. J. Am. Chem. Soc. 1987, in press. (c) Kau, L.S.; Solomon, E.I.; Hodgson, K.O. J. Physique 1986, 47, 289.
5. Solomon, E.I.; Dooley, D.M.; Wang, R.H.; Gray, H.B., Cerdonio, M.; Mogno, F.; Romani, G.L. J. Am. Chem. Soc. 1976, 98, 1029. (b) Dooley, D.; Scott, R.A.; Ellinghaus, J.; Solomon, E.I.; Gray, H.B. Proc. Nat. Acad. Sci. USA 1978, 75, 3019.
6. Eickman, N.C.; Himmelwright, R.S.; Solomon, E.I. Proc. Natl. Acad. Sci. USA 1979, 76, 2094.
7. Himmelwright, R.S.; Eickman, N.C.; Solomon, E.I. Biochem. Biophys. Res. Comm. 1978, 81, 243.
8. (a) Himmelwright, R.S.; Eickman, N.C.; Solomon, E.I. Biochem. Biophys. Res. Comm. 1978, 81, 237. (b) Himmelwright, R.S.; Eickman, N.C.; Solomon, E.I. Biochem. Biophys. Res. Comm. 1978, 84, 300. (c) Himmelwright, R.S.; Eickman, N.C.; Solomon, E.I. J. Am. Chem. Soc. 1979, 101, 1576.
9. Himmelwright, R.S.; Eickman, N.C.; Solomon, E.I. Biochem. Biophys. Res. Comm. 1979, 86, 628.
10. (a) Solomon, E.I. In "Copper Proteins"; Spiro, T.G., Ed.; Wiley: New York, 1981, 41-108. (b) Solomon, E.I.; Penfield, K.W.; Wilcox, D.E. Structure and

Bonding, Vol. 53; Springer-Verlag, 1983, 1-57. (c) Solomon, E.I. Pure and Applied Chemistry 1983, 55, 1069.

11. Himmelwright, R.S.; Eickman, N.C.; Solomon, E.I. International Symposium on Oxidases and Related Redox Systems III, 1982, 263-290.
12. (a) Jolley, R.L.; Evans, L.H.; Makino, N.; Mason, H.S. J. Biol. Chem. 1974, 249, 335. (b) Himmelwright, R.S.; Eickman, N.C.; LuBien, C.D.; Lerch, K.; Solomon, E.I. J. Am. Chem. Soc. 1980, 102, 7339.
13. Winkler, M.E.; Lerch, K.; Solomon, E.I. J. Am. Chem. Soc. 1981, 103, 7001.
14. Wilcox, D.G.; Porras, A.G.; Hwang, Y.T.; Lerch, K.; Winkler, M.E.; Solomon, E.I. J. Am. Chem. Soc. 1985, 107, 4015.
15. Graziani, M.T.; Morpurgo, L.; Rotilio, G.; Mondovi, B. FEBS Lett. 1976, 70, 87.
16. Reinhammar, B. Biochim. Biophys. Acta 1972, 275, 245.
17. LuBien, C.D.; Winkler, M.E.; Thamann, T.J.; Scott, R.A.; CO, M.S.; Hodgson, K.O.; Solomon, E.I. J. Am. Chem. Soc. 1981, 103, 7014.
18. (a) Farver, O.; Frank, P.; Pecht, I. Biochem. Biophys. Res. Comm. 1982, 108, 273. (b) Frank, P.; Farver, O.; Pecht, I. J. Biol. Chem. 1983, 258, 11112. (c) Frank, P.; Farver, O., Pecht, I. Inorg. Chim. Acta. 1984, 91, 81.
19. Powers, L.; Blumberg, W.E.; Chance, B.; Barlow, C.; Leight, J.S., Jr.; Smith, J.C.; Yonetani, T.; Vik, S., Peisach, J. Biochim. Biophys. Acta 1979, 546, 520.
20. Smith, T.A.G., Ph.D. Thesis, Stanford University, 1985.
21. (a) Blair, R.A.; Goddard, W.A. Phys. Rev. B 1980, 22, 2767. (b) Kosugi, N.; Yokoyama, T.; Asakuna, K.; Kuroda, H. Springer Proc. Phys. 1984, 2, 55. (c) Kosugi, N.; Yokoyama, T.; Asakuna, K.; Kuroda, H. Chem. Phys. 1984, 91, 249.
22. Spira, D.J.; Solomon, E.I. J. Am. Chem. Soc., in press.
23. Wilcox, D.E.; Westmoreland, T.D.; Sandusky, P.O.; Solomon, E.I., unpublished results.
24. Wilcox, D.E.; Long, J.R.; Solomon, E.I. J. Am. Chem. Soc. 1984, 106, 2186.
25. (a) Allendorf, M.D.; Spira, D.J.; Solomon, E.I. Proc. Natl. Acad. USA 1985, 82, 3063. (b) Spira, D.J.; Allendorf, M.D.; Solomon, E.I. J. Am. Chem. Soc. 1986, 108, 5318.

DISCUSSION

MASON: Why isn't magnetic dipolar interaction observed among the oxidized trinuclear Cu site?

SOLOMON: The two Cu(II) on the Type 3 site are ferromagnetically coupled with a resultant S = 0 ground state which cannot couple with the Type 2 site, S = 1/2. What we are now trying to obtain is a half reduced Type 3 site, (S = 1/2) in the presence of Type 2 Cu(II) which should be an ideal system to probe further the trinuclear copper cluster site.

LUNDEEN: (1) I think that in 1971 Byers, Williams et al., suggested that azide binds to Type 2 and Type 3 sites, in agreement with your results. (2) At that time, Williams et al., were suggesting that a disulfide bridge might be present at the Type 3 site. What do you think the bridging ligand at the Type 3 site of laccase may be, particularly in the light of the X-ray crystallography of hemocyanin which shows tyrosine to be too far away to coordinate. (3) Can you distinguish spectroscopically between a PhO- ligand and an OH- ligand?

SOLOMON: (1) All possible combinations of copper centers interacting with exogenous ligands were proposed in the early literature. Only the low temperature magnetic circular dichroism method provides a direct probe which first demonstrated the presence of a trinuclear copper cluster in laccase [Allendorf, Spiro, Solomon (1985) PNAS: 82, 306].

(2a) EXAFS studies [(1983) BBRC 112: 746] have shown that sulfur is not bound to the Type 3 site.

(2b) All of our data has indicated that the endogenous bridge (RO- in Fig. 2) is a protonatable ligand with pK > 7.0 (i.e., phenolate, alkoxide or hydroxide). In light of crystallographic results on deoxyhemocyanin, the likely candidate appears to be hydroxide.

(3) According to present literature on Cu(II) model complexes, phenolate gives rise to a CD charge transfer transition in the region of 400-450 nm; OH- gives rise to charge transfer around 330-360 nm.

PECHT: (1) Could you please comment on which reagents are capable of reoxidizing the Type 3 site in Type 2 depeleted laccase? (2) A comment: In our model for the Cu site in stellacyanin, we find that there is a carboxyl oxygen close to the copper (about 3A) which could supply the high field ligand you propose.

SOLOMON: (1) We've looked at several reagents (including ferricyanide) and find that of those studies only H_2O_2 and NO will oxidize the T3 site.

(2) The carboxyl oxygen would have to be $<$ 2.2A from the copper to have a strong enough bonding interaction to reproduce the unusual EPR spectrum of stellacyanin (rhombic splitting of $g_x + g_y$ and a large value of A_x, which requires some $d_{z}2$ mixing into the $d_{x}2{-}_{y}2$ ground state) and thus a C_{2v} distorted tetrahedral structure.

KOSMAN: We heard that the Cu-Cu distances in the tri-nuclear site in ascorbate oxidase may be 4A. Does this accommodate an azide bridge between the T2 and one T3 Cu?

SOLOMON: In Fig. 15, right, the T2 Cu(II)-N N N -T3 Cu(II) distance must be less than 5.2A. The Cu(II)...Cu(II) distance within the T3 pair was shown to be about 4A from simulation of our uncoupled T3 EPR signal in Fig. 9.

Oxidases and Related Redox Systems, pages 331–348

EVOLUTIONARY RELATIONSHIPS AMONG COPPER PROTEINS CONTAINING COUPLED BINUCLEAR COPPER SITES

K. Lerch and U.A. Germann

Biochemisches Institut der Universität Zürich, Winterthurerstrasse 190, CH-8057 Zürich, Switzerland

ABSTRACT

The amino acid sequences of different copper proteins containing coupled binuclear copper centers are compared. Hemocyanins from arthropods and molluscs and tyrosinases from three different species were found to share a highly homologous region in the C-terminal parts. This region contains three invariant histidines previously identified as ligands to Cu(B) in *Panulirus interruptus* hemocyanin by X-ray crystallography (Gaykema et al., Nature 309, 23-29 (1984)). In contrast, the ligand environment for the second copper, Cu(A), proved to be quite variable. It is proposed that hemocyanin and tyrosinase have arisen from a common mononuclear copper protein with the typical Cu(B) site. From this ancestral protein two types of binuclear proteins evolved independently into a tyrosinase and an arthropodan hemocyanin type. The amino acid sequence comparison between human ceruloplasmin and *Neurospora crassa* laccase together with the results from a preliminary X-ray structure analysis of zucchini ascorbate oxidase showed a close relationship in primary and most likely also in tertiary structure in the C-terminal parts of these enzymes. It is suggested that the multicopper oxidases have evolved from an ancestral copper protein which presumably contained all the ligands required for the binding of one binuclear and two additional mononuclear metal centers.

INTRODUCTION

Copper is an important prosthetic group of a number of different proteins and enzymes involved in redox reactions and the binding and activation of molecular oxygen. The copper proteins known to interact with dioxygen all contain a coupled binuclear metal center commonly referred to as "type 3" copper (1). As indicated in Table 1, these proteins fulfill a variety of different biological functions. Hemocyanin and tyrosinase which both contain a single coupled binuclear copper site with very similar structural features bind dioxygen reversibly leading to stable oxygenated forms (2-4). In contrast to hemocyanin whose main biological function is oxygen transport, tyrosinase catalyzes both the hydroxylation of monophenols to o-diphenols (monooxygenase) as well as the oxidation of o-diphenols to o-quinones (2é-oxidase).

Table 1. Properties of copper proteins containing coupled binuclear copper centers

Protein	Number of copper ions per molecule			Function(s)
	Type 1	Type 2	Type 3	
Hemocyanin			2	Oxygen transport
Tyrosinase			2	Monooxygenase/ Oxidase (2x2é - reduction)
Laccase	1	1	2	Oxidase (4é - reduction)
Ascorbate oxidase	3	1	4	Oxidase (4é - reduction)
Ceruloplasmin	2	1	3-4	Ferroxidase/ Oxidase (4é - reduction)

Laccase, ascorbate oxidase and ceruloplasmin contain in addition to a coupled binuclear copper site two different mononuclear copper sites designated as "type 1" or blue copper and "type 2" or normal copper (5). These so called multicopper oxidases couple one-electron oxidations of different substrates (phenols, aromatic amines, aminophenols, ascorbic acid, Fe^{2+}, etc.) to the four-electron reduction of dioxygen to water. As shown in Table 1, the three oxidases differ in their metal stoichiometries. Nevertheless, these copper proteins are structurally related in view of the striking similarities of the spectroscopic properties of their active site metal ions (2,6).

A wealth of information on the geometric and electronic structure of the coupled binuclear copper center has become available in recent years (2,6). Although the type 3 site of the multicopper oxidases is similar to the coupled binuclear copper center in hemocyanin and tyrosinase with respect to the magnetic properties (antiferromagnetic coupling between the two Cu(II) ions in the oxidized forms), there are also distinct differences. As was pointed out more than four decades ago by Keilin and Mann (7), laccase is not inhibited by carbon monoxide, whereas hemocyanin and tyrosinase bind this ligand reversibly giving rise to a peculiar luminescence (8). More recently, it was shown, that in contrast to hemocyanin and tyrosinase, exogenous ligands don't bridge the two copper ions at the binuclear copper center of type 2 depleted laccase (9). During the past years, the copper proteins containing coupled binuclear copper centers have been studied extensively also in terms of their protein and molecular structures. Thus the three-dimensional structure of the hemocyanin from the spiny lobster Panulirus interruptus has been determined by X-ray diffraction techniques (10). In addition, amino acid sequences of several arthropodan and molluscan hemocyanins and three tyrosinases have been reported (11-16). Moreover, the complete primary structures of human ceruloplasmin (17) and laccase from Neurospora crassa (18) have been elucidated. Finally, the determination of the three-dimensional structure of ascorbate oxidase from zucchini is well underway (19).

In this article, the evolution of hemocyanin and tyrosinase on one hand and that of the multicopper oxidases on the other hand will be discussed taking into account the new structural information of the copper proteins containing coupled binuclear copper centers.

RESULTS AND DISCUSSION

1. Evolution of hemocyanin and tyrosinase

Figure 1 shows a schematic representation of the polypeptide chains of tyrosinase from the bacterium Streptomyces glaucescens (Ty S.g.), the fungus Neurospora crassa (Ty N.c.) and the mouse (Ty mouse), and of hemocyanin from the mollusc Helix pomatia subunit d (Hc Helix d) and the arthropod Panulirus interruptus (Hc Pint a). From the three-dimensional structural analysis of P. interruptus hemocyanin it followed that the molecule is folded into three domains with the binuclear copper site located in the center of the second domain (10). The two copper atoms - designated as Cu(A) and Cu(B) - are each ligated by three histidines originating from a pair of α-helices.

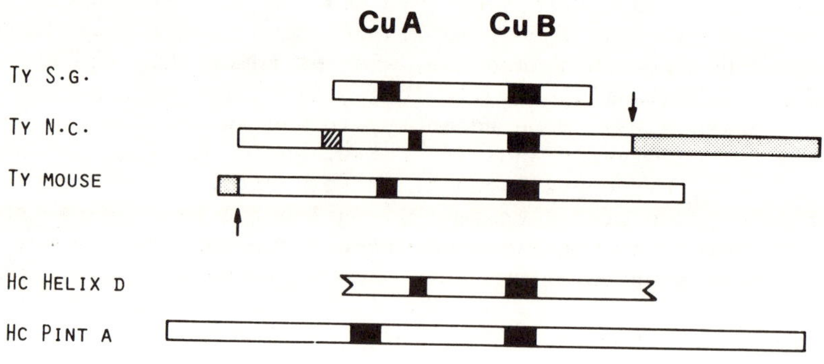

Figure 1. Schematic representation of the polypeptide chains of tyrosinase from S. glaucescens (14), N. crassa (15) and mouse (16), and hemocyanin from H. pomatia (13) and P. interruptus (12). The solid bars indicate the regions of the Cu(A) and the Cu(B) sites. The alignment of the five polypeptide chains is based on the homologous regions of the Cu(B) site. The hatched bar represents a non-functional, ancient Cu(A) site in N. crassa tyrosinase. The shaded areas indicate extensions which are post-translationally removed by limited proteolysis (site of cleavage is shown by an arrow).

A comparison of the amino acid sequences of the five proteins displayed in Fig. 1 shows them to be rather different at first sight. In the C-terminal parts (Figs. 1 and 2), however, there is a highly homologous region of 56 amino acids. The invariant and isofunctional residues comprise the three histidines known to be ligands of Cu(B) in P. interruptus hemocyanin (10). As shown in Fig. 2a two of these are separated by three other amino acids whereas the third one is located 37 residues towards the C-terminus. The flanking sequences of the third histidine show a high degree of similarity and are characterized by numerous large aliphatic and aromatic residues. To obtain the best alignment of the five sequences (Fig. 2a) major gaps had to be introduced in the three tyrosinases and H. pomatia hemocyanin. The piece which was inserted during the evolution of arthropodan hemocyanins forms part of a loop structure located between the Cu(B)-binding helical pair (10). In many proteins, such loops have been found to vary considerably in length without impairing their functional properties (20). The remarkable sequence homology beyond the loop region strongly suggests that these three histidine residues constitute the Cu(B) ligands in all five proteins quoted in Figs. 1 and 2a. This conjecture is supported both by active-site directed inactivation (21) and site-specific mutagenesis experiments (22) of N. crassa and S. glaucescens tyrosinase. From these studies the third histidine (His 306) of N. crassa tyrosinase was found to be specifically destroyed in a mechanism-based inactivation process with the concomitant loss of one copper, whereas the substitution of the first histidine (His 189) in S. glaucescens tyrosinase by asparagine resulted in an inactive enzyme containing only one copper per molecule.

In contrast to the situation of Cu(B), the sequence comparison doesn't reveal a corresponding common structure for Cu(A). The X-ray structure of P. interruptus hemocyanin (10) shows again three histidines in a helical pair with the third one separated by 27 amino acids. However, this arangement is not found in the three tyrosinases and H. pomatia hemocyanin. In N. crassa tyrosinase His 188 and His 193 have been implicated as copper ligands by photochemical oxidation (23). This segment (Fig. 2c) shows some resemblance to the Cu(A) site of two arthropodan hemocyanins from Eurypelma californicum subunit e (Eury e) and P. interruptus, although the two histidines are separated by four rather than three residues. Surprisingly, these residues are absent in S.

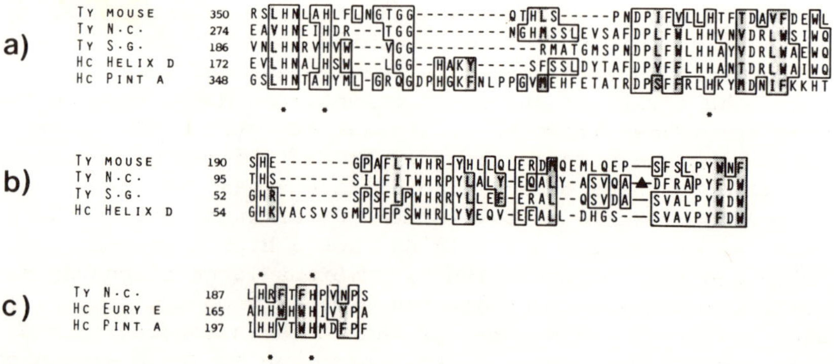

Figure 2. Alignment of amino acid sequences of tyrosinases from mouse (16), N. crassa (15), and S. glaucescens (14) and hemocyanins from H. pomatia (13), P. interruptus (12) and E. californicum (11). a) Homologous regions of 56 residues in the C-terminal part (Fig. 1) including the ligands of Cu(B) indicated by asterisks. b) Homologous regions of tyrosinases and H. pomatia hemocyanin in the N-terminal part (Fig. 1). The triangle indicates an insert of 20 residues in N. crassa tyrosinase. c) Homologous regions of two arthropodan hemocyanins and N. crassa tyrosinase including two ligands of Cu(A) designated by asterisks. The number on the left of each sequence identifies the position within the protein of the first amino acid residue shown. Positions with identical residues are boxed; those with isofunctional residues are shaded.

glaucescens and mouse tyrosinase as well as H. pomatia hemocyanin. The tyrosinase molecules and H. pomatia hemocyanin, however share a region with a rather high degree of sequence similarity (Figs. 1 and 2b). Furthermore this stretch contains an invariant histidine in a characteristic tripeptide WHR and a second histidine separated by 8 amino acids (all tyrosinases) and 15 amino acids (H. pomatia hemocyanin), respectively. It is therefore likely, that His 53 and His 62 in S. glaucescens and His 191 and His 200 in mouse tyrosinase are involved as copper ligands thus representing a new type of a Cu(A) site. This idea is supported by a recent study using site-specific mutagenesis of the S. glaucescens tyrosinase gene (24). Replacing His 62 with asparagine resulted in an inactive enzyme containing only one copper per molecule. In this context, it is of interest to note, that His 96 of N. crassa tyrosinase (corresponding to His 53 in the bacterial enzyme) is covalently linked to Cys 94 (15). This peculiar structure has been proposed earlier to be involved in the regulation of tyrosinase activity (25). Hence it is attractive to speculate that His 96 and His 105 of N. crassa tyrosinase were originally ligands to Cu(A) but during evolution were replaced by His 188 and His 193 (Fig. 2c).

Figure 3 shows an evolutionary tree of tyrosinases and hemocyanins. In view of the highly conserved Cu(B) site in tyrosinases from three very distantly related species and hemocyanins from two different phyla (molluscs and arthropods) it is likely that these proteins have arisen from a common ancestral mononuclear copper protein with the characteristic α-helical pair (10). The Cu(A) site of the arthropodan hemocyanins most probably originates from gene duplication of the Cu(B) helical pair. This hypothesis is supported by the recent finding of an internal twofold symmetry of the copper binding environment in P. interruptus hemocyanin (26). The early binuclear copper protein subsequently increased in size by addition of two new domains at the amino- and the carboxy-terminal ends to reach its present structure with three distinct domains (14). In contrast to the arthropodan hemocyanins, the generation of an early binuclear copper protein in molluscan hemocyanins and tyrosinases most probably occurred by gene fusion. As is evident from Fig. 1, the size of tyrosinase has gradually increased during the course of evolution reaching a value that is close to the average size of arthropodan hemocyanins. These changes have taken place both at the amino- and carboxy-terminal ends. Moreover, the amino-terminus of mouse tyrosinase was found to contain a signal peptide of 24 amino acids suggested to be involved in the transfer of the enzyme into the melanosomes (16).

Very recently, a structural analysis of the tyrosinase gene from N. crassa revealed the presence of a C-terminal extension (Fig. 1) of 200 amino acids (27). Although the biological significance of this structure is not clear at the present time, it is likely that it could be involved in the activation of a proenzyme by limited proteolysis. Most interrestingly, this extension shows distinct sequence similarity to the C-terminal domain of arthropodan hemocanins.

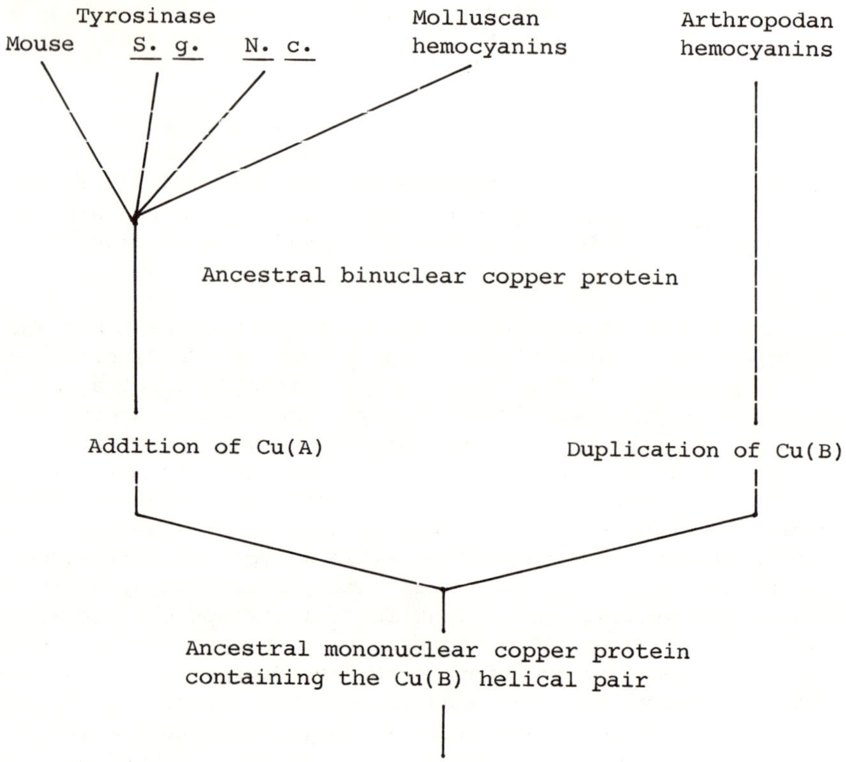

Figure 3. Schematic view of the possible evolution of tyrosinase and hemocyanin

Furthermore, secondary structure predictions of the extension revealed mostly extended structure with several β-turns. These structural elements are also very prominent in domain C of *P. interruptus* hemocyanin (10). As is evident from the X-ray structure of this hemocyanin, the amino- and the carboxy-terminal domains are shielding the binuclear copper complex very efficiently to prevent its interaction with large ligands. Hence it is quite plausible that the C-terminal extension of *N. crassa* tyrosinase may fulfill a similar function in this enzyme. Removal of this structure would then lead to the active enzyme now capable of binding and oxidizing organic substrates.

In conclusion, tyrosinases and hemocyanins have arisen from a common mononuclear copper protein containing the typical α -helical pair (10). From this ancestral protein two types of binuclear copper proteins evolved independently into a tyrosinase and an arthropodan hemocyanin type. The tyrosinase type then developed into distinctly different tyrosinase types and the molluscan hemocyanins. This pronounced variability most certainly accounts for the divergent properties and functions of the different types of hemocyanins and tyrosinases such as (i) the accessibility of the active site copper for different ligands (28), (ii) the different rates of copper removal by cyanide and copper reincorporation (2), (iii) the weak catalase activity in molluscan hemocyanins and essential absence in arthropodan hemocyanins (29), and (iv) differences in CD spectra of the oxy-forms, and in EPR spectra of half-met and dimer species, reflecting small but distinct differences in active site geometry (2).

2. Evolution of multicopper oxidases

At the present time, the complete primary structures of human ceruloplasmin (17) and *N. crassa* laccase (18) are known. Ceruloplasmin, which is a blue glycoprotein, consists of a single polypeptide chain of 1046 amino acid residues and exhibits an internal threefold homology (30). Each homology unit contains about 350 amino acids and is further subdivided into three domains designated as A1, A2, and B (Fig. 4).

In several reports (32-34) it has been suggested, that the multicopper oxidases share a common ancestry with the small blue copper proteins plastocyanin and azurin. These small proteins contain only one type 1 copper and have been studied extensively from a structural point of view. For both the poplar plastocyanin (35) and azurin from *Pseudomonas aeruginosa* (36) the three-dimensional structures are known at great precision including the ligands of the active site copper. In Fig. 5 is shown an amino acid sequence comparison of plastocyanin (37), azurin (38), the three homology units of human ceruloplasmin (17), and the C-terminal part of *N. crassa* laccase (39). It is evident, that the primary structure of *N. crassa* laccase is very similar to the C-terminal part of ceruloplasmin (CpC in Fig. 5) and to a smaller extent to the first (CpN) and second (CPM) homology units (Fig. 5) and to plastocyanin and azurin. In all six

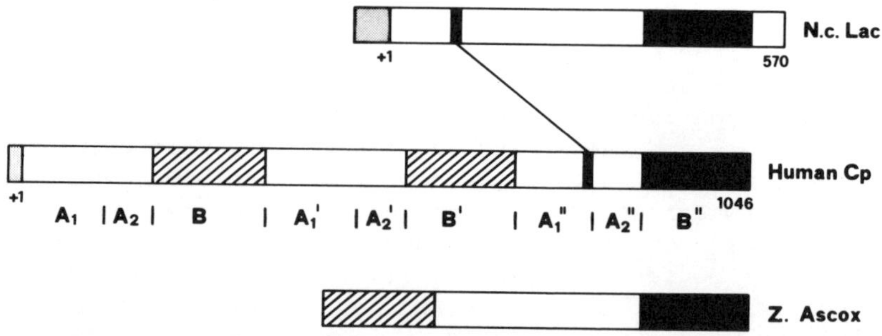

Figure 4. Schematic representation of the polypeptide chains of N. crassa laccase (18), human ceruloplasmin (30,31), and zucchini ascorbate oxidase (19). The solid bars indicate homologous regions providing ligands to the different copper types. The hatched bars represent regions containing putative type 1 copper ligands. The shaded areas indicate extensions which are posttranslationally removed. For ceruloplasmin, the subdomains of the three homology units are designated as A1, A2, and B, respectively (30). The alignment of the three proteins is based on the extensive sequence homology in the C-terminal parts of the multicopper oxidases.

polypeptides compared, five amino acids are invariant (one Cys, two Gly and two His) and four are isofunctional having a conserved aromatic (three Phe, Tyr, or Trp) or hydrophobic (one Leu or Met) character. Among the invariant ones, three represent ligands of the type 1 copper in plastocyanin (His-37, Cys-84, and His-87) and azurin (His-46, Cys-112, and His-117). The fourth blue copper ligand in plastocyanin and azurin is Met-92 and Met-121, respectively. The corresponding residues Met-690 and Met-1031 are found only in CpM and CpC and represent the putative fourth binding sites of the two type 1 copper ions of human ceruloplasmin (30). Surprisingly, however, N. crassa laccase lacks a methionine at the corresponding position. Hence, this finding implicates a type 1 copper coordination different from plastocyanin and azurin, and possibly from ceruloplasmin.

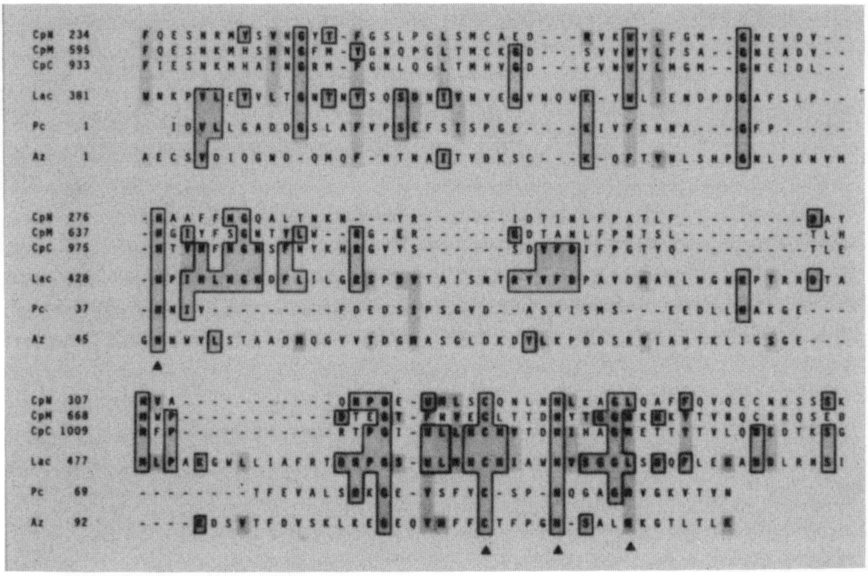

Figure 5. Amino acid sequence comparison between homologous parts of different blue copper proteins. Lac, N. crassa laccase (39), Cp, ceruloplasmin (17); CpN, first homology unit residues 234-344; CpM, second homology unit residues 595-705; CpC, third homology unit residues 933-1046; Pc, poplar plastocyanin (37), Az, P. aeruginosa azurin (38). The three homology units of ceruloplasmin were aligned according to (40). CpC and Pc were aligned according to (32) and Pc and Az according to (41) with some modifications maximizing the sequence homology of each primary structure to that of N. crassa laccase. The positions of the four copper ligands in Pc (37) and Az (38) are marked by triangles. The number on the left of each sequence identifies the position within the protein of the first amino acid residue shown. Dashes represent gaps intruduced for alignments. Identical residues are boxed, isofunctional ones are shaded.

As mentioned above, the highest sequence similarity is found when N. crassa laccase is compared with the C-terminal region of human ceruloplasmin. Thus, 114 amino acid residues of the third homology unit (CpC) of ceruloplasmin can be aligned with part of the primary structure of laccase to give 27 identical and 12 isofunctional amino acid residues (Fig. 5). Most interestingly, two short stretches show extensive sequence similarity. A computer analysis using the unitary matrix indicated maximal homology scores of 6 and 9 for the segments presented in Fig. 6 b and c, respectively (39). The remarkable similarity of these two regions suggests that they play an important role in the formation of the active site of multicopper oxidases. In particular, they could provide ligands to all three different types of copper in both enzymes. As pointed out above, at least three invariant type 1 copper binding residues are present within these two stretches. The idea that also type 2 and type 3 copper li-

```
          Lac    90   N G S I H W H G M H Q R
a)                        ● ● ●   ● ● o
          CpC   812   P Y S I H A H G V Q T E


                        •1    •3  •3  •3
          Lac   427   P H P I H L H G H D F L
b)                      ●   o ●   ● ● ●   ●
          CpC   974   L H T V H F H G H S F Q
                        •1    •3  •3  •3


                                    •3 •1 •2        •1
          Lac   492   N P G S W L M H C H I A W H V S G G L S
c)                      ● ●   ● ● o ● ● ● o     ● o     ● o
          CpC  1013   T P G I W L L H C H V T D H I H A G M E
                                    •3 •1 •2        •1  •2    •1
```

Figure 6. Amino acid sequence comparison of three regions with a high degree of similarity (a-c) of N. crassa laccase (Lac) and the C-terminal homology unit (CpC) of human ceruloplasmin (40). Numbers on the left of each sequence identify the positions within the proteins of the first residues shown. Filled circles indicate identical and open ones isofunctional residues. Potential ligands to the three different types of copper are indicated by *1,*2, and *3, respectively. The assignment of the copper ligands for N. crassa laccase is based on that proposed earlier for CpC (32).

gands could be provided by these regions (Fig. 6) is based on an earlier report (32). Using the coordinates of the known X-ray structure of poplar plastocyanin (35), a three-dimensional model of the copper active site of ceruloplasmin

plasmin was proposed (32). The model involves the C-terminal region (res. 932-1040, ref. 17) and suggests His-975, Cys-1021, His-1026, and Met-1031 as type 1; His-1022 and His-1028 as type 2; and His-978, His-980, His-982, and His-1020 as type 3 copper ligands. The amino acid sequence comparison of ceruloplasmin with laccase supports this model with respect to the type 1 (except the aforementioned lack of Met in laccase) and the type 3 copper ligands. However, no histidine residue is found in laccase corresponding to the suggested type 2 copper binding His-1028 of ceruloplasmin. Hence, these data implicate that either the structures of the type 2 site in ceruloplasmin and laccase are different or the type 2 ligands are closer to the amino terminus than suggested. As shown in Fig. 4 and 6a, there is a small segment towards the amino-terminal end of N. crassa laccase and subdomain A1" of human ceruloplasmin which shows significant sequence similarity. It is conceivable, that the two invariant histidine residues in this stretch could provide additional type 2 or type 3 ligands.

In a recent report a preliminary X-ray structure analysis of zucchini ascorbate oxidase was presented together with the amino acid sequence of the C-terminal part (19). From these data it is clear that the C-terminal part of this multicopper oxidase is homologous to those of N. crassa laccase and human ceruloplasmin. In particular, the two histidine-rich regions (Fig. 6b and c) are highly conserved. Moreover, the C-terminal part of ascorbate oxidase shows the typical folding pattern of poplar plastocyanin (35). A plastocyanin fold is also found at the N-terminal end (Fig. 4).

The finding that the C-terminal parts of the multicopper oxidases are closely related in primary and most likely also in tertiary structure together with the presence of three homology units in human ceruloplasmin (Fig. 4), suggests that the three enzymes have evolved from a common ancestral copper-binding protein. Because this protein presumably contained all the ligands required for the binding of the three types of copper, it may have been already functional as an oxidase. This hypothesis is further supported by the primary structure of N. crassa laccase (18, Fig. 4). Unlike the other multicopper oxidases, this enzyme contains only one type 1, one type 2 , and one type 3 copper site (Table 1). During the evolution of laccase, an additional domain was inserted somewhere in the middle of the ancestral copper oxidase leading to the present size of about 550 amino acids. As indicated in Fig. 4, the evolution of ceruloplasmin and ascorbate oxidase is more complex than the one of laccase. For ceruloplasmin it has been suggested (17, 32) that the molecule evolved from the primordial oxidase by tandem triplication. After this event, some of the copper sites in the two new regions were lost with the exception of the type 1 sites. In human ceruloplasmin one of these is located in CpM which is a blue type copper, whereas the second one resides in CpN which based on physical-chemical analyses may be present permanently in the monovalent state (32). In this context it is of interest to note that the blood clotting factors V (41) and VIII (42) contain regions which are homologous to the three repeat units in human ceruloplasmin. The first and the third A domains of factor VIII contain all the ligands for a typical type 1 site. It is, however, not known wether this protein actually binds copper. On the other hand, factor V was reported to contain one copper per molecule with spectroscopic properties different from those of type 1 and type 3 (43).

The evolution of ascorbate oxidase is not clear at the present time. Nevertheless, the presence of a second plastocyanin fold at the amino-terminus (19, Fig. 4) suggests the fusion of a plastocyanin gene with the one of the primordial copper oxidase rather than a gene duplication.

As shown in Fig. 4, human ceruloplasmin (31) and N. crassa laccase (19) are synthesized as preforms. The amino-terminal leader peptides are rich in hydrophobic residues and display a positively charged amino-terminus. Thus, both peptides contain the basic elements of a secretory signal sequence. This finding is in agreement with the biological function of ceruloplasmin and laccase as exoenzymes.

ACKNOWLEDGMENTS

This work was supported by Swiss National Science Foundation grants 3.285-0.82 and 3.236-0.85 and the Kanton of Zürich.

REFERENCES

1. Solomon, E.I., Penfield, K.W., and Wilcox, D.E. (1983) Structure and Bonding 53,1-57
2. Solomon, E.I. (1981) in "Copper Proteins" (Spiro, T.G., ed) pp. 41- 108, Wiley, New York
3. Lerch, K. (1981) in "Metal Ions in Biological Systems" (Sigel, H., ed) pp. 143-186, Marcel Dekker, New York
4. Lerch, K. (1983) Mol. Cell. Biochem. 52, 125-138
5. Fee, J. (1975) Structure and Bonding 23, 1-60
6. Reinhammar, B. and Malmström, B.G. (1981) in "Copper Proteins" (Spiro ,T.G., ed) pp. 109-149, Wiley, New York
7. Keilin, T. and Mann, T. (1939) Nature 143, 23-25
8. Kuiper, H.A., Lerch, K., Brunori, M., and Finazzi-Agrò (1980) FEBS Lett. 111, 232-234
9. Spira, D.J. and Solomon, E.I. (1987) J. Am. Chem. Soc., in press
10. Gaykema, W.P.J., Volbeda, A., and Hol, W.G.J. (1986) J. Mol. Biol. 187, 255-275
11. Schartau, W., Eyerle, F., Reisinger, P., Geisert, H., Storz, H., and Linzen, B. (1983) Hoppe-Seyler's Z. Physiol. Chem. 364, 1383-1409
12. Linzen, B., Soeter, N.M., Riggs, A.F., Schneider, H.-J., Schartau, W., Moore, M.D., Yokota, E., Behrens, P.Q., Nakashima, H., Takagi, T., Nemoto, T., Vereijken, J.M., Bak, H.J., Beintema, J.J., Volbeda, A., Gaykema, W.P.J., and Hol, W.G.J. (1985) Science 229, 519-524
13. Drexel, R., Siegmund, S., Schneider, H.-J., Linzen, B., Gielens, C., Préaux, G., Lontie, R., Kellermann, J., and Lottspeich, F. (1987) Hoppe-Seyler' Z. Physiol. Chem. 368, 617-635
14. Huber, M., Hintermann, G., and Lerch, K. (1985) Biochemistry 24, 6038-6044
15. Lerch, K. (1982) J. Biol. Chem. 257, 6414-6419
16. Shibahara, S., Tomita, Y., Sakahura, T., Nager, C., Chaudhuri, B., and Müller, R. (1986) Nucl. Acids Res. 14, 2413-2427
17. Takahashi, N., Ortel, T.L., and Putnam, F.W. (1984) Proc. Natl. Acad. Sci. USA 81, 390-394
18. Germann, U.A., Müller, G., Hunziker, P.E. and Lerch, K., J. Biol. Chem., in press

19. Messerschmidt, A., Rossi, A., Ladenstein, R., Bolognesi, M., Gatti, G., Marchesini, A., Petruzelli, R., and Finazzi-Agrò, A. (1987) 18th FEBS Meeting, Liubljana, Jugoslavia (Abstract)
20. Stroud, R.M., Kay, L.M., and Dickerson, R.E. (1971) Cold Spring Harbor Symp. Quant. Biol. 36, 145-140
21. Dietler, C. and Lerch, K. (1982) in "Oxidases and Related Redox Systems" (King, T.E., Morrison, M, and Mason, H.S., eds) pp. 305-317, Pergamon Press, Oxford and New York
22. Huber, M. and Lerch, K., submitted
23. Pfiffner, E. and Lerch, K. (1981) Biochemistry 20, 6030-6035
24. Huber, M. and Lerch, K. (1986) in "Invertebrate Oxygen Carriers" (Linzen, B., ed) pp. 265-276, Springer-Verlag, New York
25. Lerch, K. (1978) Proc. Natl. Acad. Sci. USA 75, 3635-3639
26. Hol, W.G.J. and Volbeda, A. (1987) Rec. Trav. Chim. Pays-Bas 106, 355
27. Niedermann, D. and Lerch, K., in preparation
28. Winkler, M., Lerch, K., and Solomon, E.I. (1981) J. Am. Chem. Soc. 103, 7001-7003
29. Ghiretti, F. (1956) Arch. Bioch. Biophys. 63, 165-176
30. Ortel, T.L., Takahashi, N., and Putnam, F.W. (1984) Proc. Natl. Acad. Sci. USA 81, 4761-4765
31. Koschinsky, M.L., Funk, W.D., vanOost B.A., and MacGillivray, R.T. (1986) Proc. Natl. Acad. Sci. USA 83, 5086-5090
32. Rydén, L. (1982) Proc. Natl. Acad. Sci. USA 79, 6767-6771
33. Dwulet, F.E. and Putnam, F. (1981) Proc. Natl. Acad. Sci. USA 78, 2805-2809
34. Rydén, L. (1984) in "Copper Proteins and Enzymes" (Lontie, R., ed) Vol. 1, pp. 157-182, CRC, Boca Raton, FL
35. Guss, M.J. and Freeman, H. (1983) J. Mol. Biol. 169, 521-563
36. Adman, E.T., Stenkamp, R.E., Sieker, L., and Jensen, L.H. (1978) J. Mol. Biol. 123, 35-47
37. Ambler, R.P. (1980) in "Poplar Plastocyanin Structure" (Guss, J.M. and Freeman, H., eds), Protein Data Bank, Brookhaven National Laboratory, Upton, NY

38. Ambler, R.P. (1971) in "Recent Developments in the Chemical Studies of Protein Structures"(Previero, A., Pechère, J.-F., and Coletti- Previero, M.-A., eds), pp. 289-305 Institut National de la Santé et de la Recherche Médicale, Paris
39. Germann, U.A. and Lerch, K. (1986) Proc. Natl. Acad. Sci. USA 83, 8854-8858
40. Takahashi, N., Baumann, R.A., Ortel, T.L., Dwulet, F.E., Wang, C.-C., and Putnam F.W. (1983) Proc. Natl. Acad. Sci. USA 80, 115-119
41. Kane, W.H. and Davie, E.W. (1986) Proc. Natl. Acad. Sci. USA 83, 6800-6804
42. Vehar, G.A., Keyt, B., Eaton, D., Rodriguez, H., O'Brien, D.P., Rotblat, F., Oppermann, H., Keck, R., Wood, W.I., Harkins, R.N., Tuddenham, E.G.D., Lawn, R.M., and Capon, D.J. (1984) Nature 312, 337-342
43. Mann, K.G., Lawler C.M., Vehar, G.A., and Church, W.R. (1984) J. Biol. Chem. 259, 12949-12951

DISCUSSION

RYDEN: Since you have determined the entire sequence of *Neurospora* laccase, how do the N-terminal 400 residues compare with the ceruloplasmin sequence?

LERCH: With the exception of a small histidine-rich region (see Fig. 6a) there is no obvious sequence homology between the two proteins.

SANDERS-LOEHR: In searching for a long axial ligand in type I sites, we may have been unduly influenced by the fact that the first two crystal structures (plastocyanin and azurin) showed an axial methionine. In the high-resolution azurin structure of Ted Baker there is an axial carbonyl group at about the same distance from the copper as the axial methionine. Perhaps we should be giving greater consideration to the possibility of carbonyl bonding in the axial location.

LERCH: I agree with you. As I pointed out, the axial methionine in *N. crassa* laccase is replaced with leucine. We have considered a nearby methionine (met 499) as a potential ligand. However, this residue can be ruled out as a ligand based on the structural comparison to plastocyanin (H. C. Freeman, personal communication). Hence, the question of the nature of the fourth ligand in *N. crassa* laccase is presently unresolved.

SOLOMON: In response to Joann Loehr's comment, carbonyl oxygen at about 3.0 A should not be considered as bonding to the copper. A methionine S at 2.9 A should be. Hence, this site should be viewed as an elongated C_{3v} tetrahedral structure as in plastocyanin and the Seattle azurin structure.

Oxidases and Related Redox Systems, pages 349–366

EVOLUTION OF BLUE COPPER PROTEINS

Lars Rydén

Department of Biochemistry, Uppsala University, Biomedical Center, Uppsala, Sweden

ABSTRACT

The evolutionary relationships of blue copper proteins are reviewed. Five homologous families of small blue proteins are recognized. Despite differences in length their peptide chains can all be accomodated into the eight-stranded fold of plastocyanin with some adjustments at three of the loops and the two termini. The C-termini of the blue oxidases ceruloplasmin and *Neurospora* laccase also fit into this fold and they are suggested to be homologous to the small blue proteins. The alignment of their amino acid sequences suggest some of the histidines to be binding active site copper. A superposition of the structures of poplar plastocyanin and bovine Cu-Zn superoxide dismutase (SOD) showed that 68 out of 99 alpha-carbons in plastocyanin overlapped with corresponding atoms in SOD with a rms distance of 2.99 Å. In addition three of the histidine residues that were proposed to be copper-binding in laccase and ceruloplasmin aligned with ligands to the Cu-Zn pair in a SOD. Thus also SOD might be related to the blue proteins.

INTRODUCTION

Many of the proteins involved in oxygen metabolism are beautifully colored in yellow, red or blue. Most spectacular are perhaps the intensely blue proteins. Some of them are simple redox proteins which can accept and deliver a single electron. These are known as the small blue proteins.

Some of them are simple redox proteins which can accept and deliver a single electron. These are known as the small blue proteins (1,2). The others are active enzymes, the blue oxidases (3-5). All blue proteins owe their color to one or several copper ions bound in a specific way.

Blue proteins are found in all types of cells, in bacteria as well as in plant and animal tissues. Physiologically they are involved in bacerial oxidative metabolism, in the evolution of oxygen in algal and plant photosynthesis as well as the consumption of oxygen in cellular oxidations.

All evidence suggests that this group of proteins have a single origin and that the original blue protein appeared on earth very early in the evolution of life, perhaps more than three billion years ago and then evolved into a number of different forms as life itself diversified. Here I will try to trace the history of these events by using the amino acid sequences and three-dimensional structures of the proteins of today. Some of the relations are very distant and can only be ascertained by using special techniques. In other cases the description of an intermediary form, the missing link, has demonstrated the relation. The contradictory hypothesis that the blue copper proteins have two or more independent origins can in theory be tested if enough data is available. There is, however, at present no evidence in favor of it.

Protein-bound copper is generally classified into one of three groups (3). First there is the intensely blue copper called type 1, with a molar absorption coefficient in the order of 5,000 at the wavelength of maximal absorption, around 600 nm. Type 2 copper is about 100-fold less blue and is often called non-blue or normal. Type 3 coppers, which are neither blue nor paramagnetic, forms diamagnetic pairs with a short distance between the two metal ions. The blue proteins all contain type 1 copper. They might be related to other proteins which lack this copper and has another one, or for tht matter, to proteins without any copper at all, as will be shown below.

SMALL BLUE PROTEINS

Occurrence and general properties

The small blue proteins have peptide chains of about 100-120 amino acid residues and a single type 1 copper per mole (1). The considerably higher molecular weights, typical for some of the plant small blue proteins, are due to bound carbohydrate.

A bacterial small blue protein was first isolated from *Pseudomonas aeruginosa* by Horio in 1958 (6). Similar proteins were later isolated from other *Pseudomonas* and *Bordetella* species and given the name azurin (7). Ambler and coworkers sequenced nine of them and proved the azurins to be a family of closely homologous proteins (8).

Several other blue proteins have since then been isolated from *Achromobacter* (9), *Paracoccus* (10), *Thiobacillus* (11) and also from other strains of *Pseudomonas* (9). Amino acid sequences determined in Ambler's laboratory (9) have allowed us to identify two new families of bacterial small blue proteins, called amicyanins and pseudoazurins. The data are summarized in Table 1. The protein from *Thiobacillus*, called rusticyanin, might belong to a sixth group. This protein has been suggested to have a role in the oxidation of iron in the bacteria. The amicyanins take part in the oxidation of C1-substrates, such as methylamine.

Also in plants several different small blue proteins are found. Best known are the plastocyanins (12) involved in oxygenic photosynthesis. They have been isolated from chloroplasts from a great number of higher plants, from algi and also from the procaryotic cyanobacteria, earlier called blue-green algae. The amino acid sequences, partial or complete, of 70 plastocyanins have been determined mostly by Boulter and coworkers for the purpose of studying plant phylogeny (1). The plastocyanins form a well defined protein family. In the chloroplasts plastocyanin receives electrons from cytochrome f at the end of the photosystem II redox chain and reduces P in photosystem I.

Small blue proteins are also found in the non-photosynthetic tissues of flowering plants. The discoverers have freely used their priority to give these proteins new names and they have thus become known as Stellacyanin (13), Ume-

TABLE 1. Small blue proteins

Name	Source	Peptide length	Function	Reference
Azurin	Pseudomonas Alcaligenes Bordetella	128-129	unknown	(1,2,6-8)
Pseudoazurin	Pseudomonas Achromobacter Alcaligenes	123	unknown	(9) (9) (17)
Amicyanin	Pseudomonas Paracoccus Methylomonas	99 129 121	oxidation of methyl-amin	(9) (10) (10)
Plastocyanin	Green plants Cyanobacteria	99 105	oxygenic photosyn-thesis	(1,12)
Phytocyanin	Lacquer tree Horse radish Cucumber	107 125 96	unknown	(13,25) (14) (18)

cyanin (14), Mavicyanin (16) etc. But the amino acid sequence data available show that they reasonably well fit into a single protein family, despite differences in peptide chain length. I thus suggest that these proteins are given the single name phytocyanin to remind of their presence in plants (1). The claim of the presence of a quite different protein, plantacyanin (16), in cucumber peelings seem doubtful in the light of its similarity to a basic blue protein from cucumber (17). The amino acid sequence of this latter protein proved it to be a phytocyanin.

The phytocyanins have no known physiological function. They differ from the plastocyanins in their basic character (plastocyanin is acidic) and the presence of covalently bound carbohydrate, which for some of them accounts for up to 40 % of their weight.

Structures of azurin and plastocyanin

The structures of poplar plastocyanin (19) and Pseudomonas aeruginosa azurin (20) were both reported in 1978. Later a much refined structure of plastocyanin (21) in both oxidized and reduced form have been reported as well as an additional azurin structure (22). These structures have allowed us to study the details of the copper-binding site in the small blue proteins and also to compare the two families.

The two structures are shown in Figure 1. Plastocyanin is made up of eight strands roughly parallel to the long axes of the elongated molecule. Seven of the strands are in correct conformation for beta structure, while the middle fifth strand is irregular and even contain a turn of helix. A structure like that of plastocyanin is often referred to as a beta barrel, but it would be more correct to talk of a double sandwich in which each slice is made up of four strands. The hydrophobic insides of the two slices are dominated by seven phenylalanine side chains.

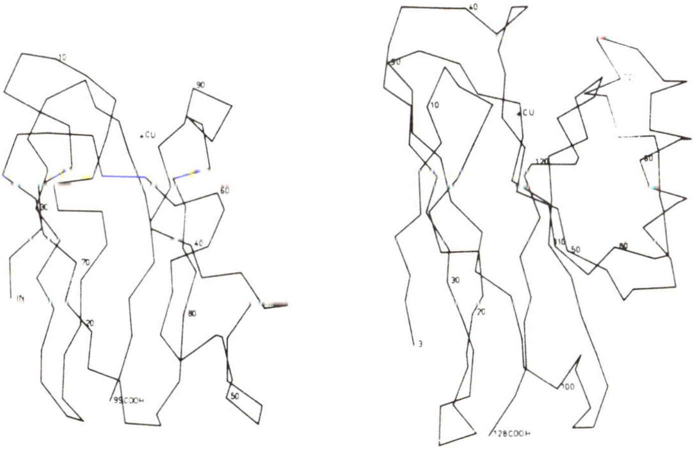

Figure 1. Structure of plastocyanin (left) (21) and azurin (right) (20).

Also azurin (20,22) contains eight strands of which the middle fifth is more irregular. In addition there is a big "external loop" in the structure inserted at the fifth strand. This loop accounts for most of the difference in length of the peptide chains in the two proteins. The plastocyanins contains 99 residues and the azurins 129 with minor variations.

The type 1 copper is bound in identical ways in the two proteins with two histidines, one cysteine and one methionine as liganding side chains. The ligands are also found in corresponding parts in the two proteins. The results thus completely confirm the previous conclusion (23) that the azurins and plastocyanins are homologous, which then was based on sequence comparisons.

Chothia (24) has analyzed the difference between the two structures in detail. He concludes that the secondary structure elements, "the two bread slices", are unchanged in spite of numerous amino acid replacements, but that they have been displaced from each other by an increased angle of 15 degrees and a translation of 3.8 Å.

Comparison between small blue proteins

The copper-binding residues provide anchor points when comparing the sequences of the small blue proteins, comparable to active site residues in enzymes. They can be easily aligned when comparing the azurins, the pseudoazurins, and the amicyanins with plastocyanin. For the phytocyanins it is not quite as simple. The two copper-binding histidine residues and the cysteine are conserved. Although it was originally claimed that the thiol group homologous to the copper-binding thiol in plastocyanin is involved in the disulphide bridge (25) that these proteins have, this was later refuted (26). In addition the methionine being the fourth ligand in plastocyanin is not present in Stellacyanin. Still the alignment scores obtained, 4 or more standard deviations, clearly indicate homology to the plastocyanins and thus the small blue proteins in general.

The alignments obtained in combination with the known three-dimensional structure of plastocyanin give an indication how the different lengths of the peptide chains in the five families of proteins can be accomodated in the eight-stranded "double sandwich" (Table 2). When using plastocyanin as reference three of them, pseudoazurin, amicyanin and the phytocyanins, have a short extension at the first strand. The pseudoazurins and Umecyanin, a phytocyanin, have a long C-terminal tail easy to identify because of the copper-binding residues close to the plastocyanin C-termi-

TABLE 2. Accomodation of peptide chains of the small blue proteins of different lengths into the plastocyanin folding. The figure refers to number of amino acid residue differences as compared to plastocyanin (99 residues). (Some gaps are not included.)

	N-terminus and/or bend 1-2	bend 3-4	strand 5 with bends	bend 7-8	C-terminus
Plastocyanin					
Azurin		+7	+17	-1	+1
Pseudoazurin	+4		-7		+31
Amicyanins	+9		-8	-2	
Phytocyanins	+(2-10)		-4	+2	+(0-21)

nus. Other differences in length are accomodated in the turns between strands 3 and 4 and, more surprisingly, the copper-binding turn between strands 7 and 8. Also the irregular fifth strand differs in length, being longest in azurin (45 residues) with its large extra loop and shortest (20 residues) in the Achromobacter amicyanin.

The now identified five families of small blue proteins provide a group of evolutionary related electron carriers just as diversified in structural and functional properties and potentially as rewarding for a more detailed study as do for example the better known cytochromes and ferredoxin--rubredoxin families.

The ragweed allergen

In 1985 the interesting discovery was published that the major ragweed allergenic protein was related to the phytocyanins (27). This protein is best known for the trouble it causes victims of heyfever but to the plant it is important for the pollen's entry into the germ. The protein does not contain copper and is not blue. It is also evident from the alignment that the copper-binding residues have been replaced for non-binding residues when this protein evolved from the phytocyanins. The allergen is only one example of how completely unexpected functions can evolve from the blue proteins. A qualitative evolutionary tree of the small blue proteins is shown in Fig. 2.

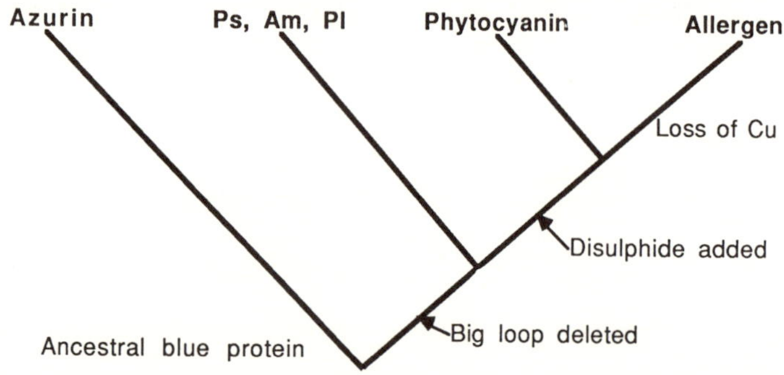

Figure 2. Evolutionary tree of small blue proteins.

THE BLUE OXIDASES

Occurrence and general properties

In the context of oxygen metabolism the blue oxidases are of special interest since they have the capacity to reduce oxygen all the way to water. For this they need four electron-carrying groups, and they also all have four or more copper atoms. This makes them much larger and considerably more complex than the small blue proteins.

The blue oxidases consist of ascorbate oxidase from plant tissue (5), laccase originally isolated from lacquer tree but later also from several fungi (3) and ceruloplasmin from sera of vertebrate animals (4). All of them contain a single peptide chain and an amount of covalently bound carbohydrate. Some properties of the blue oxidases are summarized in Table 3.

Best known of the blue oxidases is human ceruloplasmin (4), which has been studied as a serum protein and a protein involved in several diseases. The amino acid sequence of the 1046 residues long peptide chain is known since 1984 (28). In spite of considerable efforts a complete picture of the physiological relevance of ceruloplasmin has not emerged. It appears that the enzyme is important for iron metabolism and oxidizes divalent iron to promote its uptake by transferrin and liver cells. There are a great number of substrates for ceruloplasmin, and no substrate specificity in the usual sense. Rather those compounds that have a suit-

TABLE 3. The blue oxidases

Name	Source	Peptide length	Number of coppers	Function	Ref.
Ascorbate oxidase	Cucumber	1100 (dimer?)	8-10	unknown	(5)
Laccase	Lacquer tree	540	4	oxidation of urushi lac	(3)
	Fungi	500	4		
Ceruloplasmin	Mammalian serum	1046	6-7	multiple	(4)

able redox potential are oxidized by ceruloplasmin as they transfer one electron at a time.

Fungal laccases are involved in the oxidation of lignin and tree laccase in the oxidative hardening of the lacquer. The fungal laccases are enzymes with only four copper atoms and a peptide chain of no more than about 500 residues. It is therefore the least complex of the blue oxidases and has been the enzyme of choice for those studying the properties of the active site copper ions and the enzyme mechanism.

Evolution of the blue oxidases

Spectroscopic and enzymatic properties of the three blue oxidases mentioned are so similar that one would suggest they all are related. The differences in their physico--chemical properties such as the differences in peptide chain length and number of bound coppers show that the relationship still cannot be very close. A stretch of 242 amino acid residues deduced from a partial gene sequence of *Neurospora* laccase (29) and an active site peptide from *Polyporus* laccase (30) have been published. The *Neurospora* sequence was easily aligned with the C-terminal end of ceruloplasmin, and confirms the homology between the two enzymes.

The ceruloplasmin sequence of 1046 residues is composed of three closely homologous regions of about 340 residues. It is thus apparent that the ceruloplasmin ancestor was a 340 residue long protein (31). Since these regions are more similar to each other than to laccase (29) it is concluded that this smaller precursor was an active oxidase containing four copper ions. But today's ceruloplasmin does not contain 3x4=12 copper ions. Most of them have apparently been lost after the triplication. One of the homology regions must obviously contain the four coppers that make up the active site. Again the comparison with laccase is consistent with the suggestion (30) that it is the C-terminal one.

Relation to the small blue proteins

When the sequence of the C-terminal part of ceruloplasmin was available it was possible to examine a possible relationship with the small blue proteins for this part of the sequence. The potential copper-binding residues His, His, Cys and Met was present in the sequence in the proper order. A new method was deviced to determine exactly which part of the ceruloplasmin sequence should be compared to the shorter plastocyanin peptide chain. When this was applied an alignment score of 4.3 or more between the two proteins was obtained.

The alignment obtained was tested by fitting the amino acid sequence of the ceruloplasmin fragment to the known three dimensional structure of plastocyanin (31). In the model obtained deletions and insertions in all cases were positioned in the bends between the eight strands. The important copper-binding residues as well as the hydrophobic core residues were conserved all confirming that in fact this part of ceruloplasmin had the same chain-folding as plastocyanin.

But for those that consider a model built on a graphic display to be rather doubtful construction nature itself provided a further confirmation. When the amino acid sequence of the middle homology region in ceruloplasmin - expected to have the same folding - became known it was shown to contain a disulfide bridge not present in the C-terminal part. This was consistent with the model. The two residues that corresponded to the cysteines in the middle region were in close proximity in the model although they were 85 residues apart in the linear sequence. Also the deletion of three residues in the middle homology region could very easily be accomodated in the model (Fig. 3).

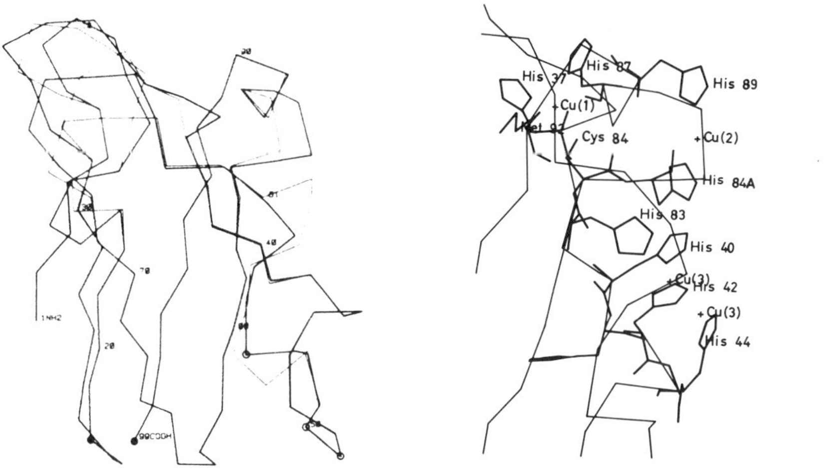

Figure 3. Model of ceruloplasmin (left) and a detailed view of a cluster of 8 histidine residues (right) on the right side of the model (31). Three copper ions, marked by crosses, were added to the histidines to form a possible active site structure. The model is based on the ceruloplasmin-plastocyanin homology, and the thin line shows the plastocyanin peptide chain as a background object. The two residues marked by solid circles are connected by a disulfide bridge and the three marked by open circles are deleted in ceruloplasmin's middle homology region.

It thus seems that the ancestral oxidase of 340 residues evolved from a small blue protein, a plastocyanin or a close relative to plastocyanin. Since the part of ceruloplasmin that is clearly homologous to plastocyanin is found at the C-terminal 120 residues the question of the origin of the rest of the 340 residues remains. There is a weak evidence for a duplication in the 340 residues long peptide. It is possible that this is the only similarity that remains from an original doubling of the plastocyanin sequence. This hypothesis can be tested when a complete structure of one of the blue oxidases is available.

Evolution of the active site

The addition of three copper ions to make up an active site in the blue oxidases requires that new liganding amino acids evolve. A suggestion for which these are could be advanced after having constructed the model of the ceruloplasmin C-terminus (31). A very conspicuous cluster of histidine residues was found on the side of the molecule where the type 1 copper site was located (Fig. 3). It was quite easy to accomodate an additional three copper ions in this region. Two of these were close to each other as expected for the type 3 copper and one further away as expected for the type 2 copper. The histidines that in this way are hypothesized to bind types 1, 2 and 3 copper are conserved in the laccase sequence - with the exception of one type 2 ligand - but not in the two additional homology regions of ceruloplasmin in accordance with the hypothesis. The ligands to type 1 copper are conserved in all of these with the important exception that methionine is substituted for leucine in the laccase sequence and the N-terminal ceruloplasmin region. One is reminded here of the same substitution in at least one of the phytocyanins (Stellacyanin) and the observation (22) that in azurin the methionine sulphur is so far away from the copper that it is just barely acceptable as a ligand atom.

Blood coagulation factor VIII

In 1984 a further surprise was presented to the biochemists interested in blue proteins, when the complete nucleotide sequence of the gene for blood coagulation factor VIII was published (32). Factor VIII turned out to have a 2332 residues long peptide chain composed of several regions. Three of these were closely homologous to the three homology regions in ceruloplasmin. These alone accounted for close to 50 % of the factor VIII structure.

When examining the new sequence it is found that the type 1 copper liganding amino acids are conserved in the first and last homology region. It is thus reasonable to conclude that factor VIII binds copper and probably also is blue. This has not yet been confirmed experimentally due to difficulties in obtaining enough protein. It is not likely that factor VIII is an oxidase. The residues postulated to bind the other active site coppers in the ceruloplasmin C--terminal region are only partly conserved in the factor VIII structure.

It should finally be added that blood coagulation factor V is closely related to factor VIII and certainly also contains the three homology regions. At present sequence information is, however, lacking. A qualitative evolutionary tree for the blue oxidases are presented in Fig. 4.

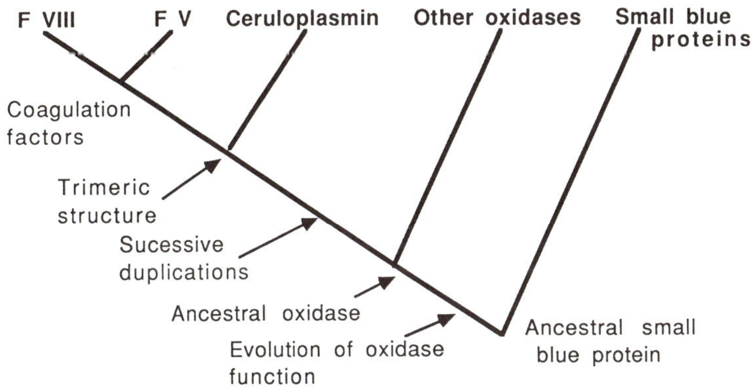

Figure 4. Evolutionary tree of the blue oxidases.

COPPER-ZINC SUPEROXIDE DISMUTASE

Relationships between SOD and the blue proteins

Copper-zinc-containing superoxide dismutase (SOD) is blue and was once isolated and characterized as a blue copper protein (33). Its zinc content and enzymatic activity was discovered much later. The blue color of the SOD is, however, much less intense than that of the blue oxidases and its metal ligand structure, with a zinc-copper pair, is completely different (34).

Some similarities between the amino acid sequences of ceruloplasmin and SOD have been observed since long and motivated speculations about similarities in function. An alignment of the bovine enzyme with human ceruloplasmin confirms the similarity but only on the level of 2.2 standard deviations in alignment score and this is to weak to allow definite conclusions. A similar comparison between the small blue proteins and SOD gave negative results.

Comparison of the folding of SOD and plastocyanin

Since structures of both SOD and plastocyanin, known to be related to ceruloplasmin, are available the possibility remained to explore the relationship on the level of peptide chain-folding. SOS is as plastocyanin made up of eight strands of approximate beta structure. In addition there are three more or less extensive loops which account for SOD's longer chain, 151 residues as compared to the 99 residues of plastocyanin. The topologies are, however, not identical since strands 1-3 have different relationships to each other in the two proteins.

Taking these differences into account a trial to fit the two structures to each other was made. A least squares fit between the alfa carbons of predetermined residues in the two proteins was obtained using one residue from each of the eight strands. After visual inspection and slight rotation several new residues were selected as a basis for a new fit of the two structures. Several iterations were done to identify all reasonable correspondences.

Out of the 99 residues in plastocyanin 64 was overlapping with residues in SOD with a rms distance of 2.77 Å. If the best 72 residues were used the distance increased to 3.28 Å. Strand 1, which points in different directions in the two proteins, did not overlap at all. The additional residues in SOD were found in loops between strands 4-5, 6-7 and 7-8.

There is no simple way to evaluate the similarities obtained, and further sequence comparisons are needed. Normally the overlap of 2/3 of the residues with a rms distance of less than 3 Å is considered to indicate homology. The comparison of the structure alignment obtained with the sequence alignment between plastocyanin and ceruloplasmin gives the startling observation that three of the ligands to the copper-zinc pair in SOD corresponds to copper ligands in ceruloplasmin. Two correspond to the first two histidines in the HFHGH sequence suggested to bind type 3 copper. The third is a zinc ligand that is homologous to the last histidine in the HCH sequence suggested to bind type 1, 2 and 3 copper. All these histidines are conserved in laccase. Again the results suggest that there is an evolutionary relationship between SOD and the blue oxidases. In fact the zinc site in SOD easily accomodates a copper ion so that a Cu-Cu SOD can be synthesized.

ACKNOWLEDGEMENTS

The least squares fit between the SOD and plastocyanin structures was carried out with a program written by Alwyn Jones at the Department of Molecular Biology of Uppsala University. The advice of Per Kraulis in handling the program and the graphic display is gratefully acknowledged.

REFERENCES

1 Rydén L (1984). In: Copper Proteins and Copper Enzymes, Vol 1 (Lontie R, Ed) CRC Press, Boca Raton, pp 157-182.
2 Adman AT (1985). In: Metalloproteins, Part I (Harrison, PM, Ed) Macmillan, London.
3 Reinhammar B (1984). In: Copper Proteins and Copper Enzymes, Vol 3 (Lontie R, Ed) CRC Press, Boca Raton, pp 157-182.
4 Rydén L (1984). In: Copper Proteins and Copper Enzymes Vol 1 (Lontie R, Ed) CRC Press, Boca Raton, pp 157-182.
5 Rydén L (1984). In: Copper Proteins and Copper Enzymes, Vol 1 (Lontie R, Ed(CRC Press, Boca Raton, pp 157-182.
6 Horio T (1958). J Biochem (Tokyo) 45:195.
7 Sutherland IV, Wilkinson JF (1963). J Gen Microbiol 30: 105.
8 Ambler RP (1971). In: Recent Developments in the Chemical Study of Protein Structures (Previero A, Pechere JF, Coletti-Previero MA, Eds) INSERM, Paris, p 289.
9 Ambler RP, Tobari J (1985). Biochem J 232:451-457.
10 Husain M, Davison L (1986). Biochemistry 25:2431-2436.
11 Cobley JG, Haddock BA (1975). FEBS Lett 60:29.
12 Katoh S, Takamiya A (1961). Nature 189:665.
13 Peisach J, Levine WG, Blumberg, WE (1967). J Biol Chem 242:2847.
14 Stigbrand T (1971). Biochim Biophys Acta 236:246.
15 Marchesini A, Minelli M, Merkle H, Kroneck PMH (1979). Eur J Biochem 101:77.
16 Markossian KA, Aikazyan VTs, Nalbandyan RM (1974). Biochim Biophys Acta 359:47.
17 Hormel S, Adman E, Walsh KA, Beppu T, Titani K (1986). FEBS Lett 197:301-304.
18 Murata M, Begg GS, Lambrou F, Leslie B, Simpson RJ, Freeman H, Morgan FJ (1982). Proc Natl Acad Sci USA 79:6434.

19 Colman PM, Freeman HC, Guss JM, Murata M, Norris VA, Ramshaw JAM, Venkatappa MP (1978). Nature 272:319.
20 Adman ET, Stenkamp RE, Sieker LC, Jensen LH (1970). J Molec Biol 123:35.
21 Guss JM, Freeman HC (1983). J Molec Biol 169:521-563.
22 Norris GE, Anderson BF, Baker EN (1983). J Molec Biol 165:501-521.
23 Rydén L, Lundgren JO (1976). Nature 261:344-346.
24 Chothia C, Lesk A (1982). J Molec Biol 160:309-323.
25 Bergman C, Gandvik EK, Nyman PO, Strid L (1977). Biochem Biophys Res Commun 77:1052.
26 Engeseth, HR, Hermodson, MA, McMillan, DR (1984). FEBS Lett 171:257-261.
27 Hunt LT, George DG, Yeh LS (1985). J Molec Evol 21:126-132.
28 Takahashi N, Ortel TL, Putnam FW (1984). Proc Natl Acad Sci USA 81:390-394.
29 Germann UA, Lerch K (1986). Proc Natl Acad Sci USA 83: 8854-8858.
30 Briving, C, Gandvik EK, Nyman PO (1980). Biochem Biophys Res Commun 93:454-461.
31 Rydén L (1982). Proc Natl Acad Sci USA 79:6767-6771.
32 Vehar GA, Keyt B, Eaton D, Rodriguez H, O'Brian DP, Rotblat F, Oppermann H, Keck R, Wood WI, Harkins RN, Tuddenham EGD, Lawn RM, Capon DJ (1984). Nature 312: 337-342.
33 Fielden EM, Rotilio G (1984). In: Copper Proteins and Copper Enzymes, Vol 2 (Lontie R, Ed) CRC Press, Boca Raton, pp 27-61.
34 Richardson JS, Thomas KA, Rubin BH, Richardson DC (1975). Proc Natl Acad Sci USA 72:1349.

DISCUSSION

MASON: It appears that blue copper protein evolved in the anaerobic era of the biosphere about 3.5 billion years ago. Heme must have evolved in the same anaerobic era because cytochromes occur in anaerobic photosynthetic organisms. Thus, ancestral proteins of aerobic metabolism and respiration seem to have evolved before the advent of an aerobic atmosphere.

RYDEN: The suggestion that blue copper proteins existed 3.5 billion years ago rests on the finding of microfossils of that age that look like cyanobacteria. They might have been oxygenic. Oxygen was produced by living cells a long time before it started to accumulate in the atmosphere. In general, one should also add that electron transport with electron acceptors other than oxygen long preceeded the advent of molecular oxygen in the atmosphere.

MANN: As regards the question of evolution of metalloproteins in general, we must bear in mind that even closely related species can differ in respect to the nature of these proteins as best shown by the differences in hemoglobin, for instance. As regards laccase, we found that it exhibits different properties according to its occurrence in latex of different species of lacquer trees.

RYDEN: I agree that there is a danger here. Sometimes the same reaction is catalyzed in different species by related but distinct so-called paralogous enzymes coded by genes at different loci. If this is not recognized one might construct false phylogenies. This is not, however, a problem until we have more sequence information on the copper containing oxidases. In the small blue proteins it seems quite clear which proteins are orthologous at the moment - they comprise the classes I mentioned.

NICHOLLS: To follow up on Dr. Mann's question, does anyone know of any polymorphisms (isozymes or allozymes) in blue copper proteins?

LERCH: We have determined the nucleotide sequences of Neurospora crassa from two wild type strains and found them to show a large divergence (German et al., JBC in press). This enzyme represents a good example of polymorphism in blue copper proteins.

MESSERSCHMIDT: I would expect ceruloplasmin to contain a trinuclear site like ascorbate oxidase. This does not fit your model for the copper site. What is your opinion about that?

RYDEN: The key finding with respect to the model is that 6 histidine residues are clustered to provide the ligands for three additional coppers to make up an active site at the C-terminus of the ceruloplasmin peptide chain. These coppers are not present in the N-terminal and middle homology regions. The precise geometry of the copper site has to be shown by structure determination and is probably the same in ceruloplasmin and ascorbate oxidase.

Session VI: Transfer of Reducing Equivalents. Heme Proteins

Oxidases and Related Redox Systems, pages 369–385

HYDRIDE VS. ELECTRON TRANSFER IN THE OXIDATION OF NADH MODEL COMPOUNDS

M. F. Powell*, T.C. Bruice

Institute of Pharmaceutical Sciences, Syntex Research, Palo Alto, CA 94304, (M.F.P.), Department of Chemistry University of California at Santa Barbara Santa Barbara, CA 93106, (T.C.B.)

INTRODUCTION

There exists a long-standing controversy regarding the detailed mechanism of hydride equivalent transfer from NADH. In principle the controversy involves the choice between a single step H^- transfer and one which involves step-wise e^-, H^+, e^- transfer (eq. 1). Early isotope

(1)

effect studies of NADH model reactions (Abeles et al., 1957; Narcross et al., 1962; Dittmer and Fouty, 1964; Shellenberg and McLean, 1966; Schellenberg et al., 1967; Huffman and Bruice, 1967; Kurtz and Frieden, 1975, 1977, 1980) supported a direct H^- transfer reaction. Steffens and Chipman (1971) reported kinetic isotope effects (k_H/k_D) that were significantly smaller than product isotope effects (Y_H/Y_D) for the reduction of trifluoroacetophenone (TFA) by N-propyl-1,4-[4,4-H,D]-dihydronicotinamide. Since this isotope effect discrepancy could be explained by invoking an intermediate the mechanistic controversy started in earnest with the suggestion that the intermediate was the radical cation $PrNADH^{\cdot +}$.

Thereafter, other dihydropyridine redox reactions were reported with k_H/k_D's less than their counterpart Y_H/Y_D's (Creighton et al.,1973; Ohno and Ohnishi, 1967; Ohno and Kito, 1972). Chipman et al. (1980) retracted their original study when they discovered that PrNADH reacted with TFA in aqueous solution to give not only the expected redox products, but also products of adduct formation (k_a) and hydration (k_{hyd}). These kinetically-significant side reactions resulted in a reduction of the observed k_H/k_D's, as shown by eq. 2 derived from the

$$k_H/k_D = \frac{2k_H + (k_{hyd} + k_a)}{k_H + k_D + (k_{hyd} + k_a)} \qquad (2)$$

expression for obtaining k_H/k_D from the rates of 1,4-[4,4-H_2]- and 1,4-[4,4-H,D]-dihydropyridines (PyHH and PyHD, respectively). Various researchers attempted to circumvent the problems of dihydronicotinamide hydration or adduct formation by carrying out their studies in aprotic solvents such as CH_3CN (Shinkai et al., 1977, 1981; Ohno et al., 1981). Discrepancies in k_H/k_D and Y_H/Y_D were also found in aprotic solvents. These were touted in favor of a stepwise mechanism until we demonstrated that comparison of isotope effects when using PyHD as reductant leads to isotope scrambling between PyHD and the products PyH and PyD. Such isotope scrambling magnifies Y_H/Y_D giving erroneous results (Powell and Bruice, 1982). This observation, and our redetermination of k_H/k_D using PyHH and PyDD, demonstrated unequivocally that all NADH isotope effect discrepancies in the literature could be explained by phenomena other than a multistep pathway. Herein, we have included some of these discrepancies and their corrections.

Although solving the isotope effect discrepancies laid to rest many claims for e^-, H^+, e^- transfer in NADH models, other indirect experiments were offered to support a multistep mechanism: (i) free energy relationships and kinetic isotope effects (Ohno et al., 1981); (ii) optical yields (Ohno et al., 1981); (iii) spectroscopic detection of charge-transfer intermediates (Ohno et al., 1981); and (iv) variable amounts of solvent exchange in the reaction of NAD(D)H models (Yasui et al., 1982).

Some of these systems were reinvestigated by us giving different experimental results consistent with single step hydride transfer (Powell and Bruice, 1983). For example, the k_H/k_D's for reaction of N-aryldihydronicotinamides were ranked differently when determined using PyHH and PyDD rather than PyHD. Also, TFA and PrNADH underwent adduct formation in CH_3CN even though only traces of water were present, contrary to earlier beliefs that the use of aprotic solvents should preclude adduct formation (Shinkai et al., 1977, 1981; Ohno et al., 1981).

Evidence for direct hydride transfer of NADH compounds has been found in both enzymatic and NADH model systems. For example, the observation that α-(hydroxyalkyl)cyclopropanes do not undergo cleavage of the cyclopropane ring when oxidized by liver alcohol dehydrogenase does not support a radical mechanism (MacInnes et al., 1982; Chung and Park, 1982; Laurie et al., 1986). Substituent effect calculations of the charge of the migrating H atom in the reduction of isoquinolinium cations by 1,4-dihydronicotinamides clearly indicate the hydridic nature of this reaction (Bunting et al., 1981, 1982). Kreevoy and co-workers (Ostovic et al., 1983) support direct hydride-transfer based on a Marcus theory correlation of relative reaction rates and equilibrium constants for hydride-equivalent reductions employing NADH model compounds. The one electron energetics for the $NADH/NADH^{\cdot +}$ couple also support single step hydride transfer (Martens et al., 1979, Carlson and Miller, 1984, 1985; Carlson et al., 1984; Verhoeven et al., 1986). Comparison of the values of ΔG^o calculated from the very high $1e^-$ oxidation potential of $NADH^{\cdot +}$ (> 0.9 V) with the corresponding experimental $\Delta G^\ddagger$ shows the latter are often less, disallowing the e^- transfer mechanism. Further experiments to probe the nature of the transition state and the contribution by quantum mechanical tunneling in these reactions support the single step mechanism (Husky and Schowen, 1983; Ostovic et al., 1983; van Gerresheim and Verhoeven, 1983; van der Kerk et al., 1984; Wu and Houk, 1987). In fact, NADH compounds prefer to react by direct hydride transfer, even when a thermodynamically more favorable $1e^-$ reaction pathway is readily available (Powell and Bruice, 1983; Powell et al., 1982), as occurs in the reaction of NADH model compounds with oxidized flavin cations and the structurally-similar flavinylium radical cations. These flavin-NADH experiments are summarized herein.

Yet, NADH compounds do not react exclusively by direct hydride transfer. High energy reactions, such as photoinduced electron transfer is an example (Fukuzumi et al., 1987). Recently we showed that $1e^-$ transfer in the NADH reduction of ferricyanide is followed by rate-determining proton transfer from the intermediate $NADH^{\cdot +}$ radical cation (Powell et al., 1984; Sinha and Bruice, 1984) even at neutral pH.

RESULTS AND DISCUSSION

For the sake of clarity, the discussion section is in three parts: (i) k_H/k_D and Y_H/Y_D discrepancies in support of a multistep mechanism are due to isotope scrambling and neglect of secondary isotope effects; while (ii) direct hydride transfer is the preferred mechanism even when a thermodynamically favorable $1e^-$ transfer is possible; and (iii) e^-, H^+, e^- occurs with obligate $1e^-$ oxidants of high potential.

Isotope Effects for NADH Model Reactions

A great deal of the early evidence for a multistep mechanism was provided by the differences in k_H/k_D and Y_H/Y_D for the reaction of methyl acridinium iodide (MAI) and various NADH compounds. We reinvestigated the reaction of MA with BzNADH, QNADH, CF_3NADH and MAH (see Scheme I for structures) and their isotopically-labeled counterparts in dry CH_3CN at 30°C.

Scheme I.

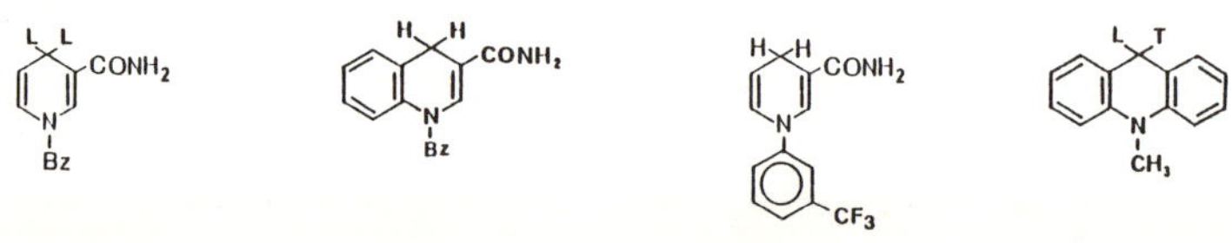

BzNADH QNADH CF_3PhNADH MAH

The symmetrical reaction of MAH and MA was followed using an isotope tracer; all others were followed spectrophotometrically. Spectral changes from 200 to 500 nm were consistent with the formation of MAH and nicotinamide. First-order kinetics were observed under our conditions of [NADH] >> [MA]; the second-order rate constants shown in Table 1

Table 1. Rate Constants, Equilibrium Constants and Kinetic Isotope Effects for the Reaction of MAI with Various NADH Model Compounds in CH_3CN at 50 °C.

Reductant[a]	$k (M^{-1} s^{-1})$	$10^4 K_{eq}$	k_H/k_D[b]	Y_H/Y_D[b]
H,H-BzNADH	79.9 ± 0.9[c]			
H,D-BzNADH			[2.76 ± 0.15][c]	[4.0 ± 0.2][c]
D,D-BzNADH	19.5 ± 0.1[c]		4.11 ± 0.05[c,d]	
H,H-QNADH	0.762 ± 0.013[c]	0.24		
H,D-QNADH	0.447 ± 0.07[c]		4.95[c]	[7.6 ± 0.3][c]
D,D-QNADH	0.149 ± 0.003[c]		5.09 ± 0.15[c,d]	
H,H-MeOPhNADH	35.16 ± 1.07	82		
H,D-MeOPhNADH	21.51 ± 0.54		3.83	[5.9]
D,D-MeOPhNADH	8.84 ± 0.08		3.98 ± 0.13[d]	
H,H-MePhNADH	26.28 ± 0.52	22		[5.9]
D,D-MePhNADH	6.56 ± 0.64		4.01 ± 0.08[d]	
H,H-CF_3PhNADH	2.903 ± 0.019	1.5		[6.2]
D,D-CF_3PhNADH	0.655 ± 0.012		4.43 ± 0.08[d]	
H,H-CNPhNADH	0.727 ± 0.005	0.24	4.77 ± 0.06[d]	[6.0]
D,D-CNPhNADH	0.152 ± 0.005			
H,H-MA	0.051 ± 0.002			
H,H-MA	0.103 ± 0.005			

[a]The isotopic variant of the NADH model compounds is shown as H,H-, H,D- or D,D-. [b]Values shown in brackets are from the literature. [c]Reaction at 30°C. [d]Includes a small secondary kinetic isotope effect.

were obtained by least squares analysis of a plot of k_{obs} vs. [NADH]. Inspection of the k_H/k_D's in Table 1 shows that those obtained from PyHH and PyDD are generally larger than those calculated from PyHH and PyHD. Kinetic isotope effects determined using PyHH and PyDD have only small secondary isotope effects incorporated into their observed values so that the 'true' kinetic isotope effect is altered by secondary isotope effects by less than 10%. Alternatively, isotope effects determined using PyHD and eqs. 3-6 contain a significantly larger correction factor brought about by the secondary isotope effect.

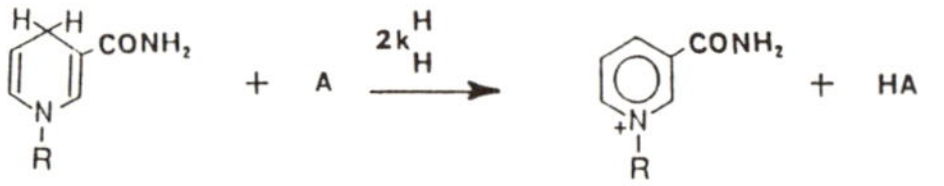

(3)

(4)

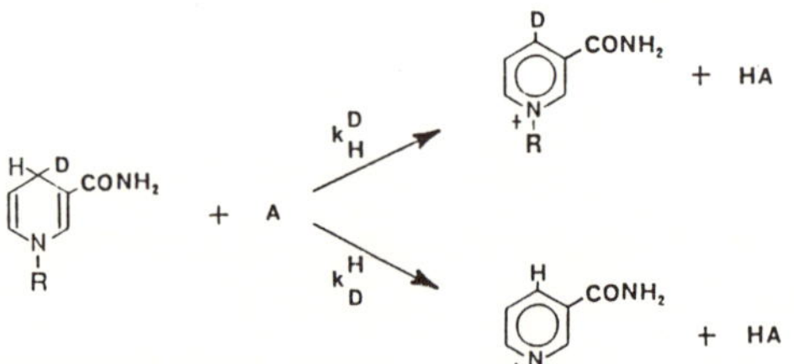

A = hydride acceptor

$$k_H/k_D = k_H^H/(k_{HD} - k_H^H) \qquad (5)$$

$$k_{HD} \simeq k_H^D + k_D^H \qquad (6)$$

For example, using eqs. 5 and 6, a primary isotope k_H/k_D of 4.0, when overlayed with varying secondary isotope effects (k^H/k^D) of 0.9-1.14 (as expected for reactions of this type), gives observed isotope effects of 2.8-8.4. This is far outside acceptable error limits for a ~10% uncertainty in secondary isotope effects. This large error in the calculated k_H/k_D (eq. 5) is due to the error in k_D calculated from eq. 6, an approximate expression at best. Although the secondary isotope effects for these reactions are close to unity (e.g. QNADH; 1.03 ± 0.03; MeOPhNADH, 1.04 ± 0.06) (see Powell and Bruice, 1983), even a 3-4% secondary isotope effect changes the calculated k_H/k_D by ~15-20%. For example, the k_H/k_D for BzNADH determined using PyHH and PyDD was found to be 4.11 ± 0.05 (Table 1), somewhat larger than the value of 2.78 ± 0.15 using PyHD. Similarly, k_H/k_D for QNADH increased from 2.0 ± 0.2 to 5.09 ± 0.15 when PyDD instead of PyHD was used as the isotopic variant. Isotope effects for the reaction of X-PhNADH and MA, when redetermined using PyHH and PyDD, also were very different than reported previously using PyHD. In our hands, we observed a slight increase in k_H/k_D with increasing electron demand of substituent X, in contrast to the sharply decreasing isotope effects observed by Ohno et al. (1981). We have determined the equilibrium constants (Table 1) and $\Delta G°$ values for these reactions, and have shown that the reaction of X-PhNADH can be easily understood in terms of a reaction profile that becomes increasingly isoendergonic with the greater electron demand of substituent X .

The product isotope effects for the reaction of MAH and NADH analogues are also complicated by isotope scrambling. For example, the observed Y_H/Y_D for QNADH (7.6 ± 0.3) is anomalously large because isotope scrambling between MAH and the product MA was unaccounted for. In this case, reaction of the H,D-MA product (decreasing Y_D) with remaining MA reactant to give additional H,H-MA (increasing Y_H) resulted in an anomalously large Y_H/Y_D. This was demonstrated by solving eqs. 7-9 by

$$\text{H,D-QNADH} + \text{H-MA} \underset{k_2}{\overset{k_1}{\rightleftharpoons}} \text{D-QNAD} + \text{H,H-MA} \quad (7)$$

$$\text{H,D-QNADH} + \text{H-MA} \underset{k_{-2}}{\overset{k_2}{\rightleftharpoons}} \text{H-QNAD} + \text{H,D-MA} \quad (8)$$

$$\text{H,D-MA} + \text{H-MA} \underset{k_{-3}}{\overset{k_3}{\rightleftharpoons}} \text{H,H-MA} + \text{D-MA} \quad (9)$$

computer simulation. The concentrations of the reaction species were calculated using the determined rate constants $k_1 = 0.37\ M^{-1}s^{-1}$, $k_{-1} = k_{-2} = 0$, $k_2 = 0.074\ M^{-1}s^{-1}$, $k_3 = 0.025\ M^{-1}s^{-1}$, $k_{-3} = 0.051\ M^{-1}s^{-1}$ and the given initial concentrations, $[\text{H,D-QNADH}]_o = 0.001$ and $[\text{MAI}]_o = 0.002$ M. A plot of the concentration of reactants and products vs. time shows that [H,H-MA]/[H,D-MA] changes during the time course of the reaction due to isotope scrambling reaching an equilibrium value of Y_H/Y_D of 7.9, close to the value of 7.3 observed experimentally.

The discrepancy between k_H/k_D and Y_H/Y_D for reaction of X-PhNADH is also misleading. By reaction scheme simulation, we have shown (1982) that the previously reported Y_H/Y_D's are incorrect and that, when isotope scrambling by the kinetically-significant back reactions is included, Y_H/Y_D's significantly smaller (not larger) than the k_H/k_D's should be observed. The largest discrepancy between k_H/k_D and Y_H/Y_D values was observed in the reaction of CF_3PhNADH and MA. For the sake of brevity the full scheme for CF_3PhNADH is not reproduced herein; those interested are referred to Powell and Bruice (1983). The calculated time courses for the reaction of MA and CF_3NADH show that the ratio of [H,H-MA]/[H,D-MA] varies with time, and that after 24 hrs Y_H/Y_D decreased from an initial value of 4.43 to 1.60. Thus, the literature values of Y_H/Y_D for x-PhNADH (0.6) (Ohno et al., 1981) are inconsistent with isotope exchange unless the "true" values of Y_H/Y_D are unusually large, i.e. > 30. This is not likely to be the case for an NADH transhydrogenation reaction. Thus, the discrepancies between k_H/k_D and Y_H/Y_D for NADH model reactions are usually of little use as a mechanistic indicator.

Hydride vs. Electron Transfer in the Reduction of Flavins

The reaction of oxidized flavin and flavin radical with various dihydropyridines has been studied. If hydride equivalent reduction of oxidized flavin by dihydropyridine involves 1e- transfer (eq. 10) then it is predicted that dihydropyridines should also reduce flavin radical to 1,5-dihydroflavin by 1e- transfer, providing the 1e- transfer is thermodynamically favorable. The reduction of 1,10-ethanolumiflavinium cation (Fl^+) by the PyHH compounds BzNADH, BzAcNADH, BzCNNADH, QNADH and MA was studied anaerobically under the pseudo-first-order conditions of [PyHH] >> [Fl^+] in 5% CH_3CN/95% t-BuOH at 30°C (see Scheme II for structures).

Scheme II.

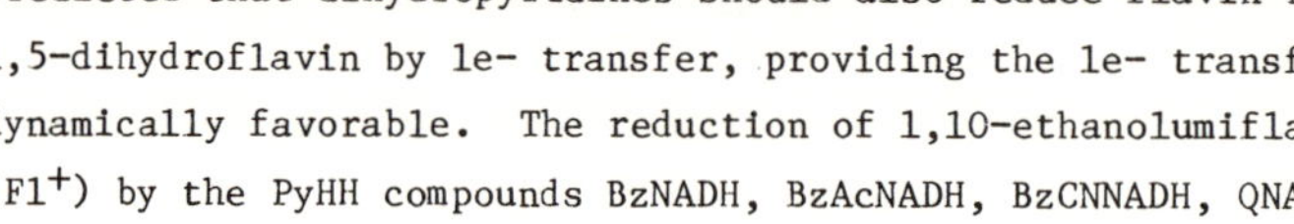

Fl^+ $FlEt^{\cdot +}$ BzAcNADH   BzCNNADH

The observed first-order rate constants (k_{obs}) were linearly dependent on [PyHH] with zero intercepts giving the second order rate constants in the middle column of Table 2.

Table 2. Rate Constants for the Reaction of Fl^+ ($M^{-1}s^{-1}$) and $FlEt^{\cdot +}$ (s^{-1}) with NADH Model Compounds at 30°C

Compound	10^5 $FlEt^{\cdot +}$(s^{-1})[a]	Fl^+ (expt'l)[b]	Fl^+ (calc)[c]
BzNADH	2.30 $\pm$ 2.0	1689 $\pm$ 32	1437
BzAcNADH	2.1 $\pm$ 1	92 $\pm$ 1	103
BzCNNADH	1.7 $\pm$ 0.1	6.34 $\pm$ 0.09	10.4
QNADH	9.6 $\pm$ 0.7	39.7 $\pm$ 0.7	60.0
MA	1.9 $\pm$ 0.2	8.4 $\pm$ 0.1	12.3

[a]5% CH_3CN/95% t-BuOH (t-butyl alcohol). [b]t-BuOH, [$FlEt^{\cdot +}$] = 2 x 10^{-5} M.
[c]See text.

The reaction of 1,10-ethano-5-ethyllumiflavinyllum radical ($FlEt^{\cdot +}$) however, was not as straightforward. Under similar reaction conditions, (100% t-BuOH, 30°C), the disappearance of $FlEt^{\cdot +}$ was zero order throughout most of the time course, changing to first order during the completion of the reaction (Figure 1).

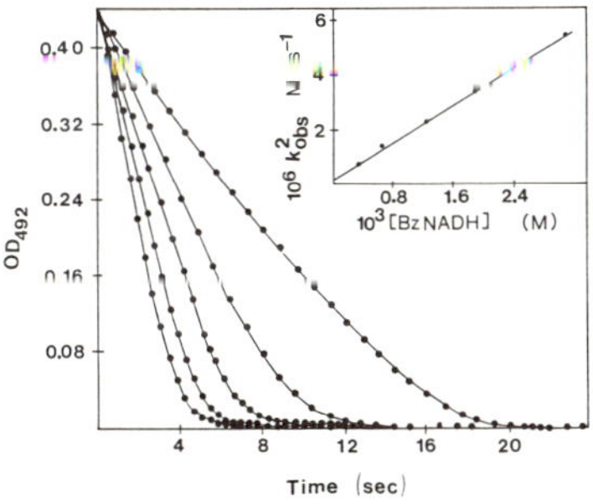

Figure 1. Time course for the reaction of $FlEt^{\cdot +}$ ([$FlEt^{\cdot +}$] = 2 x 10^{-5} M) with BzNADH (0.06 - 3.1 x 10^{-3} M) in t-BuOH at 30°C; (inset) observed pseudo-zero-order rate constants vs. the concentration of BzNADH for reaction of $FlEt^{\cdot +}$ in t-BuOH at 30°C.

The pseudo-zero-order rate constants reported in Table 2 were also linearly dependent on [$FlEt^{\cdot +}$], and so a single flavin radical concentration ([$FlEt^{\cdot +}$] = 3 x 10^{-5} M) was used in the plots of Figure 1.

The final, and perhaps the strongest piece of evidence for the reaction mechanism of Scheme III is the finding that the observed pseudo-zero-order rate constants for the reaction of $FlEt^{\cdot+}$ with BzNADH are linearly dependent upon the concentration of added Fl^{+}.

The most reasonable explanation for the reaction of $FlEt^{\cdot+}$ with PyHH is that $FlEt^{\cdot+}$ carries Fl^{+} as an impurity. By this mechanism, oxidized flavin Fl^{+} undergoes 2e- ("hydride") reduction by PyHH to yield the 1,5-reduced flavin (FlH) which then acts as a $1e^{-}$ reductant for $FlEt^{\cdot+}$ producing $FlH^{\cdot}$ and $FlEt^{-}$ (Scheme III). In other words, the flavin

Scheme III.

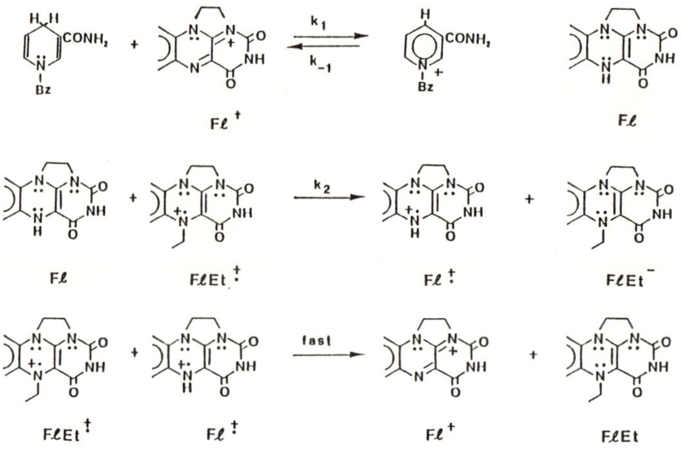

radical $FlEt^{\cdot+}$ is not reducible by BzNADH or other PyHH compounds. The flavin radical cation $Fl^{\cdot+}$ then reacts rapidly with $FlEt^{\cdot+}$ regenerating the 'catalyst', Fl^{+} and the observed product, FlEt. Kinetic competency of this scheme was demonstrated by assuming a steady state concentration of Fl^{+} and fitting the observed decay of $FlEt^{\cdot+}$ to eq. 10.

$$\frac{-d[FlEt^{\dot{+}}]}{dt} = \frac{k_1\, k_2[PyHH][Fl^+][FlEt^{\dot{+}}]}{k_{-1}[PyHH] + k_2[FlEt^{\dot{+}}]} \qquad (10)$$

The known half cell oxidation potentials of Fl^+ and $FlEt^{\cdot+}$ (Powell and Bruice, 1983) show that the reaction between FlH and $FlEt^{\cdot+}$ is exergonic and presumably fast, making $k_2[FlEt^{\cdot+}] >> k_{-1}[PyHH]$. Thus eq. 10 reduces to eq. 11, a zero-order expression in $FlEt^{\cdot+}$.

$$\frac{-d[FlEt^{\dot{+}}]}{dt} = k_1[Fl^+][PyHH] \qquad (11)$$

From eq. 12 and the kinetic data, we calculated the concentration of Fl^+ required as an impurity in $FlEt^{\cdot+}$ as $\sim 0.5\%$. The zero-order rate constants obtained for the disappearance of $FlEt^{\cdot+}$ with BzNADH, when divided by $0.05[FlEt^{\cdot+}]$, provided the independently determined second-order rate constant for reaction of Fl^+ with PyHH (Table 2). Differences in the observed and calculated values may be due in part to a small solvent effect; Fl^+ was studied in 5% CH_3CN/95% t-BuOH whereas $FlEt^{\cdot+}$ was studied in neat t-BuOH. A linear correlation of the rate constants for Fl^+ and $FlEt^{\cdot+}$ was observed in support of Scheme III wherein the rate-determining step for the consumption of $FlEt^{\cdot+}$ is actually the reaction of Fl^+ with PyHH. Finally, this mechanism is validated by the similarity of isotope effects for the reaction of H,H-BzNADH and D,D-BzNADH with Fl^+ ($k_H/k_D = 4.1 \pm 0.1$), and $FlEt^{\cdot+}$ ($k_H/k_D = 3.8 \pm 0.2$).

From Scheme III we see that hydride reduction of Fl^+ is very rapid, and that $1e^-$ transfer to $FlEt^{\cdot+}$ cannot be detected. In that $1e^-$ transfer for reduction of $FlEt^{\cdot+}$ is even more thermodynamically favored than for Fl^+ (eqs. 12 and 13), one would expect $FlEt^{\cdot+}$ to react faster than Fl^+ if $1e^-$ transfer were occurring. Clearly this is not the case. Hydride equivalent transfer from NADH analogs to oxidized flavins proceeds via a two electron transfer mechanism with concomitant motion of hydrogen; in other words, single-step hydride transfer.

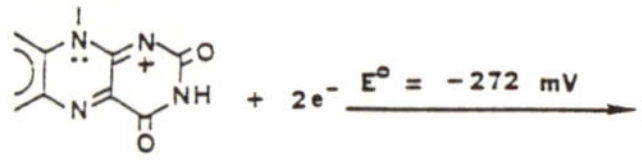

(12)

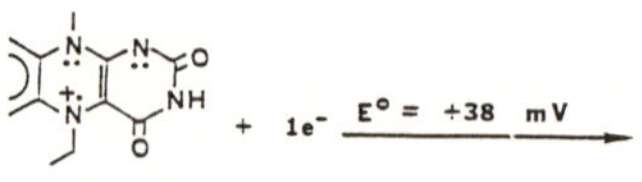

(13)

Ferricyanide Oxidation of NADH Model Compounds

The reaction of BzNADH, PrNADH, X-PhNADH, QNADH and MAH with $Fe(CN)_6{}^{3-}$ was carried out anaerobically in 20% CH_3CN-H_2O at 30°C. The $Fe(CN)_6{}^{3-}$ concentration was usually held at 2×10^{-3} M or 8×10^{-4} M, and the $Fe(CN)_6{}^{4-}$ concentration varied from $(0.2 - 6) \times 10^{-3}$ M such that both exceeded the substrate concentration by at least a factor of 50. By this method the "redox buffer ratio" ($[Fe(CN)_6{}^{3-}]/[Fe(CN)_6{}^{4-}]$) did not change appreciably during the course of the reaction. PyHH substrate disappearance followed first-order kinetics with a linear dependence upon $[Fe(CN)_6{}^{3-}]$ when $[Fe(CN)_6{}^{4-}]$ and $[K^+]$ were held constant.

In the case of constant $[Fe(CN)_6{}^{3-}]$ and varying $[Fe(CN)_6{}^{4-}]$, linear plots of $1/k_{obs}$ vs. $[Fe(CN)_6{}^{4-}]$ of positive slope were observed in most cases, indicative of $Fe(CN)_6{}^{4-}$ inhibition. These results in anaerobic solution are consistent with the sequence of reactions shown in Scheme IV.

Scheme IV.

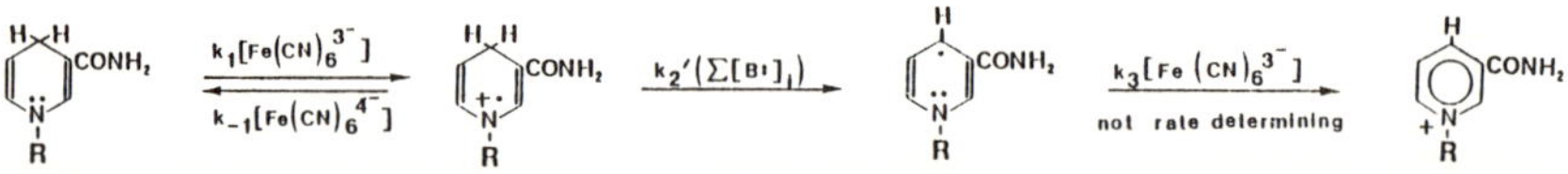

(PyH_2) ($PyH_2^{\cdot+}$) ($PyH^{\cdot}$) (PyH^+)

According to Scheme IV the pseudo-first-order rate constant (k_{obs}) for disappearance of NADH is described by eq. 16 when $NADH^{\cdot+}$ is at steady state. The rate constants k_{1y} and k_2 represent catalysis by solvent-derived lyate species (primarily H_2O) and buffer base B, respectively.

Inversion of eq. 14 provides eq. 15.

$$k_{obs} = \frac{k_1[Fe(CN_6)^{3-}](k_{1y}+k_2[B])}{k_{-1}[Fe(CN)_6^{4-}]+(k_{1y}+k_2[B])} \qquad (14)$$

$$\frac{1}{k_{obs}} = \frac{k_{-1}[Fe(CN)_6^{4-}]}{k_1[Fe(CN)_6^{3-}](k_{1y}+k_2[B])} + \frac{1}{k_1[Fe(CN)_6^{3-}]} \qquad (15)$$

The determination of k_1 and $k_{-1}/(k_{1y}+k_2[B])$ (the partitioning ratio from the NADH$\cdot^+$ intermediate) from the plots of $1/k_{obs}$ vs. $[Fe(CN)_6{}^{3-}$ and $[Fe(CN)_6{}^{4-}$ is straightforward since $[Fe(CN)_6{}^{3-}]$ and $[Fe(CN)_6{}^{4-}]$ are known and essentially invariant during the course of the reaction. The k_1 rate constants and the $k_{-1}/(k_{1y}+k_2[B])$ partitioning coefficients are given in Table 3.

By this mechanism, $Fe(CN)_6{}^{4-}$ inhibition indicates that reaction of $Fe(CN)_6{}^{4-}$ with NADH$\cdot^+$ can compete with proton transfer from NADH$\cdot^+$ to base B or H_2O, while lack of such inhibition indicates that it cannot. Thus, the lack of $Fe(CN)_6{}^{4-}$ inhibition for the reaction of CNPhNADH is indicative of slow e^- transfer followed by rapid H^+ transfer.

Table 3. Summary of Rate Constants, Partitioning Ratios, and Isotope Effects for Reaction of NADH Model Compounds with $Fe(CN)_6{}^{3-}$ at 30°C in 20% CH_3CN-H_2O Containing 10^{-2} M $CO_3{}^{2-}$.

Substrate	$k_1 (M^{-1}s^{-1})$	$k_{-1}/(k_{1y}+k_2[B])$
H,H-BzNADH	1.36 ± 0.13	$(1.29 \pm 0.18) \times 10^3$
H,H-PrNADH	$(2.01 \pm 0.13) \times 10^1$	$(1.80 \pm 0.15) \times 10^3$
D,D-PrNADH	$(1.91 \pm 0.20) \times 10^1$	$(3.59 \pm 0.00) \times 10^3$
H,H-MeOPhNADH	1.08 ± 0.13	$(3.50 \pm 0.47) \times 10^3$
H,H-MePhNADH	$(2.66 \pm 0.20) \times 10^{-1}$	$(1.68 \pm 0.17) \times 10^3$
H,H-PhNADH	$(8.17 \pm 0.21) \times 10^{-2}$	$(7.05 \pm 0.50) \times 10^2$
H,H-CNPhNADH	$(2.59 \pm 0.27) \times 10^{-3}$	$(3.57 \pm 0.37) \times 10^1$
H,H-QNADH	$(3.23 \pm 0.17) \times 10^{-3}$	$(5.19 \pm 0.30) \times 10^2$
H,T-MA	$(5.18 \pm 3.28) \times 10^{-1}$	$(2.01 \pm 1.27) \times 10^6$

[a] 0.02 M $Fe(CN)_6{}^{3-}$, no addition of $CO_3{}^{2-}$.

The partitioning ratio $k_{-1}/(k_{1y}+k_2[B])$ for X-PhNADH compounds decreases smoothly as X became more electronegative, indicating that the free energy barrier for the forward rate $(k_{1y}+k_2[B])$ falls off faster than the barrier for the reverse rate (k_{-1}).

The partitioning ratio is also sensitive to the basicity of the reaction solution. Plots of buffer dilution obey the empirical expression of eq. 16 where $[B]_t$ is the total buffer concentration and a_H is the hydronium ion activity.

$$k_{obs} = k_{1y} + \frac{k_B K_a [B]_t}{K_a + a_H} \qquad (16)$$

When 1/slope of the buffer dilution plots are plotted against a_H there is obtained as intercept $1/k_B$ and $1/k_BK_a$ as secondary slope. From eq. 14 where $[Fe(CN)_6{}^{4-}] \gg (k_{1y}+k_2[B])$ it follows that $k_B = k_1k_2[Fe(CN)_6{}^{3-}]/[Fe(CN)_6{}^{4-}]$. The known value of $[Fe(CN)_6{}^{3-}]/[Fe(CN)_6{}^{4-}] = 5.4$ in these experiments allows calculation of the k_{-1}/k_2 partitioning ratio. The values for k_{-1}/k_2 for the reaction of H,H-MA with imidazole, acetate, formate and H_2O are 670, 1190, 3330 and 2.56×10^6, respectively. If one assumes the exergonic reaction of H,H-MA$\cdot^+$ radical cation with $Fe(CN)_6{}^{4-}$ to be diffusion-controlled (10^{10} $M^{-1}s^{-1}$) then the rate constants for proton transfer from H,H-MA$\cdot^+$ to the general bases are $1\text{-}5 \times 10^7$ $M^{-1}s^{-1}$ and 10^4 for H_2O.

The isotope effects for this reaction were also determined using PyHH and PyDD. As expected, the primary isotope effect on k_1 was found to be close to unity (for example, PrNADH gives 1.05 ± 0.13) whereas the isotope effect for the proton transfer step k_2 of MAH with imidazole, acetate and formate was found to be 4.4 ± 0.1; for reaction of MAH and H_2O the isotope effect was 5.3. The smaller isotope effect for PrNADH as compared to MAH is not unexpected since the less stable species PrNADH$\cdot^+$ results in a more exothermic reaction with a reactant-like transition state. These results demonstrate unequivocally that general base catalyzed proton transfer is the rate-determining step in the NADH-ferricyanide reaction.

Acknowledgement. This work was supported by grants to TCB from the National Science Foundation and The National Institutes of Health.

REFERENCES

Abeles RH, Hutton RF, Westheimer FH (1957) J. Am. Chem. Soc. 79:712.

Bunting JW, Chew VSF, Chu GJ (1981) Org. Chem. 46:4211.

Bunting JW, Chew, VSF, Chu, GJ. (1982) Org. Chem. 47:2303, 2308.

Carlson BW, Miller LL, Neta P, Grodkowski JJ (1984) Am. Chem. Soc. 106:7233.

Carlson BW, Miller LL (1984) J. Am. Chem. Soc. 105:7454.

Carlson BW, Miller LL (1985) J. Am. Chem. Soc. 107:479.

Chipman DM, Yaniv R, van Eikeren PJ (1980) Am. Chem. Soc. 102:3244.

Chung SK, Park SU (1982) J. Org. Chem. 47:3197.

Creighton DJ, Hajdu J, Mooser G, Sigman DS (1973) J. Am. Chem. Soc. 95:6855.

Dittmer DC, Fouty RA (1964) J. Am. Chem. Soc. 86:91.

Fukuzumi S, Koumitsu S, Hironaka K, Tanaka T (1987) J. Am. Chem. Soc. 109:305.

Huffman RW, Bruice TC (1967) J. Am. Chem. Soc. 89:6243.

Husky WP, Schowen RL,(1983) J. Am. Chem. Soc. 105:5704.

Kurz LC, Frieden C (1975) J. Am. Chem. Soc. 97:677.

Kurz LC, Frieden C (1977) Biochemistry 16:5207.

Kurz LC, Frieden C (1980) J. Am. Chem. Soc. 102:4198.

Laurie D, Lucas E, Nonhebel DC, Suckling CJ (1986) Tetrahedron 42:1035.

MacInnes I, Nonhebel DC, Orszulik ST, Suckling CJ (1982) J. Chem. Soc. Chem. Commun. 121.

Martens FM, Verhoeven JW, Gase RA, Pandit UK, de Boer THJ (1978) Tetrahedron 34:443.

Norcross BE, Klinedinst PE Jr, Westheimer FH (1962) J. Am. Chem. Soc. 84:797.

Ohno A, Ohnishi Y (1967) Chem. Lett. 697

Ohno A, Kito N (1972) Chem. Lett. 369.

Ohno A, Shio T, Yamamoto H, Oka S (1981) J. Am. Chem. Soc. 103:2045.

Ohno A, Yamamoto H, Oka S J (1981) Am. Chem. Soc. 103:2041.

Ohno A, Nakai J, Nakamura K, Goto T, Oka S (1981) Bull. Chem. Soc. Jpn. 54:3486.

Ohno A, Yamamoto H, Oka S (1981) Bull. Chem. Soc. Jpn. 54:3489.

Okamoto T, Ohno A, Oka S (1980) Bull. Chem. Soc. Jpn. 53:330.

Ostovic D, Roberts RMG, Kreevoy MM (1983) J. Am. Chem. Soc. 105:7629.

Powell MF, Wong WH, Bruice TC (1982) Proc. Natl. Acad. Sci. U.S.A. 79:4604.

Powell MF, Bruice TC (1982) J. Am. Chem. Soc. 104:5834.

Powell MF, Bruice TC (1983) J. Am. Chem. Soc. 105:1014.

Powell MF, Bruice TC (1983) J. Am. Chem. Soc. 105:7139.

Powell MF, Wu JC, Bruice TC (1984) J. Am. Chem. Soc. 106:3850.

Schellenberg KA, McLean GW (1966) J. Am. Chem. Soc. 88:1077.

Schellenberg KA, McLean GW, Lipton HC, Lietman PS (1967) J. Am. Chem. Soc. 89:1948.

Shinkai S, Ide T, Hamada H, Manabe O, Kunitake TA (1977) J. Chem. Soc. Chem. Commun. 848.

Shinkai S, Tsuno T, Manabe O (1981) Chem. Lett. 1203.

Sinha A, Bruice TC (1984) J. Am. Chem. Soc. 106:7291.

Steffens JJ, Chipman DM (1971) J. Am. Chem. Soc. 93:6694.

van Gerresheim W, Verhoeven JW (1983) Recl. Trav. Chim. Pays-Bas 102:339.

van der Kerk SM, van Gerresheim W, Verhoeven JW (1984) Recl. Trav. Chim. Pays-Bas 103:143.

Verhoeven JS, van Gerresheim W, Martens FM, van der Kerk SM (1986) Tetrahedron. 42:975.

Wu YD, Houk KN (1987) J. Am. Chem. Soc. 109:2226.

Yasui S, Nakamura K, Ohno A, Oka S (1982) Bull. Chem. Soc. Jpn. 55:196.

DISCUSSION

MASON: What is your position regarding the mechanism of oxidation of NADH? Removal of two electrons and a proton, or an electron and a hydride ion?

BRUICE: If one examines the reduction of NAD^+ by NADH, the symmetry of this reaction demands a one-step H^- transfer or a stepwise $1e^-$, H^-, $1e^-$ transfer. Thus, any $1e^-$ oxidation leading to hydride equivalent transfer must involve the latter sequence of events. An example is the oxidation of a dihydronicotinamide by $Fe(CN)_6^{3-}$. We show the first step to be a reversible $1e^-$ transfer to give $NAD^{+\cdot}$ and $Fe(N)_6^{4-}$. The following step is a general base catalyzed proton transfer. This is followed by a very rapid oxidation of the resultant $NAD\cdot$ by a second $Fe(CN)_6^{3-}$. After Leslie Hellerman, this can be called an obligate $1e^-$ oxidation (i.e., $Fe(CN)_6^{3-}$ cannot accept H^- nor $2e^-$). For the more usual type of reductions which were classified by Westheimer as hydride transfer, the following may be stated. The various data (including comparisons of product and kinetic isotope effects) which have been offered in support of $1e^-$ processes have been found to be faulted. At present there is no experimental data in support of the intermediacy of dihydropyridine intermediates in the reduction of such functional groups as aldehydes, ketones, quinones, etc.

Oxidases and Related Redox Systems, pages 387–400

DETERMINANTS OF ELECTRON TRANSFER RATES: THE CYTOCHROME C: CYTOCHROME C PEROXIDASE SYSTEM

George McLendon, Jill Short Rogalsky, Ed Magner and Karen Taylor Conklin

Department of Chemistry, University of Rochester, Rochester, New York 14627

INTRODUCTION

Simple single electron transfer reactions clearly lie at the heart of the redox systems discussed in this symposium. In this paper, we briefly review the fundamentals of (outer sphere) single electron transfer reactions, in theory and in simple model systems. We then consider how these basics may be modified in biological systems, in order to insure biological specificity. In the final section we will focus on recent genetic engineering approaches to probing biological electron transfer.

The Basics: The Theory of Electron Transfer.

To a physical chemist electron transfer reactions probably represent the simplest chemical reactions, and thus can be studied at a fundamental level. The rate at which an electron is transferred from an electron donor to an acceptor depends on two basic factors: an electronic factor, which describes the "orbital overlap" between the donor and acceptor, and a nuclear term, which describes the motions of the reactant and their surroundings which are necessary to reach the transition state (where the energy of reactants and products are equivalent).[1]

The reaction coordinate can be easily visualized in a two dimensional potential surface, as in Figure 1. In this figure, the donor acceptor coupling

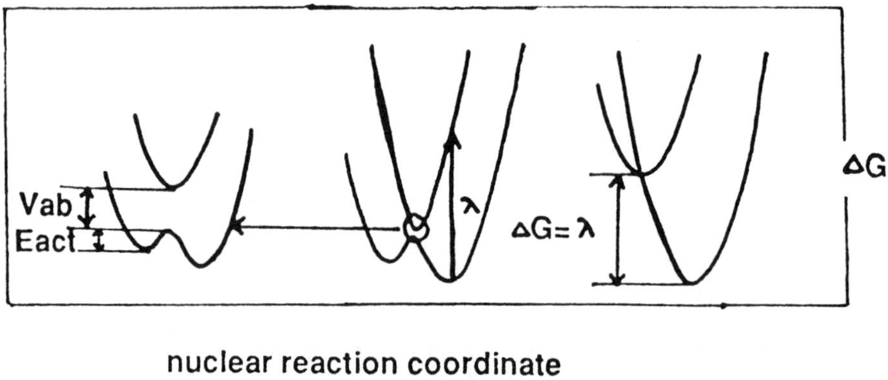

Figure 1: Simplified potential surfaces for electron transfer: parameters as defined in text.

strength (eg. orbital overlap) determines the magnitude of the splitting V of the upper and lower surfaces. If V is large, then any reaction trajectory will remain solely on the lower surface, and all trajectories which reach the transition state will lead to products. Such reactions are called "adiabatic". If V is small, then some trajectories which reach the transition state may continue along the reactant surface, and ultimately return to the reactant well, without leading to products. In such non adiabatic reactions, the transition state may be accessed many times without leading to products. V_{DA} depends on mixing of donor and acceptor states, e.g. orbital overlap, which in turn depends simply on distance: $V_{DA} = 2\pi/h \exp^{-\alpha R}$ This kinetically predicted dependence has been amply verified experimentally in cases where the donor and acceptor are held at fixed distances by an essentially insulating medium. This is the case for proteins. (For simple moleccules in fluid solution, which can undergo collisions, the orbital overlap is generally so large at a collisional distance that the reaction becomes adiabatic, and V_{DA} does not affect the rate).

Figure 1 also illustrates a simple parameterization of the nuclear terms. The activation energy required to reach the transition state depends on the shapes of the (reactant and product) potential wells, characterized by a Franck Condon transition energy, λ. In essence, λ measures how much energy is required to go from reactants to products without allowing any bond displacements, or solvent reorientation. For the electron transfer prosthetic groups found in proteins, (e.g. flavins, metalloporphyrins, the bond reorganization which accompanies electron transfer is generally small. Thus λ is dominated by the repolarization of the surrounding medium, which stabilizes a charged species in solution. In general λ will be large for high dielectric media, and small for low dielectric media. From the geometry of parabolas, λ is related to ΔG^* (when $\Delta G=0$) as $\Delta G^*=\lambda/4$. ΔG^* also depends on the overall reaction free energy. Comparing figures 1a and 1b, when ΔG increases (corresponding to a vertical displacement of the product surface) ΔG^* decreases. Thus ΔG^* depends both on λ, and ΔG, as $\Delta G^* = (\Delta G-\lambda)^2/4\lambda$.

These basic parameters can be summarized in a general rate expression for electron transfer between well separated donor acceptor pairs.

$$k_{et} = 2\pi/h \exp^{-\alpha R} \exp \frac{-(\Delta G-\lambda)^2}{4\lambda\ kT} \qquad (1)$$

Equation 1 has been dramatically successful in predicting the dependence of electron transfer on distance, free energy, and medium composition, for small molecules in colliision free matrices, or for donor acceptor assemblies.

The examples suffice for present purposes. Figure 2 shows the kinetics of electron transfer from a photoexcited donor which is which is surrounded by acceptors in a glassy collision free matrix. In such a random ensemble, since $k\alpha)\exp^{-R}$ those donors which have acceptors nearby will react rapidly, while those in which the nearest acceptors are distant will react much more slowly. Complex non exponential kinetics are expected, and, as Figure 2 shows, these exponential kinetics agree quantitatively with the theoretical predictions outlined above.

A second prediction of theory, the dependence of rate on ΔG, is beautifully illustrated by the classic work of Miller and Closs[2] who showed that k_{et} increases as ΔG increases, reaches a maximum when $\Delta G=\lambda$, and then decreases when $\Delta G>\lambda$. Furthermore, the value of λ depends markedly on solvent dielectric properties, just as theory suggests.

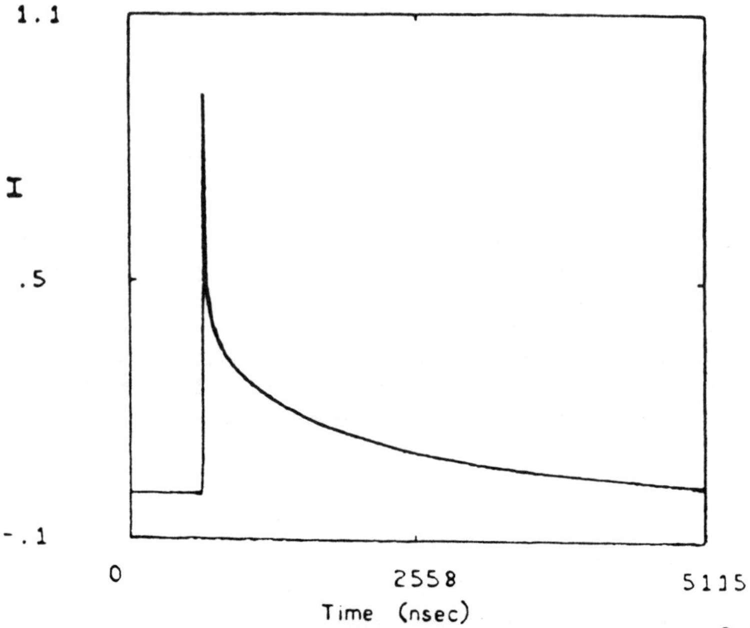

Figure 2: Electron transfer kinetics for $Ru(bpy)_3^{2+}$[4]/TMPD in a lexan matrix. Data and fit (as described in text) are shown.

These two illustrations are a small part of many experiments which independently confirm the theories suggested above. A more complete bibliography can be found in several recent reviews.[1]

Electron Transfer Proteins

Given the basic understanding of electron transfer in small molecules, the direct applicability of such understanding to proteins poses interesting questions. Although the same basic physics must hold for biological and abiological reactions, the key rate determining steps may be modified in biological systems. For example, if the "medium" which repolarizes to stabilize a charge is the protein matrix, rather than a fluid solvent, the timescale for such protein motions e.g. "conformational changes" can dominate the overall dynamics.

In this section we consider as a paradigm, the cytochrome c (cyt c)/cytochrome c peroxidase (ccp) couple, which demonstrates the application of fundamental theory to biological electron transfer, by illustrating the dependence of k_{et} on ΔG. We note that similar results have been found for the (cyt b_5) couple, and the cyt b_5/cyt c couple: Hb/met Hb reductase system.[3,4]

The cyt c/ccp System

While several protein:protein systems have now been studied in rather explicit detail (e.g. cyt c:cyt b5, cyt b_5:Hb, cyt c 1:cyt c, cyt c:ccp cyt c:cytb_2), the cyt c:ccp system has become perhaps the most important paradigm for quantitative study. This status reflects the high level of structural information available for these proteins, the relative ease of genetic manipulations for both proteins, and the ease of preparation and characterization of various metal substituted proteins which provide a powerful probe of reaction mechanism. What is missing from this system, and all the others cited above, is detailed structural information about the complex. An ingenious proposal was put forth by T. Poulos and J. Kraut, for the structure of the c:ccp complex, based on computer graphics.[5] The proposed structure is stabilized by a network of inter- protein H bonds between positively charged lysines (Lys 77, 18, 13) on cyt c and corresponding acidic residues on ccp. A host of chemical and physical evidence roughly support this proposed structure, with several significant exceptions. For example, that lys 32, which is not involved in the P–K model, is most strongly protected in differential chemical modification.[6] As discussed in the next section recent NMR and X-ray results now argue that a family of structures may exist perhaps involving rotations of the cytochromes about their charged domains. Clearly more detailed structural information is warranted. Structural questions aside, a key kinetic question in terms of the foregoing theory is, "How does rate depend on ΔG?" This question can be answered by comparing the oxidation rate by Fe(IV)ccp of Fe(II)cyt c, (ΔG≅1V) and the essentially isostructural derivatives[3] H_2 porphyrin cyt c and Zn(II) cyt c. The energetics of the reaction of ccp(ES) with these metal substituted cyt c derivatives can be assessed from electrochemical studies of these derivatives. Figure 3 compares the differential pulse voltamograms, obtained using a glassy carbon electrode, of cyt c and Zncyt c. (An excellent discussion of the application of electrochemical techniques to redox proteins, by Barber and Hill, is found elsewhere in this volume).

From such data, the oxidation potential of Zn(II)cyt c is found to be E°≅ 0.7V, and for H_2 porphyrin cyt c E°≅0.9 V. Taking E°ccp(ES)≅1.1 V, gives reaction free energies of ES/Fec ≅ 0.9 V, ES/Znc≅ 0.4V ES/porphc ≅ 0.1V.

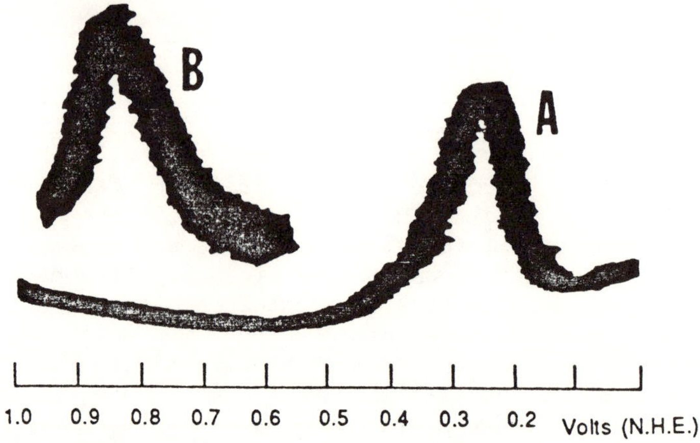

Figure 3: Differential pulse voltammogram of 3mM Fe(A) and Zn(B) cytochrome c; 3 mm graphite disc electrode; scan rate $2mVs^{-1}$; pulse amp. 50 mV.

With these free energies defined, the dependence of rate in free energy may now be addressed. As shown in figure 4, as ΔG increases, rate increases, consistent with a value of $\lambda \cong 1.5V$. Thus large configurational changes do accompany electron transfer in this system. These changes are not purely anomalies of the Fe(IV) ccp center, since equally large λ values are found for the <u>reduction</u> of ccp by cyt c^{3+}.

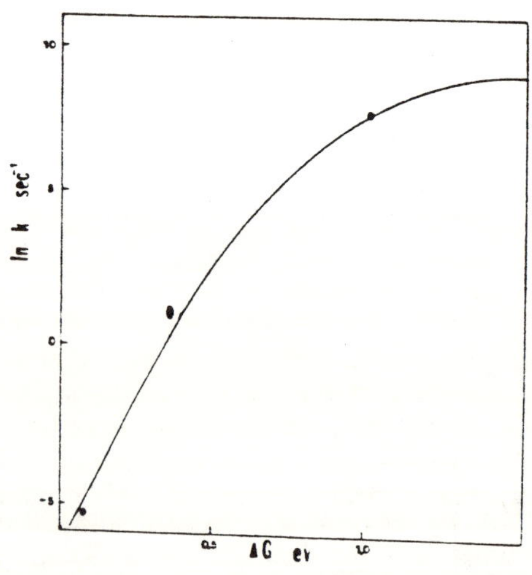

Figure 4 Oxidation rates of Fe(II) cytc, Zn(II)cytc and H_2 porph cytc by Fe(IV) ccp solid line is fitted to eqn. 1.

Similarly large values are found for other protein couples,[3,4] and suggest that large λ values, (and thus large redox dependent protein motions) may be the rule, not the exception, as previously supposed.

Reactivity and Specificity in Biological Electron Transport: an Approach involving Specific Amino Acid Replacement.

The foregoing example combined with studies of other protein complexes[3,4] suggest that protein motions might dominate the kinetics of electron transfer between proteins. What motions, in what regions, might be involved?

Elegant ongoing work by Margoliash, Bosshard and many others[6] have implicated ionic patches on cytochrome c and other proteins as important in insuring specificity between proteins, for the otherwise blatantly nonspecific single electron transfer reactions. These ideas are crucial to the elegant "model docking" work of Poulos and Kraut,[5] already mentioned. Such regions might also offer a logical region for gating of electron transfer rates by conformational matchings of the interface.

In order to study these important protein domains both in vitro and in vivo we turned to generic engineering techniques. In collaboration with Prof. Fred Sherman, we have prepared and characterized a number of single site mutants within the "binding and recognition" site of cyt c. Initial results with these mutants have been surprising. Given the crucial role of binding and recognition in the reaction of cyt c with ccp, we have used NMR to characterize this interaction in some detail. These studies are based on the fact that when ccp binds to cyt c, this binding changes the local electrostatic environment around the cyt c heme, leading to a shift in the heme methyl resonances. For example, full formation of a complex between cyt c (III) and ccp (III) leads to a ca 3 ppm shift in the frequency of the hemes 3-methyl resonance in cyt c (Fig 5). When specific lysines on cyt c are modified, some surprises occur. Mutation of cyt c Lys 13 → Ile does not decrease binding to ccp as predicted by current models of electrostatic recognition. Instead, the binding constant increases from $K_b=10^3$ for wild type cyt c to $K_b=10^5$ for Ile

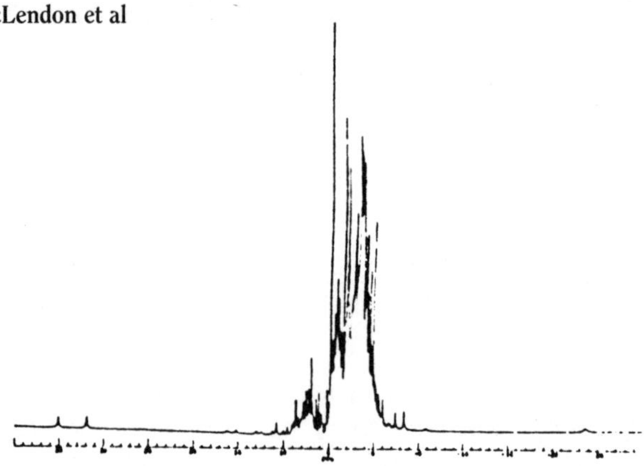

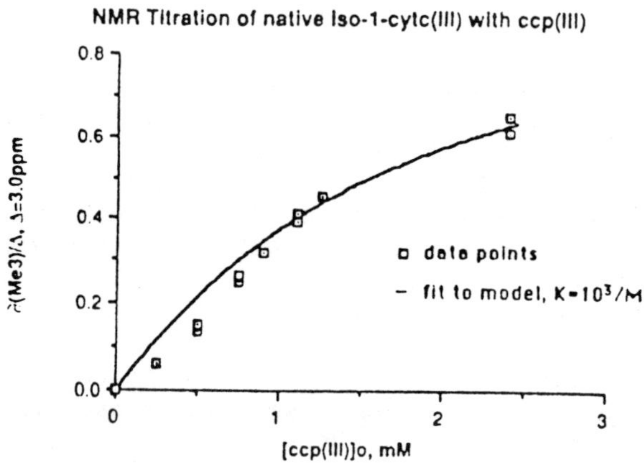

Figure 5a. ^{1}H-NMR Spectrum at 400 MHz of Saccharomyces cerevisiae iso-1 ferricytochrome c at 23°C and pH (uncorrected) = 6.1. The protein concentration was 1.1 mM contained in a 50 mM potassium phosphate buffer previously exchanged with D_2O. All chemical shifts are relative to DSS.

b. NMR titration of Saccharomyces cerevisiae iso-1 ferricytochrome c with yeast ferricytochrome c peroxidase. The ferricytochrome c was present in a 1.1 mM concentration while the ccp was varied from .25 to 2.5 mM. both proteins were contained in a 50 mM potassium phosphate buffer previously exchanged with D_2O. The uncorrected pH meter reading was 6.1 and all experiments were conducted at room temperature. The spectra were collected on a Bruker 400 MHz NMR. A typipical spectrum contained 3000-5000 transients with a sweep width of 41K Hz and a 32 KB data block. A 70° sampling pulse (7 µsec) was used at a repetition rate of .5 sec. The DSS was used as an internal standard and the residual HOD resonance was used as an internal temperature probe. A value of Δ = 3.0 ppm and a Ka of 1 X 10^3 M^{-1} were obtained upon fitting the data to the theoretical model.

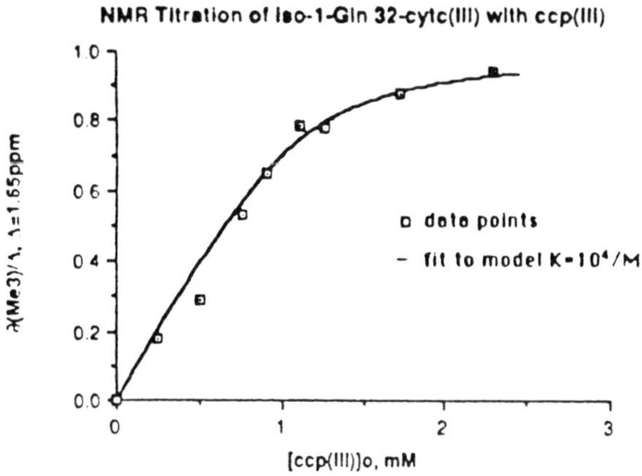

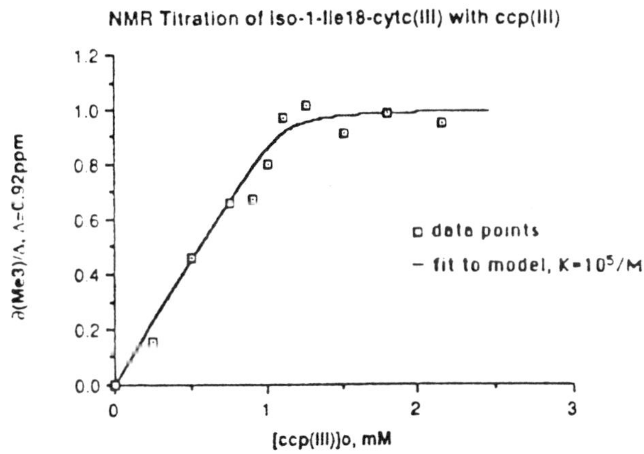

c. NMR titration of mutant iso-1-Gln 32 ferricytochrome c with yeast ferricytochrome c peroxidase. All samples and experimental conditions were the same as described in "b" above. A value of Δ=1.65 ppm and a Ka of $1 \times 10^4\ M^{-1}$ were obtained upon analyses.

d. NMR titration of mutant iso-1-Ile 18 ferricytochrome c with yeast ferricytochrome c peroxidase. All samples and experimental conditions were the same as describe in "b" above. A value of Δ=.92 ppm and a Ka of $1 \times 10^5\ M^{-1}$ were obtained upon analyses.

18! An analogous increase to K_b=10^4 is found for Lys 27 → Gln 27 cyt c. These mutant complexes also show quite different heme methyl resonance frequencies than the wild type consistent with a different electrostatic environment about the heme; ie: a different binding mode. These surprising results suggest a picture of binding and recognition which is not static, but highly plastic. This notion receives independent support from a recent x-ray study of the c-ccp complex, in which cyt c is found to be highly disordered: that is, more than one binding site on ccp is available to cyt c.

Electrostatic recognition may guide binding, but it clearly does not dominate the binding energetics (and may even impede binding). Given the result that mutation of these evolutionarily invariant lysines may enhance binding, what corresponding rate effects occur? Here the findings are equally surprising. Mutation of a given changed lysine to a neutral amino acid often accelerates the rate of intramolecular reaction with a physiological partner like ccp. In collaboration with Cusanovich, Hazzard and Tollin, we have studied intra molecular electron transfer in the cyt c/ccp (ES) complex. (The methods used in these studies are presented elsewhere in this volume). Some of our results to date are summarized in figure 6. It is worth emphasizing that in many of these mutants cytochrome c function is preserved in vivo at or near 100% of wild type levels.

Figure 6: Electron transfer rates for ccp and cytc wild type and mutants.

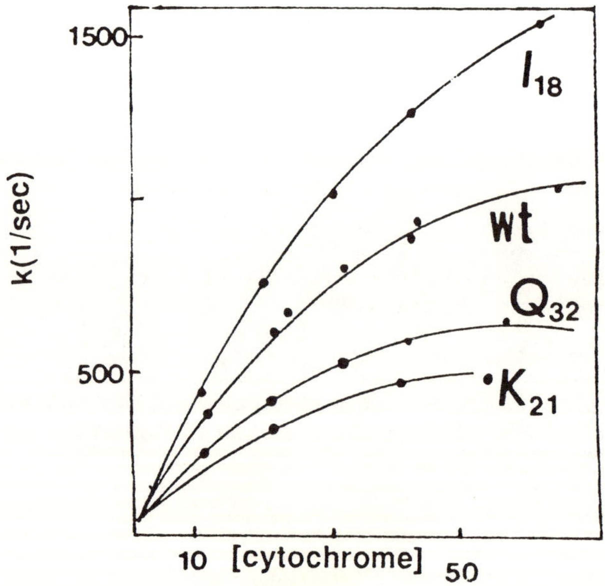

Apparently, strong electrostatic interactions at the charged interface, which help insure specificity, need not lead to the optimal configuration for electron transfer between cyt c and ccp. When the electrostatic binding requirements between cyt c and ccp are relaxed, by mutation of lysine to a neutral amino acid, the rate of electron transfer may increase. (This effect is not general. Although mutation of Lys 77, or Lys 18 increase the rate, mutation of Lys 32 tends to decrease the rate.)

A dedicated teleologist might ask "why hasn't evolution driven cyt c to a conformation which supports both maximal specificity and maximal rate?" One equally speculative answer comes from traditional Darwinism. cyt c has several cellular functions, with several potential partners. Optimizing specificity and reactivity with ccp, for example, will almost certainly insure a corresponding loss in reactivity or specificity with other proteins like cytochrome oxidase, lactate dehydrogenase, or cyt c1. Darwin's "principle of multiple utility" holds that an acceptable, not optimal, solution is found for each reaction, which may be "globally" optimal for the organism, particularly in the sense of preserving future functions yet undeveloped. This speculative note perhaps forms an appropriate postscript. It is clear that although we've learned much about biological electron transfer, there remains much left to learn. As recent theory and experimental techniques developed for small molecule reactions are increasingly applied to proteins, we hope that a clearer understanding of the principles of specificity and reactivity of these protein redox reactants will emerge.

ACKNOWLEDGEMENT

This work was supported by a grant from the NIH and, for the mutagenesis work by the NSF. Key roles in this work were played by the collaborators mentioned in the text: Fred Sherman, Mike Cusanovich, Gordon Tollin, and Jim Hazzard.

REFERENCES

1. Recent detailed reviews are available:
   a. R. Marcus and N. Sutin Biochim Biophys. Acta (1985) 811, 265-312.
   b. G. McLendon Accts. Chem. Res. (1987) in press.

2. a. J. Miller, G. Closs, L. Calcaterra J. Am. Chem. Soc. (1984) 106, 3047.
   b. J. R. Miller Science (1975) 189, 221.
   c. S. Strauch, T. Guarr, M. McGuire, G. McLendon J. Phys. Chem. (1983) 87, 3575.

3. a. G. McLendon and J. R. Miller J. Am. Chem. Soc. (1985) 107, 7811.
   b. G. McLendon, K. Simolo, A. G. Mauk J. Am. Chem. Soc. (1984) 106, 5012.
   c. K. Taylor, G. McLendon Inorganic Chem. (1986).

4. Large λ values are also suggested for electron transfer between Hb subunits in the elegant work of Hoffman, J. McGourty, N. Blough, B. Hoffman J. Am. Chem. Soc. (1983) 105, 4470, and in Ru modified proteins: H. B. Gray J. Am. Chem. Soc. (in press).

5. a. T. Poulos and J. Kraut, J. Biol. Chem. (1980) 255, 10322
   b. T. Poulos and Finzel Pept. Prot. Revs. 4 (1984) 4, 115.

6. An excellent review is: R. Bosshard and E. Margoliash TIBS (1985) 8, 316.

DISCUSSION

GARBER: Consistent with your observation that unusually high concentrations of cytochrome c, well above the reported K_D determined by gel filtration for the interaction of cytochrome c and cytochrome c peroxidase, are necessary to effect changes in the NMR spectrum, are experiments we have done in collaboration with D. Rousseau at Bell Laboratories, in which no changes in the resonance Raman spectra were observed when ferricytochrome c was bound to either cytochrome c oxidase or cytochrome c peroxidase. We interpreted these results as indicating a tethered complex between the two proteins and that a conformational change must occur subsequent to binding to produce the electron transfer permissive complex. At ionic strengths above 100 mM the monophasic kinetics of the oxidation of fungal cytochromes c of varying net charge by cytochrome oxidase displays an ionic strength dependence not representative of the net charge of the proteins, but instead a reorientation of the two proteins already in proximity to each other as being the rate limiting step. Is it possible that replacement of lysine-18 or lysine-77 favors formation of the electron transfer permissive complex while replacement of lysine-32 has the opposite effect?

McLENDON: The mutants we discussed have been fully characterized in terms of their stability. For all practical purposes their stability is identical to wild type. There is no "best" method for measuring complexation. Different methods are sensitive to different physical properties: NMR shifts, for example, provide a measure of local electrostatic fields around the heme while fluorescence energy transfer measures heme-heme distance, and ultracentrifugation measures overall binding. The fact that different techniques produce different equilibrium constants can only be explained by a multiplicity of different interactions (each of which is best sensed by different techniques). The cytochrome c - cytochrome c peroxidase interaction then is quite malleable along the binding surface. Small modulations of that surface may thus have a large effect on binding as we found for the isoleucine-18 (13) mutant. Under our conditions we measure "simple" (simple turnover) kinetics so dissocation cannot contribute to our results.

However, I am delighted to learn that the observations of the Northwestern group with cytochrome c oxidase are in many ways analogous to our results for cytochrome c - cytochrome c peroxidase. I believe we may well converge to a view first espoused by Margoliash and Koppenol of strong but relatively non-specific binding between the charge patches and hydrophobic surface of cytochrome c with its partners.

BALLOU: Do you think that systems with two redox centers in the same protein will have more specific interactions? And will the protein organization be such that reaction rates will be optimized at high rates?

McLENDON: The basic physical parameters (λ, ΔG, distance) will be important in any system, but will vary from one system to the next. In many cases, I suspect that electron transfer may be faster than associated informational re-arrangements: the reaction will be "informationally limited" providing another level of biological control. Such control is, I believe, a key evolutionary concern. The activities of most electron transfer proteins are intrinsically very high (and thus intrinsically non-specific). Many proteins are likely to be designed not to speed up reactivity of the prosthetic group, but to slow it down!

Oxidases and Related Redox Systems, pages 401–418

INTRAMOLECULAR AND INTRACOMPLEX ELECTRON TRANSFER IN REDOX PROTEINS

M. A. Cusanovich, J. T. Hazzard, T. E. Meyer and G. Tollin
Department of Biochemistry, University of Arizona, Tucson, AZ 85721

SUMMARY

The properties of intramolecular and intracomplex electron transfer are discussed for four different redox protein systems: cytochrome *c*-flavodoxin, ferredoxin $NADP^+$: reductase-ferredoxin, cytochrome *c*-cytochrome *c* oxidase and cytochrome c-cytochrome c peroxidase. It is shown that in these rather diverse systems several factors in addition to distance and thermodynamic driving force play an important role in the kinetics of electron transfer between prosthetic groups. These include dynamic motions of the reactants within the electron transfer complex, orientation of the prosthetic groups relative to each other, and the nature of the intervening media, that is, the amino acid side chains and solvent between the redox centers. It became apparent such factors were involved when distance and driving force, which play significant roles in the kinetics of electron transfer in diffusion-controlled systems, showed no obvious correlations to the kinetics of intracomplex electron transfer. Although these parameters are probably also important, they are apparently masked by the dominant contribution of the other factors (dynamics, intervening media and orientation). It can also be concluded that hypothetical computer generated protein-protein complexes are not entirely consistent with experimental data.

INTRODUCTION

The mechanism of electron transfer between redox proteins has been the subject of extensive study in recent years (see Tollin et al, 1986 for a recent review). The primary focus of these studies has been on the interaction of soluble proteins containing a single redox center, although some work exists on multicenter and membrane-associated systems. The interest in the study of soluble systems has resulted from a desire to take advantage of the substantial body of structural information available for small and relatively simple redox proteins (i.e., c-type cytochromes, copper proteins, iron-sulfur proteins, etc.). This focus has had two principal ramifications: first, the systems studied often do not use physiological redox partners, the choice being dictated by the properties of the system rather than by in vivo function; second, due to the soluble nature of the systems, much attention has been focused on the factors controlling the diffusion dependent portion of the reaction and not the direct electron transfer event itself. However, to a large extent driving force, electrostatics, and steric accessibility (distance)of prosthetic groups can be quantitated in terms of their contribution to the kinetic parameters of diffusion controlled interactions (Tollin et al, 1986).

To date, with the exception of flavocytochrome b_2 (Xia et al, 1987) and trimethylamine dehydrogenase (Lim et al, 1986), very little high resolution structural information is available concerning proteins containing multiple redox centers or complexes of interacting proteins. Nevertheless, computer graphics have been extensively used to generate structural models of interacting redox proteins when the structures of both reactants are known. Examples of this approach include cytochrome c-cytochrome b_5 (Salemme, 1976), cytochrome c-flavodoxin (Simondsen et al, 1982), cytochrome c-cytochrome c peroxidase (Poulos and Kraut, 1980) and cytochrome b_5-hemoglobin (Poulos and Mauk, 1983). Although invaluable for experimental design, the actual three dimensional structures of these hypothetical models have not been determined and the validities of the models are open to further studies. No high resolution x-ray data are presently available for many important complex redox systems, for example, flavocytochrome c, cytochrome c oxidase, cytochrome b/c_1 and xanthine oxidase, or for a variety of physiological complexes that exist at least transiently (for example, the ferredoxin $NADP^+$ reductase - ferredoxin complex).

In view of the foregoing, the purpose of this discussion is to summarize some of our recent studies on electron transfer in multicenter redox systems and transient complexes and to discuss the roles of orientation, distance, intervening media, and dynamics in these systems. Although the studies to date are not quantifiable in terms of the contribution of these factors to the observed kinetics, progress is being made in defining their role.

RESULTS AND DISCUSSION

Flavodoxin-cytochrome c. The flavodoxin-cytochrome *c* complex is an appropriate place to begin as it illustrates many of the observations and unresolved questions in intracomplex electron transfer. Although the complex is non-physiological, it has been extensively characterized. Based on the kinetic properties of the system and available structural information, a putative electron-transfer complex structure has been proposed (Simondsen et al, 1982). This predicts a strong electrostatic interaction involving four charge pairs from cytochrome *c* (cyt *c*) lysines and *C. pasteurianum* flavodoxin (Fld) carboxylates and is consistent with the high association constant (10^7 M^{-1}) measured at low ionic strength. The model requires that the heme *c* and the dimethylbenzene portion of the Fld FMN are within van der Waals contact. To test the validity of this computer model, a variety of kinetic studies have been carried out on the Fld-cyt *c* complex, utilizing the interaction of exogenous flavin semiquinones with the individual components of the Fld-cyt *c* complex (Hazzard et al, 1986).

A number of observations have been made concerning this system. First, the kinetics of reduction of Fld by 5-deazariboflavin semiquinone (5-dRfH·) were the same in the free and complexed states, indicating that no steric restrictions to the flavin moiety of Fld occurred upon complexation with cyt *c*. In contrast, the reduction of cyt *c* by lumiflavin semiquinone (LfH·) was inhibited by complex formation (∿3 fold), suggesting that Fld sterically restricted heme *c* accessibility to a small extent. Analysis of the hypothetical Fld-cyt *c* structural model using van der Waals surfaces shows that there are two pockets in the protein interface region that allow access of small flavin reductants to the FMN-heme interaction site, either lumiflavin or riboflavin when the ribityl side chain is appropriately oriented. The Fld FMN moiety is substantially more exposed to the solvent than the cyt c heme within the complex. Furthermore, these two pockets allow access to only a small fraction of the exposed cyt c heme edge. Thus, the observed kinetics do not appear to be entirely consistent with the structural model because greater steric effects would have been expected. However, the structural model and kinetic data can be rationalized by assuming dynamic motions at the protein-protein interface which allow greater access than the static model predicts (see below).

Second, it was found that the interaction with FMN semiquinone (FMNH·) yielded a nonlinear dependence of the observed rate constant on the complex concentration. This observation is consistent with a change in rate limiting step at high complex concentrations. To discriminate between a mechanism in which the reduction of complexed cyt c within a ternary complex ($complex_{ox}$-FMNH·) with FMNH· represents the rate limiting step (Eq. 1), or a mechanism in which the rate limiting reaction is controlled by a dynamic equilibrium between a closed ($complex_{ox}^{*}$) and an open ($complex_{ox}$) form of the protein complex (Eq. 2), the FMN analogue, Cl_2-FMN was used.

$$complex_{ox} + FMNH\cdot \leftrightarrow complex_{ox}\text{-}FMNH\cdot \rightarrow complex_{red} + FMN \quad (1)$$

$$complex^{*}_{ox} \leftrightarrow complex_{ox} + FMNH\cdot \rightarrow complex_{red} + FMN \quad (2)$$

Although it is much more reactive with free cyt c than is FMNH·, Cl_2-FMNH· gives kinetic results identical to those of FMNH· when cyt c is complexed with Fld. This establishes that the rate limiting process (810 s^{-1} at μ = 10 mM, 2620 s^{-1} at μ = 20 mM) is independent of the chemistry of the reductant and must be due to dynamic motions of the complex, consistent with the mechanism given in Eq. 2 (i.e., breathing). Moreover, the increase in limiting rate constant observed at higher ionic strength is consistent with a weakening of the electrostatic forces holding the complex together. Presumably, the reason why breathing is not detected with LfH· and riboflavin semiquinone (RfH·) is due to a reduced steric effect. Hence the dynamic motions required for access to the heme by these less bulky flavins were smaller and faster and not detectable in the accessible time domains.

When Fld semiquinone (Fld·) and oxidized horse cyt c are mixed, rapid reduction of cyt c occurs (Simondsen et al, 1982). However, as the cyt c concentration is increased, it is found that a change in rate limiting step occurs at high cyt c concentrations (limiting rate constant 85 s^{-1}). Unlike the reduction of cyt c within the preformed complex, this electron transfer reaction is consistent with the formation of a transient complex obeying the type of mechanism given by Eq. 1, as similar behavior was not observed with all cytochromes c. Thus, the limiting rate constant apparently reflects electron transfer from Fld· to cyt c. The relatively small value for this intracomplex electron transfer is interesting since there is a rather high thermodynamic driving force (400 mV, Table 4) and in the hypothetical complex the interprosthetic group distance is only 3.5 A and the orientations of the FMN and heme ring planes are essentially parallel.

To summarize, although the kinetic results obtained with the Fld-cyt c system are partially consistent with the computer model, they require that the complex have characteristic dynamic motions (>810 s^{-1} at μ = 10 mM) and that these dynamics affect the accessibilities of the prosthetic groups to exogenous reactants. Furthermore, the relatively small value for the intracomplex electron transfer rate constant suggests that the orientation of the prosthetic groups in the Fld-cyt c biochemical complex may not be accurately described in the hypothetical model, and/or the distance may be larger than imagined. Alternatively, the predicted orientation may be correct, but it is not conducive to rapid electron transfer.

Ferredoxin: Ferredoxin $NADP^+$ reductase. Spinach ferredoxin (Fd) and ferredoxin $NADP^+$-reductase (FNR) form a tight, and presumably physiological relevant complex at low ionic strength ($K_a \sim 10^8$ M^{-1}) which provides a useful experimental system to study intracomplex electron transport (Bhattacharyya et al, 1986). Because of the low redox potentials of FNR and Fd, this system has been studied using dRFH· ($E_o{}'$ = -650 mV). As shown in Table 1, 5-dRFH· rapidly reduces both the iron sulfur cluster of Fd_{ox} and the FAD of FNR_{ox}. Moreover, in the oxidized Fd-FNR complex, FNR is reduced with a rate constant essentially identical to that of the free protein indicating no steric hindrance of the FNR FAD moiety upon complex formation. In contrast, in the Fd_{ox}-FNR_{red} complex, it is found that reduction of Fd is slowed substantially (~30 fold) as compared to the free protein. This last result indicates that steric accessibility to the Fd iron-sulfur cluster is substantially hindered in the Fd-FNR complex and is in sharp contrast to the Fld-cyt c complex where either a small or neglible change in accessibility occurs on complex formation.

Table 1. Ferredoxin-FNR Kinetics

Free Proteins and Complexes	$kx10^{-8}$ $M^{-1}s^{-1}$	Intracomplex Electron Transfer (s^{-1})
Fd_{ox}	2.0	-
FNR_{ox}	6.3	-
Fd_{ox}-FNR_{ox}	5.6	-
Fd_{ox}-FNR_{red}	.07	-
Fd_{red}-FNR_{ox}	-	$4.0x10^3$(=310 mM) $1.6x10^3$(=460 mM)

Although intracomplex electron transfer (Fd_{red} to FNR_{ox}) cannot be measured at low ionic strength because of the slow rate of Fd reduction within the complex, at higher ionic strengths plots of the observed rate constant vs FNR_{ox} concentration (at a fixed Fd concentration) show nonlinear behavior and yield a limiting rate constant for electron transfer. These data are consistent with an intracomplex electron rate of 4×10^3 s^{-1} at μ= 310 mM and 1.6×10^3 at μ= 460 mM. Thus, we estimate that at low ionic strength ($<$100 mM) the intracomplex electron transfer rate constant must be well in excess of 10^4 s^{-1}. It is important to note that the driving force for this reaction is small (Δ E $\sim$50 mV) yet the limiting rate constant is very large. This is in contrast to the Fld-cyt c complex where the driving force is large (ΔE $\sim$400 mV) and yet the limiting rate constant is small ($\sim$85 s^{-1}).

FNR is also an interesting enzyme to study because it will strongly bind a wide variety of negatively charged non-physiological redox proteins at low ionic strength. In view of this, we have investigated intracomplex electron transfer with a variety of FNR-containing complexes (Przysiecki et al, 1985; Bhattacharyya et al, 1987). Rubredoxin binds tightly to FNR and when bound is reduced by free flavin semiquinones with the same kinetics as the free form (Przysiecki et al, 1985), in sharp contrast to FNR-bound Fd (Bhattacharyya et al, 1986). Intracomplex electron transfer (FNR semiquinone to oxidized rubredoxin) is characterized by a rate constant of 2100 s^{-1} at low ionic strength (10 mM). The driving force for this reaction (ΔE = 260 mV, Table 4) is substantially greater than that for the Fd-FNR complex but the rate constant is smaller perhaps because of a greater distance over which electron transfer occurs. Another FNR complex has been investigated using Ectothiorhodospira iso-I high potential iron-sulfur protein (HIPIP) (ΔE $\sim$440 mV, Table 4) which has an intracomplex electron transfer rate constant of 23 s^{-1} (FNR· to $HIPIP_{ox}$) which is 100-fold slower than that of the complex with rubredoxin although there is a larger driving force in the FNR-HIPIP reaction. Thus, electron transfer between ferredoxin and FNR, although having a much smaller driving force than the comparable FNR reactions with HIPIP or rubredoxin, has a substantially larger intracomplex electron transfer rate constant suggesting that orientation, intervening media and/or distance effects play significant roles in controlling the kinetics of intracomplex electron transfer in these systems.

In a somewhat different approach, we have investigated intracomplex electron transfer within FNR-cyt $\underline{c}_2$ complexes (Bhattacharyya et al, 1987). For these studies cytochromes $\underline{c}_2$ from R. rubrum and P. denitrificans were used. In both cases the cytochromes $\underline{c}_2$ have net negative charges at pH 7; however, the charge distributions are asymmetric. That is, both cytochromes $\underline{c}_2$ have a cluster of positive charges around the exposed heme edge (the presumed electron transfer site) and a negative cluster on the back side approximately 180° from the exposed heme edge. Thus, these cytochromes should bind to FNR backwards in the sense that the exposed cytochrome heme edge is presumably pointed away from the FNR electron transfer site. In both cases tight binding to FNR and intramolecular electron transfer reactions with rate constants 400-700 s^{-1} are observed. Thus moderately large rate constants, as compared to the HIPIP example, are obtained which presumably represent electron transfer through the interior of the cytochromes $\underline{c}_2$. Because the distance over which the electron would have to travel is greater in the cytochrome than in HIPIP, these examples point out possible orientational and/or intervening media effects. The results of the studies with these various nonphysiological reactants with FNR are summarized in Table 2.

In summary, FNR allows intracomplex electron transfer to be measured in a variety of biological redox complexes. It is clear from the FNR studies that orientation, distance, and/or intervening media between prosthetic groups play an important role in the intracomplex electron transfer process.

Table 2. Intracomplex Electron Transfer Between FNR· and Electron Acceptors[a]

Reaction	Intracomplex Electron Transfer (s^{-1})
$FD_{red} \rightarrow FNR_{ox}$	$>>4x10^3$
$FNR\cdot \rightarrow Rd_{ox}$	2100
$FNR\cdot \rightarrow HIPIP_{ox}$	23
FNR· → *R. rubrum* cyt. c_2	700
FNR· → *P. denitrificans* cyto. c_2	400

[a]μ = 10 mM

Cytochrome c Oxidase and Cytochrome c Peroxidase. The rather unique aspects of two physiological redox partners of mitochondrial cyt *c*, bovine cyt *c* oxidase (CcO) and yeast cyt *c* peroxidase (CcP), have permitted us to study in some detail the nature of the intracomplex electron transfer from reduced cyt *c* to either CcO or CcP within electrostatically stabilized or transiently formed 1:1 complexes. As shall be shown below, the free CcO or CcP cannot be reduced by flavin semiquinones with rate constants rapid enough to compete with disproportionation of the free flavin semiquinone to the fully reduced species. In the presence of cyt *c*, oxidation of flavin semiquinones by either CcO or CcP does occur, although the electron transfer occurs via the complexed cyt *c*.

Cytochrome c-cytochrome c oxidase. As stated above, the reaction between fully-oxidized CcO and flavin semiquinones is too slow to compete with semiquinone disproportionation (Ahmad et al, 1982); however, CcO reacts slowly with fully reduced flavins, with a change in the rate limiting step for reduction characterized by a rate constant of $\sim 20\ s^{-1}$ which is independent of the chemical nature of the flavin. The inability of flavin semiquinones to directly interact with CcO suggests that both the *a* and a_3 hemes are relatively inaccessible to exogenous reductants. Furthermore, the limiting rate constant observed with fully reduced flavins suggests that an isomerization of CcO is necessary in order for a fully reduced flavin to gain access to the prosthetic groups.

In the presence of cyt *c*, an indirect reduction of oxidized CcO by free flavin semiquinones occurs. It was found that direct reduction of the complexed cyt *c*(III) preceeded an intracomplex electron transfer to the CcO heme a, since CN^-, which reacts with cyt c and cytochrome a_3 but not cytochrome a, blocked cytochrome a reduction (Ahmad et al, 1982). Interestingly, no transient formation of reduced cyt c could be observed indicating that intracomplex electron transfer from cyt c to cytochrome a had a rate constant of $>700\ s^{-1}$ for the experimental conditions used. In contrast to the Fld-cyt c system, we found that the rates of reduction of cyt c complexed with CcO were as large or larger than free cyt c for a large number of flavin semiquinones, suggesting that the cyt c heme is fully accessible to the solvent on the time scale of these experiments.

Although we have no 3-D structural information on the cyt c-CcO complex, we assume that the cyt c heme is oriented towards and fairly close to the cytochrome a heme in the complex because of the relatively large intracomplex electron transfer rate constant. If this is the case, then the results indicate that the cyt c heme accessibility must be due to dynamic motions within the complex. The fact that flavin semiquinones are not able to reduce CcO, but that facile reduction of the oxidase by cyt c occurs suggests that cyt c binding elicits a substantial change in the conformation of CcO in the vincinity of the cyt c-CcO heme a interaction domain.

Cytochrome c-cytochrome c peroxidase. This system represents a physiologically relevant complex with the structures of both reactants known to high resolution and for which a hypothetical model has been proposed (Poulos and Kraut, 1980). Complexation with ferric CcP results in a decrease in the rate constants for reduction for all cyt c (III) species measured, relative to the free protein as shown in Table 3 (Hazzard et al, 1987). Importantly, the magnitude of the decrease in the cyt c reduction rate constant in the complex is strongly dependent upon the cytochrome species and significantly exceeds the relatively small differences in the rate contants for reduction of the free cytochromes. Thus, the data suggests that the geometric relationships of the various cytochromes within the complexes with CcP are not the same. In the complex with the H_2O_2-oxidized peroxidase, CcP(IV,$R^{+}\cdot$), the rate constant for reduction of cyt c increases, relative to the complex with CcP(III). Again, the magnitude of the change in the rate constant, and by inference the cyt c heme accessibility, varies with the species of cytochrome. These results suggest that oxidation of CcP(III) to CcP(IV,$R^{+\cdot}$) induces a conformational change in the peroxidase near the protein-protein interaction domain, thereby increasing the accessibility of the complexed cyt c heme edge. This proposal is supported by the recently published CcP(IV,$R^{+\cdot}$) structure determination (Edwards et al, 1987) which shows movement of an antiparallel beta sheet upon peroxide oxidation that lies at the interprotein interface in the hypothetical model of the complex.

Reduction of CcP(III) by 5-dRfH· is characterized by a relatively small second order rate constant, 9 x 10^6 $M^{-1}s^{-1}$, consistent with the highly buried nature of the peroxidase heme. When complexed with cyt c(II), a 45% decrease in this rate constant was observed consistent with a small extent of steric hinderance due to complexation, whereas none was expected on the basis of the computer model. When CcP(III) is oxidized to CcP(IV,$R^{+\cdot}$), reduction by either low (5-dRfH·) or high (LfH·) potential flavin semiquinones does not occur. However, in the presence of a variety of cytochromes c(III) at low ionic strength, reduction of CcP(IV,$R^{+\cdot}$) by free flavin semiquinones does occur, although the mechanism involves reduction of the complexed cyt c followed by a first-order intracomplex one-electron transfer to the peroxidase. These data are also summarized in Table 3 and establish a wide range of rate constants for intracomplex electron transfer. Interestingly, at low ionic strength the physiological substrate, yeast iso-2 cyt c, gives the lowest rate constant while tuna cytochrome yields the highest. Another interesting feature is that in the case of yeast iso-1 and iso-2 cytochromes the rate constants for the intracomplex electron transfer are markedly increased with increasing ionic strength, a point that shall be addressed below. The lack of direct reduction of CcP(IV,$R^{+\cdot}$) by free flavin semiquinones, but rapid reduction in the presence of cyt c, is similar to the results obtained with the cyt c-CcO system. This suggests that oxygen-reducing enzymes may have as a common feature the requirement for the physiological electron donor (i.e.cyt c) to establish an effective electron-transfer conduit, possibly by the reorientation of specific amino acid side chains.

In recent studies in collaboration with Drs. G. McLendon and F. Sherman (University of Rochester), we have extended the studies just described to site specific mutations in amino acid side chains of yeast iso-1 cyt c(III). When positively charged residues of cyt c (lys-27 or arg-13), which have been proposed to play a significant role in stabilization of the complex with CcP (Poulos and Kraut, 1980), are converted to electrostatically neutral residues several marked effects are observed. As shown in Table 3, there is an increase in heme c exposure relative to the wild-type cytochrome (>>14% for the wild-type reaction reflects the fact that lumiflavin semiquinone reduction was too slow to compete with disproportionation). Mutant cyt c accessibility is also substantially

Table 3. Cytochrome c Accessibility and Intracomplex Electron Transfer in the CcP-cyt c Complex

	Heme c Accessibility (% of free)		Intracomplex electron transfer (s^{-1})	
Cyt c	CcP(III) Complex	CcP(IV,$R^{+\cdot}$) Complex	μ=8 mM	μ=260 mM
horse	26%	42%	730	-
tuna	80%	-	920	-
yeast ISO-2	9%	68%	150	1850
yeast ISO-1	<<14%	69%	260	1850
Yeast ISO-1 Lys-27 → Gln	17%	85%	260	890
yeast ISO 1 Arg-13 → Ile	34%	100%	1000	3000

changed in the CcP(IV,$R^{+\cdot}$) complex. Furthermore, in the case of the Ile-13 mutant, a marked increase in the intracomplex electron transfer rate constant at low ionic strength occurs. In all cases an increase in ionic strength results in an increase in the intracomplex electron transfer rate constant. These latter two points are particularly significant since they suggest that the electrostatically-stabilized complexes (i.e., low ionic strength) are actually less favorable for rapid intracomplex electron transfer than are the high ionic strength transient complexes (Hazzard et al, 1987). Therefore, masking electrostatic interactions apparently permits the formation of more efficient electron transfer orientations. Thus, both the cyt c accessibility and the intracomplex electron transfer are very sensitive to specific amino acid interactions at the cyt c-CcP interface.

As was the case for the Fld-cyt c complex, we have been able to test the validity of the hypothetical model for the cyt c-CcP complex using the kinetic data. The model predicts that steric accessibility of the CcP and cyt c heme should not be significantly altered by complexation. While the data for

CcP and tuna cyt *c*, the structures from which the model was constructed, are in good agreement with this prediction, the data for yeast cyt *c* at low ionic strength suggests that the structure of the complex and/or the structure of the yeast cytochrome is substantially different from the structure(s) in the hypothetical model. It is important to note that whereas most hypothetical electron transfer models emphasize electrostatic stabilization of the complex, presumably important electrostatic interactions may not, in fact, facilitate the formation of an efficient electron transfer complex. Thus, only by the application of the appropriate experimental techniques can the validity of hypothetical models be tested. In sum, the available data for the cyt *c*-CcP system are consistent with a complex whose structure is quite sensitive to the specific interactions and to presumbly small structural changes which have substantial effects on accessibility and electron transfer kinetics.

CONCLUSIONS

It is clear from the data presented that much remains to be determined before we have a full understanding of biological electron transfer within multicenter systems. However, a number of features are apparent when comparing the various systems discussed. By way of summary, Table 4 is presented which attempts to relate the kinetics of intracomplex electron transfer to driving force and distance. In constructing this table, the difference in redox potential ($\Delta E_{m,7}$) was obtained from the midpoint potentials of the uncomplexed reactants. The distances used in Table 4 were obtained from either computer based models taking the distance of closest approach in the static complex (flavodoxin-cyt *c*, cyt *c*-CCP or by assuming that the flavin moiety of FNR is solvent accessible and taking the distance of the heme from the back side of the cytochromes c_2 or the maximum distance from the iron sulfur center to the surface in the iron-sulfur proteins. Flavocytochrome b_2, for which both 3-D structure and kinetics are known (Xia et al, 1987) is also included. Clearly, no obvious correlations exist between driving force or distance, or a combination of the two, and the observed kinetics. This forces one to the conclusion that other parameters must be involved. Most obvious is the intervening media, that is, the influence of the specific amino acid side chains and the solvent between interacting redox centers, and the relative orientations of

Table 4. Intracomplex Electron Transfer

Reactants	$\Delta E_{m,7}$(V)	Edge to Edge Distance (A)	$k(s^{-1})$
tuna cyt *c* → CcP(IV,$R^{+\cdot}$)	∿.75	∿18	950
Iso-1 cyt *c* → CcP (IV,$R^{+\cdot}$)	∿.75	∿18	250
Iso-1 cyt *c*(Ile 13)-CcP(IV,$R^{+\cdot}$)	∿.75	∿18	1000
Iso-1 cyt *c*(Ile 13)-CcP(IV,$R^{+\cdot}$) high ionic strength	∿.75	∿18	3000
FNR· → Paracoccus cyt c_2	0.57	<15	700
FNR· → Ecto. HIPIP	0.44	<10	23
Flavodoxin· → tuna cyt *c*	0.40	3.5	85
FNR· → Rubredoxin	0.26	<10	$2.1x10^3$
FNR· → Ferredoxin	0.05	<10	$>4x10^3$
Cyt *c* → heme *a*	-0.04	?	>700
FMN → heme b (in flavocyt. b_2)	.03	10	350

the interacting centers. In the absence of structural information, it is not possible to discuss orientation factors in detail. The redox centers in flavocytochrome b_2 are nearly coplanar and 10 A apart. Although there is essentially no driving force, kinetics are relatively fast. In the flavodoxin-cyt *c* computer generated complex the heme and flavin moieties are approximately parallel and within van der Waals contact. In this case there is ample driving force yet we observe a relatively slow rate constant for intracomplex electron transfer. A possible explanation for this is that in the flavocytochrome b_2 the N5 position of the flavin is oriented towards the heme, whereas in the putative flavodoxin-cyt *c* complex it is the dimethylbenzene part of the flavin which is in contact with the heme. Thus, the former may be more favorable than the latter. Intervening media effects are also a possibility. Thus, intracomplex electron transfer in the FNR-cyt *c* system is relatively fast. However, the HIPIP-FNR complex has a very slow rate constant even though the distance over which electron transfer occurs must be much smaller due to the size of HIPIP relative to cytochrome c_2 (M.W. ∿8,000 daltons vs. ∿14,000 daltons).

It is clear from the studies described that the static computer generated structures do not completely explain the observed kinetics. Hence a complete understanding will require consideration of additional hypothetical complexes as well as of the dynamic motions within the protein-protein complexes. Indeed, data with the flavodoxin-cyt *c* complex (using FMNH· as a reductant) clearly establish that the protein-protein complex can undergo rapid dynamic motions. Moreover, less direct evidence suggests similar dynamic properties for both the cyt c-CcP and cyt c CcO complexes. Although driving force and distance effects may be masked in some cases by the influence of intervening media, dynamics and orientation, the real situation involves a combination of factors. Having established a role for intervening media, dynamics and orientation, it is now necessary to quantitate the contribution of these factors to the electron transfer kinetics. Considering the progress being made in terms of structural data and molecular dynamics and the range of kinetic approaches now available, as well as the availability of site-directed mutants, one can be confident that very real progress in understanding biological electron transfer can be expected in the near future.

ACKNOWLEDGMENTS

This research was supported by NIH grants GM 21277 (MAC) and AM 15057 (GT)

References

Ahmad I, Cusanovich MA, Tollin, G (1982). Laser Flash Photolysis Studies of Electron Transfer Between Semiquinone and Fully Reduced Free Flavins and the Cytochrome c - Cytochrome Oxidase Complex. Biochem. 21: 3122-3128.

Bhattacharyya AK, Meyer TE, Cusanovich MA, Tollin G (1987). Laser Flash Photolysis Studies of Electron Transfer Between Ferredoxin-NADP$^+$ Reductase and Several High-Potential Redox Proteins. Biochem. 26: 758-764.

Bhattacharyya AK, Meyer TE, Tollin G (1986). Reduction Kinetics of the Ferredoxin - Ferredoxin-NADP$^+$ Reductase Complex: A Laser Flash Photolysis Study. Biochem. 25: 4655-4661.

Edwards SL, Xuong NH, Hamlin RC, Kraut J (1987). Crystal Structure of Cytochrome c Peroxidase Compound I. Biochemistr 26: 1503-1511.

Hazzard JT, Cusanovich MA, Tainer JA, Getzoff ED, Tollin G (1986). Kinetic Studies of Reduction of a 1:1 Cytochrome c-Flavodoxin Complex by Free Flavin Semiquinones and Rubredoxin. Biochem. 25: 3318-3328.

Hazzard JT, McLendon G, Cusanovich MA and Tollin G (1987) Formation of Tight Binding Complex Inhibits Interprotein Electron Transfer Between Yeast Cytochrome c and Cytochrome c Peroxidase. in preparation.

Hazzard JT, Poulos TL, Tollin G (1987). Kinetics of Reduction by Free Flavin Semiquinones of the Component of the Cyto-c - Cytochrome c Peroxidase Complex and Intracomplex Electron Transfer. Biochemistry 26: 2836-2848.

Lim LW, Shamala N, Mathews FS, Steenkamp DJ, Hamlin R, Xuong NA (1986). Three-Dimensional Structure of the Iron-Sulfur Flavoprotein Trimethylamine Dehydrogenase at 2.4-A Resolution. J. Biol. Chem. 261: 15140-15146.

Poulos TL, Kraut J (1980). A Hypothetical Model of the Cytochrome c Peroxidase · Cytochrome c Electron Transfer Complex. J. Biol. Chem. 255: 10322-10330.

Poulos TL, Mauk AG (1983). Models for the Complexes Formed Between Cytochrome b_5 and the Subunits of Methemoglobin J. Biol. Chem. 258: 7369-7373.

Przysiecki CT, Bhattacharyya AK, Tollin G, Cusanovich MA (1985). Kinetics of Reduction of Clostridium pasteurianum Rubredoxin by Laser Photoreduced Spinach Ferredoxin: $NADP^+$ Reductase and Free Flavins. J. Biol. Chem. 260: 1452-1458.

Salemme FR (1976). An Hypothetical Structure for an Intermolecular Electron Transfer Complex of Cytochromes c and b_5. J. Mol. Biol. 102: 563-568.

Simondsen RP, Weber PC, Salemme FR, Tollin G (1982) Transient Kinetics of Electron Transfer Reactions of Flavodoxin: Ionic Strength Dependence of Semiquinone Oxidation by Cytochrome c, Ferricyanide, and Ferric Ethylenediaminetetraacetic Acid and Computer Modeling of Reaction Complexes. Biochemistry 24: 6366-6375.

Tollin GT, Meyer TE, Cusanovich MA (1986). Elucidation of the Factors which Determine Reaction-Rate Constants and Biological Specificity for Electron-Transfer Proteins. Biochim. Biophys. Acta 853: 29-41.

Xia ZX, Shamala N, Bethge PH, Lim LW, Bellamy HD, Xuong NH, Lederer F. Methews FS (1987). Three-Dimensional Structure of Flavocytochrome b_2 From Baker's Yeast at 3.0-A Resolution. Proc. Natl. Acad. Sci. USA 84: 2629-2633.

DISCUSSION

HURST: Could you comment on the use of bifunctional cross-linking reagents as a means of aiding in the identification of protein interaction sites? What systems have been studied and what general patterns emerge?

CUSANOVICH: The only system for which we have data is the cross-linked cytochrome c peroxidase - horse cytochrome c complex. My co-author, Dr. James Hazzard, can comment.

HAZZARD: Covalent cross-linking techniques have been applied to two electron transfer complexes: A vinelandii flavodoxin:horse cytochrome c and horse cytochrome c:yeast cytochrome c peroxidase. To my knowledge, no kinetic data relevant to electron transfer are available for the first complex.

The cytochrome c:cytochrome c peroxidase complex has, however, been more thoroughly studied by a variety of biophysical techniques. In collaboration with Drs. S. Moench, J. Satterlee and J. Erman, we have studied the reduction kinetics of the covalent cyt c:CcP complex and found that there are some significant differences relative to the electrostatically-stabilized complex. Specifically, it was found that both the electrostatic environment near the cytochrome c heme edge and the rate constant for one-electron transfer from cyt c(II) to CcP(IV,$R^{+\cdot}$) were markedly different from those obtained with the electrostatic complex. Cyt c heme accessibility to external reductants, however, was found to be similar.

Based on the results presented by Drs. Cusanovich and McLendon at this symposium, as well as results published by Dr. Margoliash and co-workers, it is becoming clear that the processes controlling biological electron transfer can be modulated to a large degree by ionic strength conditions as well as by the presence or absence of charged residues on the surfaces of the redox proteins. Thus, dynamic processes and/or multiple orientations of the proteins within these complexes may have profound effects on the electron transfer kinetics. By virtue of covalent cross-links, dynamic processes are certainly minimized, which may lead to confusion in the interpretation of data from the study of these complexes.

Furthermore, covalent cross-linking has been shown to result in heterogeneous mixtures of complexes which have been difficult to isolate and characterize. Thus, it is debatable whether the technique answers more questions than it raises concerning the factors which control the formation of, and electron transfer within, protein:protein complexes.

KOSMAN: Have you studied the cross-linked complex which contains the Arg->Ile substitution? Specifically, can you consider the explicitly media effects of this change at interface with intracomplex electron transfer in contrast to the "loosening" to which you and McLendon attribute the enhanced rate of electron transfer?

CUSANOVICH: No. Only horse heart cytochrome c has been cross-linked. Clearly, to use site directed mutants but to define both the interaction's domain and the role of amino acid side chains in the "intervening media" is the direction to go.

BALLOU: Is it possible that in those cases where higher ionic strength increases electron transfer rates the effect might be on easing the processes involved in the reorganization terms of the rate equation?

CUSANOVICH: Yes. Assuming the strong electrostatically stabilized complex (low ionic strength) is not optimum, then a substantial reorganizational energy will be required to attain the appropriate transition state. Masking the electrostatic interactions will then facilitate the reaction. However, I should point out that the electrostatic complex is the most favorable complex in some cases (e.g., FNR- ferredoxin) so one should be careful about generalizing.

KOPPENOL: I am glad to hear the computer-generated electron transfer complexes do not always work out. It can be shown that models based only on complementary charge interactions are energetically unfavorable. The reason is that the loss of solvation energy is not compensated for by the increased electrostatic interaction energy. Clearly other interactions are important and therefore the actual structures might be different from those postulated.

CUSANOVICH: In general terms I agree. It is clear that factors other than electrostatic (e.g., topology and presumably non-polar interactions) can contribute to complex formation. Nevertheless, electrostatics must be considered as it does play an important role in many systems.

Oxidases and Related Redox Systems, pages 419–433

DIRECT ELECTROCHEMICAL PROBES OF REDOX PROTEIN AND REDOX ENZYME STRUCTURE AND FUNCTION

Paul D. Barker and H. Allen O. Hill

Inorganic Chemistry Laboratory, Oxford University, South Parks Road, Oxford, OX1 3QR, UK

ABSTRACT

The direct electrochemistry of redox proteins has been achieved at a variety of electrodes, including modified gold and pyrolytic graphite. Careful design of electrode surfaces and electrolyte conditions are required for the attainment of rapid and reversible protein-electrode interaction. Protein-protein complexes can be studied by using selective electrode surfaces. The "well-behaved" electrochemistry of redox proteins can be exploited by coupling the electrode reaction to enzymes for which the redox proteins act as cofactors. The electron-transfer reactions of more complex systems, such as redox enzymes, are now being examined.

INTRODUCTION

Now that fast, direct electron-transfer between many redox proteins and electrodes can be readily achieved (Armstrong et al., 1986), we can begin to use protein electrochemistry as a probe of the structure and function of redox proteins and the enzymes with which they interact. Many important oxidases utilise small redox proteins as electron donors or acceptors and therefore, protein electrochemistry provides a new approach to the study of these entities. Additionally, electrode surface technology realistically opens up the possibility of direct electron-transfer to, or from, redox enzymes. Protein electrochemistry can yield thermodynamic information in the form of redox potentials, as well as kinetic data on heterogeneous electron-transfer and coupled processes.

The key to direct protein electrochemistry is the design of suitable electrode surfaces. Ideally these surfaces must promote the transient binding of the protein to the electrode in a favourable orientation for electron-transfer to occur. For proteins that have a surface designed by Nature to interact specifically with another protein, the best electrode surface will be one that mimics that of the protein's partner. However, for small redox proteins, the electrode surface need not closely resemble the binding partner. A variety of electrode surfaces exist that allow rapid heterogeneous electron-tranfer between the protein and the electrode. The two most versatile types of electrode used are modified gold and edge-orientated pyrolytic graphite. The simple, yet informative, technique of cyclic voltammetry is the major experimental method used. In this experiment, the potential of the electrode is swept linearly with time and the current output is plotted against the potential. Any faradaic current due to heterogeneous electron-transfer between the electrode and solution species is seen as a wave superimposed on a background capacitative "charging" current (Figures 1-4).

Thiol-containing organic molecules (or disulphides which act as precursors) bind strongly to gold surfaces through a gold-sulphur bond thereby modifying the surface properties (Allen et al., 1984). If the organic compound is bifunctional, of the type **X-Y**, where **X** binds to the gold surface thereby presenting **Y** to the solution, the surface takes on the characteristics of **Y**. If **Y** is charged, or is dipolar in nature, then it can bind oppositely charged groups or dipoles on proteins. Pyrolytic graphite, orientated with the (b-*c*) plane parallel to the surface, presents to the solution a heterogeneous, non-specific, hydrophilic surface of carbon-oxygen functionalities. This surface alone can promote the electrochemistry of positively-charged proteins, such as cytochrome *c*, but its properties can be modulated by the presence of cations, especially multivalent cations, or by acidification and thereby protonation of the surface groups. These surfaces then promote the electrochemistry of negatively-charged proteins.

In this fashion, the direct electrochemistry of cytochromes, ferredoxins, flavodoxins, azurins and plastocyanins have been observed. This direct communication with the redox sites has been exploited to probe transient events accompanying electron-transfer at the sites. The electroactivity of plastocyanin at low pH (Armstrong et al., 1985) and the superoxidation of [4Fe-4S] clusters in ferredoxin (Armstrong et al., 1982) are good examples of the use of direct protein electrochemistry to study redox centres. A detailed study of cytochrome *c* electrochemistry at a modified gold surface (Albery et al., 1981) has derived the first order electron transfer rate of 50 s^{-1} at E^o.

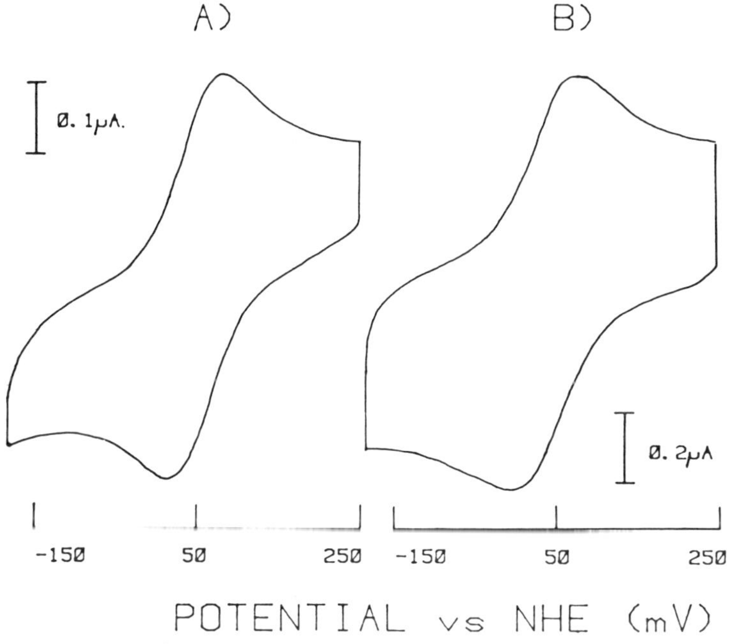

Figure 1. Cyclic voltammograms of cytochrome b_5 at a gold electrode modified with lysyl-cysteine (A) and at an edge-plane pyrolytic graphite electrode (B). Conditions: 90 μM protein, 10mM phosphate pH 7.0 20°C, 20mV s^{-1} sweep rate. A) +0.2mM $Cr(NH_3)_6{}^{3+}$ B) +1.2mM $Cr(NH_3)_6{}^{3+}$.

PROTEIN-PROTEIN COMPLEXES

Cytochromes *c* and b_5 are well-characterised proteins that mediate electron-transfer between catalytic centres. Each has a highly asymmetric surface charge distribution and a specific, electrostatic and mutually-complementary binding domain. They form a strong electrostatic complex ($K_A=10^7$ at I=0) of physiological relevance (Mauk et al., 1982). Using electrode surfaces that present an opposing charge to that of the protein, a protein-electrode complex can form allowing electron-transfer between them. Cytochrome c requires a negatively-charged surface. An edge-plane graphite electrode alone will react directly with cytochrome and this reaction is destabilised by the addition of cations. Alternatively, gold, modified with 4,4-bipyridyl or bis(4-pyridyl)bisulphide, presents a surface with the negative end of the dipole exposed to the solution thereby promoting reversible electrochemistry. More interestingly, cysteinyl peptides will modify gold surfaces and hence cysteinyl-glutamate is a good promotor of cytochrome c electrochemistry. In contrast, cytochrome b_5 electrochemistry, (Figure 1), at an edge-plane graphite electrode requires the presence of multivalent-cations and consistant with this, cytochrome b_5 is effective at a gold surface modified with lysyl-cysteine (Barker et al., 1987). The cysteinyl peptide modifiers represent a class of modifier that form surfaces which resemble natural proteinaceous ones.

While our current surface technology cannot uniquely select for one protein over another similarly charged protein, it can certainly discriminate between proteins with positively- or negatively-charged domains. Thus, at a gold electrode modified with Cys-Glu, cytochrome c electrochemistry is observed but that of cytochrome b_5 is not. However, the addition of cytochrome c to a solution of cytochrome b_5 leads to cytochrome b_5 electrochemistry at a previously unfavourable surface (Figure 2). By using zinc-cytochrome c (which is redox-inactive in normal potential ranges), this effect can be shown to be direct electron-transfer to the cytochrome b_5 heme and is not mediated by cytochrome c. This effect can be explained by complex formation at the electrode surface. Maximal response from cytochrome b_5 is achieved with less than one equivalent of cytochrome c and the response does not simply depend on the amount of complex formed in solution, (which is controlled by the ionic strength). Under conditions of very low ionic strength, where 95% of the proteins should be complexed in solution, the current observed is dependent on the cytochrome b_5 concentration alone. Cytochrome c is therefore acting as a surface promotor of cytochrome b_5 electrochemistry; cytochrome c binds to the electrode and cytochrome b_5 binds to cytochrome c. The effect is also observed with the non-physiological pair, plastocyanin and cytochrome _c_. Cytochrome _c_ promotes plastocyanin electrochemistry at a previously unfavourable surface. Therefore the redox sites of both cytochrome $\underline{b}_5$ and plastocyanin are accessible to the electrode surface when either is bound to cytochrome _c_.

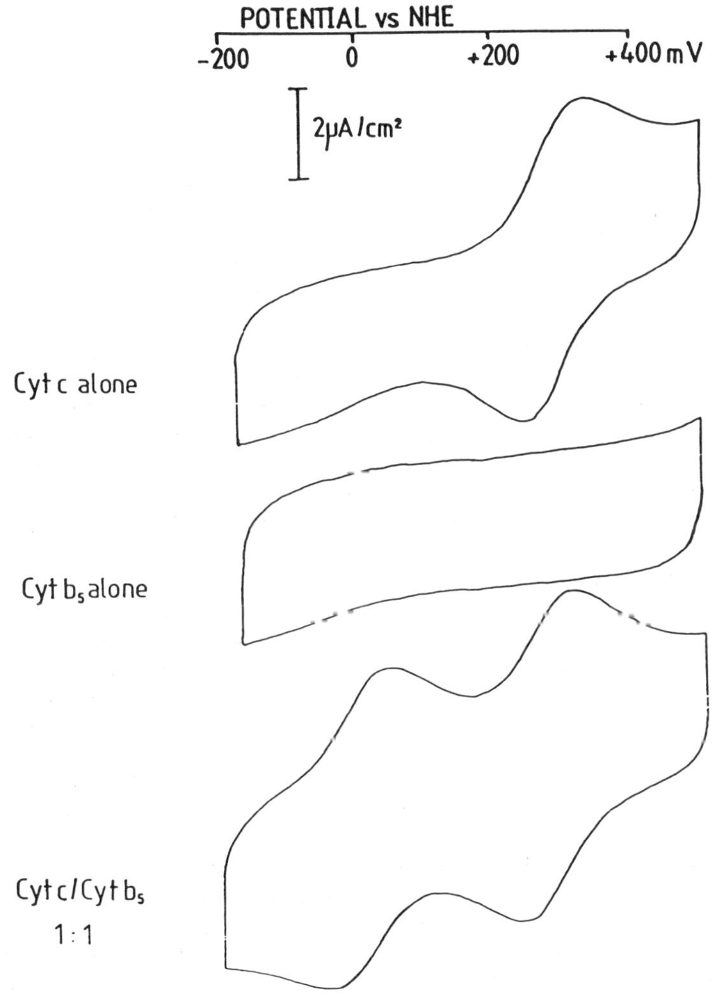

Figure 2. The promotion of cytochrome $\underline{b}_5$ electrochemistry by complex formation with cytochrome _c_ at a pyrolytic graphite electrode. Conditions: 95 μM each protein, 1mM K-HEPES pH 7.0, 1mM KCl, 20 mV s^{-1} sweep rate.

However, attempts to promote the electrochemistry of cytochrome *c* with either cytochrome b_5 or plastocyanin have so far failed. There are two possible explanations for this: firstly, when bound to the electrode, cytochrome b_5 or plastocyanin cannot then bind cytochrome *c*. Alternatively, if a ternary complex between cytochrome b_5 or plastocyanin, cytochrome *c* and the electrode can form, the heme of cytochrome *c* is inaccessible to the electrode. Cytochrome *c* complexes with plastocyanin using the face closest to the heme to bind to the "east site" acidic patch of plastocyanin. These two binding domains on the separate proteins are those that bind to the electrode surfaces. We propose that, by complexation with cytochrome *c*, the electrode binding site of plastocyanin is completely masked whereas that of cytochrome *c* is not. When bound to cytochrome *c* in the above manner, plastocyanin could present the "north site", and therefore the copper, to the electrode surface.

An important point to note is that the potential of plastocyanin (+350mV vs NHE) when bound to cytochrome *c* differs from that (+370mV) measured at a normal electrode surface (Armstrong et al., 1985) or by classical methods (Adman, 1986). Thus the potential of plastocyanin is shifted by 20mV upon binding to cytochrome *c*, in agreement with previous results. So, by introducing a third complexing surface, the electrode, we have a direct probe of protein-protein interactions.

The situation with cytochrome b_5 is less clear. The Salemme model for the b_5/*c* electrostatic complex proposes that the heme faces "cover" each other with the hemes essentially co-planar (Salemme, 1976). However, the structure of this complex is pH-dependent (Mauk et al., 1986); the pH-dependence of the complex electrochemistry is being studied, but, the ternary complex between cytochrome *c*, cytochrome b_5 and the electrode could well stabilise a different b_5/*c* orientation from that that exists in solution. We should remember that these complexes could well be dynamic. Mauk *et al.*, 1986, suggest that cytochrome *c* could well "walk around" the surface of cytochrome b_5 and that many different complexes with differing protein alignments, but comparable thermodynamic stabilities, could exist. To test the mobility of the proteins, within these complexes, at the electrode, we have rigidly constrained the complex between plastocyanin and cytochrome *c* by extensive crosslinking with 1-ethyl-3-(3-dimethylaminopropyl)carbodiimide (EDC) (Geren et al.,

1983). The resulting 1:1 linked complex can still bind to the electrode that recognises cytochrome c and electron-transfer is possible between the electrode and either centre (Figure 3). This suggests that the heme of cytochrome c is accessible to the electrode when cytochrome c is bound to another small protein.

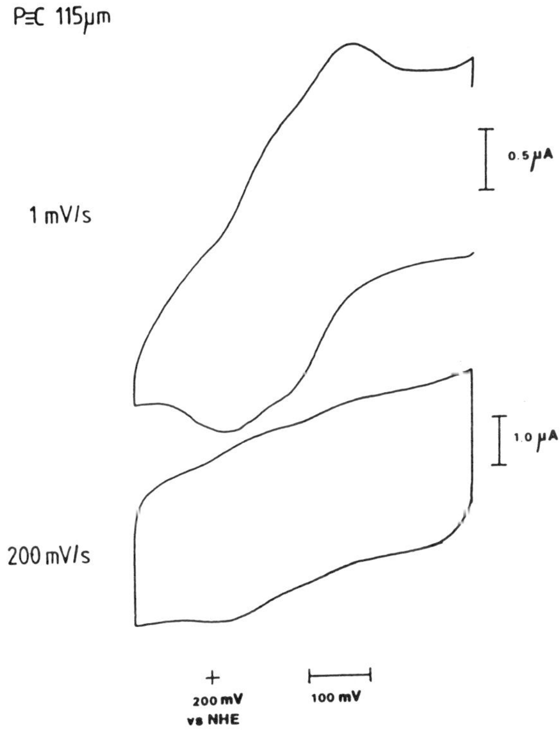

Figure 3. The direct electrochemistry at an edge-plane pyrolytic graphite surface of a 1:1 complex between cytochrome c and plastocyanin crosslinked with EDC (as described by Geren et al., 1983). 2mM K-HEPES pH 7.0, 20mM KCl, 115μM protein. The dimer was purified on Sephacryl S-200. The responses due to cytochrome c (+260mV vs NHE) and plastocyanin (+350mV) overlap.

COUPLED ENZYME ELECTROCHEMISTRY

The direct electrochemistry of redox proteins can be used to drive the enzymes, or enzymic pathways, for which these redox proteins are electron donors or acceptors. Two good examples of this involve cytochrome c-driven cytochrome c oxidase or sulphite oxidase.

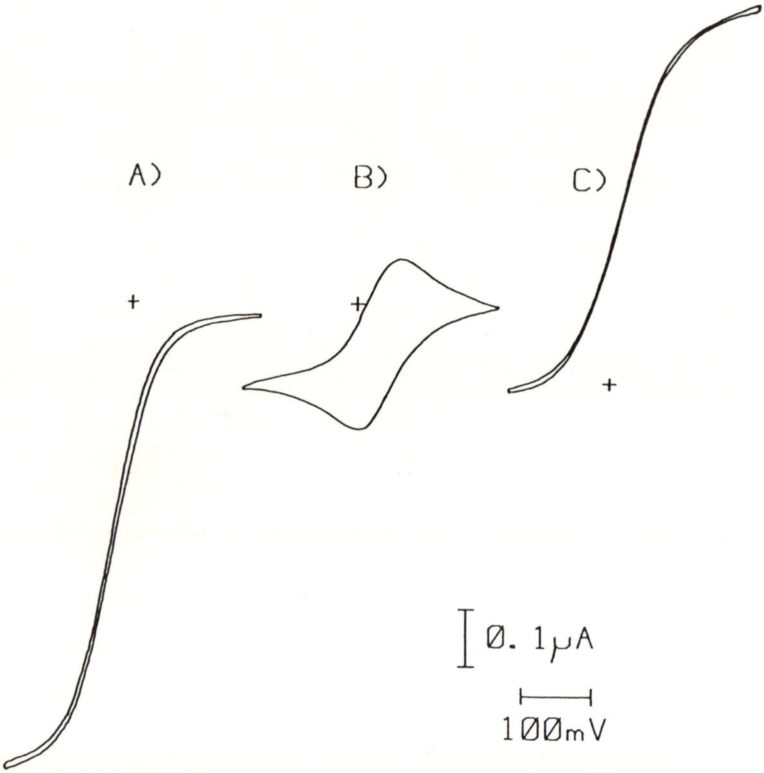

Figure 4. Coupled oxidase electrochemistry. Beef-heart cytochrome c oxidase or liver sulphite oxidase can be driven by cytochrome c electrochemistry. Conditions: A) 100μM cytochrome c, 210nM cytochrome c oxidase, 50mM K-phosphate pH6.5, 1mV s^{-1} sweep rate. B) 400μM cytochrome c, 50mM K-phosphate pH 6.5, 1mV s^{-1} sweep rate. C) 100μM cytochrome c, 585nM sulphite oxidase, 10mM Tris-Cl pH 8.0, 10mM KCl, 250uM SO_3^{2-}. + indicates +200mV vs NHE and 0μA.

As cytochrome c is reduced at an electrode it is re-oxidised by cytochrome c oxidase, in the presence of dioxygen, giving a greatly enhanced reductive current response. In the presence of the substrate (SO_3^{2-}), sulphite oxidase re-reduces electrochemically generated oxidised cytochrome c. A modified electrode can drive the steady-state turnover of these enzymes (Figure 4). Electrochemical analysis (Nicholson and Shain, 1964) of these catalytic currents can yield information about the second order rates of reaction and the rate constants derived from this method are in good agreement with results from other methods.

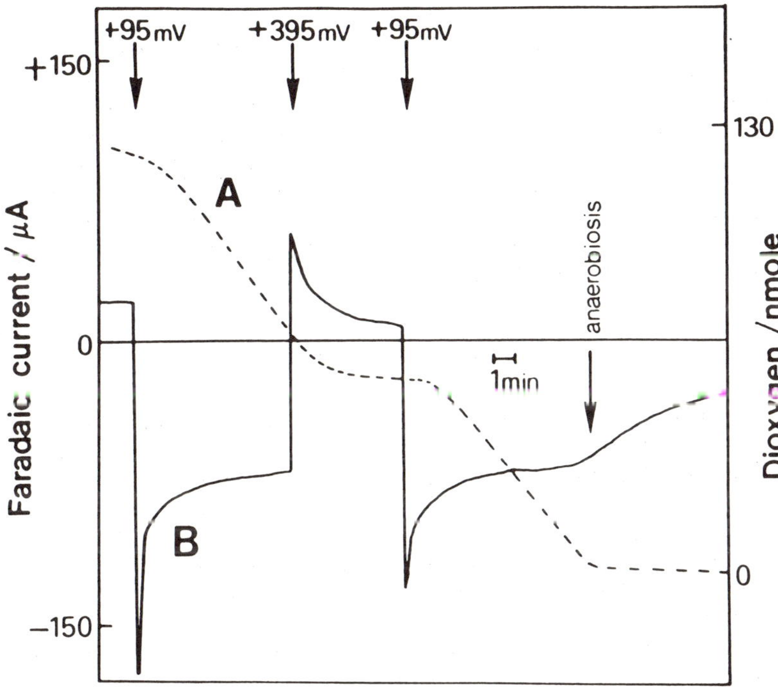

Figure 5. Electrochemically-driven respiration by rat liver mitochondria in the presence of exogenous cytochrome c. The simultaneous measurement of current and dioxygen consumption stimulated by the application of a reducing potential at a modified gold electrode.

The enzymes need not be isolated for these experiments. Cytochrome c oxidase can be driven in situ within the mitochondrion or bacterial protoplasts (Coleman et al., 1983). Respiration by mitochondria or mitoplasts is driven by cytochrome c electrochemistry (Figure 5) which can also drive the steady-state electron-transfer through complexes II and III when mitoplasts are supplied with succinate and cyanide, cytochrome c accepting electrons from cytochrome c_1 in complex III. Cytochrome c can also accept electrons from the outer mitochondrial membrane NADH/b_5 oxidoreductase. Cytochrome c is therefore a site-selective mediator between four sites within the mitochondrion and the electrode. Consequently, protein electrochemistry can be specifically coupled to many enzymic reactions. Further examples include the reduction of methemoglobin, though not an enzymic reaction, using electrochemically generated reduced cytochrome b_5 (Barker et al., 1987), the physiological electron donor in erythrocytes. Also the blue copper protein, azurin, is believed to be the natural electron acceptor for a number of hydroxylase enzymes from Pseudomonas bacteria. The direct electrochemistry of azurin can drive the turnover of p-cresol methylhydroxylase (from P. putida) with a near quantitative conversion of p-cresol to p-hydroxybenzaldehyde (Hill et al., 1985a) (Figure 6).

An advantage of electrochemical techniques is that they are often compatible with other methods of investigation. The use of transparent electrodes, for example, offers the possibility of the simultaneous measurement of changes in UV/visible spectra. This approach has been used to investigate bulk pH changes associated with mitochondrial respiration using pH-sensitive indicators (Hill et al., 1985b). Alternatively, a pH electrode can be inserted into a bulk electrolysis cell. The pH changes in mitoplasts have been investigated this way. In either system, the steady-state scalar production or consumption of protons associated with the electron transport chain can be monitored. No pH changes due to vectorial proton translocation can be detected by these methods. The problem here lies in the ability of electrochemistry to effect rapid changes in the bulk solution. It is in the region close to the electrode, the depletion layer, where the power of electrochemistry can bring about fast changes and hence where transient changes in solution properties can be monitored. The spectroelectrochemical technology exists for studying this thin layer (~10μm) close to the electrode surface but has yet to be applied to biological problems.

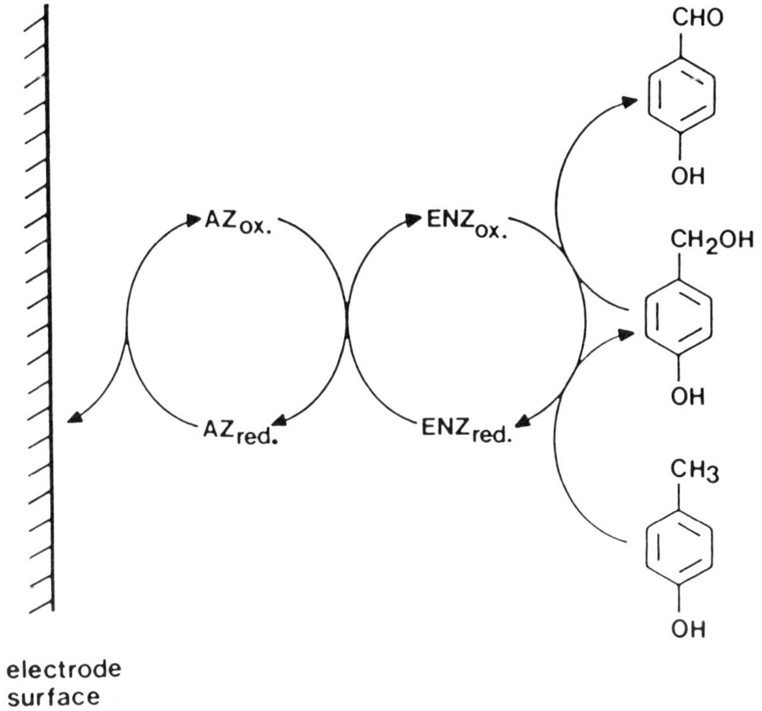

Figure 6. The catalytic scheme for the electrochemical turn-over of p-cresol methylhydroxylase, and the consequent conversion of p-cresol to p-hydroxybenzaldehyde.

DIRECT ENZYME ELECTROCHEMISTRY

Redox enzymes with binding surfaces for other redox proteins and with electron transfer pathways from their surface to their prosthetic group should, in theory, also be able to react directly with an electrode. The main problem with such enzymes is stabilising a much larger protein at the electrode surface. However, as designed electrode surfaces mimic matching protein surfaces more and more closely, the prospect for direct electronic communication to such enzymes as cytochrome *c* peroxidase, sulphite oxidase or soluble bacterial terminal oxidases becomes very real. The non-physiological one-electron reduction and reoxidation of ferric cytochrome c peroxidase has been reported (Assefa and Bowden, 1986).

The mechanisms of many more redox enzymes do not involve long range electron transfer to or from a bound redox protein mediator. Electron transfer occurs within the vicinity of the prosthetic group, between small substrates and the enzyme. Apart from increasing size, direct electrochemistry of this class of protein introduces new problems. Firstly, if these proteins only react with small organic/inorganic molecules then there will be no binding surface to recognise and hence, binding and stabilising the enzyme at a surface becomes very difficult. Secondly, the catalytic centre could be well-buried in this class of protein. Thirdly, the electron transfer mechanisms are often inner-sphere in nature and require substrates bound to the prosthetic group. An additional problem with flavoproteins is that very few of them have a covalently bound flavin moiety. Thus, of the few reports of direct electrochemistry of enzymes in this class, many involve flavoproteins and further investigation suggests that the free flavin cofactor is responsible for any electrochemistry and this can then mediate electron transfer to the enzyme. Perhaps the first real enzyme electrochemistry to be reported is that of lysyl oxidase (Govindaraju et al., 1987) - a type II copper enzyme - at a glassy carbon electrode though this system requires further investigation.

There are many oxidases in this class and a clear objective in the current drive for direct electrochemistry is to replace dioxygen with the electrode as an electron acceptor. Many oxidases specifically require O_2 as an electron acceptor whereas others are less specific and will donate electrons to many inorganic and organic acceptors and therefore could also conceivably utilise an electrode surface. Many research groups are currently engaged in achieving this goal.

SUMMARY

We have seen that the availability of electrode surfaces that permit electron transfer to redox proteins to proceed rapidly and reversibly, allows the investigation of a variety of interesting systems including protein-protein complexes, enzymes and even components of mitochondria. The attraction of electrochemical methods is that they are readily compatible with traditional methods of investigation, particularly visible spectroscopy. One can expect there to be many such investigations reported in the future. The most immediate interest centres on the investigation of protein-protein complexes but it is important to stress that it is the termolecular complex with the electrode surface that is investigated. The direct electrochemistry of enzymes poses more problems but it is likely that electron transfer to some, at any rate, will be confidently reported in the near future. Whether it will be possible to exploit such electron transfer reactions for more applied purposes remains to be seen.

REFERENCES

Adman ET (1986). Structure and Function of Small Blue Copper Proteins. In Harrison PM (ed): "Metalloproteins: Part 1," London: MacMillan, pp1-42.

Albery WJ, Eddowes MJ, Hill HAO & Hillman AR (1981). Mechanism of the reduction and oxidation reaction of cytochrome c at a modified gold electrode. J Am Chem Soc 103:3904-3910.

Allen PM, Hill HAO & Walton NJ (1984). Surface modifiers for the promotion of direct electrochemistry cytochrome c. J Electroanal Chem 170:69-86.

Armstrong FA, Hill HAO & Walton NJ (1986). Reactions of electron-transfer proteins at electrodes. Q Rev Biophysics 18:261-322.

Armstrong FA, Hill HAO, Oliver BN & Whitford D (1985). Direct electrochemistry of the photosynthetic blue copper protein plastocyanin. Electrostatic promotion of rapid charge transfer at an edge-oriented pyrolytic graphite electrode. J Am Chem Soc 107:1473-1476.

Armstrong FA, Hill HAO & Walton NJ (1982). Direct electrochemical oxidation of Clostridium pasteurianum ferredoxin. Identification of facile electron-transfer processes relevant to cluster degradation. FEBS Lett 150:214-218.

Assefa H & Bowden ED (1986). Electron-transfer reaction of cytochrome c peroxidase at tin oxide electrodes. Biochem Biophys Res Commun 139:1003-1008.

Barker PD, di Gleria K, Hill HAO & Lowe VJ (1987). In preparation.

Coleman JOD, Hill HAO, Walton NJ & Whatley FR (1983). Electrochemically driven respiration in mitochondria and Paracoccus denitrificans. FEBS Lett 154:319-322.

Geren LM, Stonehuerner J, Davis DJ & Millett F (1983). The use of a water soluble carbodiimide to cross-link cytochrome c to plastocyanin. Biochim Biophys Acta 724:62-68.

Govindaraju K, Unni Nair B, Ramasami T & Ramaswamy D (1987). Active-site chemistry of lysyl oxidase. J Inorg Biochem 29:111-118.

Hill HAO, Oliver BN, Page DJ & Hopper DJ (1985a). The enzyme-catalysed electrochemical conversion of p-cresol into p-hydroxybenzaldehyde. J Chem Soc Chem Commun 1469-1471.

Hill HAO, Walton NJ, Whitford D & Coleman JOD (1985b). The measurement of changes in pH associated with electrochemically-driven respiration in rat liver mitochondria. J Inorg Biochem 23:303-309.

Mauk MR, Mauk AG, Weber PC & Matthew JB (1986). Electrostatic analysis of the interaction of cytochrome c with native and dimethyl ester heme substituted cytochrome b_5. Biochemistry 25:7085-7091.

Mauk MR, Reid LS & Mauk AG (1982). Spectrophotometric analysis of the interaction between cytochrome b_5 and cytochrome c. Biochemistry 21:1843-1846.

Nicholson RS & Shain I (1964). Theory of stationary electrode polarography. Anal Chem 36:706-723.

Salemme FR (1976). An hypothetical structure for an intermolecular electron transfer complex of cytochromes c and b_5. J Mol Biol 102:563-568.

ACKNOWLEDGEMENTS

Our thanks go to Dr Mike Wilson, University of Essex, for porphyrin cytochrome c and also to Paul Driscoll and Val Lowe for plastocyanin and peptide modifiers respectively. One of us, P.D.B., thanks the SERC for a studentship.

DISCUSSION

KOPPENOL: It seems that cytochrome *c* is able to promote the electrochemistry of plastocyanin and cytochrome b_5, but these two cannot promote the electrochemistry of cytochrome *c*. Could it be that the charge of the electrode relative to the solution is important in the experiment?

BARKER: The charge on the electrode is almost entirely governed by that of the promoters in use. I suspect that the reasons why plastocyanin and cytochrome b_5 do not work are different. I believe plastocyanin cannot bind to both cytochrome *c* and the electrode. I think that cytochrome b_5 could bind to both. I think what you are suggesting is that this would involve bringing positively-charged cytochrome *c* to a positive electrode. However, we know that cytochrome b_5 can exchange electrons with a negatively charged electrode as long as cytochrome *c* is present.

It is possible that the relative dipole moment orientation of the two proteins depends on which protein binds to the electrode first. Then cytochrome *c* binding to cytochrome b_5, which is already at the electrode, may not be possible due to electrostatic repulsion between the electrode and cytochrome *c*. But we cannot rule out the possibility that cytochrome *c* can bind to cytochrome b_5 already at the electrode but that its heme is not favorably oriented for electron-transfer reactions with the electrode.

WEVER: Is it possible with this electrochemical method to determine the number of electrons taken up by the cytochrome *c* system? and how many electrons are transferred?

BARKER: There is an exact stoichiometric relationship of 4 moles of electrons consumed per mole of O_2 when steady state turnover of oxidase is driven by cytochrome *c* electrochemistry. In principle it should be easy to use coulometric cytochrome *c* electrochemistry to titrate the number of electrons required to reduce the enzyme fully. In practice, all traces of O_2 must be kept out of the electrochemical cell. Outside of a good glove box this is not trivial. It would, of course, be simple to titrate inhibited forms of the enzyme. This has not yet been done.

Session VII: Peroxidases

Chairman: Chen-Lu Tsou

Oxidases and Related Redox Systems, pages 437–449

THE COORDINATION AND SPIN STATES OF YEAST CYTOCHROME C PEROXIDASE AND THEIR IMPLICATION TO PEROXIDASE MECHANISM

Helen Anni and Takashi Yonetani

Department of Biochemistry and Biophysics,
School of Medicine, University of Pennsylvania,
Philadelphia, PA 19104-6089

ABSTRACT

Cytochrome c peroxidase, freshly prepared, contains a penta-coordinated heme iron and is fully reactive with hydroperoxides. On the other hand, the enzyme normally stored in frozen states invariably contains different amounts of an altered, aged species whose heme iron is hexa-coordinated. The aged enzyme reacts with hydroperoxides only after a slow conformation change leading to the formation of a reactive penta-coordinated state. Thus, the reactivity of cytochrome c peroxidase with hydroperoxides is strongly controlled by the coordination state of the heme iron. A penta-coordinated heme iron may be a prerequisite for rapid reactions of hydroperoxidases with hydroperoxides.

INTRODUCTION

The reaction of peroxidases with their oxidizing substrates, hydroperoxides, to form their peroxide intermediates, is one of the fastest biochemical reactions. Yeast cytochrome c peroxidase (CCP), reacts with hydroperoxides to form a relatively stable peroxide intermediate (Compound ES), with a second order rate constant of the order of 10^6 to 10^8 $M^{-1}sec^{-1}$ (Chance, 1951; Yonetani and Ray, 1966; Nicholls and Mochan, 1971; Loo and Erman, 1975; Ohlsson and al., 1986; Balny et al., 1987). The question is whether the ferric heme iron, the active site of a peroxidase, is penta- or hexa-coordinated to be highly reactive with hydroperoxides.

Although its structure has been most thoroughly investigated among hydroperoxidases, the coordination and spin states of CCP have not been unequivocally established. A 1.7 A-resolution X-ray structure of CCP (Finzel et al., 1984) indicates the presence of a water molecule (H_2O-595) at a distance of 2.4 A from the iron in the distal side of the heme plane. Therefore, *the heme iron in CCP is considered as penta-coordinated at ambient temperature.* On the other hand, electronic absorption, paramagnetic susceptibility (Iizuka et al., 1971), Mossbauer spectroscopy (Lang et al., 1969) and electron paramagnetic resonance (EPR) spectra show that CCP is a mixture of high- and low-spin ferric states (Fig. 1B), suggesting that *CCP is in ferric hexa-coordinated (high- and low-spin) states at 77 K.* In addition, upon prolonged storage CCP undergoes structural changes which could be partially reversible or irreversible, depending on the extent of the modifications, leading to the formation of a hexa-coordinated high-spin state (Fig. 1A). The storage-altered CCP exhibits spectroscopic and enzymic properties substantially different from the unaltered enzyme. Thus, unequivocal description of spectroscopic and enzymic characteristics of CCP has not been possible from the previously reported data.

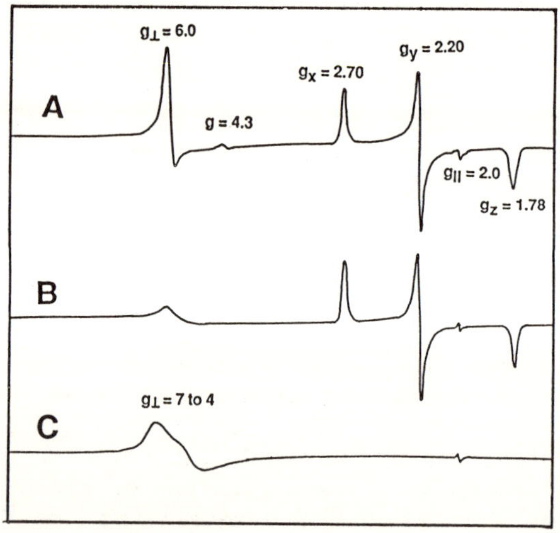

Figure 1. X-band EPR spectra at 77 K of (A) aged cytochrome c peroxidase and (B) freshly prepared cytochrome c peroxidase in 0.1 M Mes buffer, pH 7.0.in the absence and (C) presence of 66% glycerol.

This work shows that the heme crevice of CCP is highly susceptible to freezing-induced structural alterations, leading to the reversible formation of hexa-coordinated states at cryogenic temperatures by a forced coordination of strong- and weak-field ligands to the heme iron. *Upon elimination of such a freezing-induced coordination (Fig. 1C), CCP can be maintained penta-coordinated at both ambient and cryogenic temperatures*. Furthermore, upon storage, especially at acidic pHs, the penta-coordinated CCP undergoes conformational changes leading to the formation of a hexa-coordinated high-spin state, which exhibits an EPR spectrum of purely axial symmetry. The storage-modified CCP species is less reactive toward hydroperoxides. Therefore, we attempt in this work to eliminate such freezing- and aging-induced artifacts from CCP in order to characterize the spectroscopic and enzymic properties of the native enzyme. We have demonstrated that native CCP is in a penta-coordinated ferric high-spin state in both ambient and cryogenic temperatures and this is essential for its high reactivity with hydroperoxides.

EXPERIMENTAL PROCEDURES

Cytochrome c peroxidase was purified from baker's yeast (Yonetani and Ray, 1965), thrice recrystallized by dialysis against distilled water (Yonetani et al., 1966a), and stored in a polycrystalline state at -20°C. The enzyme crystals were dissolved in neutral-pH buffers and assayed spectrophotometrically using ϵ_{408nm} (CCP) = 98 $mM^{-1}cm^{-1}$ (Yonetani and Anni, 1987) and $\Delta\epsilon_{424nm}$ (Compound ES minus CCP) = 42 mM^{-1} cm^{-1} (Yonetani et al., 1966b). X-band EPR measurements were carried out using a Varian E-109 EPR spectrometer at 77 K and 10 K. Transient kinetic measurements were performed with a Union Giken rapid-scan spectrophotometer.

RESULTS AND DISCUSSION

Stoichiometric Titration of Cytochrome c Peroxidase with Hydroperoxides

As previously reported (Yonetani, 1965), a spectrophotometric titration of CCP in a concentration range from 1 to 10 μM with hydroperoxides (hydrogen peroxide or ethyl hydroperoxide) at ambient temperature gives a linear titration

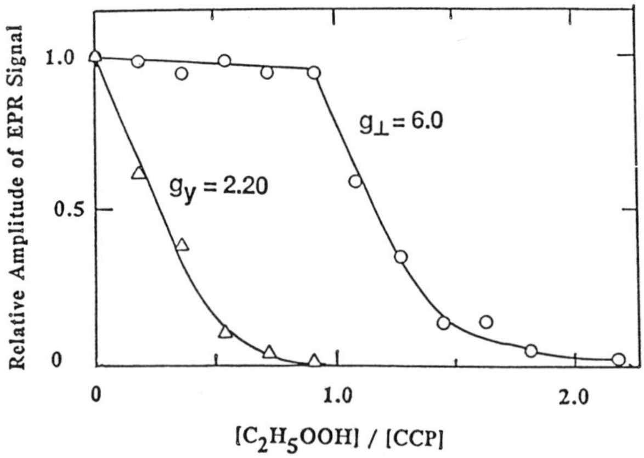

Figure 2. EPR titration of cytochrome c peroxidase (0.8 mM) with ethyl hydroperoxide in 0.1 M phosphate buffer, pH 7.0.

curve with an inflexion point at 1:1 molar stoichiometry. On the other hand, if CCP (0.1 to 1 mM) is titrated with hydroperoxides at ambient temperature and the reaction mixture is immediately frozen (within 30 sec) for EPR examination at 77 K, we observe that the low-spin component of CCP preferentially reacts with hydroperoxides (Fig. 2). Only after the low-spin component is exhausted, the g = 6 high-spin component reacts with hydroperoxides. Compound ES formed from the low-spin component acts *catalatically* to consume free hydroperoxides in competition with the slow-reacting g = 6 high-spin component, so that normally more than stoichiometric amount of hydroperoxides was required to fully convert the g = 6 high-spin component to Compound ES.

Transient Kinetics of Reaction of Cytochrome c Peroxidase with Hydroperoxides

Transient spectrophotometric measurements of the reaction of CCP with hydroperoxides show biphasic kinetics (Balny et al., 1987). The fast phase is dependent on the chemical nature and concentration of hydroperoxides with a second order rate constant of 1.38×10^7 and 3.38×10^6 M^{-1} sec^{-1} at 2°C for hydrogen peroxide and ethyl hydroperoxide respectively, and shows an activation energy of 6.3 kcal mol^{-1}. Therefore, one can conclude that the fast phase represents the bimolecular reaction between CCP and hydroperoxides. The slow phase, on the other hand, is found to be independent of the nature and concentration of hydroperoxides and to have an activation energy of 20.5 kcal mol^{-1}. This suggests that some intramolecular processes probably involving a protein conformation change in CCP may be rate-limiting. In order to probe the nature of these fast and slow phases, kinetic difference spectra were taken as shown in Fig. 3. The difference spectra of the fast phase are characterized by a broad absorbance change from 350 to 410 nm in the Soret band of CCP. The spectral characteristics of this change indicate that the reactive species of CCP in the fast phase is penta-coordinated. On the other hand, the sharp maximal absorbance change at 405 nm in the difference spectra of the slow phase shows that the reactive species of CCP in the slow phase is hexa-coordinated.

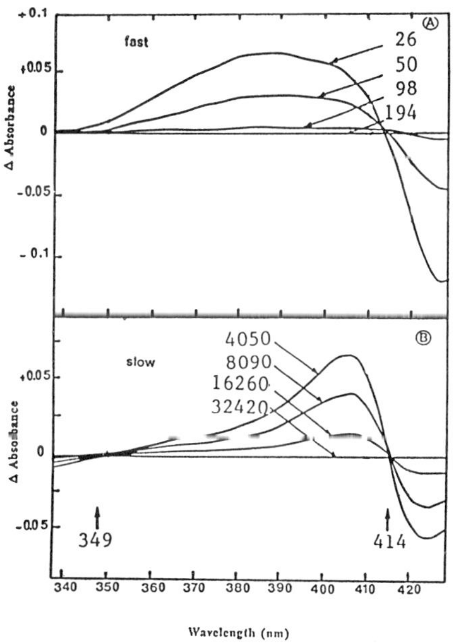

Figure 3. Transient kinetic difference spectra between cytochrome c peroxidase and Compound ES in 0.1 M phosphate buffer, pH 6.1 at 2°C. [CCP] = 10 μM, [Ethyl hydroperoxide]= 20 μM. Spectra were recorded at times (in msec) (A) the fast phase and (B) the slow phase using the 194-msec and 32-sec spectra as a baseline, respectively.

The hexa-coordinated nature of the CCP species which reacts slowly with hydroperoxides in the slow phase is supported by both EPR (cf. Fig. 2) and transient difference spectra (cf. Fig. 3). However, the coordination and spin states of the CCP species which reacts with hydroperoxides in the fast phase are ambiguous. Both a low-spin (hexa-coordinated) and a high-spin (penta-coordinated) species are indicated by EPR and transient difference spectra, respectively.

Storage- and Freezing-induced Artifacts

We have previously noted that the relative fraction of the g = 6 high-spin species of CCP (cf. Fig. 1A) tends to increase upon prolonged storage. Therefore, it is possible that the g = 6 high-spin species which is less reactive toward hydroperoxides (cf. Fig. 2 and 3) may be produced by a storage-induced modification of CCP. An EPR examination has revealed that freshly prepared CCP is in a predominantly low-spin state with the g = 6 high-spin signal replaced with a minor broad g = 7 to 4 signal (Fig. 1B).

The ambient temperature X-ray structure of CCP shows that no strong-field ligand responsible for a low-spin state is available in the distal pocket (Finzel et al., 1984). Consequenly, the low-spin species observed in 77 K EPR may be produced by a forced ligation of a strong-field ligand upon freezing. In fact, the EPR spectrum of freshly prepared CCP shows no low-spin signal in the presence of 66% glycerol (cf. Fig. 1C). Instead the low-spin signal is replaced by a

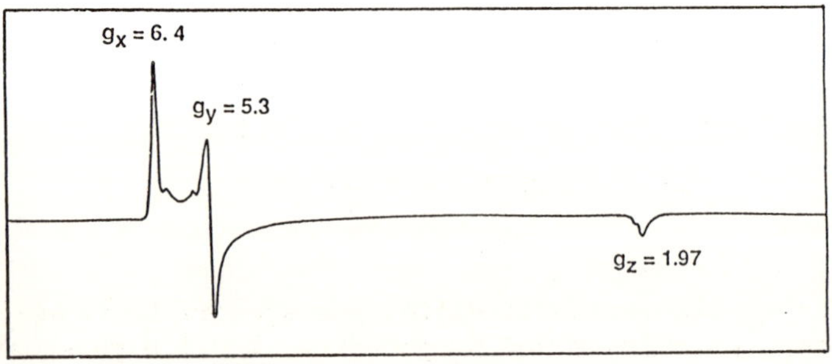

Figure 4. X-band EPR spectrum at 10 K of fresh cytochrome c peroxidase in 0.1 M Mes buffer, pH 7.0 in the presence of 66% glycerol.

broad featureless signal at g = 7 to 4, with a line shape resembling the EPR signal of the quantum-mixed spin state (Maltempo and Moss, 1976). The EPR spectrum at 10 K shows that freshly prepared CCP exhibits an EPR spectrum of rhombically-distorted high-spin signals with EPR extrema at g = 6.4, 5.3, and 1.97 in the presence of 66% glycerol, a clear indication of a penta-coordinated nature of the heme iron (Fig. 4). A rhombicity of 6.87% was calculated (Peisach and Blumberg, 1971; Yonetani and Anni, 1987).

CONCLUSIONS

Reaction of Cytochrome c Peroxidase with Hydroperoxides

On the basis of the above-mentioned results, the reaction of CCP with hydroperoxides can be schematically described as shown in Fig. 5. Native CCP (penta-coordinated) reacts with hydroperoxides in single-exponential kinetics in a bimolecular reaction. In storage-altered CCP preparations, aged CCP (hexa-coordinated) is slowly converted to native CCP which is reactive with hydroperoxides. This slow process involves a conformation change accompanying the dissociation of the ligated water (H_2O-595) from the heme iron, resulting in the formation of the reactive penta-coordinated CCP.

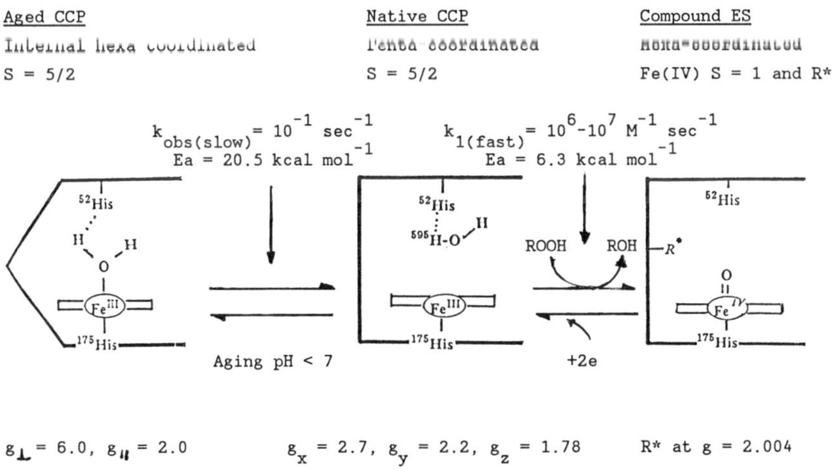

Figure 5. Schematic presentation of the reaction of cytochrome c peroxidase with hydroperoxides. EPR principal g-values at 77 K, R* = protein free radical.

The nature of the protein free-radical (R*) in Compound ES has remained a puzzle till a combination of site-directed mutagenesis and crystallographic techniques (Edwards et al., this volume) pointed to the proximal heme side and proposed the network of Trp-191, serving as a bridge to the cluster of Met-230 and Met-231. Zinc-substituted CCP can be converted to a porphyrin residing pi-cation radical with strong oxidants or illumination (Anni and Yonetani, 1986). However, there is no evidence for the transfer of the oxidizing equivalent of the pi-cation radical to the protein radical site.

Spectroscopic Characterization of Penta- and Hexa-Coordinated Cytochrome c Peroxidase

Table I summarizes EPR, electronic absorption, resonance Raman, and hyperfine-shifted proton NMR characteristics of ferric CCP in both penta- and hexa-coordinated states. A compressed Soret extinction coefficient (ϵ_{408nm} = 98 $mM^{-1}cm^{-1}$), rhombically-distorted axially symmetric EPR signals (g_x = 6.4, g_y = 5.3, and g_z = 1.97), and a +15 cm^{-1}-shifted resonance Raman ν_3 line (ν_3 = 1495 cm^{-1}) are readily recognizable characteristics of the penta-coordinated CCP (Table 1, state I). These features may be applied also to other penta-coordinated ferric hemoproteins (Schonbaum, 1982).

Temperature-dependent (25-2°C) coordination-state equilibria have been observed by resonance Raman measurements on native lignin peroxidase (Andersson et al., 1987). This transition from high-spin penta-coordinated ferric enzyme to high-spin hexa-coordinated one resembles the native ⟶ aged CCP conversion (state I ⟶ II). However, the crystal structure of CCP at -15°C shows no change in coordination or spin state of the iron (Edwards et al., 1987). Interestingly enough, there is a movement of Arg-48 side chain 1.5-2 A closer to the ligand binding site at the lower temperature ferric CCP structure and the crystal structure of CCP-fluoride (state III)(Edwards et al., 1984). In frozen samples (state IV) cytochrome c peroxidase is still high-spin hexa-coordinated, while horseradish peroxidase becomes low-spin, as judged by resonance Raman spectra (Evangelista-Kirkup et al., 1985).

In alkaline region above pH 8.0, the enzyme becomes an-

TABLE 1. Spectroscopic characteristics of several ferric states of cytochrome c peroxidase

Enzyme State	EPR (a)	Optical (b)	R.Raman (c)	Proton-NMR (d)
(I) Native CCP Penta-Coordinated High-Spin ($S = 5/2$)	Rhombic axial. symmetric HS $g_x = 6.4$ $g_y = 5.3$ $g_z = 1.97$	Penta-HS 408 nm (98) 505 nm (11) 645 nm (3.3)	Penta-HS ν_{10}=1629 cm^{-1} ν_2=1572 cm^{-1} ν_3=1495 cm^{-1} ν_4=1374 cm^{-1}	HS +80.4 ppm +70.9 ppm +67.6 ppm +58.6 ppm
(II) Aged CCP *Internal* Hexa-Coordinated High-Spin ($S = 5/2$)	Axially symmetric HS $g_\perp = 6.0$ $g_\parallel = 2.0$	Hexa-HS 408 nm (≈120) 503 nm 620 nm (≈6)	Hexa-HS ν_{10}=1619 cm^{-1} ν_2=1564 cm^{-1} ν_3=1479 cm^{-1} ν_4=1372 cm^{-1}	HS +84.3 ppm +71.2 ppm +60.8 ppm
(III) Weak-Field Complex *External* (Fluoride) Hexa-Coordinated High-Spin ($S = 5/2$)	Axially symmetric HS $g_\perp = 6.0$ $g_\parallel = 2.0$	Hexa-HS 408 nm (130) 495 nm (9.8) 620 nm (7.6)	Hexa-HS ν_{10}=1619 cm^{-1} ν_2=1564 cm^{-1} ν_3=1479 cm^{-1} ν_4=1372 cm^{-1}	HS +65.0 ppm +61.2 ppm +54.5 ppm +51.9 ppm
(IV) Frozen Native CCP *Internal* Hexa-Coordinated Low-Spin ($S = 1/2$)	LS $g_x = 2.70$ $g_y = 2.20$ $g_z = 1.78$	LS	No R. Raman at cryogenic temperature	No NMR at cryogenic temperature
(V) Alkaline CCP *Internal* Hexa-Coordinated Low-Spin ($S = 1/2$)	LS $g_x = 2.69$ $g_y = 2.19$ $g_z = 1.80$	LS	Hexa-LS ν_{10}=1635 cm^{-1} ν_2=1580 cm^{-1} ν_3=1506 cm^{-1} ν_4=1375 cm^{-1}	LS

HS, high-spin; LS, low-spin; R. Raman, Resonance Raman.

a, This paper, Principal g-values.

b, This paper, Absorption maxima (ϵ in mM^{-1} cm^{-1}).

c, Dasgupta S, Rousseau DL, Anni H, Yonetani T, unpublished results; Hashimoto et al., 1986;

d, Hyperfine-shifted porphyrin methyl proton resonances (chemical shift from sodium 3-[trimethylsilyl]propionate); Inubushi T, Anni H, Yonetani T, unpublished results.

other hexa-coordinated state (state V), by ligation of an *internal* or *external* strong-field ligand (Conroy and Erman, 1978; Dhaliwal and Erman, 1985; Dowe and Erman 1985). The diminished reactivity of alkaline peroxidases is well documented.

Penta-Coordination and Reactivity with Hydroperoxides

The coordination of hydroperoxide to the heme iron is the primary step of the hydroperoxidase-hydroperoxide reaction prior to the heterolytic cleavage of the O-O bond via an acid-base catalysis (Poulos and Kraut, 1980). In hexa-coordinated hydroperoxidases, the dissociation of the sixth ligand (either internal H_2O or exogenous ligand) is the rate-limiting step of the overall reaction. Thus, the rate of the peroxide compound formation is independent of the concentration of hydroperoxides, as in the case of aged CCP (this paper) and the horseradish peroxidase-benzohydroxamic acid complex (Schonbaum, 1982).

The electronic absorption spectrum (Yonetani and Anni, 1987) and the reactivity with hydroperoxides (Balny et al., 1987) of fresh CCP are independent of pH from pH 4 to 9, indicating that the pKa value of the participating acidic group may be unusually low (< pK 4.0). The reported acidic pKa value of 5.5 (Conroy and Erman, 1978; Shelnutt et al., 1983) appears to be related to the formation of a hexa-coordinated state in an aging-altered CCP in mildly acidic conditions. This aging-induced conformational change may have shifted the pKa value of a catalytic acidic group, such as Asp-235 (Edwards et al., 1984), to a higher value of 5.5, inducing the formation of an inactive hexa-coordinated state.

ACKNOWLEDGEMENTS

This investigation has been supported by research grants PCM83-16935 from the National Science Foundation and HL14508 from the National Heart, Lung, and Blood Institute.

REFERENCES

Anni H, Yonetani T (1986) The pi-cation radical of Zn-cytochrome c peroxidase and its interaction with ferrocytochrome c. Biophys J 49:27a.

Andersson LA, Renganathan V, Loehr TM, Gold MH (1987) Lignin peroxidase: Resonance Raman spectral evidence for compound II and for a temperature-dependent coordination-state equilibrium in the ferric enzyme. Biochemistry 26:2258-2263.

Balny C, Anni H, Yonetani T (1987). A stopped-flow study of the reaction of cytochrome c peroxidase with hydroperoxides. FEBS Lett 221:349-354.

Chance B (1951) Enzyme-substrate compounds. Adv Enzymol 12:153-190.

Conroy CW, Erman JE (1978) Proton stoichiometry of the cytochrome c peroxidase mechanism as a function of pH. Biochim Biophys Acta 527:370-378.

Dhaliwal BK, Erman JE (1985) A kinetic study of the alkaline transitions in cytochrome c peroxidase. Biochim Biophys Acta 827:174-182.

Dowe RJ, Erman JE (1985) Physicochemical characterization of the alkaline denaturation of cytochrome c peroxidase. Biochim Biophys Acta 827:183-189.

Edwards SL, Mauro JM, Fishel LA, Wang J, Miller MA, Xuong Nh, Kraut J (1987) Search for the radical in cytochrome c peroxidase compound I: Analysis by crystallography and mutagenesis. This volume.

Edwards SL, Poulos TL, Kraut J (1984) The crystal structure of fluoride-inhibited cytochrome c peroxidase. J Biol Chem 259:12984-12988.

Edwards SL, Xuong Nh, Hamlin RC, Kraut J (1987) Crystal structure of cytochrome c peroxidase compound I. Biochemistry 26:1503-1511.

Evangelista-Kirkup R, Crisanti M, Poulos TL, Spiro TG (1985) Resonance Raman spectroscopy shows different temperature-dependent coordination equilibria for native horseradish and cytochrome c peroxidase. FEBS Lett 190:221-226.

Finzel BC, Poulos TL, Kraut J (1984) Crystal structure of yeast cytochrome c peroxidase refined at 1.7-A resolution. J Biol Chem 259:13027-13036.

Hashimoto S, Teraoka J, Inubushi T, Yonetani T, Kitagawa T (1986) Resonance Raman study on cytochrome c peroxidase and its intermediate: Presence of the Fe(IV)=O bond in compound ES and heme-linked ionization. J Biol Chem 261:11110-11118.

Iizuka T, Kotani M, Yonetani T (1971) Magnetic susceptibility measurements of cytochrome c peroxidase and its complexes. J Biol Chem 246:4731-4736.

Lang G, Asakura T, Yonetani T (1969) Mossbauer spectroscopy of protohaem and mesohaem cytochrome c peroxidase and their fluorides. J Phys C (Solid St Phys) 2:2246-2261.

Loo S, Erman JE (1975) A kinetic study of the reaction between cytochrome c peroxidase and hydrogen peroxide. Depedence on pH and ionic strength. Biochemistry 14:3467-3470.

Maltempo MM, Moss TH (1976) The spin 3/2 state and quantum spin mixtures in haem proteins. Quart Rev Biophys 9:181-215.

Mathews RA, Wittenberg JB (1979) Cytochrome c peroxidase: Interconversion of chemically and enzymatically reactive and unreactive forms of the ferric protein. J Biol Chem 254:5991-5996.

Nicholls P, Mochan E (1971) Complex-formation between cytochrome c and cytochrome c peroxidase. Kinetic studies. Biochem J 121:55-67.

Ohlsson PI, Yonetani T, Wold S (1986) The formation of ES of cytochrome-c peroxidase: A comparison with lactoperoxidase and horseradish peroxidase. Biochem Biophys Acta 874:160-166.

Peisach J, Blumberg WE (1971) Departures from axial symmetry as measured by EPR for high spin ferric heme proteins. In Chance B, Yonetani T, Mildvan AS (eds): "Probes of structure and function of macromolecules and membranes" Vol II, New York: Academic Press, pp 231-239.

Poulos TL, Kraut J (1980) The stereochemistry of peroxidase catalysis. J Biol Chem 255:8199-8205.

Schonbaum GR (1982) Modulation of peroxidase-dependent reactions by acid-base catalysis. In King TE, Morrison M, Mason HS (eds): "Oxidases and Related Redox Systems", New York: Pergamon Press, pp 671-684.

Shelnutt JA, Satterlee JD, Erman JE (1983) Raman difference spectroscopy of heme-linked ionizations in cytochrome c peroxidase. J Biol Chem 258:2168-2173.

Yonetani T (1965) Studies on Cytochrome c Peroxidase. II. Stoichiometry between enzyme, H_2O_2, and ferrocytochrome c and enzymic determination of extinction coefficients of cytochrome c. J Biol Chem 240:4509-4514.

Yonetani T, Anni H (1987) Yeast Cytochrome c Peroxidase: Coordination and spin states of heme prosthetic group. J Biol Chem 262:9547-9554.

Yonetani T, Chance B, Kajiwara S (1966a) Crystalline Cytochrome c Peroxidase and complex ES. J Biol Chem 241: 2981-2982.

Yonetani T, Ray GS (1965) Studies on Cytochrome c Peroxidase I. Purification and some properties. J Biol Chem 240:4503-4508.

Yonetani T, Ray GS (1966) Studies on Cytochrome c Peroxidase III. Kinetics of the peroxidatic oxidation of ferrocytochrome c catalyzed by cytochrome c peroxidase. J Biol Chem 241:700-706.

Yonetani T, Wilson DF, Seamonds B (1966b) Studies on Cytochrome c Peroxidase. VIII. The effect of temperature on light absorptions of the enzyme and its derivatives. J Biol Chem 241:5347-5352.

DISCUSSION

YAMAZAKI: You suggested the presence of H-bonding between Fe(IV)=O and a distal base. Does the proton dissociate as the pH increases? If not, how did you confirm the H-bond?

YONETANI: This was done by resonance Raman measurement of the Fe(IV)=O lines in D_2O and H_2O. Since cytochrome c peroxidase is unstable at alkaline pH's, the pKa for the dissociation has not been measured.

YAMAZAKI: Can you separate the aged enzyme from the native one by means of crystallization or chromatography?

YONETANI: Thus far we have not been able to separate the aged enzyme from the fresh enzyme. Crystallization, particularly repeated crystallization, changes aged hexa-coordinate to penta-coordinate. In other words, the penta-coordinate is favored under the conditions of crystallization.

PAPA: Is it correct to consider the hexa-coordinate species of the peroxidase generated by aging or freezing nothing more than an artifact? Would there be some analogy between the penta- and hexa-coordinate states in peroxidase and the resting and pulsed states described by Antonini and Brunori for cytochrome c oxidase?

YONETANI: Yes, the hexa-coordinated, inactive species of cytochrome c peroxidase is a preparation- and storage-induced artifact which must be eliminated. In terms of the coordination controlled reactivity, the hexa- and penta-coordinated cytochrome c peroxidases are analogous to the resting and pulsed forms of cytochrome a_3 which are hexa- and penta-coordinated, respectively.

Oxidases and Related Redox Systems, pages 451–462

PROTONATION OF THE FERRYL AND OXY- FORMS OF HORSERADISH PEROXIDASES A AND C

Isao Yamazaki and Ryo Nakajima

Biophysics Division, Research Institute of Applied Electricity, Hokkaido University, Sapporo 060, Japan

INTRODUCTION

The redox nature of plant peroxidases is greatly influenced by acid-base groups near the heme iron. It has been suggested that protonation on such groups occurs near neutral pH and regulates catalytic activity of the enzymes (1). Theorell reviewed early studies on the pH dependence of physicochemical properties of hemoproteins under a title of "heme-linked groups and mode of action of some hemoproteins" (2).

If there is protonation in heme-linked groups, the pKa value may be measured from the pH dependence of physicochemical properties of hemoproteins. It is also reasonable to assume that the pKa value for a certain ionizing group varies with the state of hemoproteins. We reviewed a decade ago a systematic study on the pKa shift upon the redox change of the heme iron, the ligand-binding in the ferrous form, and the 2, 4-substitution in the heme porphyrin (3). Since then, there were definite advances in the study of ferryl and oxy-horseradish peroxidases.

The existence of heme-linked protonation has been detected in the ferryl form from pH dependence of absorption spectra and redox-coupled proton balance (4). Resonance Raman spectroscopy has recently afforded clear evidence that protonation occurs at a distal group and the proton is involved in a hydrogen bonding between the distal base and the O atom bound to the ferryl ion (5-7).

Oxyperoxidase has aroused unremitting attention since it was firstly found by Keilin and Mann a half century ago as a spectral species showing absorption bands at 583 and 545.5 nm (8). Oxyperoxidase, once called compound III (9) or complex III (10), is relatively unstable and its physicochemical properties of the heme-linked groups have been poorly understood. We have recently succeeded in observing absorption spectra of the protonated and the deprotonated forms of horseradish oxyperoxidases.

PROTONATION IN THE FERRYL PEROXIDASE

The presence of a heme-linked ionizing group was suggested from the pH dependence of the rate of reactions of the ferryl peroxidase with electron donors (1). Similar pH dependence is also seen in a reaction of the ferryl

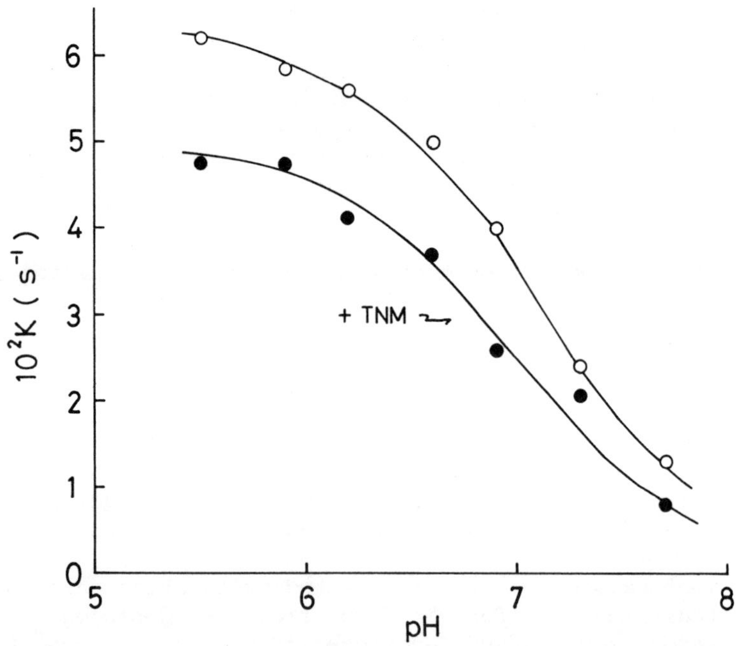

Figure 1. Effect of pH and tetranitromethane (TNM) on the rate of oxyperoxidase formation from ferryl peroxidase A in the presence of 1 mM H_2O_2 (○). In (●), 20 μM TNM was added.

peroxidase with H_2O_2, which forms the oxyperoxidase (Fig. 1). The reaction was found to consist of two independent reactions, TNM(tetranitromethane)-sensitive and -insensitive ones (11). TNM is known to be an effective scavenger for O_2^- and the reaction of TNM with O_2^- occurs as follows (12).

$$C(NO_2)_4 + O_2^- \longrightarrow C(NO_2)_3^- + NO_2 + O_2 \qquad (1)$$

The TNM-insensitive formation of the oxyperoxidase is concluded to be preceded by the dissociation of O^{2-} liganded to the ferryl ion. Therefore, the pH dependence of the rate of O^{2-} dissociation can be demonstrated as that of the rate of oxyperoxidase formation in the presence of TNM (Fig. 1). This conclusion is supported by resonance Raman data which indicate that the O^{2-} ion exchanges with solvent water only in the protonated ferryl peroxidase (6, 13). Figure 2 shows the two parallel paths of oxyperoxidase formation from the ferryl peroxidase and H_2O_2.

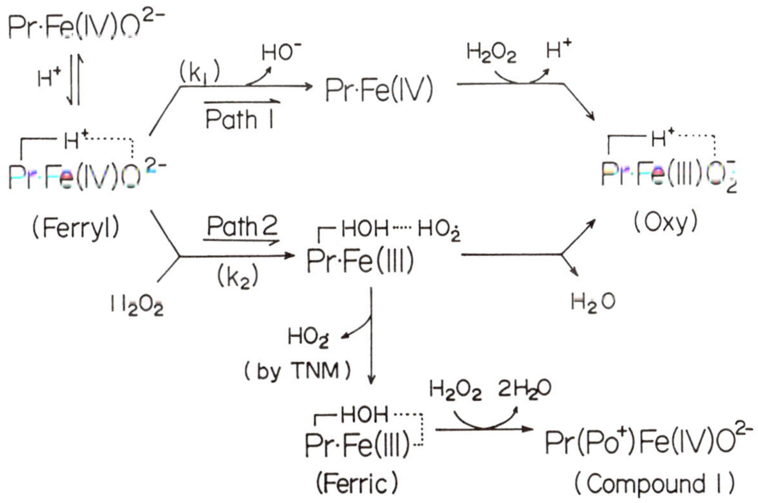

Figure 2. Two parallel paths in the reaction of ferryl peroxidase with H_2O_2 (reproduced from Nakajima and Yamazaki, J. Biol. Chem. 262, 2576-2581, 1987). Pr and Po^+ denote apoprotein and porphyrin π-cation radical, respectively.

Recently, it was found that, in the case of diacetyl deutero-peroxidase, the oxyperoxidase is formed from the ferryl peroxidase through only path 2. In this case, the oxyperoxidase formation is completely inhibited in the presence of TNM. As shown in Fig. 3, the ferryl peroxidase is converted to the oxy-form after all TNM has been reduced to $C(NO_2)_3^-$. The result, suggesting that the O^{2-} ion does not dissociate at a measureable rate even in the protonated form of the ferryl diacetyldeutero-peroxidase, does not contradict the resonance Raman data reported by Makino <u>et al</u>. (7), who have recently reported the pKa value for the protonation in ferryl diacetyldeutero-peroxidases A and C, as listed in Table I.

PROTONATION IN THE OXYPEROXIDASE

One of the most dramatic phenomena related to heme-linked ionizing groups may be that a proton is incorporated when O_2 is bound to the ferrous peroxidase (14). This may be explained in terms of the increase in the proton affinity of a distal base upon binding of O_2 to the ferrous ion. It seems likely that such a pKa shift reflects the strength of hydrogen bonding between the distal base and the negatively charged O_2 on the heme iron. A systematic study on CO complexes of the 2,4-substituted peroxidases has shown that the pKa value increases with the increase in

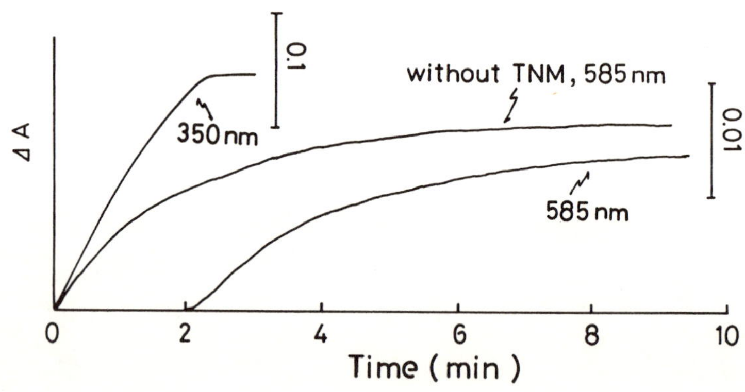

Figure 3. Effect of TNM on the reaction of ferryl diacetydeutero-peroxidase C (6.8 μM) with H_2O_2. Oxyperoxidase formation was measured at 585 nm and TNM reduction at 350 nm. $[H_2O_2]$ = 12.5 mM and [TNM] = 12 μM.

TABLE I. Proton dissociation constant for functionally heme-linked groups in ferrous, ferrous-CO, ferryl and oxy-horseradish peroxidases A and C

Enzyme / Form	Peroxidase A		Peroxidase C		Reference
	diacetyl-deutero	proto (native)	diacetyl-deutero	proto (native)	
Ferrous	5.6	5.8	7.2	7.3	3
Ferrous-CO	6.0	6.7	7.4	8.2	3, 15
Ferryl	7.4	7.0	8.5	8.6	4, 7
Oxy	8.0	8.1	9.2	10.3	15, this paper

backdonation from iron $d\pi$ to CO π^* orbitals (15). A similar study appears to be difficult with the oxyperoxidases, because the stability of the oxyperoxidases decreases as the electron withdrawing capacity of 2,4-substituents decreases (16).

We are now able to demonstrate the pH effect of optical absorption spectra of the oxy-forms of native and diacetyldeutero-horseradish peroxidases A and C. Figure 4 clearly shows that the intensity of the α-band increases with the increase of pH. It is interesting to note that the visible spectra of the deprotonated oxyperoxidases A and C resemble the spectrum of MbO_2. This reminds us the fact that no proton is incorporated on the binding of Mb with O_2. The pKa value for the protonation can be measured from pH titration at an appropriate wavelength as shown in Fig. 5. The values thus obtained are listed in Table I.

The rate constant for reactions of the oxyperoxidase with various electron donors has been reported (17, 18). Figure 6 shows pH dependence of the rate constant for two typical reactions. The results show that the effect of

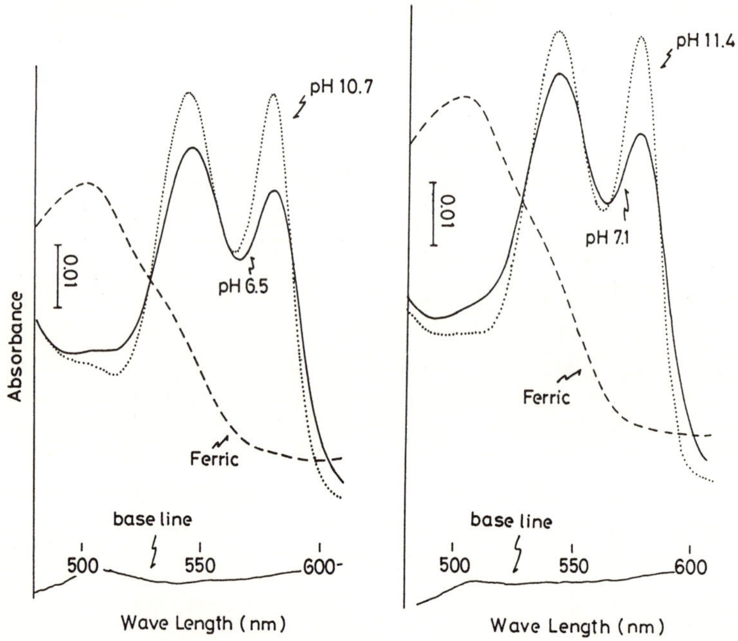

Figure 4. Effect of pH on absorption spectra of oxyperoxidases A 5.6 μM, left) and C (6.5 μM, right).

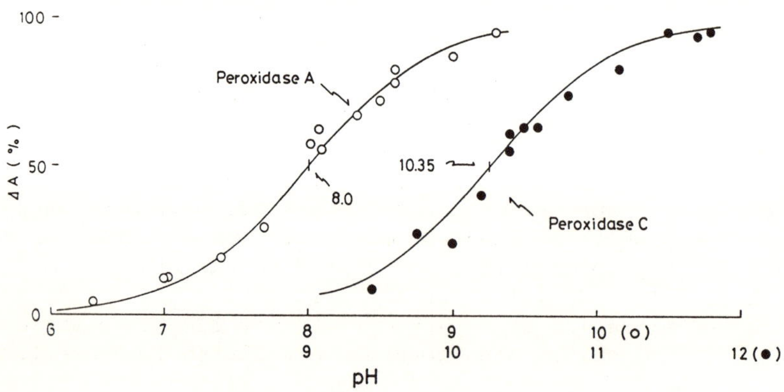

Figure 5. Spectrophotometic pH titration of oxyperoxidases A and C.

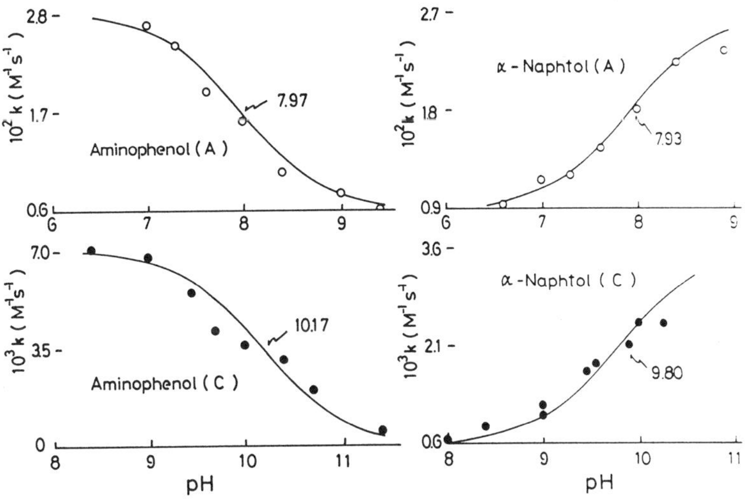

Figure 6. Effect of pH on the rate constant for reactions of oxyperoxidases A and C with electron donors.

protonation on the reactions is not simple. For instance, the kinetic pKa is not necessarily consistent with that obtained spectrophotometrically and the protonated form is not always more active than the deprotonated one. The oxyperoxidases are reduced invariably to the ferric form without any appreciable intermediate.

DISCUSSION

The pKa value of an acid-base group is determined by the local charge near the group in addition to its inherent nature. If the protonation is assumed to occur on a distal base of peroxidases A and C, the recent structural information (19) suggests that the same aminoacid residue is responsible to the protonation, irrespective of marked difference in the pKa value. The difference between the isoenzymes A and C is therefore ascribed to that of the local net charge. It may also be said that the charge polarization in the heme by 2,4-substitution further shifts the pKa value. The relation is schematically shown in Fig. 7.

LOCAL NET CHARGE

Isoenzyme	Redox state
Peroxidase A	Ferric, Compound I
↓	↓
Peroxidase C	Ferrous, Ferryl, Oxy

CHARGE POLARIZATION

Heme	Ligand
Meso	Free
↑	↓
Proto	CO
↑	↓
Diacetyl-deutero	O^{2-}
	↓
	O_2

Figure 7. Factors determining the pKa value for functionally heme-linked protonation. The arrows indicate the increase in proton affinity. We conclude that the protonation occurs on the same base in peroxidases A and C and also that there is thermodynamic additivity in these factors.

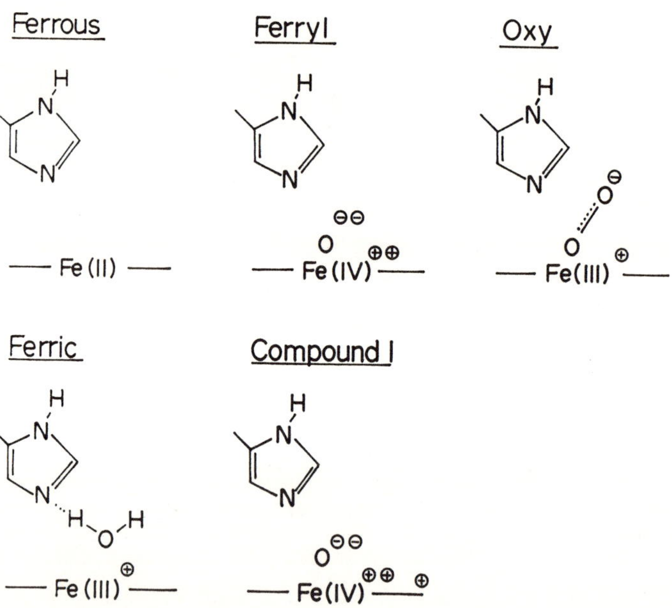

Figure 8. Demonstration of net charges around the heme iron and the distal imidazol in five redox states of horseradish peroxidases. The charge neutrality is kept in ferrous, ferryl and oxy-peroxidases, and one plus charge is excess in the ferric enzyme and Compound I.

Since the net charge of the heme is the same for the ferrous, ferrous-CO, ferryl and oxy-forms of a peroxidase (Fig. 8), the difference in their pKa values listed in Table I will be explained in terms of the nature of ligands at the 6th position. The pKa value increases in the order of ferrous, ferrous-CO, ferryl and oxy-forms. This

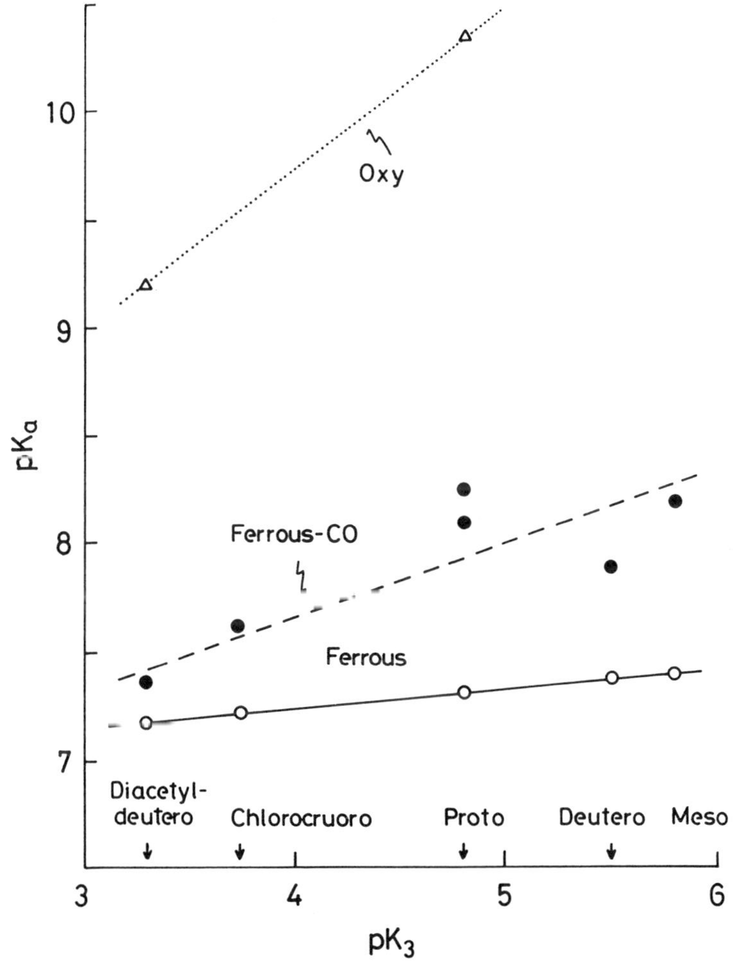

Figure 9. Effect of pK_3 on the pKa value for ferrous, ferrous-CO and oxy-peroxidases C. K_3 ; proton dissociation constant for the third protonatioon in pyrrol nitrogens of porphyrin.

tendency appears to accord with structures assumed for the ligand bindings. The effect of 2,4-substitution on the pKa values (Fig. 9) may suggest that the backdonation plays an important role in the increase of the pKa value. The fact that there is no essential change in the pKa value between the ferryl forms of native and diacetyldeutero-peroxidases will be explained by assuming that the oxygen atom bound to the ferryl ion exists in the O^{2-} state which is little affected by the 2,4-substitution.

Dunford and his coworkers have done an extensive study on the pH dependence of reactions of the ferryl peroxidase C with various electron donors (1). The reason why only the protonated form is reactive has not clearly been explained, although it is evident that the oxidation potential decreases upon deprotonation (20). It has recently been found (6, 13) that the O^{2-} ion exchanges with solvent water only in the protonated state, but the exchange rate is about a few minute, being too slow to be involved in the redox reaction. Therefore, it seems that the protonation causes a structural change to facilitate the electron transfer from donors to the heme iron.

Resonance Raman data show that the iron-oxygen stretching frequency is higher in myoglobin than in horseradish peroxidases A and C for both the ferryl forms (21) and the oxy-forms (22). Upon deprotonation, iron-oxygen frequencies of the ferryl peroxidases approach that of ferryl myoglobin and absorption spectra of the oxyperoxidases resemble that of MbO_2. The acid-base properties described in this paper appear to be common to other plant peroxidases, such as turnip peroxidases (23), but are different from those of yeast and animal peroxidases.

REFERENCES

1. Dunford, H. B., and Stillman, J. S.(1976) Coordination Chemistry Reviews 19, 187-251.
2. Theorell, H. (1947) Adv. Enzymol. 7, 265-303.
3. Yamazaki, I., Araiso, T., Hayashi, Y., Yamada, H., and Makino, R. (1978) Adv. Biophys. 11, 249-281.
4. Hayashi, Y., and Yamazaki, I. (1978) Arch. Biochem. Biophys. 190, 446-453.
5. Sitter, A. J., Reczek, C. M., and Terner, J. (1985) J. Biol. Chem. 260, 7515-7522.

6. Hashimoto, S., Tatsuno, Y., and Kitagawa, T. (1986) Proc. Natl. Acad. Sci. U. S. A. 83, 2417-2421.
7. Makino, R., Uno, T., Nishimura, Y., Iizuka, T., Tsuboi, M., and Ishimura, Y. (1986) J. Biol. Chem. 261, 8376 -8382.
8. Keilin, D., and Mann, T. (1937) Proc. Roy. Soc. B,122, 119-133.
9. Keilin, D., and Hartree, E. F. (1951) Biochem. J. 49, 88-104.
10. Chance, B. (1951) Adv. Enzymol. 12, 153-190.
11. Nakajima, R., and Yamazaki, I. (1987) J. Biol. Chem. 262, 2576-2581.
12. Rabani, J., Mulac, W. A., and Matheson, M. S. (1965) J. Phys. Chem. 65, 53-70.
13. Hashimoto, S., Nakajima, R. Yamazaki, I., Tatsuno, Y., and Kitagawa, T. (1986) FEBS Letters 208, 305-307.
14. Yamada, H., and Yamazaki, I. (1974) Arch. Biochem. Biophys. 165, 728-738.
15. Yamada, H., and Yamazaki, I. (1975) Arch. Biochem. Biophys. 171, 737-744.
16. Makino, R., Yamada, H., and Yamazaki, I. (1976) Arch. Biochem. Biophys. 173, 66-70.
17. Yokota, K., and Yamazaki, I. (1965) Biochem. Biophys. Res. Commun. 18, 48-53.
18. Tamura, M., and Yamazaki, I. (1972) J. Biochem. 71, 311-319.
19. Poulos, T. L., and Kraut, J. (1980) J. Biol. Chem. 255, 8199-8205.
20. Hayashi, Y., and Yamazaki, I. (1979) J. Biol. Chem. 254, 9101-9106.
21. Sitter, A. J., Reczek, C. M., and Terner, J. (1985) Biochim. Biophys. Acta 828, 229-235.
22. Van Wart, H. E., and Zimmer, J. (1985) J. Biol. Chem. 260, 8372-8377.
23. Ricard, J., Mazza, G., and Williams, R. J. P. (1972) Eur. J. Biochem. 28, 566-578.

DISCUSSION

<u>WEVER</u>: Comment. In our hands myleoperoxidase compound I reacts with H_2O_2 to yield compound II and O_2^-. However, O_2^- in such an oxidizing system is difficult to detect. For example, when cytochrome <u>c</u> is used as a scavenger, it is rapidly reoxidized by compound I or II.

What is the product of the reaction between oxyperoxidase and reductants like aminophenol or naphthol? You showed that native enzyme is regenerated. However, is H_2O_2 being formed under these conditions according to:

$$Fe^{3+}O_2^- \xrightarrow{e^-} Fe^{3+}O_2^{=} \longrightarrow Fe^{3+} + H_2O_2$$

or similarly with compound I?

<u>YAMAZAKI</u>: The reduction of oxyperoxidase to the ferric enzyme is expected to occur in three one electron steps, through compound I and II. But, the first step is rate-limiting and the enzyme intermediate cannot be seen when ordinary reductants are used. Only if a limited amount of hydrated electron is added to oxyperoxidase by pulse radiolysis can the formation of compound I be observed:

$$\text{Oxyperoxidase} + e^- \longrightarrow \text{compound I}$$

[Kobayashi K, Hayashi K (1981) J Biol Chem 256: 12350-12354]

Oxidases and Related Redox Systems, pages 463–475

WHERE IS THE RADICAL IN COMPOUND I OF CYTOCHROME *c* PEROXIDASE? CLUES FROM CRYSTALLOGRAPHY AND MUTAGENESIS

Steven L. Edwards, J. Matthew Mauro, Laurence A. Fishel, Jimin Wang, Mark A. Miller, Nguyen huu Xuong and Joseph Kraut

Department of Chemistry, University of California, San Diego, La Jolla, California.

ABSTRACT

A difference Fourier map shows small structural perturbations on oxidation of cytochrome *c* peroxidase (CCP) to its semi-stable intermediate, compound I. Least-squares refinement of both CCP and compound I quantifies these perturbations and suggests that the radical site may be on the distal side of the heme, since that is where most of the small movements occur.

Several engineered mutants of CCP were created in an attempt to assess the function of various side chains, among them Trp-51 → Phe, Trp-191 → Phe and His-181 → Gly. X-ray structures of the mutant CCP's confirm that only minimal changes are caused by these substitutions. Preliminary examination of the mutants' kinetic properties show that Trp-51 is not the radical site; that Trp-191 has an important enzymic function; and that His-181 is not essential for electron transfer, but probably has some more indirect role. The locus of the radical in compound I, however, remains to be established.

INTRODUCTION

At the ISOX meeting of 1979 we presented a preliminary 2.5 Å crystal structure of cytochrome *c* peroxidase (CCP). The structure was published shortly thereafter (Poulos et al., 1980) and served as the basis for a proposed catalytic mechanism that rationalized the heterolytic cleavage of the peroxide bond (Poulos and Kraut, 1980a). We also suggested that the second oxidation equivalent retained by compound I as a protein-based radical might reside on Trp-51. Additionally a model for the one-to-one electron transfer complex with cytochrome *c* was presented (Poulos and Kraut, 1980b).

Since then we have refined the structure against high-resolution x-ray data (1.7 Å) after rebuilding chain segment 178-192 and incorporating a sequence correction by including an additional asparagine residue following position 163 (Finzel et al., 1984). A high resolution crystal structure of this kind is of course a solid foundation on which to construct a theory of how the enzyme works. But additionally it also provides a sensitive experimental system for observing small structural perturbations of the molecular architecture caused by chemical transformations or alterations in amino acid sequence.

With these ideas in mind, we have tried to devise experiments that will test various aspects of the proposed mechanism. One aim has been to identify the site of oxidation associated with the free radical-type EPR signal observed in compound I (Yonetani et al., 1966). To this end we have examined the structures of the compound I intermediate (Edwards et al., 1987) and of its decay product (Anderson, 1986). We have also observed the small structural perturbations resulting from binding various heme ligands (Poulos et al., 1978; Edwards et al., 1984). But perhaps the most surprising and enlightening results have come from our experiments involving site-directed mutagenesis (Fishel et al., 1987). This communication will attempt to give a brief progress report and overview of some of that work.

Structural Perturbations When CCP Is Oxidized to Compound I

Crystals of CCP can be readily converted to the two-equivalent oxidized intermediate, compound I, simply by soaking them in a peroxide solution. Fortunately, unit-cell dimensions do not change, so that any small structural effects on the CCP molecule which may be associated with conversion to compound I can be accurately visualized in a difference Fourier map. Such a difference map is calculated from the observed F's of the compound I crystal minus the observed F's of the parent CCP crystal, using the phases of the refined parent structure. This kind of map is known to reveal subtle perturbations, which can then be quantified by continuing least-squares refinement of the compound I crystal structure.

It should be emphasized that oxidation of CCP to compound I, since it does not involve the gain or loss of non-hydrogen atoms, might conceivably have occurred without any crystallographically visible structural changes at all. Nevertheless, one could reasonably expect some movement of the heme's iron atom, which is oxidized from Fe(III) in the resting enzyme to Fe(IV) in the oxyferryl state of compound I. It was less certain, however, that the presence of a positively charged radical somewhere else in the enzyme molecule would reveal itself by causing local structural perturbations. In fact, we believe that does happen, as described below.

The most severe experimental hurdle was determining appropriate conditions to extend the compound I lifetime long enough to allow a reasonable amount of data measurement from one crystal. Since lowering the temperature stabilizes compound I, we increased the concentration of the crystallizing agent 2-methyl-2,4-pentanediol, which is also a good cryosolvent, to allow cooling of the crystals to -15 °C. At this temperature x-ray data could be collected for about 4 hours before the compound I content of the data crystal dropped below 75%. This is not enough time to collect a complete data set from a single crystal, even using the Mark II multiwire area detector diffractometer (Xuong et al., 1985), but from six crystals we were able to produce a data set complete to 2.5 Å. In order to achieve the clearest difference Fourier map, a new set of parent data was also collected at the same temperature and solvent concentration as used for the compound I data collection.

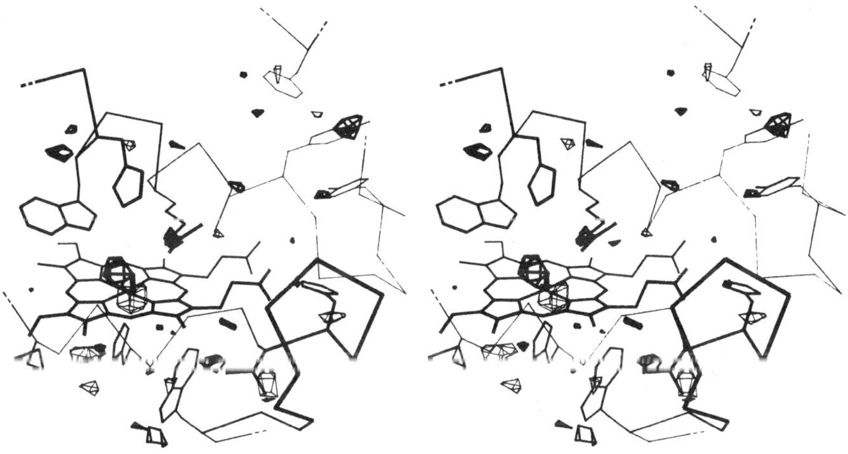

Figure 1. The difference Fourier map of compound I minus parent CCP superimposed on key residues around the heme. Positive density is shown in bold contours, negative density in light contours.

The resulting difference Fourier map contained many features, most of them in the vicinity of the heme. Although the significance of some was readily apparent, initially many were not, and were not interpreted until after least squares refinement of the compound I structure. However, one result was clear immediately: there was essentially no difference density around either of the radical candidates Trp-51 or Met-172 (Hoffman et al, 1981). At about the same time, results from mutagenesis studies on these two residues also strongly indicated that neither one is the site of the radical in compound I (Fishel et al., 1987; Goodin et al., 1986), and so our attention was redirected toward other regions of the difference map (see below).

Some structural changes that were interpretable included movement of the iron atom toward the distal side of the heme, probably due to the strong, short bond it makes with the single oxygen atom of the oxyferryl group (Chance et al., 1986). Also, difference density around the side chain of Arg-48 suggested that its guanidinium group moves toward the ferryl oxygen atom to donate a hydrogen bond. But a perplexing feature of the map was a prominent hole near Oγ of Thr-180. This general location had previously attracted our interest because it is near the proximal histidine, His-175, and because in crystals of the inhibitor complex, CCP-NO (Edwards, 1981) we had earlier observed perturbation of Trp-191, which lies between His-175 and Thr-180. The latter suggests that the exact positioning of the side chain of Trp-191 is sensitive to the chemical state of the heme iron atom.

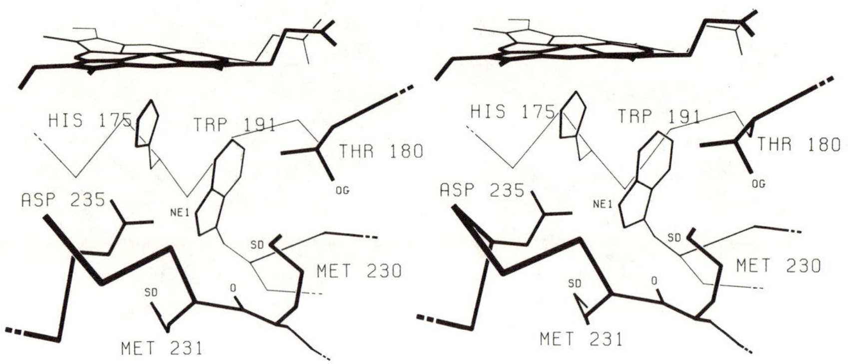

Figure 2. The locations of Met-230, Met-231 and Trp-191 on the proximal side of the heme.

As described in the following section, when Trp-51 was mutagenically replaced by phenylalanine (Fishel et al., 1987) the peroxide-oxidized mutant enzyme still displayed an EPR radical signal, although it was much less stable than the radical of compound I of parent CCP(MI). This negative result led us to consider other locations in the protein that could be the site of the radical. In particular, the difference Fourier map directed our attention to the region around the proximal His-175. Prompted by the curious difference density at nearby Thr-180, we recalled the proposal by Hoffman and coworkers

(Hoffman et al., 1979) that the radical might be supported by a pair of methionines with closely juxtaposed sulfur atoms. In fact, there is a methionine pair in the CCP sequence, Met-230 and Met-231, of which the side chains are close to Thr-180. Although the two methionine sulfurs are over 6 Å apart, both are also near Nε1 of Trp-191 (see figure 2). We were thus led to reconsider whether this methionine pair could be the radical location, with Trp-191 participating in some way. We therefore re-examined the difference Fourier map, focusing on the vicinity of the methionine pair. Although there was no strong difference density at the methionines themselves, nearby density together with positional adjustments resulting from crystallographic refinement indicated that adjacent oxygen atoms had indeed moved toward the methionine sulfurs. Are these oxygen atoms, with their partial negative charges, responding to a positive charge created at the sulfurs? Although the adjustments calculated by least-squares refinement were very small, 0.1 to 0.2 Å, these movements in fact helped to explain many small features in the difference map. The structure of the molecule shows two side chain oxygens (Thr-180 and Asp-235) and four backbone carbonyl oxygens (188,189, 190 and 230) arrayed around the methionine sulfurs, all within 4.5 Å. The closest water molecule is water-495, about 6 Å from the sulfur of Met-231. Four of these oxygen atoms in addition to the water molecule oxygen move toward the sulfurs on the formation of compound I. The environment of the methionine pair is also distinguished by the presence of numerous aromatic side chains: three tryptophans (191, 211 and 233) and three tyrosines (187, 229 and 236) combine to form a surrounding shell. We do not understand the function of each of these aromatic groups, but it is interesting that the presence of sulfur-containing side chains and aromatic groups has been cited as a common motif in proteins involved in redox chemistry (Morgan et al., 1978). The occurrence of aromatic groups around a cationic methionine radical site may explain some of the observations by Yonetani and coworkers that led them to suggest a tryptophan or a tyrosine as the site of the radical. It was proposed that changes at 241 nm in the UV absorption spectrum of compound I compared to the ground-state enzyme might be due to a tryptophan or tyrosine radical (Yonetani and Ray, 1965). And in another experiment, in which compound I was allowed to decay, tyrosines and tryptophans were the residues predominantly modified, as observed by amino acid analysis (Erman and Yonetani, 1975a). We note parenthetically that x-ray measurements on crystals of the CCP compound I decay product point to modification at Tyr-236 and Cys-128 (Anderson, 1986).

To complete this discussion of the aromatic groups around the methionine pair, we should comment on the possible role of Trp-191. The position of its indole ring is suggestive; the six membered part of the ring is parallel to and in van der Waals contact with the imidazole ring of His-175. The five membered part of indole ring is situated so that nitrogen Nε1 is about 3.5 Å from both methionine sulfurs. Simultaneously, the same nitrogen atom is hydrogen bonded to the side chain of Asp-235, which in turn also hydrogen bonds to the proximal histidine. This elaborate structural arrangement led us to wonder if the oxidation equivalent generated by cleavage of the peroxide bond could be transmitted to the Met sulfurs through the Trp-191 indole ring. Hoffman et al. (Hoffman et al., 1979) have argued on the basis of their ENDOR analysis that the radical cannot reside solely on a tryptophan ring. But perhaps the nitrogen Nϵ1 of Trp-191 plays an active role in transmitting and stabilizing a radical on the methionine sulfurs.

To summarize, then, on the basis of these observations we are now inclined to suspect that the radical of CCP compound I resides on the proximal side of the heme and probably involves Met-230 and/or Met-231. Whether it is localized to a single amino acid side chain or, alternatively, is somehow spread over several neighboring groups, remains an open question. Trp-191 is also likely to have some role in creating the radical and perhaps in its conduction.

The final idea to be gleaned from the compound I crystal structure relates to a backbone chain movement that occurs when compound I forms. The difference Fourier map and the structure refinement complement each other to suggest a coordinated positional adjustment of the loop 175-190 in response to the two-equivalent oxidation of CCP by peroxide. This loop movement may be important mechanistically because it is located near the surface of the molecule at the proposed binding site for cytochrome *c* (Poulos and Kraut, 1980b; Finzel et al., 1984). Note also that the proximal histidine residue His-175 is at one end of the loop, while Trp-191 immediately follows it. Although the two proposed sites of oxidation, the heme iron and the Met-230, Met-231 pocket, are separated by over 10 Å, it appears that the combined effect of their individual local perturbations results in the observed alteration in the structure of the 175-190 loop.

When the heme iron becomes oxidized from the ferric to the ferryl state, it forms a short bond with the single oxygen atom that remains after peroxide cleavage. The length of this bond has been measured by EXAFS to be 1.67 Å (Chance et al., 1986). The shortness of the bond probably causes the iron atom to move about 0.2 Å toward the distal side of the heme. Simultaneously the guanidinium group of Arg-48 also moves 0.3 Å toward the ferryl oxygen atom to donate a hydrogen bond. On the proximal side, His-175 moves 0.2 Å toward the heme to maintain its coordination bond with the iron. An important aspect of peroxidase chemistry has been associated with a strong hydrogen bond between the proximal histidine and the carboxyl group of Asp-235. The strength of this hydrogen bond is also manifested in the difference map and the least-squares refinement, for it is clear that the carboxyl group of Asp-235 itself moves about 0.1 Å toward His-175 to maintain the hydrogen bond. And in addition, the entire heme group slides 0.15 Å further into the heme pocket to partially compensate for the iron movement and further accomodate the Asp-His hydrogen bond. This sliding motion affects the

175-190 loop because of several hydrogen bonds between the loop and the heme propionates (Finzel et al., 1984).

The second site of oxidation in CCP compound I we propose to be the aromatic-lined cavity containing the Met pair 230-231. As described earlier, the positive charge of a cationic radical at this site appears to perturb the positions of a number of nearby oxygen atoms. A particularly prominent adjustment was noted for Oγ of Thr-180 which moves 0.2 Å toward the sulfur of Met-230.

It seems likely that the combined effect of these perturbations at the two oxidation sites in compound I causes the loop made up of residues 175 to 190 to twist slightly in a complicated manner. None of the resulting movements, however, is greater than 0.25 Å. As noted above, the 175-190 loop is at the molecular surface and constitutes part of the binding site for cytochrome *c* in the model that has been proposed for the CCP - cytochrome *c* complex (Poulos and Kraut, 1980b; Poulos and Finzel, 1984). If this model is more or less correct, twisting of the loop may play a role in the electron transfer mechanism.

The Mutants: Their Structures and Some Properties

In yeast, CCP is expressed as a pre-protein having a 68 residue leader sequence that directs transfer of the molecule from the site of synthesis in the cytoplasm into the mitochrondrial intermembrane space. There, CCP resides as the mature enzyme lacking the leader sequence (Kaput et al., 1982; Maccecchini et al., 1979).

Using oligonucleotide-directed site-specific mutagenesis we have constructed a slightly altered version of mature CCP enzyme with the leader sequence replaced by just two residues, Met-Ile. This new CCP is referred to as CCP(MI). Construction of the CCP(MI) gene involved replacement of codons for residues -2 and -1 of the leader sequence with codons specifying Met-Ile in order to create a *Bcl*I restriction site and the required Met initiation codon. The CCP(MI) gene was then substituted for the *fol* gene at the *Bcl*I site in an *E. coli* vector previously constructed for mutagenesis and overexpression of dihydrofolate reductase (Villafranca et al., 1983). Initial characterization of CCP(MI) expressed in *E. coli* revealed no significant differences from bakers' yeast CCP or *S. cerevisiae* CCP with regard to either catalytic properties or spectra. We have therefore gone on to exploit this cloning and expression system to create several variants of CCP(MI) with further mutations at specific sites of mechanistic interest, confident that we could compare them with CCP(MI) as though it were simply the wild-type *S. cerevisiae* enzyme.

Although the information afforded by characterizing the kinetic and spectroscopic properties of mutant enzymes is valuable in its own right, it is also important to know what effect each mutation has had on the molecular structure. Accordingly we have determined their crystal structures. Fortunately they all crystallize in suitable forms. However, a complicating factor is that the new parent, CCP(MI), and its mutants crystallize with different cell dimensions from the bakers' yeast enzyme whose structure was determined and refined previously (Finzel et al., 1984). Thus the structure had to be solved again.

X-ray data were measured for the new CCP(MI) crystal form on the multiwire area-detector diffractometer (Xuong et al., 1985) to 2.2 Å and the structure was solved by the molecular replacement method. Although the CCP(MI) structure is very similar to that of the bakers' yeast enzyme, substitution of isoleucine for threonine at position 53 does have a distinctly observable effect on the distal-side geometry.

Tryptophan 51 → Phenylalanine: Trp-51 Is Not the Radical Site

We have employed the mutagenesis and expression system described above to create a CCP(MI) molecule containing the mutation Trp-51 → Phe. This variant, CCP(MI,F51), was designed to test the hypothesis that Trp-51 is the locus of the characteristic free radical of compound I. That idea had been proposed (Poulos and Kraut, 1980a) simply because the crystal structure of bakers' yeast CCP showed that the indole ring of Trp-51 is in van der Waals contact with the heme, lying just 3.3 Å above it and almost parallel on the distal side. Substitution of a phenylalanine residue here should cause a minimal structural disturbance while replacing the indole ring with a much less readily oxidizable benzene ring. This expectation was confirmed by determining the crystal structure of the mutant enzyme. Since crystals of the latter have unit cell dimensions that are essentially the same as CCP(MI), we could compare the two structures quite accurately by means of a difference Fourier map (see figure 3).

The map suggests several small adjustments on the distal side of the heme. The phenylalanine ring now in position 51 occupies a position corresponding to the middle of the indole ring of Trp-51 in the wild-type enzyme. This placement pulls the alpha carbon of Phe-51 in the direction of the heme, which in turn forces the entire distal helix supporting Phe-51 0.25 Å toward the heme. A small adjustment of the distal histidine, His-52, is also consistent with this movement of the helix. His-52, moves toward the heme and its imidazole group twists slightly placing Nϵ2 0.30 Å closer to the peroxide binding site. Lastly, the side chain of Arg-48 moves 0.30 Å away from the sixth coordination site. Repositioning of Arg-48 has been noted in other CCP structures (Poulos et al., 1978; Edwards et al., 1984) that involve changes in the sixth ligand.

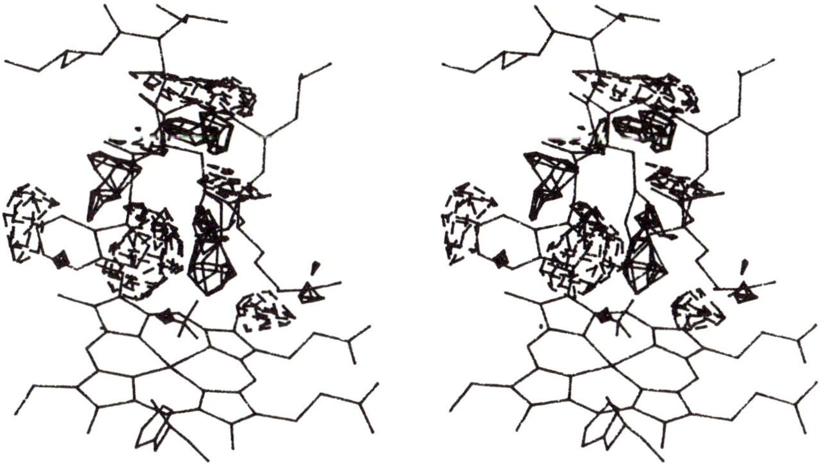

Figure 3. Difference density, CCP(MI,F51) minus CCP(MI), superimposed on selected residues on the distal side of the heme. Positive density is shown as solid lines, negative as broken lines.

The Phe-51 mutant enzyme is fully active, and the EPR spectrum of its peroxide-oxidized intermediate, corresponding to compound I, displays a strong, sharp resonance at g = 2.004 which is very similar to the signal observed for compound I of both bakers' yeast CCP and CCP(MI). However, UV-visible and EPR spectroscopy revealed that the half-life of CCP(MI,F51) compound I at 23 °C is only about 1% of that observed for compound I of CCP(MI) or bakers' yeast CCP. That is, its half-life is only minutes instead of hours. We conclude, therefore, that Trp-51 is not necessary for the formation of the free radical observed in compound I, but it does appear to exert a significant stabilizing influence.

Trp-191; a Link to the Radical Site?

Replacement of Trp-191 by a phenylalanine results in a protein with profoundly impaired catalytic ability. Steady-state assays of peroxide-dependent oxidation of cytochrome *c* reveal that the mutant, CCP(MI,F191), has only about 0.03% of the activity of the parent CCP(MI) enzyme. Adding a stoichiometric quantity of peroxide to the mutant clearly results in the formation of an oxyferryl heme, as judged by the appearance of the characteristic optical spectrum. Furthermore, this Fe(IV) species remains stable on a time scale comparable to that for CCP from bakers' yeast (Erman and Yonetani, 1975b).

Preliminary measurements of the CCP(MI,F191) Fe^{+2}/Fe^{+3} reduction potential show only a modest 10 mV decrease compared to the parent (Terence Meyers, personal communication). We expect this small difference to be negligible in affecting the electrochemistry of the higher oxidation states involved in the CCP catalytic mechanism. It appears therefore, that the mutant protein has been kinetically disabled as a result of the Trp-191 to Phe mutation. Further studies are in progress to determine which step or steps in the catalytic pathway have been affected, but in any event it is clear that Trp-191 is an important player in the kinetic mechanism of CCP.

We have also determined the structure of the Trp-191 to Phe mutant in the crystal. A difference Fourier map between the mutant and CCP(MI) shows clearly interpretable features around the 191 side chain and its neighbors (see figure 4).

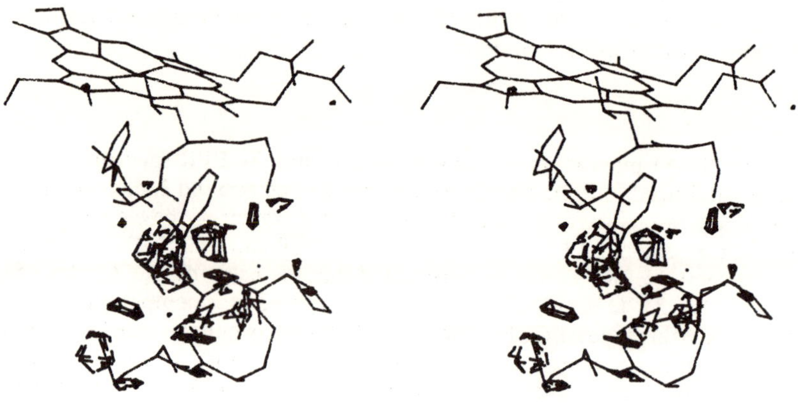

Figure 4. The mutant CCP(MI,F191) minus CCP(MI) difference density superimposed on residues on the proximal side of the heme.

In contrast to the Trp-51 to Phe mutation, an identical amino acid substitution at Trp-191 now results in the phenyl ring occupying precisely the site previously occupied by the six-membered ring portion of the Trp indole. Consequently the phenyl ring remains parallel to and in van der Waals contact with the imidazole ring of the proximal histidine, but owing to the smaller size of the phenyl ring as compared with the indole ring, the backbone chain segment containing Phe-191 has to move by about 1 Å. This perturbation is then transmitted to Met-230 by way of a hydrogen bond between the backbone carbonyl of residue 190 and the backbone amide of Met-230, which moves slightly to maintain this hydrogen bond. Perturbation of the 191-230 chain segment may be relevant to the mutation's effect on catalytic activity since the chain segment extends to the surface of the molecule near the binding site proposed for cytochrome *c* (Poulos and Kraut, 1980b).

A Role for His-181 in Electron Transfer?

An attractive feature of CCP as an object of study is that it has some obvious analogies with cytochrome oxidase. One is that CCP also receives electrons from cytochrome *c* via a one-to-one complex. Based on an observed complementarity between conserved lysine residues on the surface of cytochrome *c* and carboxylate groups around the heme crevice of CCP, a hypothetical model was presented for the complex that features a parallel arrangement of the two heme groups with an edge-to-edge separation of about 17 Å (Poulos and Kraut, 1980b; Poulos and Finzel, 1984). This large distance between the heme edges has led to speculation that electron transfer may occur through intervening pi-electron-containing side chains at the molecular interface. His-181 of CCP and the invariant Phe-87 of cytochrome *c* have been suggested as likely candidates for participation in such an electron transfer circuit, as both side-chain rings lie approximately parallel to the hemes and form a bridge with edge-to-edge distances no greater than 7 Å. Some experimental support for the proposed role of His-181 was provided by the observation that photooxidation of His-181 results in an 80% decrease in CCP activity (Bosshard et al., 1984). However, this result remains ambiguous owing to the simultaneous destruction of two additional tryptophan residues under the experimental conditions employed.

In order to more clearly define the function of His-181 in the electron transfer process, we have prepared a mutant of CCP(MI) in which His 181 is replaced by glycine. Initial characterization of the Gly-181 mutant shows that under standard assay conditions at pH 6.0 it is about 30% as active as CCP(MI). This suggests that while His-181 is not essential for electron transfer, it may play some more indirect role. Similar conclusions have been reached by others (Pielak et al., 1985) concerning Phe-87 of cytochrome *c*. We have also observed that the Gly-181 mutant is moderately unstable above pH 7, and that the spectral properties of the mutant are greatly altered between pH 6 and 7. These observations are consistent with the proposed role of His-181 in the stability of CCP at high pH (Bosshard et al., 1984).

Acknowledgements: This work was supported by NSF grant DMB-8511656.

REFERENCES

Anderson, DH, (1986) Ph D dissertation, Univ Cal San Diego.

Bosshard, HR, Banziger, J, Hasler, T, Poulos, TL (1984) *J Biol Chem. 259*: 5683-5690.

Chance, M, Powers, L, Poulos, T, Chance, B (1986) *Biochem 25*: 1266-1270.

Edwards, SL (1981) Ph D dissertation, Univ Cal San Diego.

Edwards, SL, Poulos, TL, Kraut, J (1984) *J Biol Chem 259*: 12984-12988.

Edwards, SL, Xuong, Ng-h, Hamlin, RC, Kraut, J (1987) *Biochem 26*: 1503-1511.

Erman, JE, Yonetani, T (1975a) *Biochim Biophys Acta 393*: 343-350.

Erman, JE, Yonetani, T (1975b) *Biochim Biophys Acta 393*: 350-357.

Finzel, BC, Poulos, TL, Kraut, J (1984) *J Biol Chem 259*: 13027-13036.

Fishel, LA, Villafranca, JE, Mauro, JM, Kraut, J (1987) *Biochem 26*: 351-360.

Goodin, DB, Mauk, AG, Smith, M (1986) *Proc Natl Acad Sci USA 83*: 1295-1299.

Hoffman, BM, Roberts, JE, Brown, TG, Kang, CH, Margoliash, E (1979) *Proc Natl Acad Sci USA 76*: 6132-6136.

Hoffman, BM, Roberts, JE, Kang, CH, Margoliash, E (1981) *J Biol Chem 256*: 6556-6564.

Kaput, J, Goltz, S, Blobel, G (1982) *J Biol Chem 257*: 15054-15058.

Maccecchini, M-L, Rudin, Y, Schatz, G (1979) *J Biol Chem 254*: 7468-7471.

Morgan, RS, Tatsch, CE, Gushard, RH, McAdon, RM , Warme, PK (1978) *Int J Pept Res 11*: 209-217.

Pielak, JP, Mauk, AG, Smith, M (1985) *Nature 313*: 152-154.

Poulos, TL, Finzel, BC (1984) *Pept Protein Rev 4*: 115-171.

Poulos, TL, Freer, ST, Alden, RA, Xuong, Ng-h, Edwards, SL, Hamlin, RC, Kraut, J (1978) *J Biol Chem 253*: 3730-3735.

Poulos, TL, Freer, ST, Alden, RA, Edwards, SL, Skogland, U, Takio, T, Eriksson, B, Xuong, Ng-h, Yonetani, T, Kraut, J (1980) *J Biol Chem 255*: 575-580.

Poulos, TL, Kraut, J (1980a) *J Biol Chem 255*: 8199-8205.

Poulos, TL, Kraut, J (1980b) *J Biol Chem 255*: 10322-10330

Villafranca, JE, Howell, EE, Voet, DH, Strobel, MS, Ogden, RC, Abelson, JN, Kraut, J (1983) *Science 222*: 782-788.

Yonetani, T, Schleyer, H, Ehrenberg, A (1966) *J Biol Chem 241*: 3240-3243.

Yonetani, T, Ray, GS (1965) *J Biol Chem 240*: 4503-4508.

Xuong, Ng-h, Sullivan, D, Nielsen, C, Hamlin, RC (1985) *Acta Cryst B41*: 267-269.

DISCUSSION

DUNFORD: Do you feel it is possible to have several sites of approximately the same ability to donate an electron for compound I formation and that if one site is eliminated by mutagenesis another can take its place?

EDWARDS: Yes, this has to be considered a strong possibility based on a number of experimental results. Prominent among these results is the fact that all single site mutations I am aware of have not eliminated the strong g = 2 EPR signal. Some of the mutants show subtle differences in the shape of the EPR signal at low temperature (89 K). The significance of this has been touched upon by Dr. Yonetani in his presentation.

GARBER: Is anything known about the stability and rate of formation of compound I of the trp-191 to phe-191 mutant?

EDWARDS: These properties are being investigated now. The rate of formation of phe-191 compound I is being studied by James Erman at Northern Illinois University using stop flow. Preliminary results indicate that it forms readily. The stability of this compound I appears to be like that of the phe-51 mutant, with a half-life of a few minutes.

Oxidases and Related Redox Systems, pages 477–493

STRUCTURE AND FUNCTION OF VANADIUM-CONTAINING BROMO-PEROXIDASES

R. Wever, B.E. Krenn, E. De Boer,
H. Offenberg and H. Plat

Laboratory of Biochemistry, University of Amsterdam, P.O. Box 20151, 1000 HD Amsterdam, and Biotechnological Centre, University of Amsterdam, P.O. Box 20151, 1000 HD Amsterdam (The Netherlands)

ABSTRACT

The properties of the vanadium-containing bromoperoxidases from the seaweeds Ascophyllum nodosum, Laminaria saccharina and the lichen Xanthoria parietina were studied.

Upon reduction with sodium dithionite, these bromoperoxidases show EPR spectra which are typical of a vanadyl cation (VO^{2+}). From the spectral parameters and a comparison with inorganic vanadyl complexes, we conclude that the ligand environment largely consists of oxygen donors. The data also show that the structure of the active sites in these enzymes is very similar. Since EPR spectra of vanadium(IV) bromoperoxidase are only obtained after reduction, the metal ion is present in the native enzymes in the 5+ oxidation state.

All these enzymes loose their enzymic activity upon dialysis against citrate-phosphate (PO_4^{3-}) buffer at pH 3.8, containing EDTA. The brominating activity could be reconstituted by the addition of vanadate (VO_4^{3-}). The experiments suggest that vanadate is incorporated into these enzymes. In line with the EPR data, we propose a structure of the active site in which at least 4 oxygen atoms are present as donors for the central vanadium(V) ion.

Since several inorganic peroxovanadium(V) complexes have been described, we suggest that the vanadium ion in bromoperoxidases serves as a binding site for H_2O_2. Upon subsequent binding of bromide this ion is oxidized by the peroxo-intermediate to form hypobromite. This model does not require valence state changes of the metal ion itself and indeed no changes in the EPR spectrum of reduced bromoperoxidase are observed upon addition of H_2O_2 or Br^-. Further, bromoperoxidase reduced with a small excess of sodium dithionite is not active in the bromination reaction.

The bromoperoxidases from the various sources show similarity in the amino-acid composition with a predominance of acidic amino acids. Distinct pH optima are observed in the bromination reaction catalysed by the bromoperoxidases. Despite the presence of the same prosthetic group in these enzymes with comparable vanadium ligand-field environment, the enzymic properties are very different. The specific activity as well as the $\underline{K_m}$ for bromide differ greatly. Unlike the enzymes from the seaweeds A. nodosum and L. saccharina the bromoperoxidase from the lichen X. parietina is inhibited by low concentrations (1-5 mM) of nitrate. These bromoperoxidases have a remarkable resistance towards organic solvents such as methanol, ethanol and propanol.

INTRODUCTION

The presence of vanadium in plants, mushrooms and animal tissue has been recognized for some time (for reviews see Chasteen, 1983; Lancashire, 1980; Kustin et al., 1983). Also the existence of vanadium in petroleum residuals is well documented (Dickson et al., 1972).

Vanadium has distinct effects on the physiological action of living cells and animals (Nechay et al., 1986). These effects are due to the parallelism in chemistry of orthovanadate (VO_4^{3-}) and orthophosphate (PO_4^{3-}) and thus vanadium interferes with the action of most phophatases, ATP-ases and nucleases (Chasteen, 1983). Whether vanadium is an essential element in terrestrial animals is not established (Nechay et al., 1986) and its physiological function remains elusive (Macara, 1980). However, very high concentrations of vanadium (up to 0.15 M) can be found (Dingley, 1981) in sea squirts or tunicates which are small marine animals. This has been a mystery since its discovery in 1911 (Henze, 1911) and rather detailed studies have been carried out concerning the accumulation of this transition metal in the blood cells of these marine organisms (Kustin et al., 1983). However, the question of the accumulation of vanadium in the tunicate physiology is still unanswered, although there is some evidence that vanadium may function as an antifeedant (Kustin et al., 1983).

Vanadium is an essential element for some marine macroalgae, like the brown seaweed Fucus spiralis and the green seaweed Enteromorpha compressa (Fries, 1982). Addition of vanadate, which is normally present in seawater at a concentration of $5 \cdot 10^{-8}$ M, to marine algae which were cultured in artificial seawater, enhanced the growth yield considerably. Brown, green, as well as red algae also contain vanadium in contents varying from 0.3-10 $\mu g \cdot g^{-1}$ dry weight (Yamamoto et al., 1970). This suggests some function of the element in seaweeds. Indeed, it was discovered (Vilter, 1984) that the brown seaweed Ascophyllum nodosum contained a vanadium-dependent bromoperoxidase. This peroxidase could be inactivated by dialysis at low pH against citrate-phosphate buffers containing EDTA. This was confirmed by us and we were able to demonstrate (Wever et al., 1985a,b; De Boer et al., 1986a,b; Krenn et al., 1987) that a whole range of purified bromoperoxidases from red and brown seaweeds contained vanadium as a prosthetic group.

These enzymes catalyse the bromination of many organic compounds according to the following general equation:

$$A\text{-}H + H_2O_2 + Br^- + H^+ \longrightarrow A - Br + 2H_2O$$

where A-H is a nucleophilic reagent.

Recently it was discovered (Plat et al., 1987) that vanadium-containing bromoperoxidases do not only occur in the marine environment, but are also found in the terrestrial organism Xanthoria parietina. This is a lichen that consists of the green alga Trebouxia and an Ascomycete and grows on walls and stones. That vanadium has a role in terrestrial lower organisms was also demonstrated recently (Robson et al., 1986; Hales et al., 1986). It was shown that instead of molybdenum which is present in the conventional enzyme, the alternative nitrogenases from Azotobacter vinelandii and A. chroococcum contain vanadium (for a review see Wever et al., 1987).

It is well documented (Siuda and DeBernardis, 1973; Neidleman and Geigert, 1986) that seaweeds produce a fascinating diversity of rather complex halogenated compounds. Since many of these exhibit antibiotic properties it is generally considered that these substances are produced as part of the defence system of the organisms. High amounts of rather simple, volatile halogenated compounds like CH_2Br_2, $CHBr_2Cl$ and $CHBr_3$ are also released by the marine macroalgae like the brown algae, A. nodosum, Fucus vesiculosus and the green algae Enteromorpha linza and Ulva lactua (Geschwend et al., 1985).

It has been estimated that the global production of organobromides by these macroalgae is 10^4 tons per year as compared to 5×10^4 tons per year from industrial sources. Thus the biological contribution to atmospheric pollution by volatile halogenated compounds is substantial. The biosynthetic pathways by which these halogenated compounds are produced, are still elusive. Bromoperoxidases seem to be involved since it has been demonstrated (Theiler et al., 1978; Beissner et al., 1981) that the enzymes from red and green seaweeds, respectively, acting on 3-keto octanoic acid yielded a number of brominated compounds including bromoform.

That nature has choosen the early transition metal vanadium as a prosthetic group in bromoperoxidases can be understood. As pointed out originally by Djordjevic (1982), the electron rich peroxide group will form stable complexes with metal ions of low d^n electronic configuration. She actually suggested in 1982 the possible interaction in bioprocesses of the electron rich hydrogen peroxide group and metal ions that give rise to d^o or low d^n ions like molybdenum and vanadium. As is clear from the foregoing this idea has been proven to be correct. That inorganic vanadium(V) peroxo complexes are effective oxidants is also well documented (Mimoun et al., 1983).

In this contribution, data will be presented on the mechanism of action of these novel vanadium containing bromoperoxidases and their properties will be compared.

RESULTS AND DISCUSSION

Structure of the active site

Fig. 1 shows the EPR spectra of bromoperoxidases from *A. nodosum* and *L. saccharina* reduced with sodium dithionite. The EPR spectra are typical of a vanadyl cation (VO^{2+}) showing apparent axial-symmetrical EPR signals with two sets of 8 hyperfine lines (Chasteen, 1981).

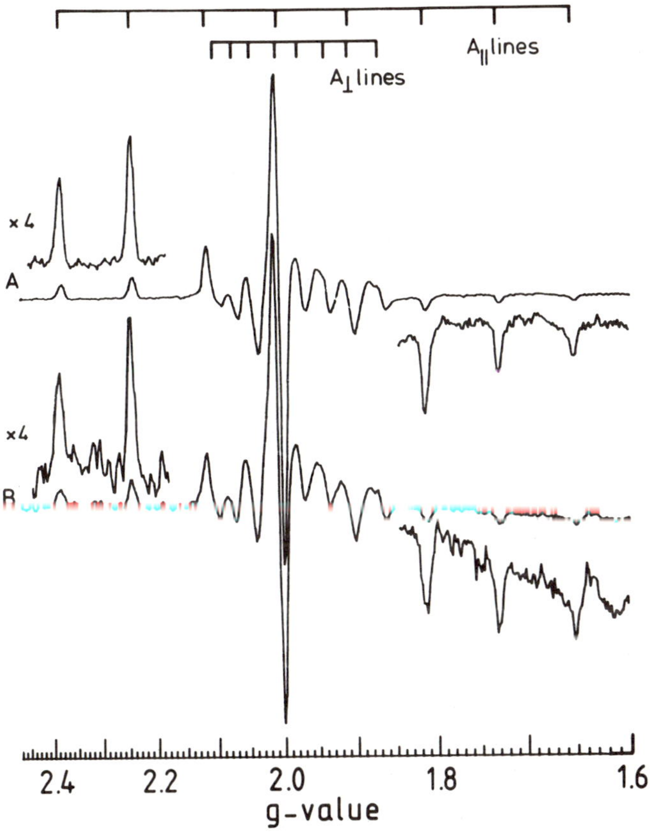

Figure 1. EPR spectra of bromoperoxidases reduced with a small amount of sodium dithionite. Trace A, bromoperoxidase from *A. nodosum*; trace B, bromoperoxidase from *L. saccharina*. Instrument settings: microwave frequency, 9.2 GHz; microwave power A and B, 10 dB and 20 dB, respectively; modulation width, 1.0 mT; temperature, 50 K.

The values for the isotropic hyperfine-coupling constants (A_o) and g_o (Table 1) are consistent with a ligand field mainly consisting of nitrogen and/or oxygen donor atoms (Boucher et al., 1969).

Table 1

EPR data of reduced bromoperoxidase

Species	$g_\parallel$	$g_\perp$	g_o	$A_\parallel$	$A_\perp$	A_o
A. nodosum	1.948	1.979	1.969	17.6	5.5	9.5
L. saccharina	1.948	1.979	1.969	17.6	5.7	9.7
X. parietina	1.945	1.979	1.967	17.7	5.8	9.8

Data obtained of reduced bromoperoxidase in 0.1 M Tris-SO_4 (pH 8.3). A in mT.

Table 1 also includes the EPR data on the bromoperoxidase from the lichen *X. parietina*. It is obvious that the structure of the active sites in these enzymes is very similar. Since as isolated, these bromoperoxidases do not show an EPR spectrum, but only after reduction, it is concluded that the metal ion is present in the native enzymes in the 5+ oxidized state.

It has been demonstrated (Vilter, 1984; Wever et al., 1985a; De Boer et al., 1986a) that these bromoperoxidases loose their enzymic activity when they are dialyzed against citrate-phosphate (PO_4^{3-}) buffer at pH 3.8 containing EDTA. The brominating activity could be reconstituted by the addition of vanadate (VO_4^{3-}) but was inhibited when phosphate, a structural analogue of vanadate was present (De Boer et al., 1986a). Since it is very unlikely that upon reincorporation of vanadate in the active site of the bromoperoxidases, vanadium-oxygen bonds are broken, we suggest that in the native enzymes at least 4 oxygen atoms are present as donors for the central vanadium(V) ion. This proposal is supported by recent ^{51}V NMR results on the enzyme from *A. nodosum* which suggest coordination of V(V) to 6-8 oxygen functions (Vilter and Rehder, 1987; Rehder et al., 1987). There is no evidence that during the turnover redox changes occur in the transition metal ion. When native bromoperoxidase is reduced with a slight excess of dithionite (2 mM) under anaerobic conditions and excess dithionite was removed by centrifugation through a column of Sephadex G-25, the enzyme lost its brominating activity. Upon aeration of the sample, the bromoperoxidase activity returned slowly ($t_{1/2}$ is 2 hours). Similarly when H_2O_2 was added to dithionite-free reduced bromoperoxidase there is no effect on the vanadium(IV) EPR signal (De Boer et al., 1987c). Also addition of bromide had no effect on this signal. It was already demonstrated (De Boer et al., 1986a) that hydrogen peroxide and bromide were unable to reduce native bromoperoxidase.

Steady-state kinetic studies (Wever et al., 1985a) do suggest a sequential type of mechanism and together with the above support the model given below.

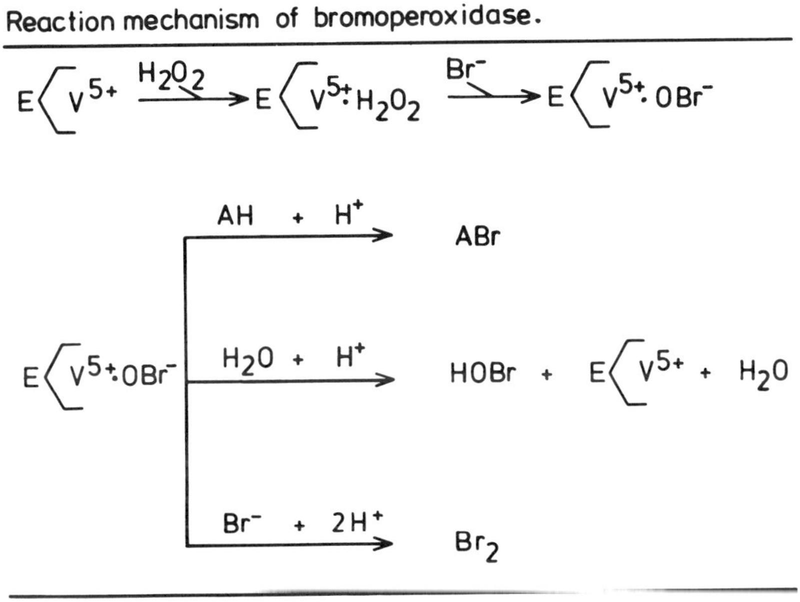

Figure 2. Reaction mechanism of bromoperoxidase from *A. nodosum*.

In this model the vanadium ion which is according to the EPR data in the 5+ oxidation state, only serves as a binding site for hydrogen peroxide. Upon subsequent binding of bromide, this halide is oxidized by the vanadium peroxo-intermediate to form an enzyme-halogenating intermediate. This enzyme intermediate may either react with a nucleophilic acceptor AH to yield the brominated compound ABr or hydrolyze to hypobromous acid. Alternatively, the intermediate may react with another Br^- ion to yield bromine

(Br_2). Such a model is supported by the well-known property of inorganic vanadium(V) compounds to form stable peroxo-vanadium(V) complexes with hydrogen peroxide (Mimoun et al., 1983). It has even been reported (Secco, 1980) that some of these peroxovanadium(V) complexes bind iodide at the metal centre with subsequent oxidation of iodide to form hypoiodite.

Amino acid composition

Table 2

Amino acid composition of bromoperoxidases.
The amino acids were determined after hydrolysis of bromo-peroxidase in 6N HCl at 110 °C in vacuo. The analyses were carried out at 24, 48 and 72 h. The values in the table were corrected for losses during hydrolysis.

Amino acid	molar percent		
	A. nodosum	*L. saccharina*	*C. pilulifera*[c]
Asx	14.0	12.9	11.4
Thr	6.9	6.4	4.9
Ser	5.0	5.8	6.7
Glx	10.2	9.1	10.2
Pro	4.9	5.2	5.3
Gly	9.8	8.9	7.8
Ala	11.1	8.0	10.0
Val	6.2	5.5	6.5
Met	1.0	2.0	1.4
Ile	4.0	2.7	5.9
Leu	8.5	9.4	9.2
Tyr	1.7	3.1	2.2
Phe	5.7	3.5	5.7
Lys	1.6	3.7	4.3
His	1.2	1.7	1.4
Arg	4.0	4.7	5.1
Cys[a]	1.7	5.5	0.8
Trp	n.d.	n.d.	0.8
Polarity[b](%)	42.9	44.3	44.0

[a] Cysteine was determined as cysteic acid after performic acid oxidation.
[b] The polarity (%) was taken to be the sum of the molar percent values for Asx, Thr, Ser, Glx, Lys, His and Arg.
[c] Itoh et al., 1986

The amino acid composition of the bromoperoxidases from the brown algae A. nodosum and L. saccharina is shown in Table 2. A comparison was also made with the bromoperoxidase from the red seaweed Corallina pilulifera. It appears to be a non-heme type bromoperoxidase (Itoh et al., 1986), but does not contain vanadium. There is reasonably close agreement between these enzymes, with a predominance of glutamic acid and aspartic acid.

Stability

As demonstrated by us (Wever et al., 1985a; De Boer et al., 1987a; De Boer et al., 1987b) the vanadium containing bromoperoxidase from the marine brown alga A. nodosum is an oxidoreductase with high operational stability in aqueous and organic media. The enzyme from the terrestrial organism X. parietina also exhibits a remarkable thermostability (Plat et al., 1987) which is probably related to the habitat of this organism. It was therefore of interest to see whether this bromoperoxidase was also resistant towards organic solvents. As Table 3 shows this enzyme is only slowly denatured by high concentrations of alcohol.

Table 3

The effect of alcohols on the brominating activity of bromoperoxidase from X. parietina

	methanol (50%)	ethanol (50%)	propanol (50%)
days		rate of bromination	
0	100%	100%	100%
1	74%	101%	111%
8	67%	71%	69%
18	44%	49%	30%

Bromoperoxidase (0.02 mg/ml) was stored in Tris-sulphate (pH 8.3) and 50% (v/v) methanol, ethanol, or propanol at 20 °C. During storage, enzymic acitivity was measured using the monochlorodimedone bromination assay as described in the legend of Fig. 2.

Enzymic activity and some properties

Like most haloperoxidases the vanadium containing bromoperoxidases showed sharp pH optima between pH 5.0 and 8.0 (Wever et al., 1985a; De Boer et al., 1986a; De Boer et al., 1986b; Krenn et al., 1987; Plat et al., 1987). Fig. 3A and B show Lineweaver-Burk plots at various bromide concentrations and at a fixed H_2O_2 concentration for the bromination of monochlorodimedone catalyzed by the bromoperoxidase from A. nodosum and from the lichen X. parietina. It is obvious that the K_m for bromide (0.02 mM) of the lichen enzyme is about 600 times smaller than that for the algal enzyme (12.7 mM) and is the smallest value ever reported for a bromoperoxidase. This high affinity for bromide may be related to the habitat of this organism which grows on rocks and stones. The essential elements for this

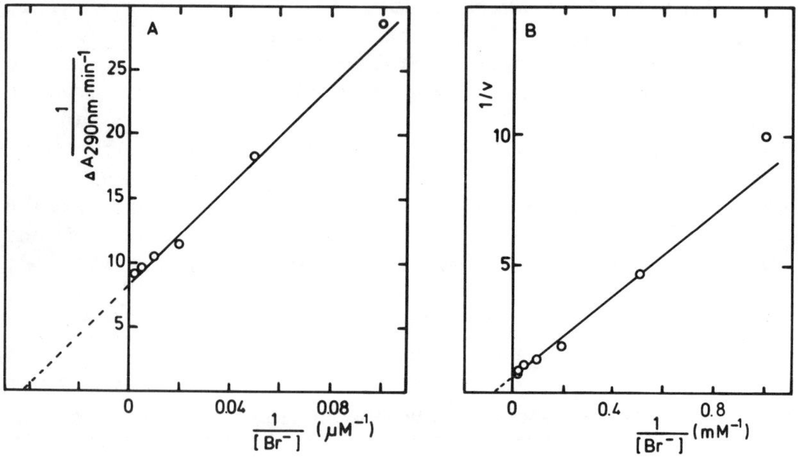

Figure 3A. Lineweaver-Burk plot of the bromination of monochlorodimedone catalyzed by the bromoperoxidase from X. parietina. Experimental conditions: 50 μM monochlorodimedone, 1 mM H_2O_2, 100 mM sodium acetate (pH 5.5), 57 ng bromoperoxidase per ml.
Figure 3B. Lineweaver-Burk plot of the bromination of monochlorodimedone catalyzed by the bromoperoxidase from A. nodosum. Experimental conditions: 50 μM monochlorodimedone, 2 mM H_2O_2, 100 mM potassium phosphate (pH 6.5), 0.6 μg bromoperoxidase per ml.

lichen apparently must come from the rocks themselves or from airborn particles. The K_m values for bromide for both enzymes were not affected by pH in the pH range studied.

Bromide is not only a substrate for the bromoperoxidase from the lichen, it is also an inhibitor at higher concentrations. When the activity was measured in the presence of 40 mM of halide ion, the enzymic activity was greatly inhibited. At these concentrations the bromoperoxidase from A. nodosum is not inhibited (Wever et al., 1985a). As was also observed for this enzyme, the K_m for H_2O_2 was affected by pH and increased considerably at lower pH values. Surprisingly, in contrast to the other vanadium bromoperoxidases the lichen enzyme was strongly inhibited by low concentrations (1-5 mM) of nitrate (NO_3^-). Both maximal rate and the K_m for H_2O_2 were affected.

Table 4

Properties of vanadium-containing bromoperoxidases

Species		pH_{opt}	$K_m^{Br^-}$ (mM)	Molecular mass (kDa) subunits[a]	Activity[b]
A. nodosum	(Wever et al., 1985a)	6.0	12.7	67	120
L. saccharina	(De Boer et al., 1986b)	6.5	1.0	66,64	600
C. rubrum	(Krenn et al., 1987)	7.4	2.0	n.d.	6
X. parietina	(Plat et al., 1987)	5.5	0.02	65	79

[a] Molecular mass was determined by SDS gelelectrophoresis under strongly denaturating conditions (100 °C in 1% SDS, 5% β-mercaptoethanol).

[b] Units per mg protein.

Table 4 gives a summary of these and other data of the various vanadium bromoperoxidases.

Mechanism of bromination

Although a number of studies have appeared on the halogenation of organic compounds by chloroperoxidase, the mechanism is still open to debate (Neidleman and Geigert, 1986; Dunford et al., 1987). Limited knowledge is available concerning the halogenation carried out by other haloperoxidases. Recently Itoh et al. reported (1987) on the bromination of aromatic heterocycles by the non-heme bromoperoxidase from the alga Corallina pilulifera. Comparison of the reactivity of the heme-containing chloroperoxidase from the fungus Caldariomyces fumago and the algal bromoperoxidase suggested that bromination catalyzed by the latter enzyme occurred at the active site of the peroxidase. In contrast, halogenation by the chloroperoxidase was found to involve the formation of molecular halogen and its release into solution (Libby et al., 1982). However, the reported differences may be due to differences in experimental conditions. The studies carried out with the algal bromoperoxidase were done at pH 6.5, whereas the studies on chloroperoxidase were performed at pH 3.0, near the pH optimum of the enzyme. As shown (Kanofsky, 1986) the rate of the reaction between HOBr and H_2O_2 at neutral pH is very high, whereas it is low at acid pH. This is the main reason why in the presence of H_2O_2, Br_3^- can only be detected at low pH.

The mechanism of halogenation by the vanadium-containing bromoperoxidase may very well differ from that of the non-heme algal peroxidase and chloroperoxidase. For example the bromoperoxidase from A. nodosum is unable to oxidize the classical peroxidase substrates (De Boer et al., 1987a). For this reason, the possible stereo- or regioselectivity of the vanadium bromoperoxidase was investigated by us (Franssen et al., 1987).

The enzyme was able to brominate barbituric acid, 1-methyl and 1,3-dimethyl barbituric acid, 5-phenyl and 4-chloro barbituric acid into the corresponding 5-bromo- or 5,5-dibromo derivatives. The structure of the examined substrates had little or no effect on the rate of bromination (Franssen et al., 1987). Thus the vanadium bromoperoxidase had no specificity regarding the structure of the organic substrates and suggests that formation of a rate-limiting ternary complex between a halogenating enzyme intermediate and an organic substrate according to Fig. 2 does not occur. It is very likely therefore that this vanadium enzyme, like chloroperoxidase from C. fumago produces free HOBr, which aspecifically reacts with organic substrates.

AKNOWLEDGEMENTS

We gratefully acknowledge the support by DSM NV, Geleen, The Netherlands. This work is part of the research programme of the Netherlands Foundation for Chemical Research (SON) and was made possible by financial support from the Netherlands Technology Foundation (STW).

REFERENCES

Beissner RS, Guilford WJ, Coates RM, Hager LP (1981). Synthesis of brominated heptones and bromoform by a bromoperoxidase of marine origin. Biochemistry 20: 3724-3731.

Boucher LJ, Tynan EC, Yenn TF (1969) "Spectral Properties of Oxovanadium(IV) Complexes. IV Correlation of ESR Spectra with Ligand Type". In Yenn TF (ed): "Electron Spin Resonance of Metal Complexes" New York: Plenum Press, pp. 111-130.

Chasteen ND (1983). The biochemistry of vanadium. Structure and Bonding, 53: 105-138.

Chasteen ND (1981) Vanadyl(IV) EPR Spin Probes, Inorganic and Biochemical aspects. In Berliner LJ, Reuben J (eds): "Biological Magnetic Resonance" New York: Plenum Press vol 3: 53-119

De Boer E, van Kooyk Y, Tromp MGM, Plat H, Wever R (1986a). Bromoperoxidase from Ascophyllum nodosum: a novel class of enzymes containing vanadium as a prosthetic group. Biochim Biophys Acta 869: 48-53.

De Boer E, Plat H, Wever R (1987a) Algal Vanadium(V)-Bromoperoxidase, a halogenating enzyme retaining full activity in apolar solvent systems. in Laane C, Tramper J, Lilly MD (eds): "Biocatalysis in Organic Media" Amsterdam: Elsevier pp. 317-322.

De Boer E, Plat H, Tromp MGM, Franssen MCR, Van der Plas MC, Meijer EM, Schoemaker HE, Wever R (1987b). Vanadium containing bromoperoxidase: an example of an oxidoreductase with high operational stability in Aqueous and Organic Media. Biotech Bioeng 30: 607-610.

De Boer E, Tromp MGM, Plat H, Krenn GE, Wever R (1986b). Vanadium(V) as an essential element for haloperoxidase activity in marine brown algae: purification and characterization of a vanadium(V)-containing bromoperoxidase from Laminaria saccharina. Biochim Biophys Acta 872: 104-115.

De Boer E, Boon K, Wever R (1987c). Electron paramagnetic resonance studies on conformational states and metal-ion exchange properties of vanadium-bromoperoxidase. Biochemistry, in press.

Dickson FE, Kunesh CJ, McGinnis, EL, Petrakis L (1972). Use of Electron Spin Resonance to characterize the vanadium(IV)-sulfur species in petroleum. Anal Chem 44: 978-981.

Dingley AL, Kustin K, Macara IG, McLeod GC (1981). Accumulation of vanadium by tunicate blood cells occurs via a specific anion transport system. Biochim Biophys Acta 649: 493-502.

Djordjevic C (1982). Heteroligand peroxo transition metal complexes. Chemistry in Britain 18: 554-557.

Dunford HB, Lambeir AM, Kashem MA, Pickard MA (1987). On the mechanism of chlorination by chloroperoxidase. Arch Biochem Biophys 252: 292-302.

Franssen MCR, Jansma JD, Van Der Plas HC, De Boer E, Wever R (1987). Enzymatic bromination of barbituric acid and some of its derivatives. Bioorganic Chemistry, submitted.

Fries L (1982). Vanadium an essential element for some marine macroalgae. Planta 154: 393-396.

Gschwend PM, MacFarlane JK, Newman KA (1985). Volatile halogenated organic compounds released to seawater from temperate marine macroalgae. Science 227: 1033-1035.

Hales BJ, Case EE, Morningstar JE, Dzeda MF, Mauterer LA (1986). Isolation of a new vanadium-containing nitrogenase from Azotobacter vinelandii. Biochemistry 25: 7251-7255.

Henze M (1911). Untersuchungen über das Blut der Ascidien. Die Vanadiumverbindung der Blutkörperchen. Hoppe-Seyler's Z Physiol Chem 72: 494-501

Itoh N, Izumi Y, Yamada H (1986). Characterization of nonheme type bromoperoxidase in Corallina pilulifera. J Biol Chem 261: 5194-5200.

Itoh N, Izumi Y, Yamada H (1987) Haloperoxidase-catalyzed halogenation of nitrogen-containing aromatic heterocycles represented by nucleic bases. Biochemistry 26: 282-289.

Kanofsky JR (1986). Catalysis of singlet oxygen production in the reaction of hydrogen peroxide and hypochlorous acid by 1,4-Diazabicyclo[2.2.2]octane (DABCO). Biochem Biophys Res Commun 134: 777-782.

Krenn BE, Plat H, Wever R (1987). The bromoperoxidase from the red alga Ceramium rubrum also contains vanadium as a prosthetic group. Biochim Biophys Acta 912: 287-291.

Kustin K, McLeod GC, Gilbert TR, Briggs Le Baron R 4th (1983). Vanadium and other metal ions in the physiological ecology of marine organisms. Structure and Bonding 53: 139-160.

Lancashire RJ (1980). The sacred mushroom. Education in Chem 17: 38-43.

Libby RD, Thomas JA, Kaiser LW, Hager LP (1982). Chloroperoxidase halogenation reactions: Chemical versus enzymic halogenating intermediates. J Biol Chem 257: 5030-5037.

Macara IG (1980). Vanadium - an element in search of a role. Trends Biochem Sci 5: 92-94.

Mimoun H, Sausinne L, Daire E, Postel M, Fischer J, Weiss R (1983). Vanadium(V) peroxo complexes. New versatile biomimetic reagents for epoxidation of olefins and hydroxylation of alkanen and aromatic hydrocarbons. J Am Chem Soc 105: 3101-3110.

Nechay BD, Nanninga LB, Nechay PSE, Post RL, Grantham JJ, Macara IG, Kubena LF, Phillips TD, Nielsen FH (1986). Role of vanadium in biology. Fed Proc 45: 123-132.

Neidleman SL, Geigert J (1986) "Biohalogenation: Principles, Basic Roles and Applications". Chichester: Ellis Horwood.

Plat H, Krenn BE, Wever R (1987). The bromoperoxidase from the lichen Xanthoria parientina is a novel vanadium enzyme. Biochem J, in press.

Rehder D, Vilter H, Duch A, Priebsch W, Weideman C (1987). A vanadate(V)-dependent peroxidase from pigweed (Ascophyllum nodosum): The ^{51}V NMR study of an unusual enzyme and simple vanadate-peptide systems. Recueil Trav Chim Pays-Bas p. 408.

Robson RL, Eady RR, Richardson TH, Miller RW, Hawkins M, Postgate JR (1986). The alternative nitrogenase of Azotobacter chroococcum is a vanadium enzyme. Nature (London) 322: 388-390.

Secco F (1980). Equilibria and kinetics of reduction by iodide ion of the vanadium(V)-hydrogen peroxide system. Inorg Chem 19: 2722-2725.

Siuda JF, DeBernardis JF (1973). Naturally occuring halogenated organic compounds. Lloydia 36: 107-143.

Theiler R, Cok JC, Hager LP, Siuda JF (1978). Halohydrocarbon synthesis by bromoperoxidase. Science 202: 1094-1096.

Vilter H (1984). Peroxidases from Phaeophyceae: a vanadium(V)-dependent peroxidase from Ascophyllum nodosum. Phytochem 23: 1387-1390.

Vilter H, Rehder D (1987). ^{51}V NMR Investigation of a vanadate(V)-dependent peroxidase from Ascophyllum nodosum. Inorg Chim Acta 136: L7-L10.

Wever R, De Boer E, Plat H, Krenn BE (1987). Vanadium - an element involved in the biosynthesis of halogenated compounds and nitrogen fixation. FEBS Lett 216: 1-3.

Wever R, Plat H, de Boer E (1985a). Isolation procedure and some properties of the bromoperoxidase from the seaweed *Ascophyllum nodosum*. Biochim Biophys Acta 830: 181-186.

Wever R, Plat H, de Boer E (1985b). Bromoperoxidases from seaweed: a novel class of enzymes containing vanadium? Rev Port Quim 27: 169-170.

Yamamoto T, Ishibashi M, Fujita T (1970). Chemical studies on the seaweeds (25). Vanadium and titanium contents in seaweeds. Rec Oceanogr Works Jpn 10: 125-135.

DISCUSSION

ESTABROOK: These are very interesting enzymes. Have antibodies been prepared and used to screen cDNA libraries for the eventual expression of the enzyme in bacteria such as *E. coli*?

WEVER: Yes, we have monoclonal antibodies against the bromoperoxidase from *Ascophyllum nodosum*. At present there is no need to express the enzyme in bacteria. Our isolation procedure yields 100 mg of protein per 10 kilograms of seaweed and this is sufficient for most purposes.

TOUSSAINT: Using bromoperoxidase is it possible to replace the bromide ion by another halide, let's say chloride or iodide, and get the same type of reactivity?

WEVER: Chloride is not oxidized by these vanadium containing bromoperoxidases. However, as can be expected, iodide is used as a substrate by them.

ANDERSSON: Do you have any information concerning the identity of the two ligands from the protein to vanadium?

WEVER: No. However, Rehder et al. (references 30,31) have done ^{51}V NMR on the enzyme from *A. nodosum*. They conclude from the large shift that six to eight oxygen atoms are around the central vanadium atom.

SMITH: Since the vanadium can be removed by dialysis, do you think this means that it is loosely bound?

WEVER: No, in fact, the vanadium is strongly bound to the enzyme. For example, you can pass it along any gel filtration column without loss of activity. Although the dissociation constant has not been exactly determined by us, we think it is less than 100 nM.

MASON: A bromoperoxidase from *Corallina pilulifera* has just been described by Itoh N, Izumi Y, Yamada, H (1987) J Biol Chem 262: 11982-11987, which has ferric non-heme iron tightly bound to cysteine, histidine, or tyrosine residues, as prosthetic group. Thus, non-heme iron, heme and vanadium all appear to act as prosthetic groups of bromoperoxidases. Would you care to comment on possible chemical or biological meanings of this diversity?

WEVER: In my opinion, it has not yet been settled that this enzyme does indeed contain iron which is essential for activity. It is obvious from the quoted paper that iron is present in less than stoichiometric amounts and further there appears to be no correlation between iron content and specific activity of the isolated enzyme. However, a novel bromoperoxidase has been discovered in *Steptomyces aureofaciens* [Van Pee KH, Sury G, Lingens F, (1987) Biol Chem Hoppe-Seyler 368: 1225-1232; Krenn BE, Plat H, Wever R (1988) Biochim Biophys Acta, in press] which contain neither heme nor vanadium. Obviously in evolution with the coming of dioxygen into the atmosphere, various metal ions entered biological systems. These enzymes apparently evolved independently but do have similar function. The question why, cannot be answered but cf. for a discussion, Williams RJP (1986) Chemica Scripta 26: 513-523.

Session VIII: Cytochrome P-450

Chairwoman: Lucile Smith

Oxidases and Related Redox Systems, pages 497–507

ROLE OF CYTOCHROME P-450 IN HYDROCARBON FORMATION FROM XENOBIOTIC AND LIPID HYDROPEROXIDES*

Minor J. Coon and Alfin D.N. Vaz

Department of Biological Chemistry, Medical School, The University of Michigan, Ann Arbor, MI 48109-0606

SUMMARY

Cytochrome P-450 has recently been found to catalyze the reductive cleavage of both xenobiotic hydroperoxides and biologically occurring lipid hydroperoxides. In a reconstituted enzyme system containing phenobarbital-inducible P-450 form 2, NADPH-cytochrome P-450 reductase, and NADPH, the corresponding alcohols are not produced, but instead carbonyl compounds and hydrocarbons. In the following equation for the reaction, X represents any of a variety of alkyl groups and R and R' are either hydrogen or alkyl groups, of which only methyl has been studied so far:

$$XRR'C\text{-}OOH + NADPH + H^{+} \rightarrow XRCO + R'H + H_2O + NADP^{+}$$

The products derived from cumyl, α-methylbenzyl, benzyl, and *t*-butyl hydroperoxides as well as 13-hydroperoxy-9,11-octadecadienoic acid indicate that the reaction involves stepwise one-electron transfer, resulting in homolysis of the peroxide oxygen-oxygen bond. The alkoxy radical thus generated undergoes β-scission to yield the carbonyl compound and an alkyl radical that is reduced to the alkane. As predicted by this radical mechanism, when the reductive cleavage of cumyl hydroperoxide was carried out in D_2O, deuteromethane was formed.

---

*This research was supported by Grant DK-10339 from the National Institutes of Health.

INTRODUCTION

As reviewed elsewhere (1-3) and illustrated in the accompanying scheme (Fig. 1), cytochrome P-450 is able to promote the monooxygenation of an organic substrate at the expense of molecular oxygen and a reductant such as NADPH, or at the expense of a peroxy compound in the absence of an external reductant. The latter reaction is of mechanistic interest because it may be studied in the absence of oxygen, a reductase capable of transferring electrons to the cytochrome, and NADPH, and is similar to and yet different from peroxidatic reactions. A difference has recently been noted between peroxide-based oxidants and their iodine counterparts. Specifically, with peroxides

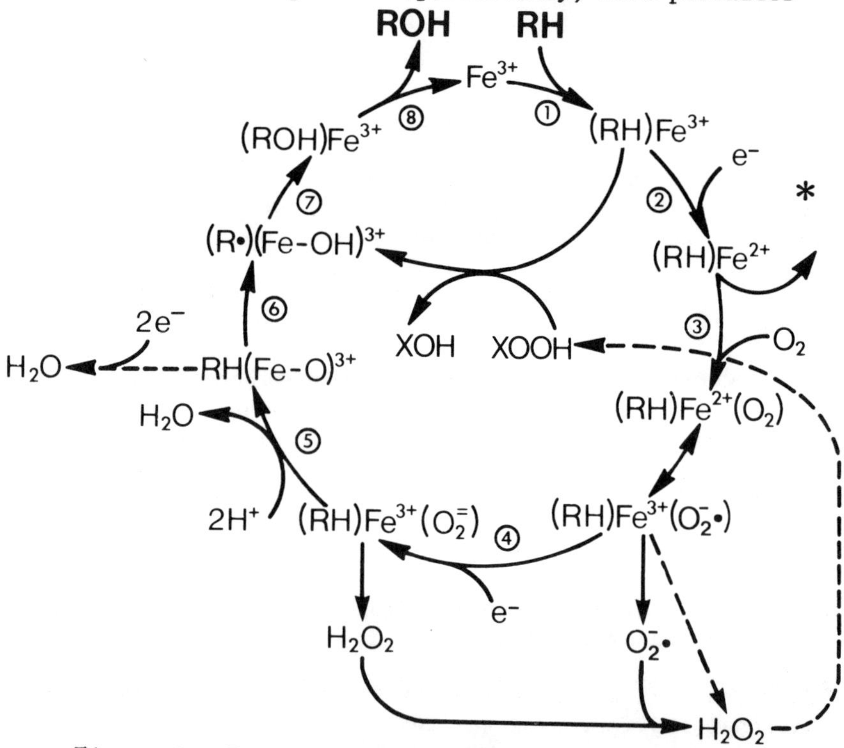

Figure 1. Proposed scheme for mechanism of action of cytochrome P-450, where RH represents the substrate, ROH the corresponding product, and Fe the heme iron atom at the active site (taken from Ref. 3). Reductive reactions, as indicated by the asterisk, involve direct electron transfer from ferrous P-450 to the substrate.

the oxygen atom incorporated into substrate comes from the peroxide (4), whereas the oxygen in alcohols formed in the presence of agents such as iodosobenzene is derived from water (5-7). Accordingly, the various oxidants may not all be equally good mechanistic models. The stoichiometry of the reaction is indicated by Eq. 1, in which XOOH is an oxidant such as cumyl hydroperoxide and RH is the substrate:

$$RH + XOOH \rightarrow ROH + XOH \qquad (1)$$

P-450 functions in such reactions as a peroxygenase, the term peroxidase already being reserved for enzymes that effect substrate oxidation without oxygen insertion:

$$RH_2 + XOOH \rightarrow R + XOH + H_2O \qquad (2)$$

Several experimental approaches were used in this laboratory in an attempt to determine whether peroxide oxygen-oxygen bond cleavage (Eq. 1) occurs by a homolytic or heterolytic mechanism. The reaction of rabbit P-450 form 2 with peroxides differs in several important respects from that of peroxidases. In the case of P-450 only, the intermediates are formed reversibly, and their spectra vary with structural differences in the peroxy compounds (8). Secondly, several P-450s were shown to bring about the decarboxylation of peroxyphenylacetic acid in a reaction presumably involving the phenylacetoxy radical as an intermediate (9). Finally, studies with a series of _m_- and _p_-substituted derivatives of cumene hydroperoxide as the oxygen donor and of toluene as the substrate showed that the hydroxylation rate constant is sensitive to alterations in both reactants (10). Additional work by McCarthy and White (11) established that peroxy acid decarboxylation is unique to P-450 enzymes. Thus, whereas all P-450s examined have been found catalytically competent in decarboxylation, all other heme proteins, including peroxidases, promote only heterolytic cleavage. In summary, these results are compatible with a homolytic mechanism of oxygen-oxygen bond cleavage but not with the heterolytic formation of a common iron-oxo intermediate from various peroxides, as generally seen with peroxidase-catalyzed reactions.

The present paper is concerned primarily with a reaction we have recently found in which P-450 catalyzes the homolytic cleavage of the oxygen-oxygen bond of peroxy compounds, but with the added requirement for an electron donor. This is an example of a P-450-requiring reductive reaction, of which several are now known, that occurs anaerobically; the reaction is indicated in Fig. 1 by an asterisk. In fact, it is such electron-transferring reactions, not the more widely studied oxygenative reactions, that justify the term "cytochrome" for this versatile catalyst. In a study of the reduction of hydroperoxides undertaken to develop a model for the four-electron reduction of dioxygen by P-450 (12), we found that simple two-electron transfer to yield the corresponding alcohol does not occur; instead acetophenone is produced (13). We then became concerned with the identity of the other product, a 1-carbon compound, as well as with the mechanism and possible biological significance of hydroperoxide reduction. This P-450-catalyzed reaction may be written as follows:

$$XRR'C\text{-}OOH + NADPH + H^+ \rightarrow XRCO + R'H + H_2O + NADP^+, \quad (3)$$

where X represents any of a variety of alkyl groups and R and R' are either hydrogen atoms or alkyl groups.

The mechanism of hydroperoxide reduction by this class of heme proteins is of biological interest because organic hydroperoxides and peroxides are used extensively in industry as initiators in polymer synthesis (14), and some of these compounds are known to produce malignant tumors (15,16) or to damage cytochrome P-450 *in vitro* (4). Subsequent to the report that ADP-activated lipid peroxidation, as measured by malonaldehyde formation, is coupled to the NADPH oxidase system of microsomes (17), numerous papers have dealt with this subject. Although NADPH-cytochrome P-450 reductase may serve as a source of superoxide (18) and hydroxyl radicals (19) for initiation of lipid peroxidation, P-450 has also been implicated (20,21). More recently, Lindstrom and Aust (22) have reported the reduction of 13-hydroperoxy-9,11-octadecadienoic acid (13-HOO-$C_{18:2}$) to the corresponding 13-hydroxy compound by a reconstituted rat liver microsomal system containing P-450, and Ekström and Ingelman-Sundberg (23) have concluded that purified rabbit P-450s may contribute to the initiation of lipid peroxidation in reconstituted membrane vesicles. Various products have been reported to arise from lipid peroxidation, including alkanes (24), and hydrocarbon exhalation is believed to be a measure of this pathophysiological process *in vivo* (25). Evidence has recently been presented for the reductive cleavage of 13-HOO-$C_{18:2}$ by purified P-450 in the reconstituted system containing NADPH, with the formation of pentane (26). This is the expected product according to Eq. 3, and we propose that lipid peroxides may be the physiological substrates for the cleavage reaction we have found using model hydroperoxides as substrates.

RESULTS AND DISCUSSION

Products Resulting From Reductive Cleavage of Model Hydroperoxides

To a reconstituted enzyme system containing P-450 form 2, NADPH-cytochrome P-450 reductase, phophatidylcholine, and NADPH under anaerobic conditions, cumene hydroperoxide was added. After a few minutes, when NADPH oxidation had ceased, the reaction mixture was analyzed by HPLC as described earlier (26); the HPLC profiles for the various products of this and other model hydroperoxides have been presented elsewhere (27). With cumyl hydroperoxide, cumyl alcohol was thought to be the likely product, but none could be detected in the petroleum ether extract of the reaction mixture as compared with the behavior of the known alcohol as a standard. Instead, acetophenone was identified. At this point it was apparent that cumyl hydroperoxide had undergone a rearrangement, probably involving radical intermediates, with the formation of an additional 1-carbon product. This was identified in the head space of the reaction mixture by GC/MS analysis as methane, and the amount of the methane was determined by GC analysis on a Carbosieve G column (26); the mass spectrometric patterns are presented in detail elsewhere (27). The results, as summarized in Table I, indicate that no cumyl alcohol or benzaldehyde was produced and the stoichiometry of the reaction, expressed as the molar ratio of hydroperoxide consumed to that of NADPH oxidized and acetophenone and methane formed, was 1.0:1.0:0.9:0.7. Ethane was not detected as a product, indicating that if a transient methyl radical had been formed (see discussion below on the mechanism of the reaction), radical dimerization did not occur. The components required for product formation are evident from the results given. Under the usual anaerobic conditions, product formation was greatly decreased in the absence of P-450 or when the P-450 had been heat-denatured. The omission of NADPH or the reductase resulted in no detectable methane formation from any of the hydroperoxides but gave acetophenone in lower but still significant amounts. Under aerobic conditions cumyl hydroperoxide yielded the expected products in comparable amounts to those found anaerobically. The oxide:NADPH:acetone:methane, was determined to be 1.0:1.1:0.9:1.0.

TABLE 1. Products resulting from reductive cleavage of cumyl hydroperoxide and components required for the reaction.

When NADPH oxidation had ceased in the complete anaerobic system (26), as judged spectrophotmetrically, the reaction mixtures were analyzed. The hydroperoxides were found to be completely consumed. The amounts of products, with the exception of methane and acetone, are corrected for efficiency of extraction from the reaction mixtures. Results are an average of three independent experiments. The data are from Ref. 26.

System	Starting amount of cumyl hydroperoxide	NADPH oxidized	Product formed			
			Cumyl alcohol	Acetophenone	Benzaldehyde	Methane
	nmol	nmol		nmol		
Complete	100	99	0	91	0	70
No NADPH				23		a
No P-450				<0.01		a
P-450 heat-inactivated				6		a
No reductase				40		a
Complete, but aerobic				88		65

[a]None could be detected.

Products Resulting From Reductive Cleavage of Lipid Hydroperoxides

In view of the results already given with model hydroperoxides, we considered that lipid hydroperoxides might undergo a similar reductive cleavage and looked for alkane formation from 13-HOO-$C_{18:2}$, with the results given in Table II. Pentane was identified as the product, and both P-450 and the reductase were found to be required. In the complete system, the stoichiometry of the reaction, peroxide consumed to NADPH oxidized to pentane formed, was 1.0:1.2:0.8. More recently, the other product has been identified as the 13-oxotridecadienoic acid, and pentane has also been identified as a product of reductive cleavage by P-450 of the arachidonic acid hydroperoxide, 15-hydroperoxy-5,8,11,13-eicosatetraenoic acid (unpublished results).

TABLE 2. Requirements for pentane formation from 13-HOO-$C_{18:2}$

The complete system contained 1.0 µM P-450 and reductase, 60 µg of phospholipid per ml, 150 µM NADPH, and 15.7 µM 13-HOO-$C_{18:2}$ in a total volume of 2.0 ml. Pentane was determined by gas chromatography, and the identity was confirmed by GC/MS. The data are from Ref. 26.

System	Hydroperoxide consumed	NADPH oxidized	Pentane formed
	nmol	nmol	nmol
Complete	31	37	25
No P-450	0	2	0
No reductase	31	5	0
P-450 and reductase heat-inactivated	0	0	0

Mechanism of the Reaction

Fragmentation of the hydroperoxide, which is a consequence of reduction of the oxygen-oxygen bond, is in our view diagnostic of homolysis. The resulting unstable alkoxy radical apparently rearranges with β-scission to form a carbonyl compound and an alkyl radical that is then reduced to an alkane. The β-scission of alkoxy radicals is a well-established process that competes with their attack on hydrogen-containing substrates to produce the corresponding alcohols. The effect of structure on the rate of β-scission of alkoxy radicals was reported in 1963 by Walling and Padwa (29), who carried out the photodecomposition of a series of t-alkyl hypochlorites and determined the relative yields of products, and Neta et al. (30) have more recently carried out radiolytic studies of cumyloxy radical β-scission in aqueous solution. Thus, ample chemical precedence exists for the reaction we have found to be catalyzed by cytochrome P-450.

The accompanying scheme (Fig. 2) shows the proposed reaction mechanism as applied specifically to cumyl hydroperoxide. This substrate is believed to undergo homolytic cleavage of the oxygen-oxygen bond with generation of a cumyloxy radical and a hydroxyl radical, with reduction of the latter by ferrous P-450 to yield a hydroxide ion. Rearrangement of the cumyloxy radical by β-scission would yield acetophenone and a transient methyl radical, which by acceptance of the second electron from ferrous P-450 and proton uptake would yield methane. That formation of the hydrocarbon occurs by reduction of the derived radical was clearly demonstrated by the incorporation of one deuteron into the methane formed when the reaction was carried out in 70% deuterium oxide as solvent (26). Ethane was not detected as a product, thus indicating that radical dimerization does not occur.

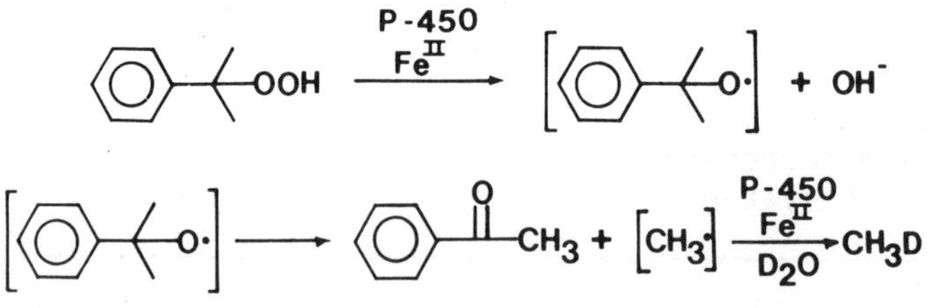

Figure 2. Proposed mechanism of P-450-catalyzed reductive cleavage of hydroperoxides, with cumyl hydroperoxide as the example.

REFERENCES

1. White, R.E., and Coon, M.J. (1980) *Annu. Rev. Biochem.* *49*, 315-356
2. Coon, M.J., White, R.E., and Blake, R.C., II (1982) in *Oxidases and Related Redox Systems*, (King, T.E., Mason, H.S., and Morrison, M., eds) pp. 857-885, Pergamon Press, Elmsford, New York
3. Black, S.D., and Coon, M.J. (1987) *Adv. Enzymol. Relat. Areas Mol.Biol.* *60*, 35-87
4. Nordblom, G.D., White, R.E., and Coon, M.J. (1976) *Arch. Biochem. Biophys.* *175*, 524-533
5. Heimbrook, D.C., and Sligar, S.G. (1981) *Biochem. Biophys. Res. Commun.* *99*, 530-535
6. White, R.E., and McCarthy, M.B. (1984) *J. Am. Chem. Soc.* *106*, 4922-4926
7. White, R.E. (1986) *Fed. Proc.* *45*, 1747
8. Blake, R.C., II, and Coon, M.J. (1980) *J. Biol. Chem.* *255*, 4100-4111
9. White, R.E., Sligar, S.G., and Coon, M.J. (1980) *J. Biol. Chem.* *255*, 11108-11111
10. Blake, R.C., II, and Coon, M.J. (1981) *J. Biol. Chem.* *256*, 12127-12133
11. McCarthy, M.B., and White, R.E. (1983) *J. Biol. Chem.* *258*, 9153-9158
12. Gorsky, L.D., Koop, D.R., and Coon, M.J. (1984) *J. Biol. Chem.* *259*, 6812-6817
13. Vaz, A.D.N., and Coon, M.J. (1985) in *Cytochrome P-450: Biochemistry, Biophysics and Induction*, (Vereczkey, L., and Magyar, K.,eds) pp. 545-548, Elsevier, Amsterdam, and Akademiai Kiado, Budapest
14. Sheppard, C.S., and Mageli, O.L. (1982) in *Encyclopedia of Chemical Technology*, (Grayson, M., ed) pp. 27-90, Wiley-Interscience, New York
15. Sittig, M. (1981) *Handbook of Toxic and Hazardous Chemicals*, Noyes, Park Ridge, NJ
16. *Commercial Organic Peroxide Toxicological Data*: Society of the Plastics Industry Bulletin (1982) Society of the Plastics Industry, New York
17. Hochstein, P., and Ernster, L. (1963) *Biochem. Biophys. Res. Commun.* *12*, 388-394
18. Aust, S.D., Roerig, D.L., and Pederson, T.C. (1972) *Biochem. Biophys. Res. Commun.* *47*, 1133-1137
19. Lai, C.-S., Grover, T.A., and Piette, L.H. (1979) *Arch. Biochem. Biophys.* *193*, 373-378
20. Hrycay, E.G., and O'Brien, P.J. (1971) *Arch. Biochem. Biophys.* *147*, 14-27

21. Svingen, B.A., Buege, J.A., O'Neal, F.O., and Aust, S.D. (1979) J. Biol. Chem. 254, 5892-5899
22. Lindstrom, T.D., and Aust, S.D. (1984) Arch. Biochem. Biophys. 233, 80-87
23. Ekström, G., and Ingelman-Sundberg, M. (1984) Biochem. Pharmacol. 33, 2521-2523
24. Poli, G., Dianzani, M.U., Cheeseman, K.H., Slater, T.F., Lang, J., and Esterbauer, H. (1985) Biochem. J. 227, 629-638
25. Wendel, A., and Dumelin, E.E. (1981) Methods Enzymol. 77, 10-15
26. Vaz, A.D.N., and Coon, M.J. (1987) Proc. Natl. Acad. Sci. USA 84, 1172-1176
27. Vaz, A.D.N., Roberts, E.S., and Coon, M.J. (1987) in Proceedings, VIIth International Symposium on Microsomes and Drug Oxidations, Adelaide, Australia, in press, Taylor & Francis, New York
28. Vaz, A.D.N., and Coon, M.J. (1987) Fed. Proc. 46, 1956
29. Walling, C., and Padwa, A. (1963) J. Am. Chem. Soc. 85, 1593-1597
30. Neta, P., Dizdaroglu, M., and Simic, M.G. (1984) Isr. J. Chem. 24, 25-28

DISCUSSION

YAMAZAKI: Do the intermediate free radicals react with external molecules such as spin trapping reagents or glutathione?

COON: To give you a general answer, the oxygen radical species that occur in P-450-catalyzed reactions appear to be tightly bound to the heme. Typical radical trapping agents have no effect, but we have not yet used spin traps.

SMITH: Is ethanol competitive?

COON: With the liver microsomal P-450, ethanol is a typical competitive inhibitor with respect to other substrates. This effect and induction of P-450 by ethanol together account for the low activity of drugs and anesthetic administered to human alcoholics or animals chronically exposed to ethanol.

PETERSON: (1) What was the stoichiometry of alkyl product formation from the fatty acid hydroperoxides? (2) Was the electron which "quenched" the alkyl radical formed transferred from NADPH via the reductase and P-450?

COON: (1) The stoichiometry of pentane formation so far as we have studied it conforms to that predicted by the general equation we have written for the reductive cleavage of organic hydroperoxides.

(2) Yes, NADPH is the primary electron donor via the reductase, and again the stoichiometry of NADPH oxidation is as predicted.

ESTABROOK: These are interesting and important studies. With regard to a role of P-450 reacting with fatty acid peroxides associated with lipid peroxidations, I wonder if you have tested fatty acid peroxides associated with phospholipid, since it has been reported that these are the intermediates formed during lipid peroxidation.

COON: We haven't yet tested hydroperoxy acids still in ester linkage in phospholipid but plan to do so. This is important with respect to the pathophysiology of alcoholism since membrane integrity might be altered with phospholipid containing shortened aldehyde acids. As to the substrate specificity of alcohol-inducible P-450, it is very broad and includes both long chain and secondary alcohols. For example, cyclohexanol is converted to cyclohexanone.

Oxidases and Related Redox Systems, pages 509–524

A UNIQUE DEUTERIUM/PROTON EXCHANGE DURING CYTOCHROME P-450 MEDIATED EPOXIDATION OF PROPENE AND BUTENE

John T. Groves

Department of Chemistry, Princeton University, Princeton, New Jersey 08544

Kenneth M. Fish, G. E. Avaria-Neisser, M. Imachi, and Robert L. Kuczkowski

Department of Chemistry, The University of Michigan, Ann Arbor, Michigan 48109

ABSTRACT

A cytochrome P-450 LM_2 reconstituted system mediated expoxidations of small terminal alkenes to the corresponding alkyloxiranes. This oxidation was accompanied by a stereoselective proton exchange of the E proton. Propene was oxidized in D_2O to *trans*-3-deuterio-2-methyloxirane, and E-1-deuteriopropene was epoxidized in H_2O to methyloxirane. Thus, oxidation occurred with complete exchange. The epoxidation of 1-butene in D_2O generated a mixture of 86% ethyloxirane and 14% E-3-deuterio-2-ethyloxirane. Reaction of E-1-deuterio-1-butene produced 10-30% ethyloxirane and 70-90% *trans*-3-deuterio-2-ethyloxirane. Larger terminal alkenes in the homologous series did not undergo such an exchange. A mechanism involving formation of organoiron porphyrin intermediates is proposed. Stereoselective deprotonation by a basic amino acid residue producing an iron carbene complex is proposed as the mechanism of exchange with water.

INTRODUCTION

Liver microsomal cytochrome P-450 catalyzes the oxygenation of alkenes (Liebman & Ortiz, 1969; Wesloki et al., 1980). Epoxides are the primary products, and several studies have been devoted to understanding the stereochemistry and mechanism of epoxidation (Watanabe & Akamatsu, 1974;

Hanzlik et al., 1978; Groves et al., 1980). Substantial amounts of other products were observed in reaction mixtures with some substrates (McCarthy & White, 1983). For example, as shown below, hydroxylation and allylic rearrangement occurred during the epoxidation of cyclohexene by cytochrome P-450 LM_2 (Groves & Subramanian, 1984; Groves et al., 1980).

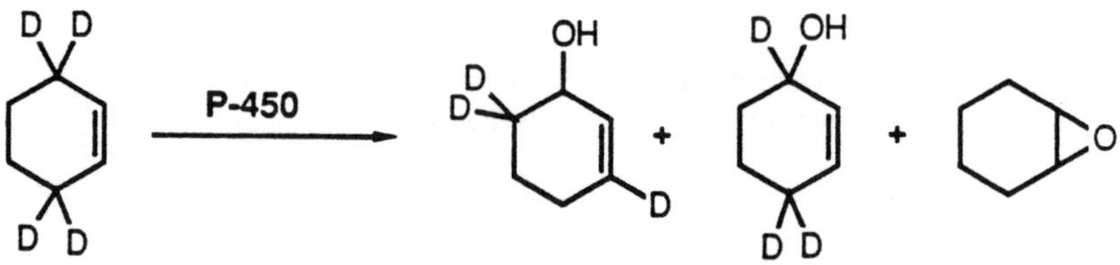

Several mechanisms have been proposed for epoxidation by cytochrome P-450. Retention of stereochemistry during epoxidation is generally observed and is accommodated by a concerted mechanism (Groves & Nemo, 1983). However, a loss of olefin configuration could indicate the formation of carbon radical intermediates which allow rotation about the carbon-carbon bond as shown below. A radical mechanism without rearrangement is conceivable if enzyme-substrate interactions are strong or if a succession of one-electron steps is very fast.

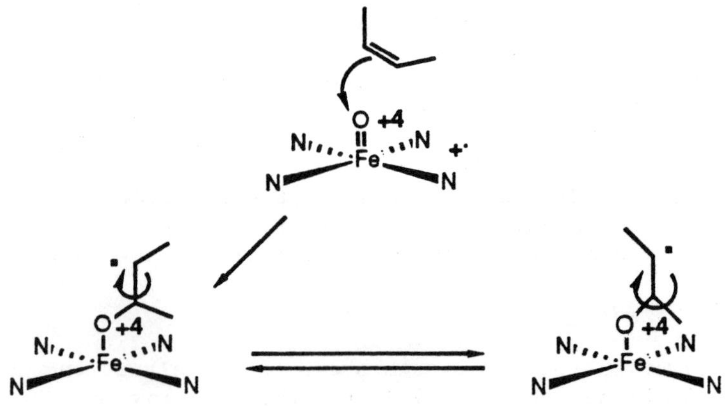

The catalytic epoxidation of alkenes by cytochrome P-450 can also result in irreversible inactivation of the enzyme (De Matters, 1978; Levin et al., 1972; Ionnides & Park, 1976; Ortiz de Montellano & Mico, 1980). Ortiz de Montellano and coworkers have demonstrated that substrate was oxidatively activated to a species which covalently bound to the heme, thus inactivating the enzyme. Isolation and characterization of hemes from inactive enzyme revealed that N-alkylation of the pyrrole nitrogen had occurred, and suggested that intermediates formed during catalysis were responsible for this suicide inhibition (Ortiz de Montellano et al., 1981). Thus, enzyme inactivation and catalytic formation of epoxides are likely to be alternate fates of a common intermediate. Only very recently has olefin epoxidation and N-alkylation been reported in model systems (Mashiko et al., 1985). Oxidation of 4,4,-dimethyl-1-pentene was epoxidized

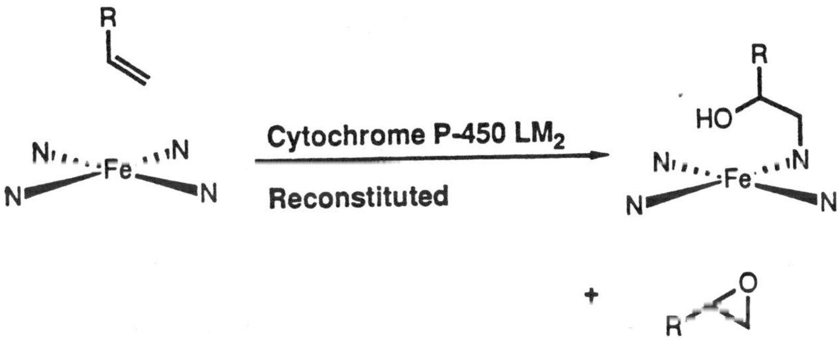

by [5,10,15,20]-tetrakis(2,6-dichlorophenyl)porphinato]-iron(III)chloride with pentafluoroiodosylbenzene as the oxidant. An N-alkyl porphyrin was also formed with this substrate.

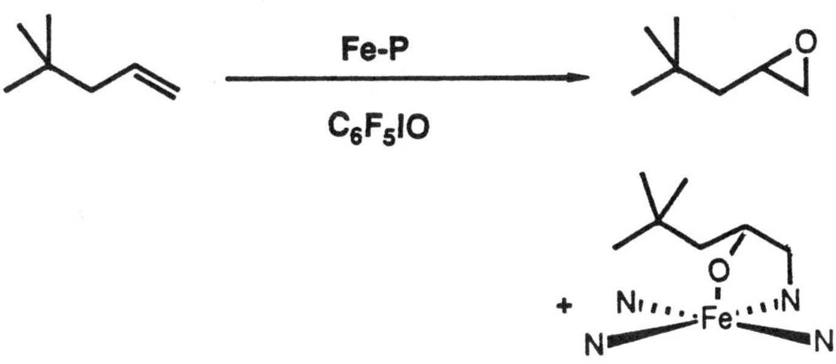

A good candidate for probing the pathway of cytochrome P-450 catalyzed epoxidations should lack steric constraints. This requirement would allow internal rotation giving stereochemical randomization if a radical species were generated. Terminal olefins fit these criteria. Selective deuteration at carbon-1 would allow stereoisomeric assignment of product epoxides. Suicide inactivation has been reported to occur once every 200-300 turnovers (Ortiz de Montellano et al., 1981); accordingly analytical sensitivities on a micromolar scale are required. Microwave spectroscopy for stereochemical assignment of small molecules fits these criteria. Mass spectrometry may be utilized to corroborate the deuterium enrichment of product epoxides and unreacted olefins.

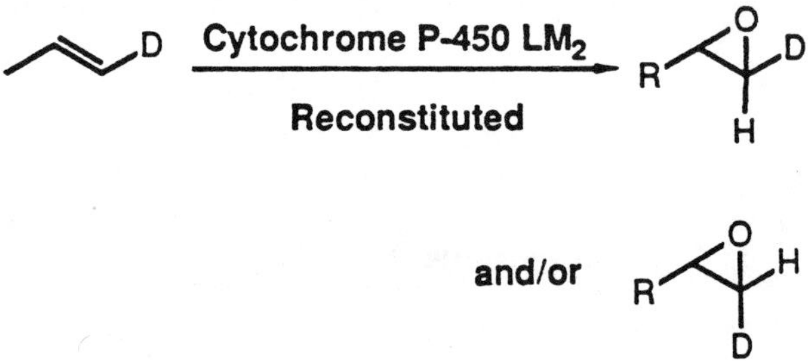

In an effort to probe the stereochenmistry of epoxidation, we have examined (Groves et al., 1986; Avaria-Neisser, 1984) the reaction of E-1-deuteriopropene with reconstituted cytochrome P-450 LM_2. Microwave spectroscopy unexpectedly showed that methyl oxirane $\underline{d}_0$ emerged as the predominant product as shown below.

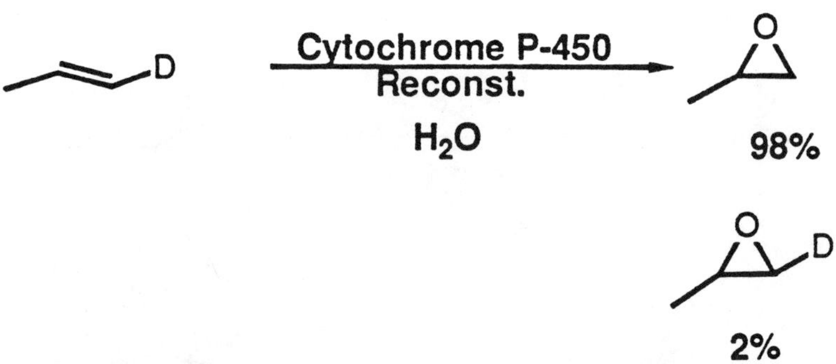

When the reaction was carried out in D_2O, propene was epoxidized to a mixture of *trans*-3-deuterio-2-methyloxirane, and methyloxirane. The relative deuterium uptake on epoxidation reflected the deuterium enrichment of the solvent.

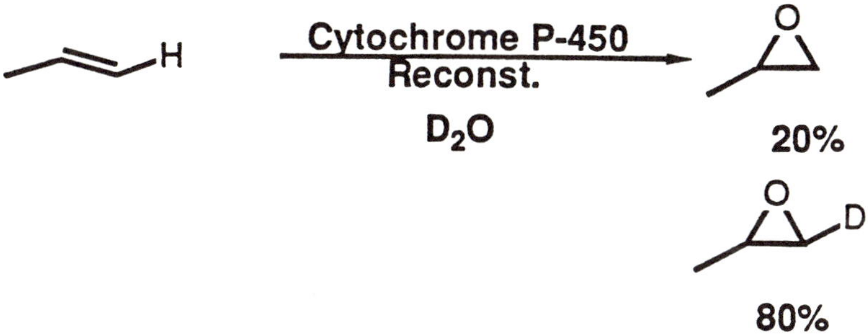

Exchange of deuterium (hydrogen) was unique to the reconstituted system containing cytochrome P-450 reductase and NADPH. Cytochrome P-450 epoxidations with oxygen donors such as iodosylbenzene led to unexchanged product. Larger substrates such as 1-hexene, 1-heptene, and 1-octene were epoxidized by reconstituted cytochrome P-450; however, no deuterium (hydrogen) exchange was observed.

Liebler and Guengerich (1983) have reported the only other terminal olefin oxidation by cytochrome P-450 which was accompanied by hydrogen/deuterium exchange. Oxidation of 1,1-dichloro-2,2-dideuterioethylene was oxidized to a variety of products including 2,2-dichloro-2-deuterioacetaldehyde by the rat liver microsomes. The formation of product was rationalized by an α-migration of deuterium.

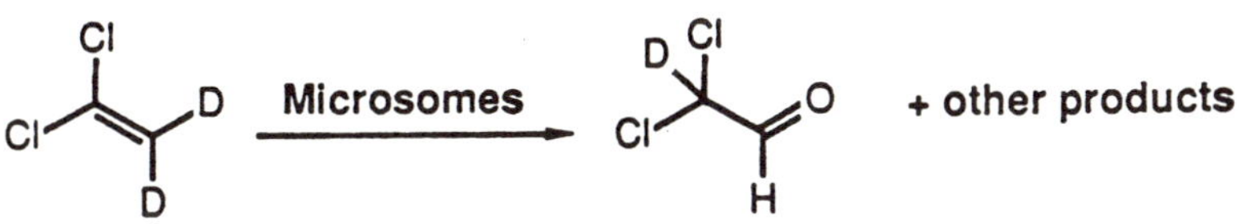

The stereoselective exchange of hydrogen during the epoxidation of propene (Groves et al., 1986; Avaria-Neisser, 1984) was not anticipated by any of the currently popular mechanisms for oxygen transfer mediated by cytochrome P-450. Accordingly, we have undertaken a more extensive examination of this unusual process. The study of the oxidation of E-1-deuteriobutene and E-1-deuteriopentene were chosen as probes of the mechanism of deuterium uptake during epoxidation by cytochrome P-450 LM_2.

RESULTS

Samples of propene, 1-butene and 1-pentene specifically labelled with deuterium *trans* at the terminal olefinic position were prepared by literature procedures (Imachi et al., 1981). All samples were determined to have >95% $\underline{d}_1$ composition by inspection of the molecular ion region of the mass spectrum. Confirmation of the deuterium composition and stereochemistry was obtained from the ^{1}H-NMR spectra of each sample (J_{H_1,H_2}(trans) = 13-14 Hz). Authentic samples of the corresponding *trans*-$\underline{d}_1$ epoxides were obtained by treating the deuterated olefins with m-chloroperoxybenzoic acid (mCPBA) in diglyme at 0°.

Enzymatic epoxidations of the alkenes were carried out in the following manner (Groves & Subramanian, 1984; Groves et al., 1980; Groves et al., 1986; Avaria-Neisser, 1984). Cytochrome P-450 LM_2 was reconstituted with dilaurylphosphatidylcholine, cytochrome P-450 reductase and equilibrated for 30 minutes with one atmosphere of alkene. Reactions carried out in D_2O were prepared identically, with buffer, dilaurylphosphatidylcholine, and NADPH prepared in 99.8% D_2O. The resulting deuterium content of the aqueous solution was calculated to be 80.6%. Oxidation was initiated by introduction of O_2 and NADPH. After 1 hour at 25°, volatile organic materials were collected and subjected to microwave analysis after bulb to bulb distillation as we have previously described (Groves et al., 1986; Avaria-Neisser, 1984; Larabee & Kuczkowski, 1978; Imachi et al., 1981; Imachi & Kuczkowski, 1982; Ljunggren et al. 1978). For the case of 1-pentene oxide deuterium content was determined by traditional NMR and GC-MS techniques. A summary of product yields, catalytic turnovers and deuterium content of each product is presented in Table 1.

DISCUSSION

The results clearly indicate the facile exchange of a single hydrogen during the epoxidation of propene, consistent with our earlier report (Groves et al., 1986; Avaria-Neisser, 1984). Significant hydrogen-deuterium exchange was observed for 1-butene but there was no detectable exchange for 1-pentene. The hydrogen exchange is shown to involve the aqueous medium. In all cases a single olefinic position *trans* to the alkyl group was susceptible to H/D exchange.

Table 1. Oxirane Yields on Reaction of Terminal Alkenes with Cytochrome P-450 LM_2

Substrate-Solvent	Turnovers	Products (%)		
		d_0	trans-d_1	cis-d_1
propene - D_2O(80.6%D)	240	20	80	0
E-1-deuterio-2-propene-H_2O	196	98	2	0
1-butene- D_2O	320	86(68)[a]	14(32)[a]	0
E-1-deuterio-1-butene -H_2O	198	11	89	0
1-pentene- D_2O	206	100[a]	0	0
E-1-deuterio-1-pentene- H_2O	198	5[a]	95[a]	0

[a] Deuterium content determined by mass spectrometry.

Suicide inactivation occurred during the epoxidation of these terminal olefins (Ortiz de Montellano, 1983, 1984; Marsh et al., 1977; Ortiz de Montellano & Mico, 1980, 1983; Ortiz de Montellano et al., 1979). Thus, only 200-300 catalytic turnovers were expected (Ortiz de Montellano & Mico, 1980, 1981; Ortiz de Montellano et al., 1979). Ortiz de Montellano and coworkers (1979, 1981) have studied π-bond interactions with cytochromes P-450, and their ability to alkylate pyrrole nitrogens during catalytic turnover. This process was used to probe the topology of liver microsomal cytochromes P-450. It is possible that any of the four pyrrole rings could be alkylated. In an effort to determine the actual site(s) of alkylation, 1-octene, 1-octyne, and 1-propyne were metabolized by microsomes. The N-alkylated protoporphyrins were purified and characterized. Results indicated that alkylation was regiospecific. Alkene oxidation resulted in carbon-nitrogen bond formation at pyrrole ring D with carbon-1 of the π bond. During acetylene oxidation, pyrrole ring A was alkylated in a similar manner.

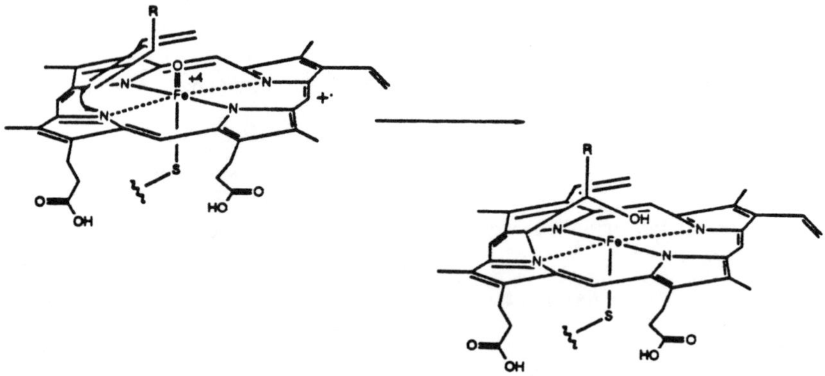

The high specificity of these reactions implied a well-defined active site topology. It is this topology which may have been responsible for the high stereoselectivity in deuterium/hydrogen exchange of propenes and butenes. It also may have contributed to the sharply decreasing exchange rate on increasing olefin size. Two mechanisms were proposed by Ortiz de Montellano for heme modification (1983; Kunze et al., 1983). Alkylation of the heme by the epoxide can be ruled out due to a decreased rate of inactivation on increased epoxide concentration. Further, nitrogen and oxygen added to the olefins from the same side of the π bond. Nucleophilic attack on the terminal carbon of the oxirane ring would have given backside attack and the oxygen and nitrogen substituents would have been found on opposite sides. Thus, it is unlikely that epoxides were involved directly with alkylation.

The observation that higher E-1-deuterioalkenes were epoxidized and caused N-alkylation without loss of configuration required a mechanism involving stringent stereocontrol. A concerted epoxidation and alkylation mechanism as alternative fates for substrates is a compelling hypothesis. However, such a mechanism does not accommodate deuterium/hydrogen exchange. A cationic mechanism is also unlikely. During oxidation of vinyl ethers, nitrogen attacked the terminal carbon (Ortiz de Montellano & Mico, 1980, 1981; Ortiz de Montellano et al., 1979). A cationic mechanism would predict the opposite regiochemistry due to stabilization of the positive charge on the oxygenated carbon. A radical mechanism is less likely because of the observed retention of stereochemistry at the π-bond.

The mechanism shown in Figure 1 is consistent with all the data presented (Groves et al., 1986; Avaria-Neisser, 1984). Oxidation of cytochrome P-450 to the oxoiron(IV) porphyrin π cation radical (Groves et al., 1981) is generally accepted as the oxygen activation process in the catalytic cycle of cytochrome P-450 as well as other heme proteins. Because reactive intermediates during the epoxidation have not been characterized, the path of oxygen transfer is unclear. The formation of an oxametallacycle is consistent with the stereochemical requirements and provides a possible mode for proton exchange. Such species have been proposed for the interaction of carbon-carbon double bonds with chromium oxo species (Sharpless et al., 1977; Collman et al., 1984) and oxidized manganese porphyrins.[22]

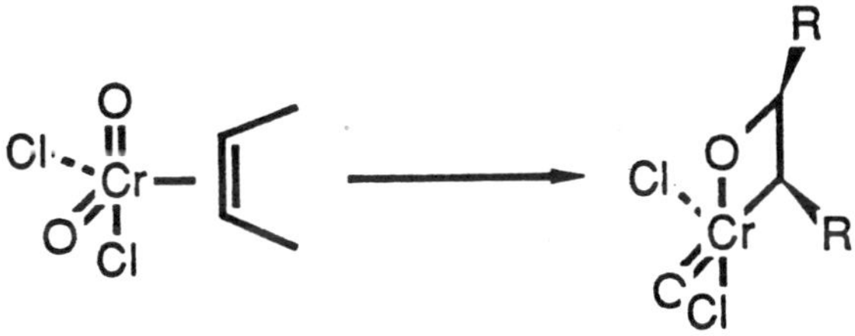

This type of cycloaddition finds precedent in the stereospecific reaction of sulfur trioxide and olefins producing cyclic sultones (Naigayama et al., 1973).

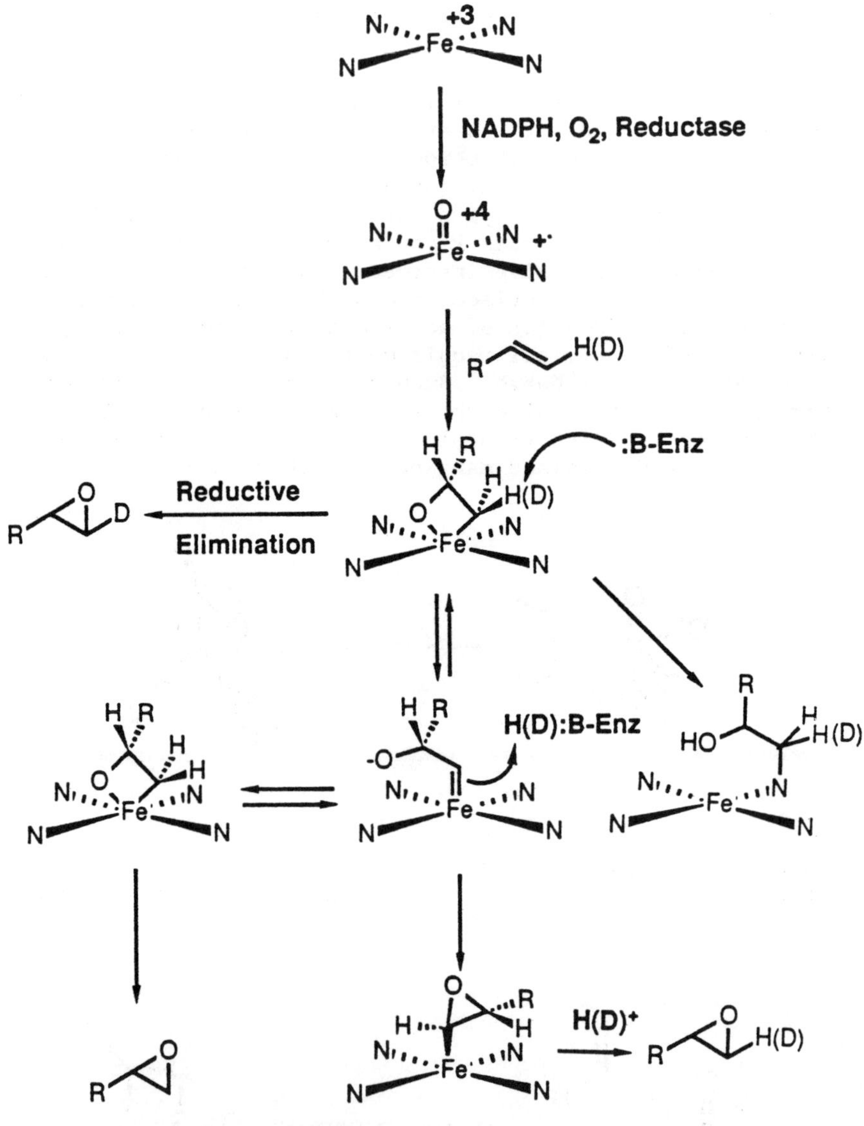

Figure 1. Proposed mechanism for stereospecific deuterium. Hydrogen exchange during the epoxidation of small olefins by cytochrome P-450.

The microscopic reverse of this process has been demonstrated by formation of cis olefins in the Wittig reaction (Vedejs & Snobel, 1973) and the deoxygenation of oxiranes by tungsten(IV) reagents (Sharpless et al., 1972; Satter et al., 1978).

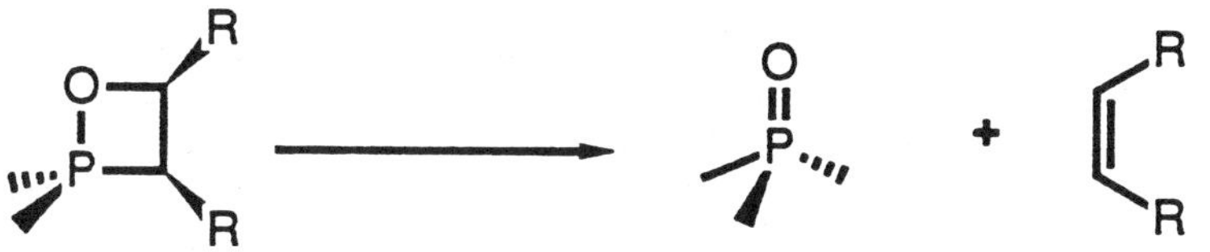

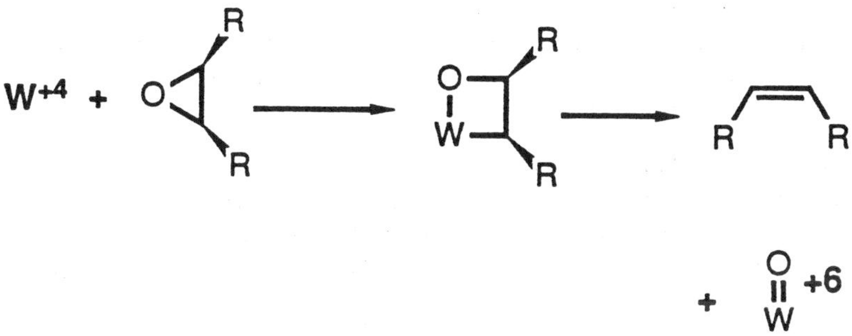

The formation of an oxametallacyclobutane may also be a plausible reaction intermediate to invoke for the unique epoxidation of terminal olefins by *Pseudomonas oleovorans* monooxygenase. This non-heme iron containing enzyme system epoxidizes E-1-deuteriooctene via a proposed Fe = 0 species with 70% inversion at carbon-1 and retentions of con figuration at carbon-2 as shown below (May et al., 1977).

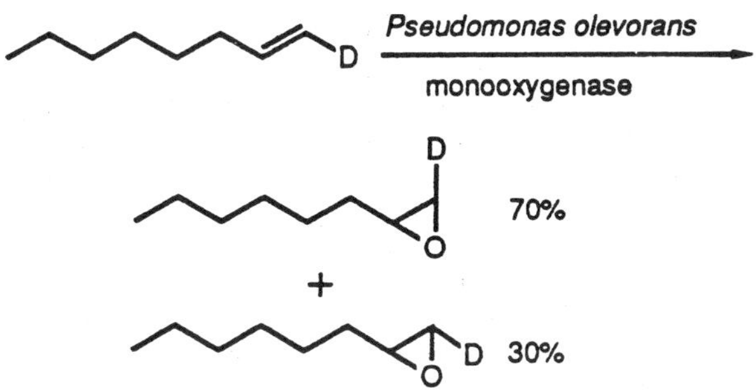

The formation of an oxametallacyclobutane by the alkene and iron oxene could be followed by ring opening to give inversion at carbon-1 while leaving the stereochemistry at carbon-2 intact.

A [2+2] cycloaddition can be envisaged for the stereospecific epoxidation of alkenes by the cytochrome P-450 active oxygen species (Sharpless & Flood, 1971). Reductive elimination would lead to epoxide while a reversible stereospecific deprotonation of the metallacycle could generate an iron carbene porphyrin complex. Removal of a proton from metal alkyls is a common route to transition metal carbene complexes (Schrock, 1979). Mansuy has reported the suicide inactivation of liver microsomal cytochrome P-450 by halomethanes and implicated the formation of a dihalocarbene complex (Wolf et al., 1977). Model compounds of this species have been generated and characterized, and later the X-ray crystal structure of dichlorocarbene-iron tetraphenylporphyrin was reported.

A stereoselective protonation of an intermediate iron-carbene can be accommodated by the microscopic reverse of the deprotonation step after equilibration with the aqueous milieu. This deprotonation/protonation equilibrium step resembles the E2 elimination reaction and would be facilitated by the stability of the iron carbene species.

Evidence from the mass spectral data suggests that exchange during 1-butene epoxidation in D_2O occurred about 2.5-3 times more often than for E-1-deuteriobutene in H_2O although the microwave data indicate a negligible difference. While it is not apparent why there should be a difference in the results obtained from the two techniques, preferential removal of hydrogen from 1-butene could result from a deuterium isotope effect. Since eliminations of this type involve C-H or C-D bond cleavage to form the proposed metallocarbene, isotope effects on β-hydrogens (deuterium) substituents are expected to be significant. In fact, the k_H/k_D during eliminations of this kind vary (usually between 1-3 but occasionally as high as 10). A generalization cannot clearly be extrapolated to the oxametallacycle intermediate. However, any isotope effect is likely to be sensitive to the H-C-C-X angle between the proton and the leaving group (X= leaving group), stability of the resulting π bond, and the steric congestion created by substituents. The nature of the base and leaving group are also important factors contributing to the magnitude of the isotope effect.

The ability of P-450 to form organoiron intermediates which stereroselectively exchange protons would also be expected to be very sensitive to steric bulk. A methyl substituent was apparently small enough to allow complete exchange. However when R=ethyl, the extent of exchange was reduced. When R=propyl or larger, this exchange phenomenon was not observed. The topology of the active site described by Ortiz de Montellano (Mansuy, 1984; Mansuy et al., 1978, 1979, 1982) would be expected to restrict the orientation of these intermediates, thus influencing the ability of an amino acid side chain base to deprotonate the metallacycle.

It is clear that an understanding of the mechanism of oxygen transfer mediated by cytochrome P-450 must account for this unusual exchange process. That this enzyme may employ organometallic intermediates during normal metabolism of hydrocarbons may open a new chapter in the rich chemistry of cytochrome P-450.

ACKNOWLEDGEMENTS

Support of this research by the National Institutes of Health is gratefully acknowledged (GM-36298 at Princeton and GM-25923 at the University of Michigan). The National Science Foundation provided funds to purchase and maintain the microwave spectrometer. We thank Prof. M. J. Coon, Department of Biological Chemistry, The University of Michigan for generous supplies of cytochrome P-450 LM_2, cytochrome P-450 reductase and for technical advice.

REFERENCES

Avaria-Neisser GE (1984) Ph.D. Thesis "Molecular Probes of the Mechanism of Oxygen Transfer by Cytochrome P-450." University of Michigan.

Collman JP, Brauman JE, Meunier B, Raybuk SR, Kodedek T (1984) Proc Natl Acad Sci 81:3245-3248.

De Matters F (1978) "Handb Exp Pharmakol." 44:129-155.

Groves JT, Krishnan S, Avaria GE, Nemo TE (1980) Adv Chem Ser (Biomimetic Chemistry) Dolphin D, Tabushi I, Murakami T. McKenna CE, eds Am Chem Soc 191:277-289.

Groves JT,Akinbote OF, Avaria GE (1980) "Microsomes, Drug Oxidations, and Chemical Carcinogenesis." Coon MJ et al eds Academic Press, New York pp 253-261.

Groves JT, Haushalter RC, Nakamura M, Nemo TE, Evans BJ (1981) J Am Chem Soc 103:2884-2885.
Groves JT, Nemo TE (1983) J Am Chem Soc 105:5786-5791.
Groves JT, Subramanian DV (1984) J Am Chem Soc 106:2177-2181.
Groves JT, Avaria-Neisser GE, Fish KM, Imachi M, Kuczkowski RL (1986) 108:3837.
Hanzlik RP, Shearer GP, Hamburg A, Gillesse T (1978) Biochem Pharmacol 27:1435-1439.
Imachi M, Kuczkowski RL (1982) J Mol Struct 96:55-60.
Ionnides C, Park DV (1976) Chem-Biol Interact 14:241-249.
Kunze KL, Mangold BLK, Wheeler C, Bielan HS, Ortiz de Montellano PR (1983) J Biol Chem 258:4202-4207.
Larabee, AL, Kuczkowski RL (1978) J Catal 52:72-80.
Levin W, Sernatinger E, Jacobson M, Kuntzman R (1972) Science 1341-1342.
Liebler DC, Guengerich RP (1983) Biochemistry 22:5482-5489.
Liebman KC, Ortiz E (1969) Biochem Pharmacol 18:552-554.
Ljunggren SO, Mjoberg PJ, Bäckvall JE (1978) Z Naturforsch 33A:1312-1322.
McCarthy M, White RE (1983) 258:9153-9158.
Mansuy D (1984) "Drug Metabolism" Proc Eur Workshop, 9th, (publ 1985) 195-203.
Mansuy D, Lang M, Chottard J-C, Battioni J-P, Chevier B, Weiss R (1978) Angew Chem Int Ed Eng 17:781-782.
Mansuy D, Lang M, Chottard J-C, (1978) J Am Chem Soc 100: 3212-3213.
Mansuy D, Battioni J-C, Chottard J-C (1979) J Am Chem Soc 101:3971-3973.
Mansuy D, Battioni F-P, Dupre D, Sartori E, Chottard G (1982) J Am Chem Soc 104: 6159-6161.
Marsh JA, Bradshaw JJ, Sapeiha GA, Lucas SA, Kaninshi LS, Ivanevitch DM (1977) Biochem Pharmacol 26:1601-1606.
Mashiko T, Dolphin N, Nakano T, Traylor TG (1985) J Am Chem Soc 107:3735.
May SW, Gordon SL, Steltenkamp MS (1977) J Am Chem Soc 99: 2017.
Naigayama M, Okumura O, Noda S, Mori A (1973) J Chem Soc Chem Commun 841.
Ortiz de Montellano PR, Yost GS, Mico BA, Cinzo SE, Corriea MA, Kombara H (1979) Arch Biochem Biophys 197:524-533.
Ortiz de Montellano PR, Mico BA (1980) Mol Pharmacol 18: 128-135.
Ortiz de Montellano PR, Mico BA (1981) Arch Biochem Biophys 206:43-50.
Ortiz de Montellano PR, Mico BA, Bielan HS, Kunze KL (1981) "Molec Basis of Drug Action." pp 151-166.

Ortiz de Montellano PR, Bielan HS, Kunze KL, Mico BA (1981) J Biol Chem 256:4395-4399.
Ortiz de Montellano PR, Mico BA (1981) Arch Biochem Biophys 206:43-50.
Ortiz de Montellano PR, Mangold BL, Wheeler C, Kunze KL, Reich ND (1983) J Biol Chem 258:4208-4213.
Ortiz de Montellano PR (1984) "Rev Biochem Pharmacol." 26: 1-26.
Satter A, Forrester J, Moir M, Roberts JS, Parker W (1978) Tett Lett 1403.
Schrock R (1979) Acc Chem Res 12:98-104.
Sharpless KB, Flood TC (1971) J Am Chem Soc 93:2316-2318.
Sharpless KB, Ubreit MA, Nieh MF, Flood TC (1972) J Am Chem Soc 94:6538-6540.
Sharpless KB, Teranishi AY, Bäckvall JE (1977) J Am Chem Soc 99:3120-3128.
Vedejs E, Snobel KA (1973) J Am Chem Soc 95:5778-6780.
Watanabe T, Akamatsu K (1974) Biochem. Pharmacol 23: 1845-1851.
Wesloki PG, Miwa GR, Lier AY (1980) Enz. Basis Detox: 135-182.
Wolf CR, Mansuy D, Nastainczyk W, Deutschman G, Ullrich V (1977) Mol Pharmacol 13:698-705.

DISCUSSION

YAMAZAKI: My question is not concerned with your elegant model, but it seems to me a nice opportunity to ask about your proposed model for activated oxygen in the P-450 reaction. Is it Fe(V)=O or $Fe(IV)=O^+$ porphyrin pi-cation radical?

GROVES: The model compound you refer to is the "green intermediate" we first reported in 1981. All of the spectroscopic evidence gathered on this species, ^1H-NMR, Mossbauer, EXAFS, Raman, and electrochemical studies, support our interpretation that the oxidized form of Fe(III) tetramethoxy porphyrin is an oxo-iron(IV)porphyrin cation radical complex. The "red intermediate" also first described in 1981 has been shown by us to be a dimethoxy iron (IV) complex.

KLINMAN: Do you have any comments on the size dependence of exchange? Also, in the propylene case, the complete exchange implies a very rapid deprotonation relative to the breakdown of your proposed intermediate. Is this compatible with the P-450 site?

GROVES: The selective regiochemistry for heme N-alkylation has been interpreted by Paul Ortiz de Montellano to require a highly congested substrate binding site in $P\text{-}450_{LM2}$. If proton exchange requires carbene formation as we have suggested, there may simply not be enough room for carbene formation when the alkyl group is longer than ethyl. One requires the relative rates of exchange and epoxide formation to vary by 100-fold. We have been induced to invoke organoiron intermediates in order to have a plausible mechanism for the terminal C-H labilization we have observed. Since rearrangements of this type have not yet been observed in model compounds, we can only speculate about the details at this time.

Oxidases and Related Redox Systems, pages 525–540

SOME ENZYMATIC VAGARIES OF A BOVINE ADRENAL MICROSOMAL CYTOCHROME P-450 INTRODUCED AND EXPRESSED IN TRANSFORMED MONKEY KIDNEY CELLS*

Ronald W. Estabrook, J. Ian Mason, Cheryl Martin-Wixtrom, Mauricio Zuber+ and Michael R. Waterman

Departments of Biochemistry, Obstetrics and Gynecology
and
The Cecil H. and Ida Green Center for Reproductive Biology
University of Texas Southwestern Medical Center
Dallas, Texas 75235, USA

INTRODUCTION

The family of hemeproteins known collectively as cytochrome P-450 serve to catalyze the oxidative transformation of a great variety of different chemicals. Interest has centered on the role of some cytochromes P-450 in the metabolism of steroids, both because of the physiological importance of such reactions and the diversity of structural derivatives of the steroid nucleus that can be tested as substrates - thereby providing some insight about the active site of the hemeprotein.

Rapid advances have been made in identifying and isolating many different P-450's. Concomitant advances in molecular biology permit the screening and identification of DNA clones unique for a specific cytochrome P-450. Such studies permit the determination of the nucleotide sequence encoding a P-450 leading to a knowledge of the amino acid sequence as well as information about possible regions involved in the regulation of expression of the hemeprotein. The availability of full-length cDNA sequences for a P-450

+ Present address: Howard Hughes Medical Institute, Massachusetts General Hospital, Boston, MA 02114.

* Supported in part by grants from the National Institutes of Health (NIGMS 16488 (RWE), AG 00306 (JIM) and DK 28350 (MRW)) and the Robert A. Welch Foundation (I-959, I-624).

allow inclusion into a plasmid that can serve as the vector for transfection of cells. Zuber et al (1985, 1985a, 1986) have successfully demonstrated the importance of this technique for the expression of different cytochromes P-450 involved in the metabolism of cholesterol and other steroids.

Most measurements of the in vivo enzymatic properties of a cytochrome P-450 are complicated by the simultaneous presence of other cytochromes P-450 in the system under study. The success of cell transfection experiments, with a single species of cDNA that encodes for a unique P-450, now affords the opportunity to carry out studies to examine the enzymologic properties of a single P-450 unencumbered by questions arising from possible contributions to the pattern of substrate metabolism by other forms of P-450 or other oxidative and/or reductive processes (enzymes). Further, the establishment of conditions for the successful cotransfection of more than one P-450 will permit, in the future, the reconstruction of more complex systems for the examination of P-450 interactions with other P-450's or other steroid and xenobiotic metabolizing enzymes.

The present report describes initial studies designed to examine the enzymatic properties of an adrenocortical microsomal P-450 which has been expressed in SV-40 transformed monkey kidney cells transfected with the cDNA for a P-450 that catalyzes the 17α -hydroxylation of pregnenolone and progesterone (termed here 17-OHase). This microsomal cytochrome P-450 was selected since it plays a pivotal role in the division of steroid metabolism for the formation of C21 gluco- or mineralocorticoids or for the generation of C19 androgens and C18 estrogens. Of interest are the reports (Nakajin et al (1981) and Shinazawa et al (1985)) that this P-450 can catalyze two reactions, i.e. the 17-hydroxylation (17-OHase) of progesterone or pregnenolone and the C-17,20-lyase reaction for the conversion of 17-OH progesterone or 17-OH pregnenolone to C19-steroids. In addition the in vivo expression of 17-OHase activity of adrenal cortex appears to be tightly regulated by the presence of ACTH (Slanina and Fevold (1982) and Fevold (1984)). Therefore studies using transfection techniques may provide an opportunity to establish methods that may provide some understanding of this type of regulation of protein synthesis. Further, the $P\text{-}450_{17\alpha}$ has been isolated and purified from testes and adrenal by Nakajin et al (1983 and

1984) and shown to possess interesting differences in the ratio of 17-hydroxylase/lyase activities when catalyzing the oxidation of different steroid substrates.

METHODS

cDNA clones for bovine adrenocortical 17-OHase were isolated and characterized as described by Zuber et al (1985, 1986). The specific cDNA's are incorporated into the vector system of Okayama and Berg (1983) and the method for growth and transfection of COS cells (Gluzman, 1981) with such a vector has been described by Zuber et al (1986). The protocol for growth, transfection and the assay of enzymatic activities using COS cells is summarized diagramatically in Figure 1. At various times following transfection, cells were exposed to fresh medium containing radiolabelled steroids and aliquots of 0.4 ml were removed at the times indicated using sterile conditions. The samples were extracted with methylene chloride and the metabolites formed were identified and quantitated using a Waters 840 computer-controlled monitor system containing a C_{18}-μBondapac reversed phase hplc column and a gradient of methanol/water. Radioactivity in the effluent from the hplc column was monitored with a Flow 1 Radiomatic radioactive detector. Figure 2 illustrates the resolution by hplc of the possible metabolites formed.

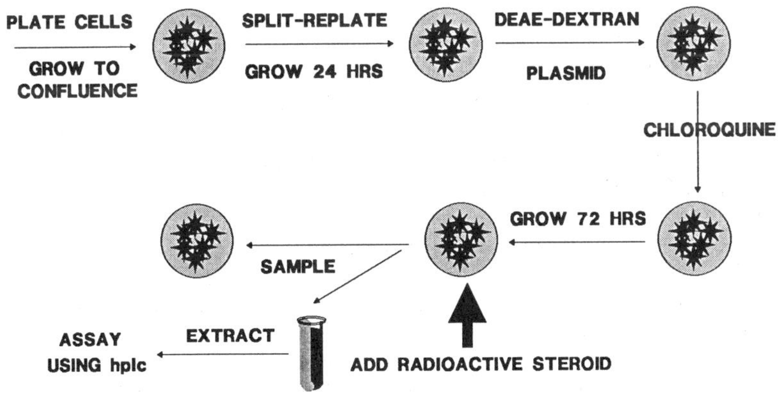

Figure 1. Diagramatic representation of the protocol for the transfection and measurement of metabolism of steroids by COS cells.

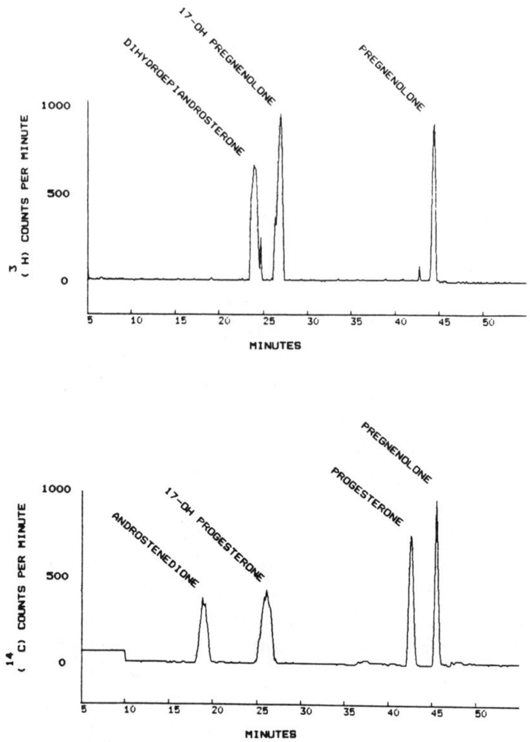

Figure 2. hplc analysis of radioactive steroids.

RESULTS

Transfection of COS cells with a plasmid containing the cDNA for bovine adrenocortical 17-OHase (termed pCD17α 2 or p17a(BOVINE) showed measureable activity when incubated with 1 μM progesterone (Figure 3). Following a brief time-delay, after exposing the cells to fresh medium containing the radioactive steroid substrate, a linear rate of progesterone disappearance was measured concomitant with the formation of the metabolite 17-OH progesterone (a rate of about 0.1 nmole of progesterone hydroxylated/hour/mg of cell protein). It should be noted that we have obtained no evidence for the formation of androstenedione during such an experiment. This result is in contrast to the initial report by Zuber et al (1986) and stands in marked contrast to the reports by Nakajin et al (1983, 1984) and Shinzawa et al (1985) who measured the metabolism of progesterone by microsomal preparations from pig or guinea pig adrenal as well as by various preparations of purified P-450's. If the COS cells were mock transfected or were transfected with the plasmid containing a different cDNA (e.g. for the cytosolic retinoic acid binding protein (pCDCRABP) or HMG-CoA reductase (pCDHMGCoARED)) no metabolism of progesterone could be detected.

In contrast, when a similar experiment was carried out using 1 μM pregnenolone as substrate (Figure 4), one notes, concomitant with the disappearance of pregnenolone, not only the formation of 17-OH pregnenolone but also the formation of a more polar metabolite which has the hplc mobility characteristics of dehydroepiandrosterone (DHEA). Of significance is the pronounced time delay before the appearance of DHEA (a result that may be attributed in part to the inhibition of the C-17,20-lyase reaction by pregnenolone and the requirement to build up a sufficiently high concentration of 17-OH pregnenolone). As with progesterone a brief initial time delay prior to the onset of metabolism was noted. The rate of 17-hydroxylation of pregnenolone was nearly the same as that obtained for the metabolism of progesterone.

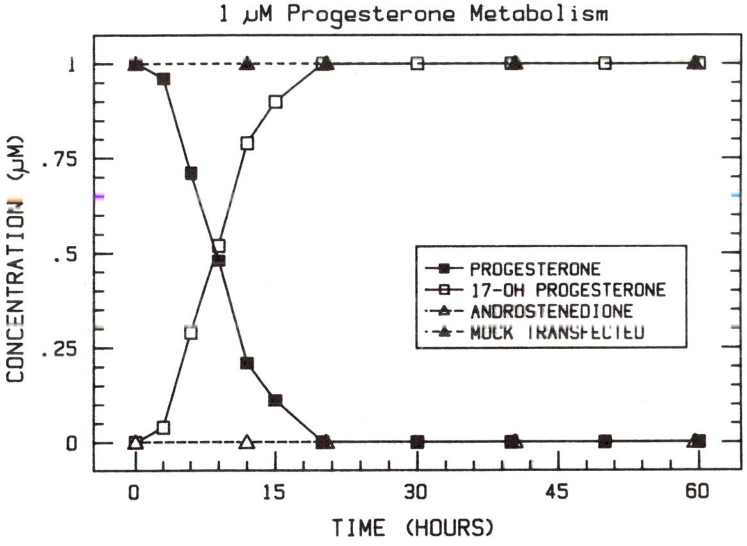

Figure 3. The metabolism of progesterone by COS cells transfected with pCD17α .

These initial experiments suggested a significant difference in the metabolism of delta-4 and delta-5 17-hydroxylated steroids by this single species of P-450,

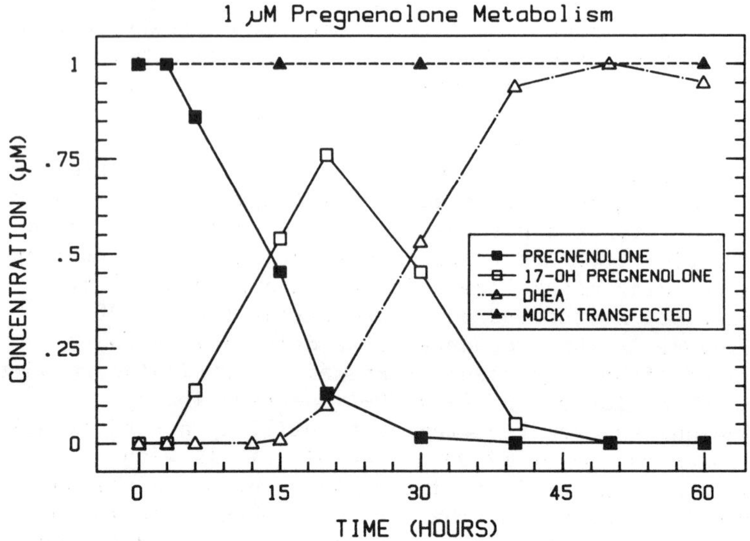

Figure 4. The metabolism of pregnenolone by COS cells transfected with pCD17α 2.

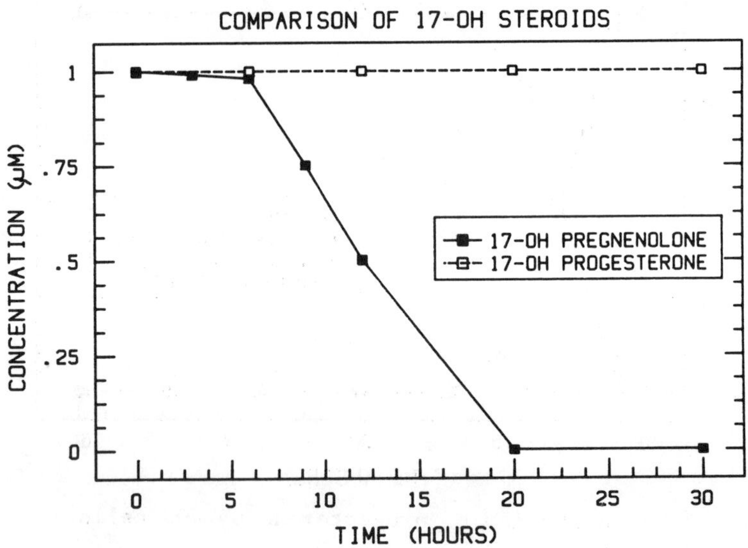

Figure 5. Comparison of the rate of C-17,20-lyase activity using 1 µM 17-OH progesterone or 1 µM 17-OH pregnenolone as substrates.

i.e. an all-or-none substrate preference for the C-17,20 lyase reaction.

To confirm this difference in substrate specificity, additional experiments were carried out using 1 μM 17-OH progesterone or 1 μM 17-OH pregnenolone as substrates. As shown in Figure 5, 17-OH pregnenolone is rather rapidly metabolised to DHEA while there is no measurable activity for the comparable lyase reaction for the conversion of 17-OH progesterone to androstenedione. It should be noted that there is a rather pronounced time delay prior to the initiation of metabolism of the 17-OH pregnenolone. Further, there was no evidence for the presence of other hydroxylated metabolites of either 17-OH progesterone or 17-OH pregnenolone when the pCD17α 2 vector was used.

Determination of the Affinity for Substrate. When a series of experiments were carried out to determine the influence of varying substrate concentrations on the initial rate of metabolism of progesterone and pregnenolone it was noted (Figures 6 and 7) that essentially the same maximal rates of steroid metabolism were obtained whether progesterone or pregnenolone was used as substrate. Likewise, the concentrations of steroid required to give half-maximal rates of metabolism were similar, i.e. about 2.5 μM. These measurements of steroid affinity, as determined from measurements of the initial rate of substrate disappearance, are within the range of those reported by Nakajin et al (1984) who measured substrate affinity using microsomal preparations from porcine adrenal.

Although the experiments illustrated in Figures 6 and 7 suggest the same Km for progesterone and pregnenolone by the expressed adrenal 17-OHase, it was noted (Figure 8) that significant differences in the pattern of substrate disappearance were apparent. For the experiment illustrated in Figure 8, duplicate samples of transfected COS cells were plated on 100 mm dishes and samples were removed at frequent intervals. It can be determined that the half-maximal rate of pregnenolone disappearance occurs when the concentration of pregnenolone is reduced to about 2 μM - a result which is consistent with the results obtained using varying initial concentrations of pregnenolone (Figure 7). In contrast, the measurement of progesterone disappearance indicates that the rate remains nearly linear for a significantly longer time and the half-maximal rate of progesterone disappearance does

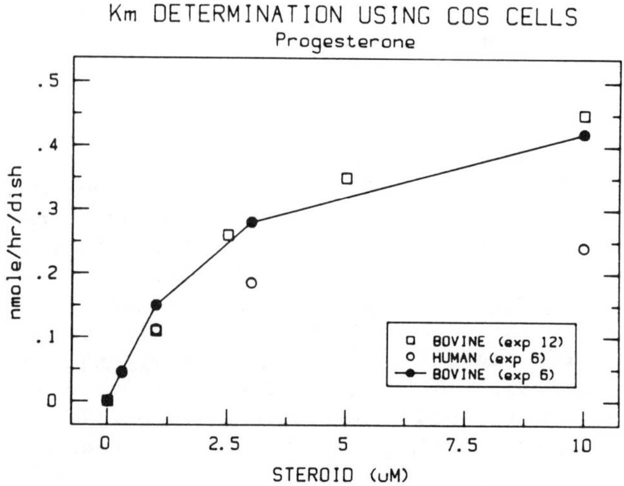

Figure 6. The effect of progesterone concentration on the initial rate of metabolism using COS cells transfected with two different preparations of pCD17α2 and a plasmid preparation containing the cDNA for human adrenal 17-OHase (pCD17α H). COS cells were plated on 60 mm dishes.

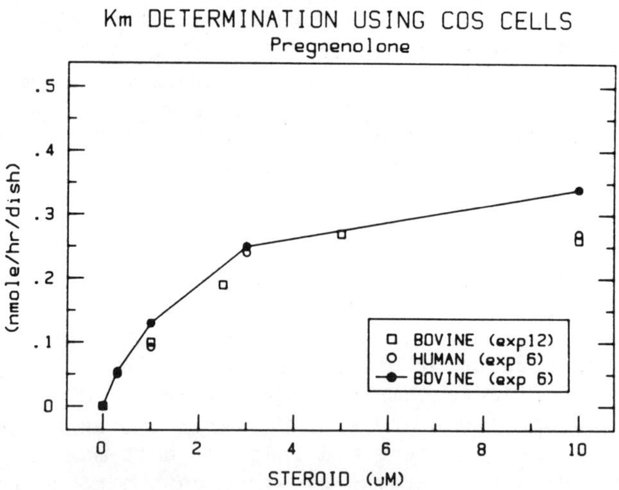

Figure 7. The effect of pregnenolone concentration on the initial rate of metabolism using COS cells transfected with two different preparations of pCD17α 2 and a plasmid preparation containing the cDNA for human adrenal 17-OHase (pCD17α H). COS cells were plated on 60 mm dishes.

not occur until the progesterone concentration is reduced to about 0.2 µM. This type of analysis would suggest a tenfold difference in the affinity of the 17-OHase P-450 for the two different substrates. Such a conclusion would be consistent with other experiments (not shown) which demonstrate the strong inhibitory effect of progesterone on pregnenolone metabolism but not the converse.

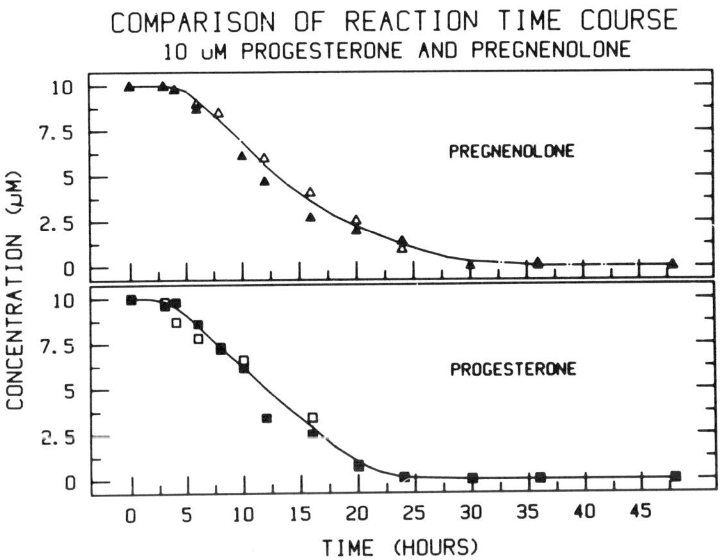

Figure 8. Measurements of the kinetics of disappearance of progesterone and pregnenolone using initial substrate concentrations of 10 µM and pCD17α 2.

<u>The Influence of Varying Concentrations of 17-OH Pregnenolone on the Rate of the $C_{17,20}$-Lyase Reaction.</u>

Measurements of the influence of varying concentrations of 17-OH pregnenolone on the maximal rate of formation of DHEA provided a surprising result. As shown in Figure 9 we were unable to introduce a saturating concentration of 17-OH pregnenolone to the cell assay system. Indeed, as the concentration of 17-OH pregnenolone increased the rate of maximal metabolism increased in essentially a linear manner. This suggests that the lyase reaction may be a first order reaction dependent solely on the concentration of the substrate present. Studies using mock transfected cells or cells transfected to express other proteins showed no measurable activity for the conversion of 17-OH pregnenolone to DHEA.

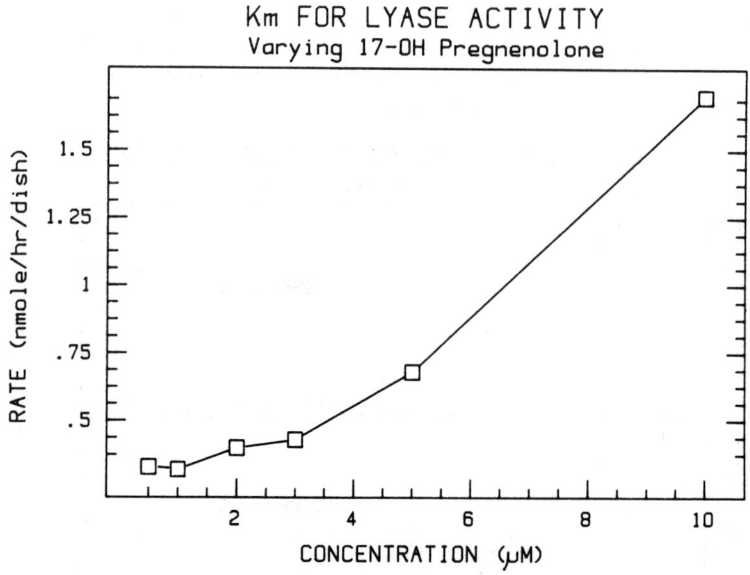

Figure 9. The influence of increasing concentrations of 17-OH pregnenolone on the rate of formation of dehydroepiandrosterone (DHEA). Experiments were carried out using pCD17α 2 and 60 mm dishes of cells.

The Influence of Cotransfection with a Vector Expressing Cytochrome b_5.

One of the explanations given for the observed differences in C-17,20-lyase activity during the metabolism of 17-OH progesterone and 17-OH pregnenolone, as seen when studying steroid metabolism by microsomal preparations derived from adrenal or testes, is the significantly higher concentration of cytochrome b_5 in testes. Recently we obtained from Dr. S. Sligar (University of Illinois, Champagne-Urbana) a cDNA for rat liver cytochrome b_5. This cDNA was incorporated into the pSVL vector system (Pharmacia) and COS cells were cotransfected with the two different plasmids to determine the influence of expressed cytochrome b_5 on the metabolism of progesterone and pregnenolone by the expressed adrenal 17-OHase. As seen in Figure 10, cotransfection with pSVLCYTb_5 caused a measurable inhibition in the observed rate of metabolism of progesterone. The reason for this inhibition remains unexplained although studies where COS cells were cotransfected with pCDCRABP or pCDHMGCoARED together with pCD17α 2 showed no evidence for the inhibition of the expressed 17-OHase activity. It was noted, however, that cotransfection with a plasmid containing a truncated form of the adrenal 17-OHase (pCD17α (TRUNC)) did inhibit the activity expressed by transfection with pCD17α 2. This type of result suggests that there may be a limitation in the amount or availability of heme for the apoenzyme expressed by pCD17α 2 when other hemeproteins are simultaneously expressed. Even so, we saw no evidence for the appearance of a C-17,20-lyase activity that would convert 17-OH progesterone to androstenedione in such cotransfection experiments. These experiments were repeated using pregnenolone as substrate and essentially the same results were obtained. As illustrated in Figure 11, cotransfection of pCD17α 2 with $pSVLCYTb_5$ inhibited both the 17-hydroxylation of pregnenolone and the subsequent C-17,20-lyase reaction. It is of interest to note that the apparent inhibitory effect of pregnenolone on the further metabolism of 17-OH pregnenolone does not appear to be as pronounced in the system cotransfected with $pSVLCYTb_5$.

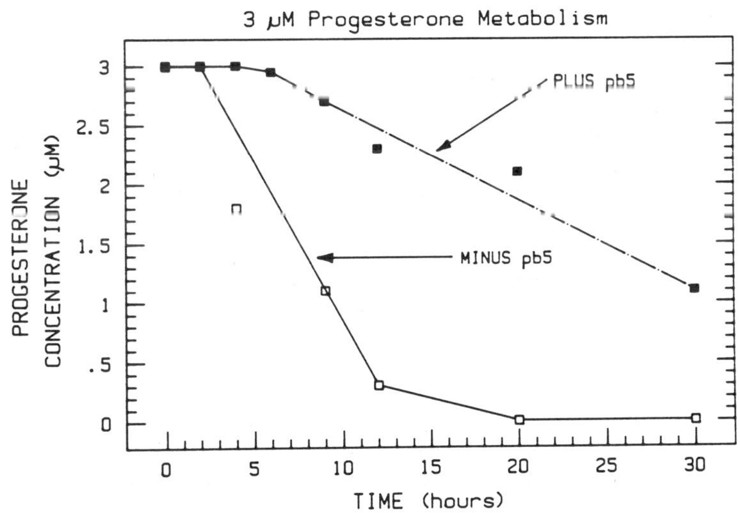

Figure 10. The Effect of Cotransfection of COS cells with $pSVLCYTb_5$ and pCD17α 2 on the 17-OHase activity expressed for the metabolism of progesterone.

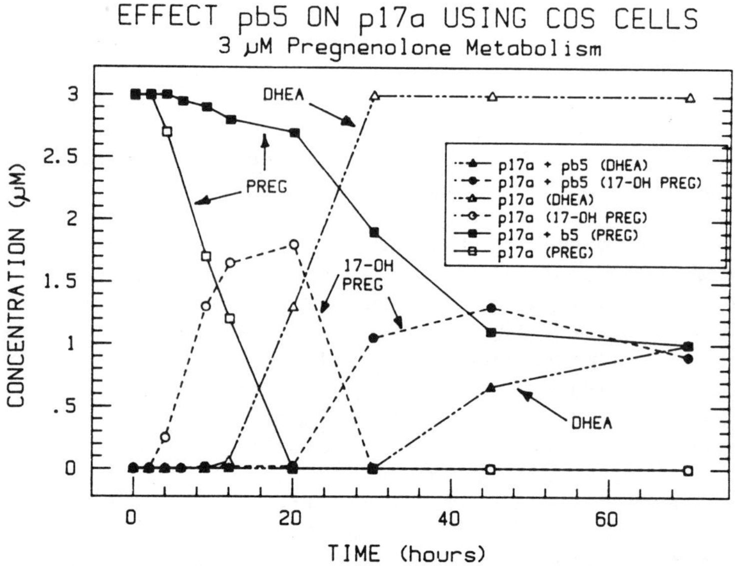

Figure 11. The Effect of Cotransfection of COS cells with pSVLCYTb_5 and pCD17α 2 on the 17-OHase activity expressed for the metabolism of pregnenolone.

SUMMARY

The present studies illustrate the ability to carry out kinetic measurements of steroid hydroxylation using a cytochrome P-450 expressed in a tissue culture cell system. For these experiments a single species of cDNA, incorporated into a suitable expression vector, has been introduced via transfection. A number of interesting preliminary observations have been made on the function of the cytochrome P-450 associated with adrenocortical microsomes which catalyses the 17-hydroxylation of progesterone and pregnenolone.

In confirmation of earlier reports the adrenal 17-OHase possesses both 17-hydroxylase as well as $C_{17,20}$-lyase activities. However, the latter is only functional with 17-OH pregnenolone and not with 17-OH progesterone as substrate. This result differs from the numerous reports that a lyase activity for both substrates is associated with this P-450. The reason for this difference between a delta 4 and a delta 5 steroid remains unresolved although initial experiments indicate that the 5-alpha reduced progesterone is a suitable substrate for both the 17-OHase as well as lyase reactions. This result suggests an inhibitory effect of the delta 4 double bond preventing the carbon-carbon cleavage of the C-17,20 bond in 17-OH progesterone. Clearly more experiments will be required to resolve this question.

Measurements of substrate affinity for the cytochrome P-450 expressed in COS cells appears to be influenced by a permeability barrier to the steroid effecting the transport of the steroid across the cell membrane into the cells. This conclusion is suggested by the presence of a time lag before the onset of metabolism as well as by the discrepancy in the concentration of substrate required to give half-maximal rates of metabolism, cf. the results obtained where the initial concentration of progesterone present in the reaction medium is altered versus those experiments measuring the kinetics of substrate depletion. The presence of such a barrier to the free movement of steroid across the membrane is interesting to contemplate when considering the build up of 17-OH pregnenolone required for the lyase reaction.

Most unexpected where the results obtained when a comparable expression vector containing the cDNA for cytochrome b_5 was cotransfected with pCD17α 2. Cytochrome b_5 has been postulated to be an electron transfer component participating in the cyclic function of some cytochromes P-450 (Hildebrandt and Estabrook, 1971). Onoda and Hall (1982) have identified the microsomal 17-OHase as one type of P-450 that is influenced by the presence of cytochrome b_5. The inhibitory effect of cytochrome b_5, as seen in the present experiments, may result from a depletion of the hemin pools in the cells - although no direct evidence is available to support this hypothesis.

Clearly much can be gained from enzymological studies using systems established by the techniques of molecular biology. The ability to express functional hemeproteins, such as cytochrome P-450, provides the opportunity to systematically construct more complex membrane systems in order to better understand the regulation of electron transport reactions.

REFERENCES

Fevold, H.R. (1984) The stimulation by ACTH of 17 alpha-hydroxylation in cultures of rabbit adrenal cells. J. Steroid Biochem. 21, 555-62.

Gluzman, Y. (1981) SV40 transformed Simian cells support the replication of early SV40 mutants. Cell 23, 175-182.

Hildebrandt, A. and Estabrook, R.W. (1971) Evidence for the participation of cytochrome b_5 in hepatic microsomal mixed-function oxidation reactions. Arch. Biochem. Biophys. 143, 66-79.

Nakajin, S., Shively, J.E., Yuan, P.M. and Hall, P.F. (1981) Microsomal cytochrome P-450 from neonatal pig testis: two enzymatic activities (17 alpha-hydroxylase and C17,20-lyase) associated with one protein. Biochemistry 20, 4037-42.

Nakajin, S., Shinoda, M. and Hall, P.F. (1983) Purification and properties of 17 alpha-hydroxylase from microsomes of pig adrenal: a second C21 side-chain cleavage system. Biochem. Biophys. Res. Commun. 111, 512-7.

Nakajin, S., Shinoda, M., Haniu, M., Shively, J.E. and Hall, P.F. (1984) C21 steroid side chain cleavage enzyme from porcine adrenal microsomes. Purification and characterization of the 17 alpha-hydroxylase/C17,20-lyase cytochrome P-450. J. Biol. Chem. 259, 3971-6.

Okayama, H. and Berg, P. (1983) A cDNA cloning vector that permits expression of cDNA inserts in mammalian cells. Mol. Cell. Biol. 3, 280-289.

Onoda, M. and Hall, P.F. (1982) Cytochrome b_5 stimulates purified testicular microsomal cytochrome P-450 (C_{21} Side-Chain Cleavage). Biochem. Biophys. Res. Commun. 108, 454-460.

Shinzawa, K., Kominami, S. and Takemori, S. (1985) Studies on cytochrome P-450 (P-450 17 alpha,lyase) from guinea pig adrenal microsomes. Dual function of a single enzyme and effect of cytochrome b5. Biochim. Biophys. Acta. 833, 151-60.

Slanina, S.M. and Fevold, H.R. (1982) The Enzyme Specificity of ACTH Stimulation of Rabbit Adrenal Microsomal 17alpha-Hydroxylase Activity. J. Steroid Biochem. 16, 93-99.

Zuber, M.X., Simpson, E.R., Hall, P.F. and Waterman, M.R. (1985) Effects of adrenocorticotropin on 17 alpha-hydroxylase activity and cytochrome P-450(17 alpha) synthesis in bovine adrenocortical cells. J.Biol.Chem. 260, 1842-8.

Zuber, M.X., Simpson, E.R. and Waterman, M.R. (1985a). Regulation of cytochrome P-450(17)alpha activity and synthesis in bovine adrenocortical cells. Ann. NY Acad. Sci. 458, 252-61.

Zuber, M.X., Simpson, E.R. and Waterman, M.R. (1986). Expression of bovine 17 alpha-hydroxylase cytochrome P-450 cDNA in nonsteroidogenic (COS 1) cells. Science 234, 1258-61.

==

A personal note: Martin Morrison was a distinguished scientist, splendid colleague and a good friend. This meeting was not the same due to his absence. We all missed Marty very much. Our thoughts and prayers are with Joyce and his family.

DISCUSSION

MANN: Among your many interesting results is one relating to the puzzling difference in behavior between 17-hydroxylases (directed toward pregnenolone and progesterone) and 17,20 lyase. I wonder if this question should be tackled by using other types of cultured cells such as, for example, the granulosa and thecal cells of the ovary?

ESTABROOK: We have attempted to express the 17-hydroxylase in a variety of cell types (placental cells, $HEPG_1$ cells, CHO cells, mouse mammary carcinoma cells) with poor success. We have used both electroporation and the DEAE-dextran technique for transfection. However, we will continue to try to obtain expression of a functional protein in other cell systems.

Session IX: Cytochrome P-450 and Related Proteins

Chairman: Ronald W. Estabrook

Oxidases and Related Redox Systems, pages 543–555

REGULATION OF STEROID HYDROXYLASE GENE EXPRESSION

Michael R. Waterman, J. Ian Mason, and Evan R. Simpson

Departments of Biochemistry and Obstetrics and Gynecology and Cecil H. and Ida Green Center for Reproductive Biology Sciences, University of Texas Health Science Center, Dallas, Texas 75235

INTRODUCTION

In the broadest sense, the cytochrome P-450 supergene family (Nebert et al., 1987) consists of two general classes of mixed-function oxidases, those which metabolize xenobiotic or exogenous substrates and those which metabolize endogenous substrates. This latter group of enzymes is involved in key biological pathways including cholesterol biosynthesis, bile acid formation, vitamin D metabolism, prostaglandin biosynthesis and steroid hormone production. The cytochromes P-450 which metabolize endogenous substrates are generally not inducible, unlike those forms which metabolize xenobiotic compounds, and are frequently found to be present in relatively low abundance. However, exceptions to this are the steroid hydroxylases involved in steroid hormone biosynthesis which are found in relatively high levels, particularly in the adrenal cortex (Hanukoglu and Hanukoglu, 1986). Consequently these forms of P-450 have been purified and studied in considerable detail, and are the first of the forms of P-450 which metabolize endogenous substrates whose regulation of gene expression has been examined.

The steroidogenic pathways in the adrenal cortex and the gonads are outlined in Figure 1. These pathways illustrate several interesting features associated with steroid hydroxylase gene expression and biosynthesis.

For example, it will be noted that tissue-specific gene expression leads to the appearance of P-450$_{11\beta}$ and P-450$_{C21}$ only in the adrenal cortex while P-450$_{scc}$ and P-450$_{17\alpha}$ are expressed in all steroidogenic tissues. Also P-450$_{11\beta}$ and P-450$_{scc}$ are localized in mitochondria while P-450$_{17\alpha}$ and P-450$_{C21}$ are found in the endoplasmic reticulum. In the adrenal cortex, steroidogenesis leads to production of glucocorticoids, mineralocorticoids, and C_{19}-precursors of sex hormones (dehydroepiandrosterone and adrostendione), while in the gonads the steroidogenic pathways lead only to production of C_{19} steroids which are then converted to the sex hormones estrogen or testosterone. However, it can be seen from Fig. 1, that the first step in all these pathways is the conversion of cholesterol to pregnenolone in the mitochondrion, catalyzed by P-450$_{scc}$.

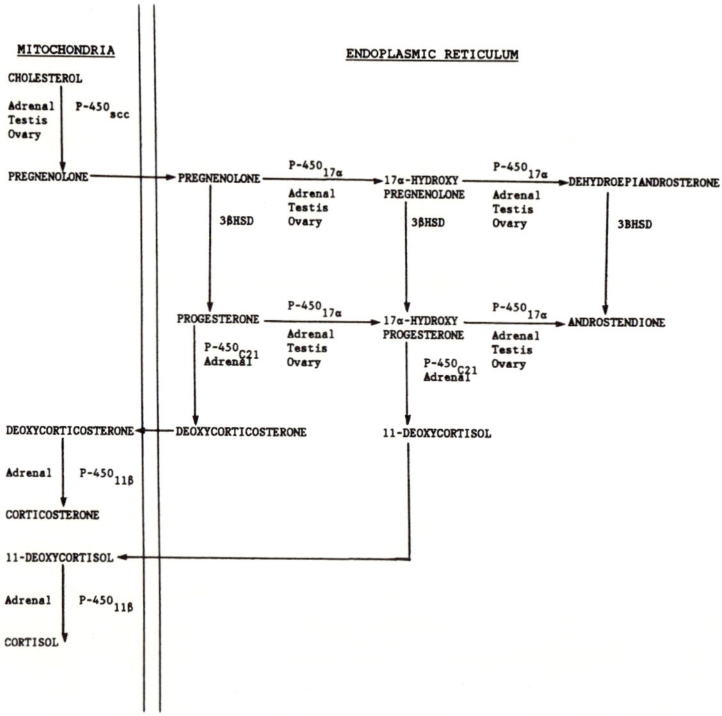

Fig. 1. Tissue localization and subcellular localization of steps in steroidogenic pathways. 3βHSD = 3β-hydroxysteroid dehydrogenase.

The steroid hydroxylases localized in the endoplasmic reticulum (P-450$_{17\alpha}$ and P-450$_{C21}$) are reduced by NADPH via the ubiquitous microsomal flavoprotein, NADPH-cytochrome P-450 reductase (Hiwatashi and Ichikawa, 1979). The mitochondrial steroid hydroxylases are reduced by NADPH via an electron transport chain localized in the mitochondrial matrix of steroidogenic tissues. Reducing equivalents are transferred from NADPH to a flavoprotein, adrenodoxin reductase (Omura et al., 1966), which in turn reduces an iron-sulfur protein adrenodoxin (Omura et al., 1966; Suzuki and Kimura, 1965). Adrenodoxin then transfers electrons to either P-450$_{scc}$ or P-450$_{11\beta}$, both of which are localized in the inner mitochondrial membrane.

While our understanding of regulation of steroid hydroxylase gene expression is not as detailed as that of TCDD regulation of P$_1$-450 gene expression (Jones et al., 1986), it will be seen that such detailed understanding is on the horizon. This is a complex process involving at least three distinct mechanisms of regulation of steroid hydroxylase gene expression; one leading to optimal steroidogenic capacity, one associated with tissue-specific expression and one required for developmental regulation.

METHODS

Primary cultures of bovine adrenocortical cells were prepared as described by Gospodarowicz et al., 1977 and manipulated as described by Kramer et al. (1984). Characterization of cDNA clones utilized in these studies has been previously described: cholesterol side chain cleavage cytochrome P-450 - P-450$_{scc}$ (John et al., 1984); 11β-hydroxylase cytochrome P-450 - P-450$_{11\beta}$ (John et al., 1985); 17α-hydroxylase cytochrome P-450 - P-450$_{17\alpha}$ (Zuber et al., 1986); steroid 21-hydroxylase cytochrome P-450 - P-450$_{C21}$ (John et al., 1986a) and adrenodoxin (Okamura et al., 1985); as has their use as probes for hybridization analysis.

RESULTS AND DISCUSSION

cAMP-Mediated Gene Expression

Treatment of primary cultures of bovine adrenocortical cells with ACTH (1 μM) or dibutyryl cAMP (1 mM) for varying periods of time leads to accumulation of mRNA specific for $P\text{-}450_{scc}$, $P\text{-}450_{11\beta}$, $P\text{-}450_{C21}$, $P\text{-}450_{17\alpha}$ and adrenodoxin. An example of such data obtained in analysis of $P\text{-}450_{17\alpha}$ mRNA is shown in Figure 2.

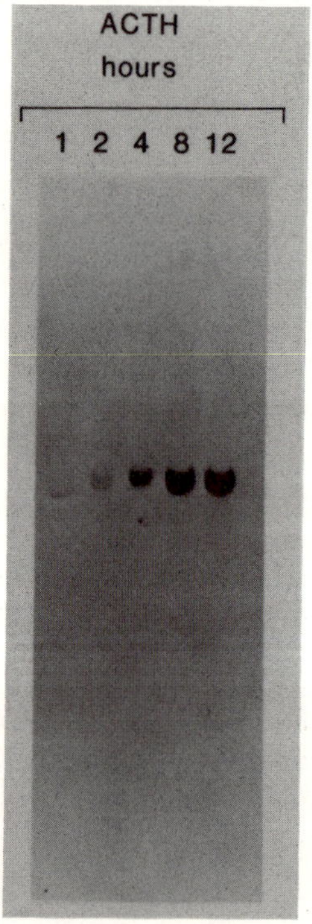

Fig. 2. Temporal pattern of enhancement of $P\text{-}450_{17\alpha}$ mRNA levels in primary cultures of bovine adrenocortical cells in response to ACTH treatment. (Zuber et al., 1986).

Utilizing nuclear ran-on techniques it was established that transcriptional activation of the genes encoding the four steroid hydroxylases and adrenodoxin plays a major role in the cAMP-mediated accumulation of mRNAs encoding these enzymes (John et al., 1986b). Moreover, this accumulation of specific mRNA encoding each of these enzymes was found to be inhibited by cycloheximide, a protein synthesis inhibitor. Thus it is concluded that labile protein factors are required for regulation of cAMP-mediated transcription of steroid hydroxylase genes. These proteins have been designated as SHIP (steroid hydroxylase inducing proteins) and a working hypothesis suggesting their transcriptional role is schematically presented in Fig. 3.

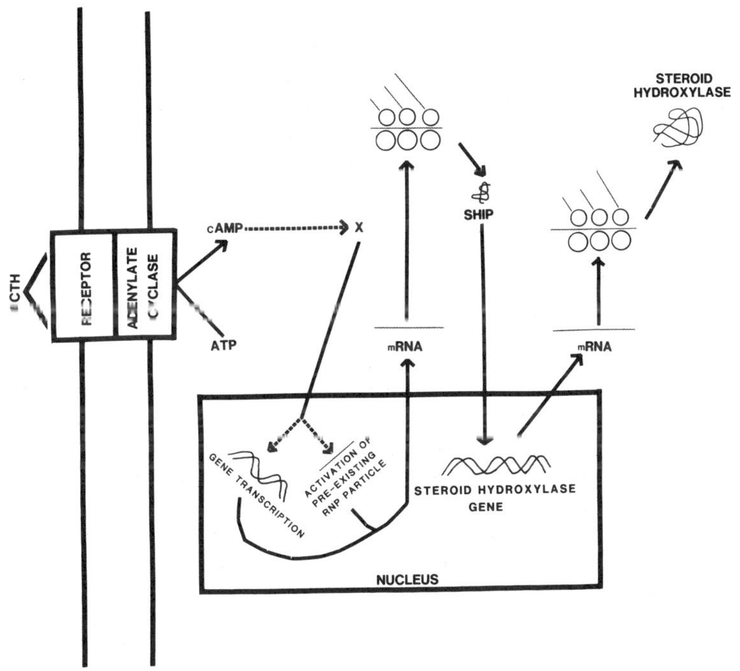

Fig. 3. Working hypothesis for cAMP-mediated transcriptional regulation of steroid hydroxylase gene expression. (Waterman & Simpson, 1985).

Studies are presently underway in our laboratories to identify the putative binding site(s) for SHIP in the 5'-flanking regions of steroid hydroxylase genes. From such studies it will initially be determined whether each P-450 requires a specific SHIP protein and ultimately the homology which may exist amongst SHIP. Of course it is also possible that a single SHIP protein might regulate transcription of all steroid hydroxylase genes.

Tissue-Specific Regulation

A second type of regulation of steroid hydroxylase gene expression is tissue-specific. The all-or-none type of tissue-specific regulation is illustrated in Fig. 1; $P\text{-}450_{scc}$ and $P\text{-}450_{17\alpha}$ being expressed in all steroidogenic tissues while $P\text{-}450_{11\beta}$ and $P\text{-}450_{C21}$ are expressed only in adrenal cortex. The molecular basis of the all-or-none type of tissue-specific regulation of steroid hydroxylase gene expression is unknown. However, Parker <u>et al</u>., (1986) have reported a DNA segment in the 5'-flanking region of the mouse $P\text{-}450_{C21}$ gene which is required for this type of regulation.

A second type of tissue-specific regulation involving steroid hydroxylase genes is evident in the ovary (Rodgers <u>et al</u>., 1987). In this case, the tissue- or cell-specific regulation is mediated by physiological factors. The ovarian follicle contains two types of steroidogenic cells, granulosa and theca interna. $P\text{-}450_{scc}$ is expressed in both cell types while $P\text{-}450_{17\alpha}$ is expressed in the theca interna. Following ovulation, these cell types differentiate into luteal cells forming the corpus luteum, a very active progesterone secreting tissue. During this differentiation process, $P\text{-}450_{scc}$ mRNA and adrenodoxin mRNA increase dramatically while $P\text{-}450_{17\alpha}$ mRNA disappears (Fig. 4). While the physiological basis of this increase in $P\text{-}450_{scc}$ gene expression and repression of $P\text{-}450_{17\alpha}$ gene expression is unknown, it is clearly different from the cAMP-dependent type of regulation where transcription of both genes is enhanced concordinately.

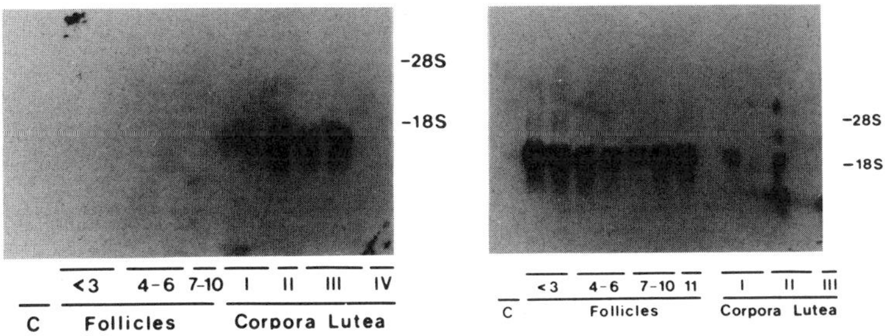

Fig. 4. Relative adrenodoxin (left) and P-450$_{17\alpha}$ (right) mRNA levels in bovine ovarian follicles and corpra lutea. The Arabic numbers refer to the diameters of the individual follicles (mm) while the Roman numerals refer to stage of development of corpra lutea. Each lane contains 5 µg poly A RNA. The pattern observed for adrenodoxin is similar to that observed for P-450$_{scc}$. (Rodgers et al., 1987).

Developmental Regulation

A third mechanism of regulation of steroid hydroxylase gene expression is developmental in nature and appears to be cAMP independent. We became aware of this mechanism during investigation of steroid hydroxylase mRNA levels in human fetal adrenals. Anencephalic human fetuses have no functional hypothalamic-pituitary axis and consequently very low or undetectable levels of ACTH in fetal blood. Also, the fetal adrenal membranes from anencephalic adrenals contain extremely low levels of adenylate cyclase activity. It was imagined, then, that RNA from such adrenals would be a good control for the cAMP-regulation described above, as undetectable levels of steroid hydroxylase mRNAs were predicted. Much to our surprise, this was not the case, as seen in Fig. 5 (John et al., 1987). The levels of steroid hydroxylase mRNA in anencephalic fetal adrenals were the same as in normal fetal adrenals. This indicates that a cAMP-independent mechanism of regulation of expression of these genes occurs during development, prior to the appearance of the cAMP-dependent mechanism. We have found the same profile in fetal sheep adrenals (John et al., 1987), which should prove to be a good model system for study of this fetal-imprinting mechanism.

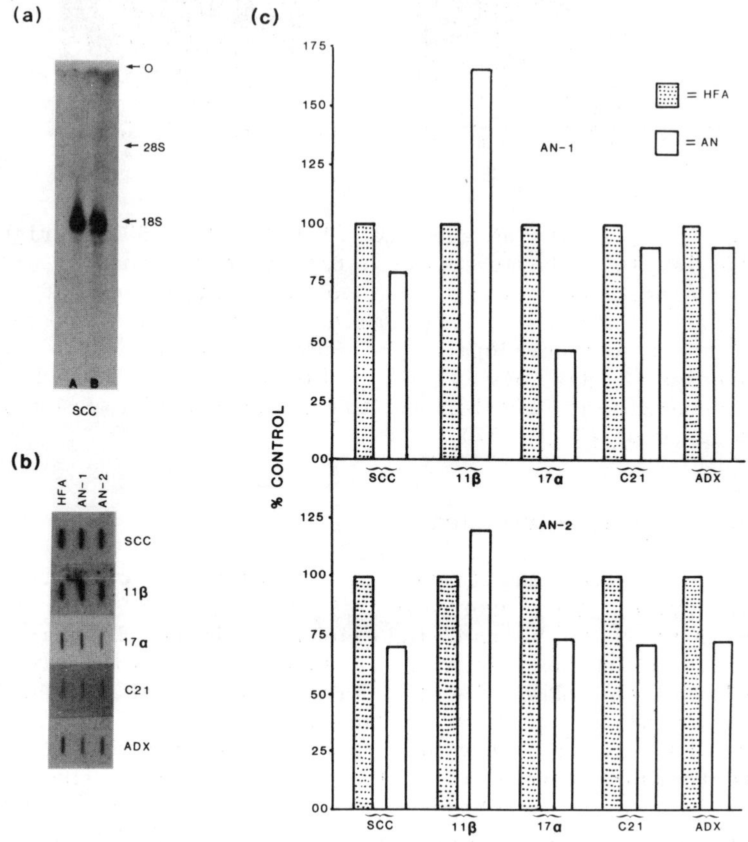

Fig. 5. Expression of steroid hydroxylase RNA in normal (HFA) and anencephalic (AN) fetal adrenals. (a) Northern blot of P-450$_{scc}$ RNA (A = AN; B = HFA); (b) Slot blot showing each RNA assayed; (c) Densitometric computations from data in (b). AN-1 and AN-2 represent data from two anencephalic fetuses. (John et al., 1987).

CONCLUSION

These studies indicate that at least 3 distinct mechanisms of regulation of steroid hydroxylase gene expression are required for steroidogenesis in the adult (Table 1). Optimal steroidogenic capacity is maintained by a cAMP-dependent mechanism involving labile protein factors (SHIP). Tissue- and cell-specific regulation of two types (all-or-none and physiologically controlled) result in the different steroidogenic pathways being localized in different steroidogenic tissues. Finally cAMP-independent fetal-imprinting leads to initial expression of these genes. A long-term goal of our laboratories is to elucidate the molecular basis of this multi-factorial regulation of steroid hydroxylase gene expression.

Table 1

Major Mechanisms of Steroid Hydroxylase Gene Expression

1. cAMP Dependent - SHIP

2. Cell-Specific

   a) All-or-none regulation
   b) Physiological regulation

3. Developmental - Fetal imprinting

ACKNOWLEDGMENTS

The authors wish to acknowledge the intellectual and experimental contributions of their colleagues to this research program. Also the support of USPHS Grants DK28350, HD11149, HD13234 and Grant I-624 from The Robert A. Welch Foundation is appreciated.

REFERENCES

Gospodrowicz, D., C.R. Ill, P.J. Hornsby and G.N. Gill (1977) Control of bovine adrenal cortical cell proliferation by fibroblast growth factor. Lack of effect of epidermal growth factor. Endocrinol. 100:1080-1089.

Hanukoglu, I. and Z. Hanukoglu (1986) Stoichiometry of mitochondrial cytochromes P-450, adrenodoxin and adrenodoxin reductase in adrenal cortex and corpus luteum. Eur. J. Biochem. 157:27-31.

Hiwatashi, A. and Y. Ichikawa (1979) Physicochemical properties of reduced nicotinamide adenine dinucleotide phosphate-cytochrome P-450 reductase from bovine adrenocortical microsomes. Biochim. Biophys. Acta 580:44-63.

John, M.E., M.C. John, P. Ashley, R.J. MacDonald, E.R. Simpson and M.R. Waterman (1984) Identification and characterization of cDNA clones specific for cholesterol side chain cleavage cytochrome P-450. Proc. Natl. Acad. Sci. USA 81:5628-5632.

John, M.E., M.C. John, E.R. Simpson and M.R. Waterman (1985) Cytochrome $P\text{-}450_{11\beta}$ gene expression: Transcriptional control by an ACTH regulated protein factor. J. Biol. Chem. 260:5760-5767.

John, M.E., T. Okamura, A. Dee, B. Adler, M.C. John, P.C. White, E.R. Simpson and M.R. Waterman (1986a) Bovine steroid 21-hydroxylase: Regulation of biosynthesis. Biochemistry 25:2846-2853.

John, M.E., M.C. John, V. Boggaram, E.R. Simpson and M.R. Waterman (1986b) Transcriptional regulation of steroid hydroxylase series by ACTH. Proc. Natl. Acad. Sci. USA 83:4715-4719.

John, M.E., E.R. Simpson, B.C. Carr, R.R. Magness, C.R. Rosenfeld, M.R. Waterman and J.I. Mason (1987) Ontogeny of adrenal steroid hydroxylases: Evidence for regulation of gene expression by an adenylate cyclase-independent mechanism. Mol. Cell. Endocrinol. 50:263-268.

Jones, D.B.C., L.K. Durrin, D.R. Galeazzi and J.P. Whitlock, Jr. (1986) Control of cytochrome P-450 gene expression: Analysis of a dioxin-responsive enhancer system. Proc. Natl. Acad. Sci. USA 83:2802-2806.

Kramer, R.E., W.E. Rainey, B. Funkenstein, A. Dee, E.R. Simpson and M.R. Waterman (1984) Induction of synthesis of mitochondrial steroidogenic enzymes of bovine adrenocortical cells by analogs of cAMP. J. Biol. Chem. 259:707-713.

Nebert, D.W., M. Adesnik, M.J. Coon, R.W. Estabrook, F.J. Gonzalez, F.P. Guengerich, I.C. Gunsalus, E.F. Johnson, B. Kemper, W. Levin, I.R. Phillips, R. Sato and M.R. Waterman (1987) The cytochrome P-450 supergene family: Recommended nomenclature. DNA 6:1-11.

Okamura, T., M.E. John, M.X. Zuber, E.R. Simpson and M.R. Waterman (1985) Molecular cloning and amino acid sequence of the precursor form of bovine adrenodoxin: Evidence of a previously unidentified C-terminal peptide. Proc. Natl. Acad. Sci. USA 82:5705-5709.

Okamura, T., M. Kagimoto, E.R. Simpson and M.R. Waterman (1987) Multiple species of bovine adrenodoxin mRNA: Occurance of two different mitochondrial precursor sequences associated with the same mature sequence. J. Biol. Chem. 262:10335-10338.

Omura, T., E. Sanders, R.W. Estabrook, D.Y. Cooper and O. Rosenthal (1966) Isolation from adrenal cortex of a nonheme iron protein and a flavoprotein functional as a triphosphopyridine nucleotide cytochrome P-450 reductase. Arch. Biochem. Biophys. 117:660-673.

Rodgers, R.J., M.R. Waterman and E.R. Simpson (1987) Levels of mRNA encoding cytochromes $P\text{-}450_{scc}$, $P\text{-}450_{17\alpha}$, adrenodoxin, and low density lipoprotein receptor in bovine follicles and corpra lutea throughout the ovarian cycle. Molec. Endocrinol. 1:274-279.

Waterman, M.R. and E.R. Simpson (1985). Mechanisms of induction of endogenous substrate oxidation. In Boobis, A.R., Caldwell, J., deMatteis, F., Elcombe, C.R. (eds): "Microsomes and Drug Oxidations", London: Taylor and Francis, Ltd., pp. 136-144.

Zuber, M.X., M.E. John, T. Okamura, E.R. Simpson and M.R. Waterman (1986) Bovine adrenocortical cytochrome $P\text{-}450_{17\alpha}$: Regulation of gene expression by ACTH and elucidation of primary sequence. J. Biol. Chem. 261:2475-2482.

DISCUSSION

NEBERT: With respect to your studies on developmental expression of steroidogenic P-450 genes in anencephalic newborns, would it not be more accurate to say that you have found an "ACTH-independent" (rather than cAMP-independent) form of gene expression?

WATERMAN: It is true that at present we cannot rule out cAMP as required for imprinting. However, it is possible that imprinting results from a cAMP-independent process such as action of a fetal growth factor.

PETERSON: In the immunological identification of the presence of proteins which cross react with P-450 antibodies, is there any enzymatic or spectroscopic indentification of these activities?

WATERMAN: There have been reports of steroidogenic activity in the brain for several years. However, the levels are very low and there is no good spectroscopic evidence for these steroid hydroxylases. Therefore, we must conclude that the immunohistochemical data suggesting the presence of $P\text{-}450_{scc}$ in brain must be confirmed by other techniques.

NEBERT: Comment. Years ago we found P-450 catalyzed metabolism (benzpyrene hydroxylation) in rodent brain and also in the pre-implantation mouse embryo. Certainly we do not expect this enzyme to exist primarily as a protective mechanism against cigarette smoke. Most likely the enzyme is important during development - if not steroid metabolism, perhaps the metabolism of prostaglandin, leukotrienes or biogenic amines.

MANN: (1) Are there any indications that the pattern of cytochromes P-450 in a given organ changes under pathological conditions? I am thinking specifically about pathways of steroidogenesis in adrenal tumors, polycystic ovary, and testicular hypogonadism. (2) Are P-450 cytochromes present in target organs for steroid hormones such as the endometrium of prostate, and if so, what is the function of these cytochromes P-450, particularly in relation to 17α-hydroxylase, but also in regard to conversions such as reduction of androstenedione to testosterone to dihydrotestosterone?

WATERMAN: (1) As the development of probes (both immunological and cDNA) specific for these enzymes is a recent event, studies on the systems that you describe have not yet been carried out. I imagine, however, that such pathological states will be examined in several laboratories in the near future.

(2) Ian Mason in Dallas has shown in sheep that following the appearance of $P\text{-}450_{17}$ in fetal adrenal, cortisol produced by this organ turns on expression of the same enzyme in placenta. As $P\text{-}450_{scc}$ and P-450 aromatase are already present in this tissue, the addition of $P\text{-}450_{17}$ leads to production of estrogen required for normal parturition. This same series of events appears to take place in bovine placenta but not in human placenta.

HORIE: Comment. We recently carried out isoelectric focusing analysis of $P\text{-}450_{scc}$-I from bovine adrenal cortex and corpus luteum. Results of protein staining, heme staining and staining with monoclonal antibody showed multiple bands and the patterns of the bands of both organs were different from each other. Although we do not know the nature of the multiple molecular forms yet, our results indicate that the multiplicity is also tissue specific.

JONES: Is there sufficient information available concerning the mechanisms of gene expression to gain any insight into evolution of gene expression for the various families of of cytochrome P-450?

WATERMAN: As Dan Nebert pointed out, only five P-450 genes have been characterized. Based on this small number, it has not been possible to evaluate evolutionary aspects of gene regulation.

Oxidases and Related Redox Systems, pages 557–576

EVOLUTION OF THE P450 GENE SUPERFAMILY

Daniel W. Nebert, John E. Jones, John Owens and Alvaro Puga

Laboratory of Developmental Pharmacology, NICHHD, NIH, Bethesda, MD 20892 (D.W.N., J.E.J., A.P.), Laboratory of Mathematical Biology, NCI, FCRF, Frederick, MD 21701 (J.O.)

ABSTRACT

A P450 gene nomenclature system has recently been proposed on the basis of divergent evolution of distinct families and subfamilies among eukaryotes and prokaryotes. At present, eleven gene families have been described, eight of which exist in all mammals. The current number of P450 genes in each species is believed to reflect, at least in part, selective advantages as animals and plants have struggled for coexistence during the last 1 billion years. It is likely that P450 genes will also be found in archaebacteria, as well as most or all eubacteria and eukaryotes. Using Wu-Kabat analysis, we propose regions of the P450 protein involved in membrane attachment, flavoprotein-binding, heme-binding, and substrate specificity. The invariant amino acid sequence (F**G***C*G in the heme-binding region) is unique to the fifty P450 proteins characterized to date and is not found in any of the more than 5,430 other proteins presently in the database. The regulatory trans-acting protein factors and their associated upstream cis-acting DNA elements of the orthologous mouse, rat and human P450IA1 gene are complex but appear to be highly conserved during the 80 million years since the mammalian radiation.

INTRODUCTION

Due to the recent advances in molecular biologic techniques, the field of P450 gene research has literally exploded. This Chapter will summarize the latest concepts in P450 gene evolution--at the level of the gene, the gene product (enzyme), and the upstream regulatory sequences controlling gene expression.

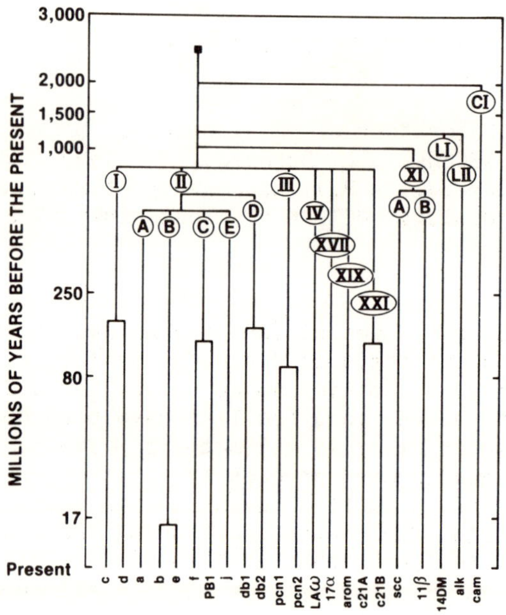

Figure 1. Evolution of the P450 gene family (Nebert and Gonzalez, 1987). Data are primarily based on rat for families I through IV, and human and cow for families XI, XVII, XIX, and XXI. Whereas these eight families exist in all mammals, the LI (Saccharomyces cerevisiae lanosterol 14-demethylase) and LII (Candida tropicalis alkane-inducible) (Sanglard et al., 1987) families have been characterized in yeast and the CI family is present in Pseudomonas. Both genes in family XI (subfamilies XIA and XIB) code for mitochondrial proteins. [Redrawn and reproduced with permission obtained from Annual Reviews, Inc.].

The Genes

Global alignment of 67 complete cDNA or protein sequences, available as of January 1987, led to a proposed nomenclature system based on evolution (Nebert et al., 1987; Nebert and Gonzalez, 1987). At present the total number of P450 sequences available exceeds 80, although more than two dozen appear to represent allelic variants of the same gene. These data are consistent with eleven families in the P450 gene superfamily, eight families of which exist in all mammals (Fig. 1). By definition (Nebert et al., 1987), the protein encoded by a gene in any one family exhibits ≦36% amino acid similarity to a protein encoded by a gene in any of the other ten families.

Within a subfamily all homologous genes so far examined appear to have diverged less than 300 million years ago (MYA), to code for proteins having ≧69% amino acid similarity, and to be chromosomally linked (*i.e.* within 50-100 kb from one another). Families II and XI contain subfamilies; in a given species, proteins within one subfamily show 37% to 68% amino acid similarity to a protein within any other subfamily of that family (Nebert and Gonzalez, 1987).

Despite ≦36% amino acid similarity between P450 proteins encoded by any two families, as examined by global alignment (Wilbur and Lipman, 1983), there is a highly conserved cysteine-containing fragment of 26 residues (*vide infra*). Between eight and eleven of the 26 residues are identical between most of the human P450 proteins and the bacterial protein (Nebert and Gonzalez, 1987). This finding argues strongly for divergent evolution, *i.e.* both eukaryotic and prokaryotic genes arose from a common ancestor.

The most popular present-day concept of ancestral genes includes the fact that exons originally were very small (6 to 15 bases), so that each gene contained dozens, or hundreds, of mini-exons divided by intron DNA (Fig. 2). We estimate that the ancestral P450 gene probably contained more than 100 exons. Examination of genes in the five P450 families studied to date (Fig. 3) provides evidence for at least 33 exons.

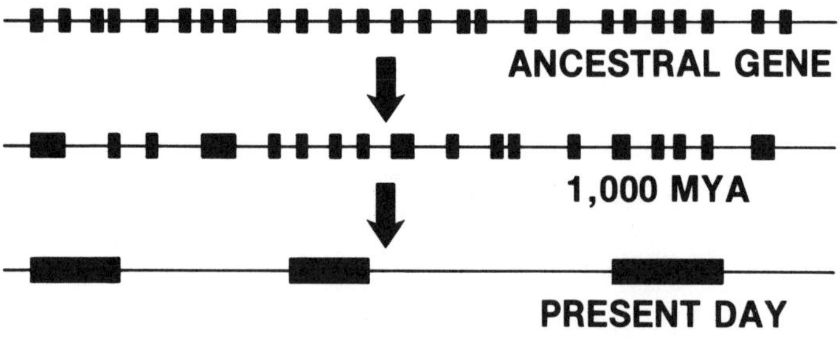

Figure 2. Illustration of the most popular theory as to how modern-day exons evolved via mini-exon fusion and intron loss. Exons are denoted as closed boxes; introns are shown as lines.

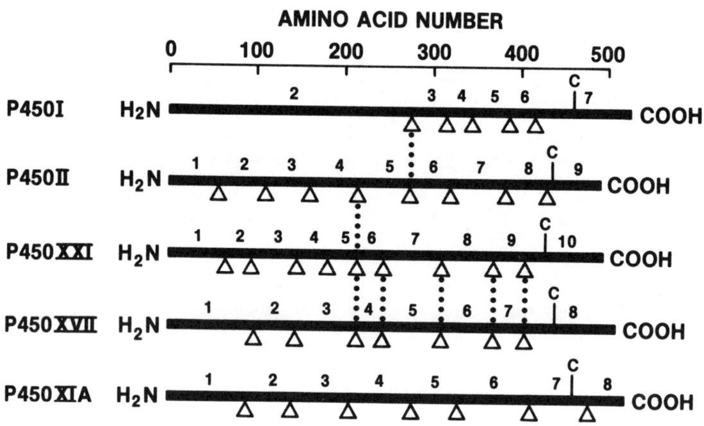

Figure 3. Exon-intron junctions (triangular arrows) of P450 proteins in each of five families. The cysteine (C) in the heme-binding site is shown in the COOH-terminal exon of all five genes. It is of interest that the 26-residue cysteine-containing conserved peptide involved in heme-binding is encoded by portions of the last two exons in the II family, but a portion of the last one exon in the other four families. P450I genes, but not the others, have a noncoding first exon. Exon-intron junctions in the IIB, IIC, IID and IIE subfamilies are identical. Except for the cases illustrated by vertical dots (in which exon-intron junctions are ≦2 residues of one another), all remaining junctions in the five families appear to be unique.

Although not appreciated by many colleagues, archaebacteria (Fig. 4) represent a third urkingdom that is almost as greatly diverged from eubacteria as eukaryotes are. An examination of the chemical structures of various archaebacterial cofactors (Fig. 5) leads one to conclude that it would be exceedingly unlikely for archaebacteria not to contain P450 enzymes needed for the synthesis and degradation of these critical life compounds. If it is true that these three urkingdoms diverged from a common ancestor between 2,500 and 3,500 MYA, we conclude that the P450 gene superfamily is extremely ancient.

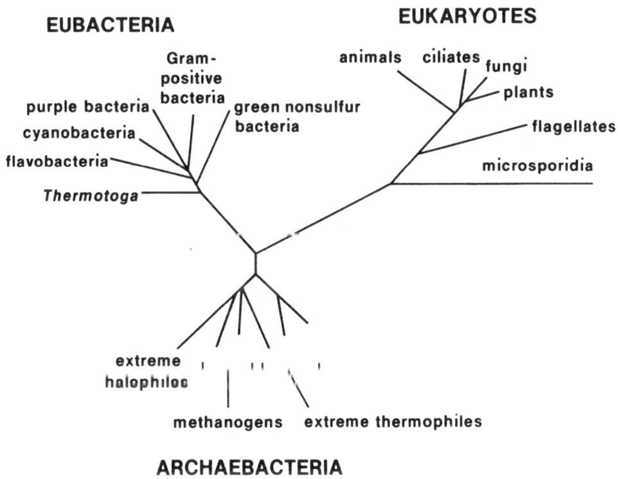

Figure 4. Universal phylogenetic tree determined from rRNA sequence comparisons (Woese, 1987). A matrix of evolutionary distances was calculated from an alignment of representative 16S rRNA sequences from each of the three urkingdoms. This was used to construct a distance tree (DeSoete, 1983), based upon those positions represented in all sequences in the alignment of homologous secondary structural elements. Line lengths on the tree are proportional to calculated distances [Reproduced with permission obtained from Academic Press, Inc.].

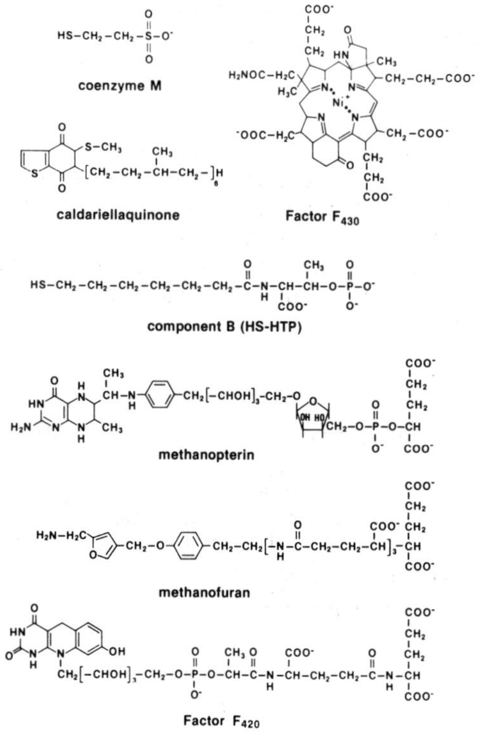

Figure 5. Structure of various archaebacterial cofactors (Woese, 1987). All are found in methanogens and utilized in methane production, except for caldariellaquinone, which occurs in Sulfolobus species [Reproduced with permission obtained from Academic Press, Inc.].

There appears to have been an accelerated rate of P450 gene duplication and divergence between 600 and 1,000 MYA (Fig. 1). This explosion of new genes might reflect the separation of the plant and animal kingdoms that occurred during this time (Nebert and Gonzalez, 1987). For the past one billion years, animals ingested plants in order to survive, plants developed new chemicals (phytoalexins) as a defense, and animals countered with new metabolic pathways in order to continue ingesting these plants. It seems reasonable that P450 genes commonly divided by duplication (Fig. 6). By means of mutations leading to one or more amino acid changes, gene products then developed new catalytic activities. In some cases the mutation would create a pseudogene. Selective pressure for a new catalytic activity could become an important evolutionary advantage for the survival of a particular plant or animal species. Present-day pharmaceutical compounds and certain industrial chemicals involved in environmental contamination are, in fact, plant metabolites or resemble closely in chemical structure such metabolites.

Species differences in drug metabolism have been appreciated for more than 30 years. Now that we have a better understanding of P450 gene evolution, it is possible to understand such observations more clearly. Since

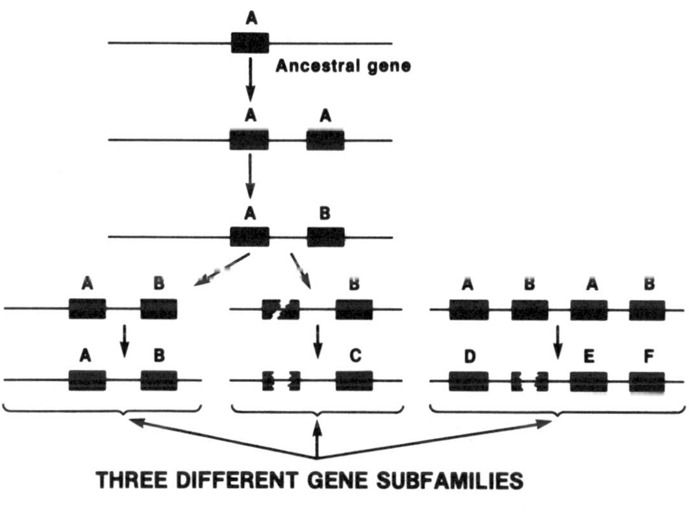

Figure 6. Illustration of the mechanism by which one ancestral gene can lead to multiple subfamilies that each contain distinctly different genes (A, B, C, D, E and F). A broken solid box denotes a nonfunctional gene. Evolutionary pressure for the selection of proteins with new catalytic activities has probably been important for animals ingesting plants, as well as for plants attempting to protect themselves from being eaten by animals.

the time of the split of human and rodents approximately 80 MYA (Fig. 7), it is likely that the number and type of P450 genes in each species has changed during the intervening period. Such changes might reflect manifestations of random mutations, the founder effect among humans (Vogel and Motulsky, 1986), bottlenecks observed among plant and animal species (O'Brien et al., 1987), compensating advantages for the heterozygote (Diamond, 1987), and survival advantages based on eating plants or being eaten by animals. Examples of P450 genes in one species but not another (Nebert and Gonzalez, 1987) include: (i) rat b and e, which have diverged since the rat-mouse split and therefore both orthologous genes will not be found in mouse (or rabbit, or human); (ii) rabbit form 1 and 88, which have separated since the rabbit-rodent split and therefore both orthologous genes will not be found in rat (or mouse, or human); and (iii) the IID subfamily, which appears to contain four functional genes in the rat yet only one functional gene in the human.

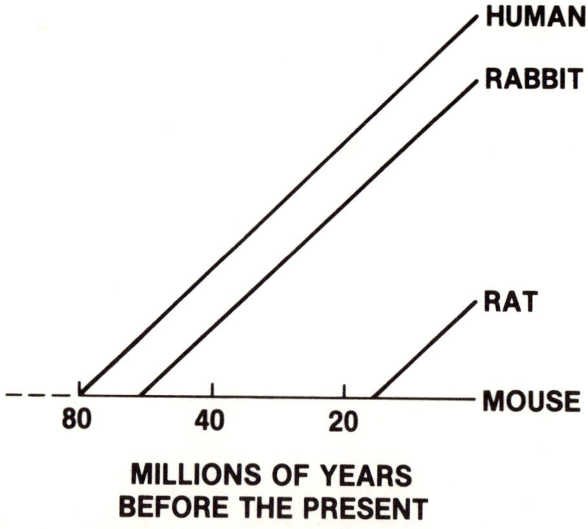

Figure 7. Diagram of the divergence of four mammalian species since the mammalian radiation, which has been estimated to have occurred about 80 MYA (Fitch and Langley, 1976; Wilson et al., 1977; Miyata et al., 1982).

The Gene Products

Proteins encoded by homologous genes within the same subfamily display strikingly similar hydropathy index profiles (Fig. 8), whereas proteins encoded by genes in other P450 gene families exhibit remarkably different hydropathy index profiles. It is particularly worth noting that the patterns of two proteins encoded by non-orthologous P450I genes in rat and mouse are very similar, in contrast to the patterns of the three proteins encoded by genes in three different rat P450 gene families (Fig. 8).

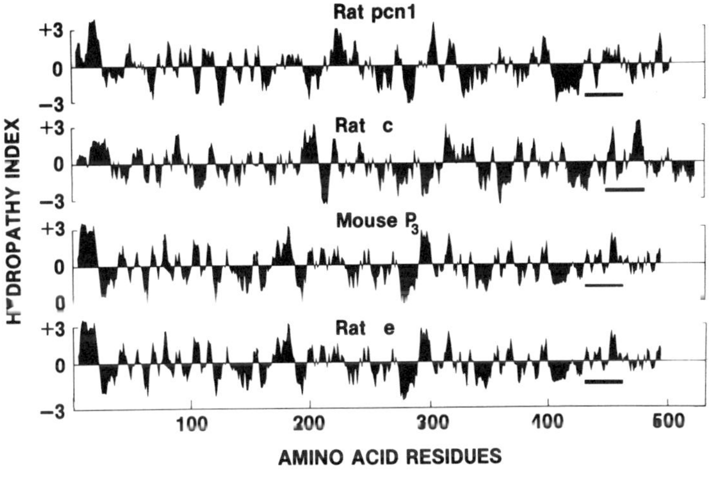

Figure 8. Comparison of the hydropathy index for four P450 proteins that are cited in Nebert and Gonzalez (1987). A sliding window of six amino acids was used (Kyte and Doolittle, 1982); positive values represent hydrophobic domains. Rat c and mouse P_3 reside in the same family (P450I), while rat pcn1 and rat e correspond to genes in the P450III and P450II families, respectively. The solid bar represents the C-terminal cysteine-containing fragment involved in the heme-binding site.

Although proteins encoded by genes within the same P450 subfamily show highly similar hydropathy index profiles, it is well known that such gene products exhibit unique substrate specificities. We therefore reasoned that any hypervariable regions found by Wu-Kabat analysis (Fig. 9) would be more likely involved in giving each enzyme its unique substrate specificity. No conclusions could be gained from the P450IA subfamily analysis (Fig. 9, top); too much background noise is probably due to the fact that only two orthologous genes among five eukaryotic species are represented. On the other hand, examination of the IIC subfamily (Fig. 9, bottom), provides several definitive regions. We suggest that the first highly conserved region (residues 1-90), which includes the signal peptide (residues 1-20), might take part in the association of P450 protein with the endoplasmic reticulum. Internal highly conserved regions (residues 110-190 and 425-455) most likely represent the flavoprotein- and heme-binding regions, respectively. The hypervariable regions, labeled substrate-binding regions 1 (residues 190-310), 2 (residues 360-420) and 3 (residues 460-end) may contribute to the diversity in catalytic activities known to exist among the P450IIC enzymes (Nebert and Gonzalez, 1987).

Future studies involving biologic mutants or genetically engineered chimeric cDNAs or hybrid proteins should benefit from Wu-Kabat analysis or similar studies. One recent report could be viewed as consistent with the Fig. 9 data. Kimura et al. (1987) examined the P_1 cDNA sequence in the mouse hepatoma Hepa-1 wild-type and the benzo[a]pyrene-resistant c37 mutant cell line and found that two mutations (L ---> R at residue 118 and R ---> P at residue 245) produced a complete absence of catalytic activity.

Conclusions from the elegant studies of Murakami and coworkers (Fig. 10) are also consistent with the Fig. 9 data. If the folding of the P450c/NADPH-P450 oxidoreductase hybrid protein in the yeast endoplasmic reticulum is anything similar to that in the Fig. 10 diagram, it seems reasonable that substrate-binding regions 1, 2 and 3 would surround the heme-binding pocket and comprise the substrate-binding site. Such a folding of the hybrid protein would also be efficient for the passage of electrons via the FAD and FMN moieties (Vermilion et al., 1981) to the substrate-Fe^{+++} and substrate-$Fe^{+++}(O_2\cdot)^-$ complexes in the enzyme active-site.

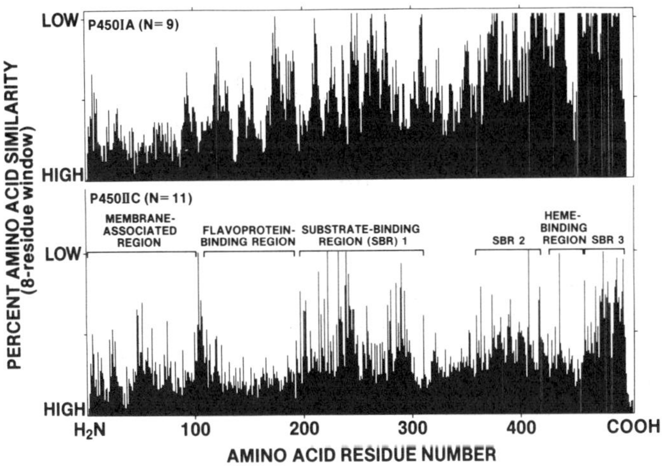

Figure 9. Wu-Kabat plot (Wu and Kabat, 1970) of P450 subfamilies IA (top) and IIC (bottom). The Wu-Kabat algorithm determines the degree of amino acid variability at each position in a set of aligned protein sequences, thus identifying the so-called hypervariable regions. The ordinate represents increasing variability values, calculated as the ratio of the number of different amino acids at a given position to the frequency of the most common amino acid at that position. A sliding window of eight residues was used. A 9-way (top) and 11-way (bottom) comparison was used to generate the data. Examination of the proteins in any other P450 subfamily was inconclusive due to the small number of sequences available, and analysis of the proteins in two or more P450 subfamilies was unsuccessful due to too much background noise. Highly conserved regions are believed to participate in membrane attachment and in binding to the flavoprotein and heme moieties, whereas highly divergent regions are believed to account for differences in substrate binding and therefore dissimilarities in catalytic activities.

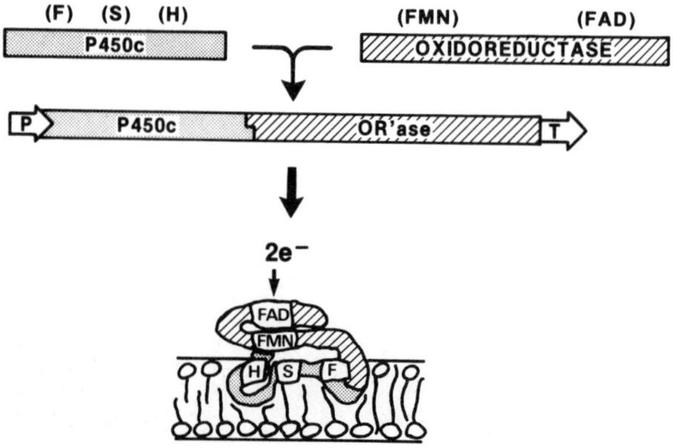

Figure 10. Illustration of the rat P450c/NADPH-P450 oxidoreductase chimeric cDNA constructed by Murakami et al. (1987). The resultant hybrid protein is believed to reside in the yeast microsomal membrane and has been shown to carry out the catalytic function of benzo[a]-pyrene hydroxylation. The 5' end of the oxidoreductase lacks the NH_2-terminal 56 amino acids, and the 3' end of the P450c lacks the five COOH-terminal amino acids. (F), (S) and (H) denote putative flavoprotein-, substrate- and heme-binding regions of P450c. (FMN) and (FAD) denote FMN- and FAD-binding regions of the oxidoreductase (Porter and Kasper, 1985). P and T inside the arrows denote the yeast ADH1 promoter and terminator, respectively.

Alignment of the cysteine-containing peptide of all fifty P450 monooxygenases in the data base (Fig. 11) demonstrates that four amino acids are invariant among all proteins in one prokaryote and ten eukaryotes. A search of 5,432 proteins currently in the data base showed that no protein other than the P450 enzymes exhibited the amino acid sequence F**G***C*G.

Two proteins having a region somewhat similar to the 26-residue P450 fragment are human cytochrome b_{245} (Royer-Pokora et al., 1986; Teahan et al., 1987) and chloroperoxidase from the fungus Caldariomyces fumago (Fang et al., 1986). Although not yet known, if the fifth ligand of the protoheme iron is indeed a thiolate ligand

Subfamily			
I	P1	MOUSE	FGLGKRKCIGETIGRSEVFLFLAILL
	c	RAT	FGLGKRKCIGETIGRLEVFLFLAILL
	P1	HUMAN	FGMGKRKCIGETVARWEVFLFLAILL
	6	RABBIT	FGLGKRKCIAETIGRLEVFLFLATLL
	P1	FISH	FGMDKRRCIGEAIGRNEVFLFLAILL
	P3	MOUSE	FGLGKRRCIGEIPAKWEVFLFLAILL
	d	RAT	FGLGKRRCIGEIPAKWEVFLFLAILL
	P3	HUMAN	FGMGKRRCIGEVLAKWEIFLFLAILL
	4	RABBIT	FGLGKRRCIGETLARWEVFLFLAILL
IIA	a	RAT	FSTGKRFCLGDGLAKMELFLLLTTIL
	2a	HUMAN	FSIGQPNCFGEGLARMELFLFFTTVM
IIB	b & e	RAT	FSTGKRICLGEGIARNELFLFFTTIL
	2	RABBIT	FSLGKRICLGEGIARTELFLFFTTIL
IIC	PB1	RAT	FSAGKRMCAGEGLARMELFLFTTILQ
	f	RAT	FSAGKRACVGEGLARMQLFLFLTTIL
	h	RAT	FSAGKRICAGEALARTELFLFFTTIL
	mp	HUMAN	FSAGKRICVGEALAGMELFLFLTSIL
	1	HUMAN	FSAGKRICAGEGLARMELFLFLTTIL
	PBc1	RABBIT	FSTGKRVCVGEALVRMELFLFLTAIL
	PBc2	RABBIT	FSTGKRVCVGEALARMELFLFLTAIL
	3b	RABBIT	FSTGKRACVGEGLARMELFLLLTTIL
	1	RABBIT	FSAGKRMCVGEGLARMELFLFLTSIL
	88	RABBIT	FSAGKRMCVGEGLARMELFLFLTSIL
	PB15	CHICKEN	FSAGKRICAGEGLARMEIFLFLTSIL
IID	db1 & 2	RAT	FSAGRRACLGEPLARMELFLFFTCLL
	db1	HUMAN	FSAGRRACLGEPLARMELFLFFTSLL
IIE	j	RAT	FSAGKRVCVGEGLARMELFLLLSAIL
	j	HUMAN	FSTGKRVCAGEGLARMELFLLLCAIL
	3a	RABBIT	FSAGKRVCVGEGLARMELFLLLSAIL
III	pcn1 & 2	RAT	FGNGPRNCIGMRFALMNMKLALTKVL
	nf (HLp)	HUMAN	FGSGPRNCIGMRFALMNMKLALIRVL
	3c	RABBIT	FGAGPRNCLGMRFALMNIKLALVRLM
IV	LA-omega	RAT	FSGGARNCIGKQFAMSEMKVIVALTL
	p-2	RABBIT	FSGGARNCIGKQFAMRELKVAVALTL
XIA	scc	HUMAN	FGWGVRQCLGRRIAELEMTIFLINML
	scc	COW	FGWGVRQCVGRRIAELEMTLFLIHIL
XIB	11-beta	COW	FGFGVRQCLGRRVAEVEMLLLLHHVL
XVII	17-alpha	HUMAN	FGAGPRSCIGEILARQELFLIMAWLL
	17-alpha	COW	FGAGPRSCVGEMLARQELFLFMSRLL
	17-alpha	PIG	FGAGPRSCVGEMLARQELFLFTAGLL
XIX	arom	HUMAN	FGFGPRGCAGKYIAMVMMKAILVTLL
XXI	c21	MOUSE	FGCGARVCLGEPLARLELFVVLARLL
	c21	HUMAN	FGCGARVCLGEPVARLELFVLLTRLL
	c21	COW	FGCGARVCLGESLARLELFVVLLRLL
	c21	COW	FGCGAPVCLGEPLARLELFVVLTRLL
LI	14dm	YEAST	FGGGRHRCIGEHFAYCQLGVLMSIFI
LII	alk	YEAST	FNGGPRICLGQQFALTEASYVTVRLL
CI	cam	Ps.putida	FGHGSHLCLGQSLARREIIVTLKEWL
		Consensus:	F**G*R*C*G***AR**L*L*L***L
Cytochrome b245		HUMAN	F Y W LCRDTHAFEWFADL L QLL
Chloroperoxidase		C.fumago	YVTGSD C GDSLVN LTL L

Figure 11. Sequence alignment of the 26-residue heme-binding fragment from the fifty P450 proteins presently known to be encoded by distinctly different genes. References to these genes, cDNAs and/or proteins can be found in Nebert et al. (1987) and Nebert and Gonzalez (1987). At bottom are the two proteins other than P450 in the current data base that contain fragments exhibiting the highest degree of similarity to the conserved P450 heme-binding peptide.

involving cysteine, it would be tempting to speculate that these peptides arose from the same ancestral peptide responsible for the P450 heme-binding region.

At present no evidence exists for an ancestral exon(s) encoding the cysteine-containing peptide responsible for the binding of iron in hemoproteins. Other residues commonly used in heme binding include histidine and methionine. It will be of interest in the future to search for cysteine-, histidine- and methionine-containing peptides, which have arisen during evolution, that participate in heme-binding.

Upstream Regulatory Sequences

Plasmids containing the mouse P_1450 upstream sequences driving the reporter chloramphenicol acetyltransferase gene (CAT gene) have been transfected into mouse and human cell lines (Gonzalez and Nebert, 1985; Jones et al., 1986; Neuhold et al., 1986; Sogawa et al., 1986; Fujisawa-Sehara et al., 1987; Jaiswal et al., 1987a,b). Stable transformants of the Hepa-1 wild-type (wt), c1 (P_1^-), and c4 (nt^-) cell lines have given particularly important information, because the degree of dependence of particular upstream regions on P_1450 metabolism and on a functional Ah receptor could be determined by using the c1 and c4 lines, respectively (Gonzalez and Nebert, 1985; Jaiswal et al., 1987a).

All of our data are consistent with a complex positive control element (TCDD-inducible enhancer) between -1140 and -920, containing at least two TCDD-responsive sites having no apparent homology to one another, and a control element responsible for augmenting constitutive P_1450 gene expression (Fig. 12). About halfway between this region and the transcription initiation site is a negative control element that responds to P_1450 metabolism by repressing constitutive P_1450 gene transcription. Loss of this negative element, and especially a defect in P_1450 metabolism, leads to an absence of this repression and, therefore, derepression of constitutive P_1450 gene transcription. It is worth noting that this negative autoregulatory loop involving the negative control element, and even the TATA box promoter, appear to be dependent upon a functional Ah receptor (Fig. 12). The TCDD-inducible

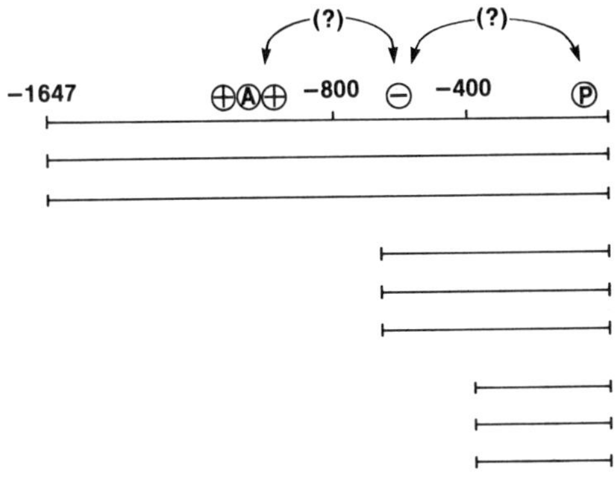

Construct	Cell type	Control	TCDD
−1647	wt	20	80
	nt^-	≤6.0	≤6.0
	P_1^-	70	70
−800	wt	≤6.0	≤6.0
	nt^-	≤6.0	≤6.0
	P_1^-	50	50
−400	wt	8.0	8.0
	nt^-	≤6.0	≤6.0
	P_1^-	20	20
pSV0-*cat*	wt	≤6.0	≤6.0

Figure 12. Summary of P_1450 upstream regulatory sequences driving the chloramphenicol acetyltransferase (CAT) gene in mouse hepatoma Hepa-1 wild-type (wt) cells and the nuclear translocation-defective (nt^-) and P_1450 metabolism-deficient (P_1^-) mutant cell lines (Nebert and Gonzalez, 1987). The -1647, -800, and -400 denote the number of bases upstream from the mRNA cap site. The positive (⊕) and negative (⊖) control regions upstream and downstream from -800, respectively, are illustrated. A DNA element within the TCDD-inducible enhancer that augments constitutive transcription (Ⓐ) and the TATA box promoter (Ⓟ) region are also shown. Interactions among the TCDD-inducible enhancer, the negative control element, and the promoter region are unknown but depicted as arrows with question marks. The pSV2-*neo* was cotransfected with a 15-fold excess of these various constructs, and stable transformants were selected on the basis of their resistance to G418. Numbers at right denote average values for CAT activity (pmol/min/mg protein) in pooled transformants treated with control medium or TCDD [Reproduced with permission obtained from Annual Reviews, Inc.].

enhancer region and the negative control region are both highly conserved between the mouse and the human P450IA1 upstream sequences (Neuhold et al., 1986). Mouse *trans*-acting protein regulatory factors can function with the human P450IA1 upstream sequences, and vice versa (Jaiswal et al., 1987a).

ACKNOWLEDGMENT

The expert secretarial assistance of Ingrid E. Jordan is greatly appreciated.

REFERENCES

Diamond JM (1987). Causes of death before birth. Nature 329:487-488.

DeSoete G (1983). A least squares algorithm for fitting additive trees to proximity data. Psychometrika 48: 621-626.

Fang G-H, Kenigsberg P, Axley MJ, Nuell M, Hager LP (1986). Cloning and sequencing of chloroperoxidase cDNA. Nucl Acids Res 14:8061-8071.

Fitch WM, Langley CH (1976). Protein evolution and the molecular clock. Fed Proc 35:2092-2097.

Fujisawa-Sehara A, Sogawa K, Yamane M, Fujii-Kuriyama Y (1987). Characterization of xenobiotic responsive elements upstream from the drug-metabolizing cytochrome P-450c gene: a similarity to glucocorticoid regulatory elements. Nucl Acids Res 15:4179-4191.

Gonzalez FJ, Nebert, DW (1985). Autoregulation plus upstream positive and negative control regions associated with transcriptional activation of the mouse P_1450 gene. Nucl Acids Res 13:7269-7288.

Jaiswal AK, Gonzalez FJ, Nebert DW (1987a). Comparison of human and mouse P_1450 upstream regulatory sequences in liver- and nonliver-derived cell lines. Mol Endocrinol 1:312-320.

Jaiswal AK, Neuhold LA, Nebert DW (1987b). Human P450IA1 upstream regulatory sequences expressing the chloramphenicol acetyltransferase gene. Effect of Ha-MSV enhancer and comparison of transient with stable transformation assays. Biochem Biophys Res Commun, in press.

Jones PBC, Durrin LK, Galeazzi DR, Whitlock Jr JP (1986). Control of cytochrome P_1-450 gene expression: Analysis of a dioxin-responsive enhancer system. Proc Natl Acad Sci USA 83:2802-2806.

Kimura S, Smith HH, Hankinson O, Nebert DW (1987). Analysis of two benzo[a]pyrene-resistant mutants of the mouse hepatoma Hepa-1 P_1450 gene via cDNA expression in yeast. EMBO J 6:1929-1933.

Kyte J, Doolittle RF (1982). A simple message for displaying the hydropathic character of a protein. J Mol Biol 157:105-132.

Miyata T, Hayashida H, Kikuno R, Hasegawa M, Kobayashi M, Koike K (1982). Molecular clock of silent substitution: At least six-fold preponderance of silent changes in mitochondrial genes over those in nuclear genes. J Mol Evol 19:28-35.

Murakami H, Yabusaki Y, Sakaki T, Shibata M, Ohkawa H (1987). A genetically engineered P450 monooxygenase: Construction of the functional fused enzyme between rat cytochrome P450c and NADPH-cytochrome P450 reductase. DNA 6:189-197.

Nebert DW, Adesnik M, Coon MJ, Estabrook RW, Gonzalez FJ, Guengerich FP, Gunsalus IC, Johnson EF, Kemper B, Levin W, Phillips IR, Sato R, Waterman MR (1987). The P450 gene superfamily. Recommended nomenclature. DNA 6: 1-11.

Nebert DW, Gonzalez FJ (1987). P450 genes. Structure, evolution and regulation. Annu Rev Biochem 56: 945-993.

Neuhold LA, Gonzalez FJ, Jaiswal AK, Nebert DW (1986). Dioxin-inducible enhancer region upstream from the mouse P_1450 gene and interaction with a heterologous SV40 promoter. DNA 5: 403-411.

O'Brien SJ, Wildt DE, Bush M, Caro TM, FitzGibbon C, Aggundey I, Leakey RE (1987). East African cheetahs: Evidence for two population bottlenecks? Proc Natl Acad Sci USA 84:508-511.

Porter TD, Kasper CB (1985). Coding nucleotide sequence of rat NADPH-cytochrome P-450 oxidoreductase cDNA and identification of flavin-binding domains. Proc Natl Acad Sci USA 82: 973-977.

Royer-Pokora B, Kunkel LM, Monaco AP, Goff SC, Newburger PE, Baehner RL, Cole FS, Curnutte JT, Orkin SH (1986). Cloning the gene for an inherited human disorder--chronic granulomatous disease--on the basis of its chromosomal location. Nature 322: 32-38.

Sanglard D, Chen C, Loper JC (1987). Isolation of the alkane inducible cytochrome P450 (P450alk) gene from the yeast Candida tropicalis. Biochem Biophys Res Commun 144:251-257.

Sogawa K, Fujisawa-Sehara A, Yamane M, Fujii-Kuriyama Y (1986). Location of regulatory elements responsible for drug induction in the rat cytochrome P-450c gene. Proc Natl Acad Sci USA 83: 8044-8048.

Teahan C, Rowe P, Parker P, Totty N, Segal AW (1987). The X-linked chronic granulomatous disease gene codes for the β-chain of cytochrome _b_-$_{245}$. Nature 327: 720-721.
Vermilion JL, Ballou DP, Massey V, Coon MJ (1981). Separate roles for FMN and FAD in catalysis by liver microsomal NADPH-cytochrome P-450 reductase. J Biol Chem 256:266-277.
Vogel F, Motulsky AG (1986). "Human Genetic Problems and Approaches," Berlin: Springer-Verlag, 807 p.
Wilbur WJ, Lipman DJ (1983). Rapid similarity searches of nucleic acid and protein data banks. Proc Natl Acad Sci USA 80: 726-730.
Wilson AC, Carlson SS, White TJ (1977). Biochemical evolution. Annu Rev Biochem 46: 473-639.
Woese CR, Gutell R, Gupta R, Noller HF (1983). Detailed analysis of the higher-order structure of 16S-like ribosomal ribonucleic acids. Microbiol Rev 47: 621-669.
Wu TT, Kabat EA (1970). An analysis of the sequences of the variable regions of Bence Jones proteins and myeloma light chains and their implications for antibody complementarity. J Exp Med 132: 211-250.

DISCUSSION

RYDEN: The cytochrome P-450 proteins now constitute one of the families where enough protein and gene sequences are available to study basic questions of evolution, such as the role of positive and negative selection, etc. Do you know if such studies are being done?

NEBERT: The fact that the rat has at least four functional IID genes whereas the human has only one functional IID gene [Gonzalez et al. (1987)] might be the first known example of selective pressure within one P-450 gene subfamily. With the advent of an increasing number of efficient cDNA expression vectors becoming available, I predict that the next decade will see a great flurry of research in the area of genetically engineered chimeric cDNAs and mutated cDNAs and studies on the catalytic activities and enzyme activities of the resultant hybrid proteins.

RYDEN: My second point is that according to more recent protein and RNA sequence phylogenies the relationships between the three kingdoms of Woese have been changed so that eukaryotes seem to have evolved from archaebacteria. Then certainly P-450 genes are expected to be found in these bacteria since they are found in the eubacteria.

NEBERT: I agree. It would be exceedingly likely to find P-450 genes in archaebacteria.

ESTABROOK: A number of years ago you predicted that the different P-450s would exceed 100 - perhaps even approach 1000 in number. Some people laughed while others violently disagreed. The current trend of new P-450s added to the inventory indicates you were correct in your prediction. You coupled your original prediction with an analogy to the diversity of antibodies. Do you still support this second prediction?

NEBERT: The review in which I raised these hypotheses [Nebert (1979) Mol Cell Biochem 27: 27-46] included a number of recently appreciated molecular mechanisms by which a finite number of genes can deal with an almost infinite number of challenges. These mechanisms include DNA rearrangements (such as those which help create antibody diversity), transpositions, alternative splicing, mRNA stabilization, and posttranslational modifications. Although I also predicted hypervariable regions and conserved regions of P-450 proteins, these can be found in virtually every protein within any given gene superfamily. I have never proposed the development of individual clonal populations of cells in liver, or any other tissue besides lymphocytes, simply because there is no physiologic or cellular evidence to support such a statement.

The observation of "as many as 15 mu-switch-like regions present in germ-line DNA" that are apparently not used by the immunoglobulin system [Kirsch et al. (1981) Nature 293: 585-587] was an interesting finding because it caused me to wonder if P-450 diversification might use the mutant switch-like regions of DNA. It therefore seemed intriguing to me in 1979 that DNA rearrangements or alternative splicing of RNA might somehow account for P-450 protein diversity in the face of chemical adversity. At present, we are aware, however, of at least 50 experimentally documented P-450 genes, where this prediction is not correct. Knowing now that each P-450 protein is derived from its own distinct gene, I believe that the animal-plant warfare hypothesis [Nebert, Gonzalez (1987) Ann Rev Biochem 56: 945-993] is most plausible. Many of the P-450 genes today represent about 1 billion years of evolution during which animals have adapted to changes in plants and vice versa. Most P-450 substrates and inducers--substances found in drug stores of the twentieth century--are plant metabolites or are chemically similar in structure to plant metabolites.

COON: In contrast to the polycyclic hydrocarbons, for which your laboratory and others have found receptors, highly water soluble inducers of P-450 like phenobarbital, ethanol, and acetone are not yet known to have receptors. Have you looked for a phenobarbital receptor?

NEBERT: Our laboratory has not looked for a phenobarbital receptor. From studies in mouse [Poland et al (1980) Mol Pharm 18: 571-580; Kende et al (1985) Mol Pharm 28: 445-453] and rat [Poland et al (1981) Mol Pharm 20: 442-450] liver with the prototype phenobarbital inducer 1,4-bis[3,5-dichloropyridoxyl)]-benzene (TCPOBOP), their data are consistent with a specific ligand-binding site (or receptor) mechanism. Inducers of the P450I family (polycyclic hydrocarbons and tetrachlorodibenzo-*p*-dioxin), IIB subfamily (phenobarbital), III family (glucocorticoids), IV family (clofibrate), and the families involved in steroidogenesis (ACTH/cyclic AMP) have all been shown to enhance P-450 gene expression via increases in the rate of transcription. I predict that each of these classes of inducers will operate by means of specific receptors. On the other hand, ethanol and acetone, inducers of the IIE subfamily, enhance P450j protein levels via a posttranslational process (*i.e.*, no increases in j mRNA are seen corresponding to induced increases in j protein). It is, therefore, quite likely that inducers of the IIE genes will not operate by means of a receptor mechanism.

Oxidases and Related Redox Systems, pages 577–587

PROSTAGLANDIN F SYNTHETASE, A DUAL FUNCTION ENZYME

Osamu Hayaishi, Kikuko Watanabe, Yutaka Fujii, Kazuhisa Nakayama*, Hiroaki Ohkubo*, Seiki Kuramitsu[¶], Hiroyuki Kagamiyama[¶] and Shigetada Nakanishi*

Hayaishi Bioinformation Transfer Project, Kyoto Laboratory, Research Development Corporation of Japan, Kyoto 601, Japan

*Institute for Immunology, Kyoto University Faculty of Medicine, Kyoto 606, Japan

[¶]Department of Medical Chemistry, Osaka Medical College, Takatsuki 569, Japan

Although the metabolism and enzymes involved in the transformation of various PG's have been delineated in detail by a number of investigators, the biological ori-

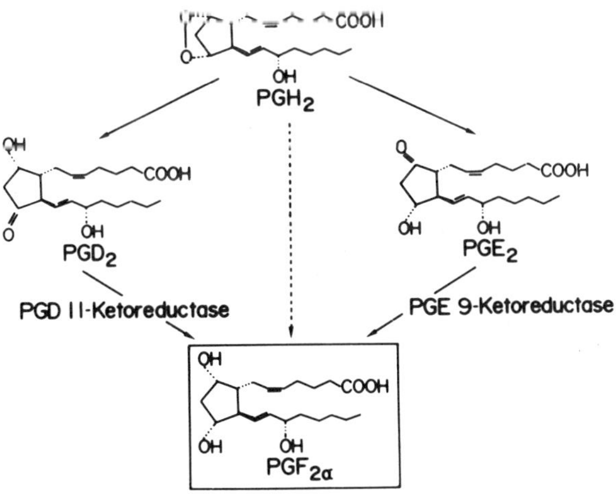

Figure 1. Putative pathways of $PGF_{2\alpha}$ biosynthesis.

gin of $PGF_{2\alpha}$, one of the oldest and most common PG's known, has long remained unclarified. Until recently, three pathways had been postulated concerning the biosynthesis of $PGF_{2\alpha}$ (Fig. 1). In 1967, Hamberg and Samuelsson presented several lines of evidence indicating that PGH_2, rather than E_2 or D_2, is the direct precursor of $PGF_{2\alpha}$(1), but the enzyme responsible for this conversion was not characterized. On the other hand, PGE_2 9-ketoreductase, which catalyzes the conversion of PGE_2 to $PGF_{2\alpha}$, has been isolated and purified from rat heart(2) and human placenta(3). However, the physiological significance of this enzyme remains to be determined because of its broad substrate specificity. In 1981, Watanabe in our laboratory(4), using rat lung, and Wong in the U.S.A.(5), using rabbit liver, independently and concurrently purified PGD_2 11-ketoreductase, which converts PGD_2 to what was then thought to be $PGF_{2\alpha}$ but the major pathway of $PGF_{2\alpha}$ biosynthesis still remained unsolved. More recently we purified this enzyme to homogeneity(6) and showed that it catalyzes the conversion of PGH_2 to $F_{2\alpha}$ and D_2 to $9\alpha,11\beta$-PGF_2, a stereoisomer of $F_{2\alpha}$, but at two different active sites on a single protein(7). The distribution of the PGD_2 11-ketoreductase activity in various rat organs is shown in Table 1.

TABLE 1. Distribution of PGD_2 11-Ketoreductase Activity in Various Tissues of Rat

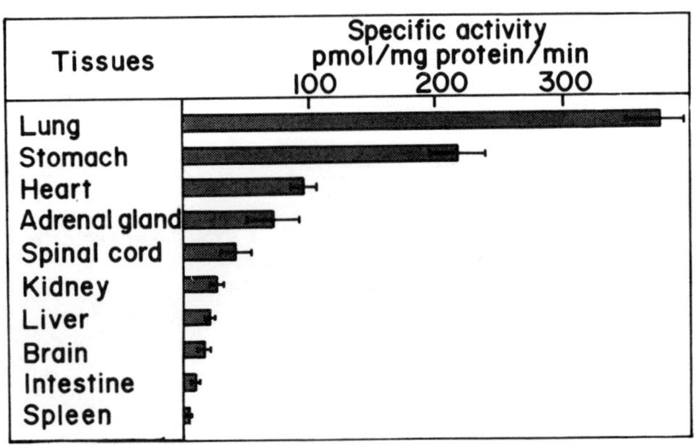

The highest specific activity was found in the lung and stomach, organs which contain high levels of $PGF_{2\alpha}$ and where $PGF_{2\alpha}$ plays important functions. Somewhat less, but still significant, enzyme activities were found in heart, adrenal gland etc.. Therefore we decided to purify this enzyme from bovine lung.

TABLE 2. Purification of PGD_2 11-Ketoreductase Activity from Bovine Lung

Step	Total protein	Specific activity	Yield
	mg	nmol/min/mg of protein	%
1. Crude extracts	14,850	0.3	100
2. Ammonium sulfate	3,150	0.7	52
3. DEAE-cellulose	1,150	1.1	40
4. P-cellulose	176	5.8	24
5. Sephadex G-100	26	35	22
6. Blue Sepharose	4	150	13

Table 2 shows the method of purification of PGD_2 11-ketoreductase from bovine lung. Through 6 steps, we were able to purify the enzyme about 500-fold with an overall yield of about 13 percent. The final preparation was apparently homogeneous as shown in Fig. 2.

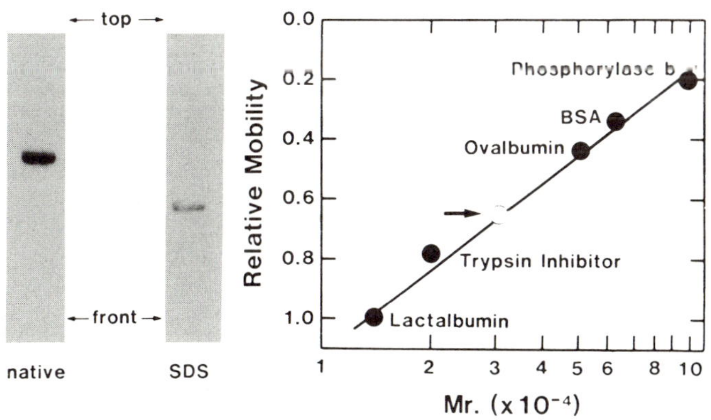

Figure 2. Polyacrylamide gel electrophoreses of the purified enzyme.

On the left are shown the native and SDS PAGE. The mobility of the enzyme, indicated by the arrow, is that of a monomeric protein with a Mr of about 31,000. In the presence of NADPH, the purified enzyme catalyzes the reduction of not only PGD_2 but also PGH_2 as well as several carbonyl compounds, including quinones.

TABLE 3. Substrate Specificity of the Purified Enzyme

Substrate	Relative activity	Km
	%	μM
PGD_2	100	120
PGH_2	44	10
PGE_2	5	-
4-Nitrobenzaldehyde	290	125
9,10-Phenanthrenequinone	290	0.7
Testosterone	3	-

Among PG's tested, PGD_2 is the best substrate in terms of maximum velocity. However, the Km value for PGD_2 is one order of magnitude greater than that for PGH_2; hence PGH_2 appears to be a preferred substrate. Various quinones and carbonyl compounds are also reduced by this enzyme, but none of the naturally occurring compounds appear to be better substrates than prostaglandins (Tab. 3). Although the major pathway of $PGF_{2\alpha}$ synthesis may differ in various organs and tissues, it was tentatively concluded that $PGF_{2\alpha}$ is probably synthesized from PGH_2 and PGD_2 by this enzyme, rather than via PGE_2 in the bovine lung. On the basis of these results, we proposed to name this enzyme "PGF synthetase". However, kinetic data indicated that PGD_2 and H_2 are metabolized at two different active sites, as shown in Fig. 3.

When the reaction was carried out in the presence of increasing amounts of phenanthrenequinone, the quinone strongly inhibited the PGD_2 reductase activity. However, the PGH_2 reductase activity was not inhibited at all, even in the presence of 100 μM quinone. This inhibition is competitive as shown by the right figure.

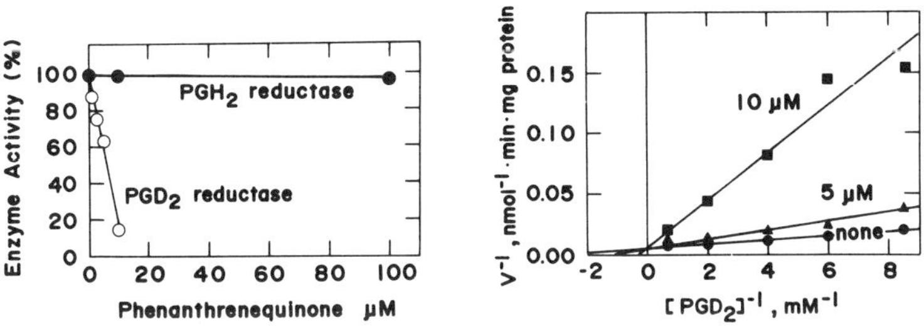

Figure 3. Effects of phenanthrenquinone on the PGD_2 and PGH_2 reductase activities.

These results indicate that PGD_2 and quinones are reduced at the same active site, while PGH_2 reduction is catalyzed at another active site on the enzyme.

In 1984, Roberts and coworkers of Vanderbilt University reported that human liver homogenate or blood converts PGD_2 to 9α, 11β-PGF_2 rather than to $PGF_{2\alpha}$(8).

TABLE 4. Biological Function of 9α, 11β-PGF_2

Contraction	Bronchial smooth muscle	Human
	Coronary artery	Human, Dog
	Pulmonary artery	Rabbit
	Tail artery	Rat
Inhibition	Platelet aggregation	Human
Increase	Arterial blood pressure	Human
	Urinary sodium excretion	Rat
	Urinary flow	Rat

Their identification was based upon the fact that this new metabolite, unlike $PGF_{2\alpha}$, did not react with butylboronic acid. They further reported that 9α, 11β-PGF_2 shows biological activities different from those of $PGF_{2\alpha}$ or PGD_2, such as contraction of bronchial smooth muscle, coronary artery, pulmonary artery, and tail artery; inhibition of platelet aggregation; and an increase in blood pressure, urinary sodium excretion, and urinary flow (Table 4).

Consequently, we synthesized 9α, 11β-PGF_2 for use as an authentic standard and carried out identification of the enzymatic products from PGH_2 and D_2 in collaboration with Dr. Roberts (7). A typical result of TLC analysis on boric acid-impregnated plates is shown in Fig. 4.

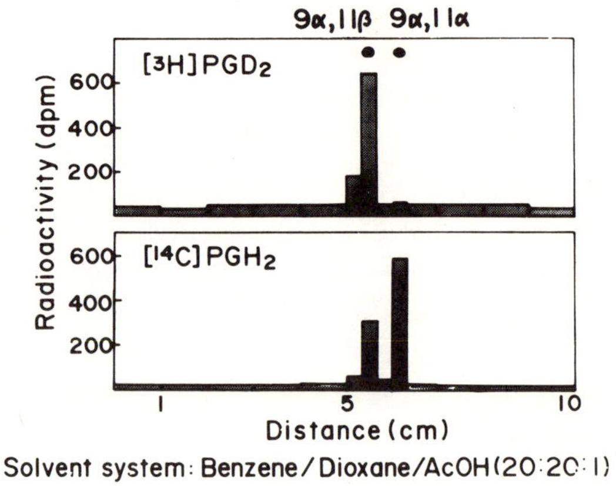

Figure 4. Separation of $PGF_{2\alpha}$ and 9α, 11β-PGF_2 on boric acid impregnated TLC plates.

As shown here, PGD_2 was converted exclusively to 9α, 11β-PGF_2. On the other hand, the enzymatic reduction of PGH_2 yielded a mixture consisting of approximately 70% $PGF_{2\alpha}$ and 30% 9α, 11β-PGF_2. When the reaction was carried out in the presence of a competitive substrate, phenanthrenequinone, or a large excess of cold PGD_2, the formation of 9α, 11β-PGF_2 from PGD_2 and also from PGH_2 in the second experiment was almost completely abolished, indicating that the formation of 9α, 11β-PGF_2 from PGH_2 is probably via PGD_2, which is nonenzymatically produced from PGH_2 during incubation.

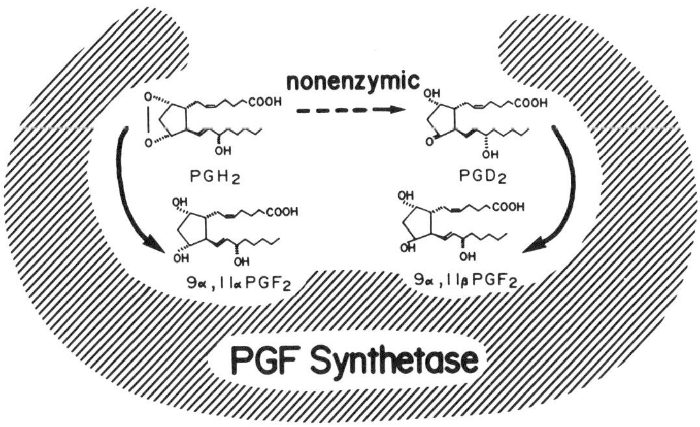

Figure 5. Two active sites of PGF synthetase.

To recapitulate, as shown in Fig. 5, PGF synthetase was purified from bovine lung to an apparently homogeneous monomeric protein with a molecular weight of about 31,000. The enzyme has two active sites and catalyzes NADPH-dependent reduction of PGH_2 to $PGF_{2\alpha}$ at one site, while PGD_2 is reduced to 9α, 11β-PGF_2, a stereoisomer of $PGF_{2\alpha}$, at the other site. Kinetic analysis and other data indicate that the formation of a small amount of the byproduct 9α, 11β-PGF_2 from PGH_2 is due to the nonenzymatic decomposition of PGH_2 to PGD_2 during incubation and subsequent enzymatic reduction to the final product.

In order to study the primary structure of this enzyme, cloned cDNA sequences specific for prostaglandin F synthetase have been isolated from a cDNA library of bovine lung mRNA sequences. Nucleotide sequence analyses of cloned cDNA inserts have revealed that prostaglandin F synthetase consists of a 966-base pair open reading frame coding for a 322-amino acid polypeptide with a Mr of 36,517 excluding the initiating methionine (9). The sequence analysis indicates that bovine lung prostaglandin F synthetase shows 39% identity and 62% homology with human liver aldehyde reductase, an enzyme similar to prostaglandin F synthetase in terms of molecular weight and substrate specificity (Fig. 6). These

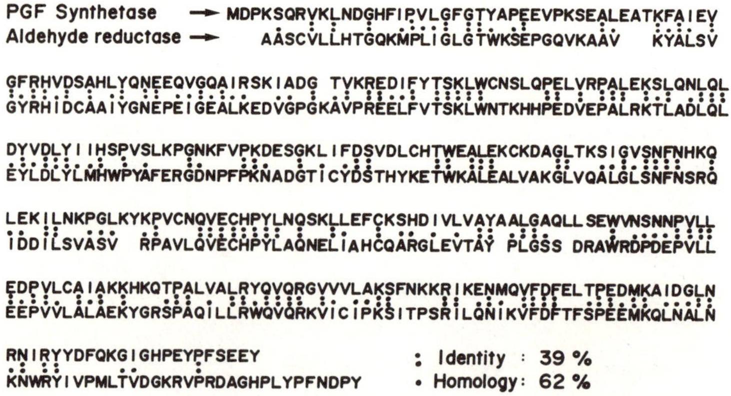

Figure 6. Amino acid sequences of PGF synthetase and human liver aldehyde reductase.

findings may provide a useful clue for the elucidation of two active sites in PGF synthetase, which are currently under investigation in our laboratory.

However, comparison of the amino acid sequence of prostaglandin F synthetase with the National Biomedical Research Foundation (NBRF) protein database reveal 58% identity and 77% homology (including conservative substitutions) with the sequence of 225 amino acids from C-terminal (about 3/4 of the full size) of the European frog lens ε-crystallin without deletions or additions (Fig. 7).

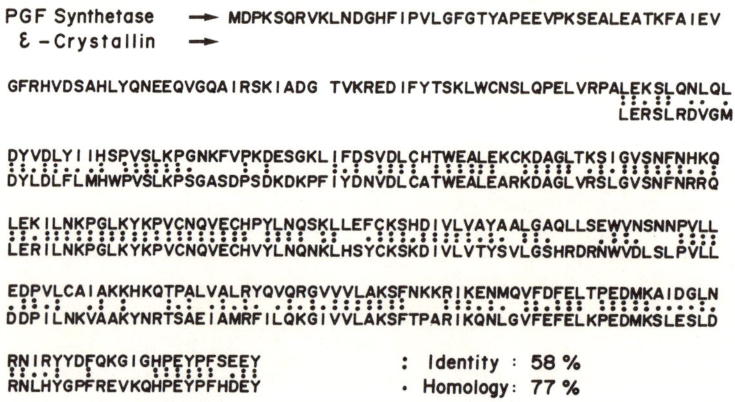

Figure 7. Amino acid sequences of PGF synthetase and ε-crystallin.

These facts suggest that European frog lens ε-crystallin may be identical to bovine lung prostaglandin F synthetase.

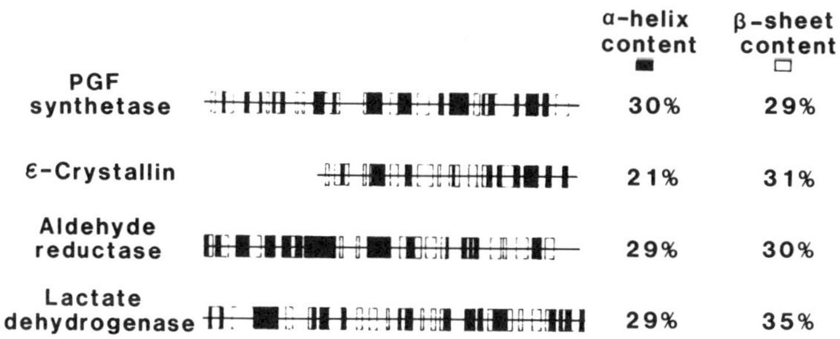

Figure 8. Secondary structures of four related proteins.

Fig. 8 shows the secondary structures of four related proteins deduced from the primary structures according to Chou and Fasman (10). Again some resemblance among PGF synthetase, ε-crystallin and aldehyde reductase is evident but LDH seems to be somewhat different.

Although the biological significance of this unexpected findings is yet to be answered, recent work from several laboratories indicate that crystallins, the major structural proteins of lens, are in many instances enzymes or proteins having other important biological functions (Tab. 5).

TABLE 5. Biological Functions of Lens Crystallins

Crystallins	
α	{ small heat shock proteins Schistosoma egg antigen p40
β, γ	Ca^{++} binding protein S
ε	lactate dehydrogenase
δ	arginosuccinate lyase
τ	enolase
SIII	glutathione S-transferase
ρ	{ aldose reductase aldehyde reductase

The α, the β and the γ crystallin belong to the super-families that include heat-shock protein, Schistosoma egg antigen and a Ca^{++} binding protein, respectively. ε, δ, and τ Crystallins were shown to have striking similarity to lactate dehydrogenase, arginosuccinate lyase, and glutathione S-transferase, respectively (11). More recently, Kinoshita and coworkers at NIH showed that rat lens aldose reductase, human liver aldehyde reductase, and ρ(rho)-crystallin (formerly called ε-crystallin) of the frog Rana pipiens show about 50% homology with one another (12).

REFERENCES

1. Hamberg, M., and Samuelsson, B. (1967) J. Biol. Chem 242, 5336-5343
2. Leslie, C. A., and Levine, L. (1973) Biochem. Biophys. Res. Commun. 52, 717-724
3. Lin, Y. M., and Jarabak, J. (1978) Biochem. Biophys. Res. Commun. 81, 1227-1234
4. Watanabe, K., Shimizu, T., and Hayaishi, O. (1981) Biochem. Int. 2, 603-610
5. Wong, P. Y.-K. (1981) Biochim. Biophys. Acta 659, 169-178
6. Watanabe, K., Yoshida, R., Shimizu, T.,and Hayaishi, O. (1985) J. Biol. Chem. 260, 7035-7041
7. Watanabe, K., Iguchi, Y., Iguchi, S., Arai, Y., Hayaishi, O., and Roberts II, L. J. (1986) Proc. Natl. Acad. Sci. USA 83, 1583-1587
8. Liston, T. E. and Roberts II, L. J. (1985) Proc. Natl. Acad. Sci. USA 82, 6030-6034
9. Watanabe, K., Fujii, Y., Nakayama, K., Ohkubo, H., Kuramitsu, S., Kagamiyama, H., Nakanishi, S., and Hayaishi, O. (1987) Proc. Natl. Acad. Sci. USA (in press)
10. Chou, P. Y., and Fasman, G. D. (1978) Ann. Rev. Biochem. 47, 25-76
11. Wistow, G. J. and Piatigorsky, J. (1987) Science 236, 1554-1556
12. Carper, D., Nishimura, C., Shinohara, T., Dietzchold, B., Wistow, G., Craft, C., Kador, P., and Kinoshita, J. H. (1987) FEBS 220, 209-213

DISCUSSION

ESTABROOK: These are very elegant experiments of great potential significance. Could you describe the possible significance of the 15-OH function as a requirement for the reaction, i.e., will the 15-keto derivative serve equally well as substrate?

HAYAISHI: As we reported previously, 15-keto PGD_2 is an unstable compound and we did not test it as substrate.

WATERMAN: Is the ε-crystallin of frog lens a prostaglandin F synthetase?

HAYAISHI: We are in the process of purifying ε-crystallin from frog lens to see if it has PGF synthetase activity.

Session X: Cytochrome *c* Oxidase

Chairman: Ronald W. Estabrook

Oxidases and Related Redox Systems, pages 591–604

BOTH HEMES ARE LOCATED IN SUBUNIT ONE OF PARACOCCUS DENITRIFICANS CYTOCHROME C OXIDASE

Michele Müller, Beatrice Schläpfer and Angelo Azzi

Institut für Biochemie und Molekularbiologie der Universität Bern, Bühlstrasse 28
CH-3012 Bern, Switzerland

INTRODUCTION

Cytochrome c oxidase, the terminal enzyme of the oxidative chains of many prokaryotes and in all eukaryotes is a polypeptide complex composed of multiple subunits (Azzi et al., 1985, Wikström et al., 1985). Two-subunit oxidases isolated from Paracoccus denitrificans (Ludwig and Schatz,1980) Rhodopseudomonas sphaeroides (Gennis et al., 1982) and other prokaryotes, have shown to be competent in electron transport and proton pumping similar to the 12 subunit mammalian enzymes. Subunit I and II of the oxidase appear to possess structural interspecies homology and immune crossreactivity.

Thermus thermophilus (Yoshida et al., 1984) contrary to a large type of bacterial oxidases contains only one subunit, the second being a cytochrome c_1 type polypeptide. The location and the sequence by which these metals participate in the catalytic process are current research matters. Due to the presence in bacterial oxidases (two subunits) of metal content, spectral and catalytic properties almost identical to those of mammalian enzymes the binding site of the metals has been assigned to the two larger subunits.

One heme and one copper (CuA) are supposed to bind to subunit II (the latter on the basis of structure similarities between a polypeptide region with blue-copper proteins (Steffens and Buse, 1979) and of labeling studies, the former on the basis of less conclusive denaturation studies (Winter et al., 1980, Corbley and Azzi, 1984)).

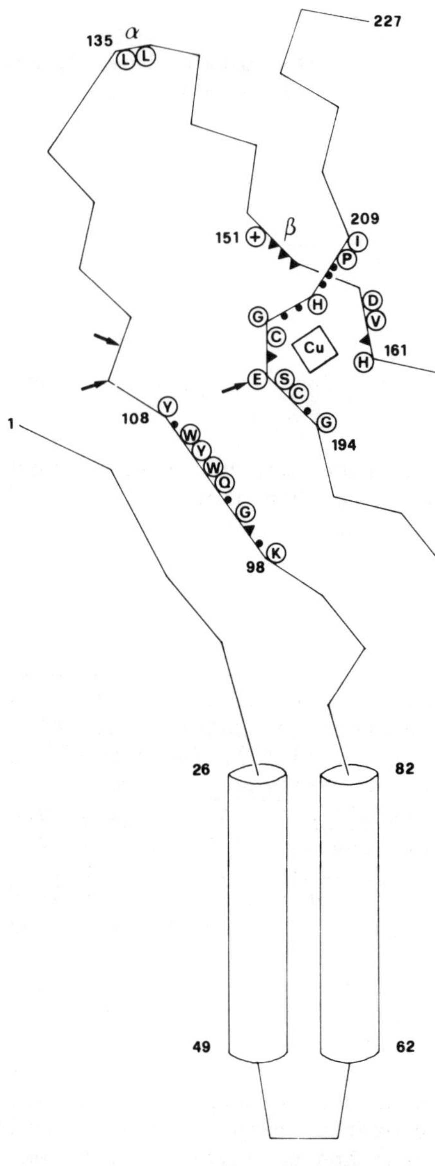

Figure 1. Model of oxidase subunit II. Numbers or symbols correspond to evolutionary conserved residues of the bovine heart enzyme sequence. Circles: amino acids or charges; triangles: hydrophobic residues; dots: residues with no apparent function. The two cylinders represent alpha-helices spanning the membrane. Arrows represent negatively charged amino acids which cytochrome c protected against modification by water soluble carbodiimides (Millet et al., 1983).

The finding that histidine 24 of subunit II (bovine heart enzyme) is not conserved in all oxidases (Benne et al., 1986) argues, however, strongly against subunit II containing the heme a binding site: in fact two imidazoles have been established to cordinate this iron while two histidines and two cysteines ligand copper A (Martin et al., 1985, Stevens et al., 1982, Scott and Schwartz, 1986) making it impossible to find in subunit II conserved residues, enough to coordinate both metals (Fig. 1). Subunit I must host two hemes and one copper, subunit II being left with the role of copper and cytochrome c binding site (Bisson and Montecucco, 1982, Millet et al., 1983, Kadenbach and Stroh, 1984). In the present study we have digested cytochrome c oxidase from P. denitrificans to a single polypeptide of Mr slightly less than subunit I which possesses two hemes. Spectral and ligand properties identical to native enzyme and significant enzymatic activity is also present.

METHODS AND MATERIALS

The oxidase was purified according to Ludwig (Ludwig, 1986) from P. denitrificans (strain ATCC 13543) cells grown on a succinate medium. To the enzyme at a concentration of 23 μM cytochrome a in 25 mM Tris/Cl, 0.5% dodecylmaltoside, at pH 8.3, chymotrypsin (treated with 0.5 mM N-p-tosyl-L-lysinechloro-methylketone) was added at a weight ratio of 1:1 and incubated for 60 min in the dark at 23°C. The proteolytic reaction was stopped by 1 mM phenyl-methyl-sulfonylfluoride. Chymotrypsin and some of the digestion products were removed by chromatography on Whatmann DE 52 cellulose while the oxidase remained bound to the column.

The enzyme was eluted with 0.5 M Tris/Cl, 0.5% dodecylmaltoside, pH 7.8. The eluate was desalted on a Sephadex G--25 column equilibrated with 10 mM Na-phosphate, 2mM EDTA, 0.1% dodecylmaltoside, pH 7.8 at a cytochrome a concentration about half that used for the chymotrypsin digestion. The proteolysis with S. aureus V8 protease was started by the addition of the enzyme at a final concentration of 100 μg/ml. After 20 min of incubation at 20°C in the dark the reaction was stopped by the addition of 1 mM diisopropylfluorophosphate. The solution was passed through a Sephadex G-100 column equilibrated with the buffer of the previous Sephadex G-25 column, or absorbed on a DE 52 and eluted with 0.5 Tris/Cl, 2 mM EDTA, 0.1% dodecylmaltoside, pH 7.8.

Further purification of the enzyme was achieved by eliminating the small digestion products using a Baker Bond PEI widebore HPLC column (4.6 x 250 mm) equilibrated with 20 mM Hepes, 0.1% Triton X-100 R, pH 7.2. The one subunit enzyme was eluted by applying a linear salt gradient (0-1 M NaCl in the same buffer) at a rate of 1.5 ml/min. Varian 5000 liquid chromatograph and Hewlett Packard HP 8541A diode array spectrophotometer equipped with an HPLC flow cell were used for elution control and detection, respectively. Pooled enzyme fractions, after ammonium sulfate precipitation (50% saturation), were centrifuged and the green floating layer was collected, resuspended in 20 mM Hepes, 0.1% dodecylmaltoside, pH 7.2, and passed through a Sephadex G-25 column equilibrated with the same buffer. Contaminating peptides of this single subunit enzyme are estimated to be less than 5%.

Cytochrome a concentration was calculated from spectra (dithionite-reduced minus air-oxidized) taken with an Aminco DW-2a or a Hewlett Packard HP 8541 A diode array spectrophotometer using a ϵ=605-630 nm ϵ=11.7 cm^{-1} mM^{-1} (4). Horse heart ferrocytochrome c was prepared by reduction with dithionite and subsequent filtration through a Sephadex G-25 column. Cytochrome c oxidase activity was measured spectrophotometrically at the wavelength couple 550-540 nm and calculated using the extinction coefficient ϵ =19.4 cm^{-1} mM^{-1}. Protein was determined according to Gornall et al. (Gornall et al., 1949), Lowry et al. (Lowry et al., 1951) or to Smith et al. (Smith et al., 1985) with bovine serum albumin standard. SDS polyacrylamide gel electrophoresis was performed according to the slightly modified version of the procedure of Laemmli (Laemmli, 1970) described by Müller and Azzi (Müller and Azzi, 1986) with no addition of already polymerized 0.5% polyacrylamide. Slab gels were conventionally stained with Coomassie blue or silver according to Bio-Rad.

RESULTS

Proteolysis of P. denitrificans oxidase with chymotrypsin results in the production of a single large fragment from each of the two subunits (called here subunit 1c and 2c, where "c" stands for chymotrypsin treated). In SDS-polyacrylamide gels (Fig. 2, lane 2) a stable pattern as a function of time was reached indicating that the chymotrypsin digestion was practically complete. This treatment changed the Vmax but not the Km for cytochrome c. Vmax was reduced by about 57% (assayed in the presence of Triton X-

100 and soy-bean phospholipid). In presence of the detergent 0.1% dodecylmaltoside, to permit protein absorbance measurements in the UV, purification of the heme binding polypeptide fragment by proteolytic treatment was also documented by the decrease of the ratio at 280 nm versus 424 nm. No spectral modifications could be detected in this preparation in comparison with the native enzyme.

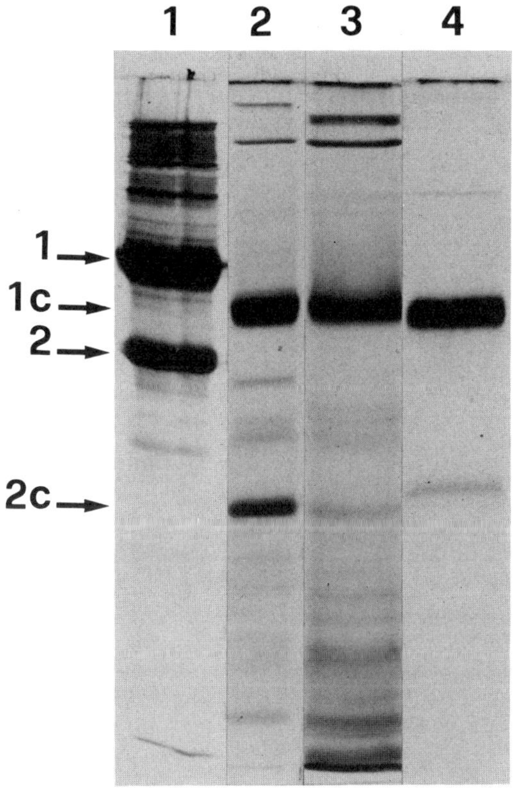

Figure 2. SDS-Polyacrylamide (15%) gel electrophoresis of P. denitrificans cytochrome c oxidase before and after proteolytic treatment. Lane 1: Native enzyme, lane 2: Chymotrypsin treated enzyme; lane 3: Enzyme after S. aureus V8 protease treatment; lane 4: same sample as lane 3 purified by HPLC. The one subunit enzyme comigrated to the same level of ovalbumin, used as standard. The gel was stained by silver.

A second proteolytic treatment was obtained by applying to the chymotrypsin digested oxidase S. aureus V8 protease. This enzyme is known to cleave specifically at aspartyl- and glutamyl-residues. In Fig. 2, lane 3 it appears that subunit 1c was not modified by this protease and that subunit 2c was almost completely reduced to smaller fragments, removable to a large extent by Sephadex G-100 filtration. By ion-exchange HPLC a better purification was achieved. This result is presented in Fig. 2, lane 4.

The small contaminant observed appears to be originated from subunit 1c. The removal of subunit 2c did not affect the spectra of the enzyme (reduced, oxidized and reduced plus CO) were not modified by the S. aureus V8 protease treatment, while the ratio of the absorption of 280 and 424 nm diminished further. Although about 30% heme was lost as a consequence of the proteolytic treatments the ratio a/a_3 remained 1. The kinetics of the single subunit enzyme showed that the apparent Km for cytochrome c remained close to the value of the native oxidase, whereas Vmax decreased to the value of 54% of the chymotrypsin treated oxidase.

DISCUSSION

The amino acid sequences cytochrome c oxidase subunit II from human, bovine or mouse heart, Drosophila melanogaster, Oenothera, maize, yeast (Rao and Argos, 1986) Lehishmania tarantolae, Trypanosoma brucei, Crithidia fasciculata (Benne et al., 1986) and Paracoccus denitrificans (only the residues from 80 to the carboxyl end (Ludwig, 1987)) have been compared and the resulting picture of the evolutionary conserved amino acids residues is shown in Fig. 1. Such a larger comparison, relative to previous similar analyses (cf ref Millet et al., 1983) has brought the following new information: (a) additional conserved histidines, besides those needed for complexing copper A are absent (b) only three negatively charged amino acids are conserved, and only one of them, aspartate 198 has been involved with cytochrome c binding (Millet et al., 1983) (c) a strongly conserved region extends only from lysine 98 to tyrosine 108 and not up to glutamate 109 and tyrosine 110, which are not conserved in L. tarantolae, T. brucei and C. fasciculata sequences; (d) the two conserved leucines 135 and 136, are located in a region where the maximal value for an alpha-helix nucleus was found; (e) a conserved beta-sheet nucleus centered at position 151 followed by three hydrophobic conserved amino acids; (f) a conserved hairpin structure from lysine 171 to arginine 178; (g) a conserved region containing three of the four proposed copper ligands (cysteines 196 and 200 and hystidine 204); (h) the residues from glycine 194 to isoleucine 209.

By using in series two mentioned proteolytic enzymes, chymotrypsin and *Staphylococcus aureus V8* protease it was possible to reduce the protein Mr of cytochrome c oxidase to 43000 without losing at least three of the metal centers of the enzyme. The spectroscopic demonstration of the presence of cytochrome a and a_3 in equal quantities, the notion that copper B is strongly associated with cytochrome a_3 an that the a_3-CO spectrum is not changed in the one-subunit enzyme are supporting the above conclusion. The suggestion that subunit II is the binding site for copper A, will be experimentally and directly tested by using its known spectroscopic and EPR (Stevens et al., 1982, Gelles and Chan, 1985) visibility in our one-subunit preparation.

The enzymatic activity left after subunit II digestion was of about 23% of the native enzyme, which compares favourably with the observation of Gelles and Chan (1985) that parahydroxymercuribenzoate modification of copper A binding site in bovine heart cytochrome c oxidase resulted in a reduction of the electron transfer activity to about 25%. The existence of a one-subunit cytochrome c oxidase (5), our data and those of Gelles and Chan (1985) strongly suggest that copper A may not be an absolute requirement for the catalytic activity of the enzyme. The one-subunit enzyme possesses a kinetically demonstratable high-affinity binding site for cytochrome c which may have its physical counterpart in the protection by cytochrome c of carbodiimide modification of subunit I (although little attention was given to such a protection, cf refs Bisson and Montecucco, 1982, Millet et al., 1983, Kadenbach and Stroh, 1984). A corollary to the results presented above, that two hemes a and one copper are located in subunit I, is that this subunit must have sufficient evolutionary conserved ligands for them.

Human, bovine, mouse, *Neurospora crassa*, *Saccharomyces cerevisiae*, *Drosophila melanogaster* (Rao and Argos, 1986), *Oenothera* (Hiesel et al., 1987) and maize (Isaac et al., 1985) oxidase subunit I were analysed for evolutionary conserved aminoacids, (charged residues, histidines and tyrosines) present in their hydrophylic parts (Fig. 3).

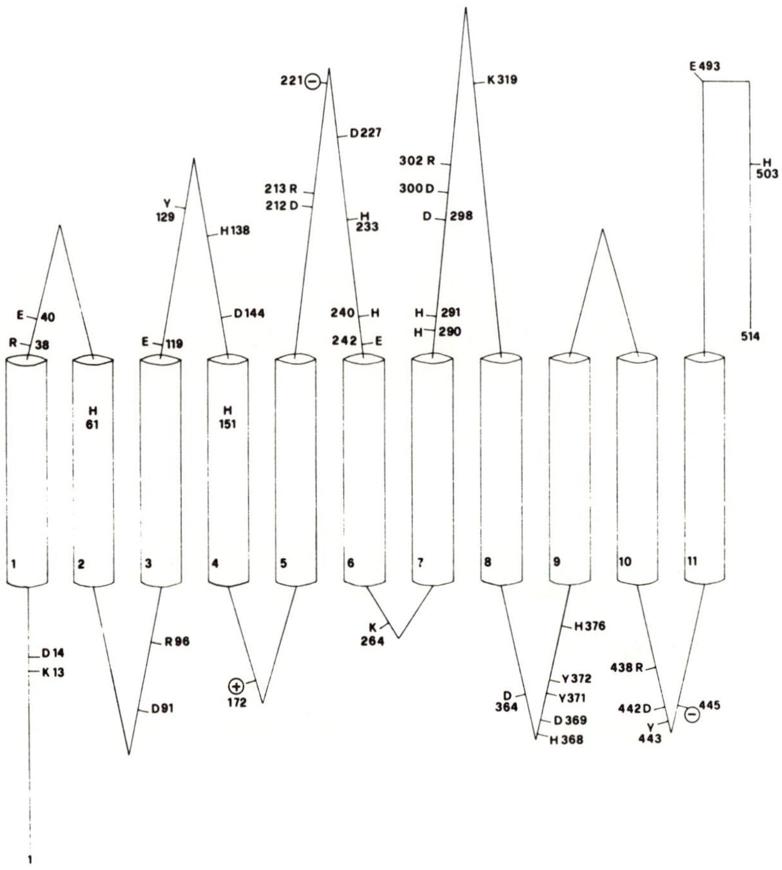

Figure 3. Schematic model of cytochrome c oxidase subunit I. The amino acids indicated are conserved in the species which were analysed until now. The numbers correspond to the sequence of the bovine enzyme. Cylinders represent alpha-helix segments of subunit I interacting with the membrane and are identified by arabic numbers. Only the charged conserved amino acids were shown in the scheme. The drawing of the protein backbone is in scale with the amino acid sequence. For further details see text.

For such an analysis the position of the hydrophobic transmembrane segments was that proposed by Lundeen et al. (Lundeen et al., 1987). In the sequence of subunit I the structural elements must be present necessary for the accommodation of at least three active centers . For one of the hemes a, (cytochrome a) two histidines, 61 of helix 2 and 151 of helix 4 may be candidates for the complex formation. The other heme a, (cytochrome a_3) should be also complexed by one histidine (Martin et al., 1985). Several alternatives can be considered in this case, such as histidines 233, 240 and 291. A remarkable feature of the above model is that the peptides linking the helices 5/6 and 7/8 are located at the cytosolic side of the enzyme and contain conserved negatively charged residues. These residues may be important in cytochrome c binding. The peptides between helices 8/9 and 10/11 (relatively hydrophobic but with clusters of charged amino acids), may be utilised as a proton-capturing device, at the inner side of the phospholipid membrane, where the pathways begin through which protons have to reach cytochrome a_3 and a; for such a path 20 conserved polar amino acids in the apolar phase of all helices (except 7) are available. Copper B which is closely associated with cytochrome a_3 may be coordinated by a serine (residues 34,15,198) a threonine (residue 31), a histidines (residue (240, 290, 291).

In conclusion the above evidence confirms previous indications already present in the literature that the catalytic centers of cytochrome c oxidases, with the exception of copper A, are located in subunit I. Subunit I has sufficient conserved residues to coordinate two heme irons and more than one copper. The proteolytically purified subunit I is competent, though at a reduced rate, in transferring electron between ferrocytochrome c and dioxygen.

ACKNOWLEDGEMENTS

This work was in part supported by Swiss National Science Foundation Grant N. 3.525-086.

REFERENCES

Argade PV, Ching V, Sassardi M, Rousseau DL (1986). Accessibility of the cytochrome a in cytochrome c oxidase to exchangeable protons. J Biol Chem 261:5969-5973.

Azzi A, Bill K, Bolli R, Casey RP, Nałęcz KA, O'Shea P (1985). Molecular aspects of the structure-function relationship in cytochrome c oxidase. In Benga G (ed): Structure and properties of cell membranes Boca Raton, Florida: CRC Press Inc, vol 2, pp 105-138.

Benne R, Van Den Burg J, Brakenhoff, JPJ, Sloof P, VanBoom JH, Tromp MC (1986). Major transcript of the frameshifted coxII gene from trypanosome mitochondria contains four nucleotides that are not encoded in the DNA. Cell 46:819-826.

Bisson R, Montecucco C (1982). Covalent binding of arylazido derivatives of cytochrome c to cytochrome oxidase. FEBS Lett 173:374-380.

Corbley M, Azzi A (1984). Resolution of bovine heart cytochrome c oxidase into smaller complexes by controlled subunit denaturation. Eur J Biochem 139:535-540.

Gelles J, Chan SI (1985). Chemical modification of the Cu_A center in c cytochrome x oxidase by sodium p-(Hydroxymercuri)benzoate. Biochemistry 24:3963-3972.

Gennis RB, Casey RP, Azzi A, Ludwig B (1982). Purification and characterization of the cytochrome c oxidase from Rhodopseudomonas sphaeroides. Eur J Biochem 125:189-195.

Gornall AG, Bardawill CJ, David MM (1949). Determination of serum proteins by means of the biuret reaction. J Biol Chem 177:751-766.

Hiesel R, Schobel W, Schuster W, Brennicke A (1987). The cytochrome oxidase subunit I and III genes in Oenothera mitochondria are transcribed from identical promotor. EMBO J 6: 29-34.

Isaac PG, Jones VP, Leaver CJ (1985). The maize cytochrome c oxidase subunit I gene: sequence, expression and rearrangement in cytoplasmic male sterile plants. EMBO J 4:1617-1623.

Kadenbach B, Stroh A (1984). Different reactivity of carboxylic groups of cytochrome c oxidase polypeptides from pig liver and heart. FEBS Lett 173:374-380.

Laemmli UK (1970). Cleavage of structural proteins during the assembly of the head of bacteriophage T4. Nature 277:680-685.

Lowry OH, Rosenbrough NJ, Farr AL, Randall RJ (1951). Protein measurement with the folin phenol reagent. J Biol Chem 193:265-275.

Ludwig B, Schatz G (1980). A two-subunit cytochrome c oxidase (cytochrome aa3) from Paracoccus denitrificans. Proc Natl Acad Sci USA 77:196-200.

Ludwig B (1986). Cytochrome c oxidase from Paracoccus denitrificans. Methods Enzymol 126:153-159.

Ludwig B (1987). Cytochrome c oxidase in prokaryotes. FEMS Microbiol Review 46:41-56.

Lundeen M, Chance B, Powers L (1987). The transmembrane helices of beef heart cytochrome oxidase. Biophys J 51:693-695.

Martin CT, Scholes CP, Chan SI (1985). The identification of histidine ligands to cytochrome a in cytochrome c oxidase. J Biol Chem 260:2857-2861.

Millet F, de Jong, C, Paulson L, Capaldi RA (1983). Identification of specific carboxylate groups on cytochrome c oxidase that are involved in binding cytochrome c. Biochemistry 22:546-552.

Müller M, Azzi A (1986). Two examples of selective fluorescent labeling of SH-groups with eosin-5-maleimide: The ADP/ATP translocator and the cytochrome c oxidase subunit III of bovine heart mitochondria. In Membrane Proteins, eds. Azzi A, Masotti L, Vecli A. (Springer-Verlag, Berlin). pp. 109-118.

Rao JKM, Argos P (1986). A conformational preference parameter to predict helices in integral membrane proteins. Biochim Biophys Acta 869:197-214.

Scott RA, Schwartz JR (1986). Structural aspects of the copper sites in cytochrome c oxidase. An X-ray absorption spectroscopic investigation of the resting-state enzyme. Biochemistry 25:5546-5555.

Smith PK, Krohn RI, Hermanson GT, Mallia AK, Gartner FK, Provenzano MD, Fujimoto EK, Goeke MM, Olson BJ, Klenk DC (1985). Measurement of protein using bicinchoninic acid. Anal Biochem 150:76-85.

Steffens G, Buse G. (1979). Studies on cytochrome c oxidase, IV(1-3). Primary structure and function of subunit II. Hoppe-Seyler's Z Physiol Chem 360:613-619.

Stevens TH, Martin CT, Wang H, Brudvig, GW, Scholes CP, Chan SI (1982). The nature of Cu_A in cytochrome c oxidase. J Biol Chem 257:12106-12113.

Vanneste WH (1966). The stoichiometry and absorption spectra of components a and a-3 in cytochrome c oxidase. Biochemistry 5:838-848.

Viola RE, Shaw RW, Ranson SC, Villafranca JJ (1983). Solvent proton relaxation studies of cytochrome c oxidase solutions. Arch Biochem Biophys 220:106-115.

Wikström M, Saraste M, Penttilä T (1985). Relationship between structure and function in cytochrome oxidase. In The enzymes of biological membranes, ed. Martonosi AN (Plenum Press, New York) vol. 4, pp. 111-148.

Winter DB, Bruynickx WJ, Foulke FG, Grinich NP, Mason HS (1980). Location of heme a on subunits I and II and copper on subunit II of cytochrome c oxidase. J Biol Chem 255:11408-11414.

Yoshida T, Lorence RM, Choc MG, Tarr GE, Findling KL, Fee JA (1984). Respiratory proteins from the extremely thermophilic aerobic bacterium, Thermus thermophilus. J Biol Chem 259:112-123.

DISCUSSION

NICHOLLS: If all redox centers are on subunit I, the same problem arises as for the ligands to cytochrome a_3 iron, namely, the absence of conserved cysteine in this subunit. Is it possible that Cu_A (as defined by the EPR spectrum) has no cysteine ligands?

AZZI: The proposal that three functional coppers exist (cf. Steffens et al., in this very meeting) complicates the issue remarkably. If Cu_A is indeed the one assumed to be bound to subunit II, still it is possible that Cu_X (the third copper) is located in subunit I. Which of the two would be functionally important is another open question. To answer all these problems in a serious way, we have to wait until the metal analysis of the one subunit enzyme is carried out together with an EPR spectrum.

MUIJSERS: I don't dispute your statement that subunit I is the catalytic core of the oxidase, but I am left with a question relevant to the proposal of Sergio Papa, the previous speaker. You state that the Paracoccus enzyme has 2 (or 3) subunits. Papa proposed the small oxidase subunits in eukaryotes have a function in the folding and assembly of the enzyme core. However, the sequence homologies for the subunits of all oxidase species are sufficient to suggest similar folds and contact surfaces between the subunits. It seems a contradiction that in Paracoccus no small subunits are essential to accomplish similar folds and contact surfaces between the subunits as in the high eukaryotic enzymes. Wouldn't it be the easiest way out to assume that the native Paracoccus enzyme also has the small subunits but these are easily lost during isolation of the enzyme, leaving a core that retains the catalytic activity?

AZZI: There is no evidence in P. denitrificans for genes coding for the small polypeptides analogous to the nuclear coded ones of the eukaryotic oxidases. A clear role, moreover, for those subunits has not yet been found, as indicated by the multiple hypotheses concerning their function, i.e., assembly, modulation of folding, regulation, etc. It may be added that many of these accessory roles may not be necessary in prokaryotes.

MASON: I have found that when the activities of isolated subunits of cytochrome c oxidase are measured by the rate of oxygen consumption in the presence of ascorbate and cytochrome c, it is essential to control for hydrogen peroxide formation by carrying out the assay in the presence of catalase. This is because displaced heme a or copper can easily catalyze the net two electron reduction of oxygen to hydrogen peroxide.

AZZI: When we carried out experiments (following your article on the controlled denaturation of bovine cytochrome c oxidase) to establish by mild denaturation followed by gel filtration the location of the coppers and heme, we found that spectral changes were always associated with the production of hydrogen peroxide. Ligand binding experiments and careful spectral studies do not support the hypothesis that we have a substantial change of the enzyme going from an oxidatic to a peroxidatic activity, which may also be associated with autoxidation of the heme a; this event we never observed.

SMITH: Do you think that the digestion effects with your two-subunit enzyme mean that subunit II covers subunit I so that it is not exposed to the digestive enzymes?

AZZI: If the order of the enzymatic proteolyses is inverted and S. aureus V8 protease is added prior to chymotrypsin no digestion of subunit II occurs under the conditions we have employed. It appears thus that the accessibility of some carboxylic amino acid residues to S. aureus V8 protease is better after a certain portion of subunit II and/or I has been digested by chymotrypsin. On the other hand, the lack of digestion of subunit I (once subunit II has been eliminated) means that no exposed glutamic and aspartic acid residues are accessible to the V8 protease. From a structure analysis it seems also that the acidic amino acid residues are very few in subunit I.

KADENBACH: You interpret your data assuming the binding site for cytochrome c in the Paracoccus enzyme to be at subunit II only. Couldn't it be that there are two binding sites for cytochrome c: one at subunit II which reacts with soluble cytochrome c and another at subunit I which interacts with cytochrome c_1, as might be concluded from the data of Dr. Lucile Smith?

AZZI: The experiments carried out by your group as well, of protection by cytochrome c from chemical modification of the oxidase, indicate that in the presence of subunit II cytochrome c shields subunit I. Such an experiment together with other published evidence would support the notion that the binding site for cytochrome c is constituted by several subunits (I, II, VI) but the fact that subunit I alone may receive electrons from cytochrome c to a considerable rate may suggest that this subunit is primarily involved in the functional interactions while the others may have rather a supporting role. As to the possibility of the existence of two cytochrome c in a monomer, no structural evidence seems to support this concept. Kinetic data, however, can be as well interpreted with negative cooperativity models involving one site per monomer or with two sites having different affinities.

Oxidases and Related Redox Systems, pages 605–617

BACTERIAL CYTOCHROME *C* OXIDASES: CLONING AND SEQUENCE DETERMINATION OF SUBUNIT II OF THE *PARACOCCUS* OXIDASE

Bernd Ludwig, Gunnar Panskus, and Peter Steinrücke

Institute of Biochemistry, Medical University,
D 2400 Lübeck, West Germany

INTRODUCTION

Interest in bacterial cytochrome *c* oxidases (E.C. 1.9.3.1.) has been stimulated by a number of recent reports showing that the structure of the prokaryotic enzymes is far less complex than that of the mitochondrial oxidases which depending on the source are composed of 6 (lower eukaryotes) to up to 13 subunits (mammalian mitochondria) (1-4).

Typically, only two to three different subunits are seen in isolated bacterial heme aa_3-type oxidases. Such preparations show high activity in electron transport; in several of them, coupling of this redox process with proton translocation could be demonstrated successfully in reconstituted systems (see Table 1).

Since evidence for homology of subunits from mitochondrial and prokaryotic oxidases has been presented (see (17) for a review), these findings have greatly supported the notion that the three largest, mitochondrially coded subunits constitute the catalytic core also in the mitochondrial enzyme.

The oxidase from the bacterium *Paracoccus denitrificans* is among the best-studied prokaryotic oxidases and may be viewed as a structurally simple model system for the mitochondrial enzyme: As commonly isolated, the *Paracoccus* enzyme consists of two different subunits of 45 and 28 kD, efficiently oxidizes reduced bacterial or mitochondrial cytochrome *c*, and acts as a redox-coupled proton pump both

Organism	heme type	subunits (no.)	copper per aa_3	proton transloc. stoichiom. H^+/e^-	ref.
Paracoccus denitrificans	aa_3	2	2	0.6	(5)
		3			(6)
			3		(7)
				1 (in cells)	(8)
PS 3	caa_3	3	2	0.45-1.4 (DCCD)	(9)
Thermus thermoph. HB8	c_1aa_3	2(1?)	2	0.8-0.9	(10)
Bacillus subtilis	aa_3	3			(11)
Nitrobacter agilis	aa_3	2	2-3	no	(12)
Thiobacillus novellus	aa_3	2	2		(13)
Rhodobacter sphaeroides	aa_3	3		no	(14)
Pseudomonas AM1	aa_3	2	2(1?)		(15)
Bacillus firmus RAB	caa_3	3	2		(16)

Table I: Some properties of purified bacterial cytochrome c oxidases
DCCD: proton translocation sensitive towards dicyclohexyl carbodiimide; for more detailed information and further references, see (17).

in proteoliposomes and in whole cells. In addition to the two subunits, the presence of a third subunit in the oxidase complex in Paracoccus, related to the mitochondrially coded subunit III, has been suggested on the basis of sequencing and DCCD labeling studies (6,18).

The cloning and DNA sequence determination for the gene of subunit II of the Paracoccus oxidase described here provide the basis for comparison of the deduced amino acid sequence with known mitochondrial sequences, but are also a prerequisite to future applications of in-vitro mutagenesis techniques to probe structure/function relationships in an attempt to learn about molecular details of electron transport and proton translocation in this enzyme.

RESULTS

Immunological screening of a gene bank of Paracoccus DNA fragments in the inducible expression vector pUC 8 (19) yielded a clone expressing a fragment of approximately 85 % of the mature size of the Paracoccus oxidase subunit II in the host, E.coli. The complete gene was isolated by hybridization and sequenced by the didesoxy nucleotide chain termination technique (20). The DNA deduced amino acid sequence and a hydropathy plot are given in Fig. 1 and 2; the sequence is compared with that of four subunits II of mitochondrial oxidases in Fig. 3.

DISCUSSION

1. Bacterial and mitochondrial polypeptides (subunit II) are homologous

Several sequence features confirm the close structural and functional relation of this Paracoccus polypeptide with the corresponding subunit of the mitochondrial enzymes as has been suggested already from earlier work (see (5)):

Two hydrophobic stretches of 22 and 23 residues within the first half of the Paracoccus subunit II sequence (Fig. 1 and 2) most likely represent two membrane spanning domains; they are interrupted by a loop of 20 residues with a positive net charge and followed by a more or less hydrophilic region for the remaining part of the molecule. Taking into account that subunit II is the site for binding of cyto-

```
____________________________
         10         20       ↓30         40         50         60
MAIATKRRGV AAVMSLGVAT MTAVPALA QD VLGDLPVIGK PVNGGMNFQP ASSPLAHDQQ
+   *+++       *    *  *        -    -    +                    -

         70         80         90        100        110        120
WLDHFVLYII TAVTIFVCLL LLICIVRFNR RANPVPARFT HNTPIEVIWT LVPVLILVAI
  - :::::: :::::::::: ::::::+  + +     +         -:::: ::::::::::

        130        140        150        160        170        180
GAFSLPILFR SQEMPNDPDL VIKAIGHQWY WSYEYPNDAF AFDALMLEKE ALADAGYSED
:::::::::+  -   - -     +           -  -      -   -+-     -   --

        190        200        210        220        230        240
EYLLATDNPV VVPVGKKVLV QVTATDVIHA WTIPAFAVKQ DAVPGRIAQL WFSVDQEGVY
-    -          ++          -               + -    +          - -

        250        260        270        280        290
FGQCSELCGI NHAYMPIVVK AVSQEKYEAW LAGAKEEFAA DASDYLPASP VKLASAE
     -              +      -+ -       +--    -  -         +    -
                                             _________________
```

Fig. 1: DNA derived amino acid sequence of the <u>Paracoccus</u> cytochrome <u>c</u> oxidase subunit II

Arrow denotes the signal sequence cleavage site; first residue in the mature sequence (pos. 29) is a pyroglutamate (see text); * : hydroxylated amino acids in the presequence; +/- : charged residues; ::: hydrophobic stretches; solid lines: N- and C- terminal extensions in the primary translation product (see Discussion)

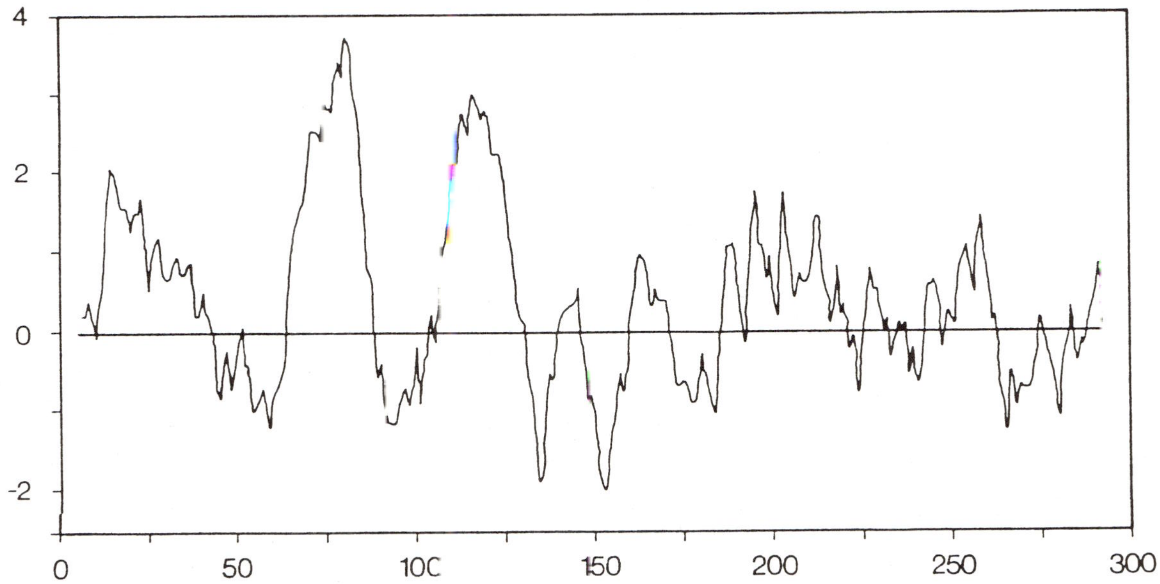

Fig. 2: Hydropathy profile of subunit II of the _Paracoccus_ oxidase (primary translation product). Values for windows of 11 residues each were calculated according to (21); positive ordinate: hydrophobic character

chrome *c* (27), which is found in the periplasmic space of the bacterial cell, this large hydrophilic center of the polypeptide, along with N- and C-terminus of the peptide, must be exposed on the periplasmic side of the cytoplasmic membrane.

This general hydropathy pattern and membrane orientation is identical to that of the mitochondrial subunit II; further homologies are seen in a more detailed sequence analysis. Fig. 3 compares the *Paracoccus* sequence with 5 mitochondrial sequences for subunits II. There is considerable homology in pairwise comparisons between the bacterial and any of the mitochondrial polypeptides (with the possible exception of the *Trypanosoma* sequence which is rather atypical anyway), whereas the inclusion of *all* sequences as given in Fig. 3 focusses on strictly conserved regions only:

a) copper binding site: Several lines of evidence suggested that subunit II of oxidase carries one of the copper atoms of the complex. Indeed, four residues are conserved in the *Paracoccus* sequence as well which could act as ligands and have previously been described in this function: his-209, cys-244, cys-248, and his-252, three of which are present in the most highly conserved domain near the C-terminus of the polypeptide. No other strictly conserved histidines are available then as ligands to heme *a* in subunit II, in agreement with more recent data on the location of both hemes on subunit I (28,7,29).

b) A second center of sequence identity in subunit II for all sequences listed in Fig. 3 is the stretch of aromatic residue around pos. 150; it has been discussed in terms of intra-molecular electron transport function (24).

c) Three acidic residues previously designated to be involved in cytochrome *c* binding are conserved in this bacterial sequence as well (pos. 206, 246, and around 160). Several other acidic residues seen in all the mitochondrial sequences (again with the exception of *Trypanosoma*) are not seen in the bacterial sequence (e.g. pos. 50, 65, 132, 185, 260). This could be a reflection of the fact that the bacterial cytochrome *c* as electron donor to oxidase must be clearly distinct from the mitochondrial cytochrome *c* in its details of binding interaction.

2. The Paracoccus subunit II precursor has both N- and C-terminal extensions

The primary translation product of subunit II carries an additional 28 residues at its N-terminus, which are missing in the mature polypeptide. This extra sequence has the typical features of signal sequences (30): a core of hydrophobic residues, several hydroxylated residues, and positively charged amino acids at its extreme N-terminus (see Fig. 1). The start of the mature sequence has unequivocally been determined by mass spectroscopy of the CNBr fragment which is N-terminally blocked (see (19)): glutamine-29 is present as pyro-glutamate as the first amino acid in the mature polypeptide. The region preceding residue 29 exactly fulfills requirements for signal peptidase recognition sites as based on empirical predictions (31).

Surprisingly, also the C-terminal region of the polypeptide is longer in the DNA deduced sequence: 17 residues are removed in an apparently distint proteolytic step.

This suggests that during biosynthesis several different forms of the polypeptide may be encountered (see Table II); the molecular weight of the mature form as isolated is close to 28 000. Expression of the gene in an E.coli host (see (19)) and immunological detection of subunit II related polypeptides indeed shows several bands in the corresponding molecular weight region.

polypeptide form	no. of residues	mol. weight
primary transl. product	297	32 473
N-terminally cleaved pp.	269	29 715
mature form as isolated (N- and C-terminus cleaved)	252	27 999

Table II: Sizes of three possible forms of the Paracoccus oxidase subunit II (for details, see Discussion)

```
    .   1o    .    2o    .    3o    .    4o    .    5o    .    6o

MAIATKRRGVAAVMSLGVATMTAVPALAQDVLGDLPVIGKPVNGGMNFQPASSPLAHDQQ   PARACOCCUS
                      MILRSLECRFLTIALCDAAEPWQLGSQDAATPMMQG-I   MAIZE
                      MLDLLRLQLTTFIMN-DVPTPYACYFQDSATPNQEG-I   YEAST
                                       MAYPMQLGFQDATSPIMEE-L   BOVINE
                                       MAHAAQVGLQDATSPIMEE-L   HUMAN
                                        MSFILTFWMIFLMDSIIVLI   TRYP.

     .   7o    .    8o    .    9o    .   1oo    .   11o    .
                     =                            =       =
WLDHFV-LYIITAVTIFVCLLLLICIVRFNRRANPVPARFT-HNTPIEVIWTLVPVL-ILV   PARACOCCUS
IDLHHDIFFFLILILVFVSWMLVRALWHFNEQTNPIPQRIV-HGTTIEIIWTIFPSV-IPL   MAIZE
LELHDNIMFYLLVILGLVSWMLYTI--VITYSKNPIAYKYIKHGQTIEVIWTIFPAV-ILL   YEAST
LHFHDHTLMIVFLIS---SLVLYIIS-LMLTTKLTHTSTMD--AQEVETIWTILPAI-ILI   BOVINE
ITFHDHALMIIFLIC---FLVLYALF-LTLTTKLTNTNISD--AQEMETVWTILPAI-ILV   HUMAN
FSFIFLSVWICALII---ATVLTVTK-INNIYCTWDFISSK----FIDTYWFVLGMMFIL-   TRYP.

     .   13o   .    14o   .    15o   .    16o    .   17o    .
         =              =  = ==== =
AIGAFSLPILFRSQEMPNDPDLVIKAIGHQWYWSYEYPNDAFAFDA-LMLEKEAL-ADAG   PARACOCCUS
FIAIPSFALLYSMDGVLVDPAITIKAIGHQWYWSYEYSDYNSSDEQSLTFDSYTIPEDDP   MAIZE
IIAFPSFILLYLCDEVI-SPAITIKAIGYQWYWKYEYSDFINDSGETVEFESYVIPDELL   YEAST
LIALPSLRILYMMDEIN-NPSLTVKTMGHQWYWSYEYTDYED-----LSFDSYMIPTSEL   BOVINE
LIALPSLRILYMTDEVN-DPSLTIKSIGHQWYWTYEYTDYGG-----LIFNSYMLPPLFL   HUMAN
CLLLRLCLLLYFSCINFVSFDL-CKVIGFQWYWVYFLFGETT------IFSN-LILESDY   TRYP.
```

```
 18o    .  19o    .  2oo    .  21o    .  22o    .  23o    .
       =                     == =         =    ===
YSEDEYLLATDNPVVVPVGKKVLVQVTATDVIHAWTIPAFAVKQDAVPGRIAQLWFSVDQ   PARACOCCUS
ELGQSRLLEVDNRVVVPAKTHLRMIVTPADVPHSWAVPSSGVKCDAVPGRSNLTSISVQR   MAIZE
EEGQLRLLDTDTSIVVPVDTHIRFVVTAADVIHDFAIPSLGIKVDATPGRLNQVSALIQR   YEAST
KPGELRLLEVDNRVVLPMEMTIRMLVSSEDVLHSWAVPSLGLKTDAIPGRLNQTTLMSSR   BOVINE
EPGDLRLLDVDNRVVLPIEAPIRMMITSQDVLHSWAVPTLGLKTDAIPGRLNQTTFTATR   HUMAN
LIGDLRILQCNHVLTLLSLVIYKLWVSAVDVIHSFTISSLGIKVEN-PGRCNEIILFATN   TRYP.

 24o    .  25o    .  26o    .  27o    .  28o    .  29o    .
     = === ==  =   ==        =
EGVYFGQCSELCGINHAYMPIVVKAVSQEKY-EAWLAGAKEEFAADASDYLPASPVKLASAE   PARACOCCUS
EGVYYGQCSEICGTNHAFTPIVVEAVTL-KDYADWVSNQLILQTN                    MAIZE
EGVFYGACSELCGTGHANMPIKIEAVSLPKFLE-WLNEQ                          YEAST
PGLYYGQCSEICGSNHSFMPIVLELVPL-KYFEKWSSML                          BOVINE
PGVYYGQCSEICGANHSFMPIVLELIPL-KIFEMGPVFTL                         HUMAN
NATLYGQCSELCGVLHGFMPIVINFI                                       TRYP.
```

Fig. 3: Sequence alignment of subunits II of oxidase from various sources. Only strict identity of residues in each position has been marked by (=); underlined regions in the Paracoccus sequence: N- and C-terminal extensions; published sequences for the maize subunit were taken from (22), yeast (23), bovine (24), human (25), and Trypanosoma (26).

CONCLUSION

The oxidase from Paracoccus has previously been shown to be functionally homologous to the mitochondrial enzyme(s) in many respects. This homology is based on a close structural relatedness as shown here for its subunit II, and, likewise, for subunit I (32) and the newly discovered subunit III gene (6). The small number of subunits makes this oxidase complex a structurally much simpler, but functionally analogous model for the mitochondrial enzyme, and it should prove useful in future studies of structure/function relationships, such as site-directed mutagenesis or 3D crystal analysis.

ACKNOWLEDGEMENTS

Supported by Deutsche Forschungsgemeinschaft (Lu 318/1-2)

REFERENCES

(1) Capaldi, R.A., Malatesta, F., and Darley-Usmar, V.M. (1983) Biochim.Biophys.Acta 726, 135-148
(2) Azzi, A., Bill, K., Bolli. R., Casey, R.P., Nalecz, K., and O'Shea, P.O. (1985) in: Structure and Properties of Cell Membranes (Benga, G., ed.) Vol.2, p. 105-137, CRC Press, Boca Raton
(3) Wikström, M., Saraste, M., and Penttilä, T. (1985) in The Enzymes of Biological Membranes (Martonosi, A.N., ed.), 2. ed., Vol. 4, p.111-148, Plenum Press, New York
(4) Kadenbach, B., Stroh, A., Ungibauer, M., Kuhn-Nentwig, L., Büge, U., and Jarausch, J. (1986) Methods Enzymol. 126, 32-45
(5) Ludwig, B. (1985) Methods Enzymol. 126, 153-159
(6) Saraste, M., Raitio, M., Jalli, T., and Perämaa, A. (1986) FEBS-L. 206, 154-156
(7) Steffens. G.C.M., Biewald, R., and Buse, G. (1987) Eur.J.Biochem. 164, 295-300
(8) van Verseveld, H.W., Krab, K., and Stouthamer, A.H. (1981) Biochim.Biophys.Acta 635, 525-534
(9) Sone, N. and Yanagita, Y. (1982) Biochim.Biophys.Acta 682, 212-226
(10) Fee, J.A., Choc, M.G., Findling, K.L., Lorence, R., and Yoshida, T. (1980) Proc.Natl.Acad.Sci.USA 77, 147-151

(11) de Vrij, W., Azzi, A., and Konings, W.N. (1983) Eur.J.Biochem. 131, 97-103

(12) Yamanaka, T., Kamita, Y., and Fukumori, Y. (1981) J. Biochem. 89, 265-273

(13) Yamanaka, T. and Fujii, K. (1980) Biochim.Biophys.Acta 591, 53-62

(14) Gennis, R.B., Ludwig. B., Casey, R.C., and Azzi, A. (1982) Eur.J.Biochem. 125, 189-195

(15) Fukumori, Y., Nakayama, K., and Yamanaka (1985) J.Biochem. 98, 493-499

(16) Kitada, M. and Krulwich, T.A. (1984) J.Bact. 158, 963-966

(17) Ludwig, B. (1987) FEMS Microbiol.Rev. 46, 41-56

(18) Haltia, T. and Finel, M. (1987) Poster Bari Meeting

(19) Steinrücke, P., Steffens, G.M.C., Panskus, G., Buse, G., and Ludwig, B. (1987) Eur.J.Biochem. 167, 431-439

(20) Sanger, F., Nicklen, S., and Coulson, A.R. (1977) Proc.Natl.Acad.Sci.USA 74, 5463-5467

(21) Kyte, J. and Doolittle, R.F. (1982) J.Mol.Biol. 157, 105-132

(22) Fox, T.D. and Leaver, C.J. (1981) Cell 26. 315-323

(23) Coruzzi, G. and Tzagoloff A. (1979) J.Biol.Chem. 254, 9324-9330

(24) Steffens, G.J. and Buse, G. (1979) Hoppe Seyler's Z. Physiol.Chem. 360, 613-619

(25) Anderson, S., Bankier, A.T., Barrell, B.G., de Bruijn, M.H.L., Coulson, A.R., Drouin, J., Eperon, I.C., Nierlich. D.P., Roe, B.A., Sanger, F., Schreier, P.H., Smith, A.J.H., Staden, R., and Young. I.G. (1981) Nature 290, 457-465

(26) Hensgens, L.A.M., Brakenhoff, J., De Vries, B.F., Sloof, P., Tromp, M.C., Van Boom, J.H., and Benne, R. (1984) Nucl.Acid Res. 12, 7327-7345

(27) Ludwig, B. and Reichardt, J. (1981) in: Vectorial Reactions in Electron and Ion Transport in Mitochondria and Bacteria (Palmieri, F. et al, eds.) p. 3-10, Elsevier, Amsterdam

(28) Holm, L., Saraste, M., and Wikström, M. (1987) EMBO-J. 6, 2819-2823

(29) Azzi, A. (1987) Bari Meeting on 'Cytochrome systems'

(30) von Heijne, G. (1985) J.Mol.Biol. 184, 99-105

(31) von Heijne, G. (1983) Eur.J.Biochem. 133, 17-21

(32) Raitio, M., Jalli, T., and Saraste, M. (1987) EMBO-J. 6, 2825-2833

DISCUSSION

ESTABROOK: Would you comment on the similarity or differences of N-terminal leader sequences proteolytically cleaved from the precursor form of Subunit II in the different species studied?

LUDWIG: These leader sequences in *Paracoccus* polypeptides precisely fulfill the requirements as deduced from a large collection of data from protein from various sources, e.g., the occurrence of positively charged residues and of hydroxylated amino acids, etc. A calculation of weight factors von Heijne assigned to individual residues in this region exactly predicts the cleavage site that we have determined experimentally for subunit II of the *Paracoccus* oxidase.

LERCH: Do you have any idea what the function of the C-terminal extension is?

LUDWIG: Only a few examples exist of such C-terminal extensions in primary translation products, such as in two of the smaller subunits of yeast cytochrome oxidase which were termed "trailer peptides" by Bob Poynton. In terms of function, I have no idea what purpose this extension could serve.

SMITH: Do you have evidence that the copper binding site is on the side towards the periplasmic space? And is there evidence that the cytochrome c binding site is near the copper site?

LUDWIG: In 1981 we showed that cytochrome c (horse heart) binds to subunit II of the *Paracoccus* oxidase. Considering that the overall topography of this subnit locates the copper site within the large hydrophilic domain facing the cytoplasmic space where bacterial c_{550} is found, it is evident that the copper site may be in close vicinity to the heme binding site.

KADENBACH: How can the *Paracoccus* subunit II have a leader sequence but the mammalian subunit II does not? Both have to be translocated through a membrane from their site of synthesis because the large N- and C-terminal domains containing the binding site for cytochrome c are oriented toward the outside?

LUDWIG: This is indeed an unresolved problem which might find an explanation in different mechanisms of assembly in different classes of organisms. I should also mention that for some mitochondrial subunits II (of fungal origin), N-terminal extensions have been observed which, however, don't follow the pattern and size of typical signal sequences.

Oxidases and Related Redox Systems, pages 619–635

KINETICS OF THE INTERACTION OF CYTOCHROME C OXIDASE OF PARACOCCUS DENITRIFICANS WITH PARACOCCUS AND MITOCHONDRIAL CYTOCHROME C

Lucile Smith, Barbara Bolgiano and Helen C. Davies

Department of Biochemistry, (L.S.) Dartmouth Medical School, Hanover, NH, 03755, Department of Microbiology, (B.B., H.C.D.) University of Pennsylvania School of Medicine, Philadelphia, PA 19104-6076

ABSTRACT

We have studied the reactions of the oxidase of *Paracoccus dentrificans* with its membrane-bound cytochrome *c* and with soluble cytochrome c_{550} of *Paracoccus* and of bovine heart. The turnover rate of *Paracoccus* oxidase with membrane-bound cytochrome *c* is high, approaching 1000/sec. at 25°. When soluble cytochrome *c* is added to the electron transport chain oxidizing NADH or succinate, no increase in O_2 uptake is observed. When the oxidase is reacting with the membrane-bound cytochrome *c*, the reaction site is not exposed for reaction with soluble cytochrome *c*. We have purified the *Paracoccus* oxidase, following relatively simple methodology. It has three major subunits similar in molecular weight to those of the larger subunits of the bovine oxidase. When reconstituted in the presence of asolectin, it is just as active as the intact membrane-bound oxidase in reaction with soluble cytochrome *c*. The soluble cytochrome *c* reacts directly with the cytochrome aa_3. We found direct evidence that the oxidase is stimulated in the presence of low concentrations of cytochrome *c*. The stimulatory effect could be the explanation for the so-called "high affinity" site for reaction with cytochrome *c*. The reaction of bovine cytochome *c* with *Paracoccus* oxidase resembles that with the bovine oxidase in every way tested. The *Paracoccus* oxidase must have a cytochrome *c* binding site equivalent to that of the bovine enzyme. The reaction of the *Paracoccus* oxidase with its own soluble cytochrome c_{550}, which has a highly negative hemisphere on the side of the molecule away from the heme crevice, has different properties from those seen in its reaction with bovine cytochrome *c*. However the properties all change to be like those with bovine cytochrome *c* on addition of poly-L-lysine. These data emphasize the importance of all of the charged groups on the cytochrome *c* in influencing binding or electron transfer reactions. The respiratory chain on the membranes of a cytochrome *c*-deficient mutant can reduce cytochrome aa_3 using NADH as substrate in a manner similar to that of the wild type, although at somewhat lower rate, suggesting diffusional encounter of the large complexes within the membrane. Our data permit speculations about the possible evolution from the bacterial to the mitochondrial electron transport system.

INTRODUCTION

Paracoccus denitrificans synthesize a respiratory chain with components similar to those of mitochondrial systems (Scholes and Smith, 1968, Albracht et al., 1980). In fact, it has been proposed that an ancestor of this type was the source of mitochondria (John and Whatley, 1980). The system of *Paracoccus* is simpler in having many less components, the oxidase having only two or three (Ludwig and Schatz, 1980, Saraste et al., 1986), in contrast with the thirteen of beef heart mitochondria. Another way in which the *Paracoccus* system is different is that it has a membrane-bound cytochrome $c(c_{552})$ associated with the cytochrome aa_3 (Berry and Trumpower, 1985) in addition to a cytochrome c_1 as well as soluble cytochrome $c(c_{550})$ in the periplasmic space (Scholes et al., 1971). This arrangement allowed a study of the *Paracoccus* oxidase with membrane-bound cytochrome *c*, with its soluble cytochrome *c* and with a mammalian cytochrome *c* with a very different distribution of charged groups on the surface.

The rate of reaction of the oxidase with the membrane-bound cytochrome c_{552} is very high (Davies et al., 1983). Soluble cytochrome *c* reacts directly with the cytochrome aa_3, but it does not react when the oxidase is interacting with the bound cytochrome c_{552} (Berry and Trumpower, 1985, Davies et al., 1983, Kuo et al., 1985). Our data give direct evidence for the stimulation of cytochrome oxidase activity in the presence of cytochrome *c* and provide an explanation for the multiphasic nature of the plots of the kinetics in the form of v/S vs. v. The *Paracoccus* oxidase is shown to have a cytochrome c binding site for bovine cytochrome c equivalent to that of the bovine oxidase. However, it reacts differently with its own cytochrome c, giving evidence of the importance of charged groups over the whole molecule in influencing the binding or electron transport reactions of this oxidation- reduction system. Experiments with a mutant of Paracoccus which cannot synthesize c-type cytochromes lead to the suggestion of diffusional encounter within the membrane of the large complexes, permitting rapid enough electron transport for the requirements of the cell. Our data permit some speculations about the reactions of the oxidase and cytochrome c, about the role of soluble cytochrome c and about the possible evolution from the bacterial to the mitochondrial electron transport system.

METHODS

Methods for the growth of Paracoccus (Scholes and Smith, 1968, Chang and Morris, 1962), isolation of fragments of cytoplasmic membrane (Scholes and Smith, 1968), purification of the oxidase with three major subunits (Bolgiano et al., 1987) have been described previously. The assays for the oxidase were devised in our laboratory (Smith and Conrad, 1956, Smith et al., 1979). The cytochrome c-deficient mutant HUUG-25 was kindly furnished by Henk van Verseveld. Cytochrome c was purified as described by Scholes et al (Scholes, et al., 1971). Experimental conditions are given in the legends to the tables and figures.

RESULTS

Reaction of Cytochrome Oxidase with Membrane-Bound and Soluble Cytchrome c:

The reaction of the oxidase with its membrane-bound cytochrome c is rapid, approaching 1000/sec at 26° and pH 7.8 with TMPD plus ascorbate (Davies et al., 1983). When it is reacting with the membrane-bound cytochrome c_{552}, addition of soluble cytochrome c results in no increase in O_2 uptake. Table 1 demonstrates this with fragments of cytoplasmic membrane oxidizing NADH or succinate by the entire electron transport chain.

TABLE 1. Effect of Bovine Cytochrome c on NADH-, Succino- and Cytochrome c- Oxidase Activities of Paracoccus membranes: Comparison of wild-type and cytochrome c-deficient strains

		O_2 Uptake (uM O_2/sec/mg)					
Prep	Deoxy-cholate	NADH oxidase	with c	Succino oxidase	with c	Cytochrome oxidase	with c
mutant	-	0.217	0.217	0.068	0.163	0	0.177
	+	0.011	0.076	0.059	0.059	0	0.669
wild-type	-	0.149	0.149	0.048	0.048	2.89	3.33
	+	1.780	1.780	0.056	0.056	2.89	3.56

The assays were done in 50 mM Tris-maleate, 2mM EDTA, pH 7.0. The following amounts of substrates were used for the appropriate assays: 1 mM NADH, 20 mM sodium succinate, 0.7 mM TMPD and 9.3 mM ascorbate, and 0.4 uM beef cytochrome c. Prior to succinoxidase assays, the membrane preparation was incubated at 26°C. for 20 min in the presence of 20 mM succinate. Membranes were incubated with an amount of deoxycholate that gave maximal stimulation (0.5 or 1 mg detergent/mg protein) and immediately diluted with cold dH_2O.

Purified Paracoccus Oxidase:

The dithionite reduced minus oxidized difference spectrum of the oxidase (Figure 1) is typical of cytochrome aa_3, with no evidence of a c-type cytochrome in the preparation. SDS-polyacrylamide gel electrophoresis of the preparation shows three major bands of molecular weight 44, 28 and 23.5 kd and several minor bands. The enzyme is stable when stored at -20° in 50% glycerol.

Reaction of Soluble Cytochrome c with the Oxidase:

Comparison of the reaction of the oxidation of soluble cytochrome c assayed spectrophotometrically with the intact system on cytoplasmic membranes showed that the soluble cytochrome reacted equally well with either oxidase (Figure 2). Thus in the spectrophotometric assay the soluble cytochrome c reacts directly with the cytochrome aa_3 rather than being mediated via the cytochrome c_{552} on the membrane.

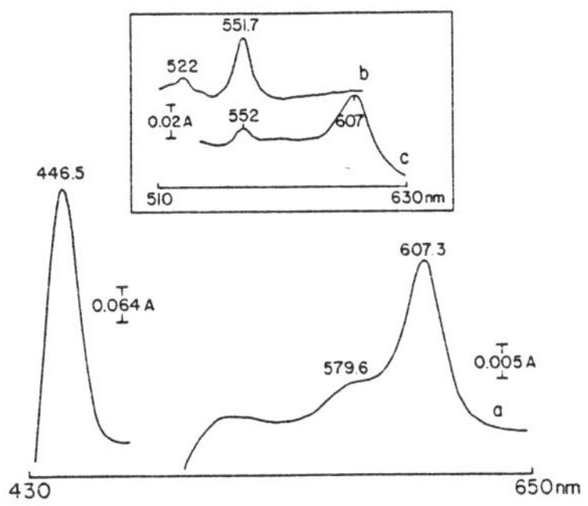

Figure 1. Difference spectra of Paracoccus cytochromes $c_{552}aa_3$, c_{552}, and aa_3. Traces a–c are dithionite-reduced minus air-oxidized spectra: trace a, purified aa_3 from wild-type Paracoccus; trace b, cytochrome c_{552}; trace c, cytochrome $c_{552}aa_3$.

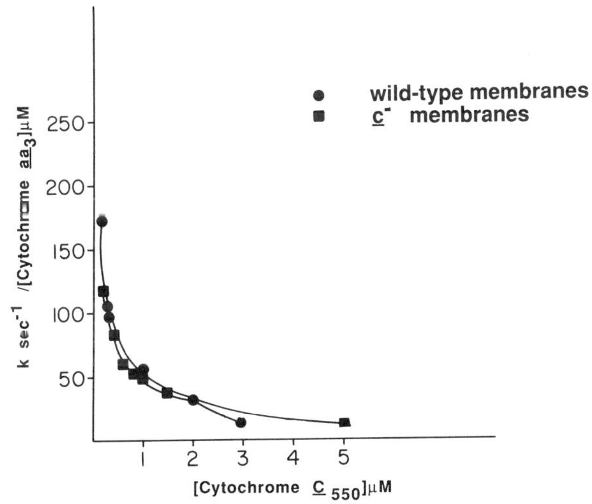

Figure 2. Rate constants (k) for oxidation of bovine cytochrome c by Paracoccus membranes of cytochrome c- deficient (■) and wild-type (●) strains measured spectrophotometrically. Data as rate k $sec^{-1}/[cyt\ aa]_3$ uM. 50 mM Tris-maleate, 2 mM EDTA, pH 7.0, 26°C. Membranes with deoxycholate (0.5 mg/mg protein for mutant and 1 mg/mg for wild-type).

Activity of the Purified Oxidase:

1) Effect of pH and ionic strength. In agreement with numerous studies, the amount of oxidase in these spectrophotometric assays was directly proportional to the first order rate constant at a fixed concentration of cytochrome c. Assays in Tris-maleate buffers showed a decrease of rate constant with increasing pH (Figure 3) with both bovine and *Paracoccus* cytochrome c. At any pH the activities with bovine cytochrome c exceeded those with *Paracoccus* cytochrome. Increasing ionic strength at pH 7 with 1 uM cytochrome c has little effect with *Paracoccus* cytochrome c, but a large effect with the bovine cytochrome c (Figure 4). The decrease in rate constants at higher ionic strengths with bovine cytochrome c is similar to that seen in the interaction of bovine oxidase with bovine cytochrome c (Davies et al., 1964). We have obtained the same results as seen in Figure 4 with Tris-cacodylate, which has "non-binding" ions (Barlow and Margoliash, 1966).

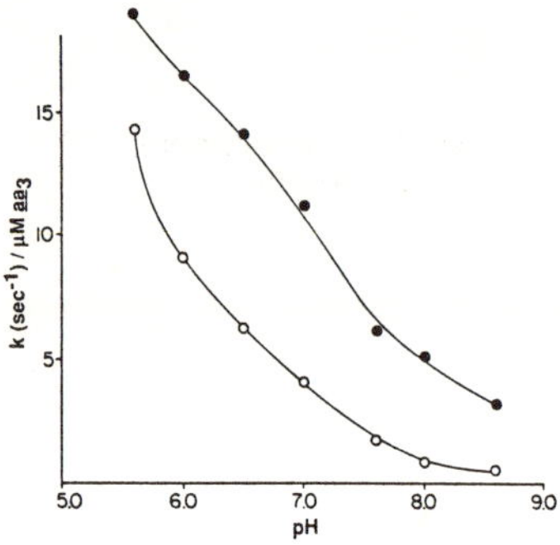

Figure 3. Effect of pH on rate constants k sec^{-1}/ [cytochrome aa_3]. Bovine (●) or *Paracoccus* (○) cytochrome c_{550} (1uM). 50 mM Tris-maleate, 2mM EDTA buffer, 26 C. Cytochrome aa_3: 7.81×10^{-3} uM in *Paracoccus* assay, and 3.91×10^{-3} in bovine assay.

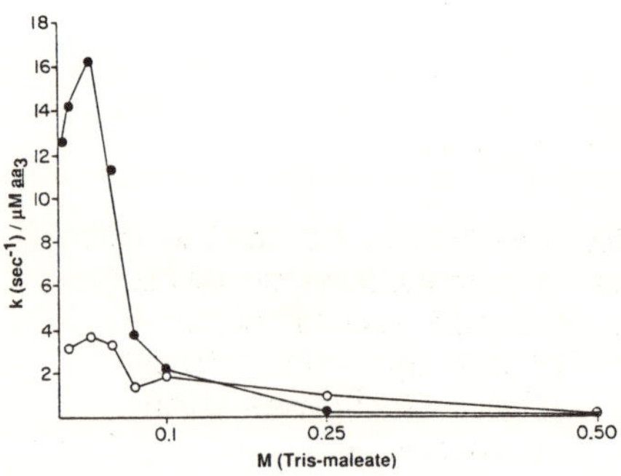

Figure 4. Effect of ionic strength on rate constant k sec^{-1}/ [cytochrome aa_3]. Bovine (●) or *Paracoccus* (○) cytochrome c_{550} (1uM). Assays at pH 7.0, 26 C. in Tris-maleate, 2mM EDTA buffer. Cytochrome aa_3: 3.91×10^{-2} uM (*Paracoccus* assay) and 1.95×10^{-3} uM aa_3 (bovine assay).

2) Effect of Cytochrome c Concentration. Figure 5 plots the oxidase activity of a partially purified cytochrome aa_3 preparation as a function of the concentration of cytochrome c at pH 7 in 50 mM Tris-maleate. With both bovine and Paracoccus cytochromes c as substrate a sharp peak in rate constant is seen at concentrations around 0.1 to 0.2 uM (with a ratio of cytochrome c to aa_3 around 30 in these experiments). The peak in activity is sometimes seen as a plateau or levelling off. As the concentration of cytochrome c is increased further, there is little or no change of rate constant with Paracoccus cytochrome c. With bovine cytochrome c the rate constant decreases in the manner previously observed in the reaction of bovine oxidase with bovine cytochrome c (Smith and Conrad, 1956). Except at even higher concentrations of cytochrome c (above 5 uM), the rate constants with bovine cytochrome c are greater than those with the Paracoccus oxidase, including membrane vesicles from wild-type and a cytochrome c-deficient mutant* as well a partially purified cytochrome c_{552} aa_3 complex and purified cytochrome aa_3.

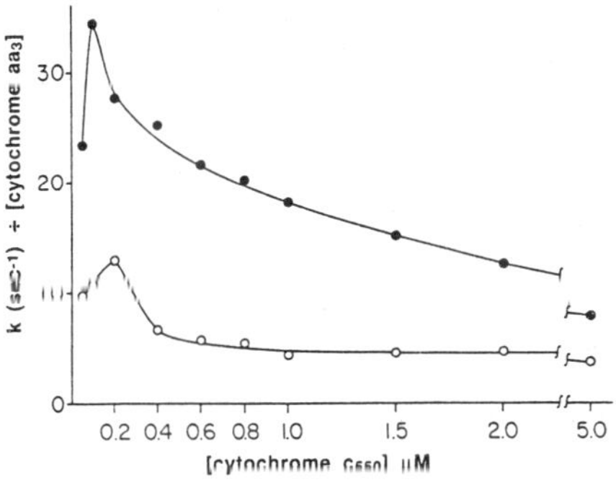

Figure 5. Rate constants of Paracoccus cytochrome c_{552} aa_3 complex on addition of either bovine (●) or Paracoccus (○) cytochromes c_{550}. Data are expressed as rate k sec^{-1}/ [cytochrome aa_3] uM. Assays at 36 C. in 50 mM Tris-maleate, 2 mM EDTA, pH 7.0. Cytochrome aa_3 9.77×10^{-4} uM/assay.

Plots of this type of data from experiments with mitochondrial enzymes represented according to Eadie-Hofstee (v/S vs v) are bi- or multi-phasic (Smith et al., 1979, Ferguson-Miller et al, 1976). This has been seen as either: (1) evidence for multiple binding sites on the oxidase for cytochrome c (Ferguson-Miller et al, 1976), or (2) indication of dimer formation by the oxidase (Nalecz et al., 1985). The results described above with Paracoccus oxidase are plotted in this way in Figure 6. The nature of the cytochrome c rather than the state of the oxidase has the predominant influence on the kinetics.

*Bolgiano, B., unpublished

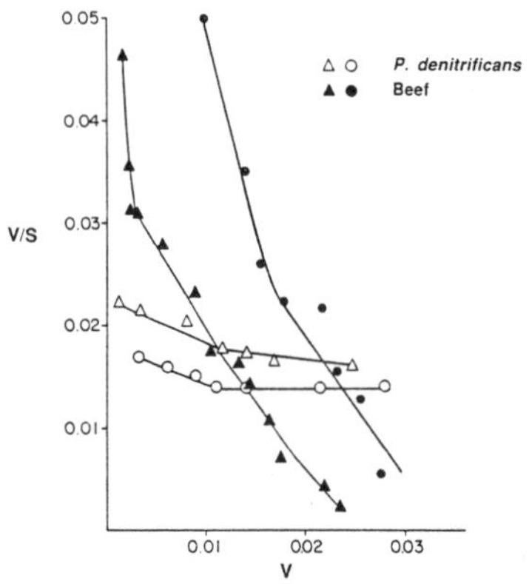

Figure 6. Oxidase activity with membranes from cytochrome c-deficient mutant (●,○) and cytochrome aa_3, (▲,△). Spectrophotometric assays, activities as v(k[cytochrome c]), plotted as v/s vs. v. Membranes with deoxycholate (0.5 mg/mg protein); purified cytochrome aa_3 with asolectin. 50 mM Tris-maleate, 2 mM EDTA, pH 7.0, 26 C with bovine (▲,●) or Paracoccus (△,○) cytochrome c_{550}.

3) Effect of poly-L-lysine. We previously showed that poly-L-lysine competitively inhibited Paracoccus oxidase when bovine cytochrome c was substrate (Smith et al., 1976), in accordance with similar data obtained with the bovine oxidase-bovine cytochrome c system (Davies et al., 1964). Surprisingly, similar concentrations of poly-L-lysine gave significant stimulation of the Paracoccus oxidase oxidizing Paracoccus cytochrome c (Smith et al., 1976). In the presence of these stimulating concentrations of poly-L-lysine the effect of the concentration of cytochrome c on the oxidase rate constants changed; the resulting plot resembles that obtained with bovine cytochrome c (Figure 7). The same change was seen in the effect of ionic strength. In the presence of poly-L-lysine the reaction of Paracoccus oxidase with Paracoccus cytochrome c became sensitive to changes in ionic strength, with a peak activity around 50 mM Tris-maleate.

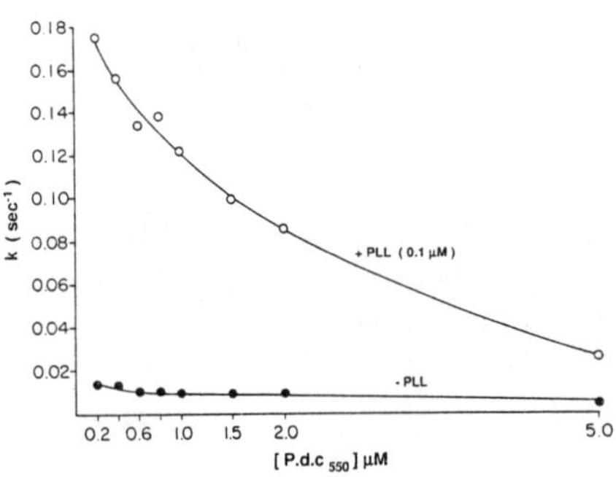

Figure 7. Effect of Poly-L-lysine on k with Paracoccus cytochrome c_{550}. The oxidation of Paracoccus cytochrome c_{550} was assayed with (○) or without (●) 0.1 uM poly L-lysine (PLL), 20,000 M.W. (added before addition of oxidase) at pH 6.0 in 50 mM Tris-maleate, 2 mM EDTA. The Paracoccus cytochrome c_{552} aa_3 contained 4.88×10^{-4} uM cytochrome aa_3.

4) Effect of Asolectin. Treatment of the purified Paracoccus cytochrome aa_3 with asolectin as a source of phospholipid, increased the rate constant with bovine cytochrome c fourfold, but had only a small effect on the reaction with Paracoccus cytochrome c (Figure 8). The turnover numbers obtained (uM cytochrome c sec^{-1}/uM cytochrome aa_3) with 5 um cytochrome c were 48.6 sec^{-1} with bovine cytochrome c and 18.7 sec^{-1} with Paracoccus cytochrome c when measured spectrophotometrically.

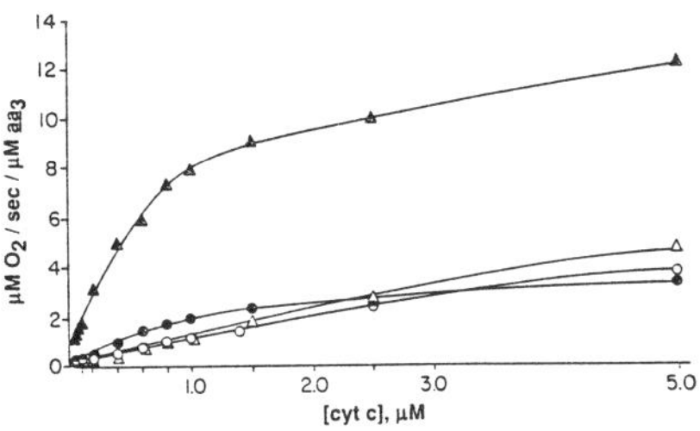

Figure 8. Effect of asolectin on cytochrome oxidase activity assayed spectrophotometrically in 50 mM Tris-maleate, 2 mM EDTA, pH 7.0, 26 C. with bovine (▲, ●) or Paracoccus (△, ○) cytochrome c_{550}. Asolectin-treated purified cytochrome aa_3 (▲, △) or untreated aa_3 (●, ○). Activities as velocities (uMO_2/sec)/ [cytochrome aa_3] uM.

Respiratory Activities of a Cytochrome c-Deficient Mutant

When grown with either glucose or succinate as substrate, the mutant and wild-type Paracoccus had nearly the same generation time. The oxidase of the mutant on cytoplasmic membrane fragments was as active as the wild-type in oxidizing cytochrome c (Figure 2). When purified preparations of cytochrome aa_3 from wild-type and cytochrome c-deficient mutants were compared, the activities of both were lower than their respective membrane preparations (at 1 uM cytochrome c, a 10-fold decrease in turnover number) and the wild-type

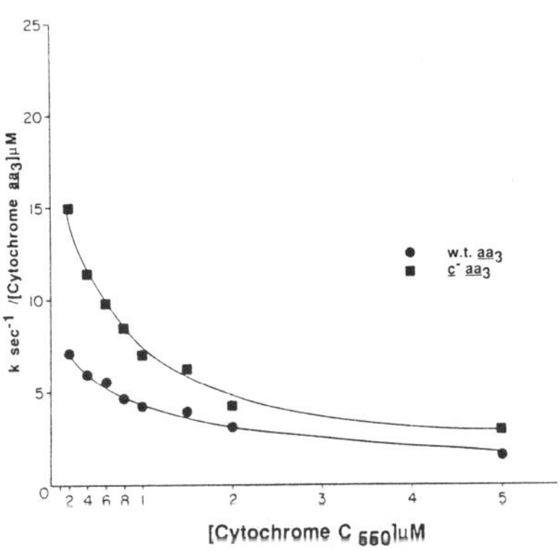

Figure 9. Rate constants (k) for oxidation of bovine cytochrome c by purified Paracoccus oxidase, of cytochrome c-deficient (■) and wild-type (●) strains measured spectrophotometrically. Data as rate k sec^{-1}/[cyt aa_3] uM. 50 mM Tris-maleate, 2 mM EDTA, pH 7.0, 26°C.

preparation was a little less active than the c-deficient mutant (Figure 9). Most important, electron transport down the chain from NADH or succinate resulted in reduction of cytochrome aa_3 when oxygen was exhausted with both wild-type and mutant, although the rate with the mutant membranes was somewhat slower (Figure 10).

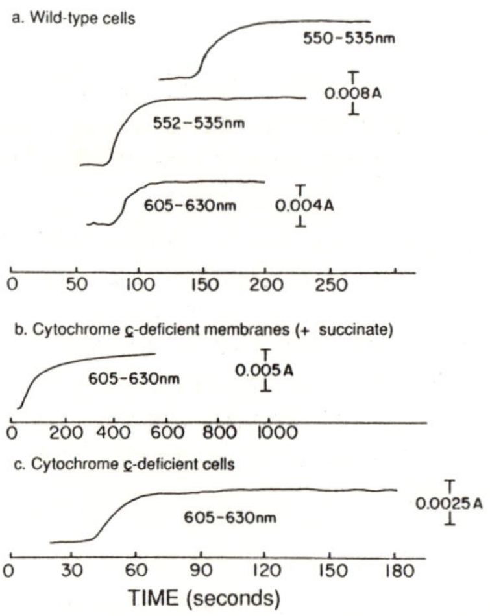

Figure 10. Steady-state reduction of cytochromes a (605-630 nm) and c (550 or 552-535 nm) from wild-type cells (a), cytochrome c-deficient membranes (b), and cytochrome c-deficient cells of Paracoccus. One ml suspension of cells (50% glycerol, 25 mM Tris-maleate, pH 7.0) or membrane fragments (25 mM sucrose, 10 mM KPi, pH 7.6) was aerated and the absorbance change monitered with a Hitatchi 557 spectrophotometer. For the cytochrome c-deficient sample, 20 mM sodium succinate was added prior to aeration.

DISCUSSION

The rate of reaction of the Paracoccus oxidase with its membrane-bound cytochrome c (TMPD plus ascorbate as reductant) is very high, exceeding the rate with even relatively high concentrations of soluble cytochrome c. This points to the importance of close association of cytochromes on membranes for rapid interaction. Soluble cytochrome c reacts directly with the cytochrome aa_3, but cannot react when the oxidase is interacting with the membrane-bound cytochrome. The oxidase must form such a tight complex with the membrane -bound cytochrome c that the reaction site is not exposed for interaction with the soluble cytochronme c which in the bacterium is primarily found in the periplasmic space.

Our method for isolating and purifying the Paracoccus oxidase is relatively simple and yields an enzyme with an absorption spectrum typical of cytochrome aa_3 with no evidence of a c-type cytochrome. The activity with bovine cytochrome c compares favorably with that of preparations obtained by more tedious methods. Assays previously reported with purified preparations were made only with mitochondrial cytochrome c and not with its own cytochrome c_{550} (Ludwig and Schatz, 1980, Nalecz et al., 1985).

The purified oxidase shows three major bands in SDS-PAGE electrophoresis. In this respect it differs from the preparation described by Ludwig and Schatz (1980), which gives evidence of two subunits resembling subunits I and II of the mitochondrial oxidase. Recently, however, Saraste et al (1986) found a DNA sequence in Paracoccus coding for protein homologous to subunit III of the mitochondrial enzyme. Perhaps we have retained the third subunit with our method of preparation.

The reaction of the Paracoccus oxidase with bovine cytochrome c resembles that of the bovine oxidase with bovine cytochrome c in every way we have investigated. Similar responses were seen in both systems to: (1) variations of pH or ionic strength (Davies et al., 1964); (2) variations of concentrations of cytochrome c (Smith and Conrad, 1956); (3) the addition of phospholipids (Vik and Capaldi, 1977); and (4) the addition of poly-L-lysine (Davies et al., 1964, Smith et al., 1976). The Paracoccus oxidase must have a cytochrome c binding site equivalent to that of the bovine enzyme and can act as an appropriate model for electron transfer reactions of the mitochondrial enzyme with mitochondrial cytochrome c.

X-ray crystallographic studies of the two species of cytochrome c showed that both possess the characteristic "cytochrome fold", and the sequence of amino acids surrounding the heme crevice is almost identical (Figure 11 from R. Timkovich (Timkovich, 1979). The method of operation must be similar. However Paracoccus cytochrome c is larger than bovine cytochrome c, and there is a vast difference in amino acid side chains on the surface on the molecules, giving the large difference in isoelectric pH (bovine 10.5 vs Paracoccus 4.5). Both cytochromes have a group of lysines around the top of the heme crevice, where the binding/reaction site appears to be localized in the mitochondrial system (Smith et al., 1977, Brautigan et al., 1978). Paracoccus cytochrome c is more of a dipole than bovine cytochrome c, with a highly negative hemisphere on the side of the molecule away from the heme crevice (Timkovich, 1979).

The interaction of the Paracoccus oxidase with its own soluble cytochrome c_{550} has very different properties from those seen in its reaction with bovine cytochrome c. Also the great sensitivity of the reaction with bovine cytochrome c to ionic strength and the decrease in rate constant with increasing concentration of cytochrome c are not apparent in the reactions with the Paracoccus pigment. However the addition of poly-L-lysine changes the properties to mimic those seen in the reaction with bovine cytochrome c. The bovine oxidase oxidizes the Paracoccus cytochrome c at a very low rate; but this rate is also increased somewhat on addition of poly-L-lysine (Smith et al., 1976).

The differences in the properties of the reaction with Paracoccus and bovine cytochromes c, with the very different distribution of charge, and the change in the reactions with Paracoccus cytochrome c in the presence of poly-L-lysine point to the importance of this charge distribution in the interaction of the oxidase with cytochrome c. Electrostatic interactions are important in orienting oxidants and reductants during an electron transfer encounter, and the ionic strength dependence of electron transfer between proteins has often been related to the surface charge environment in the vicinity of the interacting prosthetic groups. However, the present data give evidence of the role of the interactions among the charged groups around the whole surface of both molecules in determining the electrostatic stability of the complexes or in favoring electron transfer reactions, in agreement with others (Weber and Tollin, 1985, Mauk et al., 1986). Calculations of Koppenol and Margoliash (1982) point to the importance of the dipole moment in orienting molecules like cytochrome c for the formation of productive electron transfer complexes.

We previously reported (Smith et al., 1979) that with the bovine heart system the oxidase showed increasing activity, however assayed, as the cytochrome c concentration was increased up to about 0.1 to 0.2 uM, irrespective of the concentration of cytochrome aa_3 in the assays. Our observations were suggestive of a conformational form influenced by the concentration of cytochrome c. The present observations with the Paracoccus oxidase give direct evidence for an increase in oxidase activity in the presence of around 0.1 to 0.2 uM cytochrome c, since at a given cytochrome c concentration the amount of oxidase is linearly related to the first order rate constant (k sec^{-1}). We have also occasionally seen a similar stimulatory effect in the reaction of the bovine oxidase with bovine cytochrome c**. Lehninger's laboratory has reported that cytochrome c or some other metal porphyrins influenced electron transport between cytochromes a and a_3 (Bickar et al., 1985), and Kornblatt and Luu (1986) found such an effect with even porphyrin cytochrome c. The nature of the cytochrome c does not appear to be critical to this significant effect.

The stimulatory effect of cytochrome c could be the explanation for the "high affinity" site postulated from plots of v/S vs v, since in this range of cytochrome c concentration the v/S would be increased. Apparently the stimulatory effect on the oxidase is more easily seen in the reaction with Paracoccus cytochrome c due to the absence of the decrease in activity on further increasing concentration of cytochrome c seen with the mitochondrial system. This agrees with recent data suggesting that the same domain on the surface of cytochrome c is involved in both so- called "high-" and "low-affinity" reactions of cytochrome c with the oxidase (Veerman et al., 1983). The suggestion that the multiphasic nature of the plots is evidence for the dimeric state of the oxidase does not fit with the present data, where the nature of the plots depends upon the cytochrome c, rather than upon the oxidase.

The similarities of the membrane-bound respiratory chain of Paracoccus to the mitochondrial system (Scholes and Smith, 1968, Albracht et al., 1980) lead to the suggestion that an ancestor of Paracoccus was the original source of mitochondria (John and Whatley, 1977). Evolution would have involved changes in both cytochrome c and the oxidase. The ability of the Paracoccus oxidase to react rapidly with bovine cytochrome c means that there need not have been parallel evolution; the oxidase could already function efficiently with mitochondrial-type cytochrome c. Then during evolution the mitochondrial-type cytochrome c replaced both the membrane-bound and the soluble cytochrome c of Paracoccus. This introduced some new controls into the reaction with the oxidase by affecting its binding under different conditions. While the cytochrome oxidase has evolved into a complex bigenomic enzyme, the cytochrome c has become more electrostatically and sterically simple to facilitate its role in cellular reactions.

**Davies, H.C., Smith, L. and Nava, M.E. unpublished.

Our experiments with the cytochrome c-deficient mutant lead to even more radical speculation. The system of membrane particles (or intact cells) of Paracoccus is able to reduce the cytochrome aa_3 from NADH in a manner similar to that of the wild type, although at a somewhat lower rate. This suggests diffusional encounter of the large complexes within the membrane such that electron transport can proceed at a rate rapid enough for the requirements of the cell.

ACKNOWLEDGEMENT

This research was supported by the N.I.H. through Grants HL28272 and BRSG RR-0703 to Helen C. Davies.

REFERENCES

Albracht SPJ, van Verseveld HW, Hagen WR, Kalkman ML (1980). A comparison of the respiratory chain in particles from Paracoccus denitrificans and bovine heart mitochondria by epr spectroscopy. Biochim Biophys Acta 593:173-186.

Barlow GH, Margoliash E (1966). Electrophoretic behavior of mammalian-type cytochrome c. J Biol Chem 241:1473-1477.

Berry EA, Trumpower BL (1985). Isolation of ubiquinol oxidase from Paracoccus denitrificans and resolution into cytochrome bc_1 and cytochrome aa_3 complexes. J Biol Chem 260:2458-2467.

Bickar D, Lehninger AL, Turren J (1985). Cytochrome c and oxygen facilitation of electron transfer from cytochrome a to cytochrome a_3. In Quagliariello et al. (eds): "Achievements and Perspectives in Mitochondrial Research, Vol. I: Bioenergetics," New York: Elsevier, pp 367-375.

Bolgiano B, Smith L, Davies HC (1987). Submitted.

Brautigan DL, Ferguson-Miller S, Margoliash E (1978). Definition of cytochrome c binding domains by chemical modifications. J Biol Chem 253:130-159.

Chang J, Morris JG (1962). Studies on the utilization of nitrate by Micrococcus denitrificans. J Gen Microbiol 29:301-310.

Davies HC, Smith L, Nava ME (1983). Reaction of cytochrome c in the electron-transport chain of Paracoccus denitrificans. Biochim Biophys Acta 725:238-245.

Davies HC, Smith L, Wasserman R (1964). The influence of ionic strength and polycations on the oxidation of ferrocytochrome c by cytochrome c oxidase. Biochim Biophys Acta 85:238-246.

Ferguson-Miller S, Brautigan D, Margoliash E (1976). Correlation of the kinetics of electron transfer activity of various eukaryotic cytochromes c with binding to mitochondrial cytochrome c oxidase. J Biol Chem 251:1104-1115.

John P, Whatley FR (1975). Paracoccus denitrificans and the evolutionary origin of the mitochondrion. Nature 254:495-498.

John P, Whatley FR (1977). Paracoccus denitrificans Davis (Micrococcus denitrificans Beijerinck) as a mitochondrion. Advances in Botanical Research 4:51-115.

Koppenol WH, Margoliash E (1982). The asymmetric distribution of charges on the surface of horse cytochrome c. J Biol Chem 257:4426-4437.

Kornblatt JA, Luu HA (1986). The interactions of cytochromes c and porphyrin cytochrome c with cytochrome c oxidase: the resting, reduced and pulsed enzymes. Eur J Biochem 159:407- 413.

Kuo LM, Davies HC, Smith L (1985). Monoclonal antibodies to cytochrome c from Paracoccus denitrificans: effects on electron transport reactions. Biochim Biophys Acta 809:388-395.

Ludwig B, Schatz G (1980). A two-subunit cytochrome c oxidase (cytochrome aa_3) from Paracoccus denitrificans. Proc Natl Acad Sci USA 77:196-200.

Mauk MR, Mauk AG, Weber PC, Matthew JB (1986). Electrostatic analysis of the interaction of cytochrome c with native and dimethyl ester heme substituted cytochrome b_5. Biochemistry 25:7085-7091.

Nalecz KA, Bolli R, Ludwig B, Azzi A (1985). The role of subunit III in bovine cytochrome c oxidase. Comparison between native, subunit III-depleted and Paracoccus denitrificans enzymes. Biochim Biophys Acta 808:259- 272.

Saraste M, Raito M, Jalli T, Peramaa A (1986). A gene in Paracoccus for subunit III of cytochrome oxidase. FEBS Letters 206:154-156.

Scholes PB, McLain G, Smith L (1971). Purification and properties of a c-type cytochrome from Micrococcus denitrificans. Biochemistry 10:2072-2076.

Scholes PB, Smith L (1968). Composition and properties of the membrane-bound respiratory chain system of Micrococcus denitrificans. Biochim Biophys Acta 153:363-375.

Scholes PB, Smith L (1968). The isolation and properties of the cytoplasmic membrane of Micrococcus denitrificans. Biochim Biophys Acta 153:350-362.
Smith HT, Staudenmayer N, Millett F (1977). Use of specific lysine modifications to locate the reaction site of cytochrome c with cytochrome oxidase. Biochemistry 16:4971-4974.
Smith L, Conrad H (1956). A study of the kinetics of the oxidation of cytochrome c by cytochrome c oxidase. Arch Biochem Biophys 63:403-413.
Smith L, Davies HC, Nava, ME (1976). Evidence for binding sites on cytochrome c for oxidase and reductases from studies of different cytochromes c of known structure. Biochemistry 15:5827-5831.
Smith L, Davies HC, Nava ME (1979). Kinetics of reaction of cytochrome c with cytochrome c oxidase. In King TE et al. (eds): "Cytochrome Oxidase," Amsterdam: Elsevier/North-Holland Biomedical Press, pp 293-304.
Smith L, Davies HC, Nava ME (1979). Studies of the kinetics of oxidation of cytochrome c by cytochrome c oxidase: Comparison of spectrophometric and polarographic assays. Biochemistry 18:3140-3146.
Timkovich R (1979). Cytochrome c: The architecture of a protein-porphyrin complex. In Dolphin, D (ed): "The Porphyrins," Vol. 7, New York: Academic Press, pp 241-294.
Veerman ECI, Wilms J, Dekker HL, Muijers AO, van Burren KJH, van Gelder BF, Osherhoff N, Speck SH, Margoliash E (1983). The pre-steady state reaction of chemically modified cytochromes c with cytochrome oxidase. J Biol Chem 258:5739-5745.
Vik SB, Capaldi RA (1977). Conditions for optimal electron transfer activity of cytochrome c oxidase isolated from beef heart mitochondria. Biochemistry 16:5755- 5759.
Weber PC, Tollin G (1985). Electrostatic interactions during electron transfer reactions between c-type cytochromes and flavodoxin. J Biol Chem 260:5568-5573.

DISCUSSION

CUSANOVICH: Comment. It is important to note that the soluble *Paracoccus* cytochrome *c* reacts just like mammalian cytochrome *c* (at the heme edge) with small molecules (flavin semiquinones) and proteins (flavodoxin semiquinones). Combining these observations with your results leads to the suggestion that the *Paracoccus* cytochrome *c* oxidase must have a cytochrome *c* oxidation domain which interacts with a very large portion of the cytochrome surface, thus sensing both the positive and negative patches.

SMITH: I agree.

LUDWIG: (1) In your experiment using a cytochrome *c* deficient mutant did you add any cytochrome *c* when you showed reduction of aa_3 by NADH in membranes? (2) Since this mutant should also be devoid of a functional bc_1-complex, where would the aa_3 complex pick up electrons in a collision model reaction?

SMITH: (1) No cytochrome *c* was added.

(2) The aa_3 complex could react with some member of the bc_1 complex that remains, either cytochrome *c* or the Fe-S compound.

Oxidases and Related Redox Systems, pages 637–651

ANTIBODIES AS PROBES OF CYTOCHROME OXIDASE STRUCTURE AND FUNCTION

Peter Nicholls, Chris E. Cooper,
Bryan Leece & Jo A. Freedman*,

Dept. of Biol. Sciences, Brock Univ.,
St. Catharines, Ont.L2S 3A1, Canada,

and Samuel H.P. Chan, Biology Dept.,
Syracuse Univ., Syracuse, NY 13244 USA.

*Current address: Dept. of Biology,
Syracuse University.

ABSTRACT.

Antibodies have been raised in rabbits against whole beef heart cytochrome oxidase and against purified subunit V. Western blot analysis showed that antioxidase was largely composed of anti-II and anti-IV antibodies but some anti-I and antibodies against small subunits were elicited. Similar analysis of anti-V showed it to be relatively specific against subunit V. Three types of anti-V were identified by ELISA with intact enzyme: (i) weak binding and redox independent, (ii) stronger binding, redox-dependent (binding only to reduced or partially reduced enzyme), and membrane-independent, and (iii) moderate-binding, redox-dependent & membrane-dependent.

Inhibition of enzyme activity by anti-oxidase was biphasic in time, indicating populations of rapidly- and slowly- reacting molecules. Variation of cytochrome *c* concentration showed partially competitive kinetics, but the antibody also affected 'internal' enzymatic events including the turnover rate with TMPD and the spin-state change in cytochrome *a*3 that follows reduction of cytochrome a. No spectral effects could however be seen. Antioxidase also inhibits proteoliposomal respiration with external cytochrome c but not that with internally-trapped cytochrome c. No functionally significant epitopes occur on the N(matrix) side of the membrane.

The more strongly binding anti-V inhibits the isolated enzyme by at least 60%. The inhibition at high ionic strength induces a biphasic pattern with respect to cytochrome c concentration. Anti-V may thus slow the dissociation of cytochrome c from its complex with the enzyme. The redox-dependent, membrane-dependent anti-V (reported previously) gave only about 20% inhibition of the enzyme. The effective anti-V antibody had little if any influence on the activity of proteoliposomal oxidase in either orientation. Some sites on subunit V whose binding can give rise to inhibition are not accessible to anti-V when the enzyme is embedded in a functional membrane system.

INTRODUCTION.

Antibodies against membrane-bound enzymes may be used to probe structure and function in two ways. Firstly, they can be employed to determine the functionally essential subunits and domains, using either solubilized or membranous enzyme, but in essentially the same fashion as with soluble enzymes (1). Secondly, because antibodies are membrane-impermeable, they can also be employed to determine the accessibility of these same subunits and domains from the two sides of the membrane (2). Following some early studies with antibodies prepared against whole oxidase (3,4), in which only partial enzyme inhibition could be achieved, later work secured antibody against whole oxidase with strong inhibitory capability (5) and against individual subunits with partial inhibitory capability (5,6). Subunit specific antibodies permitted studies of subunit orientation and accessibility (7,8).

Subsequently, specific investigation of antibodies against subunits III and V yielded mechanistic information. Inhibition of proton pumping by anti-subunit III (9) required the intact IgG; Fab and even $F(ab')_2$ fragments were ineffective. However addition of goat $F(ab')_2$ anti-rabbit Fab restored the inhibition (10). Partial inhibition of activity by rabbit antisubunit V was associated with a redox-dependent antibody binding to membranous oxidase(6). Subunit 5 was accessible on the N (matrix) face of submitochondrial particles only in the presence of reductants.

What may be expected to happen to enzymatic activity in the presence of an enzyme-specific antibody (see ref. 1)? And will differences be expected between the inhibitory action of IgG, $F(ab')_2$ and Fab? In the case of insulin and its membrane-bound receptor, for example, either anti-insulin or anti-receptor IgG antibodies can enhance the cell's response (11). But Fab fragments, though competent to bind, are ineffectual at stimulation. Cross-linking of the receptor molecules is essential: the antibody (reacting outside the cell) promotes this and does not block the second messenger action (inside the cell).

For tight binding, some evidence of a requirement for mobility of the antigenic region has been presented (12, 13). Short peptide regions are recognized, some on the same piece of polypeptide, others on neighbouring sections (14). The cross-linking possible with intact IgG or even $F(ab')_2$ fragments inevitably increases apparent affinities. Inhibition in such cases may be expected if:

(i) the antibody binding competes with that of a substrate;

(ii) the antibody alters the chemistry of an essential reactive group (pK? E_m? etc); or

(iii) the antibody blocks an essential conformational change in the catalytic cycle or induces a conformation inconsistent with function.

However, intact IgG may induce non-specific inhibitor effects, both competitive and non-competitive in character, if it causes aggregation of the enzyme. The reduction in effective substrate concentration at the active sites in such a system will mimic competitive inhibition; blocking access of substrate to some sites will mimic non-competitive inhibition.

Apparent activations can be seen under some conditions if the antibody:

(i) reduces the 'off' constant for a substrate;
(ii) blocks the action of an inhibitor(increases its 'off' constant); or
(iii) facilitates an essential conformational change in the catalytic cycle or favours the active form in an R/T equilibrium.

As seen in the insulin case (11) activation can also be expected with IgG (but not with Fab) in systems which require dimer or oligomer formation for function (if, however, $F(ab')_2$ fragments resemble Fab in their behaviour, as with anti-subunit III (10), alternative explanations are needed).

We have examined the effects of both general anti-oxidase and specific anti-V antibodies on cytochrome oxidase behaviour and activity, in an attempt to determine something about the intersubunit relationships and subunit functional requirements.

METHODS AND MATERIALS.

Cytochrome oxidase was prepared from beef heart as previously described (15). Subunit V was isolated according to Freedman and coworkers (6, 16). Polyclonal antibodies to intact enzyme, and to isolated subunit V were raised in rabbits. Western blots and ELISA assays were carried out as previously reported (6). Activity and spectroscopic measurements were made as described in the figure legends.

RESULTS.

Figure 1 shows the results of blotting SDS-urea gels of dissociated cytochrome oxidase and challenging the blots with anti-oxidase and anti-subunit V antibodies followed by goat anti-rabbit peroxidase-conjugated antibody. Most of the anti-oxidase activity is contained in anti-subunit II and anti-smaller subunit IgG classes. Anti-subunit V is fairly free of cross-reactivity, although a small anti-subunit IV component is a contaminant in some rabbits.

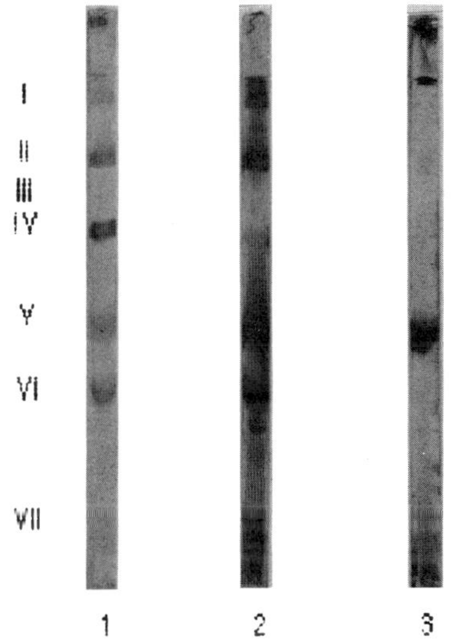

Fig. 1. Western blots of SDS-urea gels of dissociated cytochrome oxidase. Lane 1: (nitrocellulose strip): in contact with a gel-half during Western blot transfer, stained with Amido Black. Lanes 2 and 3: saturated with bovine serum albumin, allowed to bind anti-oxidase and anti-V, respectively, exposed to a second antibody-peroxidase conjugate, and then stained.

The polyclonal anti-V antibody was heterogeneous in composition and activity. The previously isolated anti-V antibody displayed redox-dependent binding to the membranous enzyme (6). The two samples of anti-V antibody were obtained from different animals in the present experiments. The first of these displayed more strongly inhibitory properties as well as stronger binding. The second was poorly inhibitory and also showed weaker binding. Figure 2 illustrates the ELISA behaviour of the inhibitory antibody towards the isolated enzyme(method of ref.5).

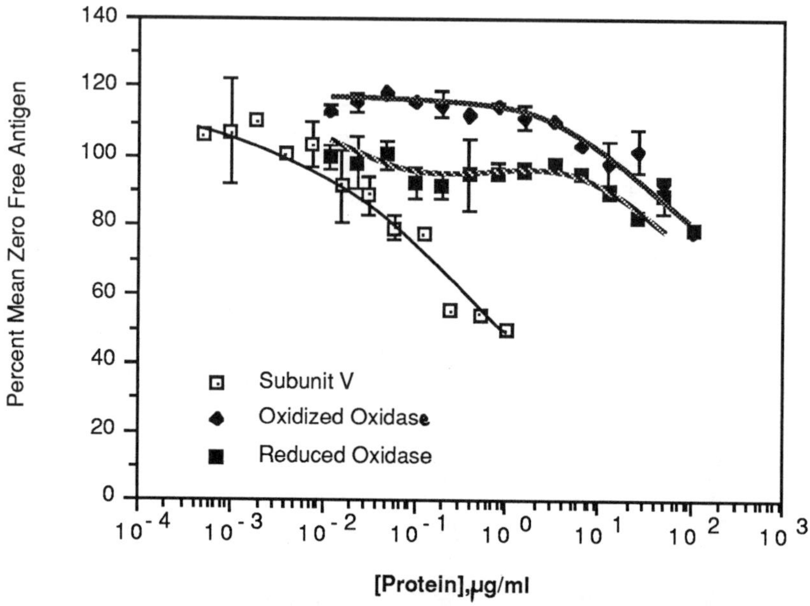

Fig. 2. Enzyme-linked immunosorbent assay of IgG fractions elicited against beef heart cytochrome oxidase subunit V. Subunit V itself, oxidized (resting) enzyme, and ascorbate-TMPD reduced enzyme were used as antigens (cf. methods in ref. 6).

The intact resting enzyme requires over 1000X as much inhibiting protein as subunit 5, or as much as 100X as much on a molar basis. Upon reduction of the inhibiting enzyme, a small proportion (25%) of the antigenic sites display an apparently higher affinity, while the remainder are unaffected. This anti-V preparation is thus heterogeneous in character, with both redox-dependent and redox-independent components. This may be compared with the membrane-dependent redox sensitivity of the previously reported anti-V preparation (6).

Anti-oxidase antibody is clearly heterogeneous in its inhibitory capability. Fig. 3 shows the time course of inhibition obtained with a typical anti-oxidase sample in an oxygen electrode assay system. Fast and slow phases of

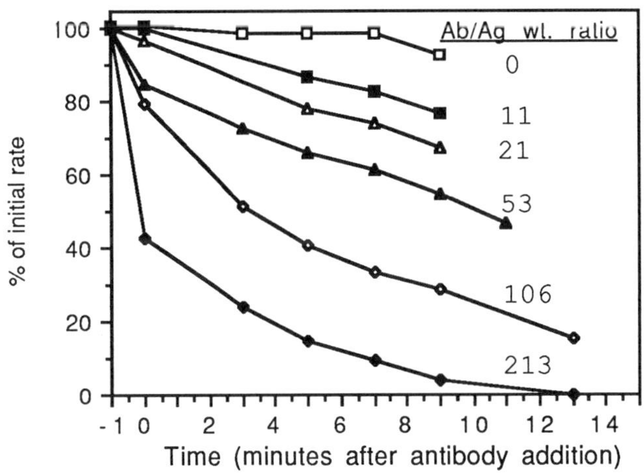

Fig. 3. Time courses of cytochrome oxidase inhibition by anti-oxidase. Oxygen uptake monitored polarographically with 18 mM cyt. c, 5 mM ascorbate, 0.4 mM TMPD, 7.3 nM cyt. aa_3 in 50 mM K^+ HEPES 0.1% lauryl maltoside at 30°C. Anti-oxidase IgG was added immediately after the reaction had commenced at the indicated weight ratio. Rates are given in terms of the initial uninhibited rate.

inhibition are seen when the antibody is added during the run. Both phases are abolished (i.e. the enzyme is fully inhibited from the beginning) if the resting enzyme is incubated with antibody prior to addition of reductant.

Cytochrome c concentration has little effect upon the observed inhibition.This is likely to be a 'pseudo non-competitive' situation, and a consequence of slow dissociation rates of antibody from its complex with the enzyme. Fig. 4 shows that the apparent non-competitive behaviour is seen in both 'high' and 'low' affinity regions of the reaction with cytochrome c.

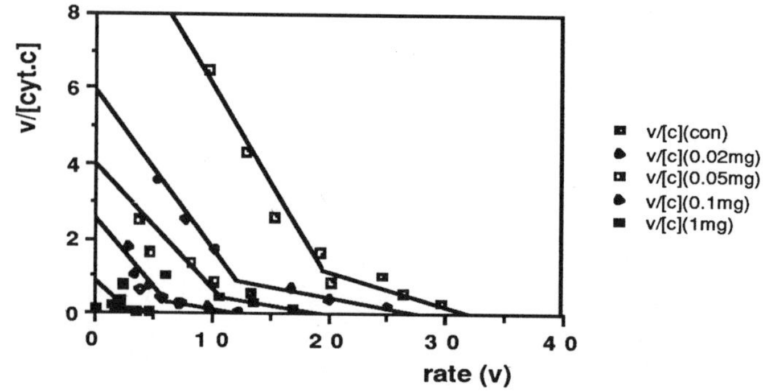

Fig. 4. Hofstee plots of cytochrome oxidase activity: effect of increasing anti-oxidase IgG concentrations. Conditions as in Fig. 3, but varying cytochrome c levels. The oxidase was preincubated for 15 minutes with the indicated level of anti-oxidase.

Even if the primary effect of the inhibitory population of antibodies is upon the cytochrome c reaction, internal enzymatic steps can also be influenced. Table 1 summarizes the effects of anti-oxidase upon three aspects of enzyme behaviour in the absence of cytochrome c, namely:(i) spectrum ;(ii) activity towards TMPD as substrate (17); and (iii) the spin state change that occurs upon cytochrome a reduction by TMPD (18). The antibody-complexed enzyme turns over more slowly and undergoes the spin state transitions more slowly than the control enzyme.

The kinetic effect of anti-V is more subtle. Under the right conditions more than 60% enzyme inhibition can be seen.The characteristic behaviour towards cytochrome c is illustrated in Fig. 5, which shows the Hofstee plots obtained in the presence and absence of antibody. An apparently competitive behaviour at high levels of cytochrome c becomes uncompetitive at low levels (in the "high affinity" region).The antibody makes the biphasicity of the Hofstee (or any other linear) plot more marked.

Table 1. Effect of anti-oxidase 1GG fraction and non-reactive serum (NRS) on TMPD-dependent turnover and spin state change.

System	$k (s^{-1})$ TMPD turnover*	cyt. a steady state reduction (605-630 nm)	$k' (s^{-1})$ spin state change**
1. BSA ($40\ mg.ml^{-1}$)	0.42	40%	1.24×10^{-2}
2. NRS ($31\ mg.ml^{-1}$)	0.7	36%	2.1×10^{-2}
3. Antioxidase ($44\ mg.ml^{-1}$)	0.15	23%	4.6×10^{-3}

* Turnover in absence of cyt. c in 8mM ascorbate, 0.3 mM TMPD, 8 μM cytochrome aa_3 in 50 mM K^+ HEPES pH 7.2, 0.1% lauryl maltoside at 30°C (cf.ref.17).
** Monitored at 433-413 nm (cf. ref. 18).

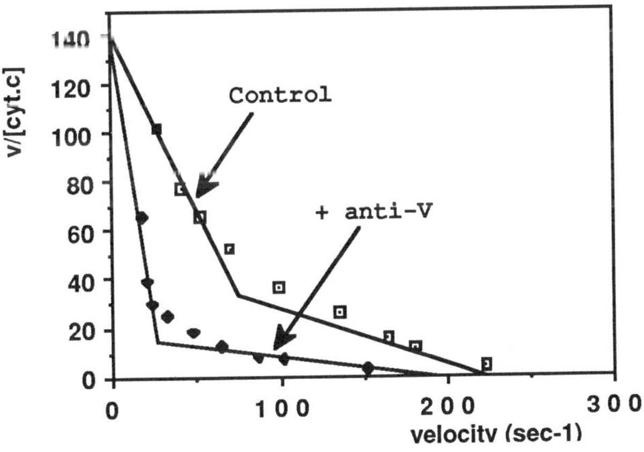

Fig. 5. Hofstee plots of cytochrome oxidase activity: effect of anti-subunit V IgG. Conditions as in Fig. 3, but with 50 mM K^+ phosphate replacing HEPES. Preincubation of oxidase and antibody as in Fig. 4.

When tested against vesicular enzyme, the antibody always has to be added from the outside. It can therefore be used to explore the inhibitable epitopes on the outside (P side) in presence of external cytochrome c and on the other side (N side) by using vesicles with entrapped cytochrome c reacting with TMPD (19, 20). Table 2 summarizes the effects on flux and on respiratory control exhibited by anti-oxidase and anti-V antibodies added to proteoliposomes.

The only effective inhibition is obtained with anti-oxidase acting on the 'P' side of the membrane. The anticipated (6) 'N' side action of anti-V antibody did not emerge. How can this behaviour be reconciled with that seen using the isolated enzyme (Fig. 5)?

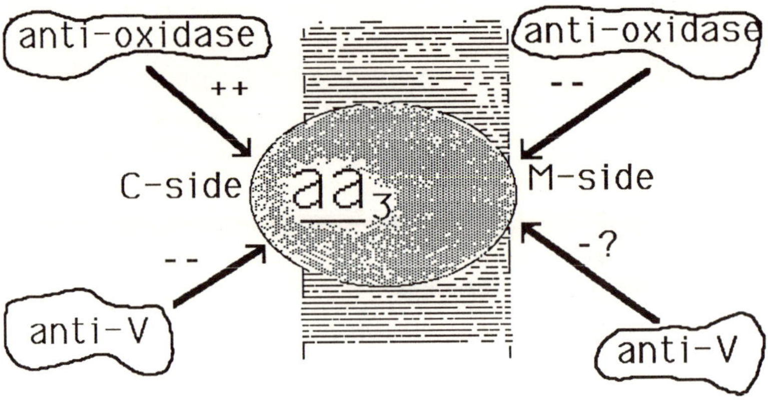

Fig. 6. Sites of antibody inhibitory action on cytochrome oxidase vesicles. P-side(C) effects determined by assay with external cyt. c; N-side(M) effects determined by assay with TMPD and internal cyt. c (see Table 2). No functional N-side epitopes were detectable with anti-oxidase, and no inhibition was seen with anti-V IgG on either side.

Table 2. Inhibition of normal and 'inverted' proteoliposomal cytochrome oxidase activity by anti-oxidase and anti-subunit V antibodies: effects on flux & respiratory control.

Antibody	Enzyme*	% inhibition Flux#	% inhibition RCR##
1. Anti-oxidase (300:1 ratio)**	Soluble	33≠	
	Normal Vesicular	25≠	50%
	Inverted vesicular	no effect (< 5)	no effect
2. Anti-subunit V (200:1 ratio)**	Soluble	50	-
	Normal vesicular	no effect (<5)	no effect
	Inverted vesicular	no+ effect (<5)	no effect

* Assayed in 50 mM K^+ phosphate buffer pH 7.4 at 30°C with ascorbate (±TMPD) + external cyt. _c_ for 'normal' activity and ascorbate + TMPD + trapped internal cyt. _c_ for 'inverted' activity. 0.1% lauryl maltoside present with soluble enzyme assays.

** antibody/protein ratio by weight

\# As % of FCCP x valinomycin 'uncontrolled' rate

\## Rate in presence of FCCP + valinomycin divided by control rate; reduction of 50% = decline of RCR from about 6 to 3.

≠ Greater than 90% when extrapolated to high Ab/protein ratios.

\+ a slight stimulation seen in some experiments.

DISCUSSION.

Figure 6 summarizes the presumed pattern of antibody interaction with liposomal cytochrome oxidase. It fails to account for the full set of observations involving subunit V. These are:

1) its redox-dependent accessibility from the matrix face (N side) of the membrane (6);
2) its redox-dependent accessibility in isolated oxidase (Fig. 2 above);
3) its effect on cyt. c kinetics when bound to antibody (Fig. 5 above);
4) its accessibility in the oxidase and only partial accessibility from both (?) sides of the membrane (5);
5) its water solubility and failure to contain an identifiable transmembranous sequence (16, 21, but cf.22).

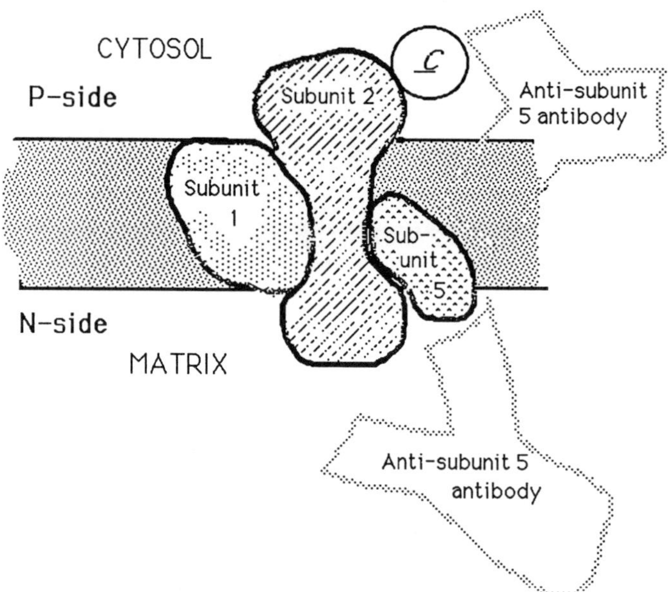

Fig. 7. Reactions of anti-subunit V antibodies with membrane-bound and soluble cytochrome oxidase (cf. Fig. 10 of ref. 6).

We conclude that subunit V of the mammalian complex may be embedded in (but not 'plugged through') the membrane sufficiently close to subunit II to enable antibody reaction with subunit V to interfere with dissociation of a neighbouring cytochrome c molecule. In the membrane the association with subunit 2 is largely at the N side, and this area is conformationally affected by reduction (largely at the P-side). Fig. 7 illustrates the kind of situation we envisage - non-inhibitory binding at the N-side, which is, however, redox-dependent, combined with inhibitory binding at a site that is blocked in the intact membrane and which interferes with cyt. c interaction with (and dissociation from) its site on subunit II.

Can we proceed further with this type of study? In order to do so, we need to increase both chemical and biochemical specificity. To achieve the former, the tendency to aggregation effects must be avoided by using Fab fragments. To achieve the latter, studies should be extended using monoclonal antibodies (23). Efforts in both directions are under way.

ACKNOWLEDGMENTS.

This work was supported by Canadian NSERC operating grant #A-0412 to PN & NIH grant #GM32246 to SHPC.We thank Mr. John Kjarsgaard, Ms. Faye Crerar & Mr. W.Z. Gai for skilled technical assistance.

REFERENCES.

1. Cinader, B. (1965) Antibodies to Enzymes in Antibodies to Biologically Active Molecules (ed. B. Cinader) Vol. 1, pp. 85-137 (Pergamon Press, Oxford).
2. Chan, S.H.P. & Schatz, G. (1979) Methods Enzymol. 50, 223-228
3. Mochan, B. S., Lang, R.W. & Elliott, W.B. (1970) Biochim. Biophys.Acta 216, 96-105.
4. Mochan, B.S., Lang, R.W. & Elliott, W.B. (1970) Biochim. Biophys. Acta 216,106-114.

5. Chan, S.H.P. & Tracy, R.P. (1978) Eur. J. Biochem. 89, 595-605.
6. Freedman, J.A. & Chan, S.H.P. (1983) J. Biol. Chem. 258, 5885-5892.
7. Frey, T.G., Chan, S.H.P. & Schatz, G. (1978) J. Biol. Chem. 253, 4389-4395.
8. Merle, P., Jarausch, J., Trapp, M., Scherka, R. & Kadenbach, B. (1981) Biochim. Biophys. Acta 669, 222-230.
9. Chan, S.H.P. & Freedman, J.A. (1983) FEBS letts 162, 344-348.
10. Freedman, J.A. Bratcher, R.L. & Chan, S.H.P. (1984) Biochem.Biophys.Res.Comm. 120, 52- 58.
11. Kahn, C.R., Baird, K.L. Flier, J.S., Grunfeld, C., Harmon, J.T., Harrison, L.C., Karlsson, F.A., Kasuga, M., King, G.L., Lang W.C., Podskalny, J.M. & van Obbershon, E. (1981) Rec. Prog. Hormone Res. 37, 477-533.
12. Tainer, J.A. Getzoff, E.D., Paterson, Y., Olson, A.J., & Lerner, R.A. (1985) Ann. Rev. Immunol. 3, 501-535.
13. Geysen, H.M., Tainer, J.A., Rodda, S.J. Mason, T.J., Alexander, H. Getzoff, E.D. & Lerner, R.A. (1987) Sciences 235, 1184-1190.
14. Berzofsky, J.A. (1985) Science 229, 932-940.
15. Kuboyama, M., Yong, F.C. & King, T.E. (1972) J. Biol. Chem 247, 6375-6383.
16. Freedman, J.A., Tracy, R.P. & Chan, S.H.P. (1979) J. Biol. Chem. 254, 4305-4308.
17. Kimelberg, H.K. & Nicholls, P. (1969) Arch. Biochem. Biophys. 133, 327-335.
18. Nicholls, P. & Hildebrandt, V.A. (1978) Biochem. J. 173, 65-72.
19. Wrigglesworth, J.M. & Nicholls, P. (1978) Biochim. Biophys. Acta 547, 36-46.
20. Cooper, C.E. & Nicholls, P. (1987) FEBS letters, in press.
21. Tanaka, M, Hamir, M., Yasonobu, K.T., Yu, C-A., Yu, L., Wei, Y-H. & King, T.E. (1979) J. Biol. Chem. 254, 3879-3885.
22. McMillen, D.A., Volwerk, J.J., Ohishi, J.-I., Erion, M., Keana, J.F.W., Jost, P.C. & Griffith, O.H. (1986) Biochem. 25, 182-193.
23. Gai, W.Z., Vanaja, B. & Chan. S.H.P. (1987) Fed. Proc. 46, 2205 (abstr.#1629).

DISCUSSION

GARBER: Since in your model you propose the effect of anti-subunit V to be on cytochrome *c* binding/dissociation, the effect on the spectrophotometric assay's kinetic parameters might be an interesting test. This is because the rate limiting step in the spectrophotometric assays is cytochrome *c* dissociation, while in the polarographic assays over the range of cytochrome *c* concentration you've employed the rate limiting step doesn't involve dissociation.

NICHOLLS: Only one type of anti-V seems to affect cytochrome *c* binding. We have kinetic values for the polarographic but not the spectrophotometric assay. However, at high ionic strengths, anti-V does inhibit the spectrophotometric assayed reaction. Competition studies with cytochrome *c* are difficult because very high total protein concentrations of the polyclonal mixture would have to be used.

Oxidases and Related Redox Systems, pages 653–668

EVOLUTION OF CYTOCHROME C OXIDASE

Bernhard Kadenbach, Achim Reimann,
Annemarie Stroh and Fritz-Joachim Hüther
Biochemie, Fachbereich Chemie der Philipps-
Universität, Hans-Meerwein-Strasse,
D-3550 Marburg, Fed. Rep. Germany

SUMMARY

The current view on the regulatory function of nuclear-encoded subunits of cytochrome c oxidase from higher evolved organisms is presented.

The activity of monomeric laurylmaltoside-dissolved, but not of reconstituted cytochrome c oxidase, is strongly affected by anions accompanied by a conformational change of the enzyme, as shown by changed visible spectra. Addition of uncoupler to proteoliposomes induces the same anion sensitivity as obtained with the soluble enzyme, suggesting dissociation of the dimeric membrane-bound enzyme by uncoupler.

Nucleotides are suggested to regulate cytochrome c oxidase activity at 3 different sites: 1) Interaction of ATP with a cytosolic site (outside) increases the K_m for cytochrome c in the enzyme from bovine heart and *Paracoccus denitrificans*; 2) binding of ADP at a matrix site decreases, and 3) binding of ATP at another matrix site increases the K_m for cytochrome c of the mammalian enzyme.

INTRODUCTION AND CURRENT STATE OF KNOWLEDGE

The energy metabolism of aerobic bacteria and of mammalian cells as presented in current textbooks is almost identical. Both contain a similar set of enzymes of glycolysis, citric acid cycle and oxidative phosphorylation. The homology concerns not only the catalytic property of enzymes but also amino acid sequences. Cytochrome c of Paracoccus denitrificans and human show 38 % sequence homology (1), indicating their evolutionary relationship. The activity of cytochrome c oxidase of Paracoccus is almost identical with that from bovine heart, and kinetic studies on electron transfer from ferrocytochrome c to oxygen could not identify basic functional differences between the two enzymes (2,3).

Thus we may ask, was there an evolution of energy metabolism, since bacteria and higher organisms evolved independently for more than a billion of years? The following results demonstrate that evolution from unicellular procaryotes to multicellular eucaryotes was accompanied by an increase of regulatory complexity of cytochrome c oxidase, the bottle neck of oxidative metabolism in all higher evolved organisms.

After a long lasting discussion on the number of protein subunits in the cytochrome c oxidase complex from different species - mainly based on the difficulty to separate and identify small polypeptides of very similar molecular weight - it is now generally accepted that mammalian cytochrome c oxidase consists of 13 (4,5), the yeast enzyme of 9 (6) and the enzyme of Paracoccus of 3 subunits (7). Thus the number of subunits increased with the evolutionary height of the organism. The N-terminal amino acid sequence of subunit VIIb, the 13th subunit of the mammalian enzyme (4) which was not found by the group of Buse (8), is shown in table 1.

TABLE 1. N-terminal amino acid sequence of subunit VIIb from rat liver and bovine liver. The rat liver sequence was determined by C. Eckerskorn and F. Lottspeich (Max-Planck-Institute, München) from gel bands (M. Ungibauer and B. Kadenbach) blotted on glass **fiber** sheets; the bovine liver sequence was taken from (5).

Rat liver	(S)H Q K K T P T F H D K Y G N A V L A G
Bovine liver	I H Q K R A P D F H D

The 3 bacterial subunits show amino acid sequence homology with the 3 largest subunits of eucaryotic enzymes (7).

These eucaryotic subunits are encoded on mitochondrial DNA whereas the small subunits are encoded on nuclear DNA. The four redox centers of the enzyme, two heme a and two copper atoms, and a third copper atom (9,10) are located within subunits I and II, and subunit III appears to participate in proton translocation (11).

Because the bacterial 3-subunit enzyme fullfills all known functional criteria of complex IV of the respiratory chain, e.g. electron transfer from ferrocytochrome c to oxygen, coupled with the translocation of protons for the formation of water, and for the generation of a μH^+ (7,11), but is lacking all nuclear-encoded subunits of the enzymes from higher evolved organisms, it was doubted that these subunits are integral parts of the eucaryotic enzyme complex and participate in the catalytic activity (for review see 11).

We have developed a hypothesis suggesting regulatory functions of nuclear-encoded subunits of eucaryotic cytochrome c oxidase (12,13). These subunits are assumed to modulate the catalytic activity of the enzyme, i.e. the rate of respiration and the H^+/e^--stoichiometry, by binding allosteric effectors like substrates, ions, nucleotides, hormones or second messengers, thus acting like receptors (13). The modulation of activity is suggested to occur via conformational change of the enzyme complex. We postulated that the rate of cellular respiration is not only controlled by the mechanism of "respiratory control", i.e. increase of mitochondrial respiration by addition of ADP, as generally assumed, but also by interaction with allosteric effectors. This variation of the rate of respiration without change of respiratory control is illustrated in fig. 1. In addition allosteric effectors are assumed to change the efficiency of energy transduction, i.e. the H^+/e^--stoichiometry.

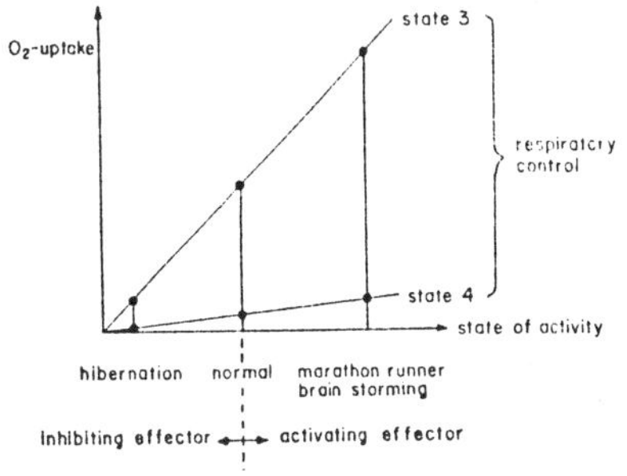

Figure 1: Schematic presentation of the regulation of mitochondrial respiration by direct interaction of allosteric effectors with respiratory chain complexes, in particular cytochrome c oxidase, without interference with "respiratory control" as postulated in (13).

The hypothesis was supported by experimental results:
1) A tight arrangement of 9 of the 10 nuclear-encoded subunits around the catalytic subunits I and II of the bovine heart enzyme was found by cross-linking studies (14).
2) A tissue-specific participation of nuclear-encoded subunits in the binding domain for cytochrome c at subunit II was found by labelling of carboxylic groups with a carbodiimide and radioactive glycine ethylester (15).
3) The sidedness and transmembranous orientation of subunits was investigated with subunit-specific antisera (16), and by protease digestion of rat liver mitoplasts (17).
4) From hydropathy plots of 9 amino acid sequences of nuclear-encoded subunits of bovine heart cytochrome c oxidase - mostly sequenced by the group of Buse - a transmembranous segment of about 25 hydrophobic amino acids was

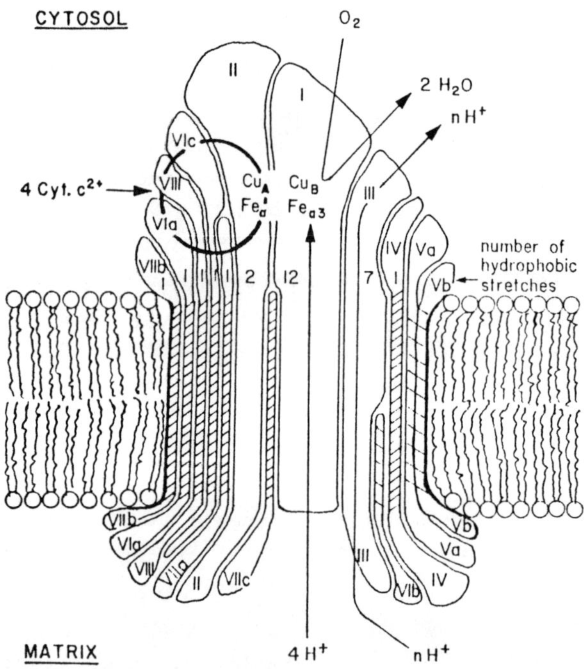

Figure 2. Scheme of the mammalian cytochrome c oxidase complex including various structural data as described in the text.

found for 6 subunits (11). Subunits Va and Vb did not show a hydrophobic segment but were also found to be accessible from both sides of the membrane (16).

By summarizing these structural data we have drawn a model of cytochrome c oxidase (fig. 2), which emphasizes the tight arrangement of hydrophilic domains on both sides of the membrane for most nuclear-encoded subunits.

A further understanding of the complexity of eucaryotic cytochrome c oxidase came from our discovery of tissue- and developmental-specific isozymes, based on identical mitochondrial-encoded "catalytic" but different nuclear-encoded "regulatory" subunits (18). Mammalian cytochrome c oxidase isozymes were found to consist always of 13 subunits, but from different N-terminal amino acid sequences, in particular of subunits VIa, VIIa and VIII, different genes were suggested to code for the corresponding subunit from liver and heart (19). These structural differences were corroborated by different catalytic properties of the isozymes from bovine heart and liver (20,21). Further support for the occurrence of isozymes came from the demonstration of gene families in the rat genome for subunits VIc and VIII (22). Isozymes of cytochrome c oxidase were also shown to occur in yeast (23) and *Dictyostelium discoideum* (24).

A strong indication for the essential role of nuclear-encoded subunits for the catalytic activity and for the occurrence of isozymes of cytochrome c oxidase came from studies of human mitochondrial myopathies (for review see 25). In one case of fatal infantile mitochondrial myopathy the specific absence of a nuclear-encoded subunit (VIIb) was demonstrated immunohistochemically in muscle cells lacking enzyme activity, but containing all other mitochondrial- and nuclear-encoded subunits of cytochrome c oxidase (J. Müller-Höcker and B. Kadenbach, in preparation).

The above described hypothesis involves a multitude of conformational changes as a basic regulatory principle of cytochrome c oxidase function. The following results demonstrate changes in activity and protein conformation by nucleotides and ions of the mammalian enzyme, but also to some extent of the enzyme from *Paracoccus*.

RESULTS

Uncoupling and the effect of anions

The influence of anions on the activity of isolated rat liver cytochrome c oxidase, dissolved in 1 mM laurylmaltoside, is presented in fig. 3. All anions tested show a biphasic effect on the activity, stimulating at low and inhibiting at high concentrations. The effectiveness is proportional to the charge of the anion, i.e. ATP acts at low and chloride at high concentration, whereas Hepes, at neutral pH a zwitterion, has very little effect. The same results were obtained with the isolated enzyme from pig heart (13), pig kidney and bovine heart (not shown).

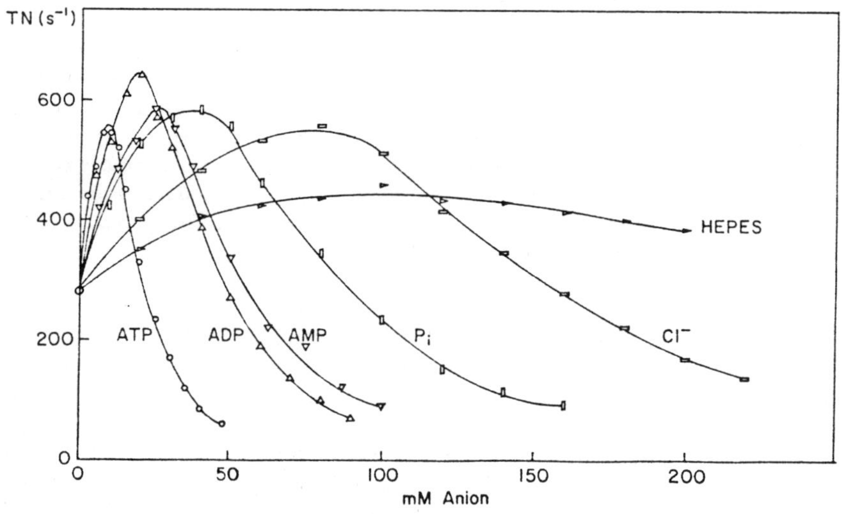

Figure 3. Effect of increasing concentrations of various anions on the activity of isolated rat liver cytochrome c oxidase. The enzyme activity was measured polarographically in 1 mM laurylmaltoside, 0.25 M sucrose, 10 mM K-Hepes, pH 7.6, 7 mM Tris-ascorbate, 0.7 mM TMPD and 40 μM horse heart cytochrome c (Sigma, Type VI). The indicated amounts of anions were added as concentrated potassium salt solutions. TN = turnover number.

A different effect of anions is obtained with cytochrome c oxidase from Paracoccus (fig. 4). No stimulation but only an inhibition of activity is found, and the anion concentrations required for half-maximal inhibition are about half from those required for the rat liver enzyme. Thus we conclude that nuclear-encoded subunits; which are lacking in Paracoccus, modulate the response of the enzyme to anions.

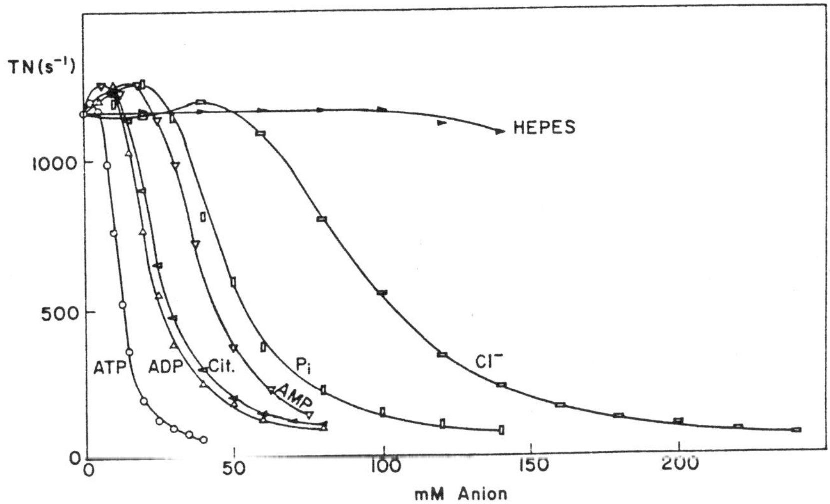

Figure 4. Effect of increasing concentrations of anions on the activity of cytochrome c oxidase of membranes from Paracoccus dissolved in 1 mM laurylmaltoside. For further details see legend to fig. 3. Cit. = citrate

How do the anions exert their effect on the activity of cytochrome c oxidase? Do they affect the electrostatic interaction with cytochrome c, or do they interact directly with the enzyme? In order to answer this question we measured the influence of anions on the visible difference spectrum of air-oxidized bovine heart cytochrome c oxidase in the absence of cytochrome c. In fig. 5 a clear change of the difference spectrum in the soret region by addition of anions can be seen. The spectral changes obtained with chloride, phosphate, ADP and ATP are different, both in respect to amplitude and location of maxima and minima. In fig. 6 is presented the anion concentration dependence of the maximal spectral change and of the enzyme activity. A clear correlation but no linear relationship is found between activity and the spectral change. From these results we conclude:
1) Anions change the visible spectrum of the soluble oxidized

enzyme, indicating a change of the heme a liganding sphere and thus a conformational change of the protein structure.
2) Different anions induce different spectral changes, indicating specific interactions of anions with the enzyme complex.

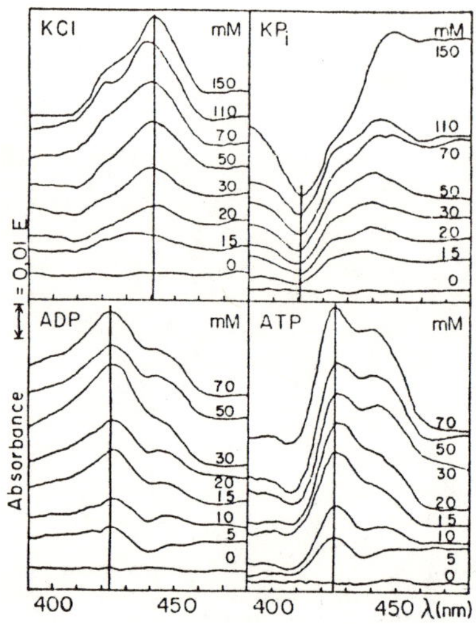

Figure 5. Effect of increasing concentrations of various anions on the visible spectrum of isolated bovine heart cytochrome c oxidase. The enzyme was dissolved in 10 mM K-Hepes, pH 7.4, 1 mM laurylmaltoside at a concentration of 17 μM. The difference spectra of the soret region were recorded at increasing concentrations of the indicated anions (potassium salts, pH 7.4) with a Uvikon 810 spectralphotometer at room temperature.

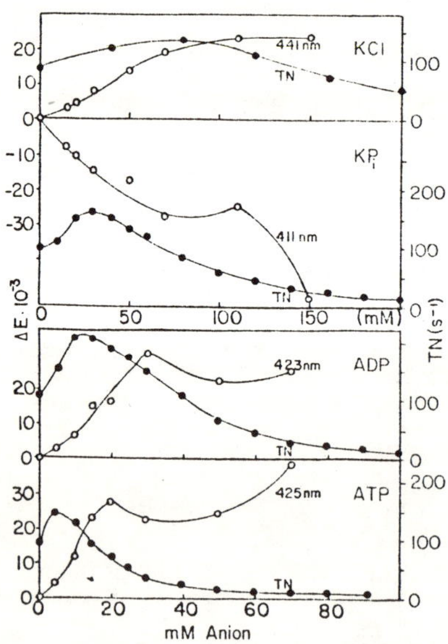

Figure 6. Relationship between the change of absorbance difference at a selected wavelength with the change of activity of isolated bovine heart cytochrome c oxidase with increasing concentrations of the indicated anions. Enzyme activity was measured polarographically as described in the legend to fig. 3 but without sucrose and with 25 mM Tris-ascorbate. The absorbance differences were taken from fig. 5.

3) The correlation between change of activity and heme a spectrum suggests that anions affect the catalytic activity via conformational changes.

In order to investigate whether the described effects of anions have a meaning on the physiological function of cytochrome c oxidase, we studied their influence on the bovine heart enzyme after reconstitution in liposomes (fig. 7). Surprisingly, phosphate does not change the activity of the reconstituted enzyme up to a concentration of 180 mM, when measured polarographically with ascorbate. However, a strong inhibition by phosphate is obtained with the uncoupler-stimulated enzyme. At 180 mM phosphate the rate is even below that measured under coupled conditions. The TMPD-stimulated activity, in the absence of uncoupler, is slightly inhibited by phosphate, which is explained by the partially uncoupling effect of TMPD, as indicated by the decreased respiratory control.

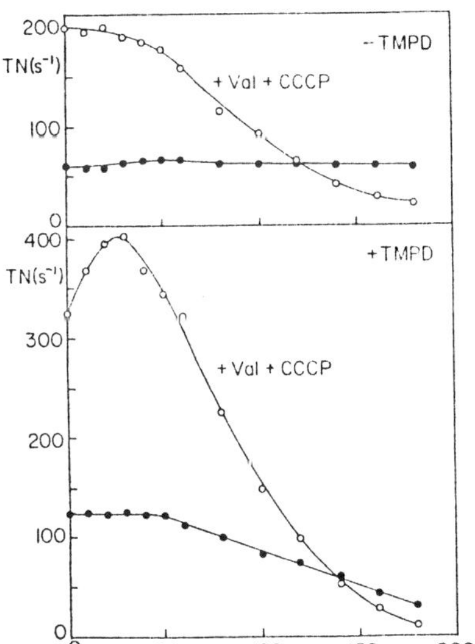

Figure 7. Influence of increasing concentrations of phosphate on the activity of reconstituted cytochrome c oxidase from bovine heart in the presence and absence of uncoupler. The isolated enzyme was reconstituted in asolectin-liposomes as described (28), and was oriented 81 % right side out. Oxygen uptake was recorded polarographically in 10 mM K-Hepes, pH 7.4, 300 mM sucrose, and the indicated amount of KP_i, pH 7.4. When indicated 0.7 mM N,N,N',N'-tetramethyl-p-phenylenediamine (TMPD) or 1 µg/ml valinomycin (val) and 3 µM carbonyl-cyanide-m-chlorphenylhydrazone (CCCP) were added.

In fig. 8 is shown the influence of increasing concentrations of ATP on the activity of reconstituted cytochrome c oxidase before and after addition of uncoupler or of laurylmaltoside. Whereas the activity without uncoupler is slightly decreased (presence of TMPD), the activity after addition of either uncoupler or of laurylmaltoside is indistinguishable inhibited by increasing concentrations of ATP. The lack of a biphasic effect of ATP is due to the presence of 50 mM KCl and 20 mM KP_i in the medium.

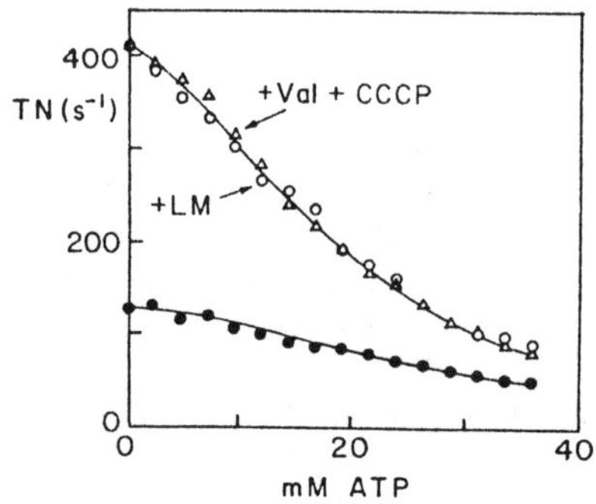

Figure 8. Influence of increasing ATP concentrations on the activity of reconstituted bovine heart cytochrome c oxidase in the presence and absence of uncoupler and laurylmaltoside. Proteoliposomes were prepared by sonication of liposomes (40 mg/ml asolectin) with the enzyme (5 nMol/ml heme aa_3) in 50 mM KCl, 20 mM KH_2PO_4, 20 mM K-Hepes,pH 7.3 for 15 min at 4°C (0.5 s pulse, 0.5 s pause) The activity was measured polarographically in the same medium containing 45 μM cytochrome c, 25 mM Tris-ascorbate, 0.7 mM TMPD and the indicated amount of K-ATP, pH 7.4. Where indicated Val + CCCP (see legend to fig. 7) or 1 mM laurylmaltoside (LM) were **added**.

From these data we conclude that uncoupling of reconstituted cytochrome c oxidase induces a conformational change of the enzyme complex leading to an anion accessibility at a site, where high concentrations inhibit the activity. Because this site is not accessible to anions under coupled conditions, the biphasic effects of anions appear not to regulate cytochrome c oxidase activity under physiological conditions. Since the laurylmaltoside dissolved bovine heart enzyme occurs as a monomer (26), but the membrane-bound enzyme as a dimer (27), uncoupling of proteoliposomes is suggested to dissociate - at least to some extent - the dimeric membrane-associated enzyme complex.

ATP and ADP modulate the kinetics of cytochrome c oxidation

A physiological regulation of cytochrome c oxidase activity by nucleotides was suggested from studies where the influence of extraliposomal and intraliposomal nucleotides was investigated on the kinetics of cytochrome c oxidation of the reconstituted enzyme. The interaction of extraliposomal ATP with the enzyme but not with cytochrome c was proven by

photolabelling of the isolated bovine heart enzyme with 8-azido-ATP (8-N_3-ATP) (28). The bound 8-nitreno-ATP increases the K_m for cytochrome c to a similar extent than free extra-liposomal ATP as shown in fig. 9, where proteoliposomes were photolabelled with 8-N_3-ATP. The presence of ATP but not of ADP during photolabelling, completely prevented the increase of K_m for cytochrome c. This increase of K_m for cytochrome c is not specific for ATP, because the following sequence of anions with increasing effectiveness was found:
P_i < AMP < ADP < PP_i < ATP < PPP_i (29). The kinetic changes were found in the presence as well as absence of uncoupler, they occur only at high millimolar concentrations of ATP, and were suggested to function under physiological conditions, because no other polyvalent anion than ATP occurs in the cytosol at sufficient high concentrations.

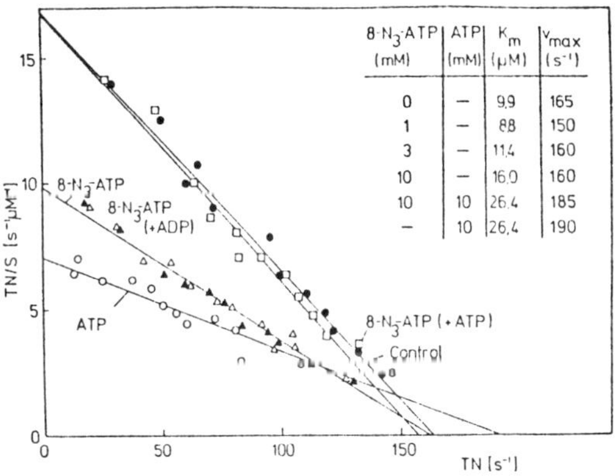

Figure 9. Change of the K_m for cytochrome c of reconstituted bovine heart cytochrome c oxidase after photolabelling with 8-N_3-ATP. Photolabeling of proteoliposomes was performed as described (28,29). The photometric assay of activity was done in 100 mM K-Hepes, pH 7.4, 50 mM KCl, 1-80 μM ferrocytochrome c, 1 μg/ml valinomycin and 3 μM CCCP. ●—●: Control, illumination of proteoliposomes without additions; o—o: illumination without additions, assay in the presence of 10 mM ATP; ▲—▲: photolabelling with 10 mM 8-N_3-ATP; □—□: photolabelling with 10 mM 8-N_3-ATP in the presence of 10 mM ATP; △—△: photolabelling with 10 mM 8-N_3-ATP in the presence of 10 mM ADP.

The increase of the K_m for cytochrome c by ATP appears not to involve an interaction with nuclear-encoded subunits, because the same results are obtained with the Paracoccus enzyme (fig. 10) (30).

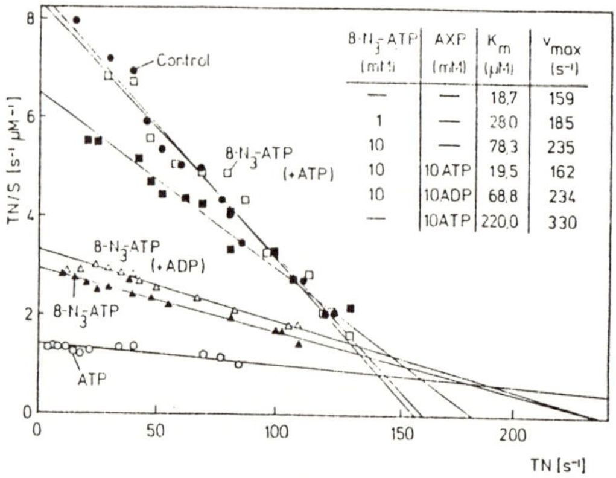

8-N$_3$-ATP (mM)	AXP (mM)	K$_m$ (µM)	v$_{max}$ (s^{-1})
—	—	18.7	159
1	—	28.0	185
10	—	78.3	235
10	10 ATP	19.5	162
10	10 ADP	68.8	234
—	10 ATP	220.0	330

Figure 10. Change of the K_m for cytochrome c of reconstituted cytochrome c oxidase from Paracoccus denitrificans after photolabelling with 8-N_3-ATP. For further details see legend to fig. 9. ■—■: Photolabeling with 1 mM 8-N_3-ATP.

A different regulation of cytochrome c oxidase activity was found by nucleotides acting from the matrix side. This was concluded from studies where the kinetics of cytochrome c oxidation were measured by the photometric method with proteoliposomes containing inside ATP or ADP (fig. 11).

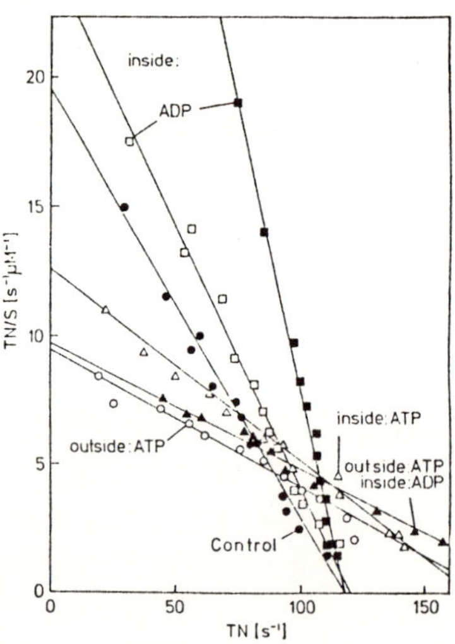

Figure 11. Influence of intraliposomal ADP and ATP and extraliposomal ATP on the kinetics of reconstituted cytochrome c oxidase. Proteoliposomes were prepared by the cholate dialysis method in the presence of the indicated nucleotides as described (31). The photometric assay of activity was performed in 100 mM K-Hepes, pH 7.4, 50 mM KCl, 1-80 µM ferrocytochrome, 1 µg/ml valinomycin, 3 µM CCCP and 10 mM ATP if indicated. ●—●: control; □—□: 5 mM ADP inside; ■—■: 10 mM ADP inside; △—△: 10 mM ATP inside; ▲—▲: 10 mM ADP inside and 10 mM ATP outside; ○—○: 10 mM ATP outside.

In contrast to extraliposomal ADP and ATP intraliposomal ADP decreased the K_m for cytochrome c, opposite to intraliposomal ATP, which acted like extraliposomal ATP. No change of the K_m for cytochrome c was obtained, however, if the effects of ADP and ATP were measured by the polarographic method (31). This result indicates that intraliposomal nucleotides change the rate of ferricytochrome c dissociation from the enzyme, the rate-limiting step in the photometric but not in the polarographic assay, as previously suggested (32).

DISCUSSION

The above described effects of anions on cytochrome c oxidase activity apparently do not reflect a physiological regulation of activity, because they are not observed with the membrane-bound enzyme under coupled conditions. However, the correlation between activity and spectral change with increasing anion concentrations indicates a conformational change as a basic principle for modulation of activity. The induction of anion-sensitivity by uncoupling of reconstituted cytochrome c oxidase also suggests a conformational change of the membrane-bound enzyme. The identical inhibition with increasing concentrations of ATP, obtained with uncoupler-treated and laurylmaltoside-dissolved proteoliposomes (fig. 6), supports the assumption that the dimeric membrane-bound enzyme is dissociated by uncoupler. We cannot distinguish between two possible causes of the conformational change: decrease of the membrane potential or direct interaction of uncoupler with the enzyme.

In previous publications it was suggested that the dimeric enzyme is essential for proton translocation (33,34). The apparent dissociation by uncoupler of membrane-bound cytochrome c oxidase supports this suggestion. Since the charged groups of the enzyme, which are involved in inhibition of activity at high anion concentrations, are only accessible after uncoupling, proton translocation in coupled proteoliposomes is suggested to involve electrostatic interactions between these charged groups at the two monomers. The different anion sensitivity of cytochrome c oxidase from Paracoccus and mammalian tissues suggests different regulatory properties of proton translocation in the enzyme of lower and higher evolved organisms.

The effect of ATP on the kinetics of cytochrome c oxidation was already described 10 years ago but not understood (35,36). The above described results clearly demonstrate the complex effects of nucleotides on the enzyme including three different categories of interactions:
1) Unspecific interaction of nucleotides and other anions with the soluble enzyme, which are absent in the membrane-bound enzyme under coupled conditions;
2) Unspecific (but possibly physiologically relevant) interaction of nucleotides and other anions with the cytosolic domain of the reconstituted enzyme, leading to an increase of the K_m for cytochrome c;
3) Specific interaction of ATP and of ADP with the matrix domain of the reconstituted enzyme, leading to an increased or decreased K_m for cytochrome c, respectively. These effects involve a conformational change of the binding domain for cytochrome c through the membrane domain of the enzyme.

Future work is required to prove the physiological relevance of the described conformational changes for the regulation of respiration in intact cells of bacteria and higher evolved organisms.

Acknowledgements: We thank Dr. B. Ludwig for membranes and isolated cytochrome c oxidase from Paracoccus, and C. Thiel for measurements of activity. This work was supported from grants of the Deutsche Forschungsgemeinschaft (SFB 103, A2) and of Fonds der Chemischen Industrie.

REFERENCES

1) Dickerson RE, Timkovich R, Almassy RJ (1976). J Mol Biol 100: 473-491.
2) Ludwig B, Gibson, QH (1981) J Biol Chem 256: 10092-10098.
3) Reichardt, JKV, Gibson QH (1983). J Biol Chem 258: 1504-1507.
4) Kadenbach B, Jarausch J, Hartmann R, Merle P (1983). Anal Biochem 129: 517-521.
5) Takamiya S, Lindorfer MA, Capaldi RA (1987). FEBS Lett 218: 277-282.
6) Power SD, Lochrie MA, Sevarino KA, Patterson TE, Poyton RO (1984). J Biol Chem 259: 6564-6570.
7) Ludwig B (1987). FEMS Microbiol Rev 46: 41-56.
8) Meinecke L, Buse G (1984). Third European Bioenergetics Short Reports 3: 177-178.

9) Steffens GCM, Biewald R, Buse G (1987). Eur J Biochem 64: 295-300.
10) Bombelka E, Richter, FW, Stroh A, Kadenbach B (1986). Biochem Biophys Res Commun 140: 1007-1014
11) Kadenbach B, Kuhn-Nentwig L, Büge U (1987). Curr Top Bioenerg 15: 113-161.
12) Kadenbach B, Merle P (1981). FEBS Lett 135: 1-11.
13) Kadenbach B (1986). J Bioenerg Biomembr 18: 39-54.
14) Jarausch J, Kadenbach B (1985). Eur J Biochem 146: 211-217.
15) Kadenbach B, Stroh A, Ungibauer M, Kuhn-Nentwig L, Büge U, Jarausch J. (1986). Methods Enzymol 126: 32-45.
16) Kuhn-Nentwig L, Kadenbach B (1985). Eur J Biochem 153: 101-104.
17) Jarausch J, Kadenbach B (1985). Eur J Biochem 146: 219-225.
18) Kuhn-Nentwig L, Kadenbach B (1985). Eur J Biochem 149: 147-158.
19) Kadenbach B, Hartmann R, Glanville R, Buse G (1982). FEBS Lett 138: 236-238.
20) Merle P, Kadenbach B (1982). Eur J Biochem : 239-244.
21) Büge U, Kadenbach B (1986). Eur J Biochem 161: 383-390.
22) Suske G, Mengel T, Cordingly M, Kadenbach B (1987). Eur J Biochem 168: 233-237.
23) Cumsky MG, Ko C, Trueblood CE, Pyton RO (1985). Proc Natl Acad Sci USA 82: 2235-2239.
24) Bisson R, Schiavo G (1986). J Biol Chem 261: 4373-4376.
25) DiMauro S, Bonilla E, Zeviani M, Nakagawa M, DeVivo DC (1985). Ann Neurol 17: 521-538.
26) Suarez MD, Revzin A, Narlock R, Kempner ES, Thompson DA, Ferguson-Miller S (1984). J Biol Chem 259: 13791-13799.
27) Henderson R, Capaldi RA, Leigh JS (1977). J Mol Biol 112: 631-648.
28) Hüther FJ, Kadenbach B (1986). FEBS Lett 207: 89-94.
29) Hüther FJ, Kadenbach B (1987a). submitted.
30) Hüther FJ, Kadenbach B (1987b). submitted.
31) Hüther FJ, Kadenbach B (1987c). Biochem Biophys Res Commun 147: 1268-1275.
32) Ferguson-Miller S, Brautigan DL, Margoliash E (1978). J Biol Chem 253: 149-159.
33) Sone N, Kosoko T (1986). EMBO J 5: 1515-1519.
34) Finel M, Wikström M (1986). Biochim Biophys Acta 851: 99-108.
35) Ferguson-Miller S, Brautigan DL, Margoliash E (1976). J Biol Chem 251: 1104-1115.
36) Roberts H, Hess B (1977). Biochim Biophys Acta 462: 215-234.

DISCUSSION

BARKER: Anions greatly increase the self-exchange rate of cytochrome c (NMR experiments). Polyanions, especially ATP, are the best at this. Is this relevant?

KADENBACH: In addition to citrate, bicarbonate and phosphate, ATP and ADP were also shown to bind cytochrome c at specific sites. The binding of ATP to cytochrome c was in fact shown to increase the electron self-exchange of cytochrome c [Concar et al. (1986) FEBS Lett 206: 15-19]. ATP increases the Km for cytochrome c to a larger extent when the exchange is measured by the polarographic as compared to the photometric assay [Zuther, Kadenbach (1986) FEBS Lett 207: 89-94]. Since 8-nitreno ATP, covalently bound to the oxidase, clearly influences the kinetics of cytochrome c oxidation, it appears that ATP and other anions interact with both cytochrome c and the oxidase, accompanied by changed kinetics of ferrocytochrome c oxidation.

NICHOLLS: At high ionic strength your results show an inhibition on addition of FCCP and valinomycin - removal of $\Delta\Psi$ or ΔpH gradient thus slows the turnover. What is the suggested explanation of this?

KADENBACH: We suggest that uncouplers change the conformation of the membrane-bound enzyme, making binding sites for anions accessible. These anions completely inhibit the activity at high concentrations. Since under coupled conditions these anion binding sites are not accessible, the coupled enzyme has a higher turnover at high anion concentrations.

SMITH: Although there are quantitative differences, there seem to be simpler changes with anions in both beef and Paracoccus oxidases, but the latter does not have all the small subunits. Also, perhaps reaction with soluble cytochrome c is not important in Paracoccus oxidase, which has the membrane bound cytochrome c_{552}.

KADENBACH: The inhibition of cytochrome c oxidase activity by anions appears to be due to the mitochondrially-coded subunits I and II because it is observed with both beef and Paracoccus enzyme as well. We suggest that these anion-sensitive sites are involved in energy transduction, which involves conformational changes of the enzyme complex.

Session XI: Cytochrome *c* Oxidase, continued
Chairman: E.C. Slater

Oxidases and Related Redox Systems, pages 671–686

HOW MANY TRANSMEMBRANE HELICES ARE THERE IN THE MITOCHONDRIALLY CODED AND OTHER SUBUNITS OF CYTOCHROME c OXIDASE? WHERE ARE THE METAL BINDING SITES?

Münime Lundeen

Atatürk Eğitim Fakültesi Kimya Bölümü, Marmara Üniversitesi, Göztepe, Istanbul, Turkey

INTRODUCTION

One of the fascinations of biochemical systems for inorganic chemists has been the presence of metal sites in these systems in special environments that are oftentimes found in proteins alone. Among such biochemical systems is the respiratory pathway of aerobic organisms that ends with cytochrome oxidase. Predictions of the secondary structures of subunits of cytochrome oxidase based on the frequencies of occurrence of side chains in alpha-helices, beta strands, reverse turns or other regions (commonly referred to as random coil) are invalidated by the special amino acid compositions of membrane proteins which do not reflect at all the frequencies of occurrence of the side chains in soluble proteins (Michel et al, 1986). Thus, hydrophobicity of a region appears to be the more reliable criterion in searching for structurally homologous regions in subunits of cytochrome oxidase from different species (Lundeen et al, 1987). It is estimated that a hydrophobic free energy gain of 20-25 kcal is required for a protein segment to traverse the thickness of the lipid bilayer (Von Heijne & Blomberg, 1979). This means that a 19-residue segment such as $A_{1.8}R_{.9}N_{.9}D_{1.1}C_{.6}Q_{.8}E_{1.2}G_{1.5}H_{.4}I_{.9}L_{1.5}K_{1.4}M_{.3}F_{.7}P_{.9}S_{1.4}$-$T_{1.2}Y_{.7}W_{.2}V_{1.4}$, for example, which shows the natural occurrence of the side chains in proteins (Klapper, 1977) would not traverse a biological membrane because the estimate of its free energy of transfer from an aqueous environment into the membrane is unfavorable ($\sim$+17 kcal) if the side chains carrying charges are included and, not favorable enough (∿-12 kcal), if arginine, aspartate, glutamate, lysine and histidine remain in contact with solvent. A transmembrane segment of Kuhn and Leigh (1985) of about the same length would have the amino acid composition $A_{2.0}R_{.2}N_{.2}D_{.3}C_{.3}Q_{.2}E_{.2}G_{1.6}H_{0.0}I_{2.5}L_{3.2}K_{.2}M_{1.0}F_{1.6}P_{.5}$-$S_{.8}T_{1.2}W_{.7}Y_{.8}V_{2.3}$. The free energy change to transfer this segment from aqueous solution to the membrane is favorable (∿-26 kcal) even including the charged side chains.

Methods

Hydrophobicity profiles of protein chains were plotted using a moving window of 10 residues by hand or of 19 residues at the midpoint by the "C" language program of Kyte and Doolittle (1982). Average hydrophobicities of the most hydrophobic 19-residue segments are reported as the hydrophobicity index, HI, of the segment. The transmembrane helices predicted for the photosynthetic reaction center from *Rhodopseudonomas viridis* are underlined below and compared to the observed transmembrane helices in the *H*, *L* and *M* subunits in the crystal structure (shown in parentheses) (Deisenhofer et al, 1985).

H helix $_{12}$(IAQLVWYAQWLVIWTVVLLYLRREDR)$_{37}$

LA helix $_{30}$FV(GFFGVSAIFFIFLGVSLIGYAASQ)GPT$_{55}$

MA helix $_{50}$YL(GASGIAAFAFGSTAILIILFNMAAEVH)$_{78}$

LB helix $_{83}$G(GFWQAITVCALGAFISWMLREVEISRKLG)$_{112}$

MB helix $_{110}$(GGWWLMAGLFMTLSLGSWWIRVYSRARALG)$_{139}$

LC helix $_{115}$(WHVPLAFCVPIFMFCVLQVFRPLLLG)$_{140}$

MC helix $_{142}$(THIAWNFAAAIFFVLCIGCIHPTLVG)$_{167}$

LD helix $_{170}$(NPGHMSSVSFLFVNAMALGLHGGLILSVAN)$_{199}$

MD helix $_{197}$(CPWHGFSIGFAYGCGLLFAAHGATILAVA-)$_{225}$

LE $_{225}$(GALSIHRLGLFLASNIFLTGAFGTIAS)$_{251}$

ME $_{259}$(TIESVHRWGWFFSLMVMVSASVGILLT)$_{285}$

The reported transmembrane helices of the photosynthetic reaction center are, on the average, 27.4 residues long and would form a regular alpha-helix just over 41 Å long which is well within the length of 48 Å estimated for the minimum packing of membrane sheets in a dehydrated specimen of bacteriorhodopsin (Engelman et al, 1982). The scale shown in Table 1 column 3 predicts the locations of all eleven transmembrane helices of the photosynthetic reaction center and overpredicts only two residues in the A helix of the L chain as being transmembrane helical residues.

Extended Huckel calculations on model copper compounds were carried out using the program FORTICON8 (J. Howell et al, QCPE 469) on a CDC Cyber 170 computer at Boğaziçi University, Istanbul, Turkey. Input atomic parameters for H, C, N, S and Cu are those reported by Nikles et al (1982).

Subunits of Cytochrome Oxidase

Altogether 23 transmembrane helices are predicted in all twelve subunits of cytochrome oxidase from beef heart so that 460 to 480 of the helical residues may be expected to be in transmembrane helices (Lundeen et al, 1987). CD and IRLD of multilamellar films of cytochrome oxidase have an overall helical tilt of 36 to 39° (Bazzi & Woody, 1985) which suggests that the transmembrane helical axis is tilted substantially from the normal to the plane of the membrane. The packing of helices together that are approximately normal to the plane of the membrane (Stamatoff et al, 1981) may maximize favorable contacts by a helical tilt (Lesk & Chothia, 1980). For example, the B and G helices in myoglobin pack in parallel in the 4n ridges-into-grooves model and have a helical tilt of ∿25° (Lesk & Chothia, 1980). The largest tilt observed in the transmembrane helices of the photosynthetic reaction center is ∿38° (Deisenhofer et al, 1985). The porphyrin of heme a appears to be tilted by ∿30° from the normal to the plane of the membrane while that of heme a_3 is approximately perpendicular (Blum et al, 1978). Based on the expected proximity of heme a to cytochrome c for initial electron transfer to occur Wikstrom and Saraste (1984) suggest that heme a may be located between the two transmembrane helices of subunit II. Another possible heme a site is in the transmembrane helices of subunit I with conserved beef heart H61 and H151 providing the proximal and distal attachment sites to a six-coordinate heme iron as in cytochromes b and b_6 (Widger et al, 1984).

Wikstrom and Saraste (1984) align the distal histidine in leghemoglobin with H61 and suggest that this is the site of heme a_3 and propose the two conserved methionines in the same transmembrane helix to be the sulfur ligands of Cu_b. On the other hand, Welinder and Mikkelsen (1983) found a long stretch of highly conserved side chains including many aromatic residues outside the transmembrane helices of subunit I and propose that this is the oxygen-binding site of cytochrome oxidase based on the homology of H233 to the distal histidine ligand of leghemoglobin. It is possible that the aromatic residues at H233-H240 serve partially to make the environment of the Cu_b ion highly hydrophobic as, unlike myoglobin and leghemoglobin, copper protein sites do not have a particular ligand (porphyrin) in a special fold (the globin fold). H138 may be close to this site and is proposed to be the proximal heme a_3 ligand with probably M208 as the distal heme a_3 ligand found in copper-depleted cytochrome oxidase (Chance et al, 1980). The alignment of myoglobin and leghemoglobin with subunit I shown below suggests this probable oxygen-binding region in subunit I is outside the transmembrane helices though the porphyrin of heme a_3 may be partially in the membrane (Fig. 1) (Lundeen, 1986).

	CD corner	D helix	E helix
Mb	FDRFKHLK	TEAEMKA	SEDLKKH_dGVTVLTAL....
legHb	FS-FLKG-	TSEVPQN	NPELQAH_dAGKVFKLV....
ppI	F--F-D-- (219)	P-AG-GG	DPILYQH LFWFFGHP....

The side chains that are underlined are strictly conserved in subunits I from ten different species.

The partial sequence of the bacterial cytochrome oxidase from *Paracoccus denitrificans* shows homology to subunits I and II of cytochrome oxidase (Steffens et al, 1983). The highly conserved segment of subunit II that

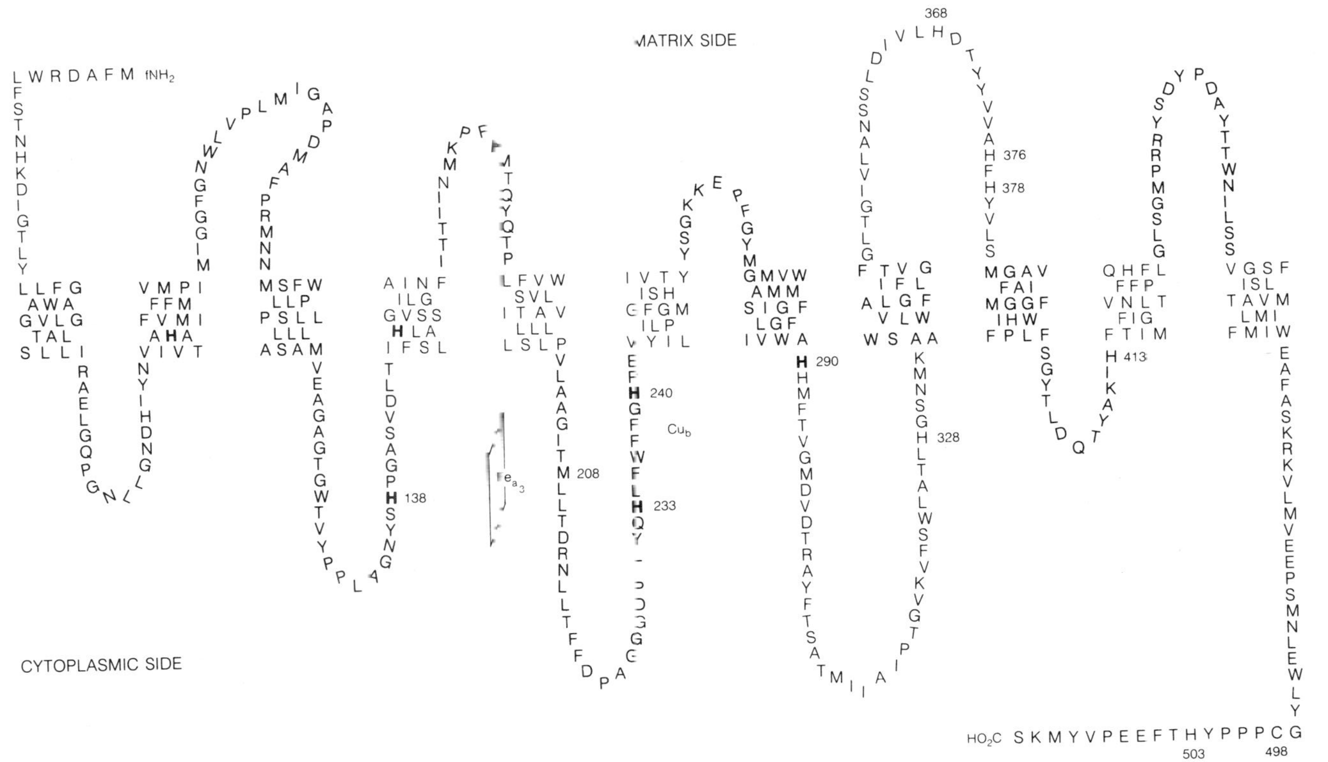

MATRIX SIDE
CYTOPLASMIC SIDE
fNH2
HO2C
138
208
233
240
Cub
Fea3
290
328
368
376
378
413
498
503

TABLE 1

HYDROPHOBICITY/HYDROPHILICITY SCALES

side chain	accessible surface area	this work	von Heijne	Engelman et al	Kyte and Doolittle
A	40 A^2	1.0	1.0	1.6	1.8
R	150	-6.5	-11.3	-12.3	-4.5
N	85	-1.8	-2.9	-4.8	-3.5
D	75	-7.1	-7.4	-9.2	-3.5
C	60	1.5	1.5	2.0	2.5
Q	105	-1.3	-2.4	-4.1	-3.5
E	115	-6.1	-5.9	-8.2	-3.5
G	0	0.0	0.0	1.0	-0.4
H	120	-0.9	-3.4	-3.0	-3.2
I	100	2.5	2.5	3.1	4.5
L	95	2.4	2.4	2.8	3.8
K	125	-5.9	-4.2	-8.8	-3.9
M	110	2.8	2.7	3.4	1.9
F	135	3.4	3.4	3.7	2.8
P	70	-0.8	-3.3	-0.2	-1.6
S	40	-0.3	-1.5	0.6	-0.8
T	65	0.3	-0.9	1.2	-0.7
W	180	3.2	2.0	1.9	-0.9
Y	155	1.3	-1.1	-0.7	-1.3
V	80	2.0	2.0	2.6	4.2

Figure 1 (previous page). The predicted transmembrane helices in subunit I of human placental cytochrome oxidase. The proposed proximal and distal histidine ligands of heme a (H61 and H151) are in helices II and IV. The proposed proximal histidine ligand of heme a_3 is H138 in the connecting link between helices III and IV. The probable histidine ligands of Cu_b are the conserved histidines that have been numbered (H233, H240, H290). The porphyrin ring of heme a may be mostly between helices II and IV while that of heme a_3 may be partially immersed in the membrane near helices IV, V and VI (Fig. 1 is reprinted with permission from Inorganic Chemistry 25: 4852-4856, copyright (1986) American Chemical Society).

TABLE 2

this work	H.I.	K&D	H.I.	K&L	W&M
19-36	1.68	14-36	2.04	12-43	14-36
54-71	1.84	56-74	2.53	53-77	56-74
100-117	1.61	99-117	2.09	97-123	-
144-161	1.35	146-164	1.65	142-168	144-168
180-197	1.93	182-210	2.25	180-210	182-209
242-259	1.31	242-260	1.62	-	-
269-286	1.32	269-287	1.73	-	266-287
-	-	-	-	304-364	-
332-349	1.78	332-350	1.70		336-357
375-392	1.61	-	-	-	369-388
412-429	1.93	-	-	-	412-434
454-471	2.27	451-469	2.10	450-470	-
					Capaldi
43-60	1.99	45-63	2.30	41-59	25-48
87-104	1.91	87-105	2.52	77-97	63-82

The predicted transmembrane segments in subunits I,II of *Saccharomyces cerevisae* cytochrome oxidase by:
This work: M. Lundeen (1986) Inorg. Chem. 25: 4852;
K&D: J. Kyte & R. F. Doolittle (1982) J. Mol. Biol. 157: 105.
W&M: K. G. Welinder & L. Mikkelsen (1983) FEBS Lett. 157: 233.
K&L: L. A. Kuhn & J. S. Leigh, Jr. (1985) Biochim. Biophys. Acta 828: 351.
Capaldi: R. A. Capaldi, F. Malatesta & V. M. Darley-Usmar (1983) Biochim. Biophys. Acta 726: 135.

contains many aromatic residues and may be a charge transfer region (Steffens & Buse, 1979) is conserved in subunit II of Paracoccus denitrificans cytochrome oxidase (Steffens et al, 1983):

	91 110
beef heart II	NNPSLTVKTMGHQWYWSYEY
P. den. II	NDPDLVIKAIGHQWYWSYEY

The side chains that are strictly conserved in subunits II from beef heart and ten other species are underlined. The segment of subunit II from beef heart that is shown follows the second transmembrane helix in this subunit (Bisson et al, 1982) and the tripeptide at 100-102 is similar to the -MGH46-tripeptide in azurin that contains one of the two histidine ligands of copper (Steffens & Buse, 1979). This histidine is not conserved in only the yeast (Saccharomyces cerevisae) subunit II sequence. Doolittle (1981) has pointed out that introducing no gaps to align beef heart subunit II with its 227 residues with azurin which has 127 residues from residue 137 to the end of the subunit II chain gives a barely better alignment score (0.1 SD) than comparing to a scrambled chain having the same amino acid composition. Introducing three gaps in aligning plastocyanin with the same segment of subunit II improves the alignment score to 1.5 SD above the mean SD for the scrambled comparisons. Remarkably, the N-terminal regions of cytochrome oxidase subunits II from eleven species contain very few identical residues up to the beginning of the second transmembrane helix (There are only four strict identities in the first sixty beef heart subunit II side chains), but the following segment from residue 61 to the end of the chain has fifty strict identities in the remaining 167 residues (One out of every three and one third residue is identical along the chains of all ten species aligned with beef heart subunit II with only four gaps between 115-116 in beef heart II for six of the species including beef heart II). This highly conserved chain segment may bind Cu_a and interact with cytochrome c (Capaldi et al, 1983). The distinct EPR spectrum of Cu_a is affected in some mercury derivatives of cytochrome oxidase after denaturation starts (Stonehuerner, 1985). The site that may bind Cu_a has some strictly conserved residues in subunits II from eleven species (Buse,

1984). This sequence differs from the blue copper sequences of azurin and plastocyanin in the number of residues in the cysteine-histidine-methionine-(copper) loops, in not having a proline and in having two cysteines. One of these cysteines is not conserved in wheat subunit II only (Bonen et al, 1984). If the alignment of beef heart subunit II with azurin and plastocyanin (Steffens & Buse, 1979) is compared position-by-position using the optimal matching hydrophobicity scale (Sweet & Eisenberg, 1983), sequence correlation is 0.25 for plastocyanin and 0.26 for azurin (For chains of 100 residues a correlation of 0.30 indicates 99.9% confidence level on the student's t-test but this may overestimate the confidence level (Sweet & Eisenberg, 1983).

RESULTS

As a first approximation to a copper ion site in a protein that has nitrogen and sulfur ligands the x-ray coordinates for the copper site in poplar plastocyanin (Guss & Freeman, 1983) were used in a Xα-SCF-SW calculation (Penfield et al, 1985). I used the same model, Copper(II)$(CH_3S)(CH_3SCH_3)(NH_3)_2^{1+}$, in a simple extended Huckel calculation to see what sulfur orbitals overlapped with the metal orbitals in $\underline{C}_s$ or $\underline{C}_1$ site symmetry for sigma- and pi-bonding.

The copper site in plastocyanin has, as ligands, two imidazole rings from histidine residues, a thiolate sulfur from cysteine and a thioether sulfur from methionine (Guss & Freeman, 1983). The very short bond distance of 2.13 Å of the cysteine sulfur from copper is indicative of the strong interaction of this ligand with copper. On the other hand, the methionine sulfur is 2.90 Å away from the copper and this bond interaction was not detectable along the bond direction by polarized x-ray absorption spectroscopy of oriented single crystals (Scott et al, 1982). The calculation used a weighted average of the atomic valence state ionization energies in the Wolfsberg-Helmholtz approximation. Future calculations may include better approximations to the resonance integrals by performing a charge iteration on the valence state ionization energies which takes more computer time.

The results of the extended Huckel calculation, unlike the SCF-Xα-SW calculations (Penfield et al, 1985), place the lone-pair electrons of methylthiolate that form π- and σ-bonds to copper in p_x and a combination of p_z,p_y orbitals that are degenerate in the free ligand (p_x and (p_z+p_y) orbitals). The other pair of the p_z,p_y combination, (p_z-p_y), has some s-orbital character and forms the C-S α-bond in methylthiolate. The combination (p_z+p_y) acquires some s character in the copper complex and forms a Cu(II)-S σ-bond which is at a higher energy than the σ-bond that forms between Cu(II) and S(thiolate) electrons in the (p_z-p_y),s orbital (Penfield et al (1985) call the former bond "pseudosigma" because the maximum electron density is along the Cu-S-C bond and not along the higher-energy Cu-S bond). The p_x lone-pair of the methylthiolate ligand is only slightly lower in energy in the copper complex and has π-interaction with d_{xy},d_{xz} orbitals on Cu(II). Penfield et al (1985) report a 1.5 ev split between the σ- and "pseudo-sigma"-lone-pair orbitals on the S in the copper complex but the extended Huckel calculation gives a 0.3 ev split. Both sets of calculations agree that the split between the σ- and "pseudosigma"-copper-sulfur orbitals is about 2 ev.

The molecular orbitals that involve lone-pair orbitals on the dimethylsulfide include contributions from the lone-pair orbitals on the methylthiolate sulfur and suggest a strongly covalent copper(II) site. Remarkably, the Cu-S (dimethylsulfide) molecular orbital is about 1 ev lower in energy than the Cu-S-C σ-bond so that it has an energy mid-way between the σ- and "pseudosigma"-Cu-S(thiolate) bonds. When the position of the dimethylsulfide ligand is moved away from the copper(II) ion so that it corresponds to a 3.13 A bond distance as in the crystal structure of <u>A. denitrificans</u> azurin (Norris et al, 1986), the molecular orbitals assigned to Cu-S(dimethylsulfide) interactions shift in energy by 0.20 and 0.53 ev. The dimethylsulfide S interacts with the d_{z^2} orbital on copper best. The σ-bond includes contributions from the $d_{yz},d_{x^2-y^2}$ Cu orbitals but the overlap is poor due to the geometry of the copper site and the long bond length. There is an increase of ∿0.1 ev in the energies of the lone-pair orbitals on the dimethyl-sulfide S upon complex formation though the C-S bond orbital is lowered in energy by ∿0.3 ev. When this ligand is removed and the two NH_3 molecules are allowed to have slightly different positions as in the plastocyanin crystal structure, the calculated energies are shifted slightly. The calculated net charges on the atoms are also not changed much.

The calculated d-level splitting of the Cu(II) ion is similar to that reported in the SCF-Xα-SW calculation of Penfield et al (1985). The odd electron is placed in a combination of $d_{x^2-y^2}$, d_{z^2} and d_{yz} orbitals. The next orbital is an almost pure d_{xy} orbital and is followed by a d_{z^2}, $d_{x^2-y^2}$, d_{yz} combination. The lowest energy d orbitals are practically degenerate. One is an almost pure d_{xz} orbital while the other is made up mostly of d_{yz} and $d_{x^2-y^2}$ orbitals. Reasonable atomic charges are calculated.

The Cu_a site in cytochrome oxidase was found to be the more covalent site in the EXAFS study of Powers, Chance et al (1979). Two trans cysteine ligands with unusual Cu-S-C angles that placed a great deal of spin density on the two S atoms were proposed as an explanation of its g_x value which is lower than that for a free electron (Hoffman et al, 1980). Chan and coworkers propose that one of the cysteine sulfurs is a free radical (Stevens et al, 1982). LEFE measurements (Mims et al, 1980) indicate a different geometry than the blue copper sites which may suggest approximately equatorial bonding of two cysteine sulfurs. A difficulty with this model is that there appears to be a single cysteine in the probable copper site of subunit II from wheat (Bonon et al, 1984). I want to explore next by extended Huckel calculations a $CuNS_3$ site with the model $(Cu(NH_3)(CH_3S)_2(CH_3SCH_3))(0)$ in which one N ligand of plastocyanin is replaced by a second thiolate so that the total charge on the copper is zero. If methionine is a ligand at the Cu_a site, this S must be replaced by a threonine O in some species. Either S or O can provide empty orbitals to the reduced copper site for back-bonding, thus stabilizing the Cu(I) state (Gray & Malmstrom, 1983).

CONCLUSIONS

There is agreement and dispute in the prediction of transmembrane helices in cytochrome oxidase. Various predictions based on different hydrophobicity scales were compared for subunits I and II in Table II. Unless evidence for a different tertiary structure exists, the ability of a scale to predict the same transmembrane locations in homologous chains from various species (using a predetermined cut-off value if possible for making the prediction) probably indicates that a hydrophobic stretch corresponds to a transmembrane helix. Though frequencies of occurrence of the side chains in membrane proteins differ greatly from their frequencies of occurrence in nature (Kuhn & Leigh, 1985), the amino acid compositions of the transmembrane regions of membrane protein chains may not differ enough from all nontransmembrane regions of the chain for reliable predictions using a statistical method. Kuhn and Leigh (1985) find that they do not predict the same locations in all species as transmembrane as a result.

In recent work on a small yeast subunit (subunit VIIa) two beta strands connected by a hairpin turn were predicted to traverse the membrane twice rather than having a single transmembrane helix (Wright et al, 1986). I find this stretch hydrophobic enough to predict a 19-residue transmembrane helix (HI= 1.69). As transmembrane helices contain many bulky side chains which have strong beta-strand preferences (Lundeen, 1986), I wonder about how to distinguish between transmembrane alpha-helices and beta strands traversing a membrane based on residue type.

ACKNOWLEDGEMENT

I want to thank the Johnson Research Foundation for a year's visit at the School of Medicine, University of Pennsylvania and the Institute for Structural and Functional Studies, University City Science Center, Philadelphia. The help of RENK YAPIMEVI is gratefully acknowledged.

REFERENCES

Bazzi, M. D., Woody, R. W. (1985). Oriented Secondary Structure in Integral Membrane Proteins I. Circular Dichroism and Infrared Spectroscopy of Cytochrome Oxidase in Multilamellar Films. Biophys J 48:957-966.

Blum, H., Harmon, H. J., Leigh, J. S., Salerno, J. C., Chance, B. (1978). The Orientation of a Heme of Cytochrome Oxidase in Submitochondrial Particles. Biochim Biophys Acta 502: 1-10.

Bonen, L., Boer, P. H., Gray, M. W. (1984). The Wheat Cytochrome Oxidase Subunit II Gene Has an Intron Insert and Three Radical Amino Acid Changes Relative to Maize. EMBO J 3: 2531-2536.

Buse, G. (1984). Cytochrome c Oxidase <u>in Copper Proteins</u>

and Copper Enzymes, R Lontie, Ed, CRC Press, Inc, 2000 Corporate Boulevard NW, Boca Raton, Florida 33431.

Capaldi, RA, Malatesta, F, Darley-Usmar, VM (1983). Structure of Cytochrome c Oxidase. Biochim Biophys Acta 726: 135-148.

Chothia, C (1974). Hydrophobic Bonding and Accessible Surface Area in Proteins. Nature 248: 338-339.

Deisenhofer, J, Epp, O, Miki, K, Huber, R, Michel, H. (1985). Structure of the Protein Subunits in the Photosynthetic Reaction Centre of *Rhodopseudomonas viridis* at 3A Resolution. Nature 318: 618-624.

Doolittle, RF (1981). Similar Amino Acid Sequences: Chance or Common Ancestry? Science 214: 149-159.

Engelman, DM, Goldman, A, Steitz, TA (1982). The Identification of Helical Segments in the Polypeptide Chain of Bacteriorhodopsin. Methods Enzymol. 88: 81-88.

Gray, HB, Malmstrom, BG (1983) On the Relationship between Protein-Forced Ligand Fields and the Properties of Blue Copper Centers. Comments Inorg. Chem. 2: 203-209.

Guss, JM, Freeman, HC (1983) Structure of Oxidized Poplar Plastocyanin at 1.6 A Resolution. J Mol Biol 169: 521-563.

Hoffman, BM, Roberts, JE, Swanson, M, Speck. SP, Margoliash, E (1980). Copper Electron-Nuclear Double Resonance of Cytochrome c Oxidase. Proc Natl Acad Sci USA 77: 1452-1456.

Howell, J, QCPE 469, Quantum Chemistry Program Exchange, Department of Chemistry, Indiana University, Bloomington, Indiana 47405.

Klapper, MH (1977) Biochem Biophys Res Commun 78:1018-1024.

Kuhn, LA, Leigh, JS Jr (1985). A Statistical Technique for Predicting Membrane Protein Structure. Biochim Biophys Acta 828: 351-361.

Kyte, J, Doolittle, RF (1982). A Simple Method for Displaying the Hydropathic Character of a Protein. J Mol Biol 157: 105-132.

Lundeen, M (1986). Where Is the Oxygen Binding Site of Cytochrome c Oxidase? Transmembrane Helices of Subunits I and II. Inorg Chem 25: 4852-4856.

Lundeen, M (1986). Preferences of the Side Chains in Proteins for Helix, Beta Strand, Turn, and Other Conformations. Secondary Structures of Copper Proteins. J Inorg Biochem 27: 151-162.

Lundeen, M, Chance, B, Powers, L (1987). The Transmembrane Helices of Beef Heart Cytochrome Oxidase. Biophys J 51:

693-695.
Lesk, AM, Chothia, C (1980). How Different Amino Acid Sequences Determine Similar Protein Structures: The Structure and Evolutionary Dynamics of the Globins. J Mol Biol 136: 225-270.
Michel, H, Weyer, KA, Gruenberg, H, Dunger, I, Oesterhelt, D, Lottspeich, F (1986). EMBO J 5: 1149-1158.
Mims, WB, Peisach, J, Shaw, RW, Beinert, H (1980). Electron Spin Echo Studies of Cytochrome c Oxidase. J Biol Chem 255: 6843-6846.
Nikles, DE, Anderson, AB, Urbach, FL (1983). A Molecular Orbital Study of Thioether Coordination to Copper Ions. In Karlin, KD, Zubieta, J (eds): "Copper Coordination Chemistry: Biochemical and Inorganic Perspectives," New York: Adenine Press, pp 203-222.
Norris, GE, Anderson, BF, Baker, EN (1986). Blue Copper Proteins. The Copper Site in Azurin from Alcaligenes denitrificans. J Am Chem Soc 108: 2784-2785.
Penfield, KW, Gewirth, AA, Solomon, EI (1985) Electronic Structure and Bonding of the Blue Copper Site in Plastocyanin. J Am Chem Soc 107: 4519-4529.
Powers, L, Chance, B, Ching, Y, Muhoberac, B, Weintraub, ST, Wharton, DC (1982). FEBS Lett 138: 245-248.
Powers, L, Chance, B (1985) Multiple Structures and Functions of Cytochrome Oxidase. J Inorg Biochem 23: 207-217.
Scott, RA, Hahn, JE, Doniach, S, Freeman, HC, Hodgson, KO (1982). Polarized X-Ray Absorption Spectra of Oriented Plastocyanin Single Crystals. Investigation of Methionine-Copper Coordination. J Am Chem Soc 104: 5364-5369.
Stamatoff, J, Blaisie, JK, Tavarmina, A, Pechance, J, Dutton, DL, Erecinska, M, Eisenberger, P, Brown, G (1981). Bull Am Phys Soc 26: 266.
Steffens, GJ, Buse, G (1979). Studies on Cytochrome c Oxidase, IV Primary Structure and Function of Subunit II. Hoppe-Seyler's Z Physiol Chem 360: 613-619.
Steffens, GCM, Buse, G, Oppliger, W, Ludwig, B (1983). Sequence Homology of Bacterial and Mitochondrial Cytochrome c Oxidases. Partial Sequence Data of Cytochrome c Oxidase from Paracoccus denitrificans. Biochem Biophys Res Commun 116: 335-340.
Steitz, TA, Goldman, A, Engelman, DM (1982). Quantitative Application of the Helical Hairpin Hypothesis to Membrane Proteins. Biophys J 37: 124-125.
Stonehuerner, J, O'brien, P, Kendrick, L, Hall, J,

Millett, F (1985). Specific Labeling and Partial Inactivation of Cytochrome Oxidase by Fluorescein Mercuric Acetate. J Biol Chem 260: 11456-11460.
Sweet, RM, Eisenberg, D (1983). Correlation of Sequence Hydrophobicities Measures Similarity in Three-dimensional Protein Structure. J Mol Biol 171: 479-488.
Von Heijne, G, Blomberg, C (1979). Trans-membrane Translocation of Proteins. The Direct Transfer Model. Eur J Biochem 97: 175-181.
Welinder, KG, Mikkelsen, L (1983). The Oxygen Binding Site of Cytochrome Oxidase. Structural Predictions on Subunit I from Amino Acid Sequences. FEBS Lett. 157: 233-239.
Widger, WR, Cramer, WA, Hermann, RG, Trebst, A (1984). Sequence Homology and Structural Similarity between Cytochrome b of Mitochondrial Complex III and the chloroplast b_6-f Complex: Position of the cytochrome b Hemes in the Membrane. Proc Natl Acad Sci USA 81: 674-678.
Wikstrom, M, Saraste, M (1984). The Mitochondrial Respiratory Chain. Bioenerg. 9: 49-94.
Wright, RM, Dircks, LK, Poyton, RO (1986). Characterization of COX9, the Nuclear Gene Encoding the Yeast Mitochondrial Protein Cytochrome c Oxidase Subunit VIIa. Subunit VIIa Lacks a Leader Peptide and Is an Essential Component of the Holoenzyme. J Biol Chem 261: 17183-17191.

DISCUSSION

LUDWIG: How do you decide whether a region is membrane-spanning? Don't you have to consider interaction of the region with other protein domains?

LUNDEEN: The criterion I'm using for whether a segment is hydrophobic enough to be transmembranal is its hydrophobicity index (average hydrophobicity per residue). If HI is equal to or greater than 1.20 the 19 residue segment is considered to be transmembranal. It is possible that some segments with lower HI values form favorable interactions with other transmembranal helices in the membrane and, therefore, form transmembranal helices anyway. Thus, hydrophobicity plots may underpredict the number of transmembranal helices that may be present.

KING: How can you place where two hemes and three copper ions are attached to the amino acid sequence of cytochrome c by computer? How do you get the hydrophobicity of the sequence from nucleotide or chemical determination? Moreover, bacteriorhodopsin seems to be more hydrophobic but, of course, does not contain metal.

LUNDEEN: Hydrophobicity by itself does not tell that a metal-binding site is present in a polypeptide chain. Hydrophobicity profiles are useful to indicate the topology of a protein disposed within a membrane. I used mostly nucleotide sequence data to calculate hydrophobicity profiles from them (as described in my paper), and to locate reasonable clusters of metal-binding ligands along the folded polypeptide chains.

I assume that metal-binding sites are possible in a protein chain when likely side-chain ligands are clustered in space even though they may be distant from each other along the linear sequence. And in spaces where there are no such clusters, metal-binding sites may be ruled out. It is generally accepted that heme- and copper-binding sites of cytochrome c oxidase are on subunits I and II [Winter et al. (1980) J Biol Chem 255: 11408-11414] with beef heart oxidase and from work on bacterial oxidases, some presented in this symposium]. For this reason, the hydrophobicity profiles of subunits I and II have been determined by myself and by others. The folding of their chains inside and outside of the membrane have been predicted, and clusters of ligands appropriate for binding heme a, heme a_3, Cu_A, and Cu_B have been sought. This is the methodology of my work.

Oxidases and Related Redox Systems, pages 687–705

INTEGRAL COMPLEXES OF CYTOCHROME C OXIDASE CONTAIN THREE COPPERS

G.C.M. Steffens and G. Buse

Abteilung Physiologische Chemie
Rheinisch-Westfälische Technische Hochschule
Aachen, Fed. Rep. Germany

ABSTRACT

Metal contents have been determined in beef heart cytochrome c oxidase by inductively coupled plasma atomic emission spectroscopy. Integral complexes of this enzyme contain three copper and two iron atoms, as well as one zinc and one magnesium atom (Cu:2,91±0,13; Fe:2,05±0,17; Zn:1,03±0,03; Mg:1,01±0,10). The combination of these results with those reported for the c_1aa_3-oxidase from <u>Thermus thermophilus</u> leads to the conclusion that subunit I is the universal two coppers and two hemes a binding catalytic unit common to all oxidases of the aa_3-type. Subunit II, which binds the third copper ion, functions as an electron conducting unit, transferring electrons from cytochrome c to the four redox centers in subunit I. Preliminary titration experiments with NADH reveal, in agreement with this catalytic organisation, the presence of five redox centers.

INTRODUCTION

If one looks at the integrity of oligomeric metalloenzymes the integral stoichiometry of all the constituting components, proteins and metals, is an important criterion which should be met. Especially in the case of the respiratory enzyme complexes from mitochondrial and bacterial membranes, solubilisa-

tion and purification procedures leave doubt, whether the obtained preparation is a faithful representation of what exists _in situ_.
The criterion of integral stoichiometries has in the case of beef heart cytochrome c oxidase often not been met. Many papers report copper stoichiometries in the range from 2,2 to 4 (Morrison et al., 1963; Yonetani, 1966; Van Gelder and Beinert, 1969; Aasa et al., 1976), the copper content of this enzyme has been a matter of controversy since its discovery. Mainly based on the work of Van Gelder (Van Gelder, 1966) and Muijsers et al. (Muijsers et al., 1972) the presence of two functional copper atoms was generally adopted (Beinert, 1966; Azzi et al., 1985). Amounts of copper present in addition to the two functional coppers were considered as "adventitious" and/or "inactive" copper (Van Gelder and Beinert, 1969; Aasa et al., 1976). In 1984 Einarsdóttir & Caughey (Einarsdóttir and Caughey, 1984 & 1985) introduced a new technique for establishing the metal contents in preparations of cytochrome c oxidase: Inductively Coupled Plasma Atomic Emission Spectroscopy (ICP-AES). Using this method they were able to demonstrate the stoichiometric presence of one zinc and one magnesium atom in bovine heart cytochrome c oxidase. They also described a Cu/Fe ratio substantially larger than one, however, no correlation was made between the protein and metal contents. We therefore started to investigate the metal stoichiometries of our preparations of cytochrome oxidase from beef heart and _Paracoccus denitrificans_ using the same ICP-AES technique. Both enzymes appeared to be characterized by integral stoichiometric relationships of Cu and Fe and the respective protein moieties, binding three coppers and two hemes a per monomer with M_r of 203000 and 84000 Da, resp. (Steffens et al., 1987). Based on these results we suggested the presence of two copper atoms in subunit I and the presence of a third copper in subunit II (Steffens et al., 1987). Combining these results with data concerning the constitution of the c_1aa_3 oxidase from _Thermus thermophilus_ (Yoshida et al., 1984; Hon-nami and Oshima, 1984) we concluded that subunit I is the universal two heme a and two copper containing unit common to all aa_3-type cytochrome oxidases. Subunit II then

functions as an electron conducting unit, transferring electrons from reduced cytochrome c to the four redox centers in subunit I (Steffens et al., 1987). This catalytic organisation requires the redox activity of all three copper and the two iron centers present in integral preparations. Therefore we decided to investigate the redox behaviour of the metal centers by titrating preparations of beef heart cytochrome c oxidase with NADH under anaerobic conditions. Initial experiments suggest, in agreement with our hypothesis, the presence of five redox active centers.

MATERIALS AND METHODS

Preparation of beef heart cytochrome c oxidase was performed as in (Steffens et al., 1987). After the final ammonium sulfate precipitation the pellet is dissolved in 0,5% Tween 80; 0,1 M phosphate pH 7,4 and dialyzed vs. this buffer for 24 h. This dialysis is followed by a 48 h dialysis against water. Cytochrome c oxidase precipitates and is, after a low speed centrifugation step, dissolved in 0,01 M ammonium hydrogen carbonate and stored at -75° C. Cytochrome c oxidase used for the anaerobic titration with NADH was not dialyzed, instead the pellet of the final ammonium sulfate precipitation was dissolved in 1% cholate; 0,1 M phosphate pH 7,4 and stored at -75° C.

Inductively Coupled Plasma Atomic Emission Spectroscopy. This technique allows for the simultaneous determination of a large number of elements and is, due to the high temperature of the plasma, not hampered by interferences from the protein moiety. Contents of Cu, Fe, Zn, Mg, Mn, P and S were determined with an ARL 34000 spectrometer after addition of a specified amount of $Ni(NO_3)_2$ as internal standard (Steffens et al., 1987). The performance of the spectrometer was tested with metalloproteins with well established metal and sulfur stoichiometries such as horse heart cytochrome c (Sigma; Type VI) and Cu,Zn-Superoxide Dismutase from bovine erythrocytes (gift of Dr. G.J. Steffens of Grünenthal GmbH, Stolberg).

Bicinchoninic Acid Assay was performed as described

in (Smith et al., 1985). Both the standard and micro protocol were used. Standard curves were obtained with bovine serum albumin (Sigma).

Amino Acid Analysis was, after acid hydrolysis of an appropriate aliquot of the cytochrome c oxidase solution (Steffens et al., 1987), carried out with a Biotronik amino acid analyzer LC 5001, using fluorescence detection of the o-phtalaldehyde derivatives.

Heme a was determined spectrophotometrically using the molar absorption coefficient ε = 12000 $l.mol^{-1}.cm^{-1}$ (Van Gelder, 1966) for the absorption difference between dithionite or NADH reduced and air oxidized cytochrome c oxidase at 605 nm.

Anaerobic titration of cytochrome c oxidase with NADH. Continuous titration of cytochrome c oxidase with NADH under anaerobic conditions was performed as follows: 2,0 ml of a cytochrome c oxidase solution (ca. 100 nmol in 1% cholate; 0,1 M phosphate buffer pH 7,4), which had been made free of oxygen by repeated evacuation and exposure to pure nitrogen, were transferred anaerobically into the sample cell, equipped with a stirrer, and kept under a continuous stream of nitrogen. Again 2,0 ml of the enzyme solution were transferred into the reference cell and both the sample and reference beam of the spectrophotometer (Perkin Elmer Lambda 5) were optically zeroed. 50 µl of a freshly prepared solution of phenazinemethosulfate (Sigma, 0,01% in cholate buffer), which had been made free of oxygen in the same way, were added anaerobically.

Titration was then started by adding 10 µl aliquots (Radiometer ABU 80 autoburette) of a 2.10^{-3}M NADH (Sigma) solution, which was free of oxygen and kept under a continuous stream of nitrogen. The exact concentration of the NADH solution was determined spectrophotometrically at 340 nm using the molar absorption coefficient of 6220 $l.mol^{-1}.cm^{-1}$ (Van Gelder, 1966). After each addition (time interval 2 minutes) the difference absorptions at 605 and 830 nm were monitored. Difference absorbances were corrected for dilution due to the titration procedure. The molar amount of cytochrome c oxidase present was obtained from the maximal difference absorbance at 605 nm after the reduction with NADH had been completed, using the molar absorption coeffi-

Table I The constitution of beef heart cytochrome c oxidase (Complex IV)

Sub-unit	Mr	Stoichio-metry	N-terminus	Sulfur atoms	Buse et al.	Capaldi et al.
I	56993	1	F-Met-Phe-Ile-Asn-	35	I	I
II	26049	1	F-Met-Ala-Tyr-Pro-	18	II	II
III	29918	1	(Met)-Thr-His-Gln-	13	III	III
IV	17153	1	Ala-His-Gly-Ser-	4	IV	IV
V	12436	1	Ser-His-Gly-Ser-	5	V	V
VI	10668	1	Ala-Ser-Gly-Gly-	2	VIa	a
VII	10063	1	Ac-Ala-Glu-Asp-Ile-	5	VII	c
VIII	9419	1	Ala-Ser-Ala-Ala-	1	VIb	b
IX	8480	1	Ser-Thr-Ala-Leu-	4	VIc	VI
X	6243	1	Phe-Glu-Asn-Arg-	3	VIIIc	
XI	ca.6000	1	Ile-His-Gln-Lys-	3	-	VIIa,b,c
XII	5441	1	Ser-His-Tyr-Gln-	2	VIIIa	
XIII	4962	1	Ile-Thr-Ala-Lys-	-	VIIIb	

Complex IV ca. 203000 Da 13 Subunits 95 Sulfur Atoms

cient $\varepsilon = 24000\ l.mol^{-1}.cm^{-1}$ for the reduced minus oxidized oxidase (Van Gelder, 1966). This determination was checked by determining the oxidase concentration in the reference cell both spectrophotometrically and by amino acid analysis.

RESULTS

Table I gives a compilation of characteristic proteinchemical data of beef heart cytochrome c oxidase (Complex IV). This oligomeric complex is built up of 13 subunits, the sequences of all 13 polypeptides have been determined either by direct protein sequencing or by deduction from the DNA-sequence of the corresponding genes (Buse et al., 1985; Takamiya et al., 1987; Capaldi, personal communication). Based on these data the total molecular mass has been calculated to ca. 203000 Da. The total number of sulfur atoms present in this complex amounts to 95.

In Table I subunits are designated according to a new nomenclature, which was recently introduced at the IUB Symposium "Cytochrome Systems: Molecular Biology and Bioenergetics" in Bari, April 14th-18th, 1987 (Buse et al., 1987). This nomenclature lists the subunits according to their molecular mass. An exception is made for the mitochondrially synthesized subunits II and III, which keep their original nomenclature. Our previous nomenclature (Buse et al., 1983) and the nomenclature of Capaldi and co-workers (Ludwig et al., 1979) are also given for comparison.

Table II gives an example of the stoichiometries of selected elements in one of our preparations of beef heart cytochrome c oxidase. It is noteworthy to mention, that unless otherwise stated, our preparations have been extensively dialyzed according to the protocol in Materials and Methods. Metal contents, as well as phosphorus and sulfur contents, have been determined with the ICP-AES technique. The data given in Table II clearly show, that this preparation is characterized by integral stoichiometries for Cu, Fe, Zn and Mg. Whereas in the past many papers have claimed the presence of non-stoichiometric amounts of copper, in most cases

Table II Stoichiometries of selected elements in a preparation of beef heart cytochrome c oxidase (heme a/protein ratio: 9,4 nmol mg^{-1})

Cu	Fe	Zn	Mg	P	S	Stoichiometries based on
2,97	2,04	0,99	1,00	23,8	94,7	Cu + Fe + Zn + Mg/7
2,99	2,05	0,99	1,01	24,0	95,4	protein determined by amino acid analysis
2,97	2,04	0,99	1,00	23,9	95,00	simultaneously determined intrinsic sulfur (S=95,00)
2,76	1,90	0,92	0,93	22,2	88,0	heme a determined spectrophotometrically
2,87	1,97	0,95	0,96	23,0	91,5	protein determined by bicinchoninic acid assay

values between 2,2 and 4 have been reported (Morrison et al., 1963; Yonetani, 1966; Van Gelder and Beinert, 1969; Aasa et al., 1976), we observe the presence of 3 copper atoms in addition to 2 irons, 1 zinc and 1 magnesium atom. The data furthermore show, that not only the mutual ratios give integral numbers, but also the protein related stoichiometries, i.e. the number of metals per monomer of cytochrome c oxidase, based on various independent parameters: thus the stoichiometries based on 1. the protein content determined by amino acid analysis, 2. the simultaneously determined intrinsic sulfur content, 3. the spectrophotometrical heme a determination and 4. the protein content determined by the BCA assay give an almost 100% match.
Similar data were found for ten other preparations of beef heart cytochrome c oxidase, which had been obtained through the same preparation procedure. All heme a/protein ratios were greater than 8,0 $nmol \cdot mg^{-1}$, with an average of $9{,}19 \pm 0{,}69$ $nmol \cdot mg^{-1}$, and all preparations were characterized by the presence of at least twelve protein bands on analytical SDS polyacrylamide gels. The data of Table III show, that our preparations generally contain 3 copper atoms per 2 hemes a, and in agreement with Einarsdóttir & Caughey (Einarsdóttir and Caughey, 1984,1985), one zinc and one magnesium atom per monomer. The mutual ratios, as well as the stoichiometries based on the protein content determined by amino acid analysis and the intrinsic sulfur content are very close to integral numbers. The stoichiometries based on the spectrophotometrical heme a determination and the protein content determined by the BCA assay are somewhat lower, which in the case of the heme a determination may be due to a slight underestimation of the molar absorption coefficient and in the case of the BCA assay to a different staining behaviour of cytochrome c oxidase and bovine serum albumin used for obtaining the standard curve.
The presence of three copper atoms, however, does not necessarily require the participation of all coppers in the redox processes. Earlier results (Van Gelder, 1966; Muijsers et al., 1972) suggest the presence of only four redox active metal centers, this observation together with the assumption of one

Table III Average stoichiometries of selected elements in eleven preparations of beef heart cytochrome c oxidase[a] ± standard deviation

Cu	Fe	Zn	Mg	S	Stoichiometries based on
2,91±0,13	2,05±0,17	1,03±0,03	1,01±0,10	97,4±9,2	Cu + Fe + Zn + Mg/7[b]
2,97±0,24	2,14±0,14	1,02±0,07	1,03±0,15	95,5±2,5	protein determined by amino acid analysis[c]
2,95±0,23	2,10±0,16	1,05±0,10	1,01±0,11	95,00	simultaneously determined intrinsic sulfur (S=95,00)[b]
2,65±0,25	1,85±0,20	0,94±0,11	0,91±0,14	87,9±10,6	heme a determined spectrophotometrically[b]
2,60±0,29	1,83±0,25	0,92±0,13	0,89±0,16	85,7±11,6	protein determined by bicinchoninic acid assay[b]

a) average heme a/protein ratio: 9,19±0,69 nmol heme a . mg^{-1}

b) data based on 11 preparations

c) data based on 5 preparations

of the two coppers being EPR-invisible, has lead to the generally adopted view of cytochrome c oxidase being a two copper, two heme a enzyme (Beinert, 1966; Azzi et al., 1985). However, some of these results are ambiguous: whereas Van Gelder (Van Gelder, 1966) and Muijsers et al. (Muijsers et al., 1972) claim the presence of exactly four electron accepting centers, Van Gelder & Beinert (Van Gelder and Beinert, 1969) observed that complete reduction required the addition of five to six electron equivalents.

In an attempt to test whether all three coppers may be involved in the redox processes we reexamined this problem with a somewhat modified technique. Figure 1 shows the curves, which we obtained in a continuous titration of beef heart cytochrome c oxidase with NADH under anaerobic conditions. This particular preparation was, according to an ICP-AES experiment, characterized by the following metal stoichiometries: Cu:2,77; Fe:2,13; Zn:0,96; Mg:1,14.

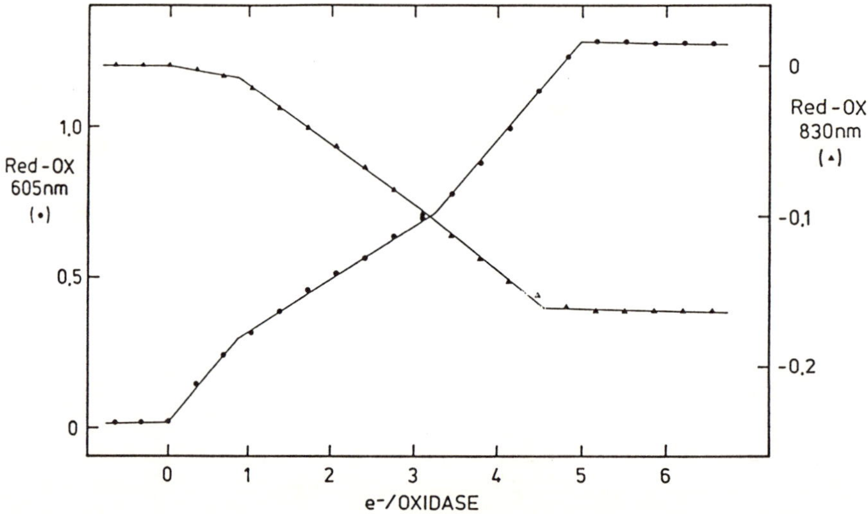

Fig. 1 Anaerobic titration of beef heart cytochrome c oxidase with NADH in the presence of phenazinmethosulfate (for details see Materials and Methods.)

The state of reduction during titration is monitored at two different wavelengths. The difference absorbance at 605 nm reflects the reduction status of the heme irons, whereas the absorbance at 830 nm mainly reflects the status of copper. The titration curves are not linear and show inflexion points after the addition of one and three electron equivalents, resp..
After the addition of the fifth electron equivalent the reduction is complete. Also the titration curve at 830 nm reveals an inflexion point after the addition of one electron equivalent, the second inflexion point, however, is much less pronounced.

DISCUSSION

The present paper describes the characterization of integral complexes of beef heart cytochrome c oxidase. Simultaneous determination of metals by ICP-AES reveals the stoichiometric presence of three copper atoms besides two irons, one zinc and one magnesium atom in beef heart cytochrome c oxidase. All the preparations used in this study had been extensively dialyzed after the final ammonium sulfate precipitation. After this dialysis procedure the sulfur content can be correlated with the number of sulfur atoms present in the polypeptides, which constitute the oligomeric complex. The reasonable fit between these two parameters furthermore supports the notion of looking at an oligomeric enzyme complex in which all the components, proteins and metals, are present with integral stoichiometric amounts. Also the average heme a to protein ratio of 9,19$\pm$0,69 nmol heme a.mg^{-1} is in good agreement with the theoretical value of 9,87 nmol heme a.mg^{-1}, which has been calculated on the basis of a molecular mass of 203000 Da (Buse et al., 1985).
More recent investigations on the metal contents of preparations, which had not been dialyzed, show that these preparations are characterized by Cu/Fe ratios in the range of 1,2-1,3 and Cu/Mg ratios in the range of 2,4-2,6. These lower ratios in our view may be caused by the presence of additional (non-heme) iron and magnesium, which apparently are removed by

the final purification procedure.
The presence of three coppers per two hemes a was also observed in preparations of cytochrome c oxidase from Paracoccus denitrificans (Steffens et al., 1987) and - in a preliminary study - in a preparation of cytochrome c oxidase from Dictyostelium discoideum (Steffens et al., unpublished results). Meanwhile Bombelka et al. showed with a different technique that their preparations of cytochrome c oxidases from various species (bovine, pig, rat) and tissues (heart, liver, diaphragm and kidney) are characterized by the presence of three coppers per two heme irons (Bombelka et al., 1986). The presence of three copper atoms thus appears to be a general feature of those cytochrome c oxidases, which have both subunits I and II.
The c_1aa_3 oxidase from Thermus thermophilus, however, according to independent investigations of Yoshida et al. (Yoshida et al., 1984) and Hon-nami and Oshima (Hon-nami and Oshima, 1984) contains only two copper atoms per three iron atoms (2 hemes a and 1 heme c).
Combining our results on beef heart and Paracoccus cytochrome c oxidase with those from the c_1aa_3 oxidase from the thermophilic bacterium, the following conclusions concerning the topology of the metal centers seem reasonable: subunit I binds two hemes a and two copper ions and is the universal aa_3-type catalytic unit common to all cytochrome c oxidases of this spectral type, subunit II binds the third copper ion. The function of this latter subunit then is to transfer electrons from cytochrome c to the four metal centers in subunit I, which represents the terminal oxidase. This conclusion asks for the redox activity of five metal centers, at least in those cytochrome c oxidases, which do have both subunits I and II.
The redox behaviour of beef heart cytochrome c oxidase was therefore reinvestigated through an anaerobic titration with NADH in the presence of PMS by a continuous technique. These experiments revealed non-linear titration curves at both 605 and 830 nm (Fig. 1). In the case of the curve monitored at 605 nm, reflecting the oxidation/reduction status of the heme irons, two clearcut inflexion points are observed after the addition of one and three

electron equivalents resp. The curve at 830 nm, monitoring the oxidation status of copper ions, shows a clear inflexion after the addition of the first electron equivalent, the second inflexion point is only weakly expressed.
Tentatively these results may be interpreted as follows: the first electron added resides, in agreement with the kinetic model of Bickar et al. (Bickar et al., 1986), to a large extent on one of the hemes a, giving rise to a relatively sharp increase of the absorbance difference at 605 nm and a slight decrease of the difference absorption at 830 nm. This interpretation fits to the observation of Van Gelder & Beinert (Van Gelder and Beinert, 1969) that the EPR-visible copper is only scarcely reduced until more than one electron equivalent has been added.
The second and third electron equivalent are preferentially used for the reduction of two copper centers, whereas the fourth and fifth electron equivalent may be required for the reduction of the remaining copper and heme a centers.
Although the presence of three functional coppers solves a number of problems e.g. the presence of the so-called "adventitious" copper (Van Gelder and Beinert, 1969) and the fact that the EPR-signal accounts for only 30-40% of the total amount of copper present (Van Gelder and Beinert, 1969), no explanation can be given for the fact that the third copper is either EPR-silent or that its EPR-signal has so far been overlooked.
The presence of two invariant cysteines (Steinrücke et al., 1987) in the presumable copper site of subunit II may keep this copper in the cuprous (+1) state, whilst this Cu-S center still would be able to accept and donate electrons (Hemmerich, 1966).

ACKNOWLEDGEMENTS

The authors are indebted to Dr. Wolff of the Nuclear Research Center (KFA) in Jülich for performing the ICP-AES determinations. The technical skills of Ursula Eberhard and Eveline Pascual are gratefully acknowledged.
This work was supported by Sonderforschungsbereich 160 "Eigenschaften Biologischer Membranen" of the Deutsche Forschungsgemeinschaft.

ABBREVIATIONS

ICP-AES	inductively coupled plasma atomic emission spectroscopy
PMS	phenazinemethosulfate
BCA	bicinchoninic acid

REFERENCES

Aasa R, Albracht SPJ, Falk KE, Lanne B, Vänngard T (1976). EPR signals from cytochrome c oxidase. Biochim. Biophys. Acta 422:260-272.

Azzi A, Bill K, Bolli R, Casey RP, Nalecz K, O'Shea P (1985). Molecular aspects of the structure-function relationship in cytochrome c oxidase. In Benga G (ed): "Structure and properties of cell membranes," Vol. II, CRC Press, Boca Raton Fl., pp 105-138.

Beinert H (1966). Cytochrome c Oxidase: Present Knowledge of the State and Function of Its Copper Components. In Peisach J, Aisen P, Blumberg WE (eds): "The Biochemistry of Copper," New York: Academic Press, pp 213-234.

Bickar D, Turrens JF, Lehninger AL (1986). The mechanism by which oxygen and cytochrome c increase the rate of electron transfer from cytrochrome a to cytochrome a_3 of cytochrome c oxidase. J. Biol. Chem. 261:14461-14466.

Bombelka E, Richter FW, Stroh A, Kadenbach B (1986). Analysis of the Cu, Fe and Zn contents in cytochrome c oxidases from different species and tissues by proton-induced X-ray emission (PIXE). Biochem. Biophys. Res. Commun. 140:1007-1014.

Buse G, Steffens GCM, Meinecke L (1983). Cytochrome oxidase: the primary structure of electron and proton translocating subunits and their hints at mechanisms. In Quagliariello E, Palmieri F (eds): "Structure and Function of Membrane Proteins," Amsterdam: Elsevier/North-Holland Biomedical Press, pp 131-138.

Buse G, Steffens GCM, Biewald R, Bruch B, Hensel S (1987). Proposal for a novel nomenclature for the subunits of cytochrome c oxidase. In "Cytochrome Systems: Molecular Biology and Bioenergetics," Pergamon press, in press.

Einarsdóttir O, Caughey WS (1984). Zinc is a constitutent of bovine heart cytochrome c oxidase preparations. Biochem. Biophys. Res. Commun. 124:836-842.

Einarsdóttir O, Caughey WS (1985). Bovine heart cytochrome c oxidase preparations contain high affinity binding sites for magnesium as well as for zinc, copper and heme iron. Biochem. Biophys. Res. Commun. 129:840-847.

Van Gelder BF (1966). On cytochrome c oxidase I. The extinction coefficients of cytochrome a and cytochrome a_3. Biochim. Biophys. Acta 118:36-46.

Van Gelder BF, Beinert H (1969). Studies of the heme components of cytochrome c oxidase by EPR spectroscopy. Biochim. Biophys. Acta 189:1-24.

Hemmerich P (1966). Model studies on the binding of univalent and redox-active copper in proteins. In Peisach J, Aisen P, Blumberg WE (eds): "The biochemistry of copper," New York: Academic Press, pp 15-32.

Hon-nami K, Oshima T (1984). Purification and characterzation of cytochrome c oxidase from <u>Thermus thermophilus</u> HB8. Biochemistry 23:454-460.

Ludwig B, Downer WK, Capaldi RA (1979). Labeling of cytochrome c oxidase with [^{35}S] diazobenzenesulfonate. Orientation of this electron transfer complex in the inner mitochondrial membrane. Biochemistry 18:1401-1407.

Morrison M, Horie S, Mason HS (1963). Cytochrome c oxidase components II. A study of the copper in cytochrome c oxidase. J. Biol. Chem. 238:2220-2224.

Muijsers AO, Tiesjema RH, Henderson RW, Van Gelder BF (1972). Biochemical and biophysical studies on cytochrome aa_3. VII. The effect of cytochrome c on the oxidation-reduction potential of isolated cytochrome aa_3. Biochim. Biophys. Acta 267:216-221.

Smith PK, Krohn RI, Hermanson GT, Mallia AK, Gartner FH, Provenzano MD, Fujimoto EK, Goeke MM, Olson BJ, Klenk DC (1985). Measurement of protein using bicinchoninic acid. Anal. Biochem. 150:76-85.

Steffens GCM, Biewald R, Buse G (1986). Fourth European Bioenergetic Conference. Prague 1986. Short reports 4:191.

Steffens GCM, Biewald R, Buse G (1987). Cytochrome c oxidase is a three-copper, two-heme-A protein. Eur. J. Biochem. 164:295-300.

Steffens GCM, Bisson R, Buse G. Unpublished results.

Steinrücke P, Steffens GCM, Panskus G, Buse G, Ludwig B (1987). Subunit II of cytochrome c oxidase from Paracoccus denitrificans. DNA sequence, gene expression and the protein. Eur. J. Biochem. 167:431-439.

Takamiya S, Lindorfer MA, Capaldi RA (1987). Purification of all thirteen polypeptides of bovine heart cytochrome c oxidase from an aliquot of enzyme. Characterization of bovine fetal heart cytochrome c oxidase. FEBS Lett. 218:277-282.

Yonetani T (1966). Copper in cytochrome oxidase. Biochem. Biophys. Res. Commun. 3:549-553.

Yoshida T, Lorence RM, Choc MG, Tarr GE, Findling KL, Fee JA (1984). Studies on cytochrome c oxidase activity of the cytochrome c_1aa_3 complex from Thermus thermophilus. J. Biol. Chem. 259:112-123.

DISCUSSION

HORIE: I understand that the near infrared peak at 830 nm was first described, if my memory is correct, by Drs. Griffith and Wharton in 1962, and is now generally attributed to copper. However, at about the same age, the late Dr. Martin Morrison and I observed that almost all ferric high spin hemoproteins had quantitatively similar absorption peaks at somewhere between 750 to 1100 nm (catalase, 775 nm; lacto-peroxidase, 955 nm; methemoglobin, 1000 nm; RHP, 1025 nm). These observations have not been published. Even P-450 seems to have this kind of peak. We studied also the spectral properties of the 830 nm peak precisely. The results suggested that heme a possibly contributed also the 830 nm peak. Therefore, I wonder how people working on the 830 nm peak can distinguish the absorption due to copper from that due to heme.

STEFFENS: The origin of the 830 nm band has been debated ever since its detection. Sunney Chan and co-workers have claimed that this absorption band is exclusively due to Cu_A, since inhibition of the Cu_B-heme a_3 center with CN^{-1} or formate did not influence the absorbance of oxidized cytochrome c oxidase at 830 nm. Now that we know that a third redox active copper might be present in bovine heart cytochrome c oxidase, this matter is open for discussion again.

WEVER: From EPR data it is known that maximally 50% of the copper is detectable. This leads to the inevitable conclusion that the third copper is reduced. This is, however, in conflict with your anaerobic titration data. How can you reconcile the observations?

STEFFENS: The initial data of Beinert and co-workers report that 30-40% of the total copper being present are EPR detectable. This implied that in the presence of three coppers only one out of three is EPR-active. Indeed our preliminary titration data suggest that three coppers are redox active and thus it is difficult to understand why this copper has not been detected by EPR spectroscopy.

RYDEN: Is it possible to study the properties of Cu in isolated subunit II, for example after having cleaved off the N-terminal hydrophobic portion?

STEFFENS: Unfortunately, experiments to isolate subunit II under native conditions have failed so far.

RYDEN: Many proteins easily pick up copper from buffer solutions, etc. How did you rule out that the third copper is not such an extraneous copper?

STEFFENS: Our cytochrome c oxidase preparations from bovine heart have been extensively dialyzed prior to the metal content determination. If a part of the copper would be bound non-specifically, the observation of integral stoichiometries would be difficult to understand. Furthermore, we have determined the copper/iron ratios after dialysis versus chelating agents such as EDTA and phenanthroline. The third copper was not extracted, in contrast we observed that, in agreement with the observations of Einarsdottir & Caughey, the copper/zinc and copper/magnesium ratios were increasing and approached the value of 3.

NICHOLLS: An "n=3" titration might be expected at equilibrium for the CN^{-1} inhibited enzyme, even where only four electron accepting centers exist, because Cu_B must eventually be reduced if the system is to show the g=3.6 signal of cytochrome a_3. The titration method therefore needs careful analysis before "n=5" values can be considered a firm possibility.

STEFFENS: I agree that the preliminary experiments, which suggest the presence of five redox active centers, have to be checked very carefully. Specifically, the anaerobic conditions and the requirement that the metals should be present with integral stoichiometries make these experiments rather difficult to perform.

Maybe other independent experiments, for instance the careful investigation of the redox properties of all the metal centers, might contribute to the solution of this problem.

KING: You may know very well your method of determination is much less precise than the atomic absorption spectroscopy or chemical methods. Moreover, how do you explain that in the past, say 20 years ago, the Cu:Fe ratio of oxidase was at the most 1.2. Do you think they are wrong? Next, do you have sufficient conserved cysteines for three coppers?

STEFFENS: I cannot agree with your statement that the ICP-AES method is much less precise than atomic absorption spectroscopy and chemical methods. The atomic absorption spectroscopy might in the case of individual elements be more sensitive, nevertheless this advantage does not match the advantages offered by ICP-AES: simultaneous determination of a large number of elements including sulfur and the elimination of the interference of the protein moiety and other metals.

Referring to your second question, I can say that there have been several reports about Cu:Fe ratios larger than 1.2, for instance Aasa and co-workers and Kadenbach and co-workers. More recently we have examined the metal contents of preparations which had not been dialyzed. From these investigations it became clear that the low Cu:Fe ratios were not due to low copper values but more likely to the presence of additional non-heme iron. Your third question addresses the number of conserved cysteines, indeed there are no conserved cysteines in subunit I. There are, however, in our opinion enough conserved histidines and methionines to accommodate two copper atoms and two heme a groups in this subunit.

Oxidases and Related Redox Systems, pages 707–730

CYTOCHROME c OXIDASE AND ITS PROTONMOTIVE ACTIVITY, AN OVERVIEW

Sergio Papa

Institute of Medical Biochemistry and Chemistry

University of Bari, Italy

Cytochrome c oxidase (E.C. 1.9.3.1.) of coupling membranes of prokaryotes and eukaryotes is an oligomeric protein which: (i) binds dioxygen with high affinity; (ii) reduces dioxygen to H_2O by the concerted delivery of four electrons from ferrocytochrome c; (iii) converts the large amount of energy thus made available (ΔE_h of 550-350 mV) into transmembrane $\Delta\mu H^+$, which acts as high-energy intermediate in oxidative phosphorylation.

This paper is intended to describe the characteristics of the protonmotive activity of the oxidase and to review structural features of the enzyme which appear relevant to the nature and mechanism of this process.

STRUCTURAL FEATURES

Electrons are delivered to the oxidase by cytochrome c which binds on a loop of subunit II protruding from the membrane in the external aqueous phase (Millet et al.,1983;Finel et al.,1987) (Fig.1). Four redox centres, common to all the oxidases, have been shown to take part in the catalytic process, organized in two pairs. Heme a and low potential Cu_A, which catalyze the entry of electrons in the oxidase, then passed to heme a_3 and high potential Cu_B involved in the reduction of dioxygen to H_2O (Nicholls et al.,1974;Malmström, 1979;Wikström et al.1981 a,b).The properties of these centres

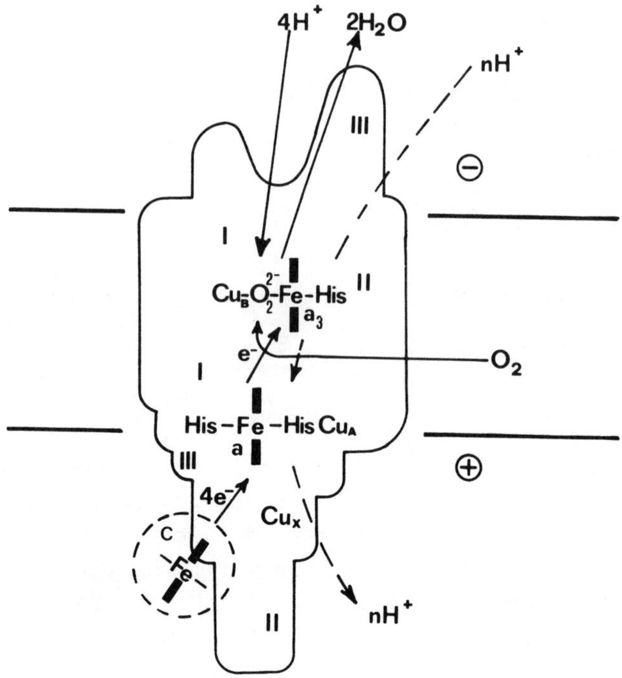

Fig.1. Some structural features of cytochrome c oxidase

have been characterized in detail (Nicholls et al.,1974; Malmström, 1979; Wikström et al.,1981;Brunori et al.,1981; Blair et al.,1985). Yet the exact catalytic mechanism is not fully understood. It has, at any event, been shown that the protons consumed in the formation of H_2O derive from the inner aqueous phase (Mitchell,1966; Mitchell et al.,1970;Papa et al.,1974; Papa,1976). This asymmetrical arrangement of the overall reaction results, as postulated by Mitchell (Mitchell, 1966), in the generation of transmembrane $\Delta\mu H^+$ with effective translocation of 1 q^+/e^- from the inner side to the outer P side (Papa, 1976; Wikström et al.,1981, 1984;Papa et al.1984).

It is now generally accepted that cytochrome c oxidase couples, in addition, electron flow with proton pumping from the N space to the P space (Wikström et al.1981a,b;Wikström and

Saraste, 1984; Papa et al.,1987a).

Definite progress has been made in the elucidation of the protein structure of the oxidase, location of the prosthetic groups and their liganding state (Papa et al.,1987a). There are two subunits (I and II) (Table 1) which are general constituents of all the cytochrome c oxidases described (Finel et al.,1987;Papa et al.1987a;Bisson et al.,1987;Buse et al.,1987), except that of bacterium <u>Thermus thermophilus</u> which apparently has only one typical oxidase subunit (I), since a second subunit of the isolated enzyme seems to be a cytochrome c_1 (Buse et al.,1987;Steffens et al.,1987). In some prokaryotes there is a third subunit(Finel et al.,1987;Sone,1987;

TABLE 1. Subunits of cytochrome c oxidase

Subunits	Mitochondrial genes
I	Binds heme a, heme a_3,Cu_B and probably also Cu_A. 10 invariant His residues. Universal subunit performing basic processes of catalysis and protonmotive activity.
II	Cu-protein essentially involved in the binding of cytochrome c. 2 invariant His, 2 invariant Cys. Contains Cu_A or Cu_X
III	Essential for folding or assembly of catalytic subunits. Possible facilitating role for protonmotive activity. Onc invariant Glu.
	Nuclear genes
IV(VI), VI(V), VIIa(XII) Va, Vb(IV)	(in parenthesis corresponding bovine subunits). Subunits shown to be essential in S.Cerevisiae for folding or assembly of catalytic subunit(s). Genetic deletion or missense mutations of these subunits produce loss of functional holoenzyme. Their removal from holoenzyme does not impair function (see text).
VII, VIII, IX, X, XI, XII, XIII.	Unknown function.Possible regulatory role of catalytic activity and protonmotive activity.(Nomenclature for beef-heart,Kadenbach et al.,1983; Buse et al.1987).

Haltia et al.,1987). In eukaryotes in addition to these three subunits, encoded by mitochondrial genes, there are 6-7 subunits in lower species (Bisson et al.,1987;Wright et al.,1987; Capaldi et al.,1987) and 10 in higher species (Kadenbach et al.,1983; Takamiya,1987) which are encoded by nuclear genes (Papa et al.,1987 Wright et al.,1987; Suske et al.;1987). Besides these structural genes more than 30 nuclear genes have been described in S.Cerevisiae, necessary for synthesis and assembly of the constituent subunits (Wright et al.,1987).

Subunit II has 2 invariant cysteines (positions 196,200) and two histidines (161,204) (Buse et al.,1987) which qualify as copper ligands (Millet et al.,1983;Finel et al.;1987). There are no additional invariant residues in subunit II which could ligate other metal centres (Buse et al.,1987). Thus the two hemes a and the other Cu atom should be associated to subunit I. This has 10 invariant histidines, that can serve as ligands, but no invariant cysteines (Buse et al., 1987). The copper associated to subunits II is generally considered to represent Cu_A (Finel et al.;1987; Wikström et al.; 1981a,b).Buse et al.(1987)(see also: Steffens et al.1987; Bomelka et al.,1986) have, however, found in bovine heart and <u>Paracoccus Denitrificans</u> oxidase that the copper atoms are three and not <u>two</u> per enzyme.

Interestingly enough, <u>Thermus thermophilus</u> oxidase which has only Subunit I (Buse et al.,1987) has only 2 copper atoms <u>per</u> enzyme and is fully competent for catalytic and protonmotive activity (Sone, 1987). It has also been observed that Paracoccus oxidase retains, after proteolytic digestion of subunit II, the capacity to oxidize ferrocytochrome c with high affinity (Müller et al.,1987). Thus subunit I probably contains not only both hemes a and a_3 but also Cu_A and Cu_B. This subunit would then be the universal catalytic unit of cytochrome oxidase (Buse et al.,1987) and the basic elements of a proton pump have to be searched for among the centres and residues in this subunit. If this were the case the role of the Cu associated to subunit II should then be re-defined.

Subunit III, which is absent in <u>Thermus thermophilus</u> and other preparations of bacterial oxidases,which still exhibit redox-linked proton ejection with good efficency (Sone,1987) when reconstituted in liposomes, does not appear to play an essential role in the catalytic process. Recently the gene for subunit III of Paracoccus oxidase, whose preparation generally lack this protein (Ludwig et al.,1980),has been identified (Saraste et al.,1986). Subunit III may, evidently be easily lost during preparation. However also when this happens the reconstituted oxidase still exhibits redox-linked proton ejection (Sone, 1987; Solioz et al.,1982). If any,subunit III may have a secondary role in optimizing the efficency of proton translocation (Finel et al.,1987;Müller et al.1987).

Some evidence has been presented in favour of the proposal (Wikström et al.,1981; Wikström et al.,1981) that dimerization of the enzyme is essential for proton pumping (Sone, 1987; Finel et al.,1986). There are, however, observations which militate against this (Müller et al.,1987; Bickar et al. 1985). One intriguing aspect which demands explanation (Papa and Lorusso 1984;Papa et al.,1987a)is the fact that various bacterial cytochrome c oxidase, with apparently the same four prosthetic groups of eukaryotic oxidases and comparable catalytic activity, do not exhibit redox-linked proton ejection, either in intact cells or in the reconstituted state,regardless of whether they have two or three subunits (Sone,1987).

HYPOTHETICAL MODELS

The presence in the oxidase only of electron carriers led Wikström (Wikström et al.,1981a,b) to propose an indirectly coupled proton pump model. An indirect proton pump model had been earlier proposed by us for cytochrome c reductase (Papa,1976; Papa et al.,1973). We postulated that co-operative, thermodynamic linkage between the redox state of metals and protolytic groups in the enzyme can result in proton pumping,provided the protonation-deprotonation steps are asimmetrically and vectorially arranged(vectorial Bohr mechanism). In the reductase, b cytochrome was proposed as the coupling centre since it displays proton electron linkage (redox Bohr effects) in the physiological pH range (Papa, 1976; Urban et al.,1969). In the oxidase Wikström considers heme a as the coupling centre (Wikström et al., 1981). Both hemes a and a_3 exhibit, in fact, redox Bohr effects in the physiological pH range, the pH dependence of Em of the low and high potential species indicating complex anti co-operativity (Wikström et al.,1981a;Blair et al.,1986). More recently two other indirect models have been proposed by Chan et al.(1979) and Malmström (1985,1987) respectively. In Chan's model protons are translocated across the Cu_A centre by redox-driven alternate protonation and Cu_A ligation of a cysteine thiolate and a tyrosine phenolate on opposite sides of the centre (Chain et al.,1979; Gelles et al.,1987). Chan et al., have also proposed that at high $\Delta\mu H^+$, coupled electron transfer through the Cu_A centre is depressed and replaced by slow uncoupled electron transfer directly from heme a to the heme a_3-Cu_B center (Blair et al.,1986). A difficulty for this specific mechanism seems to be represented by the lack of conserved cysteines in subunit I (Buse et al.,1987) (if Cu_A is associated to this subunit) where instead Cu_A may be ligated to histidine residues like Cu_B(Finel et al.1987).

In Malmström's model electrons enter the oxidase via heme a and equilibrate with Cu_A (Malmström,1987; Thörnström et al. 1987). Reduction of both these centres induces a conformation providing access and protonation of a protolytic group at the N side and electron transfer to heme a_3-Cu_B. Reduction of this pair results then in exposure and deprotonation of the acidic group at the opposite side.

Mitchell after refusal for long time of a proton pump in the oxidase (Mitchell et al.,1983;see also Papa, 1982)came to accept it and proposed a direct mechanism, the oxygen cycle (Mitchell et al.,1985), where hydrogen atoms are translocated across the membrane directly by oxygen. This model requires a second oxygen reduction site in the oxidase,in addition to the heme a_3-Cu_B centre, but no evidence for it is available (Wikström et al.,1981a;Brunori et al.,1981). Thus Mitchell has proposed a novel direct mechanism where effective proton translocation would be actuated by Cu_A mediated redox-linked hydroxide ion translocation from P to N side (Mitchell,1987).

ACTUAL CHARACTERISTICS OF THE PROTONMOTIVE ACTIVITY

The H^+/e^- stoicheiometry of the oxidase pump is generally considered to be constantly one (Wikström et al.,1981a,b;

Mitchell et al.,1985; Mitchell, 1987).
The indirect mechanisms have been implemented so as to produce optimally a **stoichiometry** of 1 H^+/e^- (Wikström et al., 1981a;Malmström, 1985). Indirect mechanism may, however,result in variable **stoichiometries** as a result of their intrinsic characteristics, influence of allosteric ligands and molecular slips (Papa, 1976;Papa et al. 987b,c,d,;Blair et al., 1986).

The protonmotive activity of the oxidase has been thoroughly investigated in intact rat-liver mitochondria and reconstituted beef-heart oxidase (Wikström et al.,1981a;Papa et al.,1987 b,c; Papa,1982;Wikström et al.1979;Lorusso et al. 1979; Casey et al.,1979).Detailed studies on the vectorial characteristics of the process (Azzi et al.,1985), charge compensation by cations (Sigel,1980) measurements of proton extraction from the N space have been carried out (Azzi et al.,1985). A close scrutiny reveals some uncertainties in the single approaches(Papa and Lorusso,1984).The data produced,when taken together, seem, however, to provide evidence indicating the process to represent electrogenic transmembrane H^+ translocation.The protonmotive activity of bacterial oxidase is,on the other hand, not equally well characterized. It should be noted that the conditions under which the protonmotive activity of preparations of cytochrome oxidase is tested in reconstituted systems are substantially different from those experienced by the enzyme in the natural membrane under physiological conditions.Thus the observations obtained in the two cases may not be completely comparable and there could be unrelated protolytic processes differently contributing to the activity of the oxidase.

In our investigations on the protonmotive activity of mammalian cytochrome c oxidase conditions have been identified where no significant proton pumping is observed either in the mitochondrial membrane (Papa et al.,1980) or in the oxidase reconstituted in liposomes(Papa et al.,1987d). Determination of the H^+/e^- stoicheiometry,based on accurate spectrophotometric measurement of the rate of oxygen consumption with hemoglobin (Papa et al.,1980) and proton release, produced with succinate pulses of aerobic mitochondria H^+/e^- ratios of 4

TABLE 2. Statistical analysis of the initial rates of respiration and proton release and of H^+/O quotients for succinate oxidation in rat-liver mitochondria

Expt.	H^+ release (ng ions·min^{-1}·mg protein^{-1})	Oxygen uptake (ng atoms·min^{-1}·mg prot^{-1}) Polarographic a) standard membrane	b) high-sensitive membrane	c) Spectrophotometric
(i)	397 ±22	46.5 ±3.0	77.4 ±8.8	
(ii)	424 ±18	53.2 ±2.7	82.0 ±7.0	106.0 ±4.5
(iii)	218 ± 4	38.3 ±1.1	50.2 ±1.3	54.2 ±1.9
		$H^+/2e^-$		
(i)		8.07±0.34	5.49±0.18	
(ii)		7.92±0.11	5.21±0.16	4.04±0.05
(iii)		5.65±0.12	4.29±0.16	4.02±0.17

(i) $|O_2|$ 250 µM; (ii) and (iii) $|O_2|$, 27 µM, hemoglobin 25 µM (50%) deoxygenated. Mitochondria were supplemented with rotenone; in (iii) rotenone was preceeded by 1 min incubation with oligomycin. From Papa et al.,(1980).

(Table 2). Higher ratios (Reynafarie et al.,1976) were shown to arise from understimation of the initial rate of oxygen consumption. Since the $H^+/2e^-$ ratio for proton ejection associated to electron flow from succinate to cytochrome c is 4 (Wikström et al.,1984;Papa and Lorusso,1984)this leaves no proton pumping for the oxidase.It is,in fact,observed that whilst under a variety of conditions a constant $H^+/2e^-$ ratio of 4 is experimentally obtained for proton ejection by cytochrome c reductase (Lorusso et al.,1987), with the oxidase the $H^+/2e^-$ ratio results experimentally to be widely variable. A few examples of this will be given below.

When rat-liver mitochondria, incubated anaerobically with succinate and rotenone were pulsed with known amounts of O_2 the $H^+/2e^-$ ratio for proton output was not higher than 4.6 independently from the amount of oxygen added which ranged

TABLE 3. $H^+/2e^-$ ratios for proton ejection elicited by oxygen pulses of rat-liver mitochondria with succinate or ascorbate + TMPD as substrates.

		O_2 added μM		
		1.98	2.98	3.97
(a)		Succinate		
$H^+/2e^-$	$\bar{M}$	4.6	4.7	4.6
	SEM	±0.02	±0.04	±0.04
	n	(32)	(10)	(6)
(b)	as (a) plus Antimycin, Myxothiazol, Ascorbate, TMPD			
$H^+/2e^-$	$\bar{M}$*	1.4	1.4	1.5
	SEM	±0.02	±0.03	±0.01
	n	(10)	(8)	(6)

*corrected for acidification arising from ascorbate.
Rat-liver mitochondria (4 mg protein/ml) were suspended in: 0.24 M sucrose, 10 mM $MgSO_4$, 1 mM K-EGTA, 0.25 mM NEM, 2 mM K-succinate, 4 μg oligomycin/ml, 0.03 μg valinomycin, 2.5 μM rotenone, pH 7.2, 25°C. Incubation in all-glass, oxygen-proof, reaction vessel under stream of argon. Oxygen content and pH changes in the medium were followed electrometrically. O_2 was pulsed as air saturated medium before and after the addition of 0.15 μg antimycin A/mg protein, 0.15 μg myxothiazol/mg protein, 19 mM K-ascorbate, 0.1 mM TMPD. Mitochondrial preparations expressing respiratory control index with succinate of ten or more were used. For other details see ref. Papa et al., 1987b.

from 2 to 4 nequivalents e^- per mg mitochondrial protein (Table 3). After 1 or 2 oxygen pulses antimycin and myxothiazol were added, which caused complete inhibition of cytochrome c reductase, followed by ascorbate plus TMPD as reductants for cytochrome c. The $H^+/2e^-$ ratio for proton output elicited by oxygen pulses was now 2.4-2.5 (Table 3). Correction for 0.5 H^+/e^- deriving from oxidation of ascorbate to dehydroascorbate left an $H^+/2e^-$ ratio of 1.4-1.5 for proton ejection by cytochrome oxidase (Papa et al.,1987b). The $H^+/2e^-$ ratios for succinate to oxygen would be expected to be equal to the sum of the $H^+/2e^-$ ratio of 4 for the span to cytochrome c, plus that measured for cytochrome c to oxygen. Two possible explanations that one can think of are: (a) $\Delta\mu H^+$ generated at their site 2 or 3 causes decoupling at the other site (cf.ref. Blair et al.,1986); (b) the oxidized form of TMPD induces a bypass of the block in the reductase, or some other redox events other than those involved in the normal oxidase activity, associated with proton ejection (Lorusso et al., 1979; Moyle et al.,1978).

Negligible H^+/e^- ratios close to zero were calculated in mitochondria supplemented with artificial reductants of cytochrome c, when contribution of vectorial and or scalar H^+ release other than proton pumping by the oxidase (see also Moyle et al.,1978) was accurately determined by rapid flow spectrophotometry (Papa et al.,1983).

In the experiments of Fig.2 succinate oxidase or cytochrome c oxidase activity were progressively changed,in the first case by addition of increasing amounts of malonate, in the second by increasing the concentration of TMPD. Depressing the rate of electron flow, the H^+/e^- ratio decreased only from 2.3 to 2 with succinate but from 0.7 to 0.2 with ascorbate plus TMPD (see also Lorusso et al.,1979). Thus it seems that the efficency of H^+/e^- coupling at site 3, but not that at site 2 (see also Lorusso et al.,1987) is impaired when the rate of electron flow is depressed.

It can be noted that in the studies from other groups measurements of the protonmotive activity of reconstituted cytochrome c oxidases were, essentially, made by pulsing with ferrocytochrome c the fully oxidized enzyme (probably in the resting state) (Brunori et al.,1979), at neutral pH in media of rather constant ionic composition. We have found that whilst under the standard conditions used in other laboratories the usual H^+/e^- ratio of around 1 was observed (Papa et al.,1987d; Papa et al.,1983), changes of possible physiological effectors like the initial state of the enzyme, pH of the medium, rate of electron flow, cyt.c^{2+}/cyt.$^{3+}$ ratio and presence of divalent cations could significantly lower the

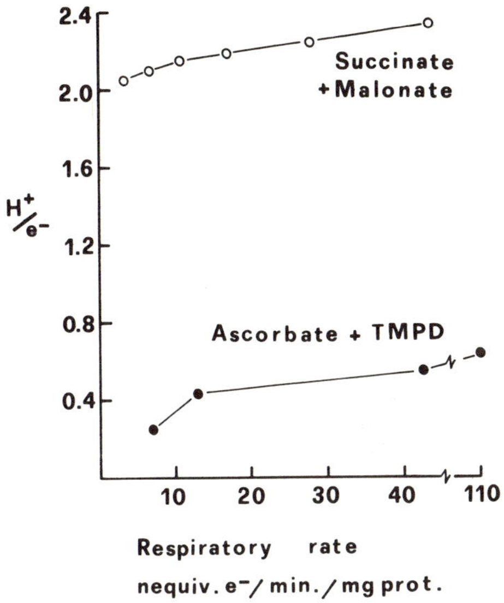

Fig.2. Dependence of the H^+/e^- ratios in rat-liver mitochondria on the rate of succinate or ascorbate respiration (see ref.Papa et al.,1987b).The experimental conditions were the same of Table 3. Oxygen in the amount of 1.62 nequivalents e^-/mg protein was pulsed as air-saturated medium. The H^+/e^- ratios obtained before and after the addition of inhibitors of the b-c_1 complex were plotted against the corresponding rate of respiration (measured polarographically). The respiratory rate with 2 mM succinate was progressively inhibited with the addition of increasing concentrations of malonate from 0.09 to 2.0 mM(o—o).The respiratory rate with 1 mM ascorbate was progressively enhanced with the addition of increasing concentrations of TMPD from 9 to 293 μM (●—●).

H^+/e^- ratio (Papa et al.,1987d). In Fig.3 experiments on reconstituted oxidase are presented which illustrate the effect of the state of the enzyme and pH of the medium on the H^+/e^- ratio. When the anaerobic reduced enzyme supplemented with ferrocytochrome c was pulsed with oxygen (probably pulsed oxidase, see ref. Brunori et al.,1979) proton ejection

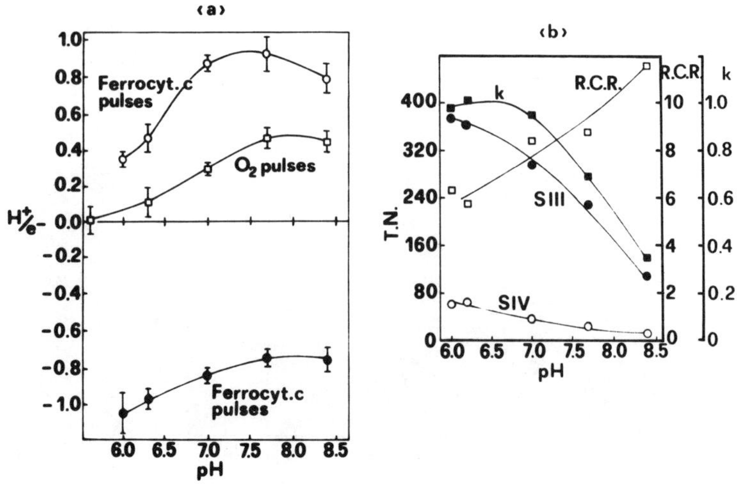

Fig.3. Effect of pH on proton translocation elicited by ferrocytochrome c and oxygen pulses, respiratory activity and rate constant of cytochrome c oxidation in cytochrome oxidase vesicles. (a) (o—o) Pulses of 4 μM ferrocytochrome c; (□—□) pulses of 1 μM O_2 in the presence of 0.2 mM N,N,N',N'-tetramethyl-p-phenylenediamine $Ph(NMe_2)_2$. The H^+/e^- ratios shown were corrected for the proton release associated with the oxidation of $Ph(NMe_2)_2$. (●—●) Pulses of 4 μM ferrocytochrome c in the presence of 3 μM CCP. Data are means of 4-7 experiments ±SEM. (b) Polarographic assay of respiratory activity elicited by the addition of vesicles (0.04 μM cytochrome oxidase) to 100 mM choline-chloride, 5 mM KCl, 5 mM choline-Hepes, 5 mM KCl, 0.1 mM choline-EDTA in the presence of 50 μM cytochrome c, 10 mM ascorbate, 0.2 mM $Ph(NM_2)_2$ in the absence (o—o) and in the presence (●—●) of 2 μg valinomycin plus 3 μM CCP. The rates of oxygen uptake were corrected for the initial baseline slope and expressed as TN (mol cytochrome $c \cdot s^{-1}$. mol cytochrome aa_3^{-1}). (□—□) Respiratory control ratio (RCR); (■—■) apparent first-order constant of the oxidation of 5 μM ferrocytochrome c measured spectrophotometrically at 4°C, to slow down the reaction. The progress curves for cytochrome oxidation when plotted as log A vs t resulted always in straight lines. The apparent k was calculated from the slopes by the regression linear procedure (from ref. Papa et al., 1987d).

which was negligible at acidic pH, increased with pH and reached a maximum H^+/e^- ratio of 0.44 at pH 7.7, with half-maximal value around pH 6.7. With the reductant pulses proton ejection also increased going from pH 6.0 to 7.7. With the reductant pulses the H^+/e^- ratios were, however, in all the pH range examined, 2-4 times larger than those measured in the oxidant pulses. A maximum H^+/e^- ratio of 0.9 was observed at pH 7.7, half-maximal H^+/e^- ratio occurred around pH 6.3. Polarographic measurements showed that the respiratory activity increased both in the coupled and the uncoupled state when the pH was decreased from 8.4 to 6.0. The extent of the increase was however larger in the coupled state than in the uncoupled state; as a consequence the respiratory control decreased from 12 to 6 between pH 8.4 and 6.0.

CROSSOVER EFFECTS OF $\Delta\psi$ AND ΔpH

A useful contribution towards elucidation of the mechanism of a protonmotive redox system is represented by identification of the individual redox steps which are under the influence of the $\Delta\psi$ and ΔpH components of the protonmotive force.

Aerobic oxidase vesicles were supplemented with cytochrome c, TMPD and ascorbate so that a respiring steady-state was established. In a few seconds the oxidase had gone through many turnovers and was conceivably all in the functional pulsed state. The steady-state redox levels of cytochrome c and that of cytochromes $a+a_3$ were monitored with a dual wavelength spectrophotometer. From the relative contributions of hemes a and a_3 to the overall absorbance changes at two wavelength couples the specific redox changes of the two hemes a were estimated (Fig.4). The addition of the reductants to the oxidized enzyme resulted in 60% reduction of cytochrome c. Heme a was reduced up to 50% whilst heme a_3 was reduced to around 10%. $\Delta\psi$ Collapse by valinomycin caused oxidation of cytochrome c. This was accompanied, during the interval in which $\Delta\psi$ was expected to be replaced by extra ΔpH, by reduction of heme a to around 70% and some oxidation of heme a_3. Further ΔpH collapse by nigericin caused definite oxidation of cytochrome c and even larger oxidation of heme a.

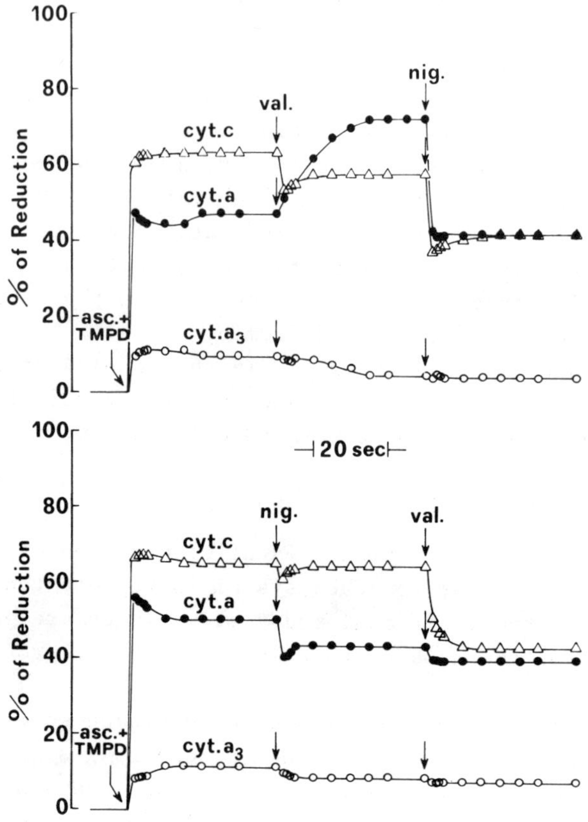

Fig.4. Redox transitions of cytochrome c, a and a_3 caused by ionophores in steady-state respiring cytochrome c oxidase vesicles. 1 µM cytochrome c oxidase vesicles were suspended in 0.180 mM sucrose, 20 mM KCl, 20 mM K-Hepes,pH 7.5 in the presence of 1 µM cytochrome c. Where indicated 10 mM ascorbate plus 0.2 mM TMPD, 2 µg/ml valinomycin, 2 µg/ml nigericin were added. Specific redox levels of hemes a and a_3 were calculated from spectrophotometric measurements at 605-630 and 443-470 nm. For estimation of the relative reduction levels of hemes a and a_3 a relative contribution of heme a of 72.5% and 35% at 605-630 and 443-470 respectively was taken (see ref.Papa et al.,1987c).

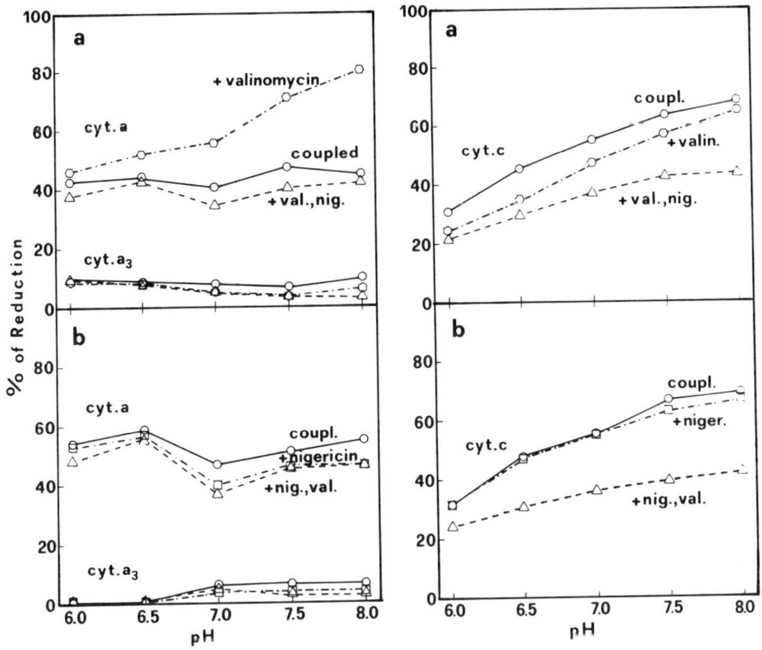

Fig.5. pH-dependence of the steady-state redox level of cytochrome c, a and a_3 in respiring cytochrome oxidase vesicles. The experimental conditions were the same of Fig.4 with the only exception of Hepes substitution with MES in the suspending medium at pH.s 6.0 and 6.5. The values reported in panels a and b are means of two different sets of experiments (see ref.Papa et al.,1987c).

Nigericin, added in the absence of valinomycin, caused oxidation of heme a with practically no change of cytochrome c and heme a. Collapse of $\Delta\psi$ by further addition of valinomycin resulted in large oxidation of cytochrome c but little if any, oxidation of heme a (there was no detectable change of heme a_3).

In Fig.5 the dependence on the pH of the suspending medium of the crossover effects exerted by valinomycin and nigericin on the steady-state redox levels of cytochromes in oxidase vesicles is shown. The following can be observed:

the extra-reduction of heme a induced by the internal extra alkalinization, produced by collapse of $\Delta\psi$ by valinomycin, increased markedly with pH. The same increase with pH was observed for the extent of oxidation induced by the subsequent addition of nigericin in the presence of valinomycin. No significant pH dependence was, on the other hand, observed in the redox changes of cytochromes caused by nigericin added before valinomycin. Thus the presence of $\Delta\psi$ obscured a pH dependence which seems to apply to the crossover effect exerted by the internal alkalinization on electron flow from heme a to a_3. The cytochrome c steady-state reduction level, which was primarily under the control of $\Delta\psi$, increased with pH in all the conditions examined. It seems possible to conclude that transmembrane $\Delta\psi$ exerts a crossover effect on electron flow from cytochrome c to heme a. There is, on the other hand, very little, if any, control by $\Delta\psi$ on internal electron flow from heme a to heme a_3. These observations agree with the reported topology of cytochrome c and hemes a and a_3 in the mitochondrial membrane. The internal alkalinization generated by respiration, and magnified by $\Delta\psi$ collapse with valinomycin, exerts a dramatic crossover effect between heme a and a_3 (see also McGovern-Moroney et al., 1984). Proton availability in the N phase does not, on the other hand, seem to be rate limiting for the reaction of reduced a_3 centre with O_2.

In respiring soluble cytochrome oxidase the reduction level of heme a was higher than that of cytochrome c and much higher that of heme a_3 (not shown, see Papa et al., 1987c). The reduction levels of heme a and cytochrome c both increased with pH. These observations, together with those obtained in the reconstituted oxidase, show that the rate of internal electron flow in the oxidase from heme a to heme a_3 is a rate limiting step. It seems that, more than transmembrane ΔpH, it is the actual pH seen by sites in the oxidase exposed to protonic equilibration with the N space to control electron flow from heme a to a_3.

We have determined the H^+/e^- coupling ratio for redox Bohr effect of heme a in the soluble oxidase with heme a_3 blocked by KCN. The pH dependence of this ratio, presented in Fig.6, shows a bell shaped curve, with maximum of 0.6 at pH 7.5

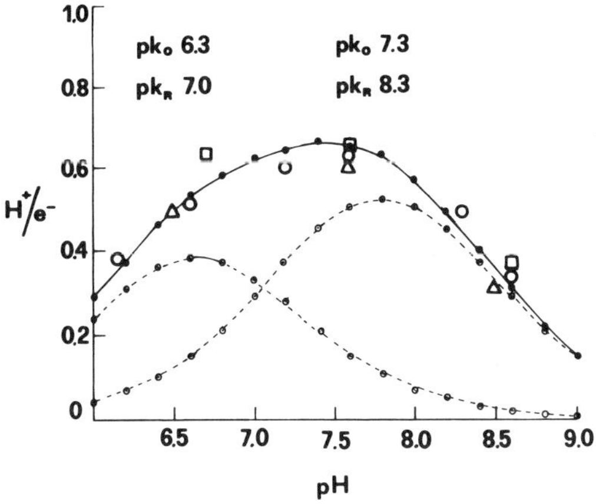

Fig.6. pH-Dependence of Bohr effects of heme a in KCN inhibited soluble cytochrome c oxidase from beef-heart.8-10 μM cytochrome c oxidase was incubated overnight at 0°C with 2 mM KCN in 150 mM KCl. 15 ml of the sample were adjusted to the indicated pH.s. The spectral changes of cytochrome a and c were followed at 605-630 and 550-540 nm respectively.
Cycles of reduction-oxidation of cytochrome a were produced by successive additions of 6-7 μM ferrocytochrome c and 10 μM ferricyanide; after this, three further additions of 10 μM ferricyanide were made to oxidize completely the cytochromes and take the spectral level, so reached, as baseline for calculations. $\Delta\varepsilon mM_{550-540}$ = 19.1 and $\Delta\varepsilon mM_{605-630}$ = 20.4 were used. pH Changes associated with the reduction-oxidation cycle of cytochrome a were followed electrometrically and quantified by comparison with additions of known amounts of titrated HCl. The extent of acidification accompanying cytochrome a oxidation was corrected for a little artifact by ferricyanide.The s/n values shown,were computed by correcting at alkaline pH.s the observed s/n ratios for the proton release associated to the oxidation of the residual ferrocytochrome c by ferricyanide.The experimental values were best fitted by a curve obtained by combining point by point two equations expressing each one the theoretical pH-dependence of the s/n ratio attributable to a single protolytic group with given different values of pKa in the two redox states of the metal center.

suggestive of contribution of more than one protolytic group (see Wikström et al.,1981). The best fit of the experimental points was,in fact, obtained by a curve resulting from combination of those constructed for two hypothetical redox-linked protolytic groups, one with pK_{ox} of 6.3 and pK_{red} of 7.0,the other with pK_{ox} of 7.3 and pK_{red} of 8.3. It is interesting to note that the pH dependence profile of Bohr effects in heme a closely resembles those of the H^+/e^- ratio for proton ejection and the oxidation of heme a induced by nigericin addition to the steady-state respiring oxidase vesicles (Papa et al., 1987c).

CONCLUSIONS

The properties exhibited by redox-linked proton ejection in mammalian cytochrome c oxidase apparently qualify the process as electrogenic transmembrane proton translocation from the inner to the outer aqueous phase of mitochondria.The process observed in lower eukaryotes and in some prokaryotes has yet to be better characterized in order to verify its nature. It may be pointed out that all cytochrome c oxidases generate transmembrane $\Delta\mu H^+$ as direct consequence of the reduction of dioxygen to H_2O, regardless whether electron flow is or not associated to proton ejection. Genetic manipulations of oxidases in prokaryotes and yeast should help to understand how his proton ejection develops in some microorganisms.

The H^+/e^- ratio of redox linked proton ejection in eukaryotic oxidases is experimentally affected by a number of physiological factors which also influence the redox activity of the enzyme (Papa et al.,1987b,d).This variability,provided it be intrinsic to the enzyme, may have important implications as far as the mechanism and the role of the process is concerned. A flexible H^+/e^- stoicheiometry would in principle favour indirect coupling models. However also direct models could be adapted to account for variable efficency (Mitchell et al.,1985).

Lowering of the H^+/e^- ratio may be the result of molecular slips, diversion of the electron current from coupled to uncoupled pathways (Blair et al.,1986) or change of the enzyme from coupled to uncoupled states (for example dimer-monomer transition: see Wikström et al.,1981a; Finel et al.,1986), either if the coupling mechanism is direct or indirect.

The observations obtained by our group would, at any event, indicate that in mitochondria under physiological conditions at the rates of electron flow supported by physiological concentrations of dehydrogenase substrates,in the presence of $\Delta\mu H^+$ generated by the other coupling sites, there is negligible if any proton pumping by cytochrome c oxidase.

Subunit I seems to be the unit of the oxidase (Buse et al.,1987) which is apparently sufficient for the catalytic process and for the basic expression of redox-linked proton ejection. This supports a direct involvement in the protonmotive activity of prosthetic groups. Heme a and Cu_A appear to be possible candidates. The present observations support a primary involvement of heme a. This heme is six-co-ordinate with two histidines (Wikström et al.,1981a).It has also been proposed that the formyl substituent of the porphyrin forms an hydrogen bond with a nearby tyrosine(Callahan et al., 1983). This interaction and other hydrogen bond interactions initiating from redox-linked displacement of the axial histidines could provide a basis for proton electron linkage and proton pumping. Heme a is in a rapid redox equilibrium with Cu_A which, as proposed by Chan et al.(1979), can also be considered a good candidate for the coupling process. A central role of Cu_A is, also, envisaged in the direct model of Mitchell (1987).

Attention to heme a and Cu_A seems promising and it may disclose the central events in the protonmotive activity of cytochrome c oxidase.

ACKNOWLEDGEMENTS

This work was supported by Grant n.86.00516.44 and Grant CT 86.02079.11 of the C.N.R.-Rome.

REFERENCES

Azzi A, Müller M, O'Shea PS and Thelen M (1985) J Inorg Biochem 23:341-350.

Bickar D, Lehninger A, Brunori M, Bonaventura J, and Bonaventura C (1985) J Inorg Biochem 23:365-372.

Bisson R and Bacci A (1987). In Papa S, Chance B and Ernster L, (eds) "Cytochrome Systems:Molecular Biology and Bioenergetics" New York, Plenum Publ.Co,in press.

Blair DF, Ellis WR, Wang H, Gray HB and Chan SI (1986) J Biol Chem 261:11524-11537.

Blair DF, Gelles J and Chan SI (1986) Biophys J 50:713-733.

Blair DF, Witt SN and Chan SI (1985) J Amer Chem Soc 107: 7379-7399.

Bomelka E, Richter FW, Stroh A and Kadenbach B (1986) Biochem Biophys Res Comm 14:1007-1014.

Brunori M, Antonini E and Wilson MT (1981. In Sigel H (ed) "Metal Ions in Biological Systems" New York, Marcel Dekker pp 210-235.

Brunori M, Colosimo A, Raimoni G, Wilson M and Antonini E (1979) J Biol Chem 254:10769-10775.

Buse G, Steffens GCM, Biewald R, Bruch B and Hensel S (1987). In Papa S, Chance B and Ernster L (eds) "Cytochrome Systems: Molecular Biology and Bioenergetics" New York Plenum Publ Co, in press.

Callahan PM and Babcock GT (1983) Biochemistry 22:452-461.

Capaldi RA, Gonzales-Halphen D and Takamiya S (1987) FEBS Lett 207:11-17.

Casey RP, Chappel JB and Azzi A (1979) Biochem J 182:149-156.

Chan SI, Bocian DF, Brudvig GW, Morse RH and Stevens TH (1979). In King, TE et al (eds) "Cytochrome Oxidase" Amsterdam Elsevier/North Holland pp 177-188.

Finel M, Haltia T, Holm L, Jalli T, Metso T, Pustinen A, Raitio M, Saraste M and Wikström M (1987).In Papa S,Chance B and Ernster L (eds) "Cytochrome Systems:Molecular Biology and Bioenergetics" New York, Plenum Publ.Co. in press.

Finel M and Wikström (1986) Biochim Biophys Acta 851:99-108.

Gelles J, Blair DF and Chan SI (1987) Biochim Biophys Acta, in press.

Haltia T and Finel M (1987). In Papa S, Chance S and Ernster L (eds) "Cytochrome Systems:Molecular Biology and Bioenergetics" New York, Plenum Publ.Co.,in press.

Kadenbach B, Jarausch J, Hartmann R and Merle P (1983) Anal Biochem 129:517-521.

Lorusso M, Boffoli D, Cocco T, Gatti D and Papa S (1987). In Papa S, Chance B and Ernster L (eds) "Cytochrome Systems Molecular Biology and Bioenergetics" New York Plenum Publ. Co. in press.

Lorusso M, Capuano F, Boffoli D, Stefanelli R and Papa S (1979) Biochem J 182:133-147.

Ludwig B and Schatz G (1980) Proc Natl Acad Sci USA 77:196-200.

Malmström Bo G (1979) Biochim Biophys Acta 549:281-303.

Malmström Bo G (1985) Biochim.Biophys Acta 811:1-12.

Malmström Bo G (1987). In Papa S, Chance B and Ernster L (eds) "Cytochrome Systems:Molecular Biology and Bioenergetics" New York Plenum Publ.Co.,in press.

McGovern-Moroney P, Scholes TA and Hinkle PC (1984) Biochemistry 23:4991-4997.

Millet F, de Jong C, Poulson L and Capaldi RA (1983) Biochemistry 22:546-552.

Mitchell P (1966) "Chemiosmotic Coupling in Oxidative Phosphorylation" Bodmin, Glynn Research Ltd.

Mitchell P (1987) FEBS Lett. 222:235-245.

Mitchell P, Mitchell R, Moody J, West J, Baum H and Wrigglesworth J (1985) FEBS Lett 188:1-7.

Mitchell P and Moyle J (1970). In Tager JM, Papa S, Quagliariello E and Slater EC (eds) "Electron Transport and Energy Conservation, Bari, Adriatica Editrice,pp 575-587.

Mitchell P and Moyle J (1983) FEBS Lett.151:1667-1678.

Moyle J and Mitchell P (1978) FEBS Lett.88:268-272.

Müller M, Labonia N, Schläpfer B and Azzi A (1978). In Papa S., Chance B and Ernster L (eds) "Cytochrome Systems: Molecular Biology and Bioenergetics" New York Plenum Publ.Co., in press.

Nicholls P and Chance B (1974). In Hayashi P (ed) "Molecular Mechanisms of Oxygen Activation" New York Academic Press pp 479-534.

Papa S (1976) Biochim Biophys Acta 456:39-84
Papa S (1982). In Martonosi (ed) "Membrane and Transport" Vol.I, New York Plenum Publ.Co.,pp363-368.
Papa S,Capitanio N and De Nitto E (1987 d) Eur J Biochem 164:507-516.
Papa S and Capitanio N (1987c). In Papa S, Chance B and Ernster L (eds) "Cytochrome Systems:Molecular Biology and Bioenergetics" New York Plenum Publ.Co.,in press.
Papa S, Capitanio N, Izzo G and De Nitto E (1987b). In Kim C et al. (eds) "Advances in Membrane Biochemistry and Bioenergetics" New York Plenum Publ.Co. in press.
Papa S, Capuano F, Markert M and Altamura N (1980) FEBS Lett 111:243-250.
Papa S, Chance B and Ernster L (1987a) "Cytochrome Systems: Molecular Biology and Bioenergetics" New York Plenum Publ. Co. in press.
Papa S, Guerrieri F and Izzo G (1983) FEBS Lett 157:15-21.
Papa S, Guerrieri F and Lorusso M (1974) Biochim Biophys Acta 357:181-190.
Papa S, Guerrieri F, Lorusso M and Simone S (1973) Biochimie 55:703-716.
Papa S and Lorusso M (1984). In Burton R and Carvalho-Guerra F (eds), New York Plenum Publ.Co.,pp 275-290.
Papa S, Lorusso M, Capitanio N and De Nitto E (1983) FEBS Lett 157:7-14.
Reynafarie B, Brand MD and Lehninger AL (1976) J Biol Chem 251:7442-7451.
Saraste M, Raitio M, Jalli T and Perämaa A (1986) FEBS Lett 206:154-156.
Sigel E and Carafoli E (1980) Eur J Biochem 111:299-306.
Solioz M, Carafoli E and Ludwig B (1982) J Biol Chem 257: 1579-1582.
Sone N (1987). In Papa S, Chance B and Ernster L (eds)"Cytochrome Systems:Molecular Biology and Bioenergetics" New York Plenum Publ.Co.,in press.
Steffens CM, Biewald R and Buse G (1987) Eur J Biochem 164: 295-300.
Suske G, Enders C, Mengel T and Kadenbach B (1987).In Papa S, Chance B and Ernster L (eds) "Cytochrome Systems:Molecular

Biology and Bioenergetics" New York Plenum Publ Co, in press.
Takamiya S, Lindorfer MA and Capaldi RA (1987) FEBS Lett 218: 277-282.
Thörnström PE, Maison-Peteri B, Brzezinski P, Arvidsson L, Fredriksson PO and Malmström BoG (1987). In Papa S, Chance B and Ernester L (eds) "Cytochrome Systems:Molecular Biology snd Bioenergetics" New York Plenum Publ.Co., in press.
Urban PF and Klingenberg M (1969) Eur J Biochem 9:519-526.
Wikström M and Krab K (1979) Biochim Biophys Acta 549:177-222.
Wikström M, Krab K and Saraste M (1981a)"Cytochrome Oxidase. A Synthesis" New York Academic Press.
Wikström M, Krab K and Saraste M (1981a)Ann Rev Biochem 50: 623-655.
Wikström M and Saraste M (1984). In Ernster L (ed) "Bioenergetis" Amsterdam Elsevier Science Publ BV, pp 49-94.
Wright RM, Trawick JD, Trueblood CE, Patterson TE and Poyton RE (1987). In Papa S, Chance B and Ernster L (eds) "Cytochrome Systems:Molecular Biology and Bioenergetics" New York Plenum Publ.Co., in press.

DISCUSSION

<u>KADENBACH</u>: You said in your talk that proton translocation in cytochrome c oxidase involves the dimeric enzyme. Is there any experimental evidence for this assumption?

<u>PAPA</u>: I was simply referring to a proposal put forward by Wikström et al. In fact there is, in my opinion, very little evidence to support it and even observations militating against it.

<u>HAZZARD</u>: It was shown by Dr. O. Einardottir, of Prof. W. Caughey's laboratory, that one zinc was tightly bound to bovine cytochrome c oxidase. (1) Has the location of this zinc been determined? (2) Has a role of this zinc in either electron transfer or proton translocation been implicated?

PAPA: The data from Dr. Buse's laboratory confirm the existence of one zinc per molecule of cytochrome c oxidase. No information is, however, yet available on the subunit location and the role of zinc.

LUDWIG: I did not understand the reason why you make the distinct discrimination between higher eukaryotes on one side and lower eukaryotes and prokaryotes on the other side in terms of their capacity to pump protons.

PAPA: I intended to point out that whilst for higher eukaryotes the observations produced on redox-linked proton ejection by cytochrome c oxidase are adequate to qualify the process as electrogenic transmembrane proton pumping, in the case of lower eukaryote and bacterial cytochrome c oxidase the observed redox-linked proton ejection is not equally well characterized.

GARBER: According to your model and data, shouldn't acidification of oxidase in vesicles result in a decrease in activity? However, Malmström and others find an increase in activity with decreasing pH. Could your results reflect lack of a titration to Vmax and hence you are observing changes in the Km and transitions between the various kinetic phases?

PAPA: Not exactly. Our data and model do, in fact, indicate in agreement with previous observations of Wilms et al., and Malmström et al., that alkalinization of the interior of the oxidase vesicles results in a depression of activity (inhibiton of electron flow from heme a to heme a_3) which is reversed by the acidification caused by nigericin addition.

The data of Malmström et al. show that Kcat (turnover) of cytochrome c oxidase decreases with pH concomitant decrease of Km. Thus, acidification seems to have two opposing effects on the oxidase: enhancement of the Vmax and decrease of the affinity for ferrocytochrome c. Evidently the oxidase is working close to saturation in our experiments.

Oxidases and Related Redox Systems, pages 731–747

A Proposal for the Site and Mechanism of Redox-linked Proton Translocation in Cytochrome *c* Oxidase*

Sunney I. Chan[§], Peter Mark Li[†], Thomas Nilsson[‡]
Jeff Gelles[†], David F. Blair[¶], and Craig T. Martin[¶]

From the Arthur Amos Noyes Laboratory of Chemical Physics, California Institute of Technology, Pasadena California 91125

Abstract

In this paper, we present a detailed mechanistic model for the redox-linked proton pump of cytochrome *c* oxidase. Using the spectroscopic properties of Cu_A, we have constructed a molecular orbital model for the electronic structure of the Cu_A site. One prediction of this model is that a ligand rearrangement should occur upon reduction of Cu_A. We further describe a detailed mechanism for proton pumping which utilizes this redox-linked Cu_A ligand rearrangement as its central feature. The turnover cycle of this proton pump model is then discussed in terms of the theoretical thermodynamic and kinetic requirements of a redox-linked proton pump.

*This paper is Contribution 7646 from the Arthur Amos Noyes Laboratory of Chemical Physics, California Institute of Technology, Pasadena, CA 91125.

Introduction

The final step in mitochondrial respiration is the transfer of electrons from cytochrome *c* to molecular oxygen, the final electron acceptor in aerobic organisms. This reaction is catalyzed by the integral membrane enzyme cytochrome *c* oxidase and constitutes one of the three coupling sites in mitochondrial energy transduction. As outlined in the chemiosmotic theory of Mitchell (1961), the energetic coupling of electron transfers to ATP synthesis and several mitchondrial transport systems is mediated by a proton gradient across the inner mitochondrial membrane. Cytochrome *c* oxidase takes part in the generation of this proton gradient via two mechanisms. First, the protons consumed in the reduction of molecular oxygen are taken from inside the mitochondrial matrix. This proton consumption results in a net proton gradient across the mitochondrial membrane. Secondly, cytochrome *c* oxidase has been demonstrated to actively transport up to one proton from the matrix side of the mitochondrial membrane to the cytosolic side for each electron transferred through the enzyme. Redox-linked proton pumping activity has been clearly demonstrated in both intact mitochondria (Wikström, 1977) as well as in isolated cytochrome *c* oxidase reconstituted into artificial lipid vesicles (Wikström et al., 1979).

---

§Fogarty Scholar-in-Residence, National Institutes of Health, Bethesda, MD 20892. Recipient of Grant 22432 from the National Institutes of General Medical Sciences, US Public Health Service.

†National Science Foundation Predoctoral Fellow. Present address: J.G., Department of Cell Biology and Physiology, Washington University Medical School, St. Louis, MO 63110.

‡Recipient of a Postdoctoral Fellowship from the Swedish Natural Science Research Council. Present address: Department of Biochemistry and Biophysics, Chalmers Tekniska Högskola, S-412 96 Göteborg, SWEDEN.

¶Recipient of National Research Service Awards 5T32GM-07616 from the National Institute of General Medical Sciences, US Public Health Service. Present addresses: D.F.B., Department of Biology, Harvard University, Boston, MA 02138, and C.T.M., Department of Molecular Biophysics and Biochemistry, Yale Univeristy, New Haven, CT 06520.

Cytochrome *c* oxidase contains four redox-active metal centers important to its catalytic activity, two coppers and two heme irons (Figure 1). These metal cofactors form two distinct groups. One iron and one copper, labelled cytochrome a_3 and Cu_B, together form the site of oxygen binding and reduction. These two centers are spatially close and magnetically coupled (Brudvig, et al., 1986). In forms of the enzyme constituting the catalytic cycle, the reduction potential of this coupled center is probably close to that of molecular oxygen. The remaining two centers, cytochrome *a* and Cu_A have lower reduction potentials (near that of cytochrome *c*), and serve to accept electrons from cytochrome *c* and pass them on to the oxygen binding site.

The largest free energy change of the redox reaction catalyzed by cytochrome *c* oxidase is associated with the internal electron transfer from the low potential centers to the high potential oxygen binding site. Since the available free energy is effectively expended once the electron has reached the oxygen binding site, the low-potential centers are the most natural candidates for the site of proton pumping. Recent experimental results on proton pumping by cytochrome *c* oxidase in whole mitochondria (Wikström and Casey, 1985) are consistent with this notion.

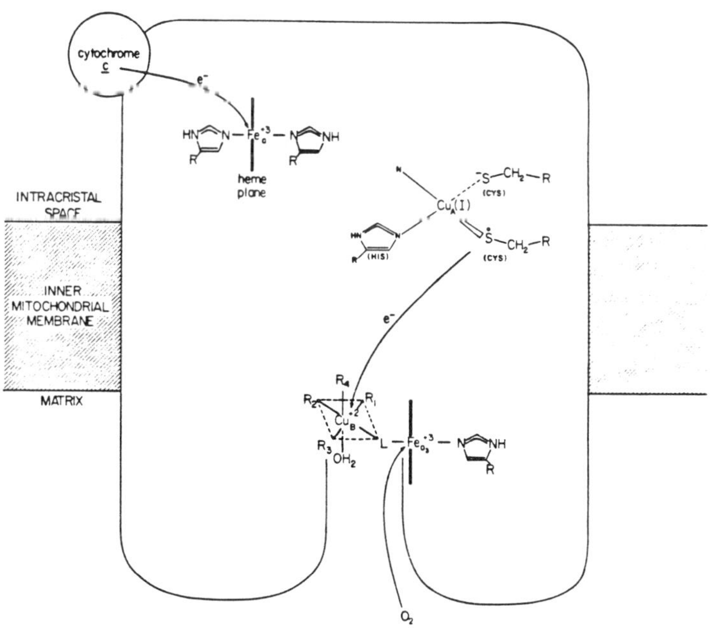

Figure 1: Redox-active metal cofactors in cytochrome c oxidase

Of the two low potential metal centers, cytochrome *a* has received the most attention as the probable site of redox- linked proton translocation. Historically, three main lines of evidence have contributed to this view. First, it is well known that cytochrome *a* exhibits a midpoint potential which is sensitive to the pH on the matrix side of the membrane (Artzatbanov, et al., 1978). Secondly, during reoxidation of the enzyme following flash photolysis of CO-inhibited enzyme, Chance et al. observed a heterogeneity in the reoxidation rate of cytochrome *a*. Thirdly, Moroney et al. (1984), observed a dependence on the steady state reduction level of cytochrome *a* on pH and potential gradients in reconstituted vesicle systems. The latter two observations have been used to implicate the existence of two or more states of cytochrome *a*, fulfilling the theoretical requirements of 'input' and 'output' states. The existence of a pH dependent midpoint potential has been generally used to imply that cytochrome *a* is the site of coupling between electron transfer and proton translocation. In a system where oxidation or reduction is linked to the stoichiometric protonation or deprotonation of a conjugate base, the expected pH dependence is 60 mV/pH unit.

Although the linkage between oxidation / reduction and protonation / deprotonation (thermodynamic linkage) that is manifested in a pH dependent midpoint potential has been considered to be an essential feature of a proton pumping site, (Wikström, 1981), several theoretical investigations show that this is not a necessity. In fact, Gelles et al. (1986) have shown that an efficient pump model may be constructed such that the redox-linked proton translocation site shows essentially no pH dependence whatsoever. Furthermore, the heterogeneous reduction and reoxidation phenomena observed for cytochrome *a* are at best circumstantial evidence for cytochrome *a* as the site for redox-linked proton translocation.

Although historically given less attention than cytochrome *a*, the Cu_A site also deserves due consideration as the possible site of redox-linked proton pumping in cytochrome *c* oxidase. In this paper, we present a model for proton pumping using Cu_A as the site of redox-linked proton translocation.

The Gating of Protons and Electrons

Any enzyme that couples two reactions must catalyze them in such a way that one reaction does not occur in the absence of the other. For the cytochrome *c* oxidase proton pump, the oxidation of cytochrome *c* by molecular oxygen is coupled to the transfer of a proton from the mitochondrial matrix to the cytosol. In this case, the *driving* reaction is the electron transfer from cytochrome *c* to the oxygen binding site; the *driven* reaction is the transfer of a proton against a electrochemical gradient.

The catalytic cycle for a proton pump must therefore include as input reactions electron transfer from the donor (ferrocytochrome *c*) and proton uptake from the matrix *to* the coupling site. Analogously, the output reactions are electron transfer to the acceptor and proton release to the cytosol *from* the coupling site.

If the pump site is not to act as a passive proton transporter in the absence of electron transfer, the proton binding site must be accessible to either the cytosol or the matrix, but not both simultaneously. Furthermore, coupling of the proton binding to electron transfer requires that the interconversion between these two states of access may occur only if the site is both reduced and protonated or oxidized and deprotonated (Wikström, et al., 1981).

Similarly, the redox element of the site should be accessible to either the donor or the acceptor (Wikström, et al., 1981). It must consequently be able to exist in two distinct states, one for electron input and one for electron output. In the input state, the site may be reduced by the donor but not oxidized by the acceptor. In the output state, it may be oxidized by the acceptor, but not reduced by the donor. For coupling to a proton transfer, the interconversion between the input and output states are subject to the same restriction as the proton site interconversion.

The aforementioned arguments imply a need for the modulation of the rate constants of both electron transfers to and from the redox element of the pump site. This is tantamount to gating the electron flow into and out of the site. Possible mechanisms for such a gating of electrons have been discussed previously by Blair et al. (1986). Using electron transfer rate theory (Marcus and Sutin, 1985), these authors have proposed that sufficient rate modulation may be obtained from changes in the effective electron-transfer distances. This requires a redox and protonation dependent structural change at the pump site.

Of the two low potential metal centers, Cu_A seems to be the most capable of undergoing structural changes that would affect rates of electron transfer during the course of reduction and oxidation. Since electron transfers to and from the hemes are likely to take place via the heme edge, gating electron flow would require substantial movement of the heme relative to the donor and acceptor sites (Blair, et al., 1986). For copper, small changes in the coordination sphere could readily modulate the rates of ligand-mediated electron transfer to and from the metal. This requires, however, that Cu_A, unlike blue copper centers, undergo a redox linked structural change. The fact that the entropy of reduction for Cu_A is significantly larger than that for most blue copper proteins is an indication that this may be the case. Since the contribution to the entropy of reduction from solvent reorganization is expected to be small in both cases, the difference in entropy of reduction is likely to reflect a larger structural change for Cu_A than for the blue coppers. In the next section paper we discuss the spectroscopic properties of Cu_A in relation to possible redox-linked ligand rearrangements.

The Cu_A Site

The Cu_A site in cytochrome *c* oxidase has been extensively compared to the blue copper proteins. The atypical spectroscopic properties of Cu_A, however, suggest an electronic structure distinct from the blue coppers and unusual for Cu(II) complexes in general. Electron paramagnetic resonance (EPR) studies have revealed an unusually small and isotropic hyperfine interaction for Cu_A of which the largest component is not coincident with the largest g-value (Hoffman, et al., 1980). Furthermore, the low field g-value of Cu_A is smaller than that found in most Cu(II) complexes while the high field g-value is reported to be located below the free electron g-value (g=2.0032) at g=1.99 (Beinert et al. (1962); Aasa et al., 1976). X-ray absorption edge measurements of Cu_A show that the coordination charge at the copper is higher than that of the blue coppers, and that little appreciable change occurs upon reduction of the site to Cu(I) (Hu et al., 1977). In addition, electron nuclear double resonance (ENDOR) measurements have revealed unusually small histidine nitrogen hyperfine coupling constants as well as two strongly coupled proton hyperfine resonances which have been assigned to cysteine methylene protons

(Martin, 1985). These spectroscopic data are consistent with the delocalization of substantial electron density onto the copper ion from one or more sulfur ligands.

In light of more recent ENDOR results (Martin, 1985) and recent x-ray absorption measurements (Li et al., 1987), we have proposed a model for the oxidized Cu_A site in cytochrome *c* oxidase (Figure 2) in which the copper is coordinated by two interacting cysteine sulfur ligands and two histidine nitrogen ligands in a distorted tetrahedral geometry (Martin et al., 1988 [in press]). Protein sequence data show that the only two highly conserved cysteines in cytochrome *c* oxidase reside three amino acids apart in subunit II. Space filling models of this region of subunit II have shown that the sulfur-sulfur distance may be as close as 3 Å. Invocation of a strong electrostatic interaction between these two cysteine sulfurs in this conformation results in a novel electronic structure that accounts for the unusual spectroscopic properties of the oxidized Cu_A site.

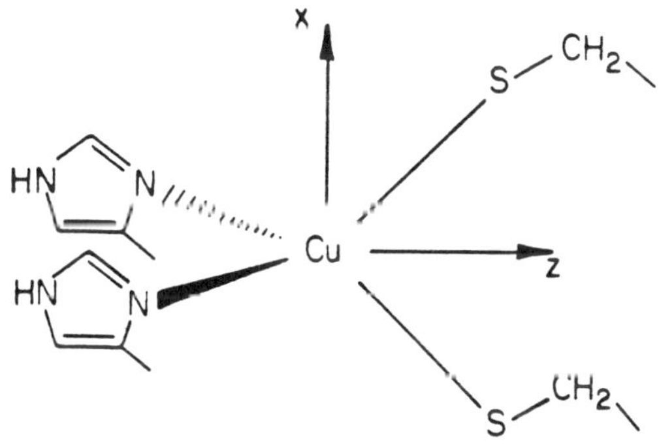

Figure 2: Ligand structure of the Cu_A site

Figure 3 shows a molecular orbital energy diagram for the Cu_A site which includes the interaction of two cysteine sulfur ligands 3.1

Å apart. The proposed distorted tetrahedral geometry results in C_{2v} symmetry, yielding five nondegenerate copper d orbitals. Because the ligand field strength of the cysteine thiolate ligands is stronger than that of the histidine nitrogens, the $3d_{xz}$ orbital of the copper will be raised in energy most by this coordination geometry. In addition, due to backbonding stabilization by the sulfur d orbitals, the manifold of copper d orbitals will be lowered in energy. At a distance of 3.1 Å, the two cysteine sulfur atoms may interact in two ways. First, the overall energies of the lone pairs will be raised due to a coulombic repulsion between the two charged atoms. Secondly, the proximity of the two sulfurs will allow for orbital overlap of the sigma lone pair orbitals. This orbital overlap will result in the formation of a symmetric (A_1) and an antisymmetric (B_1) orbital in the C_{2v} symmetry group. The antisymmetric (B_1) sigma antibonding molecular orbital is of correct energy and symmetry to overlap with the copper d_{xz} orbital. As shown in Figure 4, the resultant molecular orbital is the highest filled and will therefore contain the unpaired electron for the oxidized Cu_A site. Since this molecular orbital consists of an admixture of the copper d_{xz} orbital and sulfur sigma lone pair orbitals, this model accounts for the unusual spectroscopic characteristics of Cu_A by providing an efficient mechanism for the delocalization of the unpaired electron from the copper onto its sulfur ligands.

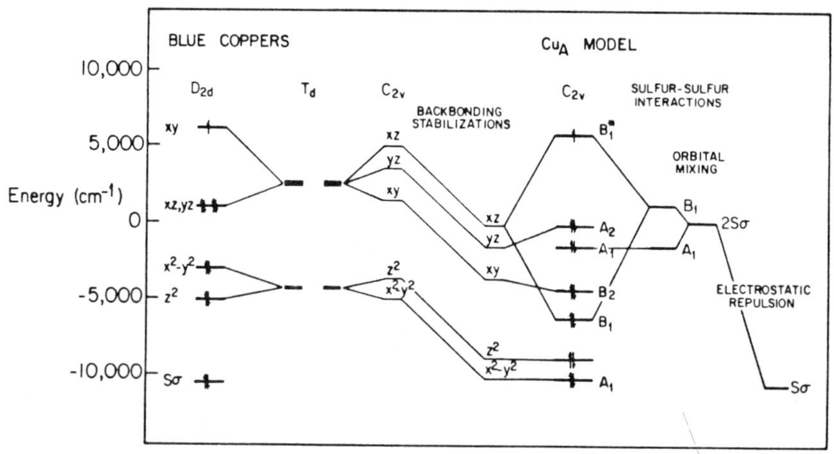

Figure 3: Molecular orbital diagram for the Cu_A site

This model for the electronic structure of Cu_A predicts a substantial rearrangement upon reduction from Cu(II) to Cu(I). The incoming electron will enter the lowest unoccupied molecular orbital, which is formally antibonding, resulting in a destabilisation of the structure. Thus, an important prediction of this electronic structural model is that one of the Cu-thiolate bonds will stretch upon reduction of the Cu_A site. This suggests that displacement of one of the sulfur ligands by another ligand is a likely redox-linked structural change. A ligand substitution is consistent with the observed large entropy of reduction for Cu_A as well as its anticooperative interaction with cytochrome *a* (Wang, et al., 1986). We will discuss how such a ligand substitution could be the salient feature in a mechanism for redox-linked proton transport at the Cu_A site.

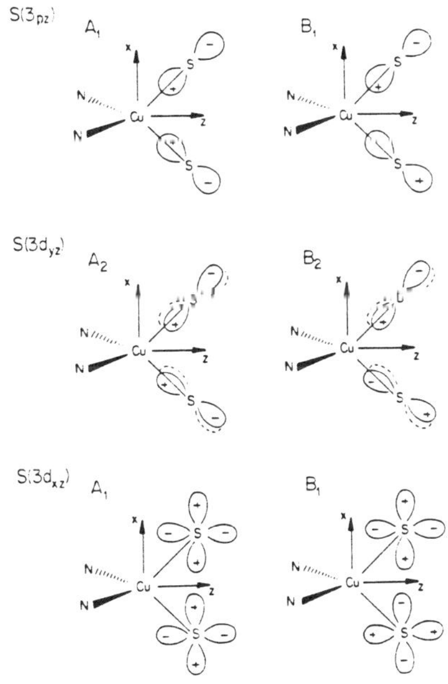

Figure 4: Molecular orbital interactions for the Cu_A site

A Model for Proton Pumping by the Cu_A Site

A ligand substitution may provide the required electron gating if electron transfer from the donor occurs through the cysteine sulfur and electron transfer to the acceptor through the ligand that replaces this cysteine in the reduced form. By allowing protonation on these two groups in their unligated states and a controlled proton transfer between them, all requirements for a pump mechanism are fulfilled. Figure 5 shows the sequence of reactions in a pumping mechanism based on these principles.

In this model, the redox-linked ligand rearrangement predicted from the electronic structure of the Cu_A site serves to interconvert the input and output states for electrons. We assume that the exchangeable cysteine is positioned between the copper and the donor, and the exchangeable tyrosine between the copper and the oxygen binding site. In this way, the effective electron transfer distance for reduction of copper by the donor will be shortened when the cysteine is coordinated. The effective electron transfer distance from the copper to the oxygen binding site will be shorter when the tyrosine is coordinated. In the state in which the cysteine is coordinated and the tyrosine is free, the copper may then be more readily reduced by the donor but not readily oxidized by the oxygen binding site. The criteria for an electron input state are then fulfilled. Similarly, the structure obtained after the ligand rearrangement has the necessary features of an output state. Here, copper may be readily oxidized by the oxygen binding site while reduction by the donor is expected to be slower.

Alternating access for protons (Tanford, 1983) is achieved by the *two* protonatable groups which are the exchangeable ligands. One is assumed to be accessible from the cytosolic side only while the other is assumed to be accessible from the matrix side only. Transfer of a proton between them alters the accessiblitiy of the proton binding site.

In the first step in the reaction cycle, copper is reduced and the coordination structure becomes unstable (Figure 5). The ligands then rearrange, giving a structure with the tyrosine coordinated and the cysteine free. This makes the tyrosine considerably more acidic and the cysteine more basic. Provided that the proton can transfer from the tyrosine to the cysteine faster than protonation of the cysteine from the cytosolic side or deprotonation of the tyrosine from the matrix side, the intermediate B will arise from intramolecular proton transfer. Proton transfer over this distance would require assistance from an additional protonatable group. This function could be filled by one of the histidine ligands. In the next step, the site is reoxidized by the oxygen binding site to form D. For oxidized copper, the coordination structure is unstable and rearrangement to form **A** will be a favorable reaction. This reaction cannot, however, take place via a reverse of the B to C conversion since the greater charge on the copper now acts as a barrier to the intramolecular proton transfer. Instead, the reaction takes place via protonation of the tyrosine from the matrix side and deprotonation of the cysteine to the cytoplasmic side. Since the Cu_A site appears to be located in a region of the protein which protrudes into the cytosol, a proton channel leading from the matrix side to the proton input side of the site is assumed. Similarly, a proton channel leading from the proton output side of the site to the cytosol may or may not be necessary.

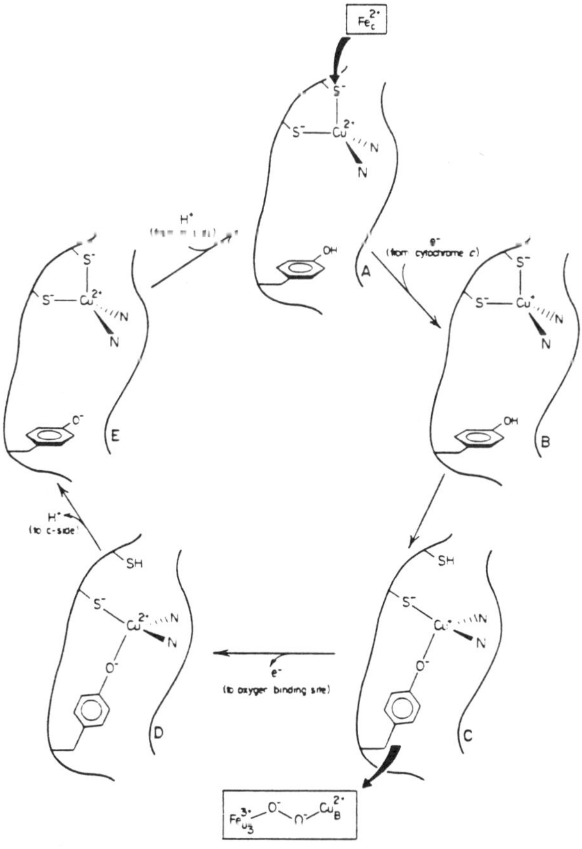

Figure 5: A model for the proton pump cycle in cytochrome c oxidase

In a discussion of the thermodynamics of active transport, Tanford (1983), has pointed out that it is a kinetic advantage for the affinity of the transported substrate to be lower on the output side than on the input side. For a proton pump, this means that the effective pK_a of the output state should be lower than the effective pK_a of the input state. One way to influence the effective pK_a is to have a strongly cooperative binding between protons and electrons in both the input and output states. The reducing conditions on the input side will then ensure a high effective pK_a while the oxidizing conditions on the output side will result in a low effective pK_a. For this reason, the cooperative binding of protons and electrons have been considered an essential feature of a redox-linked proton pump (Wikström, 1981).

The model presented here, however, achieves a kinetic effect similar to the cooperative binding of protons and electrons via a different mechanism. The pK_a in the input state is high (ca. 10); this ensures rapid protonation. To obtain rapid deprotonation of the output state, the release of the proton is obligatorily linked to the strongly exergonic conversion of the Cu_A site back to the input state. The result is a lowered effective pK_a for proton release to the cytosolic side.

This model therefore requires the input to output conversion of the Cu_A site to be strongly exergonic when the site is reduced and strongly endergonic when the site is oxidized. It follows that the reduction potential must increase when going from the input state to the output state. As a result, the potential of the site will tend to be matched to the potentials of the donor and acceptor. There must, however, be a finite potential difference between the donor and the input state to provide the necessary driving force for the electron transfers. A similar potential difference must exist between the output state and the acceptor. If this were not the case, the electron transfers would be too slow. Furthermore, overly well-matched potentials would lead to imperfect gating due to the resultant large driving force for unwanted electron transfers.

In a quantitative investigation of this model, Gelles et al. (1986) found that reduction potentials of 280 mV and 575 mV for the input and output states respectively are reasonable. These reduction potentials are well within the observed range of reduction potentials for blue copper proteins. Using these potentials and the effective electron transfer distances shown in Figure 6, the relative rates of the coupling and uncoupling reactions were calculated using the theory for long range electron transfers described by Marcus and Sutin (1985). Under the assumption that electron transfers are the rate-limiting steps in the catalytic cycle, the performance of this model was evaluated by calculating the steady-state fluxes of the electrons and protons under various conditions. Pumping efficiencies of 0.3 to 0.6 were obtained for electron fluxes of 10 to 100 s^{-1}. From these results, we conclude that electron gating by the mechanism suggested here is sufficient to make our model a viable energy transducer.

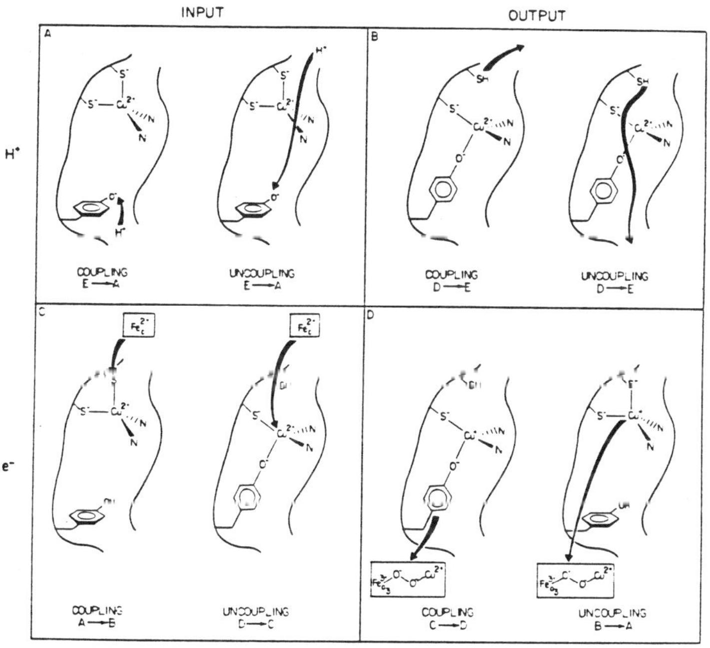

Figure 6: Input and output reactions for proton and electron transfers in the proton pumping cycle

Concluding Remarks

The experimental evidence for Cu_A being the site of redox-linked proton translocation is not strong at the present. There are, however, several results suggesting that a major portion of the electron flow through cytochrome *c* oxidase passes via Cu_A. Studies of intramolecular electron transfers at low temperature (Blair, et al, 1985). show that both cytochrome *a* and Cu_A can transfer electrons directly to the oxygen binding site. At 173 K, electron transfer from Cu_A appears to be about 3 to 5 times as fast as from cytochrome *a*. A similar branching of the electron flow at room temperature is suggested by the recent finding that chemical modification of the Cu_A site decreases the electron transfer activity to about 20% of that of the native enzyme (Gelles and Chan, 1985).

With a majority of the electrons passing through Cu_A, a pump site at cytochrome a would have to translocate at least 3 protons per electron to produce the observed average stoichiometry close to 1.0 H^+/e^-. Since the available free energy from electron transfer from cytochrome a to the oxygen binding site corresponds to the maximum transport of 1.75 H^+/e^-, we find this situation unlikely. This leaves Cu_A as the most attractive candidate for the pumping site. Experiments to determine whether the Cu_A site takes part in proton pumping are currently underway using the reconstitution of cytochrome *c* oxidase with a chemically modified Cu_A site (Gelles and Chan, 1985) into phospholipid vesicles.

An interesting consequence of imperfect gating on the input side of a proton pump is that the donor (cytochrome *c*), might bind to and reduce two different forms of the enzyme. One of these forms is the oxidized enzyme in the input state. The other is the oxidized enzyme in the output state. Reduction of the latter by the donor is an unwanted electron leak reaction (Blair, et al., 1986). Brzezinski et al., (1986) have noted that the rate equation for such a case will be quadratic in cytochrome *c* or, equivalently, contain a 'high K_m' and 'low K_m' term. It is therefore possible that the observed biphasic steady-state kinetics of cytochrome *c* oxidase arises from the imperfect gating of electron transfers within the proton pumping site.

The reaction scheme we have proposed is by necessity speculative. Our purpose, however, has not been to dwell on the specifics of a proton pump. It has been, rather, to investigate how the requirements for proton and electron gating and the thermodynamic constraints on a proton pump may be met at the mechanistic level. In doing so, we have found that the available spectroscopic information on the Cu_A site suggests a structure well suited to a function in proton transport. Although more research will be required for a definite assignment of the proton pump in cytochrome *c* oxidase, we believe that the model presented here will provide a useful framework for future thought and experimentation.

References

[1] Aasa, R., Albracht, S.P.J., Falk, K., Lanne, B., and Vängård, T., (1976) *Biochim. Biophys. Acta,* **422**, 260-272.

[2] Artzatbanov, V.Y., Konstantinov, A.A., and Skulachev, V.P. (1978) *F.E.B.S. Lett.,* **87**, 188-195.

[3] Bzezinski, P., Thörnstrom, P.-E., and Malmstöm, B.G., (1986) *F.E.B.S. Lett.,* **194**, 1-5.

[4] Beinert, H., Griffiths, D.E., Wharton, D.E., and Sands, R.H. (1962) *J. Biol. Chem.,* **237**, 2337-2346.

[5] Brudvig, G., and Chan, S.I. (1986) *J. Magnetic Res.,* **67**, 189-201.

[6] Blair, D.F., Gelles, J., and Chan, S.I. (1986) *Biophys. J.,* **50**, 713-734.

[7] Erecinska, M., and Chance, B. (1972) *Arch. Biochem. Biophys.,* **151**, 304-315.

[8] Gelles, J., Blair, D.F., and Chan, S.I., *Biochim. Biophys. Acta,* **863**, 205 236.

[9] Gelles, J., and Chan, S.I., (1985) *Biochemistry,* **24**, 3693-3698.

[10] Hoffman, B.M., Roberts, J.E., Swanson, M., Speck, S.H., and Margoliash, E. (1980) *Proc. Natl. Acad. Sci. USA,* **22**, 1452-1456.

[11] Hu, V.W., Chan, S.I., and Brown, G.S. (1977) *Proc. Natl. Acad. Sci. USA,* **74**, 3821-3825.

[12] Hu, V.W., Chan, S.I., and Brown, G.S., (1977) *Proc. Natl. Acad. Sci. USA,* **74**, 3821-3825.

[13] Li, P.M., Gelles, J., Chan, S.I., Sullivan, R.J., and Scott, R.A. (1987) *Biochemistry,* **26**, 2091-2095.

[14] Marcus, R.A., and Sutin, N. (1985) *Biochim. Biophys. Acta,* **811**, 265-322.

[15] Martin, C.T. (1988) Ph.D. Thesis, California Institute of Technology, Pasadena, CA.

[16] Martin, C.T., Scholes, C.P, and Chan, S.I. (1988) *J. Biol. Chem,* in press.

[17] Mitchell, P. (1961) *Nature,* **191**, 144-148.
[18] Moroney, P.M., Scholes, T.A., and Hinkle, P.C. (1984) *Biochemistry,* **23**, 4991-4997.
[18] Tanford, C. (1983) *Ann. Rev. Biochem.,* **52**, 379-409.
[20] Wang, H., Blair, D.F., Ellis, W.R., Gray, H.B., and Chan, S.I., (1986) *Biochemistry,* **25**, 167-171.
[21] Wikström, M. (1977) *Nature,* **266**, 271-273.
[22] Wikström, M., and Krab, K. (1979) *Biochim. Biophys. Acta,* **549**, 177-222.
[23] Wikström, M., Krab, K., and Saraste, M. (1981) *Cytochrome Oxidase, A Synthesis,* Academic Press, New York.
[24] Wikström, M., and Casey, R.P. (1985) *J. Inorg. Biochem.,* **23**, 327-334.

DISCUSSION

NICHOLLS: If, as shown by Wilson and coworkers, Cu_A is unaffected by membrane energization, and if it is also, as you say, "outside" the membrane, how can it act as a proton "gate"? Is there not some kind of problem here?

CHAN: There is no problem here. The proton "gate" can be outside of the membrane so long as there is a transmembrane proton-conducting "channel" that can direct protons to, and only to, the "input" side of the gate from the matrix.

KADENBACH: The model you have described for proton pumping in cytochrome oxidases does not necessarily involve a dimer structure of the enzyme. Could you comment on it?

CHAN: The model does not require the dimer structure of the enzyme.

MULLER: What type of device do you imagine to transfer H^+ from the M-side to the C-side, considering that Cu_A is located in the water soluble part of subunit 2?

CHAN: A proton-conducting "channel" consisting of a network of hydrogen-bonds that spans the protein across the membrane.

PECHT: Why is tyrosyl residue preferred in your model as the proton transfer agent rather than one of the ligands?

CHAN: The choice of the tyrosine as a ligand to the distorted trigonal conformation of the metal site is somewhat arbitrary and other amino acid residues with similar acid/base and ligation properties might also function as well.

KING: (1) I remember Professor Keilin's experiment that reduced cyt.c is practically not oxidized by cytochrome c oxidase in anaerobic conditions. (2) The late Professor Lemberg, among others, found that a small amount of the powerful reducing agent, $Na_2S_2O_4$, only slowly reduces cytochrome c oxidase in 30 minutes or more. Does your scheme explain these two observations? It seems to me the Em values alone are not enough.

CHAN: There are other factors beside reaction exergonicity, i.e., driving force, that can influence electron transfer rates. I alluded to these other factors in my presentation.

Oxidases and Related Redox Systems, pages 749–769

THE INTERACTION OF CYTOCHROME *c* WITH CYTOCHROME OXIDASE

Eric A. E. Garber, Thomas L. Luntz and E. Margoliash
Department of Biochemistry, Molecular Biology and Cellular Biology, Northwestern University, Evanston, Illinois 60208

INTRODUCTION

The interaction of cytochrome *c* with cytochrome *c* oxidase has been traditionally examined by either monitoring the oxidation of cytochrome *c* spectrophotometrically or the reduction of oxygen to water, manometrically or polarographically. Several disadvantages are associated with the spectrophotometric assay of the steady-state reaction. The first is that product, ferricytochrome *c*, accumulates, resulting in competitive product inhibition of the reaction of cytochrome *c* with cytochrome oxidase, so that in measuring any kinetic parameters one must analyze for the rate of the initial reaction, before any significant amount of product has accumulated. This is the case, except under the relatively high ionic strength and lower pH conditions under which K_m is close to K_S and K_P, the equilibrium binding constants of substrate and product, respectively (Minnaert, 1961; Yonetani and Ray, 1965; Speck *et al.*, 1984). In addition, because of the very low K_m for the high affinity phase of the reaction, one is limited to monitoring cytochrome *c* concentrations which are too high to yield a precise measure of the K_m for that phase.

The 1984 ***controlled binding hypothesis*** of Speck *et al.* (1984) addressed the major problem confronting any analysis of the cytochrome *c* - cytochrome oxidase steady-state reaction, namely the biphasic or multiphasic nature of the kinetics, detected many years earlier by Nicholls (1965). Speck *et al.* suggested that the non-productive binding of a molecule of cytochrome *c*, near, but not at, the electron-accepting site on the oxidase could, by electrostatic repulsion, decrease the binding affinity of the molecule of cytochrome *c* at the catalytic site. This would result in an increase in the first-order rate constant for the dissociation of the cytochrome *c* at the catalytic site, so that, as long as that dissociation was the rate-limiting step in the steady-state turnover of electrons, the presence of a nearby molecule of cytochrome *c* would lead to an increase in initial rates, as well as an increase in K_m, namely a lower affinity second kinetic phase. That the spectrophotometrically measured steady-state rate is limited by the rate at which the product, ferricytochrome *c*, dissociates from the electron-accepting site on the oxidase was most convincingly demonstrated by the Amsterdam group in 1981 (Wilms *et al.*, 1981). It was shown that ferrocytochrome c_1 reacted with the complex of ferricytochrome *c* and oxidase with a first order rate constant ($2.5s^{-1}$ in 5mM phosphate pH 7.0) that was at least 2 orders or magnitude slower than the pseudo-first-order rate of reaction of purified ferrocytochrome c_1 with ferricytochrome *c*, and which correlated well with the steady-state turnover of the high affinity phase of the spectrophotometric oxidase kinetics.

The polarographic assay under relatively low ionic strength conditions and with ascorbate and N,N,N',N' tetramethylphenylenediamine (TMPD) as reductants, behaves differently than the spectrophotometric assay. The high affinity/low turnover phase has a K_m close to that for the corresponding spectrophotometric phase and the K_D for the first molecule of cytochrome *c* bound. Furthermore, that correlation between K_m and K_D can be shown to be maintained for the reaction of a whole series of chemically-modified cytochromes *c* displaying a wide range of affinities and reactivities (Osheroff *et al.*, 1980). However, the TN_{max} for the polarographic high affinity phase is 5 to 10 times that for the corresponding spectrophotometric phase, while the first lower affinity polarographic phase shows a corrected K_m at least one order of magnitude lower than that for the spectrophotometric low affinity phase, and similarly smaller than the K_D for the binding of the second molecule of cytochrome *c* to the enzyme. Thus, as demonstrated below, there is only one molecule of cytochrome *c* bound per molecule of cytochrome oxidase throughout the first two polarographic phases, and the transition of high to lower affinity cannot be due to the binding of a second molecule of cytochrome *c*, as is the case in the spectrophotometric assay. It is suggested, from studies of the effects of viscosity, of the reactions of several different cytochromes *c* and of the role of TMPD, that upon the binding of cytochrome *c* to the electron-accepting site of cytochrome oxidase, both proteins have to go through a conformational transition before electron transfer can take place. Once the product of the reaction, ferricytochrome *c*, leaves the electron-accepting site, both proteins relax over a finite time course. If another molecule of ferrocytochrome *c* attaches itself to the oxidase before it has fully relaxed, the electron transfer permissive complex will be attained more rapidly than if the system has started from rest. Thus, as the concentration of cytochrome *c* is increased, the probability of this happening will increase, leading from the higher affinity/slower turnover phase to a lower affinity/ higher turnover phase.

EXPERIMENTAL PROCEDURES

The preparation of cytochrome *c* was according to the procedure of Margoliash and Walasek (1967), as modified by Brautigan *et al.* (1978). Prior to enzymic assay, the cytochrome *c* was reduced with minimal dithionite and chromatographed on a Sephadex G75 Superfine (Pharmacia) column, equilibrated in 25mM acetate-Tris, pH 7.9, to remove any residual reductant and any polymeric forms. The polarographic assay was according to the procedure of Ferguson-Miller *et al.* (1976), while the spectrophotometric assay was according to Speck *et al.* (1984).

To correct for distortion in the K_m value of the low affinity phase due to overlap with the high affinity reaction, the following expression was employed:

$$K_m = c_f \left[\frac{TN^L_{max} - TN^H_{max}}{TN_{obs} - TN^H_{max}} - 1 \right]$$

in which c_f is the concentration of free cytochrome *c*; TN^H_{max} and TN^L_{max} are the turnover numbers for the high affinity and low affinity phases, extrapolated to infinite cytochrome *c* concentration, while TN_{obs} is the turnover for a particular c_f. This assumes that a molecule of cytochrome oxidase cannot catalyze simultaneously both high and low affinity electron transfer. An expression derived by Sinjorgo *et al.* (1984) defines the observed low affinity V_{max} as the sum of the maximal velocities of the high and the low affinity reactions, defining a system with two independent electron transfer sites, inconsistent with the controlled binding hypothesis of Speck *et al.* (1984). Our expression also assumes that the relationship between concentrations of high and low affinity reaction complexes follows a first order dependence on the concentration of cytochrome *c*, as is clearly the case in the spectrophotometric assay and probably also valid for the polarographic assay. The validity of this expression is also dependent on the two K_m values being at least one order of magnitude apart, as was found to be the case, such that the steady state concentration of free enzyme during low affinity turnover is insignificant, a condition demonstrated to occur by the data.

The determination of the dissociation constants of ferricytochrome *c* with cytochrome oxidase using gel filtration, was carried out according to the Hummel and Dreyer procedure (1962), as further developed by Dethmers *et al.* (1979) and Osheroff *et al.* (1980). Analytical ultracentrifugation determinations of the dissociation constants were done on a Beckman model E ultracentrifuge monitoring 410nm and 420nm as described elsewhere (Garber and Margoliash, 1987).

Proton NMR, one-dimensional, longitudinal relaxation measurements and 2D-NOESY analyses of various cytochromes *c*, were carried out with a Varian XL-400 spectrometer using standard techniques (Rush *et al.*, 1987).

Site directed mutants of cytochrome *c* were obtained from a rat pseudogene (RC9) which codes for the correct amino acid sequence of rat cytochrome *c* (Scarpulla, 1984) kindly provided by Dr. Richard Scarpulla, and the *Drosophila melanogaster* major cytochrome *c* gene (DM*c* 01), which we have cloned and which codes for the thoracic muscle cytochrome *c* of the fly (Swanson *et al.*, 1985). The rat pseudogene, the sequence of which corresponds exactly to that of the 5' non-coding and coding regions of the expressed gene, was inserted in place of the yeast iso-1-cytochrome *c* coding sequence in the plasmid YCp CYC1 (2.4) (Lowry *et al.*, 1985), kindly provided by Dr. Richard Zitomer, placing it under the control of the yeast iso-1-cytochrome *c* promoter. The promoter-RC9 construct was transferred to the multicopy yeast shuttle vector YEp 13. Transformation of the cytochrome *c*-deficient yeast strain GM-3C-2 with such plasmids yielded colonies that were competent for growth on non-fermentable substrates. These yeast strains were grown and the cytochromes *c* isolated and characterized. Mutant pseudogenes were produced by the oligonucleotide method of Zoller and Smith (1984), and incorporated into expression vectors as given above. The expression of the *Drosophila* cytochrome *c* was accomplished in a similar manner.

On purifying either the wild type rat pseudogene expression products (RC9), or its tyrosine 67 to phenylalanine mutant protein [RC9(YF67)], two fractions were obtained by HPLC chromatography on a Synchropak CM300 column (Synchrom). The faster moving material was shown to have a blocked N-terminus, by sequential Edman degradation on a Model 477A protein sequencer with a Model 120A PTH Analyzer (Applied Biosystems). We presume it has an acetylmethionine at the N-terminus, as earlier observed with mutant yeast cytochromes *c* (Tsunasawa *et al.*, 1985). The slower moving peak material is not blocked and was shown to have the normal N-terminal amino acid sequence of rat cytochrome *c* for the first 10 residues. Neither cochromatographs with rat cytochrome *c* obtained from rat hearts, because, as determined by NMR, the yeast synthesized cytochromes all have trimethylated lysine 72. The effect of this post-translational modification is to decrease the isoionic point of the protein (Kim *et al.*, 1980). As expected, the blocked proteins chromatograph much closer to the normal rat protein, which has an N-terminal acetylglycine, than the unblocked rat pseudogene and rat mutant products, which carry an extra positive charge. The data presented below refer only to the N-terminally unblocked recombinant rat proteins, RC9-II and RC9(YF67)-II.

Similar wild type and mutant *Drosophila* protein preparations did not show any appreciable amounts of N-terminally blocked products and a mutant in which proline 30 was substituted by an alanine was obtained. This change eliminates the relative rigidity of the hydrogen bond between the hemochrome imidazole of histidine 18 and the backbone carbonyl group of the proline. In the native protein, this hydrogen bond serves to hold the imidazole plane at an angle which precisely bisects the porphyrin, symmetrically between Rings I and IV on one side, and Rings II and III on the other.

RESULTS

The kinetics of either ferrocytochrome *c* oxidation or oxygen consumption by cytochrome oxidase display biphasicity, with a high affinity reaction, of K_m approximately

$2 \cdot 10^{-8}$M in both systems, and with different TN_{max} values for the spectrophotometric ($5s^{-1}$) and the polarographic ($40s^{-1}$) assays. The low affinity reaction, corrected for overlap with the high affinity reaction as described in Experimental Procedures, displays a K_m and TN_{max} of approximately $1.5 \cdot 10^{-5}$M and $200s^{-1}$, respectively, in the spectrophotometric assay, and approximately $8 \cdot 10^{-7}$M and $80s^{-1}$, respectively, for the polarographic assay.

Table 1. Comparison of K_m and K_D values in the horse cytochrome *c* - purified beef heart cytochrome oxidase system.

Concentration Range	High Affinity	Low Affinity
K_m (M)		
Spectrophotometric assay	2×10^{-8}*	1.5×10^{-5}
Polarographic assay	3×10^{-8}	8×10^{-7}
K_D (M)		
Gel filtration	4×10^{-8}	ND‡
Ultracentrifugation	$>4 \times 10^{-6}$	1.1×10^{-5}

*Large spectrophotometric error
‡Not determined

These results are summarized in Table I and compared to the directly measured dissociation constants for the binding of the first and second molecules of horse cytochrome *c* with purified beef cytochrome oxidase. The determination of the binding affinity of the first molecule of cytochrome *c* to a purified preparation of cytochrome oxidase presents no difficulty by gel filtration (Ferguson-Miller *et al.*, 1976; Dethmers *et al.*, 1979; Osheroff *et al.*, 1980, 1983). Because of the low affinity of the second molecule of cytochrome *c*, it becomes difficult to determine its binding by gel filtration, and an ultracentrifugal method was used instead. Gel filtration only allowed the assignment of a lower limit of $4 \cdot 10^{-6}$M for the K_D of the second molecule of cytochrome *c*. The dissociation constants for the first and second molecules of cytochrome *c* are very close to the K_m values for the high and low affinity phases, respectively, as determined spectrophotometrically. While, for the polarographic high affinity reaction the K_m is also equal to the K_D for the binding of the first molecule of cytochrome *c*, the K_m for the low affinity phase is over one order of magnitude smaller than the K_D for the second molecule of cytochrome *c*. Hence, even though the transition from high to low affinity in the spectrophotometric assay is consistent with the binding of the second molecule of cytochrome *c* being the causative mechanism, as proposed in the controlled binding hypothesis of Speck *et al.* (1984), it is clearly irrelevant to the transition between high and low affinity kinetics in the polarographic assay. The latter must define a mechanism in which only one molecule of cytochrome *c* is bound per molecule of cytochrome oxidase.

Table 2. Polarographic K_m Values for Different Cytochromes *c**

Cytochrome *c*	High Affinity	Low Affinity
Horse	3×10^{-8}	8.4×10^{-7}
Rat	1.5×10^{-8}	12×10^{-7}
Primate		
Spider Monkey	1×10^{-8}	11×10^{-7}
Fungal		
C. krusei		
S. cerevisiae iso-1	6×10^{-9}	7.7×10^{-7}
U. sphaerogena		
Recombinant		
Rat, RC9-II	1×10^{-8}	10×10^{-7}
RC9(YF67)-II	1×10^{-8}	13×10^{-7}

* The values listed for the rat cytochrome *c* and the recombinant rat cytochromes *c* were obtained with beef heart submitochondrial particles. All other values were determined with horse heart submitochondrial particles.

Table II presents a comparison of the polarographically derived K_m values for several different eukaryotic cytochromes *c* reacting with the cytochrome oxidase of horse heart or beef heart submitochondrial particles. While for the high affinity reaction these differ for the different proteins, presumably representating variations in their binding affinities to the oxidase, all four groups of cytochromes *c* display comparable low affinity K_m values, of approximately $8 \cdot 10^{-7}$M.

To further characterize the transition from high affinity to low affinity reactions in the polarographic assay, the effects of viscosity were studied. Increasing the viscosity with polyethylene glycol-8000 (formally '6000'), resulted in an increase in the low affinity K_m with no change in the V_{max} (Figure 1). Qualitatively similar effects were also observed employing glycerol and sucrose, as had been earlier reported by Osheroff *et al.* (1979) and by Swanson *et al.* (1982).

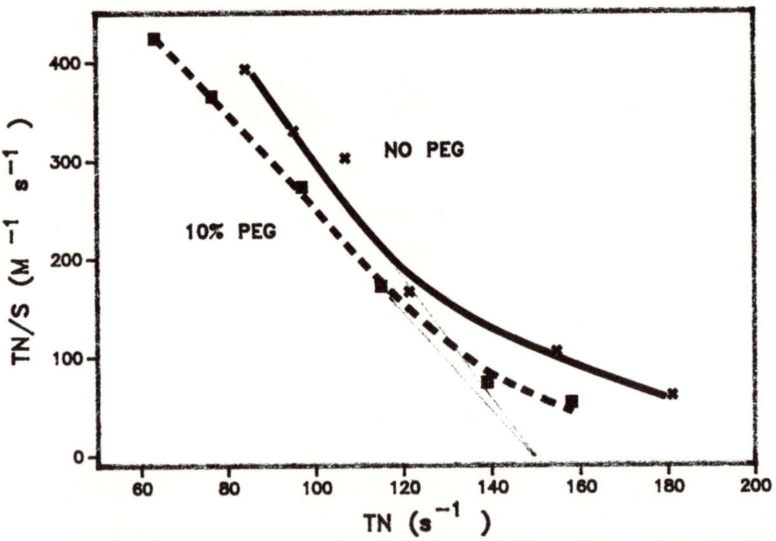

Figure 1. Effect of viscosity on the polarographically assayed low affinity reaction of horse cytochrome *c* with beef heart submitochondrial particles. The rate of oxygen consumption in the presence of 5mM ascorbate and 0.7mM TMPD in 25mM acetate-Tris, pH 7.9, was followed in the absence (■--■) and presence of 10% w/v polyethylene glycol (x—x). The concentration of cytochrome oxidase was 5nM.

Of particular interest is the observation that at the highest concentrations of cytochrome *c* tested (Figure 1), increasing viscosity does not appear to affect the K_m but changes the V_{max}. This is contrary to the effect of viscosity on the immediately preceding low affinity phase, described above, and is consistent with the occurrence of a second low affinity polarographic phase, presumably resulting from the binding of a second molecule of cytochrome *c* to the cytochrome oxidase, thereby forming the 2:1 complex. As expected for such a mechanism, the K_m for this second low affinity polarographic phase is comparable to the K_D for the second molecule of cytochrome *c* and the K_m for the low affinity spectrophotometric phase.

To clarify the mechanistic basis for the transition from high to low affinity kinetics in the polarographic assay, the roles of the various components in the assay were reexamined. Earlier work by Slater (1949) and by Yonetani (1962) had demonstrated that the reduction of oxidized cytochrome *c* in solution by ascorbate may be rate limiting in the assay of electron transfer from cytochrome *c* to cytochrome oxidase. As a result of this concern, when Ferguson-Miller *et al.* (1976) examined the use of TMPD-a functional analogue of p phenylenediamine used by Slater (1949) and ascorbate as reductants in the polarographic assay of cytochrome *c* mediated cytochrome oxidase activity, they were able to show that in the steady-state, under the assay conditions we employ throughout, the cytochrome *c* was better than 98% reduced. It should be noted that the concentration of TMPD chosen was not saturating, since increasing the concentration of TMPD results in unacceptably high background rates of oxygen consumption and inhibition of the reaction, the exact mechanistic basis of which is not know. Omission of the TMPD, results in the expected drop in oxygen consumption and kinetics which remain biphasic with K_m values of $4 \cdot 10^{-8}$M and $8 \cdot 10^{-6}$M for the high and low affinity phases, respectively, very similar to the corresponding values of the spectrophotometrically-assayed reactions (Figure 2). Had the reaction been rate limited by the reduction of cytochrome *c* by ascorbate, at a rate that was appreciably slower than the rate of oxidation of the protein by the oxidase, one would have expected monophasic kinetics. The fact that under our conditions (pH 7.9, 5mM ascorbate), a biphasic kinetic pattern is maintained, close to that observed spectrophotometrically for the initial steady state kinetics of oxidation of ferrocytochrome *c*, would indicate that the interaction with ascorbate is not rate limiting.

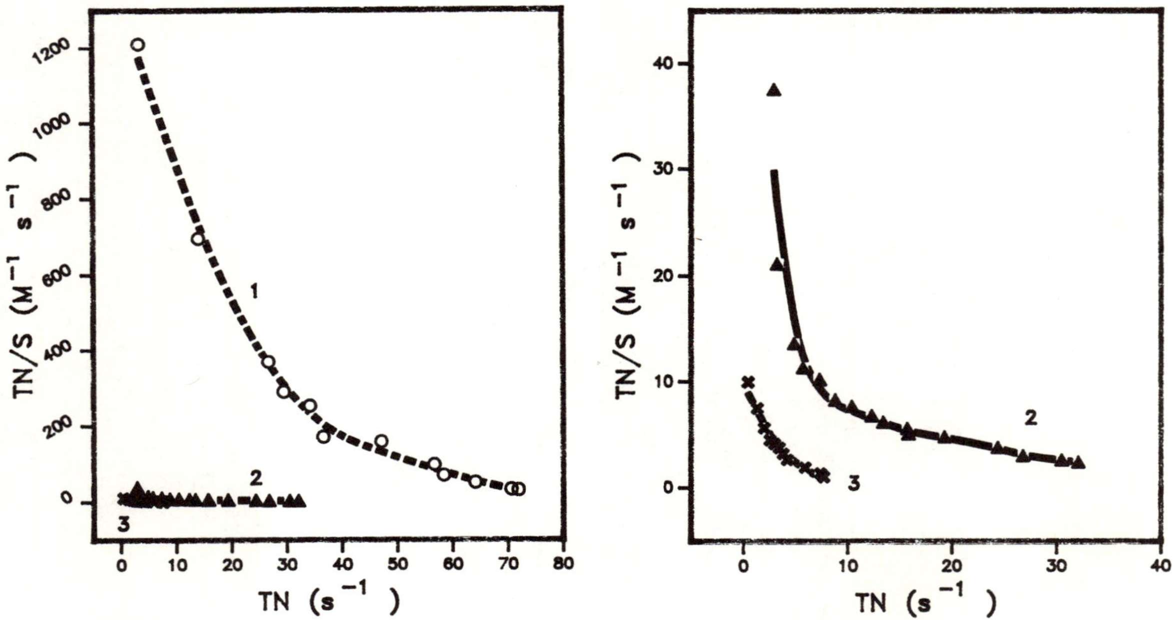

Figure 2. Comparison of steady-state assays for the interaction of horse cytochrome c with beef heart submitochondrial particles. Curve 1 represents the polarographic measurement of steady-state oxygen consumption (0.25nM cytochrome oxidase, 5mM ascorbate, 0.7mM TMPD, 25mM acetate-Tris, pH 7.9). Curve 2 represents the same conditions as for Curve 1, except that TMPD was absent. Curve 3 represents the spectrophotometric analysis of the initial steady-state rates of the oxidation of horse ferrocytochrome c by 0.2nM cytochrome oxidase, in the same buffer as that used for the two polarographic assays. Figures A and B show the same data plotted on two different scales to display the biphasic nature of the kinetics for all three assay systems.

Among the recombinant cytochromes *c* employed, the most revealing of the nature of the kinetics of reaction with cytochrome oxidase was the RC9(YF67) mutant. This protein had the same K_m as the wildtype protein, RC9, in the high affinity polarographic phase, but a V_{max} approximately 25% higher. Both the K_m and V_{max} for the low affinity phase did not differ significantly from the corresponding values yielded by the wild type recombinant rat cytochrome *c*. The larger high affinity V_{max} would appear to be best interpreted as representing a higher rate of reduction of the cytochrome *c* in its complex with the oxidase by TMPD. In attempting to determine whether any other structural parameters possibly related to the reduction by TMPD had been affected, we examined the reaction with carbon monoxide and the pK_a for methionine binding by NMR spectroscopy. It was found that RC9(YF67), in the ferrous form, reacted completely with carbon monoxide at atmospheric pressure and neutral pH, and that the initial alkaline transition between pH 9 and 10 had essentially disappeared. Thus, the kinetic properties of

Table 3. NMR determined pK_a values for Methionine 80 coordination*

Cytochrome *c*		pK_a
Rat	RC9-II	9.9
	RC9(YF67)-II	>>10.7+
Drosophila	DM *c* 01	9.3
	DM *c* 01(PA30)	8.2

* The pK_a values were determined from the relative areas of the heme methyls resonances as a function of pH for the coordinated and uncoordinated froms, as identified by Morishima *et al.* (1977). The values listed are not corrected for the ionization of D_2O as compared to H_2O.
+No detectable loss of coordination observed at pH 9.85.

the RC9(YF67) mutant were consistent with the hysteresis mechanism proposed below for the high to low affinity transition in the polarographic assay. Indeed, a single residue change in the cytochrome *c*, which causes a difference in the high affinity V_{max} without apparently affecting the binding of the two proteins, has no effect on the low affinity polarographic phase, which under the proposed mechanism is a function of the enzyme, not of the cytochrome *c*. In addition, the easy combination with carbon monoxide indicated that even though the methionine 80 sulfur was readily displaced from the heme iron when the -OH of tyrosine 67 was absent, the initial alkaline transition normally resulting in the displacement of the hemochrome sulfur was also absent. The simplest explanation of that result is that following the initial alkaline transition the heme iron ligand is the -OH of tyrosine 67, not a lysine ε-amino group, as commonly suggested and apparently supported by the results of a variety of different spectroscopies (see for example, Morishima *et al.*, 1977). Clearly, considerable more work remains to be done to substantiate this possibility.

DISCUSSION

The observation that the K_m values of the high and low affinity reactions, assayed spectrophotometrically, are equal to the dissociation constants for the binding of the first and second molecules of cytochrome *c* with cytochrome oxidase, respectively, is consistent with the controlled binding hypothesis of Speck *et al.* (1984). This model ascribes the high to low affinity transition to the binding of a modulator cytochrome *c* molecule, at a site not capable of electron transfer, which increases the rate of product dissociation, the rate limiting step. Recently, a model has been proposed by Brzezinski and Malmström (1986) which explains how, by coupling intramolecular electron transfer and proton translocation to a conformational change in cytochrome oxidase, biphasic kinetics may be generated. However, in order to explain the results indicating that under steady state turnover in the spectrophotometric assay, product dissociation appears to be rate limiting, it is proposed that the presence of bound product, ferricytochrome *c*, acts to inhibit the conformational change (Brzezinski *et al.*, 1986). This feature serves to define the rate limiting step as product dissociation, and indicates that the model represents a mechanism of how proton translocation may be coupled to turnover; nevertheless, the kinetics of steady state turnover must represent rate limiting product dissociation. As such, the basis for the transition from high to low affinity kinetics can be described by the controlled binding hypothesis.

The inability to define the role of the artificial reducing system in the polarographic assay and to elucidate the mechanistic basis for the various kinetic parameters, has hindered the development of an understanding of how cytochrome oxidase functions under steady-state conditions. The observation that the V_{max} of the high affinity reaction in the polarographic assay is 5 to 10 times greater than that observed for the high affinity spectrophotometric phase was interpreted to result from circumvention of the rate limiting step of the latter assay, namely, product dissociation (Ferguson-Miller *et al.*, 1978). In fact, stopped-flow measurements of the reduction, by TMPD, of cytochrome *c* in complex with cytochrome oxidase, have shown a faster rate constant than that observed for the reduction of cytochrome *c* free in solution (Osheroff *et al.*, 1983). By omitting TMPD from the polarographic assay and having ascorbate as the sole reductant, which apparently can only reduce cytochrome *c* free in solution, the observed steady-state kinetics are similar to those obtained by measuring spectrophotometrically the initial rates of reaction. This is consistent with TMPD being able to reduce the cytochrome *c* in complex with cytochrome oxidase. The slight difference between the results of the spectrophotometric assay and of the polarographic assay lacking TMPD, probably represents effects of ascorbate that are not yet fully understood.

Insight into the mechanistic basis for the transition from high to low affinity kinetics in the polarographic assay comes from observations that increasing the viscosity causes an increase in the K_m without affecting the V_{max} of the low affinity reaction. Viscosity is assumed to affect only the rate of cytochrome *c* diffusion, as evidenced by the compensatory effect of higher concentrations of the protein. Thus, the influence on the value of the K_m must reflect the importance of the time elapsed between successive effective bindings. A proposal consistent with this conclusion considers that cytochrome oxidase, following binding, which may or may not include turnover, is in a primed state for electron transfer with cytochrome *c* and depending on the time between turnovers, may or may not relax back to the less active form. Such a ***hysteresis mechanism***, to explain how the time between turnovers is decreased so as to trap the primed state of the enzyme, must concur with the observation that throughout the transition from high to low affinity polarographic kinetics only a 1:1 complex between the two proteins is present. This constraint implies that the TMPD reduction of cytochrome *c* complexed with the oxidase cannot provide reducing equivalents beyond those associated with the high affinity reaction. Hence, any further reducing equivalents must come by a different pathway, namely from the ferrocytochrome *c* in solution.

In this connection, it should be noted that Weber *et al.* (1987) have recently shown that upon binding to cytochrome oxidase, cytochrome *c* assumes a conformation that may be similar to that of the protein at alkaline pH values. It is thus possible that a corresponding conformation change is induced in the oxidase, a not unlikely result of the interaction of two very strongly and oppositely charged proteins.

The above model predicts that, since the K_m of the low affinity reaction is representative of the stability of the primed state of cytochrome oxidase, it should be independent of the cytochrome *c* assayed, provided that the propensity for productive binding is the same. Indeed, an examination of mammalian, both primate and non-primate, of fungal and of recombinant cytochromes *c*, show similar low affinity K_m values, despite widely varying high affinity values (see Table II). Similarly, as noted under Results, the properties of the recombinant rat cytochrome *c*, RC9(YF67), provide a test of these interpretations and turn out to be consistent with such a model.

This model also predicts that further increases in cytochrome *c* concentration will result in a change in the low affinity reaction, assayed polarographically, with a new K_m similar to the K_D for the binding of a second molecule of cytochrome *c* on the enzyme. At such high concentrations of cytochrome *c*, increasing the viscosity should not affect the K_m but lower the V_{max}. This would be consistent with the additional cytochrome *c* increasing the rate of ferricytochrome *c* dissociation, as demonstrated for the spectrophotometric assay and responsible for its low affinity phase. These characteristics are in fact indicated at the highest cytochrome *c* concentrations tested so far (Figure 1), and require further examination.

ACKNOWLEDGMENTS: The data discussed above is a summary of results presented at meetings of the American Society for Biochemistry and Molecular Biology (1986; 1987), the 70th Canadian Chemical Conference, 1987, and in papers yet to be published (Garber and Margoliash, 1987; Luntz *et al.*, 1987). This work was supported by Grants from the National

Institutes of Health (GM 19121 and GM 29001).

REFERENCES

Brautigan DL, Ferguson-Miller S, Margoliash E (1978). Mitochondrial cytochrome *c*: Preparation and activity of native and chemically modified cytochromes *cl*. In Fleischer S, Packer L (eds): "Methods in Enzymology, Volume LIII," New York: Academic Press, pp 128-164.

Brzezinski P, Malmström BG (1986). Electron-transport-driven proton pumps display nonhyperbolic kinetics: Simulation of the steady-state kinetics of cytochrome *c* oxidase. Proc Natl Acad Sci USA83:4282-4286.

Brzezinski P, Thörnström P-E, Malmström BG (1986). The rate-limiting step and nonhyperbolic kinetics in the oxidation of ferrocytochrome *c* catalyzed by cytochrome *c* oxidase. FEBS Letters 194:1-5.

Ferguson-Miller S, Brautigan DL, Margoliash E (1976). Correlation of the kinetics of electron transfer activity of various eukaryotic cytochromes *c* with binding to mitochondrial cytochrome *c* oxidase. J Biol Chem 252:1104-1115.

Ferguson-Miller S, Brautigan DL, Margoliash E (1978). Definition of cytochrome *c* binding domains by chemical modification. J Biol Chem 253:149-159.

Dethmers JK, Ferguson-Miller S, Margoliash E (1979). Comparison of yeast and beef cytochrome *c* oxidases. J Biol Chem 254:11973-11981.

Garber EAE, Margoliash E (1986). Interaction of cytochrome *c* with cytochrome oxidase: Dielectric, ionic strength and ^{1}H NMR studies. Federation Proc 45, Abstract #787.

Garber EAE, Theodorakis JL, Margoliash E (1987) The low affinity reaction of cytochrome *c* with cytochrome oxidase. 70th Canadian Chemical Conference and Exhibition, Abstract #BIE-2-A.

Garber EAE, Margoliash E (1987). The interaction of cytochrome *c* with cytochrome oxidase. In preparation.

Hummel JP, Dreyer WJ (1962). Measurement of protein-binding phenomena by gel filtration. Biochim Biophys Acta 63:530-532.

Kim C-S, Kueppers F, Dimaria P, Farouqui J., Kim S, Paik WK (1980). Enzymatic trimethylation of Residue-72 lysine in cytochrome *c*. Effect on the total structure. Biochim Biophys Acta 622:144-150.

Lowry CV, Weiss JL, Walthall DA, Zitomer RS (1983). Modulator sequences mediate oxygen regulation of CYC1 and a neighboring gene in yeast. Proc Natl Acad Sci USA 80:151-

155.
Luntz T, Koshy T, Margoliash E (1987). Production of heterologous cytochromes *c* in yeast under the control of the CYC1 promoter. Federation Proc 46, Abstract #1448.
Margoliash E, Walasek OF (1967). Cytochrome *c* from vertebrate and invertebrate sources. In Estabrook RW, Pullman ME (eds): " Methods in Enzymology, Volume X," New York: Academic Press, pp 339-348.
Minnaert K (1961). The kinetics of cytochrome *c* oxidase. I. The system: Cytochrome *c*-cytochrome oxidase-oxygen. Biochim Biophys Acta 50:23-34.
Morishima I, Ogawa S, Yonezawa T, Iizuka T (1977). Nuclear magnetic resonance studies of hemoproteins. IX. pH dependent features of horse heart ferric cytochrome *c*. Biochim Biophys Acta 495:287-298.
Nicholls P (1965). Cytochrome oxidase *in situ*: Concepts and questions. In King TE, Mason HS, Morrison M (eds) "Oxidases and Related Redox Systems," New York: Wiley, pp 764-777.
Osheroff N, Jemmerson R, Speck SH, Ferguson-Miller S, Margoliash E (1979). Site-specific anti-cytochrome *c* antibodies. J Biol Chem 254:12717-12724.
Osheroff N, Brautigan DL, Margoliash E (1980). Definition of enzymic interaction domains on cytochrome *c*. J Biol Chem 255:8245-8251.
Osheroff N, Speck SL, Margoliash E, Veerman ECI, Wilms J, König BW, Muijsers AO (1983). The reaction of primate cytochromes *c* with cytochrome *c* oxidase. J Biol Chem 258:5731-5738
Rush JD, Koppenol WH, Garber EAE, Margoliash E (1987). Electrostatic aspects in the oxidation by Tris (1,10-phenanthroline) cobalt-(III) at low ionic strength. In preparation.
Scarpulla R (1984). Processed pseudogenes for rat cytochrome *c* are preferentially derived from one of three alternate mRNAs. Mol Cell Biol 4:2279-2288.
Sinjorgo KMC, Meijling JH, Muijsers AO (1984). The concept of high- and low-affinity reactions in bovine cytochrome *c* oxidase steady-state kinetics. Biochim Biophys Acta 767:48-56.
Slater EC (1949). The measurement of the cytochrome oxidase activiity of enzyme preparations. Biochemistry 44:305-318.
Speck SH, Neu CA, Swanson MS, Margoliash E (1983). Role of phospholipid in the low affinity reactions between cytochrome *c* and cytochrome oxidase. FEBS 164:379-382.
Speck SH, Dye D, Margoliash E (1984). Single catalytic site model for the oxidation of ferrocytochrome *c* by

mitochondrial cytochrome *c* oxidase. Proc Natl Acad Sci USA 81:347-351.

Swanson M, Speck SH, Koppenol WH, Margoliash E (1982). Cytochrome *c* mobility in the eukaryotic electron transport chain. In Ho C (ed) "Electron Transport and Oxygen Utilization," Holland: Elsevier, pp 51-56.

Swanson MS, Zieminn SM, Miller DD, Garber EAE, Margoliash E (1985). Developmental expression of nuclear genes that encode mitochondrial proteins: Insect cytochromes *c*. Proc Natl Acad Sci USA 82:1964-1968.

Tsunasawa S, Stewart JW, Sherman F (1985). Amino-terminal processing of mutant forms of yeast iso-1-cytochrome *c*. The specificities of methionine aminopeptidase and acetyltransferase. J Biol Chem 260:5382-5391.

Weber C, Michel B, Bosshard HR (1987). Spectroscopic analysis of the cytochrome *c* oxidase-cytochrome *c* complex: Circular dichroism and magnetic circular dichroism measurements reveal change of cytochrome *c* heme geometry imposed by complex formation. Proc Natl Acad Sci USA 84:6687-6691.

Wilms J, Veerman ECI, König BW, Dekker HL, Van Gelder BF (1981). Ionic strength effects on cytochrome aa_3 kinetics. Biochim Biophys Acta 635:13-24.

Yonetani T (1962). Studies on cytochrome oxidase. IV. The cytochrome oxidase activity. J Biol Chem 237:550-559.

Yonetani T, Ray GS (1965). Studies on cytochrome oxidase. VI. Kinetics of the aerobic oxidation of ferrocytochrome *c* by cytochrome oxidase. J Biol Chem 240:3392-3398.

Zoller MJ, Smith M (1984). Oligonucleotide-directed mutagenesis: A simple method using two oligonucleotide primers and a single-stranded DNA template. DNA 3:479-488.

DISCUSSION

SMITH: In which buffer and at what pH did you run your assays? Recall that we showed that the relationship between the spectrophotometric and polarographic methods varies with the buffer and the pH. Also the different phases in the spectrophotometric assays can be easily explained by the increase in oxidase activity we showed yesterday in the presence of low concentrations of cytochrome *c*. This activation is seen at exactly the cytochrome *c* concentrations associated with the "high affinity phase."

MARGOLIASH: All the results presented in this talk were obtained at a pH of 7.9 in 25mM Tris acetate. Your observation that the relationship between polarographic and spectrophotometric assays varies with the pH is not surprising, since the changes in pH between 6 and 8 will alter the concentration of dianionic ascorbate, the actual reductant of cytochrome *c*, with a pK_a of 8.4, as well as affect the redox potential of TMPD. These effects are in addition to changes in the buffer anion composition. Phosphate, a well established strong ligand of cytochrome *c*, with two binding sites, most likely displays different binding affinities depending on whether it is in the monoanion or dianion state, and since the pK_a for this transition is at 7.2, its use in pH studies of oxidase reactions is probably not a good choice. Further, concern over the numerous titratable groups on the two proteins, and the lack of information on how they affect the interaction, makes the indiscriminate changing of pH, without regard for how such changes interrelate, serve only to confuse an already complex field. We therefore set out to first characterize the reactions under one set of conditions and only after this is complete, do we feel it is practical to introduce such a complex variable as pH. Your suggestion that valuable mechanistic information may be gained by varying the pH is undoubtably correct, but only after an understanding of what parameters are significant and how pH may alter them, are worthwhile results likely to be obtained.

As you mention, the activation you observe is associated with concentrations of cytochrome *c* reflective of the high affinity reaction and is replaced by a decrease in the rate constant at cytochrome *c* concentrations exceeding $2 \cdot 10^{-7}$M. The K_m of the low affinity reaction in the spectrophotometric assay is approximately 100-fold greater than the onset of a decreasing rate constant. Therefore your observed activation, while intriguing, is not relevant to the high to low affinity transition in the spectrophotometric assay.

YONETANI: Under polarographic assay conditions, cytochrome *c* is not fully reduced [Yonetani, T. (1962, *J. Biol. Chem. 237*, 550-559). Thus, the S term in the TN/S *vs* TN plot should not be the total cytochrome *c* concentration. It should be the steady state level of reduction of the cytochrome *c*. Is the "S" term in your analysis the total cytochrome *c* concentration?

MARGOLIASH: As shown in your 1962 paper, whether or not the rate of reduction of cytochrome *c* by ascorbate is the rate limiting step in the reaction with cytochrome oxidase depends on the pH. The steady state level of reduced cytochrome *c* increases with pH, consistent with the dianion of ascorbate, $pK_a = 8.4$, being the reducing agent. Your data show that at the pH 7.8, 25nM cytochrome oxidase, 20mM ascorbate and 20μM cytochrome *c*, the cytochrome *c* is fully reduced, leading to the conclusion that under these conditions the reduction of cytochrome *c* by ascorbate is not rate limiting. Even at 50nM oxidase the cytochrome *c* is better than 95% reduced, while at 100nM enzyme it is still above 90% reduced. All our assays, whether ascorbate and TMPD were used as reductants, as in our standard polarographic assay system, or ascorbate only, to examine the role of TMPD in the polarographic assay, were at pH 7.9, with up to 5nM cytochrome oxidase, when TMPD is used, and up to 10nM oxidase, when ascorbate was the sole reductant. The cytochrome *c* concentrations assayed were below 14μM throughout. Thus, under our conditions there is no problem with partial reduction of cytochrome *c* and the total concentration of cytochrome *c* is representative of the concentration of reduced cytochrome *c* in solution. In fact, we have shown directly, that under our ordinary polarographic assay conditions over 98% of the cytochrome *c* is reduced in the steady state (Ferguson-Miller *et al.*, *J. Biol. Chem.*, 1976). The problems associated with the degree of reduction of cytochrome *c* illustrate one of the major drawbacks of studying the influence of pH on a complex system without asking whether a non-enzymic event becomes rate limiting.

NICHOLLS: Lysine 79 is no longer thought to be the iron ligand in alkaline cytochrome *c*. Colin Greenwood recently suggested another lysine. Does tyrosine-67 in the 'wild type' cytochrome *c* interact with a lysine? Or is it possible that tyrosine-67 is itself the alkaline ligand?

MARGOLIASH: As you well know, the problem of what constitutes the 6th ligand in alkaline cytochrome *c* has not been settled, even though the question has been examined by numerous authors using optical spectroscopy, pH jump experiments, Raman spectroscopy, NMR and EPR spectroscopy, reactivity of lysines to chemical reagents, etc..., even old fashioned crystallographic model fitting. There is general agreement that increasing the pH results in the appearance of a transient form (pK_a = 9.1 in horse cytochrome *c*), interpreted to represent the displacement of the methionine 80 sulfur by the ε-amino group of a lysine, while at higher pH (pK_a values varying between 10.4 and 12.3), minor spectral changes are observed which have been variously interpreted, and at still higher pH (above pH 13) a hydroxide ion is supposed to occupy coordination position 6. Much of these assignments are speculative, and it is quite possible that the ionization of tyrosine 67 is what causes the initial displacement of methionine 80, though one would normally expect the pK_a of an internal hydrogen-bonded tyrosine to be greater, rather than lower than that of tyrosine in solution (pK_a = 10.4). It is of course the simplest explanation of our observations that tyrosine 67 is the long sought 6th iron ligand in the alkaline form. Obviously, much work remains to be done.

KING: How may phospholipids affect your high and low affinity reactions? How about cytochrome oxidase preparations containing only a very low phospholipid content? Cardiolipin seems to have appreciable reactivity with cytochrome *c*.

MARGOLIASH: Earlier work of Speck *et al.* (1983, *FEBS Letters 164*, 379-382) indicated that as the cardiolipin content of vesicles, in which purified oxidase was incorporated, increased, the spectrophotometric kinetics of the low affinity reaction became more complex, a mechanistic understanding of which is still lacking. Other than that, we have not examined the possible effects of phospholipids on the system.

KADENBACH: Müller and Azzi suggested a direct interaction of cytochrome *c* with subunit I of the *Paracoccus* enzyme. Do you believe that the mammalian enzyme could have two binding sites for cytochrome *c*, one at subunit I and the other at subunit II?

MARGOLIASH: The kinetic evidence and binding studies done so far appear to favor a single site for cytochrome *c* on cytochrome oxidase at which electron transfer can take place. If one accepts the evidence for that site being located mainly on subunit II in the oxidase of eukaryotes, then one should look for the possibility of electron transfer between subunits II and I. Maybe, the newly enthroned third copper atom could play the required role. As far as I am aware, there is at present no evidence in that direction.

Index